EVOLUTIONARY ECOLOGY AND ARCHAEOLOGY

EVOLUTIONARY ECOLOGY AND ARCHAEOLOGY

Applications to Problems in
Human Evolution and Prehistory

Edited by Jack M. Broughton and Michael D. Cannon

Foreword by James F. O'Connell

THE UNIVERSITY OF UTAH PRESS
Salt Lake City

© 2010 by The University of Utah Press. All rights reserved

See Acknowledgements on page 430 for a complete list of permission statements.

 The Defiance House Man colophon is a registered trademark of the University of Utah Press. It is based upon a four-foot-tall, Ancient Puebloan pictograph (late PIII) near Glen Canyon, Utah.

14 13 12 11 10   1 2 3 4 5

Publisher's Note: Because all contributions to this volume were previously published, no attempt was made to standardize the treatment of notes and references.

LIBRARY OF CONGRESS CATALOGING-IN-PUBLICATION DATA

Evolutionary ecology and archaeology : applications to problems in human evolution and prehistory / edited by Jack M. Broughton and Michael D. Cannon ; foreword by James F. O'Connell.
    p. cm.
  ISBN 978-0-87480-935-0 (pbk. : alk. paper) 1. Ecology.
2. Evolution (Biology) 3. Human evolution. 4. Archaeology and history.
I. Broughton, Jack M. II. Cannon, Michael D.
   QH541.E855 2009
   599.93'8—dc22   2009027545

Printed and bound by Sheridan Books, Inc., Ann Arbor, Michigan

Contents

List of Figures   vii
List of Tables   xi
Foreword by J. F. O'Connell   xiii

1. Evolutionary Ecology and Archaeology: An Introduction   1
   M. D. Cannon and J. M. Broughton

   Part I. Early Hominin Evolution and Behavior   13

2. The Origin of Man   17
   C. O. Lovejoy

3. Grandmothering and the Evolution of *Homo erectus*   30
   J. F. O'Connell, K. Hawkes, and N. G. Blurton Jones

4. A Theory of Human Life History Evolution: Diet, Intelligence, and Longevity   48
   H. Kaplan, K. Hill, J. Lancaster, and A. M. Hurtado

   Part II. Pleistocene Foragers and Colonists   81

5. An Optimal Foraging-Based Model of Hunter-Gatherer Population Dynamics   85
   G. E. Belovsky

6. Changing Diet Breadth in the Early Upper Palaeolithic of Southwestern France   104
   D. K. Grayson and F. Delpech

7. Paleolithic Population Growth Pulses Evidenced by Small Animal Exploitation   115
   M. C. Stiner, N. D. Munro, T. A. Surovell, E. Tchernov, and O. Bar-Yosef

8. Hunter-Gatherer Foraging and Colonization of the Western Hemisphere   122
   R. L. Kelly

9. Should We Expect Large Game Specialization in the Late Pleistocene?: An Optimal Foraging Perspective on Early Paleoindian Prey Choice   133
   D. A. Byers and A. Ugan

   Part III. Post-Glacial Adaptations   151

   Section 1. Foraging, Predator-Prey Interactions, and the Sexual Division of Labor   153

10. Effects of a Sedentary Lifestyle on the Utilization of Animals in the Prehistoric Southwest   159
    F. E. Bayham

11. Plant Utility Indices: Two Great Basin Examples   172
    K. R. Barlow and D. Metcalfe

12. Prey Spatial Structure and Behavior Affect Archeological Tests of Optimal Foraging Models: Examples from the Emeryville Shellmound Vertebrate Fauna   192
    J. M. Broughton

13. A Model of Central Place Forager Prey Choice and an Application to Faunal Remains
      from the Mimbres Valley, New Mexico   208
    M. D. Cannon

14. Sexual Division of Labor and Central Place Foraging: A Model for the Carson Desert of Western Nevada   228
    D. W. Zeanah

15. Declining Foraging Efficiency and Moa Carcass Exploitation in Southern New Zealand   257
    L. Nagaoka

    Section 2. The Intersection of Technology, Subsistence, and Mobility   271

16. A Formal Approach to the Design and Assembly of Mobile Toolkits   275
    S. L. Kuhn

17. Rocks are Heavy: Transport Costs and Paleoarchaic Quarry Behavior in the Great Basin   288
    C. Beck, A. K. Taylor, G. T. Jones, C. M. Fadem, C. R. Cook, and S. A. Millward

18. The Effect of Handling Time on Subsistence Technology   310
    J. Bright, A. Ugan, and L. Hunsaker

19. Microlithic Technology in Northern Asia: A Risk-Minimizing Strategy of the Late Paleolithic and Early Holocene   320
    R. G. Elston and P. J. Brantingham

    Part IV. Food Production Strategies: Origins, Spread, and Variation   331

20. Diffusion and Adoption of Crops in Evolutionary Perspective   335
    K. J. Gremillion

21. Bedouin Hand Harvesting of Wheat and Barley: Implications for Early Cultivation in Southwestern Asia   349
    S. R. Simms and K. W. Russell

22. Deferred Harvests: The Transition from Hunting to Animal Husbandry   356
    M. S. Alvard and L. Kuznar

23. Predicting Maize Agriculture among the Fremont: An Economic Comparison of Farming and Foraging
      in the American Southwest   372
    K. R. Barlow

    Part V. Cooperation and Competition in Complex Societies   393

24. The Calculus of Self-Interest in the Development of Cooperation: Sociopolitical Development and Risk
      among the Northern Anasazi   395
    T. A. Kohler and C. R. Van West

25. Conspicuous Consumption as Wasteful Advertising: A Darwinian Perspective on Spatial Patterns
      in Classic Maya Terminal Monument Dates   409
    F. Neiman

    Acknowledgments   430
    Contributors   431
    Index   435

Figures

2.1. Progressive prolongation of life phases and gestation in primates   19
2.2. Mechanical model of demographic variables in hominoids   20
2.3. Approximate distribution of extant Old World monkeys and pongids   24

4.1. Survival curves for forager populations   49
4.2. Daily energy acquisition data by individual among the Ache and Hiwi   52
4.3. The mean expected daily energy consumption per individual by age and sex category   53
4.4. The feeding ecology of humans and other primates   55
4.5. Ache age-sex specific daily energy acquisition   61
4.6. Hiwi age-sex specific daily energy acquisition   62
4.7. Ache fruit collection rates   63
4.8. Hiwi hourly acquisition rates by activity   64
4.9. Ache hourly acquisition rate by activity   65
4.10. Ache yearly mortality rates   68
4.11. Ecology and co-evolutionary process in the Hominid line   72

5.1. A linear programming diet model for the !Kung San   86
5.2. Changes in the population model affect the linear programming diet model's constraint equations   87
5.3. Observed relationship between the primary productivity in different environments and the cropping rates of hunter-gatherers   89
5.4. Relationship between the maximum observed biomass of large prey animals and harvestable primary productivity   89
5.5. Flow chart for the population model   90
5.6. Changes in the population's growth rate over time for an environment with a 200 g/m² harvestable primary productivity   91
5.7. Population trajectories emerging at four different harvestable primary productivities   92
5.8. Observed versus predicted relationship between densities of modern hunter-gatherers and harvestable primary productivity   93
5.9. Predicted changes in food abundances over time for different harvestable primary productivities   94
5.10. Dietary changes in environments with different harvestable primary productivities and at different stages of population growth   96
5.11. Comparison of this model's predicted chronology of colonization in the New World versus models developed to examine megafaunal overkill   97
5.A1. [A1 appendix] Plots of the population as functions of prey and human densities   100

6.1. Relationship between ungulate NISP and (Log10) ungulate NTAXA in the Le Flageolet I faunal assemblages   105
6.2. Relationship between ungulate NISP and (Log10) ungulate NTAXA with phalanges, metapodials, and radioulnae excluded   110
6.3. Red deer–reindeer dominance in the Le Flageolet I ungulate assemblages   111
6.4. The ratio of red deer to reindeer in the Le Flageolet I ungulate assemblages   111
6.5. The relationship between evenness and red deer–reindeer dominance across the Le Flageolet I ungulate assemblages   112

7.1. Middle Paleolithic fractured tortoise carapace from Hayonim Cave, Israel   115
7.2. Sources of archaeological sites in the Mediterranean Basin   117
7.3. Trend in the relative contribution of slow prey types to Pleistocene small game assemblages   117
7.4. Size reduction trend in spur-thighed tortoises from the Nahal Meged, Israel   118
7.5. Size reduction trend in limpets from Riparo Mochi, Italy   119
7.6. Comparison of hunting tolerance thresholds for tortoise, partridge, and hare populations in high and low growth models   119

8.1. Return rate experienced within a foraging area relative to that expected if the foragers moved to a new area   123

8.2. Locations of the sites used in the text to calculate potential migration rates 126
8.3. Location of reliably dated Clovis sites 127

9.1. Relationship between overall return rate and search time for a mammoth only diet 137
9.2. Frequency of radiocarbon dates for Alaskan *Equus caballus/ferrus* 140
9.3. A diet breadth model for early Paleoindians 142
9.4. Large game index, conservative estimate 145
9.5. Large game index, inclusive estimate 146

10.1. Map of Study Area 161
10.2. Correlation of prehistoric sequence of south-central Arizona with Ventana Cave 162
10.3. Change in the selective efficiency index over time at Ventana Cave 162
10.4. Schematic representation of hypothetical regional socioeconomic changes in south-central Arizona 165
10.5. Patterns of economic zonation resulting from resource depression 165
10.6. Model of relationship between travel time and prey selection 166
10.7. Proportionate representation of artiodactyls at Ventana Cave and various sites from the Picacho Reservoir and Salt-Gila study areas 169

11.1. Map of the Great Salt Lake Desert and surrounding area 173
11.2. Open pinyon cone with nuts, enlarged cone scale, and hulled nutmeat 174
11.3. Caprielle and Aaron Barlow in pickleweed patch near Grantsville, Utah. 175
11.4. Dried pickleweed plant, enlarged branchlet with flower spikes, and papery bracts encasing the pickleweed 175
11.5. Changes in the utility of pinyon with field processing time 176
11.6. Changes in the utility of pickleweed with field processing time 177
11.7. Pinyon transport decay functions 185
11.8. Pickleweed transport decay functions 185

12.1. Map of the San Francisco Bay area and the location of the Emeryville Shellmound 195
12.2. Demolition of the Emeryville Shellmound by steam shovel, 1924. 196
12.3. The distribution of the Elk Index by stratum at the Emeryville Shellmound 197
12.4. The distribution of the Sturgeon Index by stratum 197
12.5. The distribution of the Goose Index by stratum 198
12.6. The distribution of the Deer Index by stratum 198
12.7. Mean utility of artiodactyl skeletal parts across the Emeryville strata 199
12.8. Dentary widths by stratum for the Emeryville Shellmound sturgeon 200
12.9. The distribution of the Cormorant Index by stratum 201
12.10. Changes in the proportional contribution of adult cormorants across strata 201
12.11. The distribution of the Otter Index by stratum 202
12.12. Changes in the proportional contribution of adult sea otters across strata 202
12.13. The distribution of the Artiodactyl Index by stratum 203
12.14. Changes in the proportional contribution of adult artiodactyls across strata 203

13.1. The central place forager prey choice model 211
13.2. Hypothetical processing function takes the shape of a diminishing returns curve 213
13.3. The effect of search and transport time on maximum delivery rate and optimal processing time 214
13.4. Three prey types that provide different average delivery rates 215
13.5. Map of the Mimbres Valley and surrounding region 218

14.1. Map of the Carson Desert region 229
14.2. Normal year annual air-dry production of biomass in Carson Desert 232
14.3. Normal year productivity of Indian ricegrass in the Carson Desert 233
14.4. Bighorn habitat quality rankings in the Carson Desert 235
14.5. Wet regime habitat productivity of the early Late Holocene 235
14.6. Forage productivity of top five bighorn sheep habitats in the Carson Desert 236
14.7. Highest habitat return rates by season for ethnohistoric men and women in the Carson Desert 241
14.8. Highest habitat return rates by season for early Late Holocene men and women in the Carson Desert 241
14.9. Early Late Holocene women's autumn foraging returns in Carson Desert 242
14.10. Early Late Holocene men's autumn foraging returns in Carson Desert 243
14.11. Two ethnographic Toedokado central place base camps 245
14.12. Central place foraging return rate for Mount Lincoln and Stillwater Marsh camps at different proportional contribution of large game to net caloric intake of camp 247
14.13. Comparison of projectile point and ground stone tool frequencies in three regions of the Carson Desert 249

15.1. Location of the Shag River Mouth site in New Zealand   258
15.2. Mean moa utility across layers   261
15.3. Relative frequency of high utility elements (femora, tibiotarsi, fibulae, cervical vertebrae) across layers   262
15.4. Relative frequency of low utility elements (phalanges, tarsometatarsi, caudal vertebrae) across layers   262
15.5. Relative frequency of moa cervical vertebrae and tracheal rings   263
15.6. Mean moa utility for high density elements across layers   264
15.7. Percentage of unbroken moa phalanges across layers   265
15.8. Moa bone fragmentation across layers   265
15.9. Fragmentation of moa tibiotarsi across layers   266
15.10. Identifiability of moa leg elements across layers   267

16.1. Schematic illustration of tool blank   277
16.2. Examples of mass and utility calculations for tool blanks of different sizes   278
16.3. Examples of mass and utility calculations for tool blanks of different thicknesses   278
16.4. Relation between potential utility and mass for tools and cores   279
16.5. Utility/mass ratios for retouched tools plotted against length, expressed as a multiple of minimum usable size   279
16.6. Relationship between potential utility and mass for toolkits consisting of multiple small tools, and for corcs with different values of the coefficient of production efficiency   281
16.7. Utility/mass ratios for retouched tools, plotted against length, expressed as a multiple of minimum usable size   281
16.8. Hypothetical relations between artifact sizes and functional effectiveness   284

17.1. Map of locations in the text   289
17.2. Paleoarchaic long stemmed points and crescent   290
17.3. Biface reduction sequence represented in Great Basin Paleoarchaic sites   290
17.4. Central place foraging model describing the relationship between field processing and resource utility   292
17.5. Central place foraging model in which the utility function is continuous and differentiable   293
17.6. Central place foraging model describing biface reduction revised from the continuous model depicted in Fig. 17.5   294
17.7. Profile summary graphs of biface stages represented in the Little Smoky Quarry and Cowboy Rest Creek Quarry assemblages   299
17.8. Histograms of Johnson Thinning Index values for bifaces in the Little Smoky Quarry and Cowboy Rest Creek Quarry assemblages   300
17.9. Profile summary graphs of biface stages represented in the Knudtsen and Cowboy Rest Creek Quarry assemblages   301
17.10. Histograms of Johnson Thinning Index values for bifaces in the Knudtsen and Cowboy Rest Creek Quarry assemblages   302
17.11. Profile summary graphs of biface stages represented in the Little Smoky Quarry and Limestone Peak assemblages   303
17.12. Histograms of Johnson Thinning Index values for bifaces in the Little Smoky Quarry and Limestone Peak assemblages   303

18.1. Map of Nevada showing the location of the Little Boulder Basin Area   312
18.2. Temporal distribution of calibrated radiocarbon dated firehearths in the LBBA   314
18.3. Distribution of calibrated radiocarbon dates associated with brown- and grayware vessels in the LBBA   316

19.1. Organic, stone, and inset organic points compared   322
19.2. Asian microblade cores   324
19.3. Simplified model of core geometry   325
19.4. Step plot of core radius and log microblade width   326
19.5. Isovalue map for risk-sensitive Z-score model   326
19.6. Choice model for microblade production using wedge-shaped cores and boat-shaped cores   327

20.1. Graphic representation of the z score model   337
20.2. Predicted behavior with respect to novel resources of varying profitability assuming the goal of efficiency maximization   338
20.3. Predicted behavior with respect to novel resources of varying profitability assuming the goal of risk minimization   338
20.4. The opportunity cost model   339
20.5. Hypothetical change in opportunity cost and diminishing returns for maize   342

21.1. The study area, showing Bedul fields and sources of grain samples   350

22.1. Model of logistic population growth   359
22.2. Effect of herd growth on return rates to husbanding   361
22.3. Allometric relationships of body mass for mammals   362
22.4. Annual harvest as a function of body size for both the husbanding strategy and the hunting strategy   365

23.1. Map of archaeological sites in the Fremont culture area   373
23.2. Fremont material culture   374
23.3. Caloric return rates for maize agriculture with hand tools in Latin America   380
23.4. The relationship between time spent in maize fields and average annual maize harvests   381
23.5. Comparison of caloric return rates for foraging and farming in the Great Basin and American southwest   382

24.1. Good year economics   396
24.2. In good years, increasing production variance increases the relative value of pooling   398
24.3. In bad years, increasing production variance increases the relative value of not pooling   399
24.4. Smoothed annual estimates of total maize productivity in kilograms   400

25.1. Normal probability plot of the terminal monument date residuals   416
25.2. Conditioning plots of terminal date residuals   417
25.3. Contour plot of expected terminal monument dates fitted by loess   418
25.4. Contour plot of mean annual rainfall in the Maya Lowlands   419
25.5. Scatterplots of fitted terminal dates and actual terminal dates as a function of mean annual rainfall   420
25.6. Variograms of the residuals from expected terminal monument dates   422

Tables

2.1. Relative reproductive values of Old World primates   21

3.1. Post-encounter return rates from wild tubers   36

4.1. Life history parameters of human hunter-gatherers and chimpanzees   50
4.2. Production of energy by men and women in foraging societies   54
4.3. Diet of modern hunter-gatherers and chimpanzees   59

6.1. Summary of the contents and chronology of Le Flageolet I   105
6.2. The Le Flageolet I ungulates: NISP by taxon   106
6.3. The Le Flageolet I carnivores: NISP by taxon   106
6.4. Regression coefficients between ungulate NISP and (Log10) ungulate NTAXA   106
6.5. Fragmentation index and adjusted residual values for shaft specimens for the ungulate assemblages   109
6.6. NISP of major skeletal elements by high- and low-slope assemblages   109
6.7. Red deer–reindeer dominance and evenness values for the ungulate assemblages   110

7.1. Relative abundances of small prey animals in the Paleolithic faunal series from Italy   116
7.2. Relative abundances of small prey animals in the Paleolithic faunal series from Israel   116

8.1. Potential causes of migration   124
8.2. Characteristics of transient explorers and estate settlers   124
8.3. Rates of population movement, north-eastern Asia versus the western hemisphere for different possible migration dates   125

9.1. Weight of various body portions from four adult male *Loxodonta africana*   134
9.2. kcals/1000 grams of edible portion for selected prey species   135
9.3. Handling time thresholds   135
9.4. Return rates for various prey species   136
9.5. Overall return rates for mammoth hunting   138
9.6. Population density and encounter rates   139
9.7. Optimal diet breadth   141

10.1. Distribution of identified mammals and reptiles by stratigraphic level, Ventana Cave   163
10.2. Frequency of identified mammals and reptiles at the Arroyo and the Gate sites   168

11.1. Results of nutritional analysis of pinyon nutmeats   178
11.2. Pinyon processing stages and proportion of nutmeats to inedible components   179
11.3. Results of pinyon processing   179
11.4. Results of pinyon processing experiments for 3 and 15 kg loads   179
11.5. Results of nutritional analysis of pickleweed seed   180
11.6. Pickleweed processing stages and proportion of seed to inedible components   181
11.7. Results of pickleweed processing   181
11.8. Results of pickleweed processing experiments for 3 and 15 kg loads   182
11.9. Estimated energetic return rates for transporting 3 and 15 kg loads of pinyon   183
11.10. Results of pickleweed processing experiments recalculated for samples winnowed in 2 l batches   186
11.11. Values relating β to density for pickleweed samples   186

13.1. Definitions of the variables included in the central place forager prey choice model   210
13.2. Mimbres-Mogollon culture historical time periods   219
13.3. Numbers of identified specimens of artiodactyl and leporid taxa   219
13.4. Total numbers of identified artiodactyl and leporid specimens   220
13.5. Mean artiodactyl body part utility   221
13.6. Mean artiodactyl scan site volume density   221

14.1. Ethnographic plant and animal foods monitored in the Carson Desert model   234
14.2. Resource classes ranked according to caloric returns and classified as men's and women's prey sets   236

14.3. Simulated ethnohistoric women's overall foraging returns and ranks by habitat and season   237
14.4. Simulated ethnohistoric men's overall foraging returns and ranks by habitat and season   238
14.5. Simulated early Late Holocene women's overall foraging returns and ranks by habitat and season   239
14.6. Simulated early Late Holocene men's overall foraging returns and ranks by habitat and season   240
14.7. Spearman's rank correlation coefficient test results for men's and women's habitat rankings by season   242
14.8. Number of seasonal habitats falling into ordinal ranges of overall caloric return in early Late Holocene and ethnohistoric simulations   244
14.9. Constants for calculating travel and transport costs over variable grades   246
14.10. Distance, slope, and transport cost estimates for calculating central place foraging returns between Stillwater Marsh and Mount Lincoln   248
14.11. Weight, time, and load size estimates required to satisfy the net caloric requirement for 25 kg loads of dried sheep and large bulrush seeds   248
14.12. Counts of Late Holocene projectile points found in three regions of the Carson Desert   250

15.1. Summary of expectations for carcass transport and skeletal element breakage given declining foraging efficiency   261
15.2. Results of Spearman's rank correlation analysis between bone density and % survivorship for moa elements   263
15.3. The relative abundance of one-piece and composite fishhooks and lures from the Shag Mouth site   267

16.1. Equations used to calculate transport cost and potential utility   277

17.1. Total number of dacite bifaces analyzed at each of the sites   297
17.2. Biface stage determinants   298
17.3. Statistical tests between different collection samples at the two quarries and at the Knudtsen site   299
17.4. Biface stages represented in the Little Smoky Quarry and Cowboy Rest Creek Quarry assemblages   300
17.5. Johnson Thinning Index statistics for Little Smoky Quarry and Cowboy Rest Creek Quarry   301
17.6. Statistical tests between Little Smoky and Cowboy Rest Creek Quarry assemblages   301
17.7. Biface stages represented in all four assemblages   302
17.8. Johnson Thinning Index statistics for all four assemblages   302
17.9. Statistical tests between Little Smoky Quarry and Cowboy Rest Creek Quarry and their associated residential/activity sites   303

18.1. Handling times and return rates of several Australian and Great Basin prey   311
18.2. Dimensions of large and small firehearths   313
18.3. Return rate categories for LBBA prey   313
18.4. Hearth contents in the LBBA   313
18.5. Changes in groundstone technology in the LBBA   315
18.6. Investment in LBBA pottery   315

19.1. Performance characteristics of organic and flaked stone points   321
19.2. Ethnographic situational use of organic and flaked stone points   321
19.3. Experimental microblade and biface production based on data in Flenniken   322
19.4. Width and thickness of production and maintenance microblades from Ningxia, China   324

20.1. Nutritional data for selected fleshy fruits   340

21.1. Grain and straw/chaff yields for wheat and barley at Petra   351
21.2. Return rates for Bedul males hand harvesting wheat and barley near Petra   352
21.3. Return rates for alternative methods of wheat procurement   353

22.1. Symbols used in text   359
22.2. The data to calculate the maximum sustainable yield   360
22.3. Data used to iteratively calculate r_{max} from Cole's equation   361

23.1. Annual harvest yields and average field investments associated with maize farming in Latin America   377
23.2. Harvest yield data and caloric return rates for farming maize in Peru   378
23.3. Caloric return rates for farming maize in Latin America using hand tools   379

24.1. Periods with differential advantages for pooling   400
24.2. Initial test of the Pooling model   401
24.3. Performance of competing models for development of cooperative behaviors in the archaeological record   402
24.4. Differences between predictions of risk-sensitivity model and archaeological data   403

25.1. Terminal monument dates from 69 sites used in this analysis   414

Foreword

James F. O'Connell

This volume represents an important benchmark in the application of Darwin's theory of natural selection to problems in human prehistory and evolution. It includes some 24 previously published papers plus a series of original essays by the editors that introduce both the volume as a whole and the several sections into which the reprinted papers have been grouped. The theoretical framework that unifies the work is evolutionary ecology, defined by the editors as "the study of adaptive design in behavior, morphology and life history." Most of the papers are especially concerned with the subfield of evolutionary ecology known as behavioral ecology, which as the name implies is the study of adaptive behavior in relation to social and environmental circumstances.

Archaeologists have a long-standing interest in the ecological determinants of human behavior. Those raised in the American anthropological tradition often trace the roots of that interest to Julian Steward's provocative work on Native American prehistory and ethnography in the Southwest and Great Basin, published in the 1930s. The "cultural ecology" of the 1950–70s was grounded on this work. It also contributed to the "processual" approach that has dominated much of professional archaeology, at least in the English-speaking world, for the last half-century.

Contributions to this volume are certainly part of that broad tradition. What distinguishes them (and others like them in the recent literature) is their reliance on a series of formal optimality and game theoretic models developed first in economics but later elaborated in biology from the late 1950s onward. Archaeological applications of these models date from the early 1970s, but were hampered initially by the models' rigorous quantitative requirements. Background research, both ethnographic and experimental, helped meet these requirements, and a wave of archaeological studies soon followed. Early work focused on aspects of prehistoric foraging, mainly in Holocene North America, but the range of applications later broadened to include such topics as food production, technology, mating competition, the evolution of nuclear families, the development of social hierarchies, and the evolution of human life histories. Examples of these applications are now reported from all continents and across the entire time span represented by the archaeological record.

Studies in this volume reflect that florescence, and provide some of the best illustrations of the insights available from the general approach. As the editors note, there is ample room for further development. One hopes that this collection, usefully compiled and thoughtfully introduced, will be widely read, closely analyzed, and carefully critiqued. It should provide both the stimulus and guidance essential to ensure that further development.

Evolutionary Ecology and Archaeology

An Introduction

Michael D. Cannon and Jack M. Broughton

It may be asked how far I extend the doctrine of the modification of species.
The question is difficult to answer... But some arguments of the greatest weight extend very far...
In the future I see open fields for far more important researches...
Much light will be thrown on the origin of man and his history.

— CHARLES DARWIN, 1859 —

As we celebrate the sesquicentennial of Darwin's *The Origin of Species*, there is little doubt that Darwinian theory has indeed thrown much light on the origins and history of humankind. Today, the value of Darwinian theory for this purpose is fully apparent to a large and growing number of archaeologists, and at least three self-defined Darwinian approaches are highly active in the discipline, each of which anticipates ever-expanding fields of research (e.g., Bettinger et al. 1996; Bird and O'Connell 2006; O'Brien and Lyman 2002; also see the recent overview by Bentley et al. 2007). This volume is a compilation of previously published archaeological and paleoanthropological research papers that derive from one particularly productive Darwinian approach: evolutionary ecology (hereafter, EE). The purpose of the volume is to provide an overview of the ways in which EE has been used in archaeology and of the substantive advances in our understanding of human evolution and prehistory that have occurred as a result. As such, the volume should illustrate the ways in which EE can contribute to an evolutionary understanding of the human past, as well as the variety of topics to which EE can be applied. In addition, by providing archaeologists and others interested in the evolutionary study of humans with a more complete view of the structure, history, and breadth of EE in archaeology, our hope is that this volume will foster sustained development of the approach. As Darwin did nearly 150 years ago, we continue to see wide open fields ahead of us.

The papers in this volume are grouped by substantive topic and/or time period: Early Hominin Evolution and Behavior, Pleistocene Foragers and Colonists, Post-Glacial Adaptations, Food Production Strategies, and Cooperation and Competition in Complex Societies. This is done both to highlight the broad range of issues in human evolution and prehistory that have been productively studied using the EE approach and to facilitate the use of this volume in courses on world prehistory. This introductory chapter sets the stage for the substantive works that follow by discussing the structure of EE models, by exploring how these models relate to evolutionary mechanisms, and by examining issues that arise in the application of EE models to the archaeological record. We also address common criticisms and misunderstandings of the use of EE in archaeology, and we describe our rationale for selecting the papers that are included in the volume. Each section of the volume begins with an introduction that illustrates how the collective body of EE-based research into particular topics has advanced our understanding of important substantive and theoretical issues.

WHAT IS EVOLUTIONARY ECOLOGY, AND WHAT DOES IT OFFER TO ARCHAEOLOGY?

Evolutionary ecology can be defined as the application of natural selection theory to the study of adaptive design in behavior, morphology, and life history (Winterhalder and Smith 1992; Bird and O'Connell 2006). The term behavioral ecology (or HBE for human behavioral ecology) refers to the sub-area of EE that is concerned with the adaptedness of behavior, and it is this sub-area that is most directly applicable to the study of the archaeological record, though paleoanthropological applications have also addressed the evolution of life history, anatomy, and physiology. In practice, archaeological applications of EE are typically concerned with understanding how socio-ecological conditions structured the past behaviors that archaeologists are interested in studying, or with understanding variability in the archaeological record in relation to variability in the natural and social environment.

Of course, studying past human behavior as an adaptation to environmental conditions has been an explicit goal of archaeologists ever since the emergence of processual archaeology in the early 1960s (e.g., Binford 1962; see also Dunnell 1986), and anthropological interest in related issues of "cultural ecology" began well before this (e.g., Steward 1938). Thus, mere attention to the influence of the natural and social environment on culture or behavior cannot be the distinguishing characteristic of the EE approach. Rather, the distinguishing characteristic of this approach, which is evidenced in some way by all of the papers in this volume, is that it makes use of formal models—usually optimization or game theory models—that are designed to explore the fitness-related costs and benefits of behavioral alternatives in specific socio-ecological contexts. The EE approach thus builds on traditional anthropological and archaeological interest in human-environment interrelationships (for further detail, see Grayson and Cannon 1999; Winterhalder and Smith 2000), but it allows such issues to be studied in a way that is much more theoretically sound and methodologically rigorous than is typically the case in archaeological explorations of them. Theoretical soundness derives from EE's concern with explicitly specifying the parameters that are hypothesized to affect the costs and benefits of behavioral alternatives, and methodological rigor is enabled by the identification of variables that can be measured in order to test propositions derived from the theory.

Structure and Use of Evolutionary Ecology Models

The structure of the theoretical models that characterize EE has been described, and their use justified, by many other authors (e.g., Foley 1985; Parker and Maynard Smith 1990; Smith 1991; Smith and Winterhalder 1992; Stephens and Krebs 1986; Winterhalder 2002; Winterhalder and Smith 1992); we therefore give only a brief synopsis of such issues here. Simply put, EE models consist of a set of assumptions about factors that are relevant to the evolution or expression of phenotypic variants, and in HBE the focus is on the behavioral component of the human phenotype. The assumptions of an EE model are best treated as hypotheses to be tested by comparing model predictions either to direct observations of human behavior, in the case of ethnology, or to archaeological data. The purpose of such a test is *not* to determine whether human behavior is "optimal" in some sense; rather, the useful knowledge that a test produces lies in the identification of those features of the natural and social environment that have played important roles in the evolution and expression of behavior. If the predictions of a model are empirically met in a test situation, this suggests that the relevant socio-ecological factors have been appropriately captured in the model's assumptions. If predictions are not met, the process is still scientifically useful because it has shown that answers must lie elsewhere (e.g., Gremillion 2002).

The assumptions of EE models are usually categorized as follows. There is first an assumption about the *decision*, or set of behavioral alternatives, that faces an actor. This decision assumption generally incorporates some tradeoff among alternatives that each confers its own costs and benefits, and EE models can thus be thought of as analyses of the evolutionary tradeoffs that all organisms experience. For example, the decision modeled in the basic prey model of foraging theory (e.g., Stephens and Krebs 1986)—perhaps the best-known of all EE models—is whether a forager should pursue a resource upon encounter, thereby gaining the benefit of the energy provided by that resource at a cost of time spent "handling" it, or whether the forager should pass that resource by and continue searching for others, thereby incurring a cost of additional search time, which might be rewarded by the benefit of encountering a resource more profitable than the one that was passed over. (If this decision is applied to all potential resources in a forager's environment, then the model predicts the entire set of resources that the forager should pursue, and thus this model is also commonly known as the diet breadth model.) The decision assumption of a model is a major determinant of the set of research questions to which that model is applicable. For example, if a research question involves the amount of time that is spent processing resources (e.g., Metcalfe and Barlow 1992), rather than the kinds of resources that are pursued, then the prey model is obviously not the one to use.

Given a specified set of behavioral alternatives, the model then evaluates—usually through mathematical or graphical means—which of the possible alternatives is the best under a certain set of conditions, and the best possible alternative is the one that the model predicts an actor should adopt. The concept of "best possible alternative" has no practical meaning, however, without some sort of yardstick by which to measure whether one alternative is better than another. Such a yardstick is provided by the second category of model assumptions: *currency* and *strategic goal*. The currency assumed by a model is some quantity that is hypothesized to be related to fitness (we discuss the relationship between model currencies and fitness in greater detail below), and it provides the units in which the costs and benefits of behavioral alternatives are measured. For example, the currency used in many foraging theory models, including the prey model, is foraging efficiency, or net caloric energy gained from foraging per unit time spent foraging. The prey model thus evaluates alternative diets—or alternative sets of resources—in terms of the efficiency that they provide to a forager in a given environment, which depends in turn on the rates at which resource types are encountered and on the post-encounter return rates (calories gained per unit handling time) of those resource types.

If the currency assumption provides the units of measure for the yardstick, then the goal assumption determines which end of the yardstick is "up." Continuing with the example of the prey model, under some set of conditions the decision to pursue a certain low-return resource type (call it resource X) might result in lower overall foraging efficiency than would the decision to pass resource X by in order to continue searching for resources with higher post-encounter return rates. The prey model assumes a strategic goal of *maximizing* foraging efficiency, and the best possible alternative in the context of this model is thus to ignore resource X because doing so results in higher foraging efficiency. Note that the goal that is assumed is just as important as the currency: a different model might assume, for example, a goal of minimizing variability in subsistence returns (i.e., minimizing risk), and if resource X can be secured dependably, then the best possible alternative in the context of that model could be to include the low-return but reliable resource X in the diet, in contrast to the prediction of the prey model (see Gremillion 1996, reprinted here, for further discussion of risk and efficiency in foraging theory models).

Note also that the incorporation of an assumption about a goal does not necessarily amount to assuming that behavior is "goal-directed", nor does it imply that archaeological explanations based on EE models must rely on "intentionality" (e.g., Lyman and O'Brien 1998), an issue to which we return below. Rather, as is the case with all of the assumptions of EE models, the goal assumption merely provides an analytical tool that can be used to develop and test hypotheses about the evolution and expression of behavioral variants. For example, an important archaeological research question might involve learning whether resource choice in some prehistoric community had the effect of maximizing foraging efficiency or minimizing subsistence risk. The most productive approach to tackling this question would be to develop two distinguishable sets of predictions about resource choice—one set based on a model that assumes a goal of maximizing foraging efficiency and one set based on a model that assumes a goal of minimizing risk—and to then determine which set of predictions were best met by the archaeological record of that community (e.g., by considering the types of resources preserved in floral and faunal assemblages). In this way, the factors relevant to the evolution and expression of subsistence behavior might usefully be teased out.

The final category of assumptions made by EE models are *constraint* assumptions, which pertain to such things as the features of an actor's natural and social environment and the behavioral, cognitive, and technological capabilities of the actor (see, e.g., Smith and Winterhalder 1992:56–59 for further detail). For example, the prey model makes assumptions about the manner in which resources are encountered (resources must be encountered as if they were distributed in a fine-grained, or homogenous, environment) and about the capabilities of a forager (while handling one resource unit, a forager cannot simultaneously search for others). Like the decision assumption, the constraint assumptions of a model are important in determining the situations in which that model can be productively applied. If the environment of an area, for example, has features that violate constraint assumptions of a particular model, then any test of hypotheses derived from that model conducted in that region is suspect (Haccou and van der Steen 1992; Stephens and Krebs 1986), much in the same way that the results of a statistical test are suspect when that test is applied to data that violate its assumptions (assumptions such as that variables are normally distributed, etc.). Thus, it is crucial to consider whether the assumptions of an EE model are met by the situation to which it is applied. If there is reason to think that assumptions are violated, then it is necessary to evaluate what the consequences of the violation might be, and it may also be necessary to apply—or even newly develop—a model with assumptions that are more appropriate to the situation.

It should be clear from this discussion that when EE models are used—no matter whether they are relatively simple optimization models such as the prey model or more complex models such as those used in some of the papers in this volume—substantial attention must be given to the purposes for which they were designed and to the assumptions that they incorporate. Indeed, proponents of the use of these models have long warned against the dangers of applying them in cookbook fashion (e.g., Smith 1983:640).

When care is taken to use it appropriately, however, the EE approach has much to offer. The papers in this volume attest to the strengths of this approach, and we believe that these can be summarized as follows (after Broughton and O'Connell 1999). First, the EE approach is *comprehensive* in that it is applicable in principle to any fitness-related aspect of the phenotype, including behavior. It is also *integrative*, allowing variation in one aspect of the phenotype to be linked to variation in others: for example, variation in farming strategies to variation in settlement patterns (e.g., Barlow 2002, reprinted here), or changes in life history and morphology to changes in subsistence behavior (e.g., O'Connell et al. 1999, reprinted here). Finally, and most importantly, the EE approach produces hypotheses that are *testable*: it results in explicit propositions about factors relevant to the evolution and expression of behavior, life history, or morphology that can be falsified empirically.

In sum, the theoretical models of EE provide a robust means of developing and testing hypotheses about socio-ecological factors that have played important roles in shaping human morphology, life history, and behavior from prehistory into the present. We believe that this approach can greatly advance archaeology's longstanding goal of understanding human adaptations to natural and social environments in a scientific manner by providing a firmer theoretical and methodological footing, and the papers included in this volume provide strong evidence in support of this position.

Evolutionary Ecology Models and Evolutionary Processes

We have defined EE as an approach that uses formal models to understand the socio-ecological factors that are relevant to the evolution or expression of phenotypic variants, but we have not yet discussed the relationship between EE models and evolutionary processes. This is because EE models of the sort described above do not in and of themselves incorporate such processes. They are, in the words of Winterhalder and Smith (2000:52–54;

see also Winterhalder 1997), *models of circumstance*, which only explore how "socioenvironmental factors shape the costs and benefits associated with potential alternatives." A complete explanation, Winterhalder and Smith note, must incorporate not only such models of circumstance, but also *models of mechanism*, which specify the relationship between phenotypic alternatives and evolutionary processes such as natural selection.

The issue of mechanism can be dealt with, and often is, by invoking the "phenotypic gambit" (after Grafen 1984; see overview in Smith and Winterhalder 1992:33–34), or the premise that it is not necessary to understand the details of the processes by which behavioral traits are inherited in order to understand how those traits might evolve through natural selection. Researchers who explicitly adopt this approach usually justify it by noting that it is often not feasible to understand the complex genetics and environmental effects that lead to the phenotypic expression of behavior, and by pointing out that high-fitness behavioral variants should be fixed by selection regardless of the details of inheritance. Given this latter point (and assuming that it is correct: see Grafen 1984; Owens 2006), all that is necessary to link EE models to the process of natural selection is to have a strong basis for thinking that the behavioral alternatives incorporated into those models have different fitness values.

Thus, when applied to behavior, EE models have generally been used in archaeology without making specific assumptions about the mode of inheritance underlying the behaviors of interest. However, these models can be applied just as productively to behaviors that are inherited culturally, or learned socially, as to behaviors that are inherited genetically. Indeed, although Grafen's (1984) development of the phenotypic gambit discusses animal behavior as if it were strongly determined by genetics, we note that Maynard Smith (1978:35), in an earlier paper that lays the groundwork for the phenotypic gambit approach, explicitly recognizes that animal behavior can be inherited culturally as well as genetically (and see Richerson and Boyd 2004 and Avital and Jablonka 2000 for more recent discussions of cultural inheritance in the human and nonhuman realms, respectively). The phenotypic gambit makes it possible to use EE models to study the operation of natural selection in specific contexts whenever the behavior (or other phenotypic trait) of interest is inherited, in whatever manner.

In addition, some archaeologists and anthropologists using EE models have taken the phenotypic gambit a step further by invoking an assumption of "extreme phenotypic plasticity" (Smith and Winterhalder 1992:33; see also Boone and Smith 1998), which amounts to assuming that humans adopt fitness-maximizing behaviors through facultative adjustment (perhaps in conjunction with social learning, though the role that this might play in facultative adjustment is not made explicit by authors who invoke this assumption, despite their recognition of the general importance of cultural transmission; e.g., Boone and Smith 1998:S143). This assumption can be justified by pointing out that there are good reasons to think that natural selection, over the course of evolutionary history, has given humans and other organisms cognitive machinery that enables them to respond facultatively to environmental conditions in a manner that enhances fitness (e.g., Pinker 2002). Thus, when such facultative adjustment is involved—and many of the past behaviors in which archaeologists are interested certainly were the result of such facultative adjustment—the object of study is the product of natural selection, rather than the operation of natural selection, as can be the case when inheritance is involved.

The papers in this volume illustrate the range of proximate mechanisms underlying the expression of phenotypic variants that may be involved in situations to which EE models are applied. The paleoanthropological applications that begin the volume address aspects of the early hominin phenotype—such as sexual dimorphism in canine size, bipedality, and life history characteristics—that were certainly under relatively direct genetic control. Moreover, these applications argue that evolutionary changes in such aspects of the phenotype were related to changes in behavior—particularly in the realm of subsistence and mating—that may or may not have been under such direct genetic control. In a different vein, some of the papers in this volume address patterns of persistence and variability in technology, and archaeologists have long-recognized that such patterns may be largely a product of cultural transmission. Finally, many of the applications reprinted here address prehistoric subsistence behavior, which can reasonably be assumed to have been at least in part an expression of facultative adjustment, as is the case, for example, when a forager decides whether to pursue a resource based on cues to the likelihood of encountering alternative resources during a foraging trip (see Boone and Smith 1998: S145–146 for further detail). However, it is also likely that cultural transmission has played some role in subsistence behavior in many situations, as is the case, for example, when elders provide children instruction in farming, food processing, or hunting and gathering activities.

Thus, EE models have been usefully applied to questions about human evolution and prehistory in situations in which a variety of proximate mechanisms underlying phenotypic variability are likely involved. And, as we have noted, EE models can be applied to such questions without making specific assumptions about the nature of those proximate mechanisms. In this way, the phenotypic gambit has proven very successful at enabling archaeological applications of EE to produce important advances in our understanding of the human past.

For the future, however, it may be useful for archaeologists using EE to begin to consider in greater detail both the mechanisms of inheritance that underlie the evolution of phenotypic traits and the facultative mechanisms that often underlie the expression of such traits (see also Winterhalder and Smith 2000: 67; for a discussion of related issues, see Bamforth 2002). There are two reasons why it should prove productive to unpack the black box of the phenotypic gambit in this manner. First, as the recent review by Owens (2006) makes clear, behavioral ecologists in biological disciplines are beginning to explore the conditions under which the phenotypic gambit assumption is safe, spurred on by the recognition that, in at least some cases, the details of the mechanisms of inheritance can have a substantial effect on evolutionary outcomes. Applications of EE in archaeology might benefit from similar consideration of the genetic, cultural and facultative mechanisms that underlie phenotypic variability because the mechanism involved in a particular archaeological case might likewise have implications for outcomes (e.g., Richerson and Boyd 2004; Winterhalder 1997). Second, greater consideration of such issues might serve to strengthen the links between EE and other Darwinian approaches, which have traditionally placed more emphasis than EE on the role that cultural transmission plays in the evolution and expression of the human phenotype (e.g., Eerkens and Lipo 2005). Because practitioners of all Darwinian approaches in archaeology share the goal of understanding the human past through the application of evolutionary logic and scientific methods, such integration of approaches can only strengthen the general Darwinian enterprise (see also Bentley et al. 2007; Neff 2000).

Applying Evolutionary Ecology to the Archaeological Record

Archaeologists have long noted that applications of EE to the archaeological record are somewhat more difficult than ethnographic applications (e.g., Bettinger 1983). Testing the predictions of an EE model requires, among other things, measuring the costs and benefits of phenotypic alternatives in a specific empirical context, and this is easier done ethnographically than archaeologically. In an ethnographic application of the prey model, for example, it is fairly straightforward to measure the costs and benefits of behaviors in the currency specified by the model—foraging efficiency—by directly observing such things as encounter rates, handling times, and caloric returns for the resources that foragers capture or harvest (e.g., Smith 1991). In archaeological applications, on the other hand, it is not possible to measure costs and benefits through direct analysis of behavior in this way, and other means of operationalizing EE models are required. Fortunately, however, archaeologists have developed creative methods for doing so, and thus productive archaeological applications, while more difficult, are not unattainable.

This can be illustrated by applications of foraging theory, the sub-area of EE that has to date been used the most in archaeology. Making foraging theory models archaeologically operational requires, in particular, deriving estimates of, or proxy measures for, the post-encounter return rates of resources, and archaeologists have taken two approaches to doing this (see overview in Grayson and Cannon 1999). One approach has been to conduct experiments and/or to use ethnographic observations to derive estimates of return rates (e.g., Simms 1987). The other, initiated by Bayham (1979) and developed further by Broughton (e.g., 1994), has been to use body size as a proxy measure of return rate for vertebrate prey, based on a well-documented empirical correlation between these two variables (e.g., Broughton 1999; Ugan 2005).

The first of these approaches, as Bettinger (e.g., 1993) has pointed out, is problematic in that modern estimates of resource return rates bear an unknown relationship to the return rates that foragers obtained in the past due to the importance of such things as technology and skill in determining them (see also Grayson and Cannon 1999). The second approach avoids this problem, but it is limited in its applicability to animal resources and it is potentially subject to other problems. These include deviations from the general body size-return rate relationship caused by the influence of such factors as technology, capture method, or prey behavioral characteristics on return rates (e.g., Madsen and Schmitt 1998; Stiner and Munro 2002). These problems can be overcome, however, through detailed consideration, on a case-by-case basis, of the factors that might hinder the use of body size as a proxy for return rate (e.g., Cannon 2001; see also Ugan 2005). Problems associated with the first approach can likewise be overcome by treating return rate estimates as ranges rather than as point estimates (e.g., Barlow 2002; Byers and Ugan 2005, both reprinted here). Papers reprinted in this volume use both approaches to resolving the return rate issue and do so in a robust manner.

There are several other sets of issues that must also be addressed when applying EE models to the archaeological record. As noted above, one of these is determining whether the assumptions of a particular model are met in a specific application. For instance, the prey model makes a stringent assumption—known as the *fine-grained search assumption*—that relates to the manner in which a forager encounters prey, which is highly dependent on the spatial distribution of those prey. Specifically, the prey model assumes that a forager searches for all potential prey types simultaneously and that the chance of encountering any prey type is independent of previous encounters with it or any other; thus, the prey model is appropriately applied only to sets of resources that were likely derived from the same spatial patch or hunt type (e.g., Smith 1991). This assumption is frequently violated, however, which can lead to premature rejections of model predictions (or Type I errors; see, for example, McGuire and Hildebrandt 2005). Effective means of dealing with the constraints imposed by this assumption are described in a number of archaeological applications (e.g., Broughton 1999, Nagaoka 2001, Bovy 2005), one of which is reprinted here (Broughton 2002). In addition, a model of prey choice that does not require this assumption has also been developed (Cannon 2003, reprinted here), and it is suitable for use in archaeological situations in which it is not possible to determine whether the resources that foragers harvested came from a single patch or separate patches.

Finally, approximating the costs and benefits of different behavioral options and assuring that model assumptions are met is only half of the battle in archaeological applications of EE. Equally important, and no less demanding, is the task of translating model predictions cast in terms of behavior into predictions about its archaeological consequences (e.g., Broughton and Grayson 1993; Grayson and Delpech 1998, reprinted here; Jones 2004). Indeed, deriving secure predictions from EE models when the relationship between behavior and the archaeological record is so complex is a challenging obstacle, but then this is a problem facing all scientifically inclined archaeologists. Archaeologists using the EE approach have devoted considerable effort to translating model predictions into archaeologically measurable terms, and in doing so they have made valuable contributions to archaeological method more generally, as papers in this volume attest. From canine-size dimorphism in *Australopithecus* to Paleolithic stone tool manufacture to Maya monument construction, papers reprinted here illustrate how EE models can be successfully applied to the paleoanthropological and archaeological records.

Criticisms and Misunderstandings of EE in Archaeology

In an effort to further clarify the EE approach, we turn briefly to some recent criticisms and some persistent misunderstandings of the use of EE in archaeology. Such criticisms and misunderstandings have been expressed both by archaeologists who use the approach and by archaeologists who have yet to recognize its full worth.

From within the approach, some have argued that archaeological applications of EE should endeavor to become methodologically more rigorous, and these arguments have led some working outside the approach to conclude that the EE enterprise is somehow fundamentally flawed (e.g., O'Brien and Lyman 2002:34). The strongest critique to date in this vein has been that of Grayson and Cannon (1999), who noted that robust applications of EE to the archaeological record have produced major advances in our understanding of the human past, but who also argued that there remains room for improvement within the approach. For example, Grayson and Cannon pointed out that many archaeological applications of foraging theory continue to rely on ethnographic observations to a greater degree than is either desirable or necessary, and they also suggested that detailed attention to paleoenvironmental data would enable stronger tests of hypotheses derived from foraging theory models. Criticism of this sort does not imply that all archaeological applications of foraging theory are problematic, as some have concluded; rather, it merely points out ways in which the methodology of such applications can be improved and thus constitutes an important part of the scientific process. The papers reprinted in this volume, we believe, do a good job of illustrating how such methodological challenges can be met.

Also from within the approach, Bamforth (2002) has recently questioned the link between foraging theory and natural selection by observing that foraging efficiency, the most commonly used currency in foraging theory models, has never been documented to correlate with reproductive success in any human population. In responding to this we first note that, as we discussed above, we agree with Bamforth's main point, which is that archaeological applications of foraging theory may often involve situations in which individuals facultatively made subsistence decisions based on environmental conditions or cues, as opposed to situations involving an ongoing process of genetic natural selection. (Though, as we have also pointed out, some selective process may well have been operating on cultural information in situations to which foraging models are applied, and genetic selection is clearly involved in applications of EE to areas other than foraging theory, such as the early hominin examples reprinted here.) That this is the case, however, does not imply that there is no link between foraging efficiency and natural selection.

Second, we point out that, within contemporary hunter-gatherer populations for which data are available, strong correlations exist between hunting success and reproductive success for males, as discussed in a recent review by Smith (2004). Of course, the concept of "hunting success" used in such studies, which is usually determined by asking informants who the good hunters are, is not strictly the same thing as foraging efficiency as defined in EE models, and the proximate causes of the correlations between hunting success and reproductive success are unclear: they may be the result of factors such as offspring provisioning, social status acquired through hunting, or some combination thereof (Smith 2004). However, these studies come very close to providing the sort of test that Bamforth calls for. We also note that such tests have been conducted for nonhuman species and have shown a strong relationship between foraging success and reproductive success (e.g., Lemon 1991).

This brings us to our third and most important point in response to Bamforth's criticism, which is that, as we have noted several times above, the currency invoked in any EE model should always be treated as a hypothesis to be tested, rather than an a priori assertion about some factor that is relevant to human behavior or fitness. As with all model components, the currency assumption is at risk whenever a model is applied in a particular context. Predictive failures imply that one or more of the assumptions incorporated into a model, including the currency assumption, are inappropriate. Predictive successes, on the other hand, imply that model assumptions are valid, and the wide range of successful tests involving models that assume maximization of foraging efficiency as a goal and currency—many of which are included in this volume—attest to the generality of this goal and currency (see also, e.g., Kaplan and Hill 1992; Winterhalder and Smith 2000). In turn, this generality implies that foraging efficiency does have an effect on fitness and has had such an effect over the course of human evolution.

There are, of course, also cases where model predictions have failed, and in such cases currency issues have often come to the fore as the underlying cause of the failure. Most noteworthy in this context—and as the papers by O'Connell et al. (1999) and Zeanah (2005) that are reprinted here both discuss—is the observation that men's hunting behavior may be motivated not solely by concerns of foraging efficiency, but also by factors related to mating effort, costly signaling, and prestige rivalry. This observation, we must emphasize, became apparent only after careful tests conducted in several ethnographic contexts found that the predictions of standard foraging theory models failed to account for some aspects of men's foraging; in particular, men in these cases forego small-bodied prey that the prey model predicts they should pursue (e.g., Hawkes 1991, 1993; Hawkes et al. 2001; Hawkes and Bliege Bird 2002). However, in many other ethnographic cases, or when the scale of analysis is elevated to include the resources taken by both men and women collectively (thereby mirroring what is usually the case in archaeological assemblages of food remains), predictions of the prey model have fared very well (e.g., Alvard 1993; Bird and Bliege Bird 2001; Bird et al. 2002; Hawkes et al. 1982; Hill et al. 1987; Smith 1991; Winterhalder 1981).

It thus remains uncertain in what socio-ecological contexts social prestige is more relevant than foraging efficiency as a currency for men's hunting (also see discussion in Smith 2004), and we look forward to further theoretical work that might clarify this issue (see Hawkes 1990 for a good start; see also Hawkes et al. 1995). We note, however, that even if it turns out that a currency related to social prestige is generally more appropriate to men's foraging behavior, the only way that we will know this is through continued careful testing—on a case-by-case basis—of the predictions of models that assume the goal and currency of maximizing foraging efficiency (see Hildebrandt and McGuire 2002; Broughton and Bayham 2003; Byers and Broughton 2004; McGuire and Hildebrandt 2005 for lively debate on this issue). Starting simple in this way and adding complexity, *but only as needed*, is a hallmark of the EE approach, and it also makes for good science.

Turning from critique to misunderstanding, a common mistake displayed by some archaeologists who use the EE approach is evident in what we call here the *"EE predicts…" fallacy*, which surfaces in two different forms. In the first, this fallacy appears in proposed reconstructions of the past in which the predictions of EE models—and especially foraging theory models—are used in place of empirical data to support the reconstructions. For example, to support the argument that members of some prehistoric hunter-gatherer group were subsistence specialists in the absence of compelling paleodietary evidence for specialization, one might simply note that "foraging theory predicts" a relatively narrow diet breadth in an environment such as the one that the group in question occupied. In the second form of this fallacy, patterns in archaeological data are described and then casually asserted either to match or violate predictions of foraging theory.

Such reasoning fails to make the most productive use of EE models for two reasons. First, the greatest scientific value of these models lies in their ability to enable theoretically derived hypotheses to be developed and tested, and if such hypotheses are not rigorously tested, then this value is not realized. Using EE models in this manner amounts to assuming what should be demonstrated. Second, "foraging theory" does not universally predict anything; rather, a specific model from foraging theory will make a specific prediction for a specific context. Foraging theory consists of many models (and many more might be developed), each of which makes a different set of assumptions. Indeed, it is possible to derive virtually any prediction from foraging theory simply by starting from different assumptions. Merely asserting that EE predicts something is of little use without a detailed consideration of whether the assumptions of a particular model are met in a given situation, and even better yet would be to take the next step of testing model predictions: the way we actually learn something is by comparing both the predictions and the assumptions of a model to empirical data from a specific context.

As we noted above, advocates of the use of EE models have long warned against applying them in cookbook fashion without considering in detail the assumptions that they make, but those who succumb to the "EE predicts…" fallacy do just that. Fortunately, however, progress in this area can come with greater sophistication in the use of EE models, and we believe that such sophistication is effectively demonstrated by the papers included in this volume.

We now turn to criticisms and misunderstandings that have been expressed by archaeologists outside of the EE school. One of the most common criticisms of EE both within and outside of anthropological disciplines (e.g., Gould and Lewontin 1979; Joseph 2000; Martin 1983) has been an objection to the use of optimization logic, but this criticism has been successfully defended by many authors (e.g., Foley 1985; Parker and Maynard Smith 1990; Smith 1991; Smith and Winterhalder 1992; Stephens and Krebs 1986; Winterhalder 2002; Winterhalder and Smith 1992) and we do not repeat their defenses here. Rather, we refer the reader to these other works that explain and justify the use of optimization models in evolutionary studies, and we reiterate our earlier point that the purpose of these models is simply to enable the development and testing of hypotheses about factors relevant to the evolution and expression of phenotypic traits. Optimization is an analytical tool, not an empirical claim about human behavior or morphology.

Within archaeology, EE has likewise been criticized for succumbing to teleology by placing the mechanism of change in the decisions made by individuals, rather than in evolutionary processes (e.g., Lyman and O'Brien 1998). Such a charge simply reflects a fundamental misunderstanding of EE. Advocates of the EE approach have long noted that human intention is something to be explained by natural selection, rather than something that should be used in an explanatory role, a point echoed by critics of the approach (e.g., compare Smith and Winterhalder 1992:47–50 to Lyman and O'Brien 1998:620–621). Moreover, it should be pointed out that when one uses an EE model that incorporates a decision assumption, no greater primacy need be given to human intentionality or calculated decision making than must be the case in other Darwinian approaches in archaeology. For example, a selectionist analysis of evolutionary change in pottery temper (e.g., Dunnell and Feathers 1990) involves a situation in which potters made some sort of decision about the type of temper to use, just as foragers made some sort of decision about whether to pursue various resource types in a situation studied by an archaeologist using the prey model. In both cases, the decision-making apparatus involved is a product of natural selection, and archaeologists can study the operation of selective processes on the consequences of the decisions that were made. We are glad to see that practitioners of other Darwinian approaches in archaeology have recently come to agree with advocates of EE on this point (e.g., Bentley et al. 2007; O'Brien and Lyman 2002), and we concur with them that the intentionality issue does not render the various approaches incompatible.

On a related note, EE has been criticized for using models that assume rational actors who are unrealistically farsighted and who improbably possess all of the information required to make decisions in their own self-interest (e.g., Joseph 2000). Such criticism, if justified, would pertain not only to archaeological and anthropological applications of EE, but also to applications of EE in biology and to applications of optimization modeling in economics, from which many of the tools of EE derive. Such criticism is not, however, justified.

To begin with, it is important to note that in many areas of research, including those addressed by the papers in this volume (and also see Winterhalder and Smith 2000), the predictions of models that assume rational behavior (which, by the way, need not result from conscious calculation) have empirically held up well. This suggests that the assumption of rational action is not problematic in many cases.

Second, we point out that there is a large area of foraging theory that deals with issues such as learning about the environment by actors who possess incomplete information (e.g., Stephens and Krebs 1986). Such models of information acquisition are applicable to situations involving, for example, foragers who must learn new landscapes, including colonists of new lands (e.g., Kelly 2003; Meltzer 2004), an area that is ripe for further EE-based studies. Thus, EE models do not necessarily assume omniscient actors, they also can be employed to understand how individuals respond to circumstances in which they have incomplete information about their surroundings.

Third, we note that economists, who are far ahead of anthropologists on this point, have begun to respond to empirically documented shortcomings of the rational actor assumption not by abandoning the approach of developing and implementing explicit models, but by making what amount to relatively minor adjustments to this approach (e.g., Bowles 2004; Gintis 2000; see also Henrich et al. 2005). These adjustments include, for example, developing models that assume that actors make decisions based on their past experiences rather than based on "cognitively demanding forward-looking optimization processes" (Bowles 2004:11). While the specific models that have been developed by economists pursuing this area of research may or may not prove useful in archaeology, if archaeologists were to adopt some of the tools of this area of research it is likely that new lines of archaeological inquiry would be opened up to the EE approach (in the realm of economic exchange, for example, building on work such as that of Winterhalder 1997).

Our main point here is simply that the rational actor criticism of the EE approach does not hold much traction. Models that assume rational behavior have performed well empirically, and the EE approach can easily address situations involving actors who are not all-knowing or perfectly rational.

Finally, archaeologists who do not use EE have tended to equate it with a fairly narrow range of subject matter, particularly foraging behavior in hunter-gatherer societies (e.g., Schiffer 1999:167). We hope that this volume will help to dispel this misconception by demonstrating the variety of topics in human evolution and prehistory that have been addressed using the EE approach (see also Bird and O'Connell 2006), and we note in addition that the range of issues that might be addressed in the future is potentially much broader.

It is true that archaeological applications of EE initially focused mainly on questions relating to ancient diets—in both foraging and agricultural societies (e.g., Bayham 1979; Keegan 1986)—and it is in this arena that the EE approach has to date been most productive (for discussions of the history of EE in archaeology and anthropology more broadly, see Bird and O'Connell 2006 and Winterhalder and Smith 2000, respectively; see Owens 2006 for a recent review of the history and current state of EE in biology). However, since the time of the earliest archaeological applications, the scope of the EE approach in archaeology has broadened considerably to cover a diverse array of important issues, ranging from hominin life history evolution to the development of social complexity. As illustrated by the papers in this volume, EE models have been successfully applied to topics from the mating behavior of early hominins to the development of agriculture to the colonization of the Americas to the construction of Maya monuments.

Furthermore, past applications by no means define limits for the future. As we have noted, the basic EE approach of developing and testing explicit models of the fitness-related costs and benefits of phenotypic alternatives is applicable in principle to any aspect of morphology, life history, or behavior. In other words, vast expanses of the open fields that Darwin envisioned in 1859 remain to be explored using the EE approach. Just one example of a potential research avenue that has begun to be investigated but that awaits further theoretical development and systematic empirical research is provided by Broughton and O'Connell's (1999) discussion of prehistoric California. These authors link declining foraging efficiency to changes in the degree of territoriality, interpersonal violence, and overall human health that are evident in the California archaeological record. Though detailed tests of these linkages are only beginning to be conducted (e.g., Kennett and Kennett 2000), they nicely illustrate the integrative potential of the EE approach that we described earlier. Finally, we note that even in the well-trodden arena of subsistence behavior, there remains much room for further development of EE theoretical models (e.g., Barlow 2006; Cannon 2003, reprinted here). We hope that this volume makes it clear that the EE approach is very dynamic and continually developing in response to new research questions.

Archaeological Applications of Evolutionary Ecology Included in this Volume

As noted at the outset of this introduction, the primary goals of this volume are to illustrate the breadth and scope of previously published archaeological applications of EE and to demonstrate the substantive advances in our understanding of the human past that have been made by archaeologists employing this approach. In line with these goals, and to facilitate use of this volume in courses on human prehistory, the volume is organized into five sections that correspond to major developments in the human past:

Part I: Early Hominin Evolution and Behavior
Part II: Pleistocene Foragers and Colonists
Part III: Post-Glacial Adaptations
      Section 1—Foraging, Predator-Prey Interactions, and the Sexual Division of Labor
      Section 2—The Intersection of Technology, Subsistence, and Mobility
Part IV: Food Production Strategies: Origins, Spread, and Variation
Part V: Cooperation and Competition in Complex Societies

Within each of these parts and sections, we have selected papers that, in our view, address important problems relevant to the topic and represent especially well-developed or influential archaeological applications. The papers within each topical section are arranged by publication date in order to reveal the history of research and the cumulative nature of the EE enterprise.

In addition to selecting papers that help the volume meet the goals noted above, we also chose to limit the volume to papers that in some way both (1) draw on formal models of EE, and (2) incorporate empirical applications to the archaeological record. Because of this second criterion, and given space considerations, purely theoretical papers, though very important to the EE approach, had to be left out. Moreover, because the main goal of our selection process was to compile a set of papers that illustrate the breadth of the EE approach, many important papers that might have added greater depth could not be included. While we tried to include papers that provide balance in geographic and temporal coverage, applications from western North America are nonetheless overrepresented, but only because works from this region best met the selection criteria described here.

We also note that the papers reprinted in *Evolutionary Ecology and Archaeology* are merely a sample of the much larger population of archaeological and paleoanthropological applications of EE, and we hope that readers will be inspired by this sample to explore the many additional excellent applications that could not be included in this volume. The extensive references sections, together with that of Bird and O'Connell's (2006) recent survey, should provide nearly comprehensive coverage of the field.

Finally, we make three important points about this set of papers as a whole. First, although all of the papers reprinted here in some way draw on EE models, the role that these models play varies somewhat from paper to paper: some papers develop individual models and spell out specific predictions from those models, while others use one or more models to generate evolutionary scenarios or to derive inferences about important issues in prehistory (such as human population dynamics). Second, although each paper incorporates an empirical application, the nature of such applications also varies somewhat from paper to paper: some papers present detailed tests of the predictions of specific models, while others consider empirical data in a more general manner and specify additional tests that might be conducted in the future. However, despite such variability in the role that theoretical models and archaeological data play in the applications reprinted here—and perhaps even because of such variability—we believe that these applications exemplify the theoretical and methodological strengths of the EE approach, as well as the substantive advances in our understanding of human evolution and prehistory that have resulted from it.

Third, we note that that while most archaeological applications of EE make use of models originally developed in biology, several of the papers included in this volume present new models that were developed by archaeologists and anthropologists to address issues of particular interest to those who study humans. Thus, not only have archaeologists using the EE approach produced important substantive knowledge about the human past, they have also made significant contributions to the theoretical development of the general EE enterprise. We would be encouraged if this volume inspired archaeologists to continue to develop and extend EE theory and if it inspired researchers working in other fields to draw on some of the theoretical contributions of archaeologists.

Acknowledgments

We thank Doug Kennett, Hector Neff, Bruce Winterhalder and an anonymous reviewer for very helpful comments on this introduction. We also thank Peter DeLafosse and Jeff Grathwohl of the University of Utah Press for invaluable editorial assistance with the volume as a whole.

References

Alvard, M.
1993 Testing the Ecologically Noble Savage Hypothesis: Interspecific Prey Choice by Piro Hunters of Amazonian Peru. *Human Ecology* 21:355–387.

Avital, E., E. Jablonka
2000 *Animal Traditions: Behavioural Inheritance in Evolution*. Cambridge University Press, Cambridge.

Bamforth, D.
2002 Evidence and Metaphor in Evolutionary Archaeology. *American Antiquity* 67:435–452

Barlow, K. R.
2006 A Formal Model for Predicting Agriculture among the Fremont. In *Behavioral Ecology and the Transition to Agriculture*, edited by D. J. Kennett and B. Winterhalder, pp. 87–102. University of California Press, Berkeley.

Bayham, F. E.
1979 Factors Influencing the Archaic Pattern of Animal Utilization. *Kiva* 44:219–235.

Bentley, R. A., C. Lipo, H. Maschner
2007 Evolutionary Theory. In *Handbook of Archaeological Theories*. AltaMira Press, Lanham, Md.

Bettinger, R. L.
1983 Comment on Anthropological Applications of Optimal Foraging Theory: A Critical Review, by E. A. Smith. *Current Anthropology* 24:640–641.
1993 Doing Great Basin Archaeology Recently: Coping with Variability. *Journal of Archaeological Research* 1:43–66.

Bettinger, R. L., R. Boyd, P. J. Richerson
1996 Style, Function, and Cultural Evolutionary Processes. In *Darwinian Archaeologies*, edited by H. D. G. Maschner, pp. 133–164. Plenum Press, New York.

Binford, L. R.
1962 Archaeology as Anthropology. *American Antiquity* 28:217–225.

Bird, D., R. Bliege Bird
2001 The Ethnoarchaeology of Juvenile Foragers: Shellfishing Strategies among Meriam Children. *Journal of Anthropological Archaeology* 19:461–476.

Bird, D. W., J. F. O'Connell
2006 Behavioral Ecology and Archaeology. *Journal of Archaeological Research* 14:143–188.

Bird, D. W., J. L. Richardson, P. M. Veth, A. J. Barham
2002 Explaining Shellfish Variability in Middens on the Meriam Islands, Torres Strait, Australia. *Journal of Archaeological Science* 29:457–469.

Boone, J. L., E. A. Smith
1998 Is it Evolution Yet?: A Critique of Evolutionary Archaeology. *Current Anthropology* 39(Suppl.):141–173.

Bovy, K.
2005 Effects of Human Hunting, Climate Change, and Tectonic Events on the Waterbirds along the Pacific Northwest Coast during the Late Holocene. Ph.D. dissertation, Department of Anthropology, University of Washington, Seattle.

Bowles, S.
2004 *Microeconomics: Behavior, Institutions, and Evolution*. Russell Sage Foundation, New York.

Broughton, J. M.
1994 Declines in Mammalian Foraging Efficiency during the Late Holocene, San Francisco Bay, California. *Journal of Anthropological Archaeology* 13:371–401.
1999 *Resource Depression and Intensification during the Late Holocene, San Francisco Bay: Evidence from the Emeryville Shellmound Vertebrate Fauna*. University of California Anthropological Records 32. University of California Press, Berkeley.

Broughton, J. M., F. E. Bayham
2003 Showing Off, Foraging Models, and the Ascendance of Large-Game Hunting in the California Middle Archaic. *American Antiquity* 68:783–789.

Broughton, J. M., D. K. Grayson
1993 Diet Breadth, Adaptive Change, and the White Mountains Faunas. *Journal of Archaeological Science* 20:331–336.

Broughton, J. M., J. F. O'Connell
1999 On Evolutionary Ecology, Selectionist Archaeology, and Behavioral Archaeology. *American Antiquity* 64:153–165.

Byers, D. A., J. M. Broughton
2004 Holocene Environmental Change, Artiodactyl Abundances, and Human Hunting Strategies in the Great Basin. *American Antiquity* 69:235–256.

Cannon, M. D.
2001 Large Mammal Resource Depression and Agricultural Intensification: An Empirical Test in the Mimbres Valley, New Mexico. Ph.D. dissertation, Department of Anthropology, University of Washington, Seattle.

Darwin, C.
1859 *The Origin of Species*. John Murray, London.

Dunnell, R. C.
1986 Five Decades of American Archaeology. In *American Archaeology Past and Future: A Celebration of the Society for American Archaeology, 1935–1985*, edited by D. J. Meltzer, D. D. Fowler, and J. A. Sabloff, pp. 23–49. Smithsonian Institution Press, Washington, D.C.

Dunnell, R. C., J. K. Feathers
1990 Late Woodland Manifestations of the Malden Plain, Southeast Missouri. In *Stability, Transformation, and Variation: The Late Woodland Southeast*, edited by M. S. Nassaney and C. S. Cobb, pp. 21–45. Plenum Press, New York.

Eerkens, J. W., C. P. Lipo
2005 Cultural Transmission, Copying Errors, and the Generation of Variation in Material Culture and the Archaeological Record. *Journal of Anthropological Archaeology* 24:316–334.

Foley, R.
1985 Optimality Theory in Anthropology. *Man* (N.S.) 20:222–242.

Gintis, H.
2000 *Game Theory Evolving: A Problem-Centered Introduction to Modeling Strategic Interaction*. Princeton University Press, Princeton, N.J.

Gould, S. J., R. C. Lewontin
1979 The Spandrels of San Marco and the Panglossian Paradigm: A Critique of the Adaptationist Programme. *Proceedings of the Royal Society of London, Series B* 205:581–598.

Grafen, A.
1984 Natural Selection, Kin Selection and Group Selection. In *Be-

Grayson, D. K., M. D. Cannon
1999 Human Paleoecology and Foraging Theory in the Great Basin. In *Models for the Millennium: Great Basin Anthropology Today*, edited by C. Beck, pp. 141–151. University of Utah Press, Salt Lake City.

Gremillion, K. J.
2002 Foraging Theory and Hypothesis Testing in Archaeology: An Exploration of Methodological Problems and Solutions. *Journal of Anthropological Archaeology* 21:142–164.

Haccou, P., W. J. van der Steen
1992 Methodological Problems in Evolutionary Biology, IX. The Testability of Optimal Foraging Theory. *Acta Biotheoretica* 40:285–295.

Hawkes, K.
1990 Why do Men Hunt? Benefits for Risky Choices. In *Risk and Uncertainty in Tribal and Peasant Economies*, edited by E. Cashdan, pp. 145–166. Westview Press, Boulder, CO.
1991 Showing Off: Tests of an Hypothesis About Men's Hunting Goals. *Ethology and Sociobiology* 12:29–54.
1993 Why Hunter-Gatherers Work: An Ancient Version of the Problem of Public Goods. *Current Anthropology* 34:341–361.

Hawkes, K., R. Bliege Bird
2002 Showing Off, Handicap Signaling, and the Evolution of Men's Work. *Evolutionary Anthropology* 11:58–67.

Hawkes, K., K. Hill, J. O'Connell
1982 Why Hunters Gather: Optimal Foraging and the Ache of Eastern Paraguay. *American Ethnologist* 9:379–398.

Hawkes, K., J. F. O'Connell, N. G. Blurton Jones
2001 Hadza Meat Sharing. *Evolution and Human Behavior* 22:113–142.

Hawkes, K., A. R. Rogers, E. L. Charnov.
1995 The Male's Dilemma: Increased Offspring Production is More Paternity to Steal. *Evolutionary Ecology* 9:662–677.

Henrich, J., R. Boyd, S. Bowles, C. Camerer, H. Fehr, H. Gintis, R. McElreath, M. Alvard, A. Barr, J. Ensminger, K. Hill, F. Gil-White, M. Gurven, F. Marlowe, J. Patton, N. Smith, D. Tracer
2005 "Economic Man" in Cross-Cultural Perspective: Behavioral Experiments in 15 Small-Scale Societies. *Behavioral and Brain Sciences* 28:795–855.

Hildebrandt, W. R., K. R. McGuire
2002 The Ascendance of Hunting during the California Middle Archaic: An Evolutionary Perspective. *American Antiquity* 67:231–256.

Hill, K., H. Kaplan, K. Hawkes, A. M. Hurtado
1987 Foraging Decisions among the Ache Hunter-Gatherers: New Data and Implications for Optimal Foraging Models. *Ethology and Sociobiology* 8: 1–36.

Jones, E. L.
2004 Dietary Evenness, Prey Choice, and Human-Environment Interactions. *Journal of Archaeological Science* 31:307–317.

Joseph, S.
2000 Anthropological Evolutionary Ecology: A Critique. *Journal of Ecological Anthropology* 4:6–30.

Kaplan, H. S., K. Hill
1992 The Evolutionary Ecology of Food Acquisition. In *Evolutionary Ecology and Human Behavior*, edited by E. A. Smith and B. Winterhalder, pp. 167–202. Aldine de Gruyter, New York.

Keegan, W.
1986 The Optimal Foraging Analysis of Horticultural Production. *American Anthropologist* 88: 92–107.

Kelly, R. L.
2003 Colonization of New Land by Hunter-Gatherers: Expectations and Implications Based on Ethnographic Data. In *Colonization of Unfamiliar Landscapes: The Archaeology of Adaptation*, edited by M. Rockman and J. Steele, pp. 44–58. Routledge, London.

Kennett, D. J., J. P. Kennett
2000 Competitive and Cooperative Responses to Climatic Instability in Southern California. *American Antiquity* 65:379–395.

Lemon, W. C.
1991 Fitness Consequences of Foraging Behaviour in the Zebra Finch. *Nature* 352:153–155.

Lyman, R. L., M. J. O'Brien
1998 The Goals of Evolutionary Archaeology: History and Explanation. *Current Anthropology* 39:615–652.

Madsen, D. B., D. N. Schmitt
1998 Mass Collecting and the Diet Breadth Model: A Great Basin Example. *Journal of Archaeological Science* 25:445–455.

Martin, J. F.
1983 Optimal Foraging Theory: a Review of Some Models and Their Applications. *American Anthropologist* 85:612–629.

Maynard Smith, J.
1978 Optimization Theory in Evolution. *Annual Review of Ecology and Systematics* 9:31–56.

McGuire, K. R., W. R. Hildebrandt
2005 Re-Thinking Great Basin Foragers: Prestige Hunting and Costly Signaling during the Middle Archaic Period. *American Antiquity* 70:695–712.

Meltzer, D. J.
2004 Modeling the Initial Colonization of the Americas: Issues of Scale, Demography, and Landscape Learning. In *The Settlement of the American Continents: A Multidisciplinary Approach to Human Biogeography*, edited by C. M. Barton, G. A. Clark, D. R. Yesner, and G. A. Pearson, pp. 123–137. University of Arizona Press, Tucson.

Metcalfe, D., K. R. Barlow
1992 A Model for Exploring the Optimal Tradeoff between Field Processing and Transport. *American Anthropologist* 94:340–356.

Nagaoka, L.
2001 Using Diversity Indices to Measure Changes in Prey Choice at the Shag River Mouth Site, Southern New Zealand. *International Journal of Osteoarchaeology* 11:101–111.

Neff, H.
2000 On Evolutionary Ecology and Evolutionary Archaeology: Some Common Ground? *Current Anthropology* 41:427–429.

O'Brien, M. J., R. L. Lyman
2002 Evolutionary Archaeology: Current Status and Future Prospects. *Evolutionary Anthropology* 11:26–36.

Owens, I. P. F.
2006 Where is Behavioural Ecology Going? *Trends in Ecology and Evolution* 21:356–361.

Parker, G. A., J. Maynard Smith
1990 Optimality Theory in Evolutionary Biology. *Nature* 348:27–33.

Pinker, S.
2002 *The Blank Slate: The Modern Denial of Human Nature*. Viking, New York.

Richerson, P. J., R. Boyd
2004 *Not by Genes Alone: How Culture Transformed Human Evolution*. University of Chicago Press, Chicago.

Schiffer, M. B.
1999 Behavioral Archaeology: Some Clarifications. *American Antiquity* 64:166–168.

Simms, S. R.
1987 *Behavioral Ecology and Hunter-Gatherer Foraging: An Example from the Great Basin*. BAR International Series 381. British Archaeological Reports, Oxford.

Smith, E. A.
1983 Anthropological Applications of Optimal Foraging Theory: A Critical Review. *Current Anthropology* 24:625–651.
1991 *Inujjuamiut Foraging Strategies: Evolutionary Ecology of an Arctic Hunting Economy*. Aldine de Gruyter, New York.
2004 Why do Good Hunters have Higher Reproductive Success? *Human Nature* 15:343–364.

Smith, E. A., B. Winterhalder
1992 Natural Selection and Decision Making: Some Fundamental Principles. In *Evolutionary Ecology and Human Behavior*, edited by E. A. Smith and B. Winterhalder, pp. 25–60. Aldine de Gruyter, New York.

Stephens, D. W., J. R. Krebs
1986 *Foraging Theory*. Princeton University Press, Princeton, N.J.

Steward, J. H.
1938 *Basin-Plateau Aboriginal Sociopolitical Groups*. Bureau of American Ethnology Bulletin No. 120. Washington, D.C.

Stiner, M. C., N. D. Munro
2002 Approaches to Prehistoric Diet Breadth, Demography, and Prey Ranking Systems in Time and Space. *Journal of Archaeological Method and Theory* 9:181–214.

Ugan, A.
2005 Does Size Matter?: Body Size, Mass Collecting and Their Implication for Understanding Prehistoric Foraging Behavior. *American Antiquity* 70: 75–90.

Winterhalder, B.
1981 Foraging Strategies in the Boreal Forest: An Analysis of Cree Hunting and Gathering. In *Hunter-Gatherer Foraging Strategies: Ethnographic and Archaeological Analyses*, edited by B. Winterhalder and E. A. Smith, pp. 66–98. Chicago: University of Chicago Press.
1997 Gifts Given, Gifts Taken: The Behavioral Ecology of Nonmarket, Intragroup Exchange. *Journal of Archaeological Research* 5:121–168.
2002 Behavioral and Other Human Ecologies: Critique, Response, and Progress through Criticism. *Journal of Ecological Anthropology* 6:4–23.

Winterhalder, B., E. A. Smith
1992 Evolutionary Ecology and the Social Sciences. In *Evolutionary Ecology and Human Behavior*, edited by E. A. Smith and B. Winterhalder, pp. 3–23. Aldine de Gruyter, New York.
2000 Analyzing Adaptive Strategies: Human Behavioral Ecology at Twenty-Five. *Evolutionary Anthropology* 9:51–72.

PART I

Early Hominin Evolution and Behavior

The three papers in this section of *Evolutionary Ecology and Archaeology* illustrate the application of EE to human origins and address the causes and processes underlying the evolution of the hominins from hominoid ancestors. Each of these papers draws on EE to develop and test hypotheses to account for some subset of the list of widely recognized biological and behavioral differences between humans and our closest living relatives, chimpanzees. These differences include larger body and brain size, reduced sexual dimorphism, obligate bipedality, different dental and digestive anatomies, a slower life history, unique patterns of social organization and reproduction, and greater reliance on technology. The arguments in the three papers reprinted in this first section are structurally similar and point to climatic and environmental changes and consequent shifts in foraging behavior as primary catalysts; changes in morphology, life history, mating structure, and other features of the hominin phenotype are seen as evolutionary consequences. The papers in this section are perhaps the most far-reaching and ambitious, in that they deal with a broad range of questions, cover a substantial geological time depth, and integrate a considerable set of models and datasets; they are also among the most controversial.

C. O. Lovejoy's classic paper develops a scenario to account for, among other things, the emergence of bipedality and the unique sexual and reproductive behavior of *Homo sapiens*. In so doing, he pioneers the integration of a wide range of EE theory and empirical work on foraging behavior, life history, mating structure, anatomy, and demography in the context of human origins. This approach is further developed in the other, more recent papers in this section.

At the time of the publication of Lovejoy's paper in 1981, the discovery and analysis of *Australopithecus afarensis* had recently pushed back evidence for bipedality in the hominin line to four million years ago, well prior to evidence for significant expansion of the neocortex or stone tools. Lovejoy hypothesizes that the changing ecological context of the late Miocene, which involved greater seasonality and increased mosaicism, favored selection for partial separation of male and female foraging niches.

He suggests that the foraging radius of females became more restricted to a core area, while males used more distant patches and returned resources to provision mates and offspring. Such a foraging strategy, Lovejoy argues, increased the reproductive rate and the survivorship of protohominins and set in motion a series of other fundamental changes, most notably bipedal locomotion, which allowed provisions to be carried by hand, and the formation of a more monogamous mating structure, which would have enabled increased paternity confidence.

Critical tests of these ideas focus on morphological features thought to be markers of greater pair bonding: reduced sexual dimorphism in body and canine size, which appears with *Australopithecus*, and the elaborate epigamic sexual characters of modern humans (e.g., scalp manes, permanent nonlactational breast enlargement in females, etc.). Although vigorously criticized on many grounds (e.g., Cann and Wilson 1982; Harley 1982; Wood 1982; Zilman 1985; but see Lovejoy 1993 and Reno et al. 2003 for recent defenses), this influential paper represents the first comprehensive EE-based consideration of human origins, and it provides an important historical foundation for the more recent papers in this section.

The next two papers were inspired from detailed EE-based analyses of modern hunter-gatherers. The first, by J. F. O'Connell et al., stems from work with Hadza foragers of East Africa, among whom hard-working senior females have a significant influence on the nutritional welfare of their juvenile kin, whereas the meat provisions that males provide to mates and offspring are unreliable. These observations motivate O'Connell and colleagues to challenge the conventional wisdom, which holds that the evolution of *Homo ergaster* (= *H. erectus*) was driven by male's provisioning of their families with meat (either hunted or scavenged), and to suggest instead that grandmothers should be the focal point of analysis (see also Hawkes 2003, 2004; Hawkes and O'Connell 2005; Hawkes et al. 1998, 2003; O'Connell et al. 2002). In their "grandmother hypothesis," O'Connell et al. argue that late Pliocene climate change reduced the availability of foods that weanling australopiths could acquire and handle on their own, and that in such a context older females might have increased their own long-term reproductive success by provisioning their grandchildren with high-return resources, such as tubers, that they could acquire but that juveniles could not. O'Connell et al. argue that this allowed range expansion into new habitats, closer birth spacing, higher fertility, and increased survivorship. They also argue, based on principles of life history theory (e.g., Charnov 1993), that selection for greater postreproductive female longevity would have, in turn, resulted in a delayed age of maturity, allowing for an extended period of growth and the attainment of larger body size. Empirical support for these ideas involves morphological evidence of a life history shift with the appearance of *Homo ergaster* and archeological and fossil evidence for tuber use.

The paper is the first to clearly illustrate the explanatory utility for human origins of recent developments in life history theory (Charnov 1993, 2001; Charnov and Berrigan 1993); it is also novel in the pivotal role in human evolution that it gives to female foraging behavior. The grandmother hypothesis has not been without controversy, however, and critics have focused on such issues as whether grandmothers could ever help enough to have a substantial effect on the reproductive success of their descendent kin (Hill and Hurtado 1996; Kaplan et al. 2000, reprinted here), or whether extended postmenopausal longevity is a recent phenomenon (Caspari and Lee 2004; Kennedy 2003; also see Trinkaus 1995). Some have also suggested that the hypothesis fails to explain both the expansion in brain size that occurred over the course of hominin evolution and the similarity in the lifespans of males and females (Kaplan et al. 2000, reprinted here; see also Aiello and Key 2002; Key 2000). O'Connell and colleagues respond to these criticisms in a series of papers that we encourage any student of human evolution to read (Bird and O'Connell 2006; Hawkes and O'Connell 2005; O'Connell et al. 2002).

In the third and final paper in this section, H. Kaplan et al. update a version of the hunting hypothesis for the evolution of the genus *Homo*—the same hypothesis that O'Connell and colleagues challenge. Kaplan and his colleagues do so by drawing on life-history theory and the "embodied capital" model from economics (Becker 1975). Their evolutionary scenario is motivated in part by the observation that males in contemporary foraging societies provide most of the energy that supports juveniles and reproductive-aged females—a very different reading of the ethnographic data than that of O'Connell et al. Kaplan et al. propose that the emergence of African savanna habitats in the Plio-Pleistocene created the selective context for a dietary shift toward high-quality, nutrient-dense, and difficult-to-acquire food resources. Exploiting such resources, which include large and small game as well as plant resources such as nuts and tubers, would have required significant intelligence, skill, and learning, the authors argue, thereby favoring both the enlarged cranial capacity and extended period of juvenile learning and dependence that modern humans exhibit. In turn, they suggest, more time invested in learning (embodied capital) would have paid off in both higher adult productivity and lower adult mortality. Male provisioning of females and young with hunted resources underwrites much of the development in this scenario, and it coevolves with a shift toward human pair bonding and the emergence of the nuclear family.

Proposed paleontological and archaeological tests of this scenario are similar to those posited by O'Connell and associates for the grandmother hypothesis and involve evidence for changes in life history with *Homo ergaster* and a dietary shift to more easily digested foods, which is suggested by changes in torso morphology and dentition. The hunting hypothesis developed by Kaplan et al. and the grandmother hypothesis developed by O'Connell et al. are distinguishable primarily by the role of hunted foods versus tubers in early *Homo* diets, a matter on which current archaeological evidence is inconclusive (O'Connell et al. 2002). Further discussion of these EE-derived hypotheses about hominin evolution is provided by Bird and O'Connell (2006), Kaplan and Robson (2002), Kaplan et al. (2003a, 2003b), and Robson and Kaplan (2003).

It is worth noting that predictions of the theory that Kaplan et al. develop should also apply to other vertebrate taxa such as birds, carnivores, and nonhominin primates: within these taxonomic groups, species that rely on more complex, skill-intensive feeding strategies should have longer developmental periods, longer lifespans, and larger brains relative to body size. There is thus a good opportunity here for EE theory developed to understand the human past to be used to investigate evolutionary patterns in other organisms.

It should be clear from the papers in this section of *Evolutionary Ecology and Archaeology*, and from the controversies that these papers have engendered, that EE is a thriving theoretical approach in the study of human origins. Most important in the context of this volume, these papers illustrate the insights that have been gained from applications of EE in this area. Although aspects of the scenarios developed in these papers are contradictory and controversial, and more detailed empirical tests are in order, EE-based studies of early hominin evolution converge on a number of substantive points, as follows:

1. Late Miocene and Pliocene climate changes, and their effects on habitat structures and feeding ecologies, are critical catalysts to both the emergence of the hominins and the evolution of early *Homo*.
2. Those changes provide the selective context for a reduced adult mortality rate and an extended period of juvenility, key features of mammalian life history that vary in tandem with each other (and with other phenotypic characters) across taxa.
3. These life-history changes are associated with intensive intergenerational food transfers in the form of either males provisioning mates and offspring or senior females assisting younger kin.
4. The resources increasingly supplied to females with young,

such as tubers and/or meat, are high quality, nutrient dense, and easily digested.
5. Simplified digestive tracts and diminunized dental anatomies follow from the dietary shift noted above.
6. The goals of male's hunting are highlighted and are argued to relate either to social attention, mating effort, and costly signaling on the one hand, or to family provisioning and parenting effort on the other.

References

Aiello, L. C., C. Key
2002 Energetic Consequences of Being a *Homo erectus* Female. *American Journal of Human Biology* 14:551–565.

Becker, G. S.
1975 *Human Capital.* New York: Columbia University Press.

Bird, D. W., J. F. O'Connell
2006 Behavioral Ecology and Archaeology. *Journal of Archaeological Research* 14:143–188.

Cann, R. L., A. C. Wilson
1982 Models of Human Evolution. *Science* 217:303–304.

Caspari, R., Lee, S. L.
2004 Older Age Becomes Common Late in Human Evolution. *Proceedings of the National Academy of Sciences* 101:10895–10900.

Charnov, E. L.
1993 *Life History Invariants: Some Explorations of Symmetry in Evolutionary Ecology.* Oxford University Press, Oxford.
2001 Evolution of Mammal Life Histories. *Evolutionary Ecology Research* 3:521–535.

Charnov, E. L., D. Berrigan
1993 Why do Female Primates have Such Long Lifespans and so Few Babies?: Or, Life in the Slow Lane. *Evolutionary Anthropology* 1:191–194.

Harley, D.
1982 Models of Human Evolution. *Science* 217:296.

Hawkes, K.
2003 Grandmothers and the Evolution of Human Longevity. *American Journal of Human Biology* 15:380–400.
2004 The Grandmother Effect. *Nature* 428:128–129.

Hawkes, K., J. F. O'Connell
2005 How Old is Human Longevity? *Journal of Human Evolution* 49:650–653.

Hawkes, K., J. F. O'Connell, N. G. Blurton Jones
2003 Human Life Histories: Primate Tradeoffs, Grandmothering Socioecology, and the Fossil Record. In *Primate Life Histories and Socioecology,* edited by P. Kappeler and M. Pereira, pp. 204–227. University of Chicago Press, Chicago.

Hawkes, K., J. F. O'Connell, N. G. Blurton Jones, H. Alvarez, E. L. Charnov
1998 Grandmothering, Menopause, and the Evolution of Human Life Histories. *Proceedings of the National Academy of Science* 95:1336–1339.

Hill, K., A. M. Hurtado
1996 *Ache Life History: The Ecology and Demography of a Foraging People.* Aldine de Gruyter, New York.

Kaplan, H. S. A. J. Robson
2002 The Emergence of Humans: The Coevolution of Intelligence and Longevity with Intergenerational Transfers. *Proceedings of the National Academy of Science* 99:10221–10226.

Kaplan, H. S., J. B. Lancaster, A. Robson
2003a Embodied Capital and the Evolutionary Economics of the Human Lifespan. In *Lifespan: Evolutionary, Ecological, and Demographic Perspectives,* edited by J. Carey and S. Tuljapakur, pp. 152–182. Population and Development Review 29, Supplement 2003.

Kaplan, H. S., T. Mueller, S. Gangestad, J. Lancaster
2003b Neural Capital and Lifespan Evolution among Primates and Humans. In *The Brain and Longevity,* edited by C. Finch, J. Robine, and Y. Christen, pp. 69–98. Springer, New York.

Kennedy, G. E.
2003 Paleolithic Grandmothers? Life History Theory and Early *Homo. Journal of the Royal Anthropological Institute* 9:549–572.

Key, C.
2000 The Evolution of Human Life History. *World Archaeology* 31:329–350.

Lovejoy, C. O.
1993 Modeling Human Origins: Are We Sexy Because We're Smart, or Smart Because We're Sexy? In *The Origin and Evolution of Humans and Humanness,* edited by D. T. Rasmussen, pp. 1–28. Jones and Bartlett, Sudbury, MA.

O'Connell, J. F., K. Hawkes, K. D. Lupo, N. G. Blurton Jones
2002 Male Strategies and Plio-Pleistocene Archaeology. *Journal of Human Evolution* 43:831–872.

Reno, P. L., R. S. Meindl, M. A. McCollum, C. O. Lovejoy
2003 Sexual Dimorphism in *Australopithecus afarensis* was Similar to that of Modern Humans. *Proceedings of the National Academy of Sciences* 100:9404–9409.

Robson, A. J., H. S. Kaplan
2003 The Evolution of Human Life History and Intelligence in Hunter-Gatherer Economies. *American Economic Review* 93:150–169.

Trinkaus, E.
1995 Neanderthal Mortality Patterns. *Journal of Archaeological Science* 22:121–142.

Wood, J. W.
1982 Models of Human Evolution. *Science* 217:296–298.

Zilman, A. L.
1985 Gathering Stories for Hunting Human Nature. *Feminist Studies* 11:365–377.

The Origin of Man

C. Owen Lovejoy

During the last quarter-century, the study of human origins has proved remarkably successful. Crucial fossils and primate behavioral data are now available from which to reconstruct man's evolution during the last 15 million years. Equally important is the recognition of a close genetic relationship between man and the other extant hominoids (especially *Pan* and *Gorilla*).[1] Experiments on DNA hybridization indicate at least 98 percent identity in non-repeated DNA in man and chimpanzee, sufficient similarity to suggest the possibility of a viable hybrid. These data confirm studies by comparative anatomists who have emphasized the striking anatomical similarities of apes and man.[2] As a consequence of this physical similarity, models of human origin must directly address the few primary differences separating humans from apes. Clearly, the rate of acquisition of these differences, the fossil evidence bearing on their first appearance, and their underlying selection are crucial to an understanding of human evolution.

Material Culture

The most commonly cited distinction between man and apes is the former's reliance on material culture. The belief that tools were pivotal to the divergence of hominids was initiated by Darwin[3] and has remained the most popular view.[4-6] Darwin was impressed by the absence of large canines in man and attributed their reduction to tool use. As Holloway[7] and Jolly[8] have cogently argued, however, tool use is not an explanation of canine reduction since there is no behavioral contradiction in having both functional canines and tools. There is little doubt that material culture has played a role in the evolution of *Homo sapiens* and *H. erectus*, but this does not require it to have been a significant factor in the origin of hominids. In fact, the earliest recognizable tools are only about 2 million years old,[9] but there is considerable evidence placing the phyletic origin of hominids in the middle to late Miocene (12 to 6 million years ago).[10-12] Although the earliest tools will have left no record because of the use of perishable materials, there is still the necessary presumption of a 6- to 10-million-year period dominated by reliance on material culture—a view with numerous shortcomings.

The use of primitive tools by extant pongids[13] supports the contention of comparable abilities in early hominids, but it also demonstrates that tool use is a general capacity of pongids, none of which exhibit the unique characters of hominids.[14] If tools were the primary determinant of early hominization, why should their first appearance be so late in the hominid record? More importantly, what activity requiring tools was critical to early hominid survival and phyletic origin?[15] It is now clear that hunting does not qualify as such an activity.[16] From the first recognizable tools to the industrial revolution required only 2 million years, whereas if tools played a part in the origin of hominids, they must have remained primitive and unchanged for at least 5 million years. It is likely that either the earliest hominids made no use of tools at all, or that such use was comparable to that in other extant hominoids and was not critical to their survival or pivotal to their origin.

Expansion of the Neocortex

It is now clear that the marked expansion of the hominid cerebral cortex took place during the last 2 to 3 million years.[17,18] Detailed study of the Hadar crania from Ethiopia, recently attributed to *Australopithecus afarensis*,[19] has revealed that they were strikingly primitive.[20] Preliminary estimates of cranial capacity indicate a brain size well within the range of extant pongids.[21] The pelvis of the skeleton known as "Lucy" from Afar Locality (A.L.) 288 has been fully reconstructed.[22] One of its most salient features is a birth canal whose shape and dimensions show little or no effects of selection for passage of enlarged fetal crania, adaptations that so clearly dominate the form of the modern human pelvis.[23,24]

Bipedality

Bipedality is an unusual mode of mammalian locomotion. Contrary to the so-called efficiency argument, energy expenditure for bipedal walking is probably not significantly different from

*Originally published in *Science* 211(1981):341–350

that during quadrupedal locomotion.[23,25] Yet the adoption of nonsaltatory bipedal progression is disadvantageous because both speed and agility are markedly reduced.[23,24,26] All present evidence, especially that made available by the postcranium of *A. afarensis*, confirms an essentially complete adaptation to bipedal locomotion by at least 4 million years ago.[22,27] This conclusion is provided unequivocal support by the hominid footprints discovered at Laetoli in Tanzania.[28]

Dentition

Additional distinctions between hominids and pongids are found in their respective jaws and teeth. In fact, these differences have allowed the identification of possible hominids in the Miocene—there are no distinctive postcranial or cranial remains of undoubted hominid affinities before about 4 million years ago. As a result of recent field work in Mio-Pliocene deposits,[29,30] it is now possible to suggest a broad schedule of phases in the evolution of the hominoid dentition that can serve as an outline of hominoid phyletic events during the last 23 million years.

Phase I. This phase has a generalized dryopithecine dentition including a distinct Y-5 lower molar cusp pattern with bunodont crowns, thin enamel, and cheek teeth small relative to body size; incisors are broad with canine-premolar shear. This phase is associated with forest faunas and floras[31] and is shared by all hominoids before 15 million years ago (range, 23 to 15 million years).[30–32]

Phase II. This phase shows a shift toward greater molar dominance. About 14 million years ago, hominoids fall into two groups. The first retained phase I characters and may constitute ancestral populations of extant apes (*Proconsul*, "*Rangwapithecus*," and *Limnopithecus, Dryopithecus*).[30] A second group exhibits enamel thickening, increased molar wear gradient, and moderate anterior dental reduction or increased relative molar size, or both. Mandibles are more robust and prognathism is reduced. The shift toward greater molar dominance has partially been attributed to greater reliance on terrestrial food sources. This group includes genera (*Ramapithecus* and *Sivapithecus*) probably related to hominids, an extinct ape (*Gigantopithecus*), and possibly the modern orang-utan (range, 14 to 8 million years).[12,30–34]

Phase III. This phase represents a conservative period. The dentition of *A. afarensis* appears only moderately changed in morphology and proportions from phase II; the features include comparatively large incisors, frequently a unicuspid lower first premolar, canines of moderate size, molars of moderate size (relative to body size and later hominids), and loss of canine-premolar shear (range, 7 to 2.5 million years).[12,17]

Phase IV. This phase represents Plio-Pleistocene specialization. The sample in this time range is divisible into two clades or phyletic lines.[17] The first was possibly restricted to savannah and grassland. It displays extreme anterior tooth reduction and excessive molar dominance and became extinct by mid-Pleistocene (*A. africanus* → *A. robustus* → "*A. boisei*"). A second clade, ancestral to *H. erectus*, retained a more generalized dentition in the early Pleistocene but underwent dentognathic reduction in the middle and upper Pleistocene as a consequence of reliance on material culture [for example, reduced dental manipulation and greater preoral food preparation[35]]. My view is that this clade occupied more varied habitats. Both groups are probably directly descendant from *A. afarensis*.[17]

Models of Human Origin

A model of hominid origin proposed by Jolly[8] uses analogy to anatomical and behavioral characters shared by *Theropithecus gelada* and some early hominids. He suggests that early hominid populations relied on small-object feeding, that this dietary specialization led to a suite of adaptations to the grassland savannah, and that bipedality developed in response to feeding posture. Yet geladas, which do rely on small-object feeding, are not bipedal and show no significant adaptations to bipedality. Bipedal locomotion is clearly not required for extensive small-object feeding especially on grasslands where speed and agility are of great value in animals who also lack wide visual fields and sensitive olfaction.[36] Furthermore, the dental morphology of *A. afarensis* is considerably more generalized than that of later hominids. The dietary specialization seen in *A. robustus* is possibly accountable by Jolly's model, but the more generalized dentition of *A. afarensis* is not.[37] It is more likely that hominids venturing into open habitats were already bipedal and that their regular occupation of savannahs was not possible until intensified social behavior was well developed.

Other theorists have viewed hominization as the direct result of savannah occupation by prehominids. Proponents of this view believe that the selective pressures of life on grassland savannahs directly produced the human character complex. Bipedal locomotion is posited as sentinel behavior and as an adaptation allowing weapons to be used against predators. Intelligence is said to be favored because highly integrated troop behavior is necessary for predator repulsion. Differences in some behaviors of chimpanzee populations now living in woodland savannahs versus those inhabiting more forested areas are cited as evidence.[38]

There are many problems with this view. Bipedality is useless for avoidance or escape from predators. Occasional bipedality, as seen in many primates, is sufficient for the use of weapons. Most importantly, brain expansion and cultural development remotely postdate hominid divergence.

Furthermore, Miocene ecology is inconsistent with the savannah selection theory. While cooling, aridity, and increased seasonality had pronounced effects on Old World floras, the predominant effect of these climatic trends, in areas where hominids are known to have been present, appears to have been the development of diversified mosaics, rather than broad-scale forest reduction.[39–41] It would be more correct to say that hominids of the middle and late Miocene were presented with a greater variety of possible habitats than to view them as having suffered an imposed "terrestrialization." It is also clear that some Miocene sites at which possible hominids have been recovered had canopy forest conditions.[12,42] While increased seasonality would have imposed a need for larger feeding ranges, occasional use of woodlands and edaphic grasslands would not necessarily impose elevated carnivore pressure. Nor, as was pointed out above, would early hominids be required to abandon quadrupedality in order to use more orthograde positional behavior during feeding. Quite the contrary, it would appear that late Miocene habitat mosaics would allow adoption of bipedality (in forests and transition mosaics) rather than directly select for it. All present evidence therefore indicates that hominid clade evolved in forest or mosaic conditions, or both,[43] rather than only on grassland or savannahs, and that bipedal locomotion was not a response to feeding posture, material culture, or predator avoidance.

In summary, four major character complexes are usually cited as distinguishing hominids from pongids. Hominids have remarkable brain expansion, a complex material culture, anterior dental reduction and molar dominance, and bipedal locomotion. Only bipedal locomotion and partial dental modifications can be shown to have an antiquity even approximating the earliest appearance of unquestioned, developed hominids (*A. afarensis*).

Demographic Strategy and the Evolution of Hominids

The order Primates has long been recognized to display a *scala naturae* consisting of "intercalary types"—extant forms that represent earlier stages in the development of major adaptive trends. Figure 2.1 is a well-known diagram of the chronology of life phases in living primates. There is an obvious trend toward prolonged life-span, which has both physiological and demographic correlates bearing directly on the phyletic origin of hominids.

The physiological correlates (Fig. 2.1) include a longer period of infant dependency, prolonged gestation, single births, and successively greater periods between pregnancies. Cutler[44] has demonstrated that such developmental parameters are "qualitatively and sequentially similar in different mammalian species but proceed at different characteristic rates defined by the reciprocal of their MLP" (maximum life potential). The progressive slowing of life phases can in turn be accounted for by an increasingly K-type demographic strategy.[45] With each step in the *scala naturae*, populations devote a greater proportion of their reproductive energy to subadult care, with increased investment in the survival of fewer offspring. Among chimpanzee populations, this trend appears to have resulted in marginal demographic conditions. Field studies at Gombe in Tanzania show

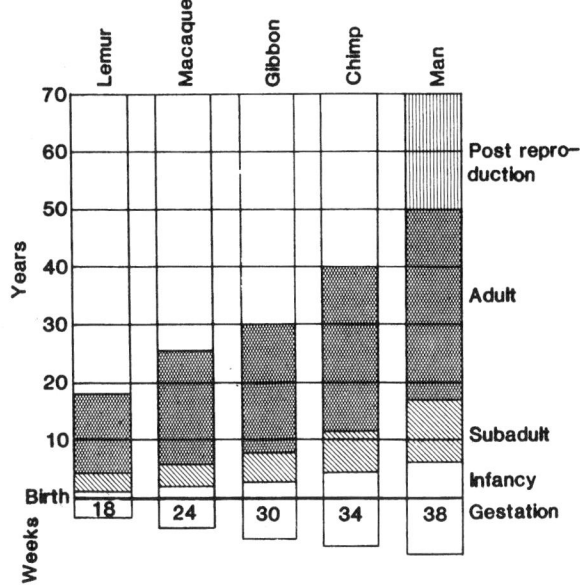

FIGURE 2.1. Progressive prolongation of life phases and gestation in primates. Note the proportionality of the four indicated phases. The post reproductive phase is restricted to man and is probably a recent development.[101, after 102]

the average period between successful births to be 5.6 years.[46] This can be attributed in part to a greatly prolonged period of subadult dependency. Van Lawick-Goodall's[47] description of the chimpanzee life phases is instructive:

> The infant does not start to walk until he is six months old, and he seldom ventures more than a few yards from his mother until he is over nine months old. He may ingest a few scraps of solid food when he is six months, but solids do not become a significant part of his diet until he is about two years of age and he continues to nurse until he is between four-and-a-half and six years old. Moreover, while he may travel short distances…when he is about four years old, he continues to make long journeys riding on his mother's back until he is five or six….

This extreme degree of parental investment has profound demographic consequences. A chimpanzee female does not reach sexual maturity until she is about 10 years old.[46] If she is to reproduce herself and her mate, that is, maintain a stable population, she must survive to an age of 21 years.[48] Whereas in rhesus macaques, the age is only about 9 years.[49,50]

Figure 2.2 shows a balance depicting the reciprocal relation between longevity and the primary demographic elements of parental investment. The two sides of this hypothetical balance are physiologically interdependent; as longevity is increased, each of the developmental stages is proportionately prolonged. The relationships between these variables, in fact, are not exactly linear, but they do have remarkably high correlations in most mammals.[44] As the scale indicates, greater longevity is

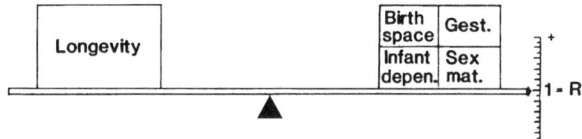

FIGURE 2.2. Mechanical model of demographic variables in hominoids. The R is the intrinsic rate of population increase (1 = static population size). An increase in the lengths of the four periods on the bar to the right (birth space, gestation, infant dependency, and sexual maturity) is accompanied by a comparable shift of longevity to the left, but without realization of that longevity, prolonged maturation reduces R and leads to extinction or replacement by populations in which life phases are chronologically shorter. Of the four variables on the right, only birth space can be significantly shortened (shifted to the left) without alteration of primate aging physiology.

accompanied by a proportionate delay in reproductive rate and therefore requires a female to survive to an older age in order to maintain the same reproductive value (measured at birth).[51] Put another way, the total reproductive rate of a primate species can remain constant with progressive increases in longevity only if the crude mortality rate is correspondingly reduced. Actual mortality rate is dependent on both maximum life potential, a genetic factor, and environmental interaction. Deaths caused by predation, accident, parasitism, infection, failure of food supply, and so forth, are at least partially stochastic events beyond the complete control of the organism. Only if mechanisms are developed to increase an organism's resistance to such factors, can the effects of increased longevity be reproductively accommodated. Strong social bonds, high levels of intelligence, intense parenting, and long periods of learning are among factors used by higher primates to depress environmentally induced mortality. It is of some interest that such factors also require greater longevity (for brain development, learning, acquisition of social and parenting skills) and that they constitute reciprocal links leading to greater longevity. This positive feedback system, however, has an absolute limit; environmentally induced mortality can never be completely under organism control, no matter how effective the mechanisms developed to resist it.

Suppose that late Miocene hominoids were approaching the effective limit of this feedback system or at least were sufficiently near the limit not to thrive in novel environments.[52] Two demographic variables could be altered to improve reproductive success—survivorship (the probability of surviving) and the time period between successive births (the birth space). All other factors are direct linear functions of mammalian developmental physiology and could not be altered. The argument is subject to the following simple quantification

$$\text{RV} = l(s) \int_{s}^{\text{MLP}} l(x)b(x)dx \qquad (1)$$

where RV is reproductive value of a cohort measured at birth, that is, the expected number of offspring produced by a unit radix; $b(x)$ is fertility at age x; $l(x)$ is survivorship at age x; s is age at sexual maturation; and MLP = maximum life potential. Assuming that a female gives birth at age s years and subsequently every β (birth space) years until reaching MLP, her total offspring would be given by

$$\frac{\text{MLP} - s}{\beta} \qquad (2)$$

Fertility is then seen to be dependent on birth space β according to

$$\int_{s}^{\text{MLP}} b(x)dx = \frac{\text{MLP} - s}{\beta} \qquad (3)$$

A simple solution (but one which is fully acceptable because of the proportionate relation between MLP and s) is $b(x) = 1/\beta$. The expression for RV then becomes

$$\text{RV} = \frac{1}{\beta}\left[l(s)\int_{s}^{\text{MLP}} l(x)dx\right] \qquad (4)$$

Because the term in brackets is independent of β, RV is inversely proportional to β, and RV is increased by a shorter birth space, by greater values of $l(x)$ for any age, or by both. Table 2.1 provides reproductive values for chimpanzees, Old World monkeys, and man from estimated values of β, s, and MLP under the simplifying assumption of $l(x) = l^x$. It can be seen from this table that both chimpanzees and humans have considerably lower reproductive values than Old World monkeys for low values of $l(x)$. As the values used for calculation are conservative, the existence of successful hominid clades in Pliocene mosaics suggests that both birth space reduction and elevation of survivorship had probably been accomplished. This is without explanation unless a major change in reproductive strategy accompanied occupation of novel environments by these hominids. Yet neither brain expansion nor significant material culture appear at this time level and were therefore not responsible for this shift.

A Behavioral Model for Early Hominid Evolution

Any behavioral change that increases reproductive rate, survivorship, or both, is under selection of maximum intensity. Higher primates rely on social behavioral mechanisms to promote survivorship during all phases of the life cycle, and one could cite numerous methods by which it theoretically could be increased. Avoidance of dietary toxins, use of more reliable food sources, and increased competence in arboreal locomotion are obvious examples. Yet these are among many that have remained under strong selection throughout much of the course of primate evolution, and it is therefore unlikely that early hominid adaptation was a product of intensified selection for adaptations almost universal to anthropoid primates. For early hominids we must look beyond such common variables to novel forms of behav-

TABLE 2.1. Relative reproductive values of Old World primates calculated from Equation 2.4 (see text) and multiplied by 10 for clarity.

	REPRODUCTIVE VALUES		
ANNUAL SURVIVORSHIP	OLD WORLD MONKEYS[a]	CHIMPANZEES[b]	MAN[c]
.90	17	4	2
.92	23	7	4
.94	31	13	9
.96	42	25	24
.98	58	50	64

[a] Maximum life potential = 20; sexual maturity = 4; birth space = 2 [49,50,103]
[b] Maximum life potential = 40; sexual maturity = 10; birth space = 3 [46,54,103]
[c] Maximum life potential = 60; sexual maturity = 15; birth space = 2.5

ioral change. The tendency has been to concentrate on singular, extraordinary traits of later human evolution such as intense technology, organized hunting, and the massive human brain. Yet these adaptations were not likely to have arisen de novo from elemental behaviors seen in extant nonhuman primates, such as the primitive tool using of the chimpanzee, in the absence of a broad selective milieu. It is more probable that significant preadaptations were present in early hominids that served as a behavioral base from which the "breakthrough" adaptations[53] of later hominids could progressively develop. We are therefore in search of a novel behavioral pattern in Miocene hominoids that could evolve from typical primate survival strategies, but that might also include important elements of other mammalian strategies, that is, a behavioral pattern that arose by recombination of common mammalian behavioral elements and that increased survivorship and birthrate.

In her essay on mother-infant relationships among chimpanzees, van Lawick-Goodall[54] noted two primary causes of mortality among infants: "inadequacy" of the mother-infant relationship and "injuries caused by falling from the mother." An intensification of both the quality and quantity of parenting would unquestionably improve survivorship of the altricial chimpanzee infant. The feeding and reproductive strategies of higher primates, however, largely prevent such an advancement. The mother must both care for the infant and forage for herself. A common method of altricial infant care in other mammals is sequestration of offspring at locations of maximum safety. Nests, lodges, setts, warrens, dreys, dens, lairs, and burrows are examples of this strategy. A similar adaptation in primates is usually not possible, however, because the need to forage requires both mother and infant to remain mobile. The requirement of mother-infant mobility is a significant cause of mortality and is at the same time the most important restriction on primate birth spacing.

Many primates display significant sex differences in foraging. Diet composition, selection of food items, feeding time, and canopy levels and sites differ in some species.[55] In at least *Pongo pygmaeus* and *Colobus badius*, males often feed at lower canopy levels than females.[56–58] In the gelada baboon, all-male groups "tended not to exploit quite the same areas as the reproductive units thus reducing indirect competition for food."[59] Clutton-Brock[55] notes that an increased separation of males from female-offspring foraging sites is advantageous where (i) animals feed outward from a fixed base, (ii) the adult sex ratio is close to parity, and (iii) feeding rate is limited by search time rather than by handling time, which is the time spent both preparing and consuming food. Similar feeding differences by sex are found in birds and other mammals.[60]

It is reasonable to assume that Miocene hominoids traveled between food sources on the ground and that these primates would be best characterized as omnivores.[12] These are ecologically sound assumptions. Increased seasonality coupled with already occurring local biotic variation (edaphic grasslands, savannah, woodland, forest)[8,12,39–41] would have presented variable and mosaic conditions. Occupation of heterogeneous ("patchy") environments and use of variable food sources favors a generalist strategy, whereas reliance on a homogeneous diet requires high food concentrations.[61,62] The time spent searching for food is greatest among generalists who live in food-sparse environments.[63] In short, Miocene ecological conditions support the view that feeding rate would have been more dependent on search time than handling time.

Greater seasonality and the need to increase both birthrate and survivorship would also favor at least partial separation of male and female day ranges since this strategy would increase carrying capacity and improve the protein and calorie supply of females and their offspring. Terrestriality, however, would require a centrifugal or linear displacement of males, as opposed to vertical stratification in canopy feeding. Given the Miocene conditions described above, such separation could become marked especially in the dry season. If such separation were primarily due only to an increase in the male day range, moreover, the range of the female-offspring group could be proportionately reduced by progressive elimination of male competition for local resources. This separation would be under strong positive selection. Lowered mobility of females would reduce accident rate during travel, maximize familiarity with the core area, reduce exposure to predators, and allow intensification of parenting behavior, thus elevating survivorship.[64] Such a division of feeding areas, however, would not genetically favor males unless it specifically reduced competition with their own biological offspring and did not reduce their opportunities for consort relationships. Polygynous mating would not be favored by this adaptive strategy because the advantage of feeding divergence is reduced as the number of males is reduced. Conversely, a sex ratio close to parity would select for the proposed feeding strategy. Such a ratio would obtain if the mating pattern were monogamous pair bonding. In this case, males would avoid competition with their bonded mates and biological offspring (by using alternative feeding sites) and not be disadvantaged by physical separation, that is, there would be no loss of consort opportunity. In short, monogamous pair bonding would favor feeding divergence by "assuring" males of biological paternity and by reducing feeding competition with their own offspring and mates.

Such a system would increase survivorship and would also favor any increase in the reproductive rate of a monogamous pair so long as the feeding strategy was sufficient to meet the increased load on the sources of protein and calories. One element of feeding among forest chimpanzees is the "food call" sometimes made by males upon discovery of a new food source.[65] In the proposed system, however, selection would not favor this behavior; instead, selection would favor a behavior that would benefit only the male's own reproductive unit. The simple alternative to the food call would involve collecting the available food item or items and returning them to the mate and offspring. Contrary to the opinion that such behavior would be altruistic, it would not be so in the proposed system, because it would only benefit the biological offspring of the male carrying out the provisioning and thus would be under powerful, direct selection. If this behavior were to become a regular component of the male's behavioral repertoire, it would directly increase his reproductive rate by correspondingly improving the protein and calorie supply of the female who could then accommodate greater gestational and lactation loads and intensify parenting.[66] The behavior would thus achieve both an increase in survivorship and a reduction in birth space. It would allow a progressive increase in the number of dependent offspring because their nutritional and supervisory requirements could be met more adequately.

Behaviors associated with similar reproductive strategies are in fact present in other primates. In both the Callitrichidae and Aotinae, extensive paternal care of the young constitutes a critical part of reproductive strategy in some species.[67,68] Among callitrichids, the social unit is usually an adult male and female, plus one to several subadults. Maternal care is largely restricted to suckling and grooming, the male being responsible for subadults at all other times. The modal birth is dizygotic twins.[55,68] It is likely that this system is a partitioning of care in response to the high protein and calorie requirements of these small species. Male care during foraging tends to equilibrate the high caloric load imposed on females by lactation and gestation of two (and sometimes three) offspring—the process of twinning being an obvious demographic adaptation of elevated birthrate. As Hershkovitz[68] notes: "survival of a population [of callitrichids] in the wild depends on close synchronization between cyclical nutritional requirements for young and old and the seasonal changes in the quality and quantity of available food." This same statement could be as well applied to early hominids, especially given increased Miocene seasonality and the need for a decrease in birth spacing. The altricial infants of Miocene hominoids, however, would have required reduced mobility and therefore prevented a callitrichid strategy of male care, with the simplest solution being the male provisioning model proposed above.

THE ORIGIN OF BIPEDALITY

Provisioning is, of course, the primary parental care strategy of most canids and birds.[69–71] Both groups exhibit direct male involvement similar to that described for callitrichids. Their offspring are normally immature at birth, immobile, and require constant provisioning and parenting. In some species, a sexual division of labor, like that posited here for early hominids, is observed. Female hornbills (Bucerotidae), for example, depend totally on male provisioning for their survival and that of their offspring. Monogamous pair bonding is characteristic of 90 percent of bird species[70,72] and is the most common mating system in provisioning canids.[69] Both groups, as a fundamental feature of reproductive strategy, commonly sequester their offspring at home bases.[73]

One critical difference separates provisioning in birds and canids from that suggested for early hominids. Birds and canids can carry in their mouths or regurgitate (or both) a significant proportion of their body weight. Oral carrying would have been inadequate for early hominids, however, and a strong selection for bipedality, which would allow provisions to be carried "by hand," would thus accompany provisioning behavior.[74]

Chimpanzees are fully capable of short-range bipedal walking and a variety of hindlimb stances,[75] but because they lack the pelvic and lower limb adaptations characteristic of hominids, bipedal walking leads to rapid fatigue.[23] It appears likely that the skeletal alterations for bipedality would be under strong selection only by consistent, extended periods of upright walking, and not by either occasional bipedality or upright posture. While primitive material culture does not impose this kind of selection, carrying behavior of the type suggested above, does. It is likely that the need to carry significant amounts of food was a strong selection factor in favor of primitive material culture.[76] Although it is not a significant shift from primitive tools of the type used by chimpanzees today, such as "termite sticks" and "leaf sponges," to simple and readily available natural articles that could be used to enhance carrying ability, it is a significant shift from such primitive and occasional tool use to the stone tools of the basal Pleistocene. Development of such tools is most likely to have followed an extended period of more primitive material culture, which was not critical to survival. It has been suggested frequently that the earliest tools were weapons. However, the progressive development of more advanced stone tools from rudimentary weapons is unlikely. A prolonged and extensive period of regular and habitual use of simple (primitive) carrying devices could eventually allow the coordination and pattern recognition necessary for a more advanced reliance on material culture.

The sequential evolution of behavior proposed in this article has a high probability of mirroring actual behavioral events during the Miocene. In most higher primates, male fitness is largely determined by consort success of one sort or another.[77] Male enhancement of offspring survival is for the most part indirect and is expressed more in terms of demic or kin selection by general behaviors such as territory defense or predator recognition and repulsion.[78] Females are solely responsible for true parenting and their ability in this is under strong selection. However, progressive intensification of higher primate K strategy elevates parenting requirements and lowers reproductive rate. The most obvious, and perhaps only, additional mechanism available with which to meet this "demographic dilemma" is an increase in the direct and continuous participation by males in the reproductive process. Whatever the actual sequence of events, whether as posed above or by some alternative order, such additional investment would improve survivorship and favor a mating structure that intensified energy apportionment to the male's biological offspring. Two mating patterns satisfy this latter requirement: polygyny (one male and several females) or monogamy. The former, however, requires male energy to continue to be devoted to maintaining consorts, and a pool of competing males is ensured by polygynous structure itself, thereby directing it away from direct enhancement of survivorship.

In their synthesis of the evolution of mating systems, Emlen and Oring[72] stress three factors common to polygynous mating structure. (i) One sex is predisposed to assume most, or all, of the parental care. (ii) Parental care requirements are minimal. (iii) A super-abundant food resource enables a single parent to provide full parental care. As has been noted above, however, survivorship of offspring must have been critical to Miocene hominoids; further female parenting is negated by the mobile feeding strategy; hominoid males may be considered an "untapped pool" of reproductive energy; and Miocene ecological conditions required a generalist feeding strategy. Conditions were prime for the establishment of male parental investment and a monogamous mating structure. Finally, it should be pointed out that only among primates in which the male is clearly and directly involved in the parenting process should monogamy be found. This is exactly the case, as this mating structure is found only in gibbons, siamangs, and the New World taxa discussed above.[55]

Human Sexual Behavior and Anatomy

The highly unusual sexual behavior of man may now be brought into focus. Human females are continually sexually receptive[79] and have essentially no externally recognizable estrous cycle; male approach may be considered equally stable. Copulation shows little or no synchronization with ovulation.[80] As was pointed out above, the selective emergence of a monogamous mating structure and male provisioning would require that males not be disadvantaged in obtaining consorts. Provisioning in birds and canids is normally made possible by highly restricted breeding seasons and discrete generations—the female normally is impregnable for only brief periods during which parental care is not required. The menstrual cycle of higher primates,[81] however, requires regular male proximity for reproductive success. The progressive elimination of external manifestations of ovulation and the establishment of continual receptivity would require copulatory vigilance in both sexes in order to ensure fertilization. Moreover, copulation would increase pair-bond adhesion and serve as a social display asserting that bond. Indeed, any sequestration of ovulation[82] would seem to directly imply both regular copulatory behavior and monogamous mating structure. It establishes mathematical parity between males restricted to a single mate and those practicing complete promiscuity, and the balance of selection falls to the offspring of pair-bonded males, since their energetic capacity for provisioning (and improved survivorship and reproductive rate) is maximized.

Man displays a greater elaboration of epigamic characters than any other primate.[8,59,83,84] Frequently, our sexual dimorphism is tacitly accepted as evidence for a polygynous mating structure because marked sexual dimorphism is most often a product of elaboration of characters of attraction, display, and agonistic behavior in males of polygynous species. Among primates, the degree of sexual dimorphism corresponds closely to the degree of male competition for mates.[56,59,83] Yet human sexual dimorphism is clearly not typical as is even made clear by the fossil record. In their discussion of *A. afarensis*, Johanson and White[17,85] note that although this species shows "marked body size dimorphism, the metric and morphological dimorphism of the canine teeth is not as pronounced as in other extant, ground-dwelling primates. This implies a functional pattern different from that seen in other primates and may have significant behavioral implications." There can be no doubt that large male canines are part of the "whole anatomy of bluff, threat, and fighting."[6] The reduction and effective loss of canine dimorphism in early hominids therefore serves as primary evidence in favor of the proposed behavioral model.[86] But it is important to stress that while canine dimorphism was undergoing reduction, other forms of dimorphism were apparently being accentuated, as judged from their expression in modern man, who remains the most epigamically adorned primate.

Since man displays a highly unusual mating structure, it is perhaps not surprising that his epigamic, or perhaps parasexual, anatomy is equally unusual and fully explicable by that mating structure. If pair bonding was fundamental and crucial to early hominid reproductive strategy, the anatomical characters that could reinforce pair bonds would also be under strong positive selection. Thus the body and facial hair, distinctive somatotype, the conspicuous penis of human males, and the prominent and permanently enlarged mammae of human females are not surprising in light of Mayr's[87] observation that in "monogamous species such as herons (egrets) in which the pair bond is continuously tested and strengthened by mutual displays, there has been a 'transference' of the display characters from the males to the females with the result that both sexes have elaborate display plumes." In man, however, marked epigamic dimorphism is achieved by elaboration of parasexual characters in both males

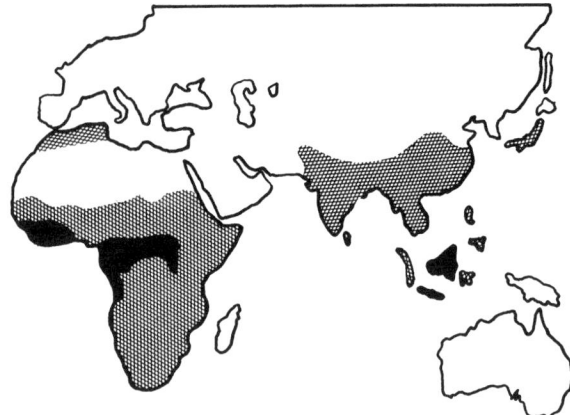

FIGURE 2.3. Approximate distribution of extant Old World monkeys (hatched) and pongids (gorilla, chimpanzee, orangutan) (solid).[38,58,102,103]

and females, rather than in males alone. Their display value is clearly cross-sexual and not intrasexual as in other primates. It should be stressed that these epigamic characters are highly variable and can thus be viewed as a mechanism for establishing and displaying individual sexual uniqueness, and that such uniqueness would play a major role in the maintenance of pair bonds.[59] This is especially important when other epigamic features of man (pubic, axillary, and scalp hair), which have been elaborated in both sexes, are considered. Such characters may also contribute to individual sexual uniqueness.[88] Redolent individuality is clearly the most probable role of axillary and urogenital scent "organs" (eccrine and apocrine glands plus hair), which are unique among mammals.[89] An objection that might be voiced in response to these suggestions is that such auxiliary pair-bond "enhancers" are eclipsed by the paramount role of culture in the mating practices of nontechnological societies. Quite the contrary, the more that culture can be shown to dominate the mating structure and process of recent man, the more ancient must be the anatomical-physiological mechanisms involved in the formation and maintainence of pair bonds.[90]

Higher Primate Paleogeography

The present-day geographic distributions of Old World monkeys and apes are shown in Figure 2.3. The great apes are markedly restricted and occupy only minor areas where minimal environmental changes have taken place since the early Miocene. Yet the fossil record shows that their lineal ancestors (dryopithecines, *sensu lato*) spread throughout the Old World following the establishment of a land bridge and forest corridor between Africa and Eurasia about 16 to 17 million years ago, and that they enjoyed considerable success after their colonization of Europe and Asia.[41,91] Old World monkeys, on the other hand, were much less abundant during this period.[92] After the middle and late Miocene, however, a marked reduction in dryopithecine numbers occurred. While this cannot be deduced from the sparse fossil record of the late Miocene and early Pliocene, the distribution of extant descendants of the dryopithecines is ample evidence of their relict status. Today, Old World monkeys are clearly the dominant and successful group, having replaced the dryopithecines and their descendants during the last 12 million years.[91] One hominoid group did survive and remain relatively abundant—the Hominidae. It is probable that the hominoid trends of prolonged longevity and increased parental investment are the key to the replacement of most pongid taxa by Old World monkeys, which are reproductively more prosperous. If only a portion of Miocene hominoids made the adaptations described above, two distinct groups would subsequently result. One group might counter the "demographic dilemma" according to the model suggested in this article; a second group could survive by occupying habitats with minimal environmental hazards. Hominids, being more demographically resistant to environmentally induced mortality, would be more capable of expanding into novel and varied habitats, especially mosaics, and of competing with the radiating Old World monkeys. Conversely, the extant pongids are by implication descendant of populations progressively more restricted to highly favorable forest conditions, where minimal seasonality in food supply, low predation pressure, and limited size of the home range would be in effect. These differences in habitat preference would result in a more extensive fossil record for hominids than pongids, both by virtue of the geographic expansion of hominids and as a consequence of the occupation of habitats with more favorable conditions of fossilization. It is therefore quite possible that the sivapithecines (*sensu lato*) of the middle and late Miocene, which already evince dental modifications adumbrating those of late Pliocene hominids,[12] may have contained primitive emergent hominids, at least behaviorally, if not phylogenetically.

The Nuclear Family

Man's most unique character is without question his enormous intelligence, and its evolutionary pathway has fascinated all who have attempted to explain the human career. Hunting and toolmaking are most frequently cited as "primal causes" for the Pleistocene acceleration in hominid brain development. Yet have these not figured so prominently because they leave ubiquitous evidence—the archeological record? Other human behaviors at least as critical to survival (especially reproductive behavior) are not "fossilized." It is now clear that man probably remained an omnivore throughout the Pleistocene and that hunting may have always been an auxiliary food source.[93]

As Reynolds[94] stressed, intense social behavior would seem the most likely single cause of the origin of human intelligence if one origin must be isolated. Tools are used to manipulate the environment and are thus a vehicle of intelligence, not necessarily a cause. Chimpanzees occasionally use tools (a behavior that has fascinated many early hominid theorists), but tools are not critical to their survival. Primates, which are the most intelligent mammals, have achieved evolutionary success primarily by their social and reproductive behavior, which is their most developed ordinal character. It seems reasonable therefore to propose that a further elaboration of this adaptive strategy is the most likely "cause" of early hominid success and the further development of intelligence.

It is of interest to explore one further effect of the proposed model on early hominid social structure. The strong maternal and sibling ties of higher primates are now well documented.[47,54,94,95] The matrifocal unit of chimpanzees continues throughout the life of the mother, as do sibling ties. In the proposed hominid reproductive strategy, the process of pair bonding would not only lead to the direct involvement of males in the survivorship of offspring, in primates as intelligent as extant hominoids, it would establish paternity, and thus lead to a gradual replacement of the matrifocal group by a "bifocal" one—the primitive nuclear family.[84,96] The effects of such a social unit on survivorship and species success could be profound. It could lead to a further shortening of birth space, which would accelerate the reproductive rate and amplify sibling bonds. Reduction of birth space would allow coincident protraction of the subadult (learning) period.[97] Behaviors that in other primates are common causes of infant death (for example, agonistic buffering)[78] would be largely eliminated, while those that might improve survivorship (for example, adoption)[46,98] would be facilitated. The age until which an orphaned chimpanzee does not survive the death of its mother is "around 5 years of age, but may stretch another 3 to 4 in special circumstances."[46] Survival of a second parent may have been a crucial reproductive advance in early hominids.[99] Primiparous females are much less adept than multiparous mothers. Drickamer[50] found that in free-ranging *Macaca mullata* "between 40 and 50% of the infants born first or second to a female did not survive their first year, but by the fourth infant born to the same female only 9% died during the first 12 months." Lancaster[100] notes that: "Recent field and laboratory workers have shown that in many species of mammals, and especially in monkeys and apes, learning and experience play vital roles in the development of the behavior patterns used in mating and maternal care." The effect of intensified parenting, protracted learning, and enhanced sibling relationships would have a markedly beneficial effect upon survivorship. Such projections of the behavior of developing hominids are certainly not new, but they have not received their due emphasis. Can the nuclear family not be viewed as a prodigious adaptation central to the success of early hominids? It may certainly be considered as being within the behavior repertoire of hominoid primates, provided that the reproductive and feeding strategies commensurate to its development were themselves under strong selection. This brief review of the fossil record and some primate behavioral and ecological adaptations would seem to strongly favor the correctness of this view.

Conclusion

It is a truism to say that even late Pliocene hominids must have been unusual mammals, both behaviorally and anatomically. As was pointed out above, emphasis in models of human origin has traditionally been on singular, extraordinary traits of later human evolution. The model proposed in this article has placed greater emphasis on a fundamental behavioral base from which these unusual adaptations could be directionally selected.

The proposed model accounts for the early origin of bipedality as a locomotor behavior directly enhancing reproductive fitness, not as a behavior resulting from occasional upright feeding posture. It accounts for the origin of the home base in the same fashion as it has been acquired by numerous other mammals. It accounts for the human nuclear family, for the distinctive human sexual epigamic features, and the species' unique sexual behavior. It accounts for a functional, rudimentary material culture of long-standing, and it accounts for the greater proportion of r-selected[45] characters in hominids relative to other hominoids. It accounts for these characters with simple behavioral changes common to both primates and other mammals and in relatively favorable environments, rather than by rapid or forced occupation of habitats for which early hominoids were clearly not adaptively or demographically equipped. It is fully consistent with primate paleogeography, present knowledge of higher primate behavior patterns (as well as those of other mammals), and the hominid fossil record.

If the model is correct, the conventional concept that material culture is pivotal to the differentiation and origin of the primary characters of the Hominidae is probably incorrect. Rather, both advanced material culture and the Pleistocene acceleration in brain development are sequelae to an already established hominid character system, which included intensified parenting and social relationships, monogamous pair bonding, specialized sexual-reproductive behavior, and bipedality. It implies that the nuclear family and human sexual behavior may have their ultimate origin long before the dawn of the Pleistocene.

Acknowledgments

I thank G. J. Armelagos, T. Barton, B. Campbell, T. Gray, F. C. Howell, K. Jacobs, D. C. Johanson, B. Kimbel, A. E. Mann, R. S. Meindl, R. P. Mensforth, M. H. Wolpoff, P. Shipman, A. C. Walker, T. D. White, and S. Ward, who read earlier versions of this paper and provided valuable comments. I thank D. C. Johanson, C. J. Jolly, J. Lancaster, R. S. Meindl, and T. D. White for valuable discussions about its content. I thank T. Barton for discussions and advice with respect to the quantitative approach used, L. Don Carlos and R. P. Mensforth for research assistance, and R. S. Meindl for listening to endless anecdotes about the behavior of canids, rodents, and birds.

References and Notes

1. M. Goodman, in *Molecular Anthropology,* M. Goodman and R. E. Tashian, Eds. (Plenum, New York, 1977), pp. 321–353; M.-C. King and A. C. Wilson, *Science* 188, 107 (1975); V. M. Sarich, in *Per-*

spectives on Human Evolution, S. L. Washburn and P. C. Jay, Eds. (Holt, Rinehart & Winston, New York, 1968), vol. 1, pp. 94–121; in *Background to Man,* V. M. Sarich and P. Dolhinow, Eds. (Little, Brown, Boston, 1971), pp. 182–191; D. E. Kohne, in *Perspectives on Human Evolution,* S. L. Washburn and P. Dolhinow, Eds. (Holt, Rinehart & Winston, New York, 1972), vol. 2, pp. 166 and 168; _____, J. A. Chiscon, B. H. Hoyer, *Carnegie Inst. Washington Yearb.* 69, 488 (1970); C. O. Lovejoy and R. S. Meindl, *Yearb. Phys. Anthropol.* 16, 18 (1972).

2. T. H. Huxley, *Evidence as to Man's Place in Nature* (Williams & Norgate, London, 1863); W. K. Gregory, *Science* 71, 645 (1930); A. Keith, *Br. Med. J.* 1, 451 (1923); *The Antiquity of Man* (Williams & Norgate, London, 1929), vols. 1 and 2; W. E. Le Gros Clark, *The Antecedents of Man* (Edinburgh Univ. Press, Edinburgh, ed. 2, 1962).

3. C. Darwin, *The Descent of Man* (John Murray, London, 1871).

4. G. A. Bartholomew and J. B. Birdsell, *Am. Anthropol.* 55, 481 (1953); A. E. Mann, *Man* 7, 379 (1972); S. L. Washburn, *The Study of Human Evolution* (Oregon State System Higher Education, Eugene, 1968); in *Classification and Human Evolution,* S. L. Washburn, Ed. (Aldine, Chicago, 1963), pp. 190–203; *Sci. Am.* 203, 3 (September 1960); I. DeVore, in *Horizons in Anthropology,* S. Tax, Ed. (Aldine, Chicago, 1965), pp. 25–6; C. F. Hockett and R. Ascher, *Curr. Anthropol.* 5, 135 (1964).

5. P. V. Tobias, *The Brain in Hominid Evolution* (Columbia Univ. Press, New York, 1971).

6. S. L. Washburn and R. L. Ciochon, *Am. Anthropol.* 76, 765 (1974).

7. R. L. Holloway, *ibid.* 69, 63 (1967).

8. C. J. Jolly, *Man* 5, 5 (1970).

9. Artifacts have been found in situ in the Gona region of the Hadar formation by the International Afar Research Expedition. Stratification of the Gona region is at present under investigation and correlations with the KH member are as yet uncertain. At present the artifacts are thought to overlie deposits equivalent to BKT$_2$ tuff of the KH member, which has a potassium-argon age determination of about 2.6 million years. The artifacts may therefore be older than 2 million years [D. C. Johanson, personal communication].

10. E. L. Simons, *Primate Evolution: An Introduction to Man's Place in Nature* (Macmillan, New York, 1972); in *(11),* pp. 543–566; D. Pilbeam, *The Ascent of Man: An Introduction to Human Evolution* (Macmillan, New York, 1972).

11. C. J. Jolly, Ed., *Early Hominids of Africa* (St. Martin's Press, London, 1978).

12. L. O. Greenfield, *Am. J. Phys. Anthropol.* 50, 527 (1979).

13. A. Kortlandt, in *Progress in Primatology,* D. Starck, R. Schneider, H.-J. Kuhn, Eds. (Fischer, Stuttgart, 1967); J. M. Goodall, in *Primate Behavior: Field Studies of Monkeys and Apes,* I. DeVore, Ed. (Holt, Rinehart & Winston, 1965), pp. 425–473; R. V. S. Wright, in *Human Evolution: Biosocial Perspectives,* S. L. Washburn and E. R. McCown, Eds. (Benjamin Cummings, Menlo Park, Calif., 1978), pp. 215–238.

14. A convincing and more detailed argument is provided by Jolly (8).

15. Whether or not the early evolution of material culture did in fact proceed in a gradualistic manner is difficult to establish. A punctuated equilibrium model is equally applicable to the early artifact record, and it is not unlikely that material culture proceeded at variable rates.

16. Contrary to popular opinion, there is no evidence whatsoever that early hominids hunted. Bipedality is probably the mode of locomotion least adapted to hunting, unless sophisticated technology is available (or unusually high levels of intelligence, or both). The evidence made available by *A. afarensis* is particularly striking. Further australopithecine evolution from that species is documented by a reduction of the anterior dentition and further enlargement of the grinding teeth. Artifacts do not appear until 2 million years ago, and when they do appear it is difficult to interpret them as hunting implements. In short, if the evidence made available by the fossil record is to be used in reconstructing early hominid evolution, one of its clearest implications is that hunting was not a dietarily significant behavior. See also *(8, 41, 62, 91, 102).*

17. D. C. Johanson and T. D. White, *Science* 203, 321 (1979).

18. L. Radinsky, *Am. J. Phys. Anthropol.* 41, 15 (1974); *Am. Sci.* 63, 656 (1975).

19. D. C. Johanson, T. D. White, Y. Coppens, *Kirtlandia* 28, 1 (1978).

20. Among the salient features of the *A. afarensis* cranium are a convex nasal clivus, a uniformly shallow palate anterior to the incisal foramen, an independent juga with individual sharp lateral margins of the nasal aperture, a true canine fossa, a rounded tympanic plate with an inferiorly directed surface, a compound temporal-nuchal crest in large and small individuals, an occipital plane that is short relative to the nuchal plane, and a relatively vertical nuchal plane and posterior positioned foramen magnum [W. H. Kimbel, in preparation]. See also *(17).*

21. R. L. Holloway, personal communication.

22. C. O. Lovejoy, *Am. J. Phys. Anthropol.* 50, 460 (1979).

23. _____ in *(11),* pp. 403–429.

24. _____, K. G. Heiple, A. H. Burstein, *Am. J. Phys. Anthropol.* 38, 757 (1973); C. O. Lovejoy, *Yearb. Phys. Anthropol.* 17, 147 (1974); in *Primate Morphology and Evolution,* R. H. Tuttle, Ed. (Mouton, The Hague, 1975), pp. 291–326.

25. C. R. Taylor and V. J. Rowntree, *Science* 179, 186 (1973); G. A. Gavagna, F. P. Saibene, R. Margaria, *J. Appl. Phys.* 19, 249 (1964).

26. I have elsewhere pointed out (23, 24) that the resting lengths of the major propulsive muscles about the hip and knee in quadrupedal primates are so substantially altered by the adoption of erect posture that the regular effective use of both quadrupedal and bipedal locomotion is not possible. Thus the transition to bipedality as a habitual mode of locomotion must have been relatively rapid.

27. D. C. Johanson, C. O. Lovejoy, A. H. Burstein, K. G. Heiple, *Am. J. Phys. Anthropol.* 44, 188 (1976); C. O. Lovejoy and D. C. Johanson, in preparation.

28. D. C. Johanson and T. D. White, *Science* 207, 1104 (1980).

29. D. C. Johanson and M. Taieb, *Nature (London)* 260, 293 (1976); D. C. Johanson and Y. Coppens, *Am. J. Phys. Anthropol.* 45, 217 (1976); M. D. Leakey, R. L. Hay, G. H. Curtis, R. E. Drake, M. K. Jakes, T. D. White, *Nature (London)* 262, 460 (1976); T. D. White, *Am. J. Phys. Anthropol.* 46, 197 (1977); Y. Coppens, F. C. Howell, G. L. Isaac, R. E. Leakey, Eds., *Earliest Man and Environments in the Lake Rudolf Basin* (Univ. of Chicago Press, Chicago, 1976); P. V. Tobias, *Annu. Rev. Anthropol.* 2, 311 (1973); R. E. Leakey, J. M. Mungai, A. C. Walker, *Am. J. Phys. Anthropol.* 35, 175 (1971); *ibid.* 36, 235 (1972); M.H. Day and R. E. Leakey, *ibid.* 39, 341 (1973); *ibid.* 41, 367 (1974); R. E. Leakey and B. A. Wood, *ibid.* 39, 255 (1973); M. H. Day, R. E. Leakey, A. C. Walker, B. A. Wood, *ibid.* 42, 461 (1974); *ibid.* 45, 369 (1976); F. C. Howell, *Nature (London)* 223, 234 (1969); _____, and B. A. Wood, *ibid.* 249, 174 (1974).

30. D. Pilbeam, G. E. Mayer, C. Badgley, M. D. Rose, M. H. L. Pick-

ford, A. K. Behrensmeyer, S. M. Ibrahim Shah, *Nature (London)* 270, 689 (1977).
31. D. R. Pilbeam, *Peabody Mus. Nat. His. Yale Univ. Bull.* 31, 1 (1969); E. L. Simons and D. R. Pilbeam, *Folia Primatol.* 3, 81 (1965); in *The Functional and Evolutionary Biology of Primates*, R. Tuttle, Ed. (Aldine, Chicago, 1972), pp. 36–62; D. R. Pilbeam, in (*32*), pp. 39–59; E. L. Simons, in (*32*), pp. 60–72.
32. P. V. Tobias and Y. Coppens, Eds., *Le Plus Anciens Hominides* (Centre National de la Recherche Scientifique, Paris, 1976).
33. While there is general agreement as to those morphological features directly associated with molar dominance in the group of hominoids referred to as phase II, there is some disagreement as to the distinctiveness of other dentognathic characters. See P. Andrews and I. Tekkaya, in (*32*), pp. 7–25.
34. L. O. Greenfield, thesis, University of Michigan(1976).
35. C. L. Brace and A. Montagu, *Human Evolution* (Macmillan, New York, ed. 2, 1977).
36. This is especially true of *Erythrocebus patas,* which is both a small-object feeder and the fastest ground-living primate [K. R. L. Hall, *J. Zool.* 148, 15 (1965)].
37. It should also be pointed out that changes in the masticatory apparatus of hominids are a reflection of changes in habitat and are not necessarily the initial cause of Clade differentiation. Care must be taken not to view those characters that allow identification of early hominids as synonymous with acutal forces of divergence.
38. A. Kortlandt, *Sci. Am.* 206, 128 (May 1962); *New Perspectives on Ape and Human Evolution* (Stichting Voor Psychobiologie, Amsterdam, 1972).
39. R. E. Moreau, *Proc. Zool. Soc. London* 121, 869 (1951); K. W. Butzer, *Am. Sci.* 65, 572 (1977); B. Campbell, in (*40*), pp. 40–58.
40. B. Campbell, Ed.; *Sexual Selection and the Descent of Man 1871–1971* (Aldine, Chicago, 1972).
41. P. Andrews and J. A. H. van Couvering, in *Approaches to Primate Paleobiology,* F. S. Szalay, Ed. (Karger, Basel, 1975), pp. 62–105.
42. M. Kretzoi, *Nature (London)* 257, 578 (1975); G. E. Kennedy, ibid. 271, 11 (1978).
43. Greenfield (*12*) concludes that "*Sivapithecus* utilized a broad range of zones including tropical rain forests (Chinji of India), subtropics (Rudabanya and Chinji-Nagri of India and Pakistan), and woodland and bush habitats (Late Nagri and Early Dhok Pathan of India and Pakistan, Fort Ternan)."
44. R. G. Cutler, *J. Hum. Evol.* 5, 169 (1976), and references therein.
45. R. H. MacArthur and E. O. Wilson, *The Theory of Island Biogeography* (Princeton Univ. Press, Princeton, N.J., 1967). The K and r are opposite ends of the continuum of reproductive strategy. In the r strategy, the number of offspring is maximized at the expense of parental care; at the K end (the effective limit of which is 1), parental care is maximized.
46. G. E. Teleki, E. Hunt, J. H. Pfifferling, *J. Hum. Evol.* 5, 559 (1976).
47. J. Goodall, in *Human Origins: Louis Leakey and the East African Evidence*, G. L. Isaac and E. R. McCown, Eds. (Benjamin, Menlo Park, Calif., 1976), p. 86.
48. A female chimpanzee reaches sexual maturity at about age 10 years. Using the average span of 5.6 years between successful births (*46*) gives a required life expectancy of about 21 years (a chimpanzee infant usually dies after his mother's death if it is not at least 4 years old). The authors (*46*, p. 577) conclude similarly that "The mean generation span, or elapsed time between birth of a female and birth of her median offspring, is about 19.6 years for a sample of ten Gombe females with three or more recorded births." Their most realistic estimate of achieved reproduction is three to four "offspring that are successfully raised to sexual maturity" (*46*, p. 580).
49. Demographic studies of Old World monkey populations are at present insufficient to provide accurate data for unprotected and undisturbed populations [D. S. Sade et al., *Yearb. Phys. Anthropol.* 20, 253 (1976)], but approximations can be made adequately from observations of protected or introduced populations. Ninety percent of adult females in the Chhatari population studied by Southwick and Siddiqi [ibid. 20, 242 (1976)] gave birth yearly during the 14-year study period. Infant mortality averaged 16.3 percent, and juvenile mortality was judged very low (the actual figure was 33 percent but most of this loss was attributed to trapping). A reasonable figure with this data is 7 to 9 years, from methods similar to those used in (*48*). L. C. Drickamer (*50*) found first birth to occur in females of 4 years on the average in free-ranging rhesus at La Parguera, Puerto Rico, and in animals of this age, 68 to 77 percent produced infants each year. Although first-year mortality was high for first and second born offspring (40 to 50 percent), it was only 9 percent for third born. These data support the above conclusion as do those of C. B. Koford [in *Primate Behavior: Field Studies of Monkeys and Apes,* I. DeVore, Ed. (Holt, Rinehart & Winston, New York, 1965, pp. 160–174] from rhesus on Cayo Santiago.
50. L. C. Drickamer, *Folia Primatol.* 21, 61 (1974).
51. R. A. Fisher, *Genetical Theory of Natural Selection* (Clarendon, Oxford, 1930).
52. It is in fact more likely that they were not, or at least not in the extreme forms seen in extant hominoids. The hominid adaptations proposed in this article are more likely to have been developed to prevent the "demographic dilemma." Modern pongids probably represent terminal phases of extreme parental investment only because of long-term occupation of particularly favorable environments, hence their very restricted present-day distribution.
53. By breakthrough adaptation is meant that which allows or precedes an adaptive radiation. While the Old World monkeys replaced all other hominoids, the hominids were successful in many of the same environments occupied by monkeys, including the Pleistocene savannahs.
54. J. Van Lawick-Goodall, in *Primate Ethology,* D. Morris, Ed. (Doubleday, New York, 1969), p. 424. It should also be pointed out that falls as a consequence of mother-infant travel would be a more critical selection factor in early hominids than other primates. Van Lawick-Goodall points out that "in striking contrast to most other primate species, the small chimpanzee infant often appears unable to remain securely attached to its mother if she makes a sudden movement. For several months after the birth of her infant the mother may have to support it, thus hindering her movements.…" Mortality as a consequence of falling may thus be viewed as one selection factor in favor of terrestrial care, but more importantly, the adoption of bipedality would mean the loss of most or all of the prehensibility of the infant foot, which is an important grasping organ in the chimpanzee infant. This, in turn, selects for a more secure infant carrying ability in the mother and thus bipedality. This form of selection can clearly not be viewed as the initial selective force for bipedal locomotion, but in conjunction with others, would certainly contribute to the total selective pattern.
55. T. H. Clutton-Brock, in *Primate Ecology,* T. H. Clutton-Brock, Ed. (Academic Press, New York, 1977), pp. 539–556.

56. _____, *J. Zool. London* 183, 1 (1977).
57. Clutton-Brock (*56*) notes that "although sex differences in feeding behavior are common among primates there is little evidence to suggest that they have evolved to minimize feeding competition between the sexes." Yet for the two species just cited this may indeed be the case. *Colobus badius* uses a more generalist feeding strategy than its sympatric congenerics, has a larger number of adult males within the troop [T. T. Struhsaker and J. F. Oates, in (*58*), pp. 165–186] and has the highest population density of any Old World monkey [T. H. Clutton-Brock, in (*58*), pp. 503–512]. It is clearly true of the orang (*64*).
58. R. W. Sussman, Ed., *Primate Ecology: Problem Oriented Field Studies* (Wiley, New York, 1979).
59. J. H. Crook, in (*40*), pp. 231–281.
60. R. K. Selander, in (*40*), pp. 180–230; *Condor* 68, 113 (1966); G. E. Hutchinson, *The Ecological Theater and the Evolutionary Play* (Yale Univ. Press, New Haven, Conn., 1965).
61. This forms the basis for an additional criticism of the hominid model proposed by Jolly (*8*). Gelada baboons are highly specialized feeders whose feeding rate is largely limited by handling time.
62. "Although nutritional factors alone would not preclude the possibility that early hominids were dietarily quite specialized, [from] the available archaeological evidence and [from] what is known of the dietary patterns of living gatherer-hunters and chimpanzees, it appears unlikely that all early hominids were almost exclusively carnivorous or herbivorous. It is more reasonable to suggest that the diet fell within the broad range of today's gatherer-hunter diets, but that within the wide spectrum of this adaptation, local environmental resources and seasonal scarcity may have forced some individual populations to become more dependent on vegetable or animal tissue foods than others" [A. E. Mann, personal communication: see "The evolution of hominid dietary patterns" in *Primate Dietary Patterns*, R. S. O. Harding and G. Teleki, Eds. (Columbia Univ. Press, New York, in press)].
63. E. R. Pianka, *Evolutionary Ecology* (Harper & Row, New York, 1974), and references therein.
64. D. A. Horr [in (*58*), p. 320] comments with respect to the orangutan: "…orang social organization might easily be explained as follows: In order not to overload the food supply, orangs disperse themselves in the jungles. Females carrying infants or tending young juveniles can best survive if they don't have to move far. Young orangs could also best learn the jungle in a restricted, familiar area…. Adult males are unencumbered by young and can more easily move over wider areas. This means that they compete with females for food only for short periods of time, and thus do not overload her food supply and force her to move over wider areas."
65. V. Reynolds and F. Reynolds, in *Primate Behavior*, I. DeVore, Ed. (Holt, Rinehart & Winston, New York, 1965), pp. 368–424; J. Van Lawick-Goodall, *Anim. Behav. Mono.* 1, 161 (1968); Y. Sugiyama, *Primates* 9, 225 (1968); ibid. 10, 197 (1969). See R. W. Wrangham [in *Primate Ecology*, T. H. Clutton-Brock, Ed. (Academic Press, New York, 1977), pp. 504–538] for further discussion on the possible functions of the "food call."
66. Such loads can become intense (up to 1.5 times normal resting basal metabolic rate in females) [O. W. Portman, in *Feeding and Nutrition of Nonhuman Primates*, R. S. Harris, Ed. (Academic Press, New York, 1970), pp. 87–116]. The modern human preparation for lactation is an average accumulation of 9 pounds subcutaneous fat [D. B. Jelliffe and E. F. P. Jelliffe, *Human Milk in the Modern World* (Oxford Univ. Press, Oxford, 1978)]. A major birth interval limitation in hominoid primates may well be the lactational loads placed on the mother (in contradistinction to the needs of the infant) and any improvement in feeding strategy could "support" a reduction in birth space on this basis.
67. G. Mitchell and E. M. Brandt, in *Primate Socialization*, F. E. Poirier, Ed. (Random House, New York, 1972), pp. 173–206; H. Kummer, *Primate Societies* (Aldine, Chicago, 1971).
68. P. Hershkovitz, *Living New World Monkeys* (Univ. of Chicago Press, Chicago, 1977), vol. 1.
69. H. C. B. Grzimek, Ed., *Animal Life Encyclopedia* (Van Nostrand Reinhold, New York, 1975), vol. 12.
70. D. Lack, *Ecological Adaptations for Breeding in Birds* (Metheun, London, 1968). Monogamy is especially characteristic of long-lived (K-selected) birds [J. W. F. Davis, *J. Anim. Ecol.* 45, 531 (1976); A. Mills, ibid. 42, 147 (1973); J. C. Coulson, ibid. 35, 269 (1966)].
71. J. F. Eisenberg, *Handb. Zool.* 10, 1 (1966).
72. S. T. Emlen and L. W. Oring, *Science* 197, 215 (1977).
73. Such sequestration is common among rodents as well with perhaps its most classic expression in castorids, which are comparatively K-selected (requiring 3 to 4 years to mature sexually) and live in stable family groups.
74. A second, also important, element of food-handling behavior may have been premastication. Reduction of birth space would have required an earlier reinitiation of ovulation, which would in turn have required a reduction of mechanical stimuli to the mechano-receptors of the nipple and areola (and thereby a reduction in prolactin levels) [R. C. Kolodny, L. S. Jacobs, W. H. Daughaday, *Nature (London)* 238, 284 (1972)], and hence an earlier age of weaning. Parental premastication would have facilitated such behavior, at the same time enhancing parental bonds (see discussion of the nuclear family). This could have also increased the rate of dental wear and have been an auxiliary selection component of the dentognathic changes characteristic of early hominids.
75. H. R. Bauer, *Primates* 18, 913 (1977).
76. R. E. Leakey and R. Lewin, *People of the Lake* (Doubleday, New York, 1978); G. Hewes, *Am. Anthropol.* 63, 687 (1961).
77. R. L. Trivers, in (*40*), pp. 136–179; R. P. Michael and D. Zumpe, *J. Reprod. Fertil.* 21, 199 (1970); T. E. Rowell, in *Social Communication Among Primates*, S. Altman, Ed. (Univ. of Chicago Press, Chicago, 1967), pp. 15–32; I. Devore, in *Sex and Behavior*, F. Beach, Ed. (Wiley, New York, 1965), pp. 266–289; *see* (*36, 67*).
78. S. H. Hrdy, in *Advances in the Study of Behavior*, J. S. Rosenblatt, R. A. Hinde, E. Shaw, C. Beer, Eds. (Academic Press, New York, 1976), vol. 6, pp. 101–158.
79. D. C. Johanson, personal communication.
80. R. A. McChance, M. C. Luff, E. E. Widdowson, *J. Hyg.* 37, 571 (1937); J. R. Udry and N. M. Morris [*Nature (London)* 220, 593 (1968)] did find such a relationship, but it is a moot point since with sequestration of ovulation and its external manifestations, copulation would require female initiation.
81. F. A. Beach, *Psychol. Rev.* 54, 297 (1947); M. A. Chance, in *Culture and the Evolution of Man*, M. F. Ashley Montagu, Ed. (Oxford Univ. Press, New York, 1962), pp. 84–130.
82. J. Lancaster, personal communication.
83. D. Morris, *The Naked Ape* (Cape, London. 1967).
84. F. A. Beach, in *Human Evolution: Biosocial Perspectives*, S. L. Washburn and E. R. McCown, Eds. (Benjamin-Cummings, Menlo Park, Calif., 1978), pp. 123–154.
85. In other primates that are monogamous, there is little sexual di-

morphism. All of these, however, live in territorial family groups [A. Jolly, *The Evolution of Primate Behavior* (Macmillan, New York, 1972)] and there is therefore no intragroup competition for mates. Strong sexual dimorphism is usually a consequence of either differential competition for mates or differential exploitation of resources (*56, 60*). In *A. afarensis*, according to the proposed model, dimorphism would be favored on the latter basis. Small female body size would reduce caloric-protein requirements, while large body size would increase male mobility and predator resistance.

86. This is not meant to imply that it was a cause of canine reduction, but only that the process could occur by a combination of relaxation of selection on large male canine size and a positive selective mechanism for reduced canines. The latter is most likely to be found in the concurrent dentognathic changes of greater molar dominance and general anterior tooth reduction (*8, 17*).
87. E. Mayr, in (*40*), p. 97.
88. Modern man displays a remarkable number of morphological traits that may be considered epigamic (hair color and type, lip size and form, corporal hair patterning, eyebrows, facial countenance, and so forth). Attempts have been made to correlate some of these with geographic variables, but they have been largely unsuccessful. An alternative explanation is that disruptive selection acts to maximize the variability of these features within populations, thereby enhancing the distinctiveness of potential and actual mates in establishing and maintaining pair bonds. The subsequent geographic isolation, whether partial or complete, of a population could then have resulted in a truncation of expression and apparent uniqueness of some features that maintained their epigamic significance in the population (for example, the epicanthic eye fold). The obvious polygenic basis of such traits and their reappearance in unrelated populations (Bushman, Lapps, infant Euroamericans) indicate that their expression is a consequence of elevated frequencies of genes that may be universal in *H. sapiens,* but below an expressive threshold in some populations.
89. W. Montagna, in *Biological Anthropology*, S. H. Katz, Ed. (Freeman, San Francisco, 1975), pp. 341–351.
90. Further evidence of the age of pair bonding is provided by the absence of strong canine dimorphism in *A. afarensis* (*17*). The only other Old World higher primates without canine dimorphism are the gibbon and siamang, which are monogamous (*6*).
91. E. Delson, in *Approaches to Primate Palaeobiology*, F. S. Szalay, Ed. (Karger, Basel, 1975), pp. 167–217.
92. _____, and P. Andrews, in *Phylogeny of the Primates*, W. P. Luckett and F. S. Szalay, Eds. (Plenum, New York, 1976), pp. 405–446.
93. R. B. Lee, in *Man the Hunter*, R. B. Lee and I. DeVore, Eds. (Aldine, Chicago, 1968), pp. 30–48. The provisioning model proposed here effectively accounts for the origin of hunting by means of a progressive elaboration of provisioning behavior (that is, collecting → scavenging + collecting → hunting + scavenging + collecting) without the requirement that hunting be critical to human evolution at any point. The similarity in social behavior between canids and early humans has often been cited and attributed to hunting. It is more likely that such similarities take origin in reproductive strategy (pair bonding, intratroop cooperation, provisioning, male involvement in subadult care, and so forth) and that hunting merely represents one food procurement method that satisfies the economic requirements of the social system. There are numerous carnivores that do not display this form of reproductive strategy, and there are some rodents that do but, of course, do not hunt (*73*).
94. V. Reynolds, *The Biology of Human Action* (Freeman, San Francisco, 1976).
95. J. H. Kaufmann, *Ecology* 40, 500 (1965); M. Yamada, *Primates* 4, 43 (1963); C. B. Koford, *Science* 141, 356 (1963).
96. The term bifocal is preferable to nuclear family because the latter carries manifest connotations from its application to Western and non-Western modern human cultures, none of which are implied by its use here.
97. A. E. Mann's extensive studies of dental development and wear in australopithecines ["Some palaeodemographic aspects of the South African australopithecines" (*Univ. Penn. Publ. Anthropol.* 1, 1975)] indicate a prolonged period of development was established by about 2.5 million years ago. K. R. McKinley's survivorship calculations based upon Mann's data [*Am. J. Phys. Anthropol.* 34, 417 (1971)] led him to conclude that australopithecines show a hominid rather than a nonhuman primate "birth spacing pattern." While these calculations require a number of assumptions about the origin and nature of the death assemblages at Swartkrans and Sterkfontein, they are strong evidence that a major demographic shift was fully developed by 2.0 to 2.5 million years ago which included an extended period of subadult dependency.
98. H. Kummer, in *Social Communication Among Primates*, S. Altman, Ed. (Univ. of Chicago Press, Chicago, 1967), pp. 63–72; D. S. Sade, in ibid., pp. 99–115; see (*47, 54*).
99. M. H. Wolpoff, cited in *Mosaic* 10, 28 (1979).
100. J. B. Lancaster, in *Primate Socialization,* F. E. Poirier, Ed. (Random House, New York, 1972), pp. 83–104.
101. C. O. Lovejoy, R. S. Meindl, T. R. Pryzbeck, T. S. Barton, K. G. Heiple, D. Kotting, *Science* 198, 291 (1977).
102. A. H. Schultz, *The Life of Primates* (Universe, New York, 1969).
103. J. R. Napier and P. H. Napier, *A Handbook of Living Primates* (Academic Press, New York, ed. 3, 1970).

3

Grandmothering and the Evolution of *Homo erectus*

James F. O'Connell, Kristen Hawkes, Nicholas G. Blurton Jones

Homo erectus presents an important and challenging problem to students of human evolution. A distinctive, well-defined fossil form, it was far more widely distributed than any previous hominid, extant for more than a million years, and behaviorally different from both apes and modern humans, though in ways that are not yet entirely clear. Conventional wisdom has long associated its evolution and dispersal with big game hunting and paternal provisioning. These practices are thought to have stimulated the development of other modern human-like patterns of behavior, including extended periods of juvenile dependence, central place foraging, a sexual division of labor, and the nuclear family. Until recently, strong support for this argument came primarily from the archaeological record, particularly from the common association between stone artifacts made by *H. erectus* and the remains of large animals allegedly butchered and transported from kill sites to distant base camps for further processing, sharing, and consumption. The apparent match between this archaeological "signature" and those produced by modern hunter-gatherers led to the inference that other behaviors associated with the latter were also present in *H. erectus*. Key elements of the modern pattern, minus language and certain cognitive capabilities, were projected well back into the Pleistocene; arguments about the subsequent emergence of *Homo sapiens* were structured accordingly.

Appealing as this argument once was, the results of recent primatological, ethnographic, and archaeological research undercut it. Hunting is now known to be far more common among non-human hominoids than previously appreciated; yet it does not involve paternal investment. Big game hunting among modern tropical foragers, though sometimes productive, has been shown to be a poor strategy for feeding a family. Even the significance of archaeological assemblages formerly seen as providing clear evidence for ancient hunting is contested, some analysts arguing that they indicate little more than passive scavenging at kills made by other predators. If big game hunting, central place foraging, and paternal provisioning are eliminated from the *H. erectus* mix, then support for contingent inferences about modern human-like social organization and mating arrangements disappears. Reconstructions of its behavior are reduced to intriguing but somewhat disconnected inferences based on skeletal anatomy and archaeology; questions about its evolutionary origins and persistence are largely begged.

Here we offer a different model, one that has been foreshadowed in several ways over the past two decades, but only now developed in comprehensive form. We consider the proposition that *H. erectus* evolved as a result of climate-driven changes in *female* foraging and food sharing practices, possibly involving the exploitation of tubers. These changes may not only have had important effects on ancestral human ecology and physiology, but could also have provoked the first fundamental move away from hominoid life history patterns.

In the following pages, we briefly review the conventional wisdom on *H. erectus* ecology and the current critique thereof. We then turn to our alternative model, organizing empirical and theoretical elements developed elsewhere in terms of a series of predictions about relationships between past environmental change, adjustments in female foraging and food sharing, and their implications for hominid life history and ecology. We draw attention to recently proposed inferences about *H. erectus* life history; then review the evidence for anticipated climatic, environmental, and economic correlates. Though critical data are patchy, results are generally consistent with expectations. Potentially productive avenues for further research are clearly indicated. We comment on the implications of these results for widely held ideas about a much later date for the first appearance of long post-reproductive human lifespans, and for the notion that male philopatry and female dispersal are basic features of social organization among all members of the African ape clade, including ancestral hominids. We conclude with some remarks on the archaeological evidence that continues to shape conventional wisdom on *H. erectus*.

---

*Originally published in *Journal of Human Evolution* 36(1999):461–485

Homo erectus: A Critique of Conventional Wisdom

Homo erectus is a large-bodied hominid, similar in size and overall form to modern humans, but with a different cranium and a smaller brain (Rightmire 1990, 1998; Walker & Leakey 1993). The earliest examples (sometimes called *H. ergaster*, Wood 1992) are from Africa, where they date to about 1.8 Ma (Feibel et al. 1989; White 1995). Near-contemporary specimens may be represented from the Caucasus, south China, and Indonesia (Brauer & Schultz 1996; Gabunia & Vekua 1995; Huang et al. 1995; Swisher et al. 1994), suggesting an early spread throughout the Old World tropics and into the temperate zone, a proposition consistent with emerging archaeological data (e.g., Dennell & Roebroeks 1996; Dennell et al. 1988; Gibert et al. 1998; Tchernov 1989). Once dispersed, *H. erectus* persisted, with little change in either physical form or geographical range, into the late Middle Pleistocene (<500 ka), when various populations evolved to or were displaced by *H. sapiens*.

Important inferences about *H. erectus* ecology and life history are drawn from its morphology, especially in contrast with that of contemporary and earlier hominids. Its modern humanlike limb proportions are read to indicate fully terrestrial bipedality; its larger body size and more linear form a broader foraging range and higher tolerance for aridity; its thoracic cavity a simpler gut and correspondingly higher quality diet; its reduced sexual dimorphism the presence of multi-male, multi-female social groups; its dental eruption sequence and brain size and age at maturity and average adult lifespan intermediate between those of australopithecines and modern humans (Aiello & Wheeler 1995; McHenry 1994; Ruff 1994; Smith 1993; Walker & Leakey 1993).

Until recently, equally important inferences about its behavior were drawn from the contemporary archaeological record. Generally speaking, early Pleistocene archaeological sites are larger in size, more diverse in terms of assemblage composition, and found in a broader range of habitats, both locally and globally, than those of the late Pliocene. Their associated faunal assemblages are often especially striking. Many include the remains of one of more (sometimes many more) large animals, mainly ungulates, often of several species. Many of the bones have been damaged, some by stone tools, apparently in connection with processing for consumption of associated meat and marrow.

These data were formerly interpreted by reference to presumed differences in the behavior of great apes and modern human foragers (e.g., Isaac 1978; Leakey 1971; Washburn & Lancaster 1968; see also Binford 1981; Fedigan 1986; Hill 1982; Isaac 1984; Sept, 1992 for critical review). Special attention was drawn to nuclear families, central place foraging, and a sexual division of labor in which men hunt (mainly big game) to provision their wives and children; all thought to be typical of modern hunter-gatherers and pre-agricultural humans in general, but unrepresented among the great apes. Since some of the earliest faunal assemblages contained elements that had arguably been transported from distant kill sites, presumably by males who gave the meat to dependent mates and offspring, then other aspects of modern human behavior might be, and often were, inferred accordingly (see especially Isaac 1978).

The evolution of this pattern was explained by appeal to a long-term trend toward cooler, drier climate that reduced the availability of previously important plant foods while favoring the spread of game-rich savannas. Ancestral hominid males were thought to have responded by adding large animals to their diets, thereby producing a potentially sharable resource. Hominid females paired with hunters to ensure access to the new food, which in turn enabled them to reduce their own foraging efforts. Nuclear families, a sexual division of labor, and paternal provisioning were established as a result. Female fertility and offspring survivorship were enhanced; an extended period of juvenile dependence, larger brain size, increased learning, and greater behavioral flexibility were all underwritten. Greater ability to cope with environmental variation, significant increases in geographical range, and long-term evolutionary success followed.

Over the past 20 years, this "hunting hypothesis" has been undercut in three important ways:

- Primates, particularly chimpanzees, are now known to hunt often (Boesch & Boesch 1989; Stanford 1996; Uehara et al. 1992; Wrangham & Bergman-Riss 1990). Most hunting is done by males; the meat obtained is widely shared; yet there is no evidence for central place foraging or paternal provisioning.

- Evidence for pervasive paternal provisioning among modern human hunters has also been challenged (Hawkes 1990, 1993; Hawkes et al. 1991, 1998). In the best known tropical foraging populations, men consistently pursue large game rather than other resources, despite the fact that returns are *highly* variable in the short run and impossible to defend from other claimants once in hand. If paternal provisioning were truly an important goal, they would do better by spending more time on small game and plant foods, both of which produce more reliable income and are often easily secured for family consumption. The fact that they target either less regularly than would meet this goal strongly suggests that big game hunting serves some other end, unrelated to provisioning wives and children.

- Archaeological evidence for big game hunting and paternal provisioning by *H. erectus* has also been re-evaluated. At some important sites, analysis of damage patterns on bones shows *no* evidence of hominid involvement (e.g., Binford 1981; Klein 1987). At others, the data can be read to indicate little more than "passive" hominid scavenging, mainly of long bone marrow and brain cavity contents (Binford 1981; Blumenschine 1991; Marean et al. 1992).[1] If accurate, this means that the amounts of edible tissue so acquired were probably too small and obtained too irregularly to provision juveniles (O'Connell et al. 1988b). Hunting and aggressive scavenging may also have been practised (Bunn & Ezzo 1993; Capaldo 1997; Dominguez-Rodrigo 1997); but even if they were, it is unlikely that they were reliable enough to meet the *daily* nutritional needs of younger offspring, especially since active hunters with better weaponry (bows and poisoned arrows) living in similar habitats today cannot do so consistently (Hawkes et al. 1991, 1998). Finally, there is no good evidence for the transport of meat by early hominids to distant "central places." The archaeological criteria formerly used to support this inference are now seen to be themselves unsupported (O'Connell et al. 1988a, 1990). Large animals represented in early archaeological assemblages may well have been processed and consumed by hominids at or near the point of initial acquisition (e.g., Marean et al. 1992; O'Connell 1997).

These observations have very important implications: they eliminate all standard justifications for inferences about nuclear families and a modern human-like sexual division of labor in *H. erectus*. This raises key questions about *H. erectus* ecology and evolution. Current approaches to answering them fall into two categories: (1) those that entail continued insistence on the importance of big game hunting, central place foraging, and the prevalence of near-modern human patterns of social organization and reproduction in *H. erectus, despite* compelling challenge (e.g., Gowlett 1993; Leakey & Lewin 1992; Stanley 1996); (2) those that respond to the challenge by treating the archaeological record much more inductively, using recent advances in taphonomy to reconstruct whatever possible about past hominid behavior, but with few exceptions about the shape it might take, apart from being "not modern" (e.g., Bunn & Ezzo 1993; Potts 1988; Rogers et al. 1994).

However commonly adopted, the first approach is indefensible. The second attends to archaeological evidence simply because it is easily recognized and relatively easily interpreted. The goal of accounting for it—of identifying and describing all the processes, human and non-human, involved in its creation—takes center stage. Though there are good historical reasons for the change in focus (compare Binford 1981; Isaac 1983), its net effect has been to beg the larger, ultimately more important ecological and evolutionary questions Pleistocene archaeology was once seen to address. The long-running dispute about whether early Pleistocene zooarchaeological assemblages represent hunting or some form of scavenging illustrates the problem. As currently phrased, and unlike the argument originally developed by Isaac (1978), *no other interesting questions about hominid behavior are resolved by the answer,* mainly because they are no longer asked. The only way of identifying such questions, and, by extension, the data pertinent to addressing them, is by developing and evaluating comprehensive models of early Pleistocene hominid ecology and evolution comparable to, but better-warranted than, the now-discredited hunting hypothesis.[2]

THE GRANDMOTHER HYPOTHESIS[3]

One such model, the "grandmother hypothesis," has so far been developed as follows:

- Observations among modern hunter-gatherers show the importance of older women's foraging when mothers of childbearing age rely on resources that weaned children cannot handle on their own. This suggests that mother-child food sharing could have favored the evolution of increased post-menopausal longevity that distinguishes humans from other hominoids.

- Several key life history attributes vary systematically with adult lifespans across the mammals generally and among primates in particular. If the underlying tradeoffs hold for humans, and longer adult lifespans are due to grandmothering, then the other departures from "typical ape" life histories that characterize our species may be (unexpectedly) explained as well.

- Since the ecological circumstances that would have favored increased mother-child food sharing and related changes in life history can be specified, the grandmother hypothesis provides the basis for an evolutionary scenario that can be evaluated in light of pertinent paleoclimatic, environmental, fossil and archaeological evidence.

Hadza Women's Foraging and Food Sharing

Our model is grounded on the results of fieldwork with the Hadza, a small population of traditional foragers living in the arid savanna woodlands of the Eastern Rift, northern Tanzania (Blurton Jones et al. 1996).

Apart from the very old and very young, Hadza of all ages are active, productive foragers. Time allocation and foraging returns are particularly striking for senior females and younger children. Women in their 60s and early 70s work long hours in all seasons, often with return rates equal to (sometimes greater than) those of their reproductive-age female kin (Hawkes et al. 1989, 1995, 1997). Hadza children are involved in the food quest virtually from the time they can walk, and by the age of five can and do supply, in some seasons, up to 50% of their daily nutritional requirements by their own efforts (Blurton Jones et al. 1989, 1994a, 1997; Hawkes et al. 1995, 1997).

Hadza mothers and grandmothers routinely capitalize on children's foraging capabilities by targeting resources that youngsters can take at high rates, notably fruit. Sometimes this involves bypassing items from which women earn better returns, but that children cannot handle. These choices mark an effort to maximize "team" returns, those earned by women and children together. In the wet season, when fruit is widely available, children's foraging opportunities largely determine adult female foraging strategies (Hawkes et al. 1995).

When resources easily taken by children are unavailable (especially in the dry season), Hadza women provision their offspring with foods they can procure reliably and efficiently. A good example is the woody rootstock, *Vigna frutescens* (Hadza:

//*ekwa*), which favors deep stony soils and requires both substantial upper body strength and endurance to collect and the ability to make and control fire to process. Adult women, including seniors, take it often in all seasons, routinely earning up to 2000 kcal/h as a result (Hawkes et al. 1989, 1995, 1997; Vincent 1985a,b). Pre-adolescents seldom pursue it, and rarely gain more than about 200 kcal/h when they do (Blurton Jones et al. 1989). Youngsters under 8 years old ignore it entirely.

This provisioning has an important ecological implication: it allows the Hadza to operate in habitats from which they would otherwise be excluded if, as among other primates, weanlings were responsible for their own subsistence.

It also creates the opportunity for another adult to influence a mother's birth-spacing: if someone else supplies food for her weaned but still dependent child, she can have the next baby sooner. Under these circumstances, grandmother, whose fertility has declined, can have a large impact on her own fitness by feeding the weaned children of her younger kin. Analyses of time allocation, foraging returns, and children's nutritional status (measured by seasonal changes in weight) provide a compelling measure of her effect (Hawkes et al. 1997). In families where mother is *not* nursing, children's nutritional status varies in accordance with mother's own foraging effort. At the arrival of a newborn, however, mother's foraging time drops and the correlation between her foraging effort and her weaned children's weight changes disappears. Instead, those weight changes vary closely with the effort of a related senior female, usually grandmother.

This suggests an hypothesis to account for the differences in average adult lifespans among the hominoids. Child-bearing careers in humans and apes are similar in length, but humans survive far longer after menopause. The dependence of weaned children on food from adults would have allowed ancestral human grandmothers to affect their fitness in ways that other apes could not, increasing the strength of selection against senescence, lowering adult mortality rates, and so lengthening average adult lifespans.

Female Food Sharing and the Evolution of Human Life Histories

Life histories differ widely among the mammals: some grow fast and die young; others mature slowly and live long adult lives (e.g., Harvey & Read 1988). Although this variation correlates with body size, relationships among life history features persist even when the effect of body size is removed. Some are strong enough to be labeled "approximately invariant." Charnov (1993) has developed a model to account for them in which adult mortality rates set the tradeoffs that determine optimal age at maturity. Annual fecundity varies with both. Though very simple, the model also captures and accounts for important differences between primates and other mammals (Charnov & Berrigan 1993).

If the grandmother hypothesis explains extended human lifespans, and if human life histories maintain the broad patterning apparent across primates, then other human life history traits should be adjusted to predictable values relative to those observed in other living hominoids. Specifically, age at maturity should be delayed as a function of reduced adult mortality rates, but instead of the lower annual fecundity that normally goes with later age at maturity in other primates, grandmother's help pays off in higher fertility. Comparison of averages for modern human foragers and wild populations of chimpanzees, gorillas, and orangutans yields results consistent with these expectations (Hawkes et al. 1998a). Not only do we have longer lifespans, but, as predicted by the combination of the grandmother hypothesis and Charnov's life history model, we also mature later and produce offspring at a higher rate.

An Evolutionary Scenario

Our analysis of Hadza women's foraging and food sharing leads us to propose a set of closely related hypotheses about the evolution of these distinctive features of human life history (Hawkes et al. 1997, 1998a,b). Imagine an ancestral hominid with life history characteristics and foraging patterns comparable to those of the modern chimpanzee. In particular (and unlike modern humans), age at maturity was about 10–12 years and fecundity was relatively low. Children were sometimes fed by mothers and older siblings, particularly with items they themselves could not handle, but the overall importance of these foods was marginal. The fertility of older females declined sharply in tandem with other aspects of physiology; maximum lifespan was about 50 years.

Imagine further a significant change in environment that reduced the availability of resources that younger juveniles could take on their own. Under these circumstances, local populations might have adjusted their foraging ranges, perhaps abandoning some areas entirely. Alternatively, they might have invested more in provisioning, especially with resources that may have been avoided before because although adults could handle them effectively, young children could not. For the strategy to be effective, returns must have been high enough to support the collector and at least one other individual. They must also have been available on a daily basis, with relatively low variance in returns between collecting bouts. Otherwise, their utility to small, growing youngsters would have been limited.

As provisioning became established, older females who were slightly more vigorous, despite declining fertility, could have assisted in the process, enhancing the survivorship of youngsters they helped while allowing the mothers of those offspring to begin a new pregnancy sooner. Less vigorous menopausal females would have provided less help. Higher reproductive success for the junior kin of more vigorous older females would have reduced the relative frequency of deleterious alleles expressed around menopause. Higher reproductive success for young adults with older helpers would also alter the tradeoffs between allocation to current reproduction in early adulthood versus allocation to maintenance for later adaptive performance. The help of vigorous oldsters could more than compensate for reduced allocation to current reproduction by the junior kin themselves. Selection against senescence would be strengthened by both these pathways, decreasing adult mortality rates so that more would live to peri-menopausal, then postmenopausal ages. Longer adult lifespans would in turn have an effect on age at maturity. Lower adult mortalities increase the likelihood of reproducing before dying. Consistent with general mammalian (including primate) patterns (Charnov 1993), delayed maturity, a longer period of growth, larger adult body size, and later age at maturity would have followed as a result. Extended fertility would *not* have been favored as it would have interfered with assistance to grandchildren and the enhanced fecundity at younger ages enjoyed by the daughters of older helpers. Instead, a fertile span similar to that of the other apes would have been conserved, the derived feature being extended post-menopausal longevity.

Increased offspring provisioning and related changes in fitness would also have had important ecological implications (Hawkes et al. 1997, 1998b). High juvenile mortality rates in modern primates are often attributable to feeding competition (van Schaik 1989). If this were the case in the ancestral hominid population, then any increase in offspring provisioning should have reduced juvenile mortality. If the resources involved occurred in dense patches, with returns limited by handling requirements rather than by abundance, then their use should also have allowed the formation of larger foraging groups (Janson & Goldsmith 1995; Wrangham et al. 1993). These would have been strongly favored by the requirements that grandmothers be near enough to daughters and grandchildren to support them. Where juvenile foraging capabilities previously limited habitat use, sharp increases in geographical range should also have been facilitated. To the degree that handling costs constrained adult returns from newly adopted resources, innovations in handling efficiency, including new technology, should also have been favored (Hawkes & O'Connell 1992).

Applying the Argument to *Homo erectus*

Pursuing this hypothesis into the fossil and archaeological record has so far involved five steps: (1) marking the points in the hominid past at which life history changes are indicated, (2) examining the fit between changes inferred by paleoanthropologists and those predicted by the grandmother hypothesis in combination with Charnov's life history model; (3) assessing the evidence for coincident changes in climate and environment that might have reduced access to "children's" resources; (4) nominating resources previously unused but likely to have been adopted in response to these changes; (5) assessing the evidence that these resources were actually exploited more heavily coincident with changes in climate and hominid life history.

Homo erectus Life History

Significant changes in hominid life history are currently identified at two, possibly three points in the fossil record, one associated with the appearance of *H. erectus*, another with archaic *H. sapiens*, a third (least certainly, see below) with fully modern humans (Smith & Tompkins 1995). Here we are concerned only with those changes associated with *H. erectus*. The appearance of this form is marked by shifts in brain size, dental eruption schedules and adult body weight, all read to indicate increased longevity and delayed maturity. Age at weaning should also have been adjusted, though the data needed to test this prediction have yet to be assessed:

- Estimates of longevity in fossil taxa are based on the correlation between brain size and longevity in living primates, including modern humans (Austad & Fisher 1992; Sacher 1959). Australopithecine brains were about the same size as those of modern chimpanzees (400–500 cc), suggesting similar adult mortality rates and, by this index, lifespans of about 50 years. Modern humans, with brain sizes of 1100–1700 cc, have much lower adult mortalities, with maximum lifespans estimated at 90–100 years. Brain sizes in *H. erectus* range from about 800–1100 cc, intermediate between values for australopithecines and modern humans, indicating similarly intermediate rates of adult mortality, and so intermediate maximum lifespans (Sather 1975).

- Dental eruption schedules provide an index of age at maturity (e.g., Beynon & Dean 1988; Bromage & Dean 1985; Smith 1986, 1989, 1993). In australopithecines, M1, an important developmental marker closely correlated with other features of life history, including age at maturity, erupted at about age 3–3.5 years, the same age as in chimpanzees, but short of the 5.5–6 year figure for modern humans. In *H. erectus*, age at M1 eruption is estimated at about 4.5 years. These data suggest that australopithecines matured at about age 10, as do modern chimps, while *H. erectus* reached that threshold at about age 15.

- Adult body size is also an index of age at maturity. In Charnov's life history model, reduced adult mortality favors growing longer before switching production from growth into offspring. Maternal size is thus expected to increase with delayed maturity. Estimates of fossil hominid body weight are based on various postcranial indicators, notably the correlation between femoral head diameter and body weight in modern humans. This relationship suggests average adult weights of 35–40 kg for australopithecines, 55–60 kg for *H. erectus*, an increase of about 55% (McHenry 1994; Ruff & Walker 1993). The difference across females is especially striking: 30–35 kg for australopithecines, 50–55 kg for *H. erectus*, an increase of roughly 70%.

- Age at weaning in *H. erectus* should be no later than in apes and australopithecines. Although in mammals later age at maturity is usually correlated with lower annual fecundity, grandmothering raises the rate of baby production. Shorter interbirth intervals should be indicated by age at weaning, which in turn should be marked by changes in the chemical composition of permanent teeth that formed across the weaning period, specifically by lower post-weaning values for O^{18}, N^{15} and Sr/Ca, and higher values for C^{13}, all associated with the shift in trophic level and the adoption of solid foods (e.g., Wright & Schwarcz 1998). It may also be indicated by an increase in the incidence of stress-related enamel hypoplasia (e.g., Goodman et al. 1984; cf. Hillson & Bond 1997). These changes can be tracked relative to crown formation schedules (e.g., Wright & Schwarcz 1998), or on the incremental growth features of individual teeth (Cerling & Sharp 1996), both of which may in future allow estimates of age at weaning in *H. erectus*.

On the basis of general correlations, available brain and body weight and dental eruption data, though limited, can be read to indicate that longevity was increased and maturity delayed in *H. erectus* relative to the broader hominoid (including earlier hominid) pattern. In australopithecines, values for both were apparently similar to those in modern chimpanzees; in *H. erectus,* intermediate between those of australopithecines and modern humans.[4] Our model assumes that the length of the fertile period did not differ between *H. erectus* and australopithecines. This is difficult to assess in the fossil record, but given the apparent conservativeness of this attribute [fertile periods in chimpanzees and humans are essentially the same (Hill & Hurtado 1996:463; Schultz 1969)], it seems simplest to assume the same period for all ancestral hominids, including *H. erectus*.

Climate Change and "Children's" Resources

Our hypothesis leads us to expect that life history changes in *H. erectus* were prompted by a decline in the availability of resources easily taken by children (e.g., fruit). Generally speaking, such declines should have been associated with shifts toward cooler, drier, more seasonal climates. In tropical Africa, cooler, drier winters would have been especially critical. Plant foods accessible to humans are very limited in this season. Those that are available (e.g., seeds, nuts, underground storage organs) typically have relatively heavy handling costs (Peters & O'Brien 1981; Peters et al. 1984).

Data from deep marine sediments indicate a general trend toward cooler climates world-wide over the past three million years, with marked steps in this direction at 2.8–2.5, 1.9–1.7, and 0.9–0.8 Ma (e.g., deMenocal 1995). Terrestrial data (e.g., soil chemistry, pollen, fossil faunas) show progressive increases in aridity and seasonality and related expansion of open habitats in tropical Africa from 2.5–1.7 Ma (Behrensmeyer et al. 1997; Cerling 1992; Cerling et al. 1988; Reed 1997; Spencer 1997; Vrba et al. 1995). These changes were evidently reinforced by continental uplift (Partridge et al. 1995), and a long term trend toward lower levels of atmospheric CO_2 (e.g., Street-Perrott et al. 1997).

At least three lines of evidence mark the 1.9–1.7 Ma period bracketing the earliest dates for African *H. erectus* (Feibel et al. 1989; White 1995) as especially critical from the perspective of our model:

- Soil carbonates indicate a sharp increase in the abundance of C_4 biomass (an index of aridity and seasonality) in both the Turkana and Olduvai regions at about this time (Cerling 1992). Prior to ca. 1.7 Ma, neither area had more than about 50% C_4 biomass present; thereafter values jump to 60–80%.
- Feeding and habitat preferences of animals represented in early East African hominid sites show a complementary trend: arboreal and (more notably here) frugivorous animals formerly common in these localities are much less so after 1.8 Ma (Reed 1997).
- Indicators of seasonal dietary stress are common in the teeth of fossil theropiths from Koobi Fora after 2.0 Ma (Macho et al. 1996).

"Tubers" as the Newly Exploited Resource

Resources adopted to provision juveniles in response to these changes must have been: (1) generally available, especially in the dry season, (2) capable of yielding returns high enough to support the collector and at least one other person, (3) reliable enough to provide those returns with little or no daily variance, and (4) open to exploitation by adults but not younger children. Many resources meet these criteria, notably certain varieties of small game, shellfish, nuts, seeds and the underground storage organs of plants. Here we restrict our attention to underground storage organs (hereafter "USOs" or, loosely, "tubers"), primarily because their availability and exploitation costs are relatively well understood, and because it has often been suggested that tubers were important in early hominid diets (e.g., Hatley & Kappelman 1980; Isaac 1980; McGrew 1992; Peters & O'Brien 1981; Stahl 1984; Vincent 1985a,b). Parallel treatment of other potential provisioning resources is clearly in order.[5]

USOs store water and carbohydrates (e.g., Anderson 1987; Chapin et al. 1990). They take many forms, including bulbs, corms, rhizomes, taproots, tubers, and woody rootstocks, and are especially well-represented among the Liliaceae, Dioscoreaceae, Araceae, Taccaceae, and Icacinaceae (Raunkiaer 1934; Thoms 1989). Consistent with function, they are common in seasonally dry and/or cold habitats, often representing up to 20% of local species, sometimes occurring at densities of more than 1000 kg/hectare (e.g., Thoms 1989; Vincent 1985b). Edible carbohydrate content varies but generally represents 50–90% of dry weight in most species.

Wild forms are heavily exploited by modern humans in tropical through cool temperate latitudes on all continents (e.g., Bahuchet et al. 1991; Coursey 1967; Endicott & Bellwood 1991;

Gott 1982; Hladik & Dounias 1993; Hurtado & Hill 1990; Johns 1990; Lee 1979; Malaisse & Parent 1985; O'Connell et al. 1983; Thoms 1989; Turner & Davis 1993; Vainshtein 1980; Vincent 1985a,b; Watanabe 1973). By contrast, they are rarely eaten by other primates except in arid, highly seasonal habitats, and even then only if they are found close to the ground surface (McGrew 1992; McGrew et al. 1988; Moore 1992; Peters & O'Brien 1981; Whiten et al. 1992).

Although attractive as a potential energy source, tubers can present certain problems to human consumers: they may be heavily defended, either mechanically or chemically (Anderson 1987; Coursey 1973), and their carbohydrate content may be difficult to digest without pre-consumption processing (Thoms 1989; Wandsnider 1997). Mechanical defenses can often be countered by simple technology (e.g., digging sticks), chemical ones by a variety of techniques including maceration, leaching, boiling, baking, or roasting (Johns & Kubo 1988; Lancaster et al. 1982; Spenneman 1994; Stahl 1984; Wandsnider 1997).

Cooking also has an important effect on digestibility. The principle storage carbohydrates in tubers are starch, sucrose, and fructan (Banks & Greenwood 1975; French 1973; Lewis 1984; Macdonald 1980). Each occurs in a variety of molecular forms. Simpler types are water-soluble and easily handled raw by human digestive systems, though cooking usually improves nutrient yield. More complex forms definitely require cooking (French 1973; Gaillard 1987; Macdonald 1980; Stahl 1984; Wandsnider 1997). Cooking also softens structural cellulose, which reduces intestinal "hurry", the speed with which high-fiber foods otherwise move through the gut. Slower passage generally increases nutrient yield (Macdonald 1980; Stahl 1984).

Two USOs favored by Native Americans, biscuit root (*Lomatium cous*) and camas (*Camassia quamash*), illustrate the effect of carbohydrate form on processing requirements. The primary storage medium in biscuit root is starch (dry weight fraction 40%; Yanovsky & Kingsbury 1938). Under traditional conditions, this tuber was eaten raw, dried, or lightly boiled. It was also occasionally ground into flour and pressed into small cakes (Couture 1978). No roasting or other extensive cooking was required. In contrast, the principal storage carbohydrate in camas is inulin (35–45% dry weight), a molecularly complex form of fructan only marginally digestible in raw form by humans. Traditional processing involved steaming the roots for 24–72 h in large rock-lined earth ovens, hydrolyzing the fructan to easily digested fructose (Konlande & Robson 1972; Thoms 1989). (See also Gott 1983; Incoll et al. 1989; Turner & Kuhnlein 1983; Turner et al. 1992; Wandsnider 1997 for additional examples.)

The difficulties of coping with USO defenses and managing any required cooking are probably great enough to prevent pre-adolescent human children from exploiting many, perhaps most, USOs effectively. In the Hadza case, children as young as five often take shallow-growing *makalita* (*Eminia antenuliera*), but cannot cook the starchy, fibrous roots for themselves if fire

TABLE 3.1. Some post-encounter return rates from wild tubers.

Location	Resource	Type	Return (kcal/hr)
Central Australia	*Cyperus* sp.	Corm	≥4500[1]
	Ipomoea costata	Rhizome*	6200[1]
	Vigna lanceolata	Rootstock	1700[1]
East Africa	*Vigna frutescens*	Rootstock*	1000–3500[2,3]
	V. macrorhyncha	Rootstock*	3000[2]
	Vigna sp.	Rootstock*	900[2]
	Vatovaea pseudolablab	Rootstock*	2000[2]
South Africa	*Coccinea rehmannii*	Rootstock*	2900[4]
	Vigna diteri	Rootstock	3000[4]
Western North America	*Camassia quamash*	Bulb*	2000–4000[5]
	Lomatium spp.	Rootstock	1000–4000[6]
	Lewisia redivivia	Rootstock*	1200–1400[6,7]

* Item requires roasting, baking, or boiling.
[1] O'Connell & Hawkes 1981; O'Connell et al. 1983.
[2] Vincent 1984a,b.
[3] Hawkes et al. 1995.
[4] Blurton Jones et al. 1994b.
[5] Thoms 1989.
[6] Couture 1978; Couture et al. 1986.
[7] Simms 1987.

kindled by an elder is unavailable. Thus, their returns may often be relatively low. The hard, sustained effort entailed in acquiring deeply buried //*ekwa* (the tuber favored by adults) prevents even older children from digging it efficiently, long after they can handle the necessary roasting.

These difficulties also constrain tuber use in chimpanzees. McGrew (1992:146) observes that the only USOs exploited by chimps are "either small bulbs simply pulled up by hand or surface roots directly gnawed." Though chimpanzees can make simple tools that might be suitable for collecting deeply buried tubers, and in some circumstances are even able to maintain fire [e.g., in connection with cigarette smoking (Brink 1957)], they apparently never use either skill to take tubers in the wild, probably because other resources, easily taken by juveniles, are readily available in the habitats they occupy.

Quantitative data on return rates from wild tuber collecting are limited, but sufficient to show that they are often high enough to support the collector and one or more dependents, even where significant processing is required (Table 3.1). Values for a sample of tropical African and Australian and temperate North American forms range from 1000–6000 kcal/h (in patch). Assuming collectors spend about 4–6 h/day at the task (e.g., Hawkes et al. 1997; Thoms 1989), daily returns from these resources would vary from roughly 4000–36,000 kcal/collector. Short-term variance in the best controlled case (Hawkes et al. 1989, 1995) is low, «50%.[6] Observations among the Hadza, !Kung, and Australian Alyawarra (Blurton Jones et al. 1994a,b; Hawkes et al. 1989, 1995; O'Connell et al. 1993) indicate that returns of 8000–12,000 kcal/collector-day may be common in tropical savanna habitats. Thoms' (1989) summary of historical data on various Columbia Plateau groups suggests that in cool temperate steppe situations returns may be 2–3 times that high. Because tubers often occur at relatively high densities, such returns can be sustained for long periods of time (weeks or months) within daily foraging distance of a single residential base, even under intense collecting pressure (Hawkes et al. 1989, 1997; Thoms 1989; Vincent 1985a,b). Heavy culling may actually improve return rates in successive seasons (e.g., Anderson 1987; Gott 1983; Thoms 1989).

The potential importance of tubers to ancestral hominids, specifically *H. erectus*, depended in part on their availability. Two lines of evidence point to greater abundance over the last 2.5 Ma, and especially after 1.8 Ma. One is overall pattern of tuber density in modern habitats. Surveys of African tropical forest communities show that tubers useful to humans are present at densities of about 1–10 kg/hectare (Hladik & Dounias 1993). Similar assessments in African savanna and North American steppe situations indicate values in the range 1–100 T/hectare, *up to five orders of magnitude higher* (Thoms 1989; Vincent 1985a, b). Since these and other open habitats have become more common over the past 2–3 Ma, it seems reasonable to think that tuber abundance, diversity and distribution have increased accordingly, particularly after 1.8 Ma. The African paleontological record provides striking confirmation: suids, which rely heavily on USOs as a food resource, show a sharp increase in taxonomic diversity at ca. 1.8 Ma (White 1995).

Archaeological and Fossil Evidence for Tuber Use

Archaeological evidence of tuber exploitation is often limited and indirect. Nevertheless, we can identify at least four patterns in the record consistent with the use of USOs beginning with the appearance of *H. erectus*:

- *Geographical range.* Prior to ca. 1.8 Ma, hominids were confined to relatively well-watered parts of tropical and subtropical Africa (Reed 1997). *H. erectus* was far more widespread, both within Africa and beyond, though never further north than about latitude 45–50 degrees (Dennell & Roebroeks 1996; Gabunia & Vekua 1995; Gibert et al. 1998; Roebroeks et al. 1992). Though the increase in range is often read to mean heavy reliance on hunting, this was unlikely to have been a productive strategy in many of the habitats newly occupied, particularly the more arid ones. Large animal biomass is typically low in such settings, implying high variance and low reliability in prey acquisition rates. USOs would have been much more dependable targets. Interestingly, latitude 50 marks not only the northern boundary of *H. erectus*, but also the approximate limit of reliance on tubers as a staple among ethnographically known hunter-gatherers in continental habitats (Thoms 1989).[7]
- *Digging tools.* Shallow-growing USOs can often be gathered by hand, but efficient acquisition of deeply buried forms requires, at minimum, a digging tool. Nearly a dozen pointed long-bone fragments, all showing damage to the tip said to be consistent with such use, are reported from Member 1 at Swartkrans Cave, dated at about 1.7 Ma (Brain 1988). Stone tools suitable for the manufacture of wooden digging sticks (unifacial choppers and heavy scrapers) are common in the Oldowan Industry, which dates to ca. 2.5 Ma (Harris 1983; Harris et al. 1987; Howell et al. 1987), but are especially well-known and widely encountered after 2.0 Ma (Isaac & Harris 1978; Leakey 1971). Keeley & Toth (1981) report damage consistent with woodworking on at least some early specimens. Later Acheulean and so-called "chopper-chopping tool" industries also include implements applicable to this task.
- *Evidence of fire.* Some tubers may be eaten without preparation, but, as noted above, cooking typically improves the nutritional yield of even the simplest starches. Where storage carbohydrates are more complex and/or chemically defended, cooking is essential. Where cooking is practised, the likelihood of encountering archaeological evidence depends on the particular techniques employed. Some will be obvious [e.g., rock-filled earth ovens used in connection with camas processing (Thoms 1989)]; others less so. Among the Hadza, for example, //ekwa roots are typically roasted for 5–15 min in large fires kindled on unprepared ground surfaces. Even where cooking sites are used repeatedly, archaeological evidence of this practice is likely to be ephemeral. The earliest unambiguous evidence for the use of fire by humans dates to the late Middle Pleistocene, 250–400 ka (Clark & Harris 1985; James 1989). Earlier indications include burned animal bones from Swartkrans, dated 1.0–1.5 Ma (Brain & Sillen 1988), and small patches of reddened earth associated with stone tools and large animals bones at Chesowanja and East Turkana, dated 1.4–1.6 Ma (Bellomo 1994; Gowlett et al. 1981; Isaac & Harris 1978). Fire is indicated in all three cases; the question is whether hominids, specifically *H. erectus,* were involved in creating or maintaining it. In at least the eastern African cases, the associations with stone tools might be read to suggest that they were.
- *H. erectus digestive anatomy.* Milton & Demment (1988) report that modern human digestive tracts are smaller, relative to body size, than those of chimpanzees, probably as a function of differences in diet. The foods humans eat generally require less digestive processing than those favored by chimpanzees, partly because of pre-consumption processing, including cooking. Aiello & Wheeler (1995) argue on grounds of thoracic morphology that modern (or near-modern) human digestive systems first appeared with *H. erectus*. Earlier hominids display the funnel-shaped thorax typical of modern chimpanzees; *H. erectus* and later *H. sapiens* fossils all show the barrel shape found in modern humans. If Aiello and Wheeler are right, then beginning with *H. erectus,* humans either (1) narrowed the range of resources commonly

exploited, focusing on those most readily digested, or (2) invested more effort in pre-consumption processing as means of reducing digestive costs. Tuber cooking is a good example of the latter strategy.

We can also point to at least two other potential tests for the importance of tubers in *H. erectus* diets. One involves analyses of dentition. Suwa et al. (1996) summarize evidence for reduced molar surface areas in early *H. erectus* relative to australopithecines, a pattern consistent with the idea of higher food quality and/or increased pre-consumption food processing just described. If tuber consumption produces distinctive damage patterns on teeth, then it might be implicated in these changes by inspection of the teeth in question.

The second test involves trace element analysis. Sillen & Lee-Thorp (1994) review research indicating differential concentration of strontium relative to calcium in plant storage organs, and discrimination against strontium across trophic levels. They further report that dietary fiber binds calcium more effectively than it does strontium; thus, all else equal, high-fiber eaters should display higher skeletal Sr/Ca than low-fiber eaters. These observations collectively suggest that if *H. erectus* relied heavily on tubers, their skeletons should display relatively high Sr/Ca ratios. If, on the other hand, and as generally believed, meat was the critical new element in *H. erectus* diets, their skeletal Sr/Ca ratios should be relatively low.

An Evolutionary Scenario Grounded in the Plio-Pleistocene

Data on *H. erectus* life history, climate, and resource use developed and integrated so far can be summarized along the following lines. By ca. 1.8 Ma, a long-term trend toward cooler, drier climate led to sharp reductions, at least seasonally, in the availability of plant foods previously exploited by hominids, especially juveniles. In some populations, adults and older juveniles increased a previously infrequent practice of using resources that younger children could not acquire on their own, with mothers and older siblings providing shares to the younger ones. Without weanlings of their own, aging females were able to feed their daughter's youngsters. Those more vigorous could support a weanling fully, allowing their daughters to wean early and begin their next pregnancies sooner, with less impact on their weanlings' welfare. This increased selection against senescence, thereby lowering adult mortality rates and in turn favoring later maturity. Extended lifespans did not favor delaying menopause since females who continued to have babies of their own were unable to enhance their daughters' fertility. Lineages with higher fertility rates were those with post-menopausal helpers. Relaxation of the limits previously imposed on adult foraging by children's resource handling capabilities opened a broader range of habitats to exploitation. Longer-lived, and so later-maturing, larger-bodied, bigger-brained hominids, identifiable as *H. erectus,* quickly spread throughout the Old World tropics and into temperate latitudes.

Tubers may have been among the newly or increasingly used resources that enabled these changes. The same shift toward more open habitats that probably reduced access to "children's" foods almost certainly increased the availability of USOs. Whether they were exploited and, if so, how effectively depended on the handling problems they posed, and on the technology available to cope with them. Digging sticks could have been fashioned with then-extant stone tools and used to exploit deeply buried but chemically-undefended tubers, particularly those with simpler forms of carbohydrate storage. If fire could be made on demand, or at least maintained once "captured", then chemically more challenging forms might also have been accessible. If not, the returns potentially available from USOs, the advantages associated with provisioning, and related adjustments in life history would have been constrained accordingly. Apparent changes in hominid gut anatomy and evidence for fire at Lower Pleistocene sites might be read to indicate that tubers were cooked and consumed extensively. If so, they may well have played a critical role in the emergence and subsequent evolutionary success of *Homo erectus.*

Discussion

Although this argument raises a wide range of issues, we comment here on just three: (1) a likely objection to the notion of long lifespans for *H. erectus,* (2) the implications of the grandmother hypothesis for current ideas about early human social organization, and (3) the importance of meat-eating in early human evolution.

Long Lifespans Among Pre-modern Humans

As applied to *H. erectus,* our model runs counter to the notion, widely held in some quarters, that long post-menopausal lifespans are a recent phenomenon, possibly the product of advances in medicine and basic sanitation made just in the last century or so, but definitely dating no earlier than the appearance of modern *H. sapiens* sometime in the early Upper Pleistocene (50–100 ka). There are good reasons to be skeptical of this proposition.

The idea that long lifespans are a very recent development stems at least in part from confusion over the implications of well-documented changes in *average life expectancy at birth*. This has increased substantially in some contemporary populations, largely due to sharp drops in infant and juvenile mortality rates. Low survivorship in the early years has a large effect on average lifespans. If, for example, half of those born die in the first year of life while everyone else lives to 100, life expectancy at birth must be near 50, even though all adults live well past that age. As it happens, old people, at least in their late 70s, are encountered everywhere, even in small populations, far from scientific medical care. Among the Ache, !Kung, and Hadza, for example, average life expectancy at birth is about 35 years, yet average female life expectancy at age 45 is about 20 additional years (Blurton Jones et al. 1992; Hill & Hurtado 1996; Howell 1979). Again, this is an average: many live much longer. In all these cases, nearly 40% of adult women are post-menopausal.

The notion that long lifespans are restricted to modern *H. sapiens* is based on the fact that individuals identified as "old" (aged >50 yrs) are uncommon in the fossil record. Trinkaus (1995) and others appeal to a variety of skeletal indicators (e.g., epiphysial fusion, dental attrition, and long bone histomorphometry) in arguing that archaic *sapiens* routinely sustained high mortalities in young adulthood and seldom lived past age forty (see also Abbott et al. 1996; Bermudez de Castro & Nicolas 1997). Though comparable analyses have yet to be undertaken on *H. erectus*, a similar argument might well be anticipated: long post-reproductive lifespans, predictable in theory, were seldom if ever actually achieved.

There are two important problems with this argument, either as applied to archaic *H. sapiens* or anticipated with respect to *H. erectus*:

- Older adults will *always* be underrepresented in archaeological samples, simply because their remains are more susceptible to decay due to bone mineral depletion (e.g., Buikstra & Konigsberg 1985; Galloway et al. 1997). The absence of elderly individuals in excavated cemetery populations of which they are *known*, on the basis of historical records, to have once been a part, illustrates the effect (e.g., Walker et al. 1988).
- Skeletally-based age estimates on adults over age 25 are notoriously inaccurate and commonly underestimate true age, sometimes by decades (e.g., Aiello & Molleson 1992; Bocquet-Appel & Masset 1982, 1996; Jackes 1992; Konigsberg & Frankenberg 1992; Paine 1997).

Failure to appreciate the pervasive effects of these factors has sometimes led analysts to infer high young adult mortality rates and short lifespans even for modern human populations known archaeologically (e.g., Lovejoy et al. 1977). It is now generally recognized that such inferences are inappropriate (Aiello & Molleson 1992; Bocquet-Appel & Masset 1982, 1996; Howell 1982; Jackes 1992; Konigsberg & Frankenberg 1992; Paine 1997). Similar arguments about earlier hominid mortality and lifespan should be treated with equal skepticism.

Correlations between brain weight, maximum lifespan, and age at maturity documented across the primates (e.g., Austad & Fisher 1992; Sacher 1975) lead us to expect that long lifespans and late maturity have been typical of most humans since the late Middle Pleistocene (200–500 ka), when brain sizes in archaic *H. sapiens* reached the modern range (Leigh 1992; Ru et al. 1997). Though skeletal data enabling estimates of age at maturity in early *H. sapiens* are limited, the few available are consistent with the idea that this threshold was achieved late, at roughly the same age as in modern humans (e.g., Dean et al. 1986; Mann & Vandermeersch 1997; Stringer & Dean 1997; Stringer et al. 1990; Tompkins 1996). The same reasoning leads us to expect earlier increases in average adult lifespan and age at maturity, beyond the hominoid range but short of those associated with *H. sapiens*, coincident with the first appearance of *H. erectus*. As indicated above, data on brain size, dental eruption schedules, and body weight are all consistent with this proposition.

Implications for *H. erectus* Social Organization

The grandmother hypothesis has important implications for current opinion about hominid social organization; in particular, for the suggestion, commonly made (e.g., Foley & Lee 1989; Ghiglieri 1987; Rodseth et al. 1991; Wrangham 1987) and now widely echoed (e.g., Mellars 1996), that evidence of male philopatry and female dispersal among chimpanzees and modern human hunter-gatherers implies that both were characteristic of *all* ancestral hominids, including *H. erectus*. Elsewhere (Hawkes et al. 1997), we have detailed reasons to be skeptical of this suggestion; among them, that patterns of residence, alliance, and dispersal vary widely among both chimps and modern hunters, and that all are evidently sensitive to local ecological conditions, especially as they affect female subsistence.

The grandmother hypothesis allows us to build on these observations more pointedly, specifically with reference to *H. erectus* social organization. Heavy reliance on high cost/high yield resources in connection with offspring provisioning should have given daughters a strong incentive to remain with their natal group. As daughters grew, they acquired the strength and skill needed to feed younger siblings. When they matured, the assistance of aging mothers continued to enhance the benefits of proximity. From this perspective, long post-menopausal lifespans, late age at maturity, and high fertility suggest a pattern of co-residence among related females. The stronger the pattern, the greater the incentive for males to leave their natal group.

Grandmothers could certainly have improved their fitness by aiding sons, but the benefits associated with helping daughters are likely to have been much greater. Mothers and daughters face similar reproductive tradeoffs: both do better by attending to offspring survivorship. Sons generally do better by investing in mating (Anderson 1994; Hawkes et al. 1995). A food-sharing mother might attract females to her son's group, but this would not assure her son paternity of those females' offspring. His fitness would depend on his success in competing with other males. Winners of that competition would enjoy higher reproductive success whether or not their mothers contributed to the fertility of their mates. Even if a grandmother could identify her son's offspring and single out grandchildren to feed, her potential fitness gains through increased fertility of "daughters-in-law" would be devalued by the uncertain paternity of subsequent children more quickly born to the mother of those grandchildren.

What About the Archaeological Evidence for Big Game Hunting and Scavenging?

This brings us to a third issue, the common association between stone tools and the remains of large animals at sites of Lower and early Middle Pleistocene age that many continue to see as strong support for the hunting hypothesis. Though comprehensive consideration of these remains is well beyond us here, a brief comment grounded on our general argument seems pertinent.

As indicated above, heavy reliance on resources like tubers that occur at high densities, with returns limited primarily by handling requirements, should have favored larger group sizes. The associated predator-defense advantages should have reinforced the pattern, especially in more open habitats. Along with larger body size, larger group size should also have provided an important edge in "aggressive" or "confrontational" scavenging, where the kill is seized from the initial predator while still substantially intact (O'Connell et al. 1988b). Increased consumption of meat and marrow, and correspondingly increased evidence of such consumption in the archaeological record, should have been among the outcomes. There is little indication of meat consumption by hominids in the late Pliocene (Kibunjia 1994). Instead, the earliest sites implicated in recent arguments about the hunting hypothesis date to the Plio-Pleistocene boundary (e.g., Bunn 1994; Bunn & Kroll 1986), coincident with the first evidence for African *H. erectus*. Recent analyses of cut- and tooth-mark distribution in the assemblage from the well-known "Zinjanthropus" site at Olduvai (Capaldo 1997; Dominguez-Rodrigo 1997) suggest that the sequence of carcass access there may have been carnivore-hominid-carnivore, consistent with the notion that hominids were successful in aggressive confrontations over large animal carcasses killed by other predators.

Success at acquiring carcasses need not have implied restricted mating access ("pair bonding") or paternal provisioning, any more than it does among chimpanzees. Neither would transport of parts to "central places" be indicated (cf. Bunn & Ezzo 1993; Potts 1988; Rose & Marshall 1996): individuals or groups may simply have called attention to any carcass they encountered or acquired, just as do modern human hunters (O'Connell et al. 1988a,b, 1992). If the carcass had not yet been taken, the crowd so drawn could have done so, then consumed it on or near the spot, again just as modern hunters sometimes do (O'Connell et al. 1988b). The same advantages that helped secure the carcass initially—large body size and large group size—would often have deterred any counter-attack, either by the original predators or others arriving later. Repeated successes at the same spot, perhaps a dry season water source in a stream channel, would create archaeological sites very like those often identified as characteristic of the Lower Pleistocene record, particularly in East Africa. Subsequent density-dependent attrition of the bone assemblage would have sealed the match (Marean et al. 1992). Archaeological visibility notwithstanding, the sites so created need not indicate that large animal prey were either commonly acquired or an important part of *H. erectus* diets. On the contrary, their appearance might simply reflect changes in hominid group and body size stimulated largely if not entirely by prior changes in female foraging, food sharing, and life history.

Summary

Clear-cut, probably widespread patterns in women's foraging and food sharing among modern tropical hunter-gatherers have an important effect on children's nutritional welfare. This observation is the basis for an hypothesis about the evolution of extended lifespans typical of all living humans. Because elements of mammalian life histories vary with each other systematically, other aspects of ancestral hominid life histories should have been entrained simultaneously. Longer adult lifespans favored by the payoffs for grandmothering when mothers provision their offspring should account for the delayed maturity, relatively high fertility and mid-life menopause that collectively distinguish humans from other living hominoids. Here we have used this foundation to develop a scenario for the evolution of *Homo erectus*; then assessed it in light of the available data on *H. erectus* life history and anatomy, Plio-Pleistocene environment, the economics of tuber exploitation, and Lower Paleolithic archaeology. Results show that these lines of evidence are consistent with the proposition that grandmothering played a central role in the evolution and spread of this long-successful taxon. Widely held ideas about the recent development of long human lifespans, the prevalence of male philopatry among ancestral hominids, and the catalytic role of hunting and scavenging in early human evolution are challenged accordingly.

Acknowledgments

Research reported here was supported by the University of Utah, University of California (Los Angeles), and by the Division of Archaeology and Natural History, Research School of Pacific and Asian Studies, Australian National University. Useful assistance and advice (not always taken) was generously provided by Leslie Aiello, Jim Allen, Helen Alvarez, Margaret Avery, Thure Cerling, Ric Charnov, Joan Coltrain, Mike Cannon, Steve Donnelly, Jim Ehleringer, Jennifer Graves, Don Grayson, Geo Hope, Richard Klein, Julia Lee-Thorp, Henry McHenry, Ric Paine, Doug Price, Alston Thoms, Eric Trinkaus, LuAnn Wandsnider, Tim White, Polly Wiessner and Richard Wrangham. We dedicate the paper to the late Bettina Bancroft, whose generous assistance made our initial fieldwork with the Hadza possible.

Notes

1. Though it has long been assumed that FLK "Zinj" and other sites of similar age (~1.75 Ma) were produced in part by "early *Homo*" (cf. *habilis*, or more narrowly, *rudolfensis* [e.g., Bunn & Ezzo 1993]), the earliest dates for African *H. erectus* (*ergaster*) (Feibel et al. 1989; White 1995) and the coincident change in the archaeo-

logical record, marked by a sharp increase in assemblage diversity (compare Bunn 1994; Kibunjia 1994), now make it a better candidate for that role.

2. Blumenschine and associates (e.g., Blumenschine & Peters 1998; Blumenschine et al. 1994) and Rose & Marshall (1996) both take steps in this direction; but in each case the potential impact is limited by the narrow goal of accounting for certain features of the archaeological record (mainly faunal assemblage composition), rather than the larger evolutionary phenomenon of which it is a part.

3. Two very different versions of the grandmother hypothesis are discussed in the recent literature, one focusing on factors that might favor an "early" end to fertility, the other on the evolution of long post-reproductive lifespans. Here we are concerned only with the latter (see Hawkes et al. 1997; Hawkes et al. 1998b; Kaplan 1997; Peccei 1995 for discussion).

4. On brain and body weight criteria (e.g., McHenry 1994), it might be argued that *H. habilis,* not *H. erectus,* is the earliest hominid to display distinctively nonpongid life history characteristics. Attempts to confirm this through analysis of dental eruption schedules have so far proven inconclusive (e.g., Dean 1995; B. H. Smith 1991; R. J. Smith et al. 1995). The issue is complicated by small sample size and continuing uncertainty about the taxonomy of key specimens (e.g., White 1995). Even if *H. habilis* life history differed from the general hominoid pattern, available data indicate that *H. erectus* marked a more pronounced departure.

5. For data on age-related handling costs and nutrient returns for other potential provisioning resources, see (for example) Blurton Jones et al. (1994a,b), Boesch & Boesch (1984) and Peters (1987) on nuts, Bird & Bliege Bird (1997), Bliege Bird et al. (1995), and Meehan (1982) on shellfish.

6. Most of the return rate variance indicated for this case probably reflects differences in collector effort, not encounter rate. It is our impression that day-to-day variance in return rates for older women is *very low,* perhaps negligible.

7. An extensive literature review leads Thoms (1989:94) to observe: "Although there is considerable variation in the use of geophytes across [northern Eurasia], the overall pattern is one of minor use in the tundra and the northern part of the taiga zone, moderate use in the taiga zone, and comparatively heavy use along the southern margins of the taiga and the northern part of the steppe." The same review indicates a similar pattern in western North America. In both cases, the steppe/forest boundary falls at about latitude 50 degrees. Clearly, it would be useful to know more about the determinants of this pattern, as well as about whether and in what ways they may have limited the distribution of *H. erectus* in the past.

References

Abbott, S., E. Trinkaus, D. B. Been
1996 Dynamic bone remodeling in later Pleistocene fossil hominids. *Am. J. Phys. Anthrop.* 99, 585–601.

Aiello, L. C., C. Dean
1990 *An Introduction to Human Evolutionary Anatomy.* London: Academic Press.

Aiello, L. C., T. Molleson
1992 Are microscopic aging techniques more accurate than macroscopic aging techniques? *J. Archaeol. Sci.* 20, 689–704.

Aiello, L. C., P. Wheeler
1995 The expensive–tissue hypothesis: The brain and the digestive system in human and primate evolution. *Curr. Anthrop.* 36, 199–221.

Anderson, D. C.
1987 Below-ground herbivory in natural communities: a review emphasizing fossorial animals. *Q. Rev. Biol.* 62, 261–286.

Anderson, M.
1994 *Sexual Selection.* Princeton: Princeton University Press.

Austad, S. N., K. E. Fisher
1992 Primate longevity: its place in the mammalian scheme. *Am. J. Phys. Anthrop.* 28, 251–261.

Bahuchet, S., D. McKey, I. de Garine
1991 Wild yams revisited: is independence from agriculture possible for rain forest horticulturists? *Hum. Ecol.* 19, 213–244.

Banks, W., C. T. Greenwood
1975 *Starch and Its Components.* Edinburgh: Edinburgh University Press.

Behrensmeyer, A. K., N. E. Todd, R. Potts, G. B. McBinn
1997 Late Pliocene faunal turnover in the Turkana Basin, Kenya and Ethiopia. *Science* 278, 1589–1594.

Bellomo, R. V.
1994 Methods of determining early hominid behavioral activities associated with the controlled use of fire at FxJj 20 Main, Koobi Fora, Kenya. *J. Hum. Evol.* 27, 173–195.

Bermudez de Castro, J. M., M. E. Nicolas
1997 Paleodemography of the Atapuerca-Sima de los Huesos Middle Pleistocene hominid sample. *J. Hum. Evol.* 33, 333–355.

Beynon, A. D., M. C. Dean
1988 Distinct dental development patterns in early fossil hominids. *Nature* 335, 509–514.

Binford, L. R.
1981 *Bones: Ancient Men and Modern Myths.* New York: Academic Press.

Bird, D. W., R. L. Bliege Bird
1997 Contemporary shellfish gathering strategies among the Meriam of the Torres Strait Islands, Australia: testing predictions of a central place foraging model. *J. Archaeol. Sci.* 24, 39–63.

Bliege Bird, R. W., D. W. Bird, J. M. Beaton
1995 Children and traditional subsistence on Mer (Murray Island), Torres Strait. *Australian Aboriginal Studies* 1995, 2–17.

Blumenschine, R. J.
1991 Hominid carnivory and foraging strategies and the socio-economic function of early archaeological sites. *Phil. Trans. R. Soc.,* series B 334, 211–221.

Blumenschine, R. J., C. Peters
1998 Archaeological predictions for hominid land use in the paleo-Olduvai Basin, Tanzania, during lowermost Bed II times. *J. Hum. Evol.* 34, 565–608.

Blumenschine, R. J., J. A. Cavallo, S. D. Capaldo
1994 Competition for carcasses and early hominid behavioral ecology: A case study and conceptual framework. *J. Hum. Evol.* 27, 197–213.

Blurton Jones, N. G., K. Hawkes, P. Draper
1994a Differences between Hadza and !Kung children's work: Original affluence or practical reason? In (E. S. Burch, Ed.) *Key Issues in Hunter-Gatherer Research,* pp. 189–215. Oxford: Berg.

1994b Foraging returns of !Kung adults and children: Why didn't !Kung children forage? *J. Anthrop. Res.* 50, 217–248.

Blurton Jones, N. G., K. Hawkes, J. F. O'Connell
1989 Studying costs of children in two foraging societies: implications for schedules of reproduction. In (V. Standon & R. Foley,

Eds.) *Comparative Socioecology of Mammals and Man*, pp. 365–390. London: Blackwell.

1996 The global process and local ecology: how should we explain differences between the Hadza and !Kung? In (S. Kent, Ed.) *Cultural Diversity Among Twentieth Century Foragers: An African Perspective*, pp. 159–187. Cambridge: Cambridge University Press.

1997 Why do Hadza children forage? In (N. Segal, G. E. Weisfeld & C. C. Weisfeld, Eds.) *Uniting Psychology and Biology: Integrative Perspectives on Human Development*, pp. 164–183. Washington DC: American Psychological Association.

Blurton Jones, N. G., L. C. Smith, J. F. O'Connell, K. Hawkes, C. Kamuzora,

1992 Demography of the Hadza, an increasing and high density population of savanna foragers. *Am. J. Phys. Anthrop.* 89, 159–181.

Bocquet-Appel, J. P., C. Masset

1982 Farewell to paleodemography. *J. Hum. Evol.* 11, 321–333.

1996 Paleodemography: expectancy and false hope. *Am. J. Phys. Anthrop.* 99, 571–583.

Boesch, C., H. Boesch

1984 Possible cause of sex differences in the use of natural hammers by wild chimpanzees. *J. Hum. Evol.* 13, 415–440.

1989 Hunting behavior of wild chimpanzees in Tai National Park. *Am. J. Phys. Anthrop.* 78, 547–573.

Brain, C. K.

1988 New information from Swartkrans Cave of relevance to "robust" astralopithecines. In (F. Grine, Ed.) *Evolutionary History of the "Robust" Australopithecines*, pp. 311–316. Hawthorne, NY: Aldine.

Brain, C. K., A. Sillen

1988 Evidence from the Swartkrans cave for the earliest use of fire. *Nature* 336, 464–466.

Brauer, G., M. Schultz

1996 The morphological affinities of the Plio-Pleistocene mandible from Dmanisi, Georgia. *J. Hum. Evol.* 30, 445–481.

Brink, A. S.

1957 The spontaneous fire-controlling reactions of two chimpanzee smoking addicts. *S. Afr. J. Sci.* 53, 241–247.

Bromage, T. G., M. C. Dean

1985 Reevaluation of age at death of immature fossil hominids. *Nature* 317, 525–527.

Buikstra, J., L. Konigsberg

1985 Paleodemography: critiques and controversies. *Am. Anthrop.* 87, 316–333.

Bunn, H. T.

1994 Early Pleistocene hominid foraging strategies along the ancestral Omo River at Koobi Fora, Kenya. *J. Hum. Evol.* 27, 247–266.

Bunn, H. T., J. A. Ezzo

1993 Hunting and scavenging by Plio-Pleistocene hominids: Nutritional constraints, archaeological patterns, and behavioural implications. *J. Archaeol. Sci.* 20, 365–398.

Bunn, H. T., E. M. Kroll

1986 Systematic butchery by Plio/Pleistocene hominids at Olduvai Gorge, Tanzania. *Curr. Anthrop.* 27, 413–452.

Capaldo, S. D.

1997 Experimental determinations of carcass processing by Plio-Pleistocene hominids and carnivores at FLK 22 (*Zinjanthropus*), Olduvai Gorge, Tanzania. *J. Hum. Evol.* 33, 555–597.

Cerling, T. E.

1992 Development of grasslands and savannas in East Africa during the Neogene. *Palaeogeog., Palaeoclimatol., Palaeoecol.* 97, 241–247.

Cerling, T. E., Z. D. Sharp

1996 Stable carbon and oxygen isotope analysis of fossil tooth enamel using laser ablation. *Palaeogeog., Palaeoclimatol., Palaeoecol.* 126, 173–186.

Cerling, T. E., J. R. Bowman, J. R. O'Neil

1988 An isotopic study of a fluvial-lacustrine sequence: The Plio-Pleistocene Koobi-Fora Sequence, East Africa. *Palaeogeog., Palaeoclimatol., Palaeoecol.* 63, 335–356.

Chapin, F. S. III, E-D. Schulze, H. A. Mooney

1990 The ecology and economics of storage in plants. *Ann. Rev. Ecol. Syst.* 21, 423–448.

Charnov, E. L.

1993 *Life History Invariants: Some Explorations of Symmetry in Evolutionary Ecology*. Oxford: Oxford University Press.

Charnov, E. L., D. Berrigan

1993 Why do female primates have such long lifespans and so few babies? Or life in the slow lane. *Evol. Anthrop.* 1, 191–194.

Clark, J. D., J. W. K. Harris

1985 Fire and its roles in early hominid lifeways. *Afr. Archaeol. Rev.* 3, 3–27.

Coursey, D. G.

1967 *Yams: An Account of the Nature, Origins, Cultivation and Utilization of the Useful Members of the Dioscoreaceae*. London: Longmans.

1973 Hominid evolution and hypogeous plant foods. *Man* 8, 634–635.

Couture, M. D.

1978 Recent and contemporary foraging practices of the Harney Valley Paiute. Master's Dissertation, Department of Anthropology, Portland State University.

Couture, M. D., M. R. Ricks, L. Housley

1986 Foraging behavior of a contemporary northern Great Basin population. *J. Calif. Great Basin Anthrop.* 8, 150–160.

Dean, M. C.

1995 The nature and periodicity of incremental lines in primate dentine and their relationship to periradicular bands in OH 16 (*Homo habilis*). In Q. Moggi Cecchi, Ed.) *Aspects of Dental Biology: Paleontology, Anthropology and Evolution*, pp. 239–265. Florence: International Institute for the Study of Man.

Dean, M. C., C. B. Stringer, T. G. Bromage

1986 Age at death of the neanderthal child from Devil's Tower, Gibraltar, and the implications for studies of general growth and development in neanderthals. *Am. J. Phys. Anthrop.* 70, 301–309.

deMenocal, P. B.

1995 Plio-Pleistocene African climate. *Science* 270, 53–59.

Dennell, R., W. Roebroeks

1996 The earliest colonization of Europe: the short chronology revisited. *Antiquity* 70, 535–542.

Dennell, R., H. Rendell, E. Hailwood

1988 Late Pliocene artifacts from Northern Pakistan. *Curr. Anthrop.* 29, 495–498.

Dominguez-Rodrigo, M.

1997 Meat-eating by early hominids at the FLK Zinjanthropus site, Olduvai Gorge (Tanzania): An experimental approach using cut-mark data. *J. Hum. Evol.* 33, 669–690.

Endicott, K., P. Bellwood
1991 The possibility of independent foraging in the rain forest of Peninsular Malaysia. *Hum. Ecol.* 19, 151–185.

Fedigan, L. M.
1986 The changing role of women in models of human evolution. *Ann. Rev. Anthropol.* 15, 25–66.

Feibel, C. S., F. Brown, I. McDougal
1989 Stratigraphic context of fossil hominids from the Omo Group deposits: Northern Turkana Basin, Kenya and Ethiopia. *Am. J. Phys. Anthrop.* 78, 595–622.

Foley, R. A., P. C. Lee
1989 Finite social space, evolutionary pathways, and reconstructing hominid behaviour. *Science* 243, 901–906.

French, D.
1973 Chemical and physical properties of starch. *J. Anim. Sci.* 37, 1048–1061.

Gabunia, L., A. Vekua
1995 A Plio-Pleistocene hominid mandible from Dmanisi, East Georgia, Caucasus. *Nature* 373, 509–512.

Galliard, T. (Ed.)
1987 *Starch: Properties and Potential.* New York: John Wiley and Sons.

Galloway, A., P. Willey, L. Snyder
1997 Human bone mineral densities and survival of bone elements: A contemporary sample. In (W. D. Haglund & M. H. Sorg, Eds.) *Forensic Taphonomy: The Postmortem Fate of Human Remains,* pp. 279–317. Boca Raton: CRC Press.

Ghiglieri, M.
1987 Sociobiology of the great apes and the hominid ancestor. *J. Hum. Evol.* 16, 319–358.

Gibert, J., Ll. Gibert, A. Iglesias, E. Maestro
1988 Two "Oldowan" assemblages in the Plio-Pleistocene deposits of the Orce region, southwest Spain. *Antiquity* 72, 17–25.

Goodman, A. H., G. J. Arnelogos, J. C. Rose
1984 The chronological distribution of enamel hypoplasias from prehistoric Dickson Mounds. *Am. J. Phys. Anthrop.* 65, 259–266.

Gott, B.
1982 The ecology of root use by Aborigines in southern Australia. *Archaeology in Oceania* 17, 59–67.
1983 Murnong-*Microseris scapigera*: A study of a staple food of Victorian Aborigines. *Australian Aboriginal Studies* 1983/2, 2–18.

Gowlett, J. A. J.
1993 *Ascent to Civilization: The Archaeology of Early Humans* (2nd ed.). New York: McGraw-Hill.

Gowlett, J. A. J., J. W. K. Harris, D. Walton, B. A. Wood,
1981 Early archaeological sites, hominid remains, and traces of fire from Chesowanja, Kenya. *Nature* 294, 125–129.

Harris, J. W. K.
1983 Cultural beginnings: Plio–Pleistocene archaeological occurrences from the Afar, Ethiopia. *Afr. Archaeol. Rev.* 1, 3–31.

Harris, J. W. K., P. G. Williamson, J. Veniers, M. J. Tappen, K. Stewart, D. Helgren, J. de Heinzelin, N. T. Boaz, R. V. Bellomo
1987 Late Pliocene hominid occupation of the Senga 5A site, Zaire. *J. Hum. Evol.* 16, 701–728.

Harvey, P. H., A. F. Reed
1988 How and why do mammalian life histories vary? In (M. S. Boyce, Ed.) *Evolution of Life Histories: Patterns and Process from Mammals,* pp. 213–232. New Haven: Yale University Press.

Hatley, T., J. Kappelman
1980 Bears, pigs, and Plio-Pleistocene hominids: a case for the exploitation of below ground food resources. *Hum. Ecol.* 8, 371–387.

Hawkes, K.
1990 Why do men hunt? Some benefits for risky strategies. In (E. Cashdan, Ed.) *Risk and Uncertainty in Tribal and Peasant Economies,* pp. 145–166. Boulder, CO: Westview Press.
1993 Why hunter-gatherers work: An ancient version of the problem of public goods. *Curr. Anthropol.* 34, 341–361.

Hawkes, K., J. F. O'Connell
1992 On optimal foraging models and subsistence transitions. *Curr. Anthropol.* 33, 63–65.

Hawkes, K., J. F. O'Connell, N. G. Blurton Jones
1989 Hardworking Hadza grandmothers. In (V. Standen & R. Foley, Eds.) *Comparative Socioecology of Mammals and Man,* pp. 341–366. London: Blackwell.
1991 Hunting income patterns among the Hadza: Big game, common goods, foraging goals, and the evolution of the human diet. *Phil. Trans. R. Soc.,* series B 334, 243–251.
1995 Hadza children's foraging: Juvenile dependency, social arrangements and mobility among hunter-gatherers. *Curr. Anthropol.* 36, 688–700.
1997 Hadza women's time allocation, offspring provisioning, and the evolution of post-menopausal lifespans. *Curr. Anthropol.* 38, 551–578.
1998 Why do Hadza men hunt big animals? Unpublished ms.

Hawkes, K., J. F. O'Connell, N. G. Blurton Jones, E. L. Charnov, H. Alvarez
1998a Grandmothering, menopause, and the evolution of human life histories. *Proc. Nat. Acad. Sci. U.S.A.* 95, 1336–1339.
1998b The grandmother hypothesis and human evolution. In (L. Cronk, N. Chagnon & W. Irons, Eds.) *Evolutionary Biology and Human Behavior: 20 Years Later.* Hawthorne, NY: Aldine de Gruyter.

Hawkes, K., A. R. Rogers, E. L. Charnov
1995 The male's dilemma: increased offspring production is more paternity to steal. *Evol. Ecology* 9, 662–677.

Hill, K.
1982 Hunting and human evolution. *J. Hum. Evol.* 11, 521–544.

Hill, K., A. M. Hurtado
1996 *Ache Life History: The Ecology and Demography of a Foraging People.* Hawthorne, NY: Aldine de Gruyter.

Hillson, S., S. Bond
1997 Relationship of enamel hypoplasia to the pattern of tooth crown growth: A discussion. *Am. J. Phys. Anthrop.* 104, 89–104.

Hladik, A., E. Dounias
1993 Wild yams of the African forest as potential food resources. In (C. N. Hladik, A. Hladik, O. F. Linares, H. Pagezy, A. Semple & M. Hadley, Eds.) *Tropical Forests, Food and People: Biocultural Interactions and Applications to Development,* pp. 163–176. Paris: Parthenon Publishing Group.

Howell, F. C., P. Haesaerts, J. de Heinzelin
1987 Depositional environments, archaeological occurrences and hominids from Members E and F of the Shungura Formation (Omo Basin, Ethiopia). *J. Hum. Evol.* 16, 665–700.

Howell, N.
1979 *Demography of the Dobe !Kung.* New York: Academic Press.
1982 Village composition implied by a paleodemographic life table: The Libben Site. *Am. J. Phys. Anthrop.* 59, 263–269.

Huang, W., R. Ciochon, Y. Gu, R. Larick, Q. Fang, H. Schwarcz, C. Yonge, J. De Vos, W. Rink
1995 Early *Homo* and associated artifacts from Asia. *Nature* 378, 275–278.

Hurtado, A. M., K. Hill
1990 Seasonality in a foraging society: variation in diet, work effort, fertility and the sexual division of labor among the Hiwi of Venezuala. *J. Anthrop. Res.* 46, 293–345.

Incoll, L. D., G. D. Bonnett, B. Gott
1989 Fructans in the underground storage organs of some Australian plants used for food by Aborigines. *Journal of Plant Physiology* 134, 196–202.

Isaac, G. Ll.
1978 The food sharing behavior of protohuman hominids. *Scientific American* 238(4), 90–108.
1980 Casting the net wide: A review of archaeological evidence for early hominid land-use and ecological relations. In (L-K. Konigsson, Ed.) *Current Argument on Early Man: Proceedings of a Nobel Symposium Organized by the Royal Swedish Academy of Sciences and Held at Bjorkborns Herrgard, Karlskoga, Sweden, 21–27 May 1978, Commemorating the 200th Anniversary of the Death of Carolus Linnaeus*, pp. 226–251. Oxford: Pergamon.
1983 Bones in contention: competing explanations for the juxtaposition of early Pleistocene artifacts and faunal remains. In (J. Clutton-Brock & C. Grigson, Eds.) *Animals and Archaeology*, vol. 1, pp. 1–20. British Archaeological Reports 163.
1984 The archaeology of human origins: studies of the Lower Pleistocene in East Africa, 1971–1981. *Advances in World Archaeology* 3, 1–87.

Isaac, G. Ll., J. W. K. Harris
1978 Archaeology. In (M. G. Leakey & R. E. F. Leakey, Eds.) *Koobi Fora Research Project*, vol. 1, pp. 64–85. Oxford: Clarendon.

Jackes, M.
1992 Paleodemography: problems and technique. In (S. R. Saunders & M. A. Katzenberg, Eds.) *Skeletal Biology of Past Peoples: Theory and Methods*, pp. 189–224. New York: Wiley-Liss.

James, S.
1989 Hominid uses of fire in the Lower and Middle Pleistocene: a review of the evidence. *Curr. Anthrop.* 30, 1–26.

Janson, C. H., M. L. Goldsmith
1995 Predicting group size in primates: foraging costs and predation risks. *Behavioral Ecology* 6, 326–336.

Johns, T.
1990 *With Bitter Herbs They Shall Eat It: Chemical Ecology and the Origins of Human Diet and Medicine.* Tucson: University of Arizona Press.

Johns, T., S. Kubo
1988 A survey of traditional methods employed for the detoxification of plant foods. *Journal of Ethnobiology* 8, 81–129.

Kaplan, H.
1997 The evolution of the human life course. In (K. W. Wachter & C. E. Finch, Eds.) *Between Zeus and the Salmon: The Biodemography of Longevity*, pp. 175–211. Washington, D.C.: National Academy Press.

Keeley, L. H., N. Toth
1981 Microwear polishes on early stone tools from Koobi Fora, Kenya. *Nature* 293, 464–465.

Kibunjia, M.
1994 Pliocene archaeological occurrences in the Lake Turkana Basin. *J. hum. Evol.* 27, 159–171.

Klein, R. G.
1987 Problems and prospects in understanding how early people exploited animals. In (M. H. Nitecki & D. V. Nitecki, Eds.) *The Evolution of Human Hunting*, pp. 11–45. New York: Plenum Press.
1999 *The Human Career: Human Biological and Cultural Origins* (2nd ed.). Chicago: University of Chicago Press.

Konigsberg, L. W., S. R. Frankenberg
1992 Estimation of age structure in anthropological demography. *Am. J. Phys. Anthrop.* 89, 235–256.

Konande, J. E., J. R. K. Robson
1972 The nutritive value of cooked camas as consumed by Flathead Indians. *Ecology of Food and Nutrition* 1, 193–195.

Lancaster, P. A., J. S. Ingram, M. Y. Lim, D. G. Coursey
1982 Traditional cassava-based foods: A survey of processing techniques. *Economic Botany* 36, 12–45.

Leakey, M. D.
1971 *Olduvai Gorge, Volume 3: Excavations in Beds I and II 1960–1963.* Cambridge: Cambridge University Press.

Leakey, R., R. Lewin
1992 *Origins Reconsidered.* New York: Little, Brown.

Lee, R. B.
1979 *The !Kung San: Men, Women, and Work in a Foraging Society.* Cambridge: Cambridge University Press.

Leigh, S. R.
1992 Cranial capacity evolution in *Homo erectus* and early *Homo sapiens*. *Am. J. Phys. Anthrop.* 87, 1–13.

Lewis, D. H.
1984 Occurrence and distribution of storage carbohydrates in vascular pants. In (D. H. Lewis, Ed.) *Storage Carbohydrates in Vascular Plants*, pp. 1–52. Cambridge: Cambridge University Press.

Lovejoy, C. O., R. S. Meindl, T. R. Pryzbeck, T. S. Barton, K. G. Heiple, D. Kotting
1977 Paleo-demography of the Libben Site, Ottowa County, Ohio. *Science* 198, 291–293.

Macdonald, I.
1980 Suppliers of energy: carbohydrates. In (R. B. Alfin-Slater & D. Kritchevsky, Eds.) *Nutrition and the Adult: Macronutrients*, pp. 97–116. New York: Plenum Press.

Macho, G. A., D. J. Reid, M. O. Leakey, N. Jablonski, A. D. Beynon,
1996 Climatic effects on dental development of *Theropithecus oswaldi* from Koobi Fora and Olorgesailie. *J. Hum. Evol.* 30, 57–70.

Malaisse, F., G. Parent
1985 Edible wild vegetable products in the Zambezian woodland areas: A nutritional and ecological approach. *Ecology of Food and Nutrition* 18, 43–82.

Mann, A., B. Vandermeersch
1997 An adolescent female neanderthal mandible from Montgaudier Cave, Charente, France. *Am. J. Phys. Anthrop.* 103, 507–527.

Marean, C. W., L. M. Spencer, R. J. Blumenschine, S. D. Capaldo
1992 Captive hyena bone choice and destruction, the Schlepp Effect, and Olduvai archaeofaunas. *J. Archaeol. Sci.* 16, 101–121.

McGrew, W. C.
1992 *Chimpanzee Material Culture: Implications for Human Evolution.* Cambridge: Cambridge University Press.

McGrew, W. C., P. J. Baldwin, C. Tutin
1988 Diet of wild chimpanzees (*Pan troglodytes verus*) at Mt. Assirik, Senegal: I. Composition. *Am. J. Primatol.* 16, 213–226.

McHenry, H. M.
1994 Behavioral ecological implications of early hominid body size. *J. Hum. Evol.* 27, 77–87.

Meehan, B.
1982 *Shell Bed to Shell Midden.* Canberra: Australian Institute of Aboriginal Studies.

Mellars, P.
1996 *The Neanderthal Legacy: An Archaeological Perspective from Western Europe.* Princeton: Princeton University Press.

Milton, K., M. W. Demment
1988 Digestion and passage kinetics of chimpanzees fed high and low fiber diets and comparison with human data. *Journal of Nutrition* 118, 1082–1088.

Moore, J.
1992 Savanna chimpanzees. In (T. Nishida, W. McGrew et al., Eds.) *Topics in Primatology, Volume 1; Human Origins,* pp. 99–118. Tokyo: University of Tokyo Press.

O'Connell, J. F.
1997 On Plio/Pleistocene archaeological sites and central places. *Curr. Anthrop.* 38, 86–88.

O'Connell, J. F., K. Hawkes
1981 Alyawara plant use and optimal foraging theory. In (B. Winterhalder & E. A. Smith, Eds.) *Hunter-Gatherer Foraging Strategies: Ethnographic and Archaeological Analyses,* pp. 99–125. Chicago: University of Chicago Press.

O'Connell, J. F., K. Hawkes, N. G. Blurton Jones
1988a Hadza hunting, butchering, and bone transport and their archaeological implications. *J. Anthrop. Res.* 44, 113–162.
1988b Hadza scavenging: Implications for Plio-Pleistocene hominid subsistence. *Curr. Anthrop.* 29, 356–363.
1990 Reanalysis of large mammal body part transport among the Hadza. *J. Archaeol. Sci.* 17, 301–316.
1992 Patterns in the distribution, site structure, and assemblage composition of Hadza kill-butchering sites. *J. Archaeol. Sci.* 19, 319–345.

O'Connell, J. F., P. K. Latz, P. Barnett
1983 Traditional and modern uses of native plants among the Alyawara of Central Australia. *Economic Botany* 37, 83–112.

Paine, R. R.
1997 The need for a multidisciplinary approach to prehistoric demography. In (R. R. Paine, Ed.) *Integrating Archaeological Demography: Multidisciplinary Approaches to Prehistoric Population,* pp. 1–18. Occasional Paper 24, Center for Archaeological Investigations. Carbondale: Southern Illinois University.

Partridge, T. C., B. A. Wood, P. S. deMenocal
1995 The influence of global climatic change and regional uplift on large mammalian evolution in East and Southern Africa. In (E. Vrba, G. Denton, T. Patridge & L. Burckle, Eds.) *Paleoclimate and Evolution, with Emphasis on Human Origins,* pp. 331–355. New Haven: Yale University Press.

Peccei, J. S.
1995 The origin and evolution of menopause: The altriciality-lifespan hypothesis. *Ethology and Sociobiology* 16, 425–449.

Peters, C. R.
1987 Nut-like oil seeds: Food for monkeys, chimpanzees, humans and probably ape-man. *Am. J. Phys. Anthrop.* 73, 333–363.

Peters, C. R., E. M. O'Brien
1981 The early hominid plant food niche: Insights from an analysis of plant exploitation by *Homo, Pan,* and *Papio* in eastern and southern Africa. *Curr. Anthrop.* 22, 127–140.

Peters, C. R., E. M. O'Brien, E. O. Box
1984 Plant types and seasonality of wild plant foods, Tanzania to southwestern Africa: Resources for models of the natural environment. *J. Hum. Evol.* 13, 397–414.

Potts, R.
1988 *Early Hominid Activities at Olduvai.* Hawthorne, NY: Aldine de Gruyter.

Raunkiaer, C.
1934 *The Life Forms of Plants and Statistical Plant Geography.* Oxford: Clarendon Press.

Reed, K. E.
1997 Early hominid evolution and ecological changes through the African Plio-Pleistocene. *J. Hum. Evol.* 32, 289–322.

Rightmire, G. P.
1990 *The Evolution of* Homo erectus: *Comparative Anatomical Studies of the Extinct Human Species.* Cambridge: Cambridge University Press.
1998 Human evolution in the Middle Pleistocene: the role of *Homo heidelbergensis. Evol. Anthropology* 6, 218–227.

Rodseth, L., R. Wrangham, A. Harrigan, B. Smuts
1991 The human community as primate society. *Curr. Anthrop.* 32, 221–254.

Roebroeks, W., N. Conard, T. van Kolfschoten
1992 Dense forests, cold steppes and the paleolithic settlement of northern Europe. *Curr. Anthrop.* 33, 551–586.

Rogers, M., C. S. Feibel, J. W. K. Harris
1994 Changing patterns of land use by Plio-Pleistocene hominids in the Lake Turkana Basin. *J. Hum. Evol.* 27, 139–158.

Rose, L., F. Marshall
1996 Meat eating, hominid sociality, and home bases revisited. *Curr. Anthrop.* 37, 307–338.

Ruff, C.
1994 Morphological adaptation to climate in modern and fossil hominids. *Yearb. of Phys. Anthrop.* 37, 65–107.

Ruff, C. B., E. Trinkaus, T. W. Holliday
1997 Body mass and encephalization in Pleistocene *Homo. Nature* 387, 173–176.

Ruff, C. B., A. Walker
1993 Body size and body shape. In (A. Walker & R. Leakey, Eds.) *The Nariokotome* Homo erectus *Skeleton,* pp. 234–265. Cambridge: Harvard University Press.

Sacher, G. A.
1959 Relation of lifespan to brain weight and body weight in mammals. In (G. E. W. Wolstenholme & M. O'Connor, Eds.) *Ciba Foundation Colloquia on Aging, Vol. 5, The Lifespan of Animals,* pp. 115–133. London: Churchill.
1975 Maturation and longevity in relation to cranial capacity in hominid evolution. In (R. Tuttle, Ed.) *Primate Functional Morphology and Evolution,* pp. 417–441. The Hague: Mouton.

Schultz, A. H.
1969 *The Life of Primates.* New York: Universe Books.

Sept, J. M.
1992 Archaeological evidence and ecological perspectives for reconstructing early hominid subsistence behavior. In (M. B. Schiffer, Ed.) *Archaeological Method and Therapy,* pp. 1–56. Tucson: University of Arizona Press.

Sillen, A., J. Lee-Thorp
1994 Trace element and isotope aspects of predator-prey relationships in terrestrial food webs. *Palaeog., Palaeoclimatol., Palaeoecol.* 107, 243–255.

Simms, S. R.
1987 *Behavioral Ecology and Hunter-Gatherer Foraging: An Example from the Great Basin.* British Archaeological Reports, International Series 381.

Smith, B. H.
1986 Dental development in *Australopithecus* and early *Homo*. *Nature* 323, 327–330.
1989 Dental development as a measure of life history in primates. *Evolution* 43, 683–688.
1991 Dental development and the evolution of life history in the Hominidae. *Am. J. Phys. Anthrop.* 86, 157–174.
1993 The physiological age of KNM-WT 15000. In (A. Walker & R. Leakey, Eds.) *The Nariokotome* Homo erectus *Skeleton,* pp. 195–220. Cambridge: Harvard University Press.

Smith, B. H., R. L. Tompkins
1995 Toward a life history of the Hominidae. *Ann. Rev. Anthrop.* 24, 257–279.

Smith, R. J., P. J. Gammon, B. H. Smith
1995 Ontogeny of austalopithecines and early *Homo*: Evidence from cranial capacity and dental eruption. *J. Hum. Evol.* 29, 155–168.

Spencer, L. M.
1997 Dietary adaptations of Plio-Pleistocene Bovidae: implications for hominid habitat use. *J. Hum. Evol.* 32, 201–228.

Spennemann, D. H. R.
1994 Traditional arrowroot production and utilization in the Marshall Islands. *Journal of Ethnobiology* 14, 211–234.

Stahl, A. B.
1984 Hominid dietary selection before fire. *Curr. Anthrop.* 25, 151–168.

Stanford, C.
1996 The hunting ecology of wild chimpanzees: Implications for the evolutionary ecology of Pliocene hominids. *American Anthropologist* 98, 96–113.

Stanley, S. M.
1996 *Children of the Ice Age.* New York: Random House.

Street-Perrott, F. A., Y. Huang, R. A. Perrott, G. Eglinton, P. Barker, L. B. Khelifa, D. D. Harkness, D. O. Olago
1997 Impact of atmospheric carbon dioxide on tropical mountain ecosystems. *Science* 278, 1422–1426.

Stringer, C. B., M. C. Dean
1997 Age at death of Gibraltar 2: a reply. *J. Hum. Evol.* 32, 471–472.

Stringer, C. B., M. C. Dean, R. D. Martin
1990 A comparative study of cranial and dental development in a recent British population and neanderthals. In (C. J. DeRousseau, Ed.) *Primate Life History and Evolution* 14, pp. 115–152.

Suwa, G., T. D. White, F. C. Howell
1996 Mandibular postcanine dentition from the Shungura Formation, Ethiopia: Crown morphology, taxonomic allocations, and Plio-Pleistocene hominid evolution. *Am. J. Phys. Anthrop.* 101, 247–282.

Swisher, C., G. Curtis, T. Jacob, A. Getty, A. Suprijo, Widiasmoro
1994 Age of the earliest known hominids in Java. *Science* 263, 1118–1121.

Tchernov, E.
1989 The age of the Ubiediya Formation. *Israel Journal of Earth Sciences* 36, 3–30.

Thoms, A.
1989 The northern roots of hunter-gatherer intensification: Camas and the Pacific Northwest. Ph.D. dissertation, Department of Anthropology, Washington State University, Pullman.

Tompkins, R. L.
1996 Relative dental development of Upper Pleistocene hominids compared to human population variation. *Am. J. Phys. Anthrop.* 99, 103–116.

Trinkaus, E.
1995 Neanderthal mortality patterns. *J. Archaeol. Sci.* 22, 121–142.

Turner, N. J., A. Davis
1993 "When everything was scarce": The role of plants as famine foods in northwestern North America. *Journal of Ethnobiology* 13, 171–202.

Turner, N. J., L. M. Johnson Gottesfeld, H. V. Kuhnlein, A. Ceska
1992 Edible wood fern rootstocks of Western North America: Solving an ethnobotanical puzzle. *Journal of Ethnobiology* 12, 1–37.

Turner, N. J., H. V. Kuhnlein
1983 Camas (*Camassia* spp.) and riceroot (*Fritillaria* spp.): Two liliaceous "root" foods of the Northwest Coast Indians. *Ecology of Food and Nutrition* 13, 199–219.

Uehara, S., T. Nishida, M. Hainai, T. Hasagawa, H. Hazaki, M. Hu man, K. Kawanaka, S. Kobayashi, J. Mitani, Y. Takahata, H. Takasaki, T. Tsukahara
1992 Characteristics of predation by chimpanzees in the Mahale Mountains National Park. In (T. Nishida, W. C. McGrew, P. Marker, M. Pickford & F. B. M. deWaal, Eds.) *Topics in Primatology, Volume 1, Human Origins,* pp. 143–150.

Vainshtein, S.
1980 *Nomads of South Siberia* (Translated from the original [1972] Russian by M. Colenso). Cambridge: Cambridge University Press.

Van Schaik, C. P.
1989 The ecology of social relationships amongst female primates. In (V. Standen & R. Foley, Eds.) *Comparative Socioecology of Mammals and Man,* pp. 195–218. London: Blackwell.

Vincent, A. S.
1985a Plant foods in savanna environments: a preliminary report of tubers eaten by the Hadza of northern Tanzania. *World Archaeology* 17, 131–148.
1985b Wild tubers as a harvestable resource in the East African Savannas: ecological and ethnographic studies. Ph.D. dissertation, Department of Anthropology, University of California, Berkeley.

Vrba, E., G. Denton, T. Partridge, L. Burckle (Eds.)
1995 *Paleoclimate and Evolution, with Emphasis on Human Origins.* New Haven: Yale University Press.

Walker, A., R. E. F. Leakey (Eds.)
1993 *The Nariokotome* Homo erectus *Skeleton.* Cambridge: Harvard University Press.

Walker, P. L., J. R. Johnson, P. M. Lambert
1988 Age and sex bias in the preservation of human skeletal remains. *Am. J. Phys. Anthrop.* 76, 183–188.

Wandsnider, L. A.
1997 The roasted and the boiled: food composition and heat treatment with special emphasis on pit-hearth cooking. *J. Anthrop. Archaeol.* 16, 1–48.

Washburn, S. L., C. S. Lancaster
1968 The evolution of hunting. In (R. B. Lee & I. DeVore, Eds.) *Man the Hunter,* pp. 293–303. Chicago: Aldine.

Watanabe, H.
1973 *The Ainu Ecosystem: Environment and Group Structure.* Seattle: University of Washington.

White, T. D.

1995 African omnivores: global climatic change and Plio-Pleistocene hominids and suids. In (E. Vrba, G. Denton, T. Partridge & L. Burckle, Eds.) *Paleoclimate and Evolution, with Emphasis on Human Origins*, pp. 369–384. New Haven: Yale University Press.

Whiten, A., R. W. Byrne, R. A. Barton, P. G. Waterman, S. P. Henzie

1992 Dietary and foraging strategies of baboons. *Phil. Trans. R. Soc.*, series B 334, 187–197.

Wood, B.

1992 Origin and evolution of the genus *Homo*. *Nature* 355, 783–790.

Wrangham, R.

1987 The significance of African apes for reconstructing human social evolution. In (W. G. Kinzey, Ed.) *The Evolution of Human Behavior: Primate Models*, pp. 51–71. Albany: State University of New York Press.

Wrangham, R., E. Bergmann-Riss

1990 Rates of predation on mammals by Gombe Chimpanzees, 1972–1975. *Primates* 31, 157–170.

Wrangham, R. W., J. L. Gittleman, C. A. Chapman

1993 Constraints on group size in primates and carnivores: Population density and day range as assays of exploitation competition. *Behavioral Ecology and Sociobiology* 32, 199–209.

Wright, L. E., H. P. Schwarcz

1998 Stable carbon and oxygen isotopes in human tooth enamel: Identifying breast feeding and weaning in prehistory. *Am. J. phys. Anthrop.* 106, 1–18.

Yanovsky, E., R. M. Kingsbury

1938 *Analyses of Some Indian food plants.* Contribution 138, Carbohydrate Research Division, Bureau of Chemistry and Soils. Washington, D.C.: United States Department of Agriculture.

A Theory of Human Life History Evolution

Diet, Intelligence, and Longevity

Hillard Kaplan, Kim Hill, Jane Lancaster, A. Magdalena Hurtado

Human life histories, as compared to those of other primates and mammals, have at least four distinctive characteristics: an exceptionally long lifespan, an extended period of juvenile dependence, support of reproduction by older post-reproductive individuals, and male support of reproduction through the provisioning of females and their offspring. Another distinctive feature of our species is a large brain, with its associated psychological attributes: increased capacities for learning, cognition, and insight. In this paper, we propose a theory that unites and organizes these observations and generates many theoretical and empirical predictions. We present some tests of those predictions and outline new predictions that can be tested in future research by comparative biologists, archeologists, paleontologists, biological anthropologists, demographers, geneticists, and cultural anthropologists.

Our theory is that those four life history characteristics and extreme intelligence are co-evolved responses to a dietary shift toward high-quality, nutrient-dense, and difficult-to-acquire food resources.

The following logic underlies our proposal. First, high levels of knowledge, skill, coordination, and strength are required to exploit the suite of high-quality, difficult-to-acquire resources humans consume. The attainment of those abilities requires time and a significant commitment to development. This extended learning phase, during which productivity is low, is compensated for by higher productivity during the adult period and an intergenerational flow of food from old to young. Because productivity increases with age, the investment of time in acquiring skill and knowledge leads to selection for lowered mortality rates and greater longevity. The returns on investments in development occur at older ages. This, in turn, favors a longer juvenile period if there are important gains in productive ability with body size and growth ceases at sexual maturity.

Second, we believe that the feeding niche that involves specializing on large, valuable food packages promotes food sharing, provisioning of juveniles, and increased grouping, all of which act to lower mortality during the juvenile and early adult periods. Food sharing and provisioning assist recovery in times of illness and reduce risk by limiting juvenile time allocation to foraging. Grouping also lowers predation risks. These buffers against mortality also favor a longer juvenile period and higher investment in other mechanisms to increase the life span.

Thus, we propose that the long human life span co-evolved with lengthening of the juvenile period, increased brain capacities for information processing and storage, and intergenerational resource flows, all as a result of an important dietary shift. Humans are specialists in that they consume only the highest-quality plant and animal resources in their local ecology and rely on creative, skill-intensive techniques to exploit them. Yet the capacity to develop new techniques for extractive foraging and hunting allows them to exploit a wide variety of different foods and to colonize all of earth's terrestrial and coastal ecosystems.

We begin with an overview of the data on which the theory is based: a comparative examination of hunter-gatherer and chimpanzee life-history traits and age profiles of energy acquisition and consumption. The data show that hunter-gatherers have a longer juvenile period, a longer adult lifespan, and higher fertility than chimpanzees do. Hunter-gatherer children are energetically dependent on older individuals until they reach sexual maturity. Energy acquisition rates increase dramatically, especially for males, until mid-adulthood and stay high until late in life.

We then present both theoretical and empirical tests of our theory. For the theory to be correct, a model of natural selection must show that mortality rates, the length of the juvenile period, and investments in learning co-evolve in the ways predicted by the theory. Building on existing models of life-history evolution,[1-5] we develop such a model. The results of our analysis confirm the theory's predictions. Those theoretical tests are

*Originally published in *Evolutionary Anthropology* 9(2000):156–185

Survival of Hunter-Gatherers and Chimpanzees

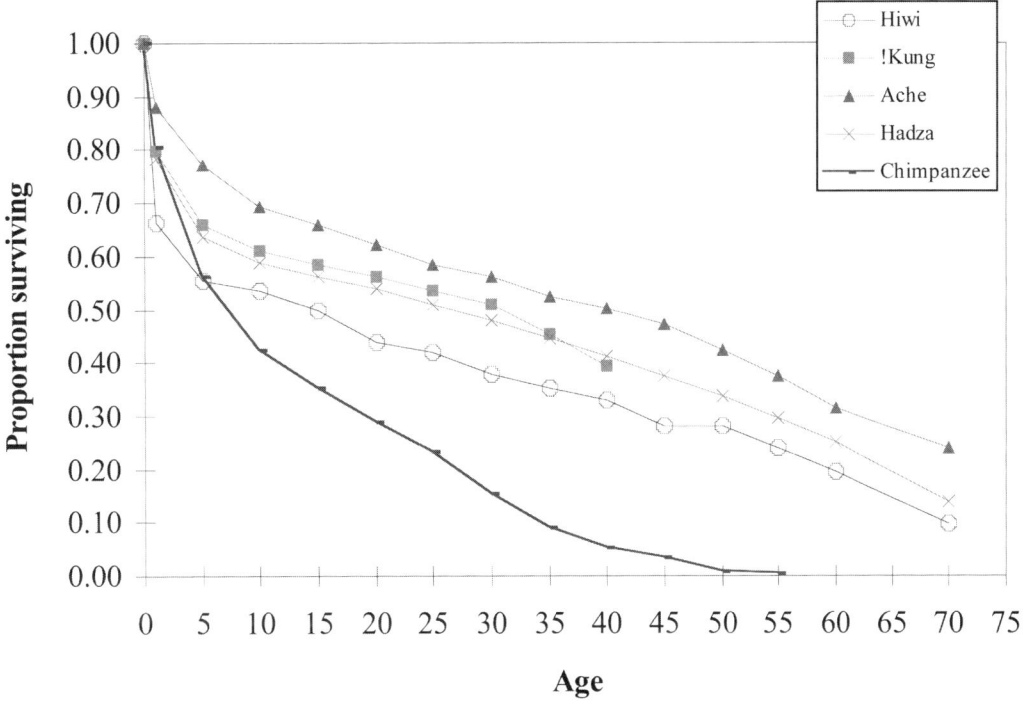

FIGURE 4.1. Survival curves for forager populations were derived from sources listed in notes for Table 4.1. Chimpanzee mortality is from a synthetic life table combining all mortality data from Bossou, Gombe, Kibale, Mahale and Tai.[8]

followed by empirical tests. In order for our theory to be correct, we must demonstrate that: Humans do, in fact, consume more skill-intensive, difficult-to-acquire, high-quality foods than do chimpanzees and other nonhuman primates; Difficulty of acquisition explains the age profile of production for both humans and chimpanzees; Men play a large role in supporting human reproduction; The foraging niche occupied by humans lowers mortality rates among juveniles and adults relative to corresponding rates among chimpanzees and other nonhuman primates. We present strong evidence in support of the first three propositions and suggestive evidence in support of the fourth.

We then examine the evolution of the primate order to determine whether the same principles invoked in our theory of hominid evolution explain the major primate radiations. We then consider the fundamental differences between our theory and the "grandmother hypothesis" recently proposed by Hawkes, Blurton Jones and O'Connell.[6,7] We conclude with a listing of the new and unique predictions derived from our theory.

Our theory is not the first to propose that high-quality foods, extractive foraging, and hunting are fundamental to human evolution. However, it is the first to do so with a specific model of natural selection that unifies the evolution of life history, brain and intelligence, diet, and age profiles of food production and consumption. As a result, it organizes existing data in a new way and leads to a novel set of predictions.

DIFFERENCES BETWEEN THE LIFE-HISTORY TRAITS OF HUNTER-GATHERERS AND CHIMPANZEES

Mortality, Fertility, and Growth

Figure 4.1 shows the differences between the life spans of traditional human foragers and chimpanzees; Table 4.1 compares a variety of life-history traits of the two species. The hunter-gatherer data come from studies on populations during periods when they were almost completely dependent on wild foods, having little modern technology and no firearms, no significant outside interference in interpersonal violence or fertility rates, and no significant access to modern medicine. The chimpanzee data are compiled from all published and unpublished sources that we are aware of. Because of small sample sizes at individual sites, mortality data were combined to create a single synthetic life table and survival function that combines all data for wild chimpanzees.[8]

The data suggest that hunter-gatherer children have a higher rate of survival to age 15 (60% versus 35%) and higher growth rates during the first 5 years of life (2.6 kg/yr versus 1.6 kg/yr) than do juvenile chimpanzees. Chimpanzees, however, grow faster between ages 5 and 10, both in absolute weight gain (2.5 kg/yr for chimps versus 2.1 kg/yr for humans) and proportional weight gain (16% per year for chimps versus 10% per year for humans) (Table 4.1). The early higher weight gain for humans may be due to an earlier weaning age (approximately

TABLE 4.1. Life history parameters of human hunter-gatherers and chimpanzees.

Group	Probability of Survival to Age 15	Expected Age of Death at 15 (years)	Mean Age at First Reproduction (years)	Mean Age at Last Reproduction[b] (years)	Interbirth Interval[a] (months)	Mean Weight Age 5 (kg)	Mean Weight Age 10 (kg)
Humans							
Ache female[d]	0.61	58.3	19.5	42.1	37.6	15.7	25.9
Ache male	0.71	51.8				15.5	27
Hadza female[e]	0.58	54.7				15.5	20
Hadza male	0.55	52.4				14.2	21.2
Hiwi female[f]	0.58	51.3	20.5	37.8	45.1	18	29.8
Hiwi male	0.58	51.3				16.4	33.6
!Kung female[g]	0.6	56.5	19.2	37	41.3	14	19.5
!Kung male	0.56	56.5				16	22.5
Forager mean[c]	0.60	54.1	19.7	39.0	41.3	15.7	24.9
Chimpanzees							
Bossou female[h]					51		
Bossou male							
Gombe female[i]	0.545	32.7	14.1		64.6	10	21
Gombe male	0.439	28.6				10	24
Kibale female[j]	0.805	35.6			68		
Kibale male	0.408	40.6					
Mahale female[k]			14.6		72		
Mahale male							
Tai female	0.193	23.8	14.3		69.1		
Tai male[l]	0.094	24					
Chimpanzee mean	0.35	29.7	14.3	27.7**	66.7	10	22.5

[a] Mean interbirth interval following a surviving infant.
[b] Age of last reproduction for chimpanzee females was estimated as two years prior to the mean adult life expectancy.
[c] The forager mean values were calculated by weighting each forager study equally. The chimpanzee mean mortality is from a synthetic life table using data from all five sites listed.[8,138]
[d] Ache: Demographic and weight data from Hill and Hurtado.[10]
[e] Hadza: Demographic data from Blurton Jones and colleagues.[16] Weight data from Blurton Jones (personal communitcation).
[f] Hiwi: Demographic data from Hill and Hurtado unpublished database collected on the Hiwi foragers from reproductive-history interviews conducted between 1982 and 1991 using the same methodology published in Hill and Hurtado.[10]
[g] !Kung: Demographic and weight data from Howell.[83]
[h] Bossou: Data from Sugiyama.[139]
[i] Gombe Data on mortality from Hill and coworkers,[8] and Pusey and Willams (personal communication). Gombe data on fertility from Pusey,[140] Tutin,[141] and Wallis.[142] Weights from Pusey.[140]
[j] Kibale: All data from Wrangham (personal communication). Mortality data in Hill and coworkers.[8]
[k] Mahale: Data from Nishida, Takasaki, and Takahata.[143]
[l] Tai: Data from Boesch and Boesch.[20]

2.5 years for hunter-gatherers versus 5 years for chimpanzees) and parental provisioning of highly processed foods. Among humans, the slow growth during middle childhood is intriguing. According to the allometric growth law, mammalian growth can be described by the equation $dw/dt = Aw^{0.75}$ (where change in weight per unit of time is expressed as a function of a growth constant, A, and weight, w, to the 0.75 power). Most mammals show a yearly growth constant of about 1, whereas the mean primate value for A is about 0.4.[9] Hunter-gatherer children between the ages of 5 to 10 years are characterized by extremely slow growth, with A being approximately 0.2.

Chimpanzees spend less time as juveniles than humans do: Female chimpanzees give birth for the first time about 5 years earlier than do hunter-gatherer women. In natural habitats, chimpanzees also have a much shorter adult life span than humans do. At age 15, chimpanzee life expectancy is an additional 15 years, as compared to 39 more years for human foragers. Importantly, women spend more than a third of their adult life in a postreproductive phase, whereas very few chimpanzee females survive to the postreproductive phase. The differences in overall survival and life span are striking (Fig. 4.1). Less than 10% of chimpanzees survive to age 40, but more than 15% of hunter-gatherers survive to age 70. These naturalistic observations are also consistent with data on maximum life spans. The maximum life span of humans is between 100 and 120 years, depending upon how it is calculated, which is about two times longer than the maximum adult chimpanzee life span (approximately 60 years for captive populations).

Despite the fact that the human juvenile and adult periods are longer than those of chimpanzees and that human infants are larger than chimpanzee infants at birth (about 3 kg versus 2 kg), hunter-gatherer women characteristically have higher fertility than do chimpanzee females. The mean interbirth interval between offspring when the first survives to the birth of the second is more than 1.5 times longer among wild chimpanzees than among modern hunter-gatherer populations. These numbers lead to an interesting paradox. Life tables from modern human foragers always imply positive growth (see Hill and Hurtado,[10] chapter 14), whereas the chimpanzee numbers presented here imply slightly negative population growth rates. Chimpanzee negative population growth may be a real feature of recent habitat destruction and other human intrusion, or "natural" mortality rates may have been overestimated due to the inclusion of deaths from viral epidemics such as ebola and polio (see Hill and coworkers[8] for a discussion).

To summarize, hunter-gatherers have a juvenile period that is 1.4 times longer than that of chimpanzees and a mean adult life span that is 2.5 times longer than that of chimpanzees. They show higher survival at all ages after weaning, but lower growth rates during middle childhood. Despite a longer juvenile period, slower growth, and a longer life span, hunter-gatherer women achieve higher fertility rates than do chimpanzee females.

The Age and Sex Profile of Energy Acquisition

Data on food acquisition by age and sex category exist for only three modern foraging populations. Ache and Hiwi food production was directly monitored by weighing all food produced by those in different age and sex categories throughout most months of various years. (See Hill and coworkers[11] and Hurtado and Hill[12] for definitions, methodology, and sampling plan). Hadza women's and children's plant-food acquisition was estimated indirectly from samples of in-patch return rates for different fruit and root resources over various age or sex classes during part of the wet season and part of the dry season of various years. (For details, see Hawkes and coworkers, and Blurton Jones, Hawkes, and O'Connell[13,14]). These data were combined with sample estimates of time spent foraging and frequency across days to estimate daily food acquisition.[13,14] Hadza men's food acquisition from hunting was measured directly by weighing all large game brought to camp.[15]

Although there is some cross-cultural variation, all three societies show similar patterns. Hunter-gatherer children produce little food compared to adults (Fig. 4.2). In the late juvenile period, daily food acquisition rates rise dramatically, especially for males. These rates continue to increase until mid-adulthood for males in all three groups and even longer for Hadza and Hiwi females. Adult men acquire much more food than do those in any other age-sex category. Although the patterns for men seem consistent for all three societies, Hadza children and postreproductive women appear to acquire substantially more food than do their Ache and Hiwi counterparts. But total food-consumption estimates for the Hadza may be unrealistically high, since the data suggest per capita consumption of about 3,400 calories per day (Tables 4.2, 4.3). That is 126% of the mean daily caloric consumption of the Ache, despite the fact that 10-year-old Hadza children weigh only 78% of the weight of 10-year-old Ache children, and adult Hadza women weigh only 89% of the weight of Ache women. Those estimates are derived by assuming an age-structure consistent with the Hadza life-table.[16] If the dependency ratio in the camps studied by the Hadza researchers was greater than expected by the life table, as Blurton Jones (personal communication) believes to be the case, the per-capita consumption estimates would be reduced accordingly and would be more realistic.

Figure 4.3 shows the mean daily energy consumption and acquisition rates for all three hunter-gatherer societies as compared to the rates for chimpanzees of the same age and sex. The food-consumption rates of forager children and adults is estimated from body weight and total group production.[17] Chimpanzee energy acquisition, while not measured directly, can be estimated from body size and caloric requirements, since very little food is transferred between age-sex categories after weaning. Daily food acquisition and consumption are virtually the same for chimpanzees from the juvenile period onward. The human consumption-acquisition profile is strikingly different from that of chimpanzees, with chimpanzee juveniles acquiring considerably more energy than forager children do until about the age of sexual maturity. No children in any forager society produced as much as they consumed until they reached their mid- to late teens. Thus, human juveniles, unlike chimpanzee juveniles, have an evolutionary history of dependency on adults to provide their daily energy needs. This can be appreciated by realizing that by age 15 the children in our forager sample had consumed over 25% of their expected life-time energy consumption but had acquired less than 5% of their life-time energy acquisition.

The area in Figure 4.3 where food acquisition is greater than consumption (where the solid line for each species is above the dotted line) represents surplus energy provided during the later part of the life span. These averaged data imply that hunter-gatherer men provide most of the energy surplus that is used to subsidize juveniles and reproductive-aged women. Although based on averaging only three societies, this trend can be confirmed by comparing the food-acquisition rates of adult males and females from a sample of ten hunter-gatherer societies in which food acquisition has been measured with a systematic sample (Table 4.2).

A Theoretical Test: Would Natural Selection Actually Produce the Co-Evolutionary Effects Proposed by the Theory?

Our proposal is that the shift to calorie-dense, large-package, skill-intensive food resources (Fig. 4.4) is responsible for the unique evolutionary trajectory of the genus *Homo*. The key element in our theory is that this shift produced co-evolutionary selection pressures, which, in turn, operated to produce the

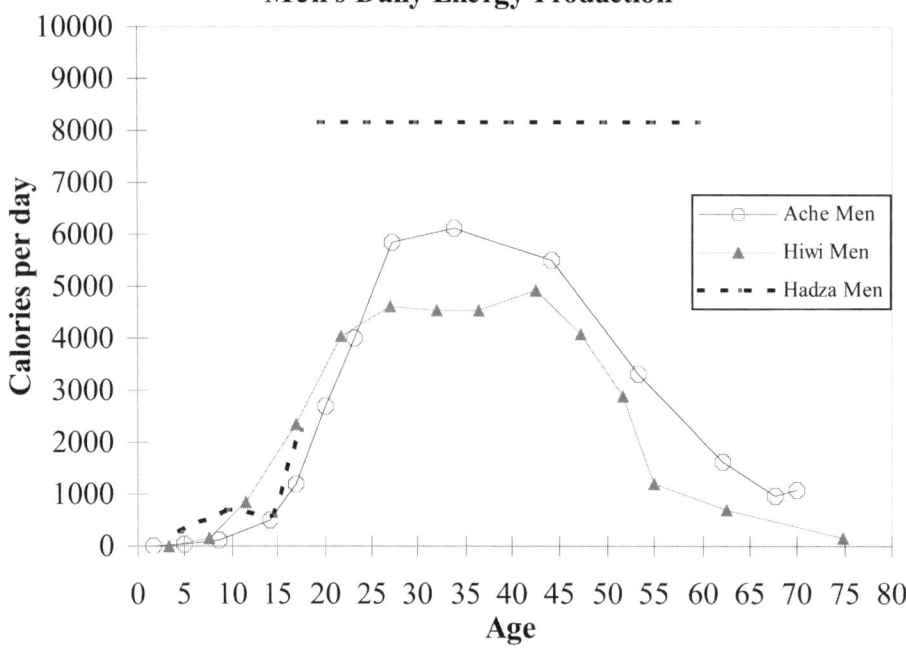

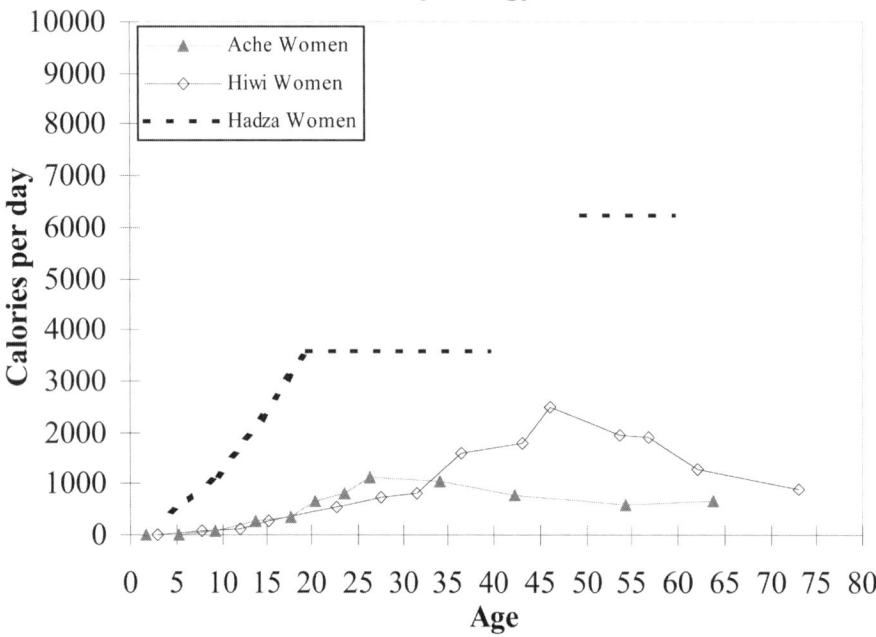

FIGURE 4.2. Daily energy acquisition data are recorded by individual among the Ache and Hiwi. Thus, the age and sex of each acquirer is known for every day sampled. Mean production for 5- or 10-year age intervals (y value) was calculated from raw data by summing all calories produced over the sample period by individuals in that age-sex class and dividing by the total sample of person days monitored for individuals in that category. This was plotted at the mean age of person days sampled (x value) in the category analyzed. Hadza production levels are given for various juvenile age categories, for all adult men combined (no age breakdown), and for all reproductive women and all women of postreproductive age combined (no age breakdown). All values are calculated as described in the notes for Tables 4.2 and 4.3.

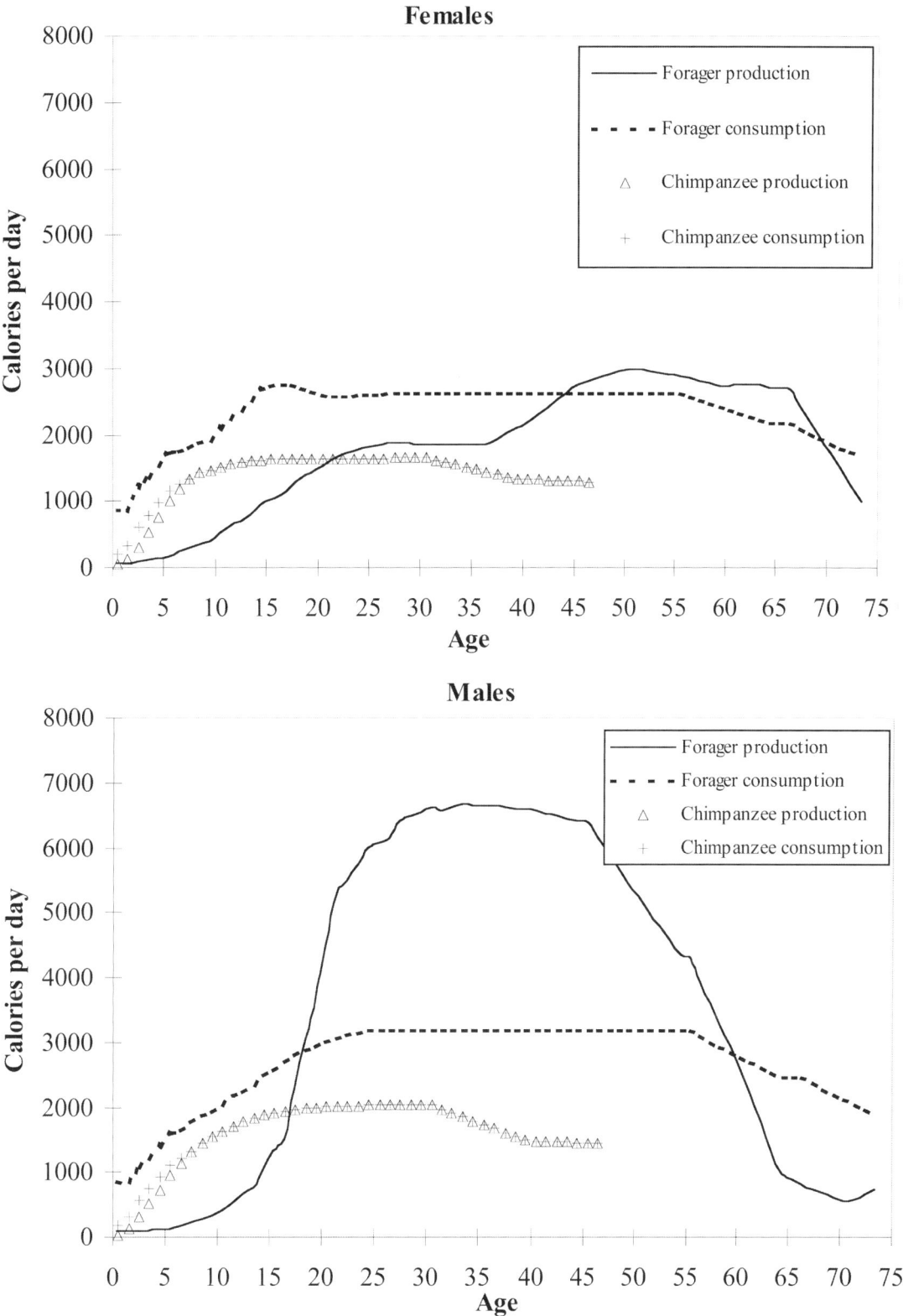

FIGURE 4.3. The mean expected daily energy consumption per individual for each age-sex category of each foraging group was estimated by first multiplying all the age-sex specific production rates for that foraging group times the proportional representation of that age-sex category as expected from the survival curves and summing expected production across all age categories. This total expected production for the group was then divided by the expected total number of individuals (as determined by the survival curve) at all ages times their proportion of a standard consumer. This procedure assumes that all populations are in steady state and that the proportional representation of each age category is determined by the probability of surviving to that age, and gives consumption per standard consumer in each group. This gives the mean consumption for a standard consumer in the group. Daily expected consumption for individuals in various age-sex categories is estimated by multiplying the proportion of a consumer represented by each age class times the mean consumption of a standard consumer. Kaplan[17] provides a detailed description of these calculations and how proportional standard consumers were determined for each age-sex category. A standard consumption rate of 1 is assigned to young adult males and females; children begin at a consumption level of 0.3 that of a standard consumer. Daily energy acquisition for age-sex categories is calculated as described for Figure 4.2 and averaged across the Ache, Hiwi, and Hadza, weighting each group equally.

TABLE 4.2. Production of energy by men and women in foraging societies.

		Daily Adult Production in Calories[a]						
		Meat	Roots	Fruits	Other	Mean Daily Total	% Total Adult Calories	% Total Adult Protein
Onge[b]	men	3919	0	0	81	4000	79.7	94.8
	women	0	968	1	52	1021	20.3	5.2
Anbarra[c]	men	2662	0	0	79	2742	70.0	71.8
	women	301	337	157	379	1174	30.0	28.1
Arnhem[d]	men	4570	0	0	8	4578	69.5	93.0
	women	0	1724	37	251	2012	30.5	7.0
Ache[e]	men	4947	0	6	636	5590	84.1	97.1
	women	32	0	47	976	1055	15.9	2.9
Nukak[j]	men	3056	0	0	1500	4556	60.4	98.6
	women	0	0	2988	0	2988	39.6	1.4
Hiwi[g]	men	3211	2	121	156	3489	79.2	93.4
	women	38	713	83	82	916	20.8	6.6
!Kung[1,h]	men	2247			974	3221	45.5	44.7
	women	0	348	348	3169	3864	54.5	55.3
!Kung[2,i]	men	6409				6409	>>50	
	women							
Gwi[f]	men	1612	800	0	0	2412	43.0	78.7
	women	0	0	0	3200	3200	57.0	21.3
Hadza[k]	men	7248	0	0	841	8089	64.8	94.1
	women	0	3093	1304	0	4397	35.2	5.9

[a] Edible portion and caloric values were taken from individual studies when available. Otherwise we assumed vertebrate meat at 85% edible, the Ache measured average for animals, and used the following conventions for calories/100 g edible: mammals 150; roots 150; fruits 70; fish 120. When not specified, protein was assumed at 20% by weight for meat and 2% for roots and fruits.

[b] Onge: Data come from Bose.[144] We assumed that all food is produced by adults and that men and women make up equal percentages of the reported population. Caloric values (p. 156) and edible portions are taken from Meehan.[145] We assumed that males got all pigs, turtles, fish, and honey, whereas females acquired all crabs, bivalves, and plant products. Total caloric intake seems very low, but the Onge are the smallest foragers in this sample and had very low fertility.

[c] Anbarra: Data come from Meehan[145]; diet is found in Tables 29–32. It is assumed that women collected 85% of shellfish (p. 125) and that men obtained only birds, fish, mammals, and some shellfish (p. 149). Total person days of consumption are in each table. Women's production days come from Table 27. We assumed an equal number of production days for men.

[d] Arnhem: Arnhem land data are from McArthur[150] (pp. 127–128 and p. 138). It is assumed that adults acquired all food, that men obtained only vertebrate meat and honey, and that women acquired all other resources.

[e] Ache: Data come from all observed foraging trips between 1980 and 1996 on which KH, HK, and/or MH were present. Data prior to 1984 were published in Hill and coworkers.[11] Subsequent data come from forest trips between 3 and 15 days long when nearly all foods consumed were acquired from the forest. All foods were weighed on site and the edible portion was calculated from refuse samples collected after consumption. Caloric values were determined as previously published. Total production of fruits was estimated by multiplying measured collection rates for different age categories times the time spent collecting by each individual. We have made two important modifications of 1984 data because of new field measures: 1) We now estimate the edible portion of wild honeycomb to be only 35% by weight; 2) The edible portion of palm starch is estimated at only 6% by weight, with the caloric value of the edible portion being 3,920 cal/kg. These corrections and new production data have lowered previously published estimates of daily caloric intake.

[f] Gwi: Meat production per hunter day is averaged from Silberbauer's[146] one-year observations of a band including 20 men and Tanaka's[147] 180-day observations (p.111) of 10 men. We estimate Silberbauer's band to contain 20 men and 24 women because there were 80 individuals, 46.5% of whom were male and 55% were adult (p. 286, 287). For live weight meat, we assume 85% edible weight containing 1,500 cal/kg. Plant production for adult women is estimated at the observed per-capita consumption, 800 g/consumer day times 80/24 (the ratio of the total population to adult women) times 1,500 cal/kg raw plant, times 80% collected by women[147] (p.70). Men are assumed to have produced 20% of the plant calories. Meat consumption per capita is the average from Tanaka[147] (p.70) and Silberbauer[146] (p. 446). We assumed that Tanaka's raw weights are 85% edible; we also assume 1,500 cal/kg edible meat for both studies. Plant consumption is reported to be 800 g/person in both studies[147] (p.70),[146] (p.199). We assume that this is equally split between roots and melons, with a mean caloric value of 1,500 cal/kg raw weight. Man days hunting are reported for both studies, but calculations of the sample size of women's production and per-capita consumption are not specified in either study.

[g] Hiwi: Data come from a sample of days between 1985 and 1988 when KH and MH resided with the Hiwi and weighed all food produced by a sample of camp members. Details of calculations of edible portion and food value are published in Hurtado and Hill.[12,148]

[h] !Kung[1]: All data on adult production and per-capita consumption are from Lee[42] (pp. 260–271). Women's plant production (non-mongongo) was assumed to be evenly split between roots and fruits.

[i] !Kung[2]: Data are from Yellen[43] as calculated in Hill[149] (pp. 182-183). Edible portion and caloric value are the same as in Lee.[42] Only hunting data are recorded. In order to estimate per-capita consumption, adult men and women are assumed to comprise equal percentages of the band members. The percentage of the diet from meat is calculated assuming total consumption of 2,355 calories per person day, as per Lee.[42]

[j] Nukak: Data come from Politis[151] (chapter 4). We assume that all food was produced by adults and that men and women make up equal percentages of the population. Edible portions and caloric values for foods come from similar Ache resources. Fruits show edible portions varying from 21% for fruits brought in and weighed with the stalk to 40% for fruits without the stalk collected in baskets. Caloric values of fruits ranged from 600 cal/kg for sweet pulpy fruits to 1,430 cal/kg for oily palm fruits. Other resources were equivalent to common Ache and Hiwi resources.

[k] Hadza: Data on the daily caloric production of children are from Blurton Jones, Hawkes, and O'Connell.[14] We assumed that 61% of the calories produced come from fruit and the remainder from roots, as for youngest girls.[6] Daily production of women taken from Hawkes, O'Connell, and Blurton Jones.[6] We multiplied in-patch rates by time foraging for each season (both in Table 1), and the proportion of time in patch (60% root, 66% berry) (Hawkes personal communication and

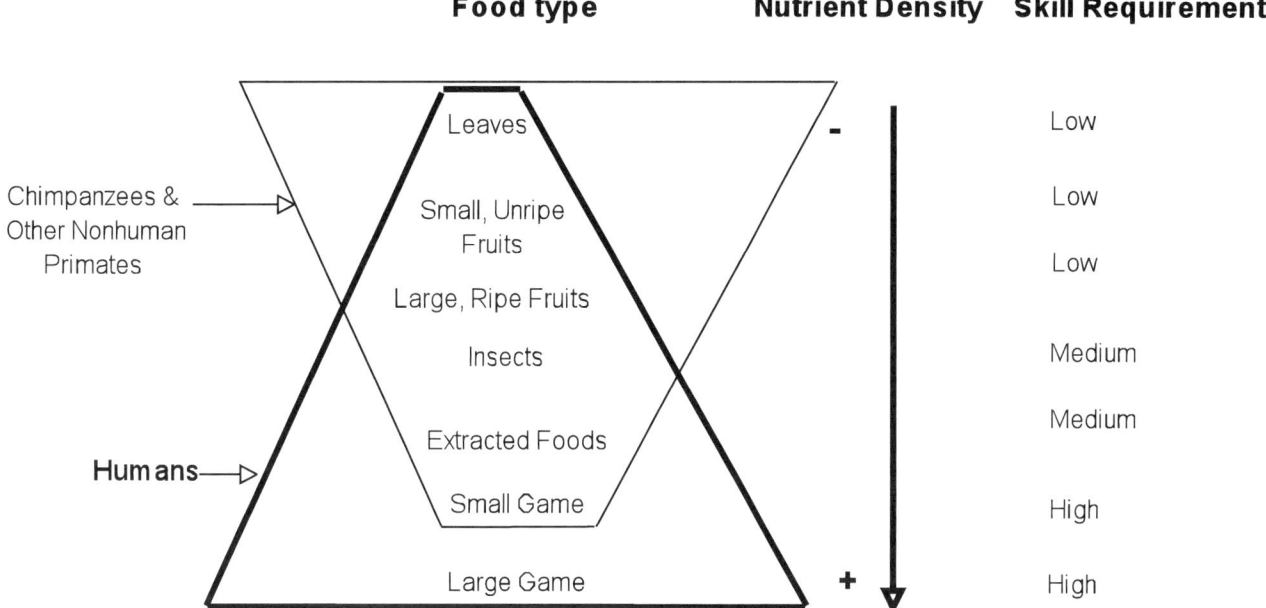

FIGURE 4.4. The feeding ecology of humans and other primates.

extreme intelligence, long developmental period, three-generational system of resource flows, and exceptionally long adult life characteristic of our species. We envision two important effects of the change in feeding niche.

First, a long developmental period, parental provisioning, and a large brain are necessary foundations of the skill-intensive feeding niche, and therefore are products of selection as a result of entry into that niche. Our view is that human childhood is elongated by including a period of very slow physical growth, during which the brain is growing, learning is rapid, and little work is done. This is followed by adolescence, during which growth is accelerated so that the brain and body can function together in the food quest. Early adulthood is a time for vigorous work during which resource acquisition rates increase through on-the-job training. Thus, investment in this life history involves three important costs: low productivity early in life, delayed reproduction, and a very expensive brain to grow and maintain. The return from those investments is delayed, with extremely high productivity occurring in the middle and latter portions of the adult period. That return increases with lengthening of the adult life span because the return is realized over a greater period of time. Second, we propose that the shift in the human feeding niche operated directly to lower mortality rates because it increased food package size, which, in turn, favored food sharing, provisioning, and larger group size. Another indirect effect was that the added intelligence and use of tools associated with the feeding niche also lowered predation rates.

Underlying our theory is the hypothesis that these two effects produce co-evolutionary processes of large magnitude. Holding all else constant, ecological changes that increase the benefits of a long developmental period and a concomitant increase in later adult productivity not only produce selection pressures to delay the onset of reproduction, but also produce selection pressures to invest more in survival during both the juvenile and adult periods. At the same time, ecological changes that lower mortality rates during the juvenile and adult periods also produce selection pressures that favor a longer juvenile period if it results in higher adult productivity. If both types of change occur (increased payoffs for time spent in development and lower mortality rates), great changes in both mortality rates and time spent in development may result. Furthermore, if those changes are accompanied by large increases in productivity after adulthood

---

Hawkes, O'Connell, and Blurton Jones,[6] p. 350), equally weighting production in dry and wet seasons. Ekwa roots are calculated as 88% edible (Hawkes, personal communication) and 850 cal/kg edible (Hawkes, O'Connell, and Blurton Jones,[6] p. 691). Fruits are assumed to be 50% edible and have a caloric value of 2,500 cal/kg edible[42] (pp. 481, 484 for grewia sp. berries). Meat acquisition is 4.89 kg/day for adult men (over age 18) and assumed to have a caloric value of 1,500 cal/kg with the discounting for the edible portion.[7] Honey production was assumed to be 0.78 kg/man-day for males over age 18[15] (p.86), with 35% edible and 3,060 cal/kg (as for the Ache). All food production by age and sex category was weighted by the probability of survival to that age for the Hadza,[16] then divided by the total of all survival probabilities to obtain the expected per-capita consumption. The estimate of total per-capita consumption is very high, probably in part because actually sampled camps contained more juvenile consumers than the life table implies (Hawkes, personal communication). However, we cannot correct the estimate of daily consumption without a complete age-sex breakdown of the sampled camps, which currently is not available.

[l] Chimpanzee diet: We use the Gombe diet from Goodall[76] (Fig. 10.1). The absolute amount of meat in the diet is from Wrangham and Riss.[22] Kibale plant percentages are from Wrangham, Conklin-Brittain, and Hunt.[97] The Kibale meat percentage is from Wrangham and coworkers.[152] The Mahale diet is taken from Hiraiwa-Hasegawa.[25] The absolute amount of meat in the diet was calculated from Uehara,[153] assuming adult prey at 13 kg and juvenile prey at 6 kg, on average, the percentage of adult prey was taken from Stanford.[154] For Tai forest chimpanzees, the absolute amount of meat in the diet was calculated from Boesch and Boesch[20] (Table 7.4).

is reached, we expect additional increases in time spent in development and in survival rates. Our proposal is that the skill-intensive feeding niche, coupled with a large brain, is associated with a significant amount of learning during the adult period.

To test this hypothesis, we developed a model to determine whether or not natural selection would actually result in co-evolution of the developmental period and the life span. This model builds on two bodies of theory, life-history theory in biology and human-capital theory in economics. Life-history theory is based on the premise that organisms face trade-offs in the allocation of their time and effort. Gadgil and Bossert[2] offered the first explicit treatment of allocations trade-offs with respect to reproduction and longevity. They postulated that during the life course selection acts on the allocation of energy to each of three competing functions: reproduction, maintenance, and growth. Energy allocated to reproduction will necessarily reduce the quantity available for maintenance and growth. Maintenance and growth may be seen as investments in future reproduction, for they affect both the probability that an organism will survive to reproduce in the future and the amount of energy it will be able to harvest and transform into reproduction. Thus, one fundamental trade-off is between current and future reproduction.

Human-capital theory in economics[18,19] is designed to analyze investments in education and training through the course of life. Central to this theory is the notion of foregone earnings: Time spent in education and training reduces current earnings in return for increased earnings in the future. The economic trade-off between current and future earnings is directly analogous to the trade-off between current and future reproduction in biology.

Charnov,[1,9] building on earlier work on optimal age at first reproduction, developed a mathematical model of the trade-off between growth and reproduction for mammals. His model is designed to capture determinate growth, in which an organism has two life-history phases after attaining independence from its parents. These are a pre-reproductive growth phase in which all excess energy, remaining after maintenance requirements have been met, is allocated to growth and a reproductive phase in which all excess energy is allocated to reproduction. By growing, an organism increases its energy capture rate, and thus increases reproductive rate. During adulthood the fundamental trade-off is between the expected length of the reproductive span (which is shorter with each additional year spent growing, because of the increased probability of dying before reproducing) and the adult reproductive rate (which is higher with every year spent growing because of increased energy stored in the form of adult body mass). The model predicts the amount of time mammals will grow before switching to the reproductive phase by selecting the time that maximizes expected energy for reproduction over the life course.

Here we extend Charnov's model in three ways. First, we broaden the concept of growth from body size alone to include all investments in development. Development can be seen as a process in which individuals and their parents invest in a stock of embodied capital, a term that generalizes the concept of human capital to all organisms. In a physical sense, embodied capital is organized somatic tissue. In a functional sense, embodied capital includes strength, immune function, coordination, skill, knowledge, and social networks, all of which affect the profitability of allocating time and other resources to alternative activities such as resource acquisition, defense from predators and parasites, mating competition, parenting, and social dominance. Because such stocks tend to depreciate with time due to physical entropic forces and direct assaults by parasites, predators, and conspecifics, allocations to maintenance efforts, such as feeding, cell repair, and vigilance, can also be seen as investments in embodied capital. In our model, the energy capture rate increases with embodied capital (that is, time spent in development).

Second, in Charnov's model, organisms have no control over mortality rates, which are exogenously determined; the only variable of choice for the organism is age at first reproduction, or time spent in development. In our model, the organism can exercise control over mortality rates by allocating energy to mortality reduction. We include this second choice variable to determine if time invested in development and energy allocated to mortality reduction co-evolve. Our theory predicts that increased investment of time in development due to the exploitation of difficult-to-acquire, high-quality resources selects for increased longevity or lower mortality rates. Third, our model allows learning to continue after physical growth has ceased to analyse the effects on the increase in return rates from foraging during the adult period (see Figs. 4.2 and 4.3) as a consequence of "on-the-job" training.

Thus, the model has two choice variables upon which selection can act: age at first reproduction, a proxy for time spent in development and physical growth, and allocation of energy to lowering mortality. It also has three ecological parameters: factors affecting the pay-offs to investments in development, factors affecting mortality rate, and the growth in productivity after adulthood due to learning. The formal model is presented in Box 4.1.

The six main results of the mathematical model confirm the co-evolutionary selection pressures predicted by the theory. Ecological factors increasing the productivity of investments in developmental embodied capital (in the context of the present theory, a skill-intensive foraging niche) increase both time spent as a juvenile and investments in mortality reduction. Ecological factors that lower mortality rates increase both time spent in development and investment in mortality reduction. Finally, the greater the growth rate in production during the adult period, due to large brains and "on-the-job" training, the more it pays to invest in development and mortality reduction. These results all show that investments in development and investments in mortality reduction and longevity co-evolve.

Box 4.1. A Formal Model of Natural Selection on Age at First Reproduction and Investments in Mortality Reduction

The model treats two phases of the life course, the juvenile and adult periods. The juvenile period begins after the high mortality phase associated with infancy and weaning, is dedicated to growth and development, and lasts a variable amount of time, t, upon which selection acts. During this time, all energy is invested in either embodied capital (growth and learning) affecting future energy production, P, or in reducing mortality rate, μ. The two choice variables during the juvenile period are its length, t, and the proportion of energy invested in mortality reduction, λ (implying that $\{1-\lambda\}$ is the proportion allocated to growth and learning). The adult production of energy at the end of the juvenile period, P_a, is determined by t, λ, and the combined ecological effects of the environment and the technology of production, captured by the vector, ϵ, which is assumed to increase energy production (i.e. $\partial P_a / \partial \epsilon > 0$). We can think of this as composed of two functions $\pi(t, \epsilon)$, which captures the growth in production due to learning, growth, and development, and $\Psi(\lambda)$, which represents the proportional loss in production due to investments in mortality reduction during the juvenile period. Thus, we have $P_a = p(t, \lambda; \epsilon) = \pi(t, \epsilon) \psi(\lambda)$.

The second period is reproductive. During this period, growth in body size ceases and all excess energy is allocated to reproduction. Production grows at some constant rate, g, due to the effects of learning. Thus, production at some age, x, after adulthood, P_x, is $P_a e^{g(x-t)}$.

During both phases, the instantaneous mortality rate, μ, remains constant, at a level determined by the amount of energy production diverted to mortality reduction, λ, and by ecology factors affecting mortality (such as density of predators and diseases), θ, which is assumed to increase mortality rates (i.e., $\partial \mu / \partial \theta > 0$). Thus, $\mu = \mu(\lambda, \theta)$. The net energy available for reproduction at age x during the adult period, $P_{r,x}$, will be equal to total energy production times the proportion allocated to reproduction, $P_{r,x} = (1-\lambda)P_a e^{g(x-t)}$.

The basic logic of the model is that an organism will maximize fitness by maximizing its lifetime energy allocated to reproduction. By increasing the length of the growth and development phase, t, the adult rate of energy capture increases, but the expected length of the reproductive period, R, decreases. This decrease results from the fact that no energy is allocated to reproduction during the growth and development phase, though the organism is still exposed to mortality. Allocations to mortality reduction also have opposing effects. Allocations to growth during the juvenile period and to reproduction during the adult period are reduced by allocations to mortality reduction. Yet lowered mortality increases the length of the reproductive period by increasing the probability of reaching reproductive age and by increasing the expected time from reproductive age to death. Thus, R, is determined by both t and λ. If we consider only individuals who survive infancy, R is equal to expectation of adult reproductive years lived, given the probability of dying at each age, x. Thus, the expected number of adult years during which energy is allocated to reproduction is $R = r(t, \mu(\lambda, \theta)) = \int_{x=t}^{\infty} (x-t) \mu e^{-\mu x} dx = 1/\mu e^{-\mu t}$, where $1/\mu$ is the expected adult life span conditional on reaching age t and $e^{-\mu t}$ is the probability of reaching age t, conditional on having survived infancy. The expected energy production during the adult period is $e^{-\mu t}(1-\lambda) P_a \int_{x=t}^{\infty} e^{(g-\mu)(x-t)} dx$, which is equal to $e^{-\mu t}(1-\lambda) P_a (\mu - g)^{-1}$ for $\mu > g$ (which we assume to hold true, because expected lifetime income otherwise would be infinite).

Selection is expected to optimize t and λ so as to maximize lifetime energy allocated to reproduction. Thus, we have the following maximization problem:

$$\max_{t, \lambda, \mu} W = e^{-\mu(\lambda, \theta)t}[1-\lambda]P_a(t, \lambda, \epsilon)(\mu(\lambda, \theta) - g)^{-1} \quad (4.1)$$

Partially differentiating the fitness function, W, with respect to t and λ, respectively, the following first-order conditions for an optimum are obtained:

$$\frac{\partial P_a}{\partial t} P_a^{-1} = \mu \quad (4.2)$$

And

$$-\frac{\partial \mu}{\partial \lambda}\left(t + (\mu - g)^{-1}\right) = -\frac{\partial P}{\partial \lambda} P_a^{-1} + (1-\lambda)^{-1} \quad (4.3)$$

Equation 4.2 for optimal t, taking μ as given, replicates Charnov's[1] result. Optimal t occurs at the age when the proportional increase in adult production due to a small increase in time spent in development (the left hand side of equation 4.2) is equal to the proportional loss in the probability of reaching adulthood (the right-hand side). The benefits and costs are measured in terms of proportions because fitness is a product of the probability of reaching adulthood and the production rate as an adult.

Equation 4.3 concerns optimal allocations to mortality reduction given time spent in development. The left-hand side of the equation is the benefit of a small increase in investment in mortality reduction. It is the proportional increase in reaching adulthood, plus the proportional increase in the adult lifespan, adjusted for the growth in income due to learning, g. The right-hand side is the cost, the proportional loss in production due to increased investment in mortality reduction. This proportional cost is twofold because allocations during the juvenile period reduce the growth rate and therefore reduce P_a (the first term), while allocations to mortality reduction during the adult period reduce the proportion of adult production allocated to reproduction (the second term).

Differentiating equations 4.2 and 4.3 with respect to the ecological parameters, ε, θ, and g, we derive our six main analytical results, confirming the co-evolutionary effects predicted by our theory. $\partial t/\partial \varepsilon$ and $\partial \lambda/\partial \varepsilon$ are both positive. This means that ecological factors increasing the productivity of investments in developmental embodied capital, as indexed by ε, not only increase time spent in development but also increase investments in mortality reduction. $\partial t/\partial \theta$ and $\partial \lambda/\partial \theta$ are both negative, meaning that exogenous or extrinsic increases in mortality, as indexed by θ, reduce both time spent in development and investments in mortality reduction. Conversely, ecological factors that lower mortality rates increase both time spent in development and investment in mortality reduction. Finally, $\partial t/\partial g$ and $\partial \lambda/\partial g$ are both positive, showing that the greater the growth rate in production during the adult period due to "on-the-job" training, as indexed by g, the more it pays to invest in development and mortality reduction. (A formal proof of the six results was developed by Arthur Robson and is available from the authors). These results all show that investments in development and investments in mortality reduction/longevity co-evolve.

Empirical Tests of the Theory

Composition of the Diet

Figure 4.4 illustrates our proposal about the differences between the diets of nonhuman primates and humans. While the diets of nonhuman primates vary considerably by species and by local ecology, the inverted triangle in Figure 4.4 represents the greater importance of lower quality, easier-to-acquire foods in the diets of most nonhuman primates (excluding insectivores). The upward-pointing triangle in Figure 4.4 represents the greater importance of large-package-size, nutrient-dense, difficult-to-acquire foods in human diets.

Table 4.3 presents data on the diets of ten foraging societies and four chimpanzee communities for which caloric production or time spent feeding were monitored systematically. As far as we are aware, this is a complete sample of the available data. The diet is subdivided into vertebrates, roots, nuts and seeds, other plant parts (such as leaves, flowers, and pith) and invertebrate resources. The diets of all modern foragers differ considerably from that of chimpanzees. Measured in calories, the major component of forager diets is vertebrate meat. This ranges from about 30% to 80% of the diet in the sampled societies, with most diets consisting of more than 50% vertebrate meat (equally weighted mean = 60%). The emphasis on vertebrate meat would be even more clear if any high-latitude foraging societies were included in the sample. In contrast, chimpanzees spend only about 2% of their feeding time eating meat. Unfortunately, the diet of wild primates is not usually expressed in calories, as is that of human foragers. Field workers studying nonhuman primates use time spent feeding on specific foods as the closest approximation of energy acquired and rarely either measure ingestion rate or calculate calorie intake. The absolute intake of meat per day also varies tremendously: Chimpanzee per capita meat intake is estimated at about 10 to 40 g per day, while human meat intake ranges from about 270 to 1,400 g per person per day. Although it is true that chimpanzee males eat much more meat than do females and juveniles,[20–22] we conclude that, in general, members of foraging societies eat more than ten times as much meat as do chimpanzees.

The next most important food category in our forager sample is roots, which make up an average of about 15% of the energy in the diet and were important in about half the societies in our sample. In contrast, the chimpanzee diet is primarily composed of ripe fruit, which accounts for over 60% of feeding time. Only two foraging societies ate large amounts of ripe fruit, the Gwi San of the Kalahari desert, who consume melons for water and nutrients during much of the year, and the Nukak of Colombia, who extensively exploit tropical palm fruits, which, however, are difficult to acquire. Other plant products are an important secondary food for chimpanzees, making up about 25% of observed feeding time. This category is unimportant for the foragers in our sample.

The data suggest that humans specialize in rare but nutrient-dense resource packages or patches (meat, roots, and nuts), whereas chimpanzees specialize in ripe fruit and plant parts with low nutrient density. These differences in the nutrient density of foods ingested are also reflected in human and chimpanzee gut morphology and food passage time. The chimpanzee's gut is specialized for rapid processing of large quantities and low-nutrient, bulky, fibrous meals.[23] However, a stronger contrast is apparent when we consider how the resources are obtained. We have categorized all foods into three types. Collected foods are those that can be obtained and eaten simply by gathering them from the environment. Extracted foods are non-mobile but are embedded in a protective context from which they must be removed. Such foods may be underground, in hard shells or associated with toxins. Hunted foods include mobile resources that must also be extracted and processed before consumption. Collected resources include fruits, leaves, flowers, and other easily accessible plant parts. Extracted resources include roots, nuts and seeds, most invertebrate products, and plant parts that are

TABLE 4.3. Diet of modern hunter-gatherers and chimpanzees.[a]

	HUNTER-GATHERERS										CHIMPANZEES			
	ONGE	ANBARRA	ARNHEM	ACHE	NUKAK	HIWI	!KUNG[1]	!KUNG[2]	GWI	HADZA		GOMBE	KIBALE	MAHALE	TAI
Sample[b]	1256	3654	276	3645	941	4756	866	928	?	?	Sample[b]				
kg/person day[c]											kg/person day[c]				
meat	0.59	0.58	1.34	1.36	0.62	0.97	0.26	0.59	0.30	1.10	meat	0.03		0.01	0.04
roots	0.21	0.07	0.43	0.00	0.00	0.44	0.15		0.40?	1.62	roots				
seeds, nuts	0.00	0.00	0.00	0.00	0.00	0.00	0.21		0.00	0.00	seeds, nuts				
fruits	0.00	0.15	0.01	0.61	0.86	0.29	0.15		0.40?	0.54	fruits				
other plants	0.00	0.00	0.00	0.96	0.00	0.04	0.00		0.00	0.00	other plants				
invertebrate	0.01	1.01	0.09	0.11	0.35	0.02	0.00		0.00	0.24	invertebrate				
calories/person day															
meat	980	822	1821	2126	764	1350	690	1602	417	1940					
roots	242	93	456	0	0	268	150		600?	1214					
seeds, nuts	0	0	0	0	0	0	1365		0	0					
fruits	0	44	10	22	747	82	150		600?	621					
other plant	0	0	3	255	0	36	0		0	0					
invertebrate	20	127	67	308	375	57	0		0	255					
Total	1243	1085	2357	2712	1886	1793	2355		1617?	4030					
Non-foraged[d]	0	1116	0	trace	378	626	trace	trace	0	trace					
Dietary percentage of foraged foods											Feeding time percentage				
meat	79	75	77	78	41	75	29	68?	26?	48	meat	1.5	0.9	2.5	2.1
roots	19	8	19	0	0	15	6		37?	30	roots	0.0	0.1	0	1.4
seed, nuts	0	0	0	0	0	0	58		0	0	seed, nuts	5.1	0	0	7
fruits	0	4	0	1	40	5	6		37?	15	fruits	60.2	78.5	57.7	67.6
other plant	0	0	0	9	0	2	0		0	0	other plant	29.3	21.3	33.4	15.5
invertebrate	2	12	3	11	20	3	0		0	6	invertebrate	3.9	0	6.4	5.6
collected	0.0	4.0	0.6	0.8	20?	4.6	4.9	?	37?	15	collected	94.2	99.1	91.1	92.3
extracted	21.9	20.3	30.1	24.3	40?	21.6	63.4	?	37?	36	extracted	3.8	0	6.4	5.6
hunted	78.0	75.7	69.4	74.9	40	73.7	31.7	68.0	26.0	48	hunted	2	0.9	2.5	2.1

[a] For information on methods of calculation and data sources, see footnotes, Table 4.2.
[b] Person days sampled, including all men women and children as equal consumers.
[c] This is the weight of the edible portion for meat and field weight for all other resources.
[d] Intake of nonforaged foods in calories/person day when measured or reported.

difficult to extract, such as palm fiber or growing shoots. Hunted resources include all vertebrates and some mobile invertebrates.

Table 4.3 shows a breakdown of forager and chimpanzee foods according to our three acquisition categories. Chimpanzees obtain an average of about 95% of their diet from collected foods, whereas the foragers in our sample obtain an average of 8% of their food energy from collected resources. On the other hand, foragers obtain about 60% of their food energy from hunted resources and about 32% from extracted resources, whereas chimpanzees obtain about 2% of their food energy from hunted foods and about 3% from extracted resources. While these categories may be somewhat rough, it is clear that humans are much more dependent on resources that can be obtained only by complicated techniques. Thus, the dietary data are consistent with our theoretical model. Humans appear to be more dependent on resources that require skill and learning to acquire.

The Age Profile of Acquisition for Collected and Extracted Resources

The proposition that difficulty of acquisition predicts the age profile of food production can be tested in two ways, by looking at the daily amount of different resource types produced by individuals of different ages and by observational and experimental measures of hourly rates of acquisition of different resource types by individuals of different ages. The daily data are determined by both time allocation and rates of return per unit of time spent on a resource, whereas the hourly data are based exclusively on rates of return. Both sources of data support the proposition that juveniles cannot easily obtain extracted and hunted foods.

Beginning with the daily data, Figures 4.5 and 4.6 show, respectively, the daily caloric contribution of various food types acquired by Ache and Hiwi males and females as a function of age. The upper panels represent the calorically less important foods (those for which daily production is less than 200 kcal/day); the lower panels represent the more important foods (7200 kcal/day). Among young Ache, both males and females acquire only fruits. Both sexes reach their peak rates of daily fruit acquisition in their mid- to late teens. Extracting palm hearts requires strength and some skill (about three minutes of chopping in the right spot); daily palm heart acquisition is asymptotic for both sexes by the age of 20 years. More skill and learning are required to extract honey or palm starch (knowing how to open a "window" to the resource and then extract it). The daily acquisition of these resources does not peak until individuals are in their late 20s. Daily returns from hunting do not peak until individuals are about 35 years old.

Hiwi foragers show similar production patterns with age, except that daily fruit acquisition becomes asymptotic at later ages. Hiwi fruit collection is more complicated because many trips entail walking through the night to distant groves, followed by a return trip of 10 to 20 km with a heavy load of fruit.

Daily honey extraction rate also reaches its peak level when these foragers are in their early 30s (these are very small nests of native bees). Female root-production rates increase four-fold from age 20 to age 40, but men's meat production does not peak until they are about age 35.

The fact that forager children, like chimpanzees, primarily acquire ripe fruits is supported by additional data. Among the Ache, children acquire five times as many calories per day during the fruit season as they do during other seasons of the year.[24] Among the Hadza, adolescent girls acquire 1,650 calories per day during the wet season, when fruits were available, and only 610 calories per day during the dry season, when fruits are not available. If we weight the data for the wet and dry season equally, teenage Hadza girls acquire 53% of their calories from fruits, compared to 37% and 19%, respectively, for reproductive-aged women and postreproductive women (all calculated from Hawkes and coworkers).[6] Hadza boys, like Ache and Hiwi boys, switch from easier tasks, such as fruit collection, shallow tuber extraction, and baobab processing to honey extraction and hunting in their mid- to late teens.[13,14] Chimpanzee juveniles also focus on more easily acquired resources than do adult chimpanzees. Juvenile chimpanzees practice difficult extraction activities such as collecting termites, fishing for ants, or nut cracking less than adults do.[20,26] Hunting is strictly an adult or subadult activity.[20,21,27]

Hourly return rates provide further evidence that important human food resources require long periods of learning and observations of Ache fruit collection show that foragers generally acquire the maximum observed rate by about age 20 (Fig. 4.7). Some fruits (for example, pretylla) that are simply picked from the ground are collected by children as young as one-and-a-half to three years at 30% of the adult maximum rate. For fruits such as vijulla, which must be picked off branches, children do not reach 50% of the adult maximum rate until age 15 (Fig. 4.7). As mentioned earlier, fruit collection by the Hiwi (Fig. 4.8), unlike that done by others, is labor-intensive and requires travel to distant food sites. Both males and females reach maximum return rates by about age 25.

Hadza data also show competent fruit collection by children. The Kongoro berry collection rate of young married girls is equal to the adult women's rate.[6] Baobab collecting and processing seems to reach 50% of the adult rate by about age 12.[14] Baobabs are an interesting food resource because they are both collected and extracted. While they can easily be picked off the ground, much more food energy is obtained when the pith is extracted with pounding and water. Young children do not practice these activities, whereas older children do.[14]

In contrast to the hourly acquisition rate of fruits, that of extracted resources often increases through early adulthood as foragers acquire necessary skills. Data on Hiwi women show that their root-acquisition rates do not become asymptotic until the women are about age 35 to 45 years (Fig. 4.8). The root-acquisition rate of 10-year-old girls is only 15% of the adult

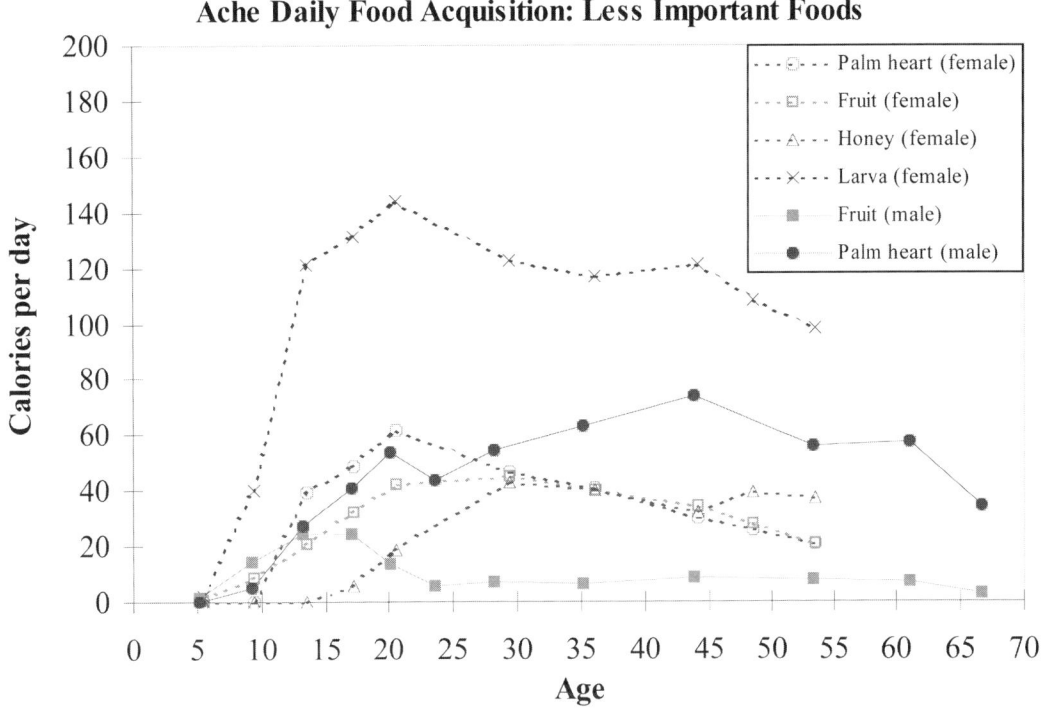

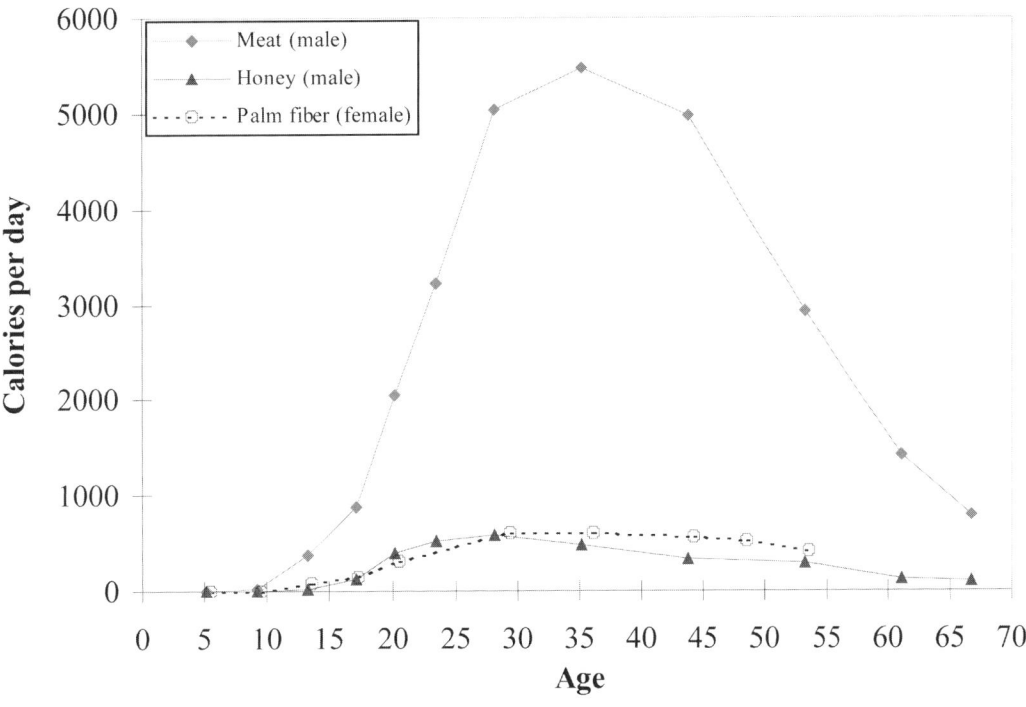

FIGURE 4.5. Age-sex specific daily energy acquisition is calculated as described for Figure 4.2 and Tables 4.2 and 4.3. Mean daily acquisition for each resource class was calculated by summing all the calories acquired for individuals of that age-sex category and that resource class, then dividing by all person days sampled in the relevant age category. Y values are plotted at the mean age for each class analyzed.

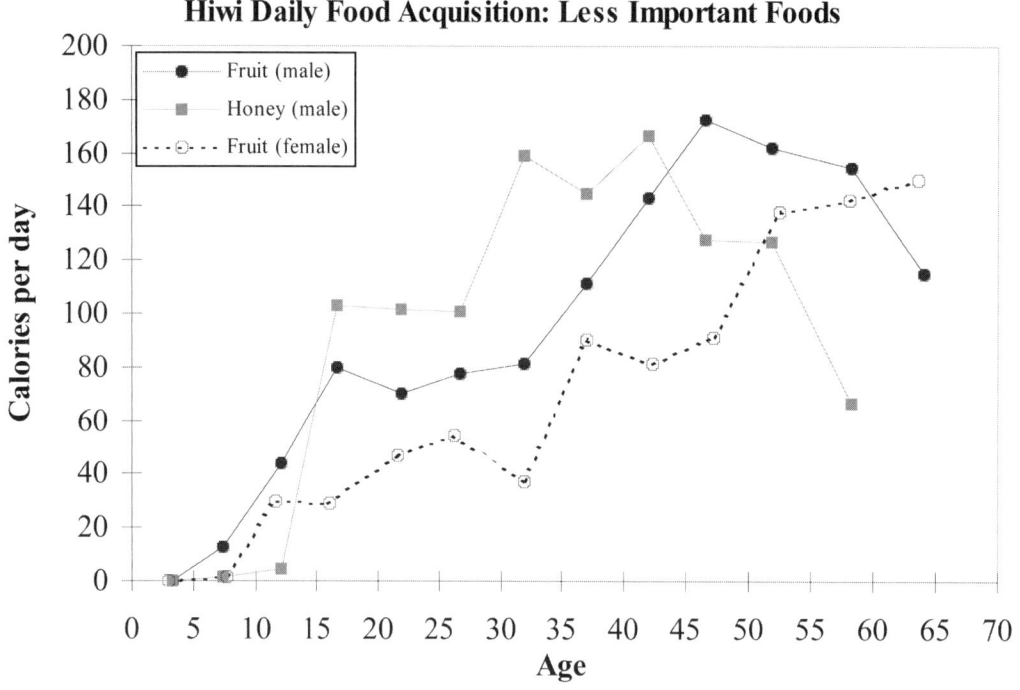

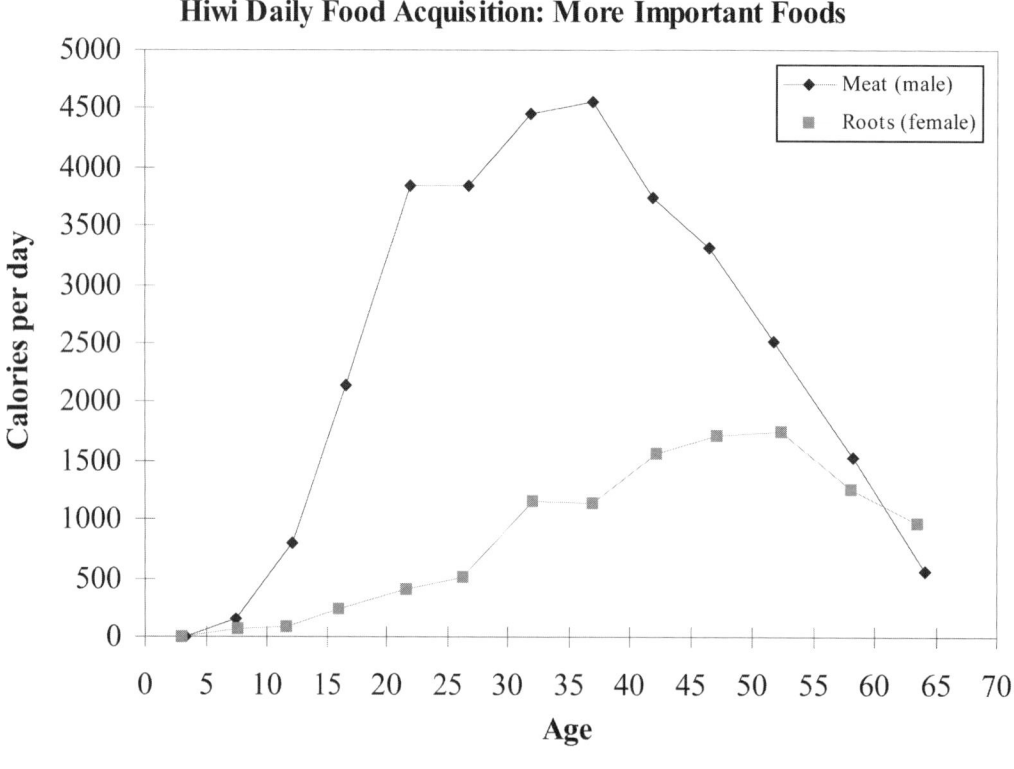

FIGURE 4.6. Age-sex specific daily energy acquisition is calculated as described for Figure 4.2 and Tables 4.2 and 4.3. Mean daily acquisition for each resource class was calculated by summing all the calories acquired for individuals of that age-sex category and that resource class, then dividing by all person days sampled in the relevant age category. Y values are plotted at the mean age for each class analyzed.

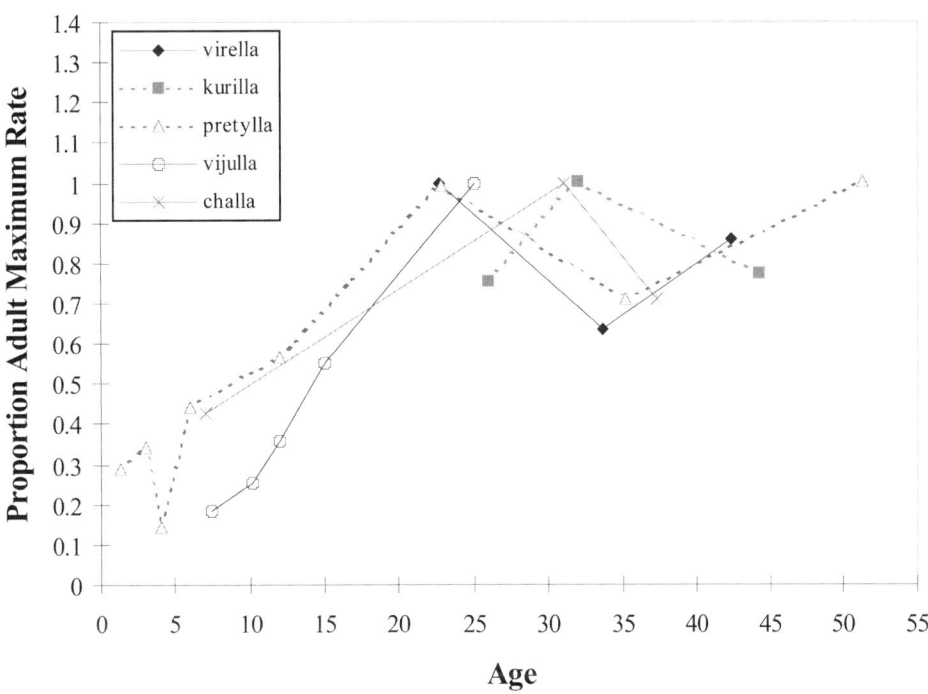

FIGURE 4.7. All Ache data, including timed counts of fruits acquired per minute, were analyzed for each age category. The mean plotted is the average from each independent monitored count for a given age class of acquirers. Fruits other than those that are simply collected from the ground or low branches were not included in the analyses. Y values are plotted at the mean age for each class analyzed.

maximum. For Hiwi males, the honey-extraction rates peak at about age 25. Again, the extraction rate of 10-year-olds is less than 10% of the adult maximum. Experiments done with Ache women and girls clearly show that the young adults are not capable of extracting palm products at the rate obtained by older Ache women (Fig. 4.9). Girls take longer than women to cut palms because they lack strength and because they cannot judge whether a palm will fall to the ground or get stuck in nearby tree branches. Girls take longer than women to extract the growing shoot from the palm after it is on the ground because this task requires strength and knowing where to cut across the palm leaf stalks. Girls take longer to extract the starchy fiber from the trunk of a downed palm because they do not know how to cut open a window nor how most efficiently to pound the fiber away from the hard outer trunk wood. When these component activities of palm extraction are combined, Ache women do not reach peak return rates until their early 20s (Fig. 4.9).

Supporting data are also available from other groups. !Kung (Ju/'hoansi) children crack mongongo nuts at a much slower rate than adults do.[28] Bock[29] has shown that nut-cracking rates among the neighboring Hambukushu do not peak until about age 35. Hadza women, however, appear to obtain maximum root-digging rates by early adulthood,[6] perhaps because they obtain a good deal of practice throughout childhood[13,14] and thus require only adult strength in order to produce at adult rates.

Casual ethnographic observation supports the generalization that fruit collection is easily learned, extraction skills require more time to develop, and hunting is the most difficult foraging behavior. Anthropologists working with modern foragers often participate in fruit collection and can rapidly achieve aboriginal return rates. Some types of extraction can also be mastered with practice. One of us (KH) chops down palms as fast as Ache women do and can extract the heart at a slightly slower rate. However, despite some practice, KH has not achieved the rate of palm-fiber extraction of Ache women. We know of no ethnographers who successfully dig roots or crack nuts at the rate of members of their study populations.

Indeed, the same patterns are seen among captive animals released into the wild. Animals that have grown to adulthood in captivity can be successfully released into the wild if their feeding niche is simple (for example, if they are herbivores). However, the success rate for carnivores and apes with complex feeding strategies is much lower.

Hunting and the Role of Men in Human Reproduction

Male behavior plays a distinctive role in human life histories in two ways. First, hunting, the primary subsistence activity of adult men, is the most learning-intensive foraging strategy practiced by humans. Second, unlike most higher primates, men play a major role in the energetics of human reproduction.

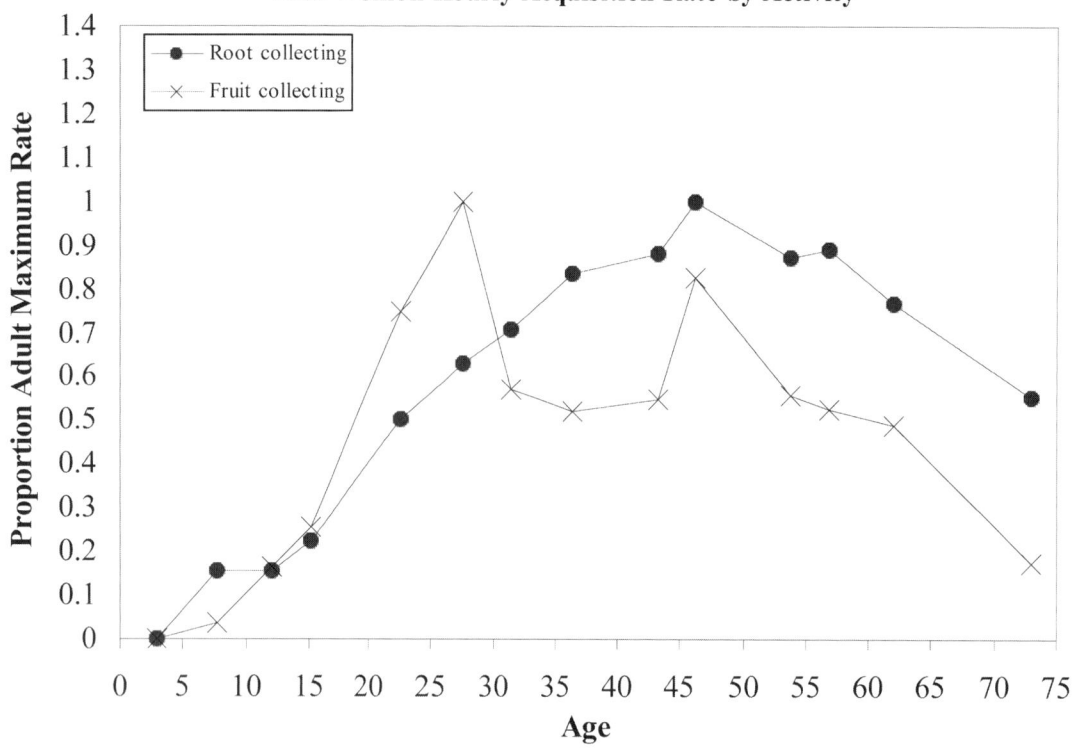

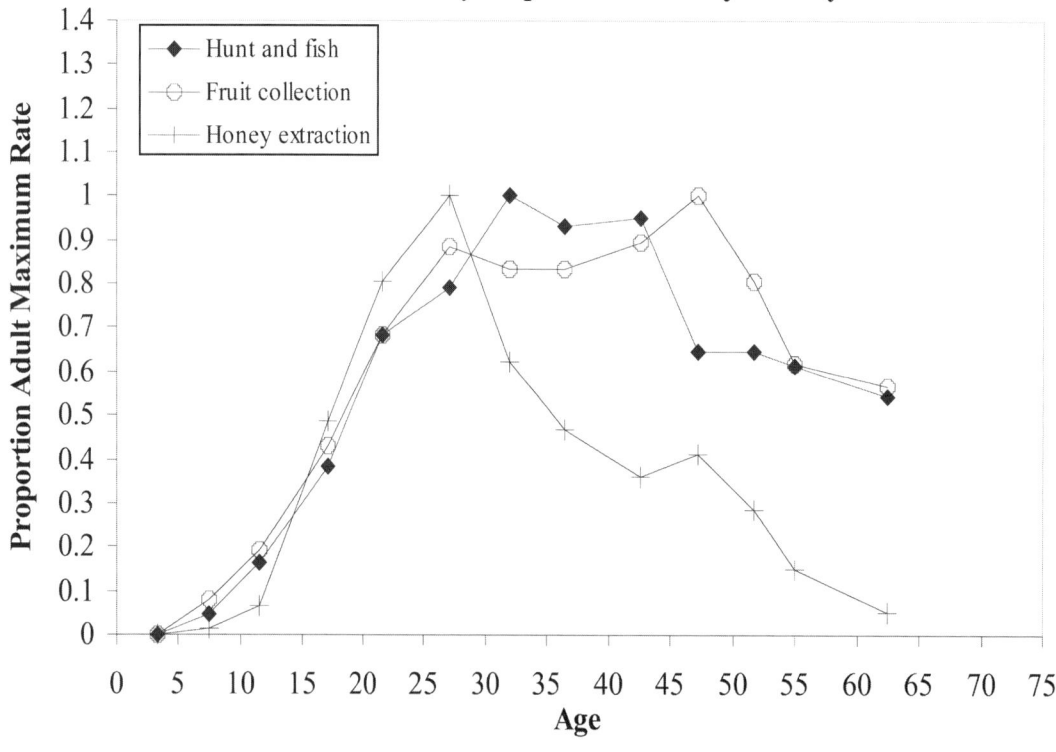

FIGURE 4.8. Because Hiwi foragers target specific resources when they leave their central camp, we recorded all time dedicated to foraging for different resource types over the sample period. Total energy production for each resource type and age-sex class was divided by the total number of hours reported to be out-of-camp foraging for that resource type to obtain the hourly return rate from foraging for that resource type. The time foraging included in-patch pursuit of resources as well as walking time to and from the patch. Y values are plotted at the mean age for each class analyzed.

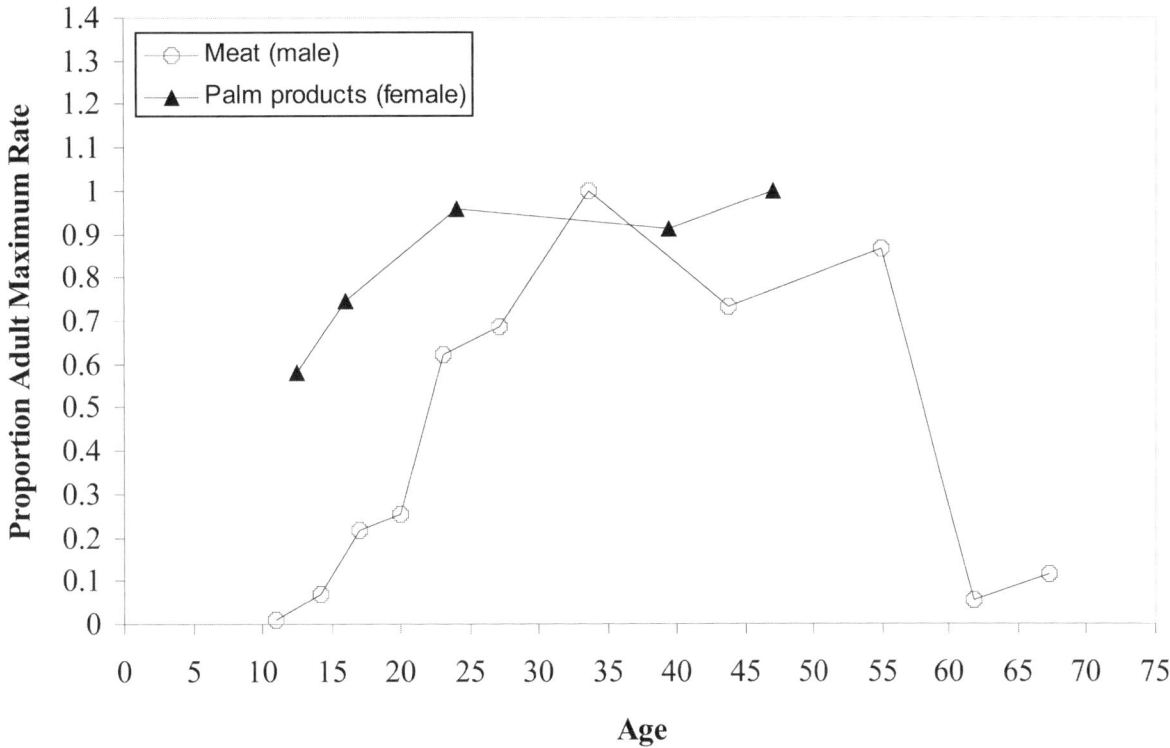

FIGURE 4.9. Data from experiments in which Ache females of different ages were asked to chop down a palm tree, extract its growing shoot (heart), and extract palm starch from the palm for at least ½ hour. Y values are plotted at the mean age for each class analyzed. Age-specific hunting returns are based on measured kg of live weight of game acquired by Ache males between 1980 and 1984 and measured out-of-camp foraging time for males during the same time period. Y values are plotted at the mean age for each class analyzed. Differences between age categories may be due to differences in encounter rates with game or success in pursuits of game after encounter.

Although a detailed quantitative analysis of the learning process involved in human hunting has not yet been conducted, it is clear that human hunting differs qualitatively from hunting by other animals. Unlike most animals, which either sit and wait to ambush prey or use stealth and pursuit techniques, human hunters use a wealth of information to make context-specific decisions, both during the search phase of hunting and then after prey is encountered. Specifically, information on ecology, seasonality, current weather, expected animal behavior, and fresh animal signs are all integrated to form multivariate mental models of encounter probabilities that guide the search and are continually updated as conditions change.[30] Various alternative courses of action are constantly compared and referenced to spatial and temporal mental maps of resource availability.[30] This information is collected, memorized, and processed over much larger spatial areas than chimpanzees ever cover. For example, interviews with Ache men show that fully adult men (those over the age of 35 years) had hunted in nearly 12,000 km² of tropical forest during their lives. Almost all foragers surveyed use more than 200 km² in a single year, and many cover more than 1,000 km² in a year (Table 4.1 in Kelly[31]). Male chimpanzees, on the other hand, cover only about 10 km² in a lifetime.[32,33]

After potential prey are encountered, humans employ a wide variety of techniques to obtain them, using astounding creativity. Here are just some examples that Hill, Hurtado and Kaplan have seen among the Ache, Hiwi, Machiguenga, and Yora: Arboreal animals have been shot with arrows from the ground or a tree, driven by climbing, shaken down from branches, frightened into jumping to the ground, brought down by felling a tree with an axe, lured by imitated calls, lured by making captured infants emit distress calls, captured by the spreading of sticky resin on branches to trap them, and captured by scaffolding constructed from tree branches and vines. Ground-dwelling prey are shot with arrows, driven to other hunters or capture devices, run down upon encounter, slammed to death against the ground, strangled at the neck, or suffocated by stepping on them while they are trapped in a tight spot. Burrowing prey are dug out, chopped out of tree trunks, stabbed through the ground with spears, frightened to the point at which they bolt from the burrow, smoked out, and captured by introducing a lasso through a small hole. Aquatic prey are shot on the surface or below it, driven into traps, poisoned, discovered on muddy bottoms by systematically poking the bottom of a pond, and speared underwater by random thrusts in drying lakes. The widely varied kill techniques are tailored to a wide variety of prey under a wide variety of conditions. Although all groups probably specialize in the most abundant and vulnerable prey in their area, the total array of species taken is impressive, and probably is much larger than that of most, if not all, other vertebrate predators. For example, from 1980 to 1996 our sample of weighed prey taken by the Ache included a minimum of 78 different mammal species, at least 21 species of reptiles and amphibians, probably more than 150 species of birds (more than we have been able to identify) and more than 14 species of fish. Moreover, human hunters tend to select prey that is in prime condition from the perspective of human nutritional needs rather than prey made vulnerable by youth, old age, or disease, as do many carnivorous animals.[34,35]

The skill-intensive nature of human hunting and the long learning process involved are demonstrated dramatically by data on hunting return rates by age. Hunting return rates among the Hiwi do not peak until the men reach the age of 30 to 35 (Fig. 4.8). The acquisition rates of 10-year-old and 20-year-old boys reach, respectively, only 16% and 50% of the adult maximum. In the 1980s, the hourly return rate for Ache men peaked when they were in their mid-30s (Fig. 4.9). The return rate of 10-year-old boys is now about 1% of the adult maximum, and that of 20-year-olds is still only 25% of the adult maximum.

It is not surprising that no ethnographer has ever described being able to hunt at a rate equivalent to that of study subjects. Indeed, most who hunt make kills only after being led to the game animal by a competent hunter, and then generally with a firearm (as is the case with recreational big-game hunters). These patterns also mirror the effects of acculturation in most groups. Foragers who grow up in settled communities can often collect fruits and other plant resources at rates equivalent to those obtained by older individuals who grew up in the bush (see, for example, Blurton Jones and colleagues[28] on the !Kung). But most young Ache men who have grown up on reservation settlements cannot hunt using traditional technology nearly as successfully as older men who grew up in the forest (unpublished data).

Chimpanzees too, appear to require many years to learn successful hunting techniques. Older males are more likely than younger adult males to ambush prey during a group hunt and perform more complicated maneuvers during the hunt.[20] (But see Stanford.)[21]

Although the learning process is long, investments in hunting ability by human males allow them to be highly productive as adults. The comparative analysis of diets and productivity among foragers shows that men play a major role in the energetics of human reproduction. For example, among the Ache the total expected net caloric production (food produced minus food consumed) from age 18 to death is +21,638,000 calories for males and −924,000 calories for females. The corresponding figures for the Hiwi are +11,151,000 for males and −3,096,000 for females. Even among the Hadza, where women play a much greater role in subsistence production, males provide at least as much support for reproduction as females do, if not more. While the estimates may require revision when researchers complete their analysis of age-specific production among the Hadza, our initial approximation, based on published data, is that over the entire expected life course (including the probability of survival to each age), net production for Hadza males is +16,671,000 calories, while that of females is only +3,352,000 calories. Table 4.2 shows that men provide more food energy per day than women do in all but one or two of the ten foraging societies for which there is quantitative data. Men also provide the vast majority of protein in the diet. This is critical, since higher daily protein intake increases weight gain,[36,37] immune function response,[38] and survival.[39,40]

The fact that men produce more food than women do in most low-latitude foraging societies is not conventional wisdom. This is a result of the influence of Richard Lee's[41,42] pioneering study of the !Kung (Ju/'ho-ansi), which showed that women provide more food than men do. There has been a tendency to generalize those results to all foraging societies. Table 4.2 shows that those results are not general; it also shows the possibility that weaknesses in Lee's study have been misleading for the !Kung as well. Lee's sample covered only 28 days of one month of 1964, and on two of those days he took women out collecting mongongo nuts in his vehicle, thus inflating the collected portion of the diet and female return rates. During much of the year, mongongo nuts are not abundant, nor are they as abundant at other !Kung study sites as at the Dobe site. Another study of !Kung food production shows much higher hunting success than Lee reported. Yellen[43] provided data showing that !Kung men acquired twice as much meat per day when they were in bush camps than they did in the permanent dry-season waterhole settlement where Herero raised cattle, which probably had a depressive effect on game densities. Yellen's sample of person consumption days was larger than Lee's and covered all months of the year. The daily meat consumption in Yellen's sample was about 1,600 calories per capita, which would represent 68% of all calories if total food consumption was the same as Lee reported. In addition, both Lee and Draper (p. 262 in Lee)[42] found that !Kung men spent more hours per week on food acquisition than did !Kung women.

The fact that humans were successful in colonizing high-latitude ecologies where plant foods are not abundant and are available only for short periods also demonstrates the important role that men play in the energetics of reproduction. It is also interesting that many Neanderthal and early *Homo sapiens* are found at high latitudes where plant consumption was minimal and, in some cases, where even fuel and residential construction were provided by animal products (see Hoffecker[44] for a review of east European sites). Moreover, some archeological sites beginning with *Homo ergaster* (for example, Boxgrove, England)[45] contain super-abundant animal remains and evidence of spears and other hunting tools, but no evidence of plant consumption. More recent low-latitude archeological assemblages also often have extremely dense bone scatter, which suggests high meat consumption. At Kutikina, Tasmania, for example, bone fragments comprise nearly 20% of the weight per cubic meter of some archeological strata.[46] On the other hand, complete assessment of plant pollen at some assemblages, among them Tamar Hat on the north coast of Africa,[47] suggests no edible plant species, despite dense faunal scatters. Such evidence suggests that adult males have often been the main, and sometimes the only food providers in foraging societies.

In another vein, the Hadza research team[15,48,49] has argued that men do not play a major role in the energetics of human reproduction because the vagaries of hunting luck render meat an unreliable and indefensible resource. Two separate issues must be distinguished in evaluating this argument. The first concerns the sources of the caloric and nutritional subsidization of human reproduction. The second issue is whether or not the proceeds from hunting preferentially support the spouse and children of the hunter.

With respect to the first issue, the data we present clearly demonstrate that hunted foods provide a substantial proportion of the energy and essential nutrients consumed by women and children. The fact that humans in some places depend on hunting for almost 100% of their energy needs suggests that this can be a reliable means of subsistence because they have developed cultural solutions to the variability problem. The main solutions are food sharing and, to a lesser extent, food storage. The second issue is more complex and, as yet, unresolved. Hawkes and associates[49] claim that prey items are not controlled or owned because they are impossible to defend. Thus, males do not provision their families, but simply provide equally for everyone in a band by hunting. There are, however, no data to support this interpretation, other than the observation of sharing itself. In fact, there is considerable evidence that carcasses can be defended when conditions do not favor food sharing. The Ache of Paraguay, who supplied the initial data for the indefensibility view of sharing share game resources widely when on long foraging treks, yet withhold even large game, sharing mainly with preferred partners when on reservation settlements.[50] Thus, the same resources are shared differently in the forest and in the reservation. The same is true in Africa, where the large game items taken by the Hadza are treated as private property by other ethnic groups who trade or sell them in the bush meat market.

The indefensibility hypothesis also asserts that shares given up by successful hunters are never repaid in any useful currency (meat, other goods, or services) and that shares are given to everyone equally, regardless of whether they have done or will do anything to pay them back. However, recent research shows that contingent giving is typical in all societies where it has been examined. Ache and Hiwi foragers share more with those who share with them and, when they are sick or injured, receive help from those with whom they have shared and in relation to how generously they have shared.[50,51] Yanomamo gardeners work more hours in the gardens of non-kin who work more hours in their gardens.[52] We believe that this type of reciprocity, often in different currencies, is the basis for all human economies, divisions of labor, and specialization, and that its critical development in the hominid line distinguishes us from our ape relatives.

Taken together, currently available evidence suggests that men generally provide a considerable portion of the energy consumed by juveniles and reproductive-aged women. This does not mean that men contribute more to society or reproduction than do women. Women process food (part of providing nutrients), care for vulnerable children, and do a variety of other important tasks in all societies. It is the partnership of men and women that allows long-term juvenile dependence and learning and high rates of survival. Indeed, analyses show that among both the Ache and the Hiwi, individual women produce less food if their husband is a high producer.[53] Divorce or paternal death leads to higher child mortality among the Ache,[10,54] the Hiwi,[55] and the !Kung,[56] but not the Hadza.[57]

In our view, human pair bonding and male parental investment is the result of complementarity between males and females. The commitment to caring for and carrying vulnerable young, common to primate females in general, together with the long period required to learn human hunting strategies, renders hunting unprofitable for women. The fact that human males can acquire very large packages of nutrient-dense food means that they can make a great difference to female reproductive success. That is not true of most other primates (the exception being callithricid males, who provide other necessary assistance). This difference creates a major discontinuity between humans and apes, and results in a partnership between men and women. That partnership is ecologically variable, in that the roles of men, women, and children vary with the availability of food resources in the environment and the risks posed to children.[28,55,58,60]

When plants are abundant and game is scarce (as in the Gwi environment), men specialize in providing the rarest nutrients in the environment, protein and lipid, because these nutrients are critical for growth and good health,[61,62] while women provide more food energy. When plant foods are scarce, men's hunting provides the bulk of energy and women concentrate on the processing of food and other raw materials, and on child care. There is much to be learned about the determinants of male and female roles in foraging societies, but the primary activity of adult males is hunting to provide nutrients for others.

In sum, hunting is the human activity that requires the longest period before maximum return rates are achieved, but it also provides the highest overall return rate once maximum skill levels are reached. As a result, hunting provides the greatest energy component of human diets in many foraging societies (Table 4.3) and is a fundamental feature of the human life-history adaptation. Resolution of the debate about whether hunting is primarily direct parental investment, as we contend, or mating effort, as proposed by Hawkes and coworkers,[63] awaits further data and sophisticated tests. Nevertheless, the

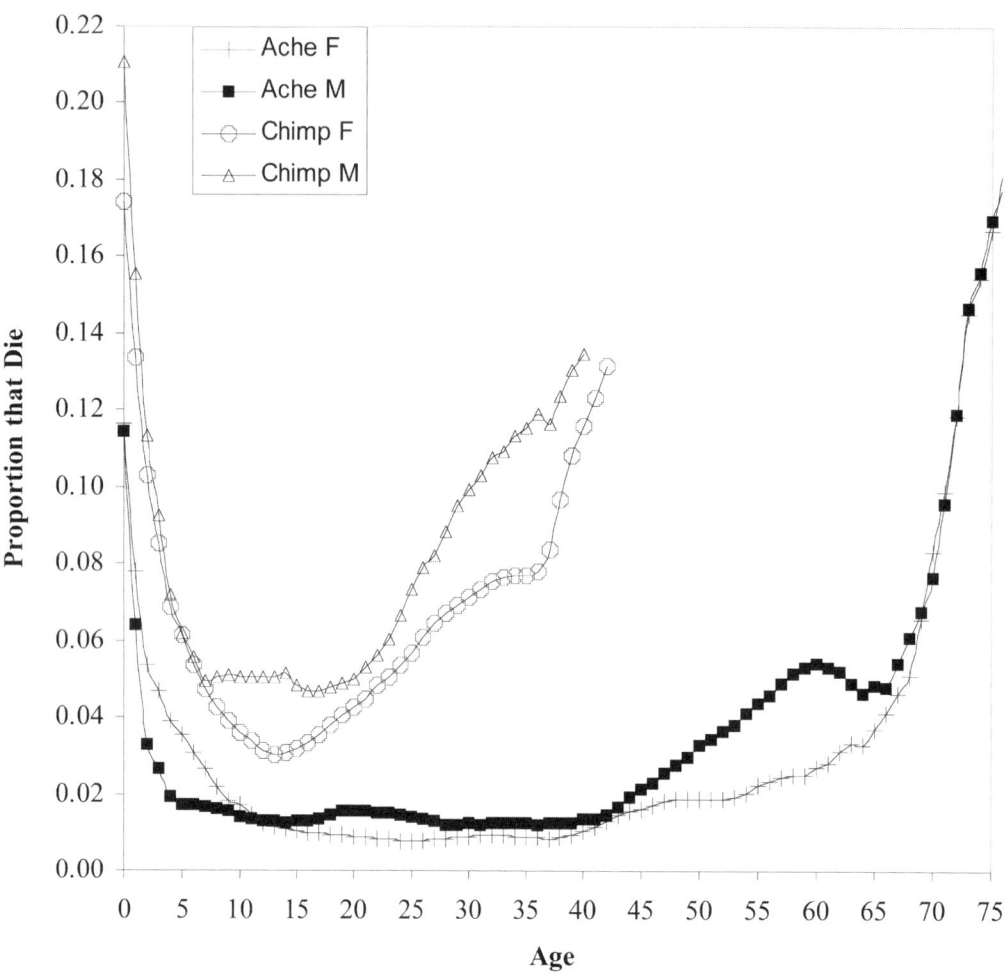

FIGURE 4.10. Yearly mortality rates for the Ache are from Hill and Hurtado[10] and for chimpanzees from Hill and coworkers[8] as described for Figure 4.1. Raw mortality data were smoothed using a double running average. Ages 1–5, 3 pt; ages 5–10, 5 pt; ages >10, 9 pt with truncation at the end of the life table.

nutritional support of reproduction by human males is a fundamental feature of our species.

Reduction in Juvenile and Early Adult Mortality Rates

Human foragers have longer maximum life spans than chimpanzees do, suggesting that they may have lower mortality rates over much of the life span.[64–67] Both groups experience minimum mortality rates in the late juvenile and early adult period, a pattern typical of many living organisms.[68] However, the minimum mortality rate of foragers is about 1% per year, whereas the minimum rate for chimpanzees is about 3.5% per year (Fig. 4.10). We propose that the character of food resources taken by humans ultimately lowers the mortality rate of juveniles and young adults in three ways. First, the hunted and extracted foods taken by humans generally come in large packages. Large, valuable packages favor food sharing, which reduces fluctuations in daily food intake, allows sick and injured individuals to recover at higher rates, and facilitates provisioning of children. Second, the food types taken favor larger foraging parties and residential groups, which offer protection, particularly for juveniles who associate closely with adults. Third, the food niche has led to the development of tools and an understanding of animal behavior, which can be used effectively to repel predators.

Large package size has been positively associated with the probability of food transfers or the degree of food transfer in virtually every study in which it has been examined.[50,63,69–75] Among the Ache and Hiwi, the percentage of foods not eaten by the acquirer's nuclear family is directly related to the mean package size of different food categories[72] or package size within food categories.[50] Daily variability in the acquisition of different food categories also positively correlates with the percent shared in both societies. Importantly, meat comes in larger packages and is more variable than extracted resources, which, in turn, come in larger, more variable packages than do fruits. Meat is transferred most between non-kin, followed by extracted products, then fruits.[72] Chimpanzees also share meat more than they do any other food resource.[76,77] Thus, we propose that as humans moved into a hunting-extraction feeding niche, levels of food sharing increased dramatically over that seen among chimpanzees.

Among the Ache, individuals who are sick or injured are frequently fed by other individuals, often ones who are not kin. Those who share a greater percentage of their production are provisioned by more individuals when they are sick or injured.[51] Illness and injury are common among those in foraging societies, as we might imagine, given their frequent exposure to dangers, parasites, and pathogens, and the lack of modern medical care. A systematic health survey in one reservation Ache community in 1997 showed that adults required care at the community clinic on 6% of all person days, whereas children required care on 3.5% of all person days. Sugiyama and Chacon[78] have calculated that Yora men of Peru were unable to hunt on about 10% of all person days monitored. Bailey[79] reported that Efe men came to ask for medical treatment on 21% of all man days and often had problems that precluded foraging. Most relevant to our hypothesis is how often men, women, or children are sick or injured for many days in a way that would preclude food acquisition. No data are yet available on this topic. However, we have seen a variety of serious medical problems including snakebites, injuries sustained in piranha and jaguar attacks, broken bones, large punctures, and animal bites, as well as, arrow wounds, massive infections, and occasional illness that have kept people from foraging for more than a week at a time. Indeed, we researchers have all experienced medical problems that lasted for a week or more and prevented us from working in the field. If people were not food-subsidized during such periods, mortality rates would undoubtedly be higher.

Large, widely dispersed food patches may promote grouping among animals.[80,81] Human women and children generally spend the day in parties that contain many individuals of the foraging band, whereas chimpanzee females and juveniles are often dispersed into parties of one or two individuals.[76,82] This appears to have two important consequences. First human males experience higher mortality from predators than do females. Eight of the nine Ache who died from jaguar attacks in the twentieth century were men, as were 14 of the 18 who died from snakebite.[10] Second, human juveniles are rarely killed by predators. In recent memory and historical mythology, no Ache child has been killed by a jaguar. The only !Kung reportedly killed by a predator in Howell's[83] demographic study was an older man. Nevertheless, there is good evidence that chimpanzee juveniles are killed by predators[20,76] and that adult females are probably killed more often than adult males.

Human tool use probably provides a good deal of protection from predators. Fire appears to frighten many predators and may provide a good deal of nocturnal protection. Hunting tools can inflict serious injury or death on predators, thus providing protection for women and children, particularly when multiple males are present. The knowledge and analysis of animal behavior, fundamental to human hunting, can also be employed to fend off predators. Ache often know when the group is being stalked by a jaguar by analyzing its footprints and movement patterns. When they become aware of a predatory jaguar, they build brush walls around the camp and take turns acting as sentinels through the night.

These quantitative and qualitative data about food sharing, tools, and knowledge provide suggestive evidence that the shift to large-package, high-quality foods indirectly acted to lower mortality rates. Research on causes of death and risks of predation and illness among both humans and chimpanzees is necessary to determine the factors responsible for the more than three-fold differences in juvenile and adult mortality rates.

Discussion

Primate Life-History Evolution

The life-history traits and large brains of fully modern *Homo sapiens* may be seen as the extreme manifestation of a process that defines the primate order as a whole. Our theory organizes the major evolutionary events in the primate order and the specific changes that occurred in the hominid line.

The early evolution of the primate order (60 mya to 35 mya) was characterized by small increases in encephalization. Relatively little is known about early life-history evolution except that it appears that even the more "primitive" prosimian primates were long-lived and delayed in reaching reproductive maturity as compared to mammals of similar body size. Austad and Fischer[84,85] have related this evolutionary trend in the primates to the safety provided by the arboreal habitat. They compare primates to birds and bats, which are also slow in developing and long-lived for their body sizes. Thus, the first major grade shift that separated the primate order from other mammalian orders was a change to a lowered mortality rate and the subsequent evolution of slower senescence rates.

The second major grade shift occurred with the evolution of the anthropoids, the lineage containing monkeys, apes, and humans, beginning about 35 mya. This was characterized by an increasing emphasis on plant foods as opposed to insects, and by more rapid increases in brain size relative to body size.[80] The major defining characteristic of the evolution of the anthropoids was reorganization of the sensory system from one in which olfaction and hearing were relatively dominant to one completely dominated by binocular color vision.[80] This grade shift is almost certainly tied to a dietary shift toward a diverse array of plant parts, particularly fruits and leaves.

The diet of the anthropoids has been characterized as both "broad and selective."[86,87] The diet is broad precisely because it is so selective. Anthropoid primates tend to select foods on the basis of the ripeness, fiber content, nutrients, and toxicity of foods consumed early in the day.[88–90] This selectivity requires the allocation of increased brain tissue to visual processing.[91,92] It also requires that many different species of plants and animals be included in the diet, increasing the demands for memory and learning.[93,94]

This grade shift is reflected in brain size. Regressions of log brain size on log body size show that the intercept is significantly lower for strepsirhine primates (including most prosimians) than for the haplorhine primates (including all anthropoids and a few prosimians).

The third major grade shift in primates occurred with the evolution of the hominoid lineage, the branch leading to apes and humans. This shift included further encephalization. The intercept in a regression of log brain size on log body size is significantly higher for apes than monkeys, and apes clearly perform better on most tasks reflecting higher intelligence.[93,95] Evidence on teeth and tooth wear among early hominoids suggests a diet composed mainly of soft, ripe fruits.[96] The frugivorous emphasis of the hominoid lineage is evident in later hominoid species as well.[96] Wrangham and colleagues[97,98] also show that the chimpanzee diet is based on a much greater percentage of ripe fruits than is the diet of other sympatric primate frugivores. (This is probably true of orangutans as well.) The cognitive demands of a diet that emphasizes ripe fruits are likely to be much greater than simple frugivory. For one thing, there are greater perceptual demands in detecting the state of fruit from visual cues against a background.[91] For another, the lower abundance of ripe fruits and the short time in which they are available (ripe, but not yet eaten by competitors) is likely to impose greater demands with respect to monitoring the environment, remembering the state of individual trees, and predicting when the fruits will become ripe on the basis of their current state.

In addition to eating ripe fruits, chimpanzees and gorillas also use complex techniques for extracting foods from protected substrates, such as nut-cracking, termite and ant fishing, and removal of bark to get at pith.[93] Gibson[99,100] identified extractive foraging in primates as an important selective force in primate intelligence and presented evidence in support of this idea. Furthermore, these behaviors vary between groups as social traditions. In a comprehensive review of chimpanzee cultures using 151 years of chimpanzee observations from seven long-term studies, 39 behavior patterns were found to be customary or habitual in some communities but absent from others where ecological explanations could be discounted.[101] Of these, 19 were patterns of extractive foraging. An additional 14 extractive foraging behaviors were identified but failed to achieve habitual status in any one community. Furthermore, chimpanzee males are avid hunters. Boesch and Boesch[20] present data indicating that older chimpanzee males are capable of predicting escape patterns of their prey and of predicting how prey will respond to the behaviors of other chimpanzees. This appears to require even greater levels of cognitive processing than are required for extractive foraging. Although, as compared to humans, chimpanzees engage in relatively little extractive foraging and hunting, they do much more than monkeys. In this sense, their superior intelligence and greater encephalization than occurs in monkeys illustrates the same evolutionary forces that separate humans from apes. (See Lancaster and associates[102] for a review of chimpanzee behavior and cognition.)

There is some debate, however, about the relative importance of diet versus group living in the evolution of primate intelligence and brain size. According to one view, the increase in brain size was largely driven by the complexities of the primate diet. Jerison[103,104] suggested that brain tissue evolves in response to two kinds of demands: One depends on body size, based on monitoring and supporting an animals body tissue and particularly its surface area; the other is the ability to assimilate, integrate, and remember environmental information. He therefore predicted that differences in brain size, after controlling for body mass, would be associated with an animal's ecological niche and its demands for information processing. Jerison[103] hypothesized that the need to process information in a complex three-dimensional environment was the cause of the large brain of primates relative to the brains of other mammals.

Clutton-Brock and Harvey[105] tested a version of this hypothesis with intergenera comparisons within families of primates. They reasoned that frugivores need to assimilate and retain more environmental information than folivores do because fruits are more scarcely distributed than leaves, requiring more specific locational memory. They found that, after controlling for body size, both dietary emphasis (leaves versus fruits and insects) and the size of the home range predicted brain size. Milton[23,89,106] extended their work, focusing on gut specialization and brain size as alternative routes to energetic efficiency. Leaves, while abundant, tend to contain high amounts of fiber and often toxins as well. The ability to extract nutrients from leaves depends on the size of the gut and other specializations designed to facilitate fermentation for nutrient extraction. Fruits, on the other hand, are ephemeral resources, patchily distributed but offering a higher density of easily processed energy. Milton showed in paired interspecific comparisons that gut size and brain size were inversely correlated, corresponding to the dietary emphasis on fruits versus leaves.

Another view is that brain-size evolution was driven primarily by the complexities of social life in primate groups.[107,108] Many species of primates exhibit complex dominance hierarchies that are mediated by political alliances and relations among relatives in genetic lineages.[109–112] It is not clear why higher primates have such complex social relations, but it appears that group living is at least partially a response to predation that significantly lowers predation. Among species that tend to eat higher-quality foods that are easy to monopolize, social relationships also mediate access to foods within groups.[113]

The most recently published analyses with the largest samples show that both group living and diet are associated with brain size and the size of the neocortex in primates.[91,92] In a separate set of analyses, Allman and colleagues[66] have shown that group living and diet are positively associated with maximum life span in primates after controlling for body size. They also have shown a high positive correlation between brain size and life span, again controlling for body size. Smith[114] showed that after controlling for body size, brain size in primates is positively associated with the age of first molar eruption, an indicator of the age at which individuals begin to consume adult diets, as well as with age of first reproduction and longevity. The relationship is strongest for age of first molar eruption, probably because brain growth and postcranial morphological development compete.[115]

Kaplan and colleagues[116] have recently conducted multivariate analyses specifically designed to test the present theory. They found that when brain size is regressed on body size, age at first reproduction, maximum life span, percent of fruit in the diet, range size, and group size in a multivariate model, all but group size were significant predictors of brain size. In addition, the frugivory and range size variables, meant to capture the cognitive demands of the diet, also predict the life-history variables (age of first reproduction and maximum lifespan) after controlling for body weight.

Regardless of the importance of group size in determining brain size in monkeys, the grade shift in brain size between monkeys and apes is almost surely related to diet and not to group size.[95] Apes show dietary specialization, but do not live in particularly large groups. In fact, gibbons, orangutans, and to a lesser extent, gorillas, live in relatively small groups while chimpanzees and bonobos form groups of the same size as do baboons and vervets.

It seems likely that the cognitive ability associated with foraging in a complex three-dimensional environment is an important pre-adaptation for social intelligence and complex social relations. Of particular interest here are the specializations inherent in the primate visual system, including binocular vision, high visual acuity, and associated increases in the size of the lateral geniculate nucleus.[64, 65, 91] Barton[91] has shown that the size of the parvocellular system involved in the processing of visual stimuli increases with increasing group size among nonhuman primates. Thus, it may be that selection favoring social intelligence increases once the necessary cognitive pre-adaptations exist. This may be one reason why primates exhibit such high levels of social complexity relative to other group-living mammals, such as many herbivores. Adept social manipulation that leads to a higher position in a dominance hierarchy may be favored by natural selection, for dominance has correlated positively with measures of fitness in a multitude of primate studies.[109–113, 117] Yet this would probably be true for other social mammals as well. Many of the same abilities to store and analyze information may be employed to solve dietary and social problems.[94] Once social intelligence and social complexity evolve, the cognitive adaptations may be maintained even when dietary complexity is secondarily reduced in some species, as is the case with the derived simplicity of the diet of colobine monkeys.[105] Apes show remarkably sophisticated social intelligence[94, 112, 118, 119] and stand out in comparison to monkeys. Although this cannot be a result of greater group size per se, it is again possible that selection on social applications of intelligence increased when the cognitive adaptations to feeding acted as a pre-adaptation.

It is useful to think of primate evolution as both a branching and a directional process. It is branching in the sense that ecological variation, intraniche competition, and segregation generate variable selection pressures that lead to the evolution of multiple species with different traits. Some are selected to rely on more easily acquired foods and travel less, to invest more in gut physiology and less in brains, and to mature more rapidly and live shorter lives. Others are selected to rely on more complex feeding strategies and exhibit their respective, correlated life histories. It is a directional process in the sense that cognitive and life-history adaptations that evolved previously act as pre-adaptations for further selection in the same direction in response to niche competition and ecological change. *Homo sapiens* is an extreme in one of those directions.

The Hominid Line

The fourth major grade shift in primate evolution occurred with divergence of the hominid line, particularly the evolution of the genus *Homo*. The brain and life span of modern humans are clearly "outliers" compared to those of other mammals, and even as compared to the relatively large-brained, slow-living primates. The evolution of these extreme adaptations in the hominid line is built on a hominoid base that already showed a significant tendency toward large brains, long lives, and exploitation of high-quality foods. Available evidence on Australopithecines suggests that bipedalism preceded changes in brain size and life history.[114, 115]

Although the record is still incomplete, it appears that brain enlargement and life-history shifts co-occurred. Early *Homo ergaster* shows both significant brain expansion and an elongated developmental period,[120] but much less so than modern humans. Neanderthals display both brain sizes and dental development that are in the same range as those of modern humans. Bipedalism can be thought of as a pre-adaptation that evolved to facilitate terrestrial locomotion, allowing for long day ranges through energetic efficiency.[121] Ache men walk an average of 15 km per day, about three times the distance covered by chimpanzee males,[32, 33] even though both inhabit tropical forests. Bipedality also had the secondary effect of freeing the hands for specialization in manipulative activities. Environmental conditions in the early Pleistocene interacted with bipedality to favor an increased emphasis on extractive foraging, hunting, learning, prolonged development, and long lives for at least one line of the hominid family. It may be that this unique suite of conditions is responsible for the extreme differences between humans and other primates in intelligence, development, longevity, and resource flows across generations. The complexities of a highly

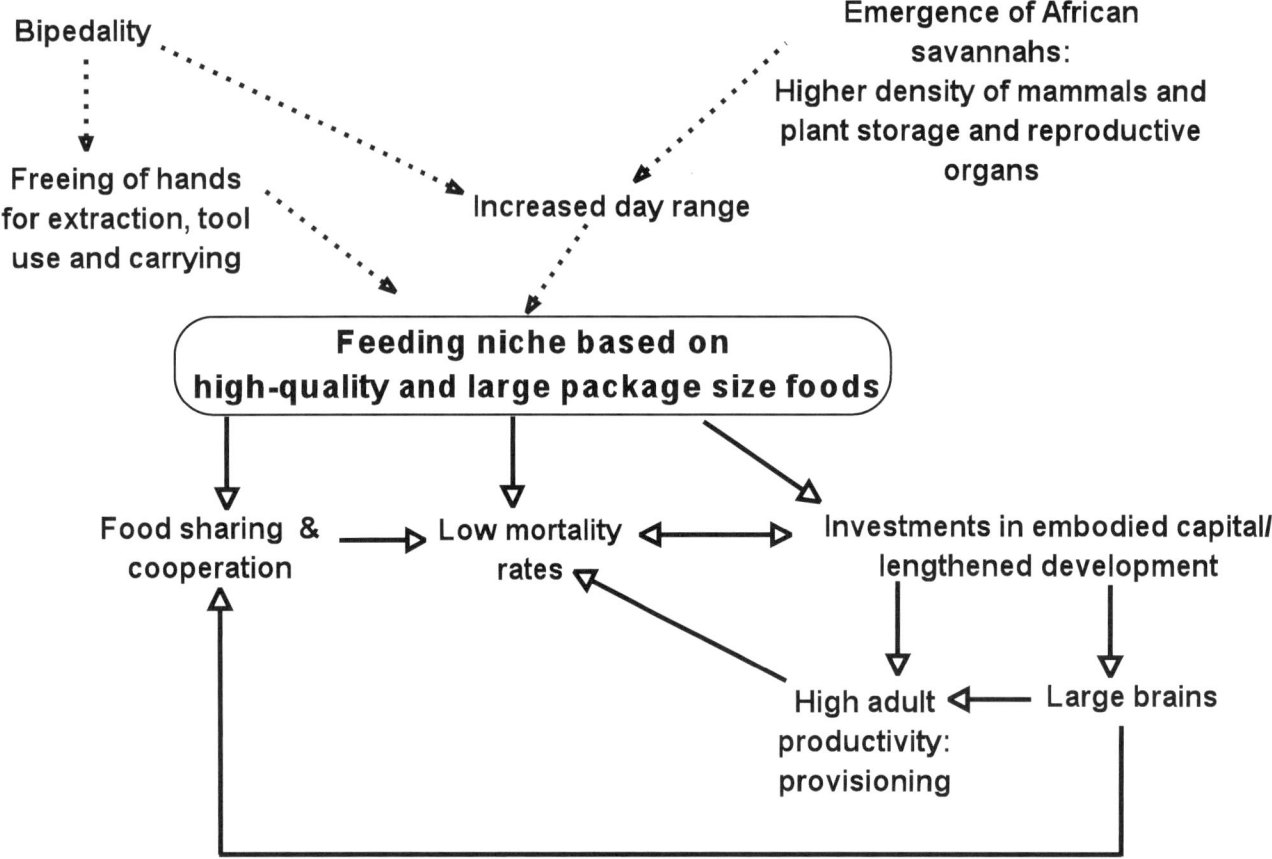

FIGURE 4.11 Ecology and co-evolutionary process in the Hominid line.

variable climate also would have favored a trend toward cognitive solutions that led to new foraging opportunities. Hominids with such solutions were probably better able to withstand rapid climate and habitat change than those who were more inflexible.[122]

The shift toward a high-quality diet based on learned foraging techniques also implies another co-evolutionary process. Just as the juvenile period and the life span co-evolved, so too did brain and gut sizes co-evolve with life-history traits and with each other.[23,123] Many other physiological traits are probably related to this evolved complex. For example, it is likely that hidden estrus among humans is related to the male energy subsidization of adult female reproduction,[62] and that the postreproductive life span of both sexes is related to high food production by older individuals. Almost certainly, the reason that human females have shorter interbirth intervals than apes do[124] (and probably different physiological sensitivity to nursing stimulation) is because reproductive-aged human women are able to decrease rather than increase their food production during lactation due to subsidies by other age and sex classes.[102] This commitment to food sharing is evident in human physiology and behavior. Women obtain more of the extra energy required for pregnancy and lactation by increases in energy intake[125] and re-

ductions in energy expenditure[12,126] rather than by an increase in fat mobilization or metabolic economy.[125] Reductions in basal metabolic rate during early pregnancy[127] are not sufficient to make up for the increased energy requirements. It is likely that this response evolved in the context of provisioning.

Figure 4.11 illustrates our historical hypothesis regarding the hominid diet and its impact on both pay-offs to learning and development and on mortality rates. The figure begins with two important exogenous changes. The first is a change in the distribution of foods with the emergence of African savannas in the Pleistocene, which increased the abundance of high-quality but protected plant foods (nuts and tubers, in particular) and animal foods. The second change is the pre-adaptation of bipedality, emerging in the Australopithecines as a locomotor adaptation. This adaptation frees the hands for tool use, which allows more efficient extraction and hunting and frees the hands for carrying large packages of food suitable for sharing.[128,129] It also results in efficient terrestrial locomotion and higher daily mobility that would increase the daily encounter rate with rare but energetically rich resources. These changes led to an increased emphasis on large, high-quality, but difficult-to-acquire foods. Our hypothesis is that this feeding niche had multiple effects: It increased the premium on learning and intelligence, delaying growth and maturation; increased nutritional status and decreased mortality rates through food sharing (predicted by the provisioning of young, sick or injured individuals); and released selection against larger group size, which lowers predation mortality.

The Grandmother Hypothesis

The present theory shares some features with a model of menopause and human life history recently proposed by Hawkes and colleagues,[130] often referred to as the grandmother hypothesis. This hypothesis proposes that humans have a long life span relative to that of the other primates because of the assistance that older postreproductive women contribute to descendant kin through the provision of difficult-to-acquire plant foods. Women, therefore, are selected to invest in maintaining their bodies longer than chimpanzee females do. Thus, both theories focus on the exceptionally long human life span. Rather than regarding the cessation of reproduction in the fifth decade of life as the critical adaptation (a feature shared with apes), both theories attempt to explain the extension of the human life span beyond menopause.[24,130] This implies that some fertility earlier in life is given up in order to prolong the adult life span.[24] The theories differ in important ways, however.

First, the grandmother theory focuses on the productivity of older women, but not on investments in learning and development. The present theory specifically links high productivity later in life to investments in a large brain and to the unique features of growth in human children. The grandmother hypothesis is silent about the expansion of the costly human brain. It also offers no explanation of why human children grow so slowly and take so long to mature, except for the fact that adult mortality rates are low. In fact, Hawkes and colleagues[130] suggest that learning is a secondary effect of a long juvenile period (determined by the long lifespan) rather than the cause of the long juvenile period. Second, the grandmother theory fails to account for why men live to about the same age as women. If the benefits of living longer derive, for women, from provisioning descendant kin and, for men from direct reproduction, there is no reason why their life spans should be so similar. Third, the grandmother model fails to capture the important role that human males play in supporting women's reproduction. It also fails to explain the age profile of production among males and why men take so long to reach their productive peaks.

The evidentiary basis of the grandmother hypothesis is also very weak. There is no direct evidence that postreproductive women are the major food providers in any society. While Hadza data do suggest that older Hadza women produce more food than younger women do, postreproductive women produce less than Hadza men (Fig. 4.2). Furthermore, it is unlikely that the Hadza pattern of high food production by postreproductive women is common among other foragers. Ache and Hiwi postreproductive women do not acquire even half the daily food energy acquired by adult men. !Kung data also suggest that older women produce very little, even leading the Hadza research team to ask "Why don't elderly !Kung women work harder to feed their grandchildren?" (Blurton Jones and coworkers,[13] p. 388). And certainly none of the high-meat-consuming societies of Table 4.3 nor any of the societies in human history dwelling at high latitudes could have been mostly dependent on the food production of postreproductive women. While we agree that the age profile of production is shifted toward older ages among women as well as men, and that postreproductive women provide many important services in foraging societies, among them child care, tool making, food processing, and camp maintenance, there is no evidence to support the hypothesis that they have been the "breadwinners" in most societies during human history.

The present theory organizes all of these facts. Both males and females exploit high-quality, difficult-to-acquire foods (females extracting plant foods and males hunting animal foods), sacrificing early productivity for later productivity, with a life-history composed of an extended juvenile period in which growth is slow, a large brain is programmed, and a high investment is made in mortality reduction and maintenance to reap the rewards of those investments.

PREDICTIONS OF THE THEORY TO BE TESTED IN FUTURE RESEARCH

Comparative Biology

The co-evolutionary selection pressures posited by our theory and supported by the quantitative theoretical model should be generally applicable. Holding pay-offs to time spent as a juvenile constant, increased survival rates during the juvenile period should select for delayed reproduction, as predicted by the models of Kozlowski and Weigert[5] and Charnov.[1,9] Holding constant extrinsic mortality hazards and the pay-offs to investments in mortality reduction, increased pay-offs to time spent in development as a juvenile should select for increased allocations to survival and lower mortality rates. The latter is a novel prediction of our theory. The testing of our predictions will require careful comparative research within taxonomic groups at different levels of analysis (for example, among birds, mammals, and reptiles, and among orders and genera within those higher taxonomic levels). It will be necessary to distinguish pay-off functions and mortality hazards from the actual levels achieved, given the observed allocations. For example, it will be necessary to distinguish the risk of dying from different causes, such as predation and infectious disease, from observed mortality rates due to those causes, because the observed mortality rate will be affected by allocations to predator avoidance and feeding, and by allocations to immune function.

A principal innovation of the proposed theory is the incorporation of co-evolutionary selection among life-history traits without requiring any trait to be treated as extrinsic. This is more realistic than previous approaches, but makes for increasingly demanding empirical analysis. There are many possible pay-offs to time spent in development. The most general benefit of increased time spent in development is increased body size and its effects on energetic turnover, survival, and mating success. The co-evolutionary selection measures posited here should apply to those effects, but of primary interest in our theory are the benefits associated with learning and information storage. Variation within birds, carnivores, pinnipeds, and primates should be particularly fertile ground for testing predictions generated by the theory with respect to learning, brain size, and life-history traits. The general prediction is that within those taxonomic groups, species that rely on more learning-intensive, complex feeding strategies will have longer developmental periods, longer life spans, and larger brains relative to body size. With respect to brain size, the correlations of brain size with life-history characteristics and feeding strategies should be reflected in those parts of the brain associated with learning and information storage, and not with raw perceptual processing, such as sonar and visual acuity (unless, of course, those functions are more developed in organisms that learn and store information). Suggestive evidence in support of the theory is available for birds[131] and primates,[116] but rigorous testing of those predictions awaits further research and analysis.

Hominid Evolution

Following this line of reasoning, the first major increases in brain size in the hominid line should be accompanied by extensions of the juvenile period, increased longevity, and increased complexity of learned foraging strategies. Smith's analyses of tooth eruption among early *Homo ergaster* provide suggestive evidence of extension of the juvenile period beyond that of chimpanzees and Australopithecines, but to a lesser extent than that of fully modern humans. So far, the record is silent with respect to longevity and mortality rates. Our ability to test the prediction that longevity increases with brain size during hominid evolution may await new developments such as advances in the extraction of DNA from fossil remains and in understanding the genetics of aging. Nevertheless, this extension of the expected life span with increased emphasis on learning in development and brain size is a firm prediction of our theory.

Although the early evolution of the genus *Homo* could have been accompanied by increased complexity of foraging strategies with respect to either hunting or gathering, we strongly suspect that future research will demonstrate increases in both the importance of hunted foods and complex extractive technologies for gathering.[23,121,128,129,132,133] Recently, O'Connell and associates[49] have rejected hunting as being important in early hominid evolution on two grounds. First, some archeological assemblages that include large accumulations of animal bone, thought to illustrate the hunting lifestyle, have now been reinterpreted as possibly resulting from natural processes. Second, recent data show that chimpanzees hunt considerably more than was previously thought. Therefore, hunting cannot be the cause of the changes in hominid life histories and social systems.

With respect to the first point, it is premature to reject the "hunting hypothesis" simply because the causes of bone accumulation at hominid sites are not well understood and are open to various interpretations. It has never been demonstrated that early hominids did not rely heavily on hunted foods. It is also the case that too much emphasis has been placed on hunting large game (Jones presents suggestive evidence that bone accumulations at Olduvai reflect reliance on small game hunting).[134] Chimpanzees hunt small game, as do many modern hunter gatherers. For example, the Ache rely heavily on small game: About 50% of the animal food in the Ache diet is small game acquired using hand-hunting techniques (no projectile weapons). In fact, it may be that the learning demands of a diet based on small game may be greatest because it can require killing many different species at regular intervals. Thus, much encounter-specific and species-specific knowledge and creativity may also be required.

With respect to the second point, Tables 4.1 and 4.2 show that humans and chimpanzees exploit very different food niches. While chimpanzees do hunt, the most successful chimpanzee hunters obtain less than 10% of the daily per capita energy intake from meat reported for any human foraging group. To claim that humans and chimpanzees both must have the same life history and social system because they both hunt is equivalent to asserting that chimpanzees must have the same social system as black and white colobus monkeys because they both eat some ripe fruit. We predict that early hominids will show a major increase in the consumption of hunted foods. Analyses of vitamins and minerals in hominid bone remains may provide new avenues for testing that hypothesis, in addition to archeological evidence.[135]

It is clear that the process of hominization occurred over a long time. Our theory is silent about the dates of important events and evolutionary changes in the hominid line. Whether the process will look ratchet-like, in that small increases in longevity will precede small increases in brain size, which, in turn, lead to further increases in longevity, is also an open question. Whether the process is gradual or punctuated, our theory predicts that changes in brain size, the dietary importance of meat and other difficult-to-acquire foods, gut size, and life span will be seen to co-evolve in the archeological and paleontological record. It is important to recognize that many of those changes may be quantitative as well as qualitative. For example, although chimpanzees hunt monkeys, they only pursue them in a very small proportion of their encounters with them, presumably because most of those encounters do not occur under conditions likely to result in a successful kill. Ache foragers only rarely ignore an encounter with monkeys, even when they hear them at a great distance, because most such encounters do result in successful kills, presumably because of the Ache's skill and effective weaponry. The quantitative changes in brain size and life-history traits during hominid evolution should be accompanied by quantitative changes in diet and skill-intensive foraging techniques.

Modern Humans Versus Chimpanzees

There are at least two major avenues of future research with extant hunter-gatherers and chimpanzees for testing the predictions of our theory. First, we have provided only suggestive evidence that food sharing and provisioning lower both juvenile and adult mortality rates in humans and that this is a major cause of the difference between human and chimpanzee age-specific survival probabilities. We need to know a great deal more about the frequency of illness in both species and the relationship between morbidity and mortality. What is the relationship between illness and food intake rates in both species? What is the relationship between reduced food intake and mortality? Our theory predicts that weight loss accompanying illness will be more frequent and severe, and that illness will more frequently result in mortality among chimpanzees than humans. Our theory also predicts that chimpanzee juveniles will spend more time than human juveniles in contexts that expose them to predation, and that this difference will be due to both group size and time spent in the food quest. Our theory also predicts absolutely greater frequency of predation on chimpanzees than on humans.

A second avenue of research is on the components of foraging success. A principal difference between our theory and the grandmother hypothesis is that we propose that the human life course and large brain are the results of a long learning process that is necessary for successful foraging, and that both males and females engage in this process. Recently, on the basis of some experiments among Hadza juveniles, Blurton Jones and Marlowe[136] have suggested that human foraging is not very difficult to learn. We propose that hunting, as practiced by humans, but not necessarily by other predators, is exceedingly difficult to learn and requires many years of experience. Our observations of hunters in six different groups suggests to us that it is not marksmanship, but the knowledge of prey behavior and remote signs of that behavior such as tracks and vocalizations that are the most difficult features of human hunting.[30]

This impression is testable. We should find that marksmanship is acquired relatively early in the learning process and that naive individuals such as anthropology graduate students could be trained to be effective marksmen with relatively little practice. In contrast, we should find that knowledge of animal behavior and signs of behavior are learned only gradually over many years. We should also find that encounter rates with prey are much more strongly age-dependent than are kill rates upon encounter. It should also be the case that kills that rely on marksmanship are less age-dependent than kills that rely on finesse and skill, such as luring animals out of holes. In addition, it should be much more difficult to train naive individuals to find prey than to shoot projectile weapons accurately. Similarly, we should find that age effects on chimpanzee hunting should be a result of differences in knowledge of prey behavior. Boesch and Boesch[20] provide suggestive evidence of such effects, but more research is necessary for a definitive test. An analogous set of predictions also could be tested with extractive gathering techniques and the specific knowledge required for each (Bock[29,137] presents a series of experiments with grain processing as a model of how such research could be conducted).

Conclusions

The human adaptation is broad and flexible in one sense, but narrow and specialized in another sense. It is broad in the sense that, as hunter-gatherers, humans have existed successfully in virtually all of the world's major habitats. This has entailed eating a very wide variety of foods, both plant and animal, both within and among environments. It also has entailed a great deal of flexibility in the contributions of individuals of different ages and sex. The relative contributions of men and women to food production appear to vary from group to group. Even the contributions of children and teens to food production vary predictably with the abundance of easy-to-acquire foods.

Our adaptation is narrow and specialized in that it is based on a diet composed of large, nutrient-dense, difficult-to-acquire packages and a life history with a long, slow development, a large commitment to learning and intelligence, and an age profile of production shifted toward older individuals. We do not expect to find any human population that subsists on leaves or other low-quality foods. Indeed, we expect humans to remain at the very top of the food hierarchy in every environment they live in (for example, humans often exterminate all other top predators in their habitat). Humans ingest foods that are already high in quality and do not require much digestive work or detoxification. And if a food contains toxins, they are generally removed prior to ingestion by processing techniques. This dietary commitment is reflected in the extremely reduced size of the human hindgut.[23] Humans use their great intelligence to extract and hunt those foods. In order to achieve this diet, humans also engage in extensive food sharing both within and among age and sex classes of individuals. Finally, the effect of the commitment to food sharing is evident in the reproductive physiology of human women. Provisioning permits human women, in contrast to other female primates, to reduce rather than increase their rate of energy production during their reproductive years, when they have both infant and juvenile nutritional dependents and a greatly reduced spacing between births.

The model and the data we have presented suggest that the human life course is based on a complex set of interconnected, time-dependent processes and the co-evolution of physiology, psychology, and behavior. There appears to be a tight linkage among the ordering of major psychological milestones (language learning, understanding and mastering the physical, biological, and social environment); the timing of brain growth; growth rates during childhood and adolescence; developmental changes in survivorship; behavioral, psychological, and physiological changes with the transition to adulthood; profiles of risk with age; and rates of senescence and aging. It is very likely that a species-typical life course evolved in response to the demands of a hunting and gathering lifestyle that was broad and flexible enough to allow successful exploitation of the world's environments, but specialized toward the acquisition of learned skills and knowledge to obtain very high rates of productivity later in life.

Acknowledgments

We thank K. G. Anderson, Theodore Bergstrom, John Bock, Robert Boyd, Pat Draper, Peter Ellison, Michael Gurven, Sarah Hrdy, Monique Borgerhoff Mulder, David Lam, and Peter Richerson for helpful discussions of these issues as well as Anne Pusey and Janette Wallis for sharing data on chimpanzee development and birth-spacing. We especially thank Nicholas Blurton Jones, Eric Charnov, Kristen Hawkes, and James O'Connell whose seminal work in this area both challenged us and helped stimulate our thinking, and who took time to critique various versions of this manuscript. Arthur Robson deserves special recognition for performing the mathematical analysis to generate the results outlined in Box 4.1 and writing the Mathematical Appendix, which is available from the authors.

Field research on the Ache, Hiwi, and Machiguenga has been supported by grants to Kim Hill and A. M. Hurtado by the National Science Foundation (BNS-8613215, BNS-538228, BNS-8309834, BNS-8121209, BNS-9617692), the National Institutes of Health (RO1HD16221-01A2), Fundacion Gran Mariscal de Ayacucho, Caracas, Venezuela, and two grants from the L. S. B. Leakey Foundation. Further support has gone to H. Kaplan for field research on the Machiguenga and Yora by the National Science Foundation (BNS-8718886) and the L. S. B. Leakey Foundation, and for life history theory and life course evolution by the National Institute of Aging (1R01AG15906-01).

References

1. Charnov, E. L. 1993. *Life History Invariants: Some Explanations of Symmetry in Evolutionary Ecology*. Oxford: Oxford University Press.
2. Gadgil, M., W. H. Bossert. 1970. Life historical consequences of natural selection. *Am Nat* 104:1–24.
3. Janson, C. H., C. P. Van Schaik. 1993. Ecological risk aversion in juvenile primates: slow and steady wins the race. In: Pereira, M., L. Fairbanks, editors. *Juvenile Primates: Life History, Development and Behavior*. New York: Oxford University Press. p. 57–76.
4. Kozlowski, J., R. G. Wiegert. 1986. Optimal allocation to growth and reproduction. *Theoret Popul* 29:16–37.
5. Kozlowski, J., R. G. Weigert. 1987. Optimal age and size at maturity in the annuals and perennials with determinante growth. *Evol Ecol* 1:231–244.
6. Hawkes, K., J. F. O'Connell, N. Blurton Jones. 1989. Hardworking Hadza grandmothers. In: Standen, V., R. A. Foley, editors. *Comparative Socio-Ecology of Humans and Other Mammals*. London: Basil Blackwell. p. 341–366.
7. Hawkes, K., J. F. O'Connell, N. Blurton Jones. 1997. Hadza women's time allocation, offspring provisioning, and the evolution of long postmenopausal life spans. *Curr Anthropol* 38:551–577.
8. Hill, K., C. Boesch, A. Pusey, J. Goodall, J. Williams, R. Wrangham. 2000. *Chimpanzee Mortality in the Wild*. University of New Mexico.
9. Charnov, E. 1993. Why do female primates have such long lifespans and so few babies? *Evol Anthropol* 1:191–194.
10. Hill, K., A. M. Hurtado. 1996. *Ache Life History: The Ecology and Demography of a Foraging People*. Hawthorne, N. Y: Aldine de Gruyter.
11. Hill, K., K. Hawkes, A. Hurtado, H. Kaplan. 1984. Seasonal variance in the diet of Ache hunter-gatherers in eastern Paraguay. *Hum Ecol* 12:145–180.
12. Hurtado, A. M., K. Hill. 1990. Seasonality in a foraging society: Variation in diet, work effort, fertility, and the sexual division of labor among the Hiwi of Venezuela. *J Anthropol Res* 46:293–345.
13. Blurton Jones, N., K. Hawkes, J. O'Connell. 1989. Modeling and measuring the costs of children in two foraging societies. In: Standen, V., R. A. Foley, editors. *Comparative Socioecology of Humans and Other Mammals*. London: Basil Blackwell. p. 367–390.
14. Blurton Jones, N. G., K. Hawkes, J. O'Connell. 1997. Why do Hadza children forage? In: Segal, N. L., G. E. Weisfeld, C. C. Weisfeld, editors. *Uniting Psychology and Biology: Integrative Perspectives on Human Development*. New York: American Psychological Association. p. 297–331.
15. Hawkes, K., J. F. O'Connell, N. G. Blurton Jones. 1991. Hunting income patterns among the Hadza: big game, common goods, foraging goals and the evolution of the human diet. *Philos Trans R Soc London* (B) 334:243–251.
16. Blurton Jones, N., L. Smith, J. O'Connell, K. Hawkes, C. L. Samuzora. 1992. Demography of the Hadza, an increasing and high density population of savanna foragers. *Am J Phys Anthropol* 89:159–181.
17. Kaplan, H. K. 1994. Evolutionary and wealth flows theories of fertility: empirical tests and new models. *Popul Dev Rev* 20:753–791.
18. Becker, G. S. 1975. *Human Capital*. New York: Columbia University Press.
19. Mincer, J. 1974. *Schooling, Experience, and Earnings*. Chicago: National Bureau of Economic Research.
20. Boesch, C., H. Boesch. 1999. *The Chimpanzees of the Tai Forest: Behavioural Ecology and Evolution*. Oxford: Oxford University Press.
21. Stanford, C. G. 1999. *The Hunting Apes: Meat Eating and the Origins of Human Behavior*. Princeton: Princeton University Press.
22. Wrangham, R. W., E. Van, Z. B. Riss. 1990. Rates of predation on mammals by Gombe chimpanzees, 1972–1975. *Primates* 3:157–170.
23. Milton, K. 1999. A hypothesis to explain the role of meat-eating in human evolution. *Evol Anthropol* 8:11–21.
24. Kaplan, H. K. 1997. The evolution of the human life course. In: Wachter, K., C. E. Finch, editors. *Between Zeus and Salmon: The Biodemography of Aging*. Washington, D.C.: National Academy of Sciences. p. 175–211.
25. Hiraiwa-Hasegawa, M. 1990. The role of food sharing between mother and infant in the ontogeny of feeding behavior. In: Nishida, T., editor. *The Chimpanzees of the Mahale Mountains: Sexual and Life History Strategies*. Tokyo: Tokyo University Press. p. 267–276.
26. Silk, J. B. 1979. Feeding, foraging, and food-sharing behavior in immature chimpanzees. *Folia Primatol* 31:12–42.
27. Teleki, G. 1973. *The Predatory Behavior of Wild Chimpanzees*. Lewisburg, P. A: Bucknell University Press.

28. Blurton Jones, N. G., K. Hawkes, P. Draper. 1994. Foraging returns of !Kung adults and children: why didn't !Kung children forage? *J Anthropol Res* 50:217–248.
29. Bock, J. A. 1995. The determinants of variation in children's activities in a Southern African community. Ph.D. dissertation, University of New Mexico.
30. Leibenberg, L. 1990. *The Art of Tracking: The Origin of Science.* Cape Town: David Phillip.
31. Kelly, R. 1995. *The Foraging Spectrum: Diversity in Hunter-Gatherer Lifeways.* Washington, D.C.: Smithsonian Institution Press.
32. Wrangham, W. 1975. The behavioral ecology of chimpanzees in Gombe National Park, Tanzania. Ph.D. dissertation, Cambridge University.
33. Wrangham, R. W., B. Smuts. 1980. Sex differences in behavioral ecology of chimpanzees in Gombe National Park, Tanzania. *J Reprod Fertil* (Suppl) 28:13–31.
34. Alvard, M. 1995. Intraspecific prey choice by Amazonian hunters. *Curr Anthropol* 36:789–818.
35. Stiner, M. 1991. An interspecific perspective on the emergence of the modern human predatory niche. In: Stiner, M., editor. *Human Predators and Prey Mortality.* Boulder: Westview Press. p. 149–185.
36. Martorell, R., A. Lechtig, C. Yarbrough, H. Delgado, R. E. Klein. 1976. Protein-calorie supplementation and postnatal physical growth: a review of findings from developing countries. *Arch Latinoam Nutr* 26:115–128.
37. Mora, J. O., M. G. Herrera, J. Suescun, L. Denavarro, M. Wagner. 1981. The effects of nutritional supplementation on physical growth of children at risk of malnutrition. *Am J Clin Nutr* 34:1885–1892.
38. Coop, R., P. Holmes. 1996. Nutrition and parasite interaction. *Int J Parasitol* 26:951–962.
39. Baertl, J. M., G. Morales, G. Verastegui, G. G. Graham. 1970. Supplementation for entire communities: growth and mortality of infants and children. *Am J Clin Nutr* 23:707–715.
40. Kielmann, A. A., C. E. Taylor, R. L. Parker. 1978. The Narangwal nutrition study: a summary review. *Am J Clin Nutr* 31:2040–2052.
41. Lee, R. B., I. De Vore. 1968. editors. *Man the Hunter.* Aldine, Chicago.
42. Lee, R. B. 1979. *The !Kung San: Men, Women, and Work in a Foraging Society.* Cambridge: Cambridge University Press.
43. Yellen, J. 1977. *Archaeological Approaches to the Present: Models for Reconstructing the Past.* New York: Academic Press.
44. Hoffecker, J. E. 1999. Neanderthals and modern humans in Eastern Europe. *Evol Anthropol* 7:129–141.
45. Pitts, M., M. Roberts. 1997. *Fairweather Eden: Life in Britain Half a Million Years Ago as Revealed by the Excavations at Boxgrove.* London: Century.
46. Jones, R. 1990. From Kakadu to Kutikina: the southern continent at 18,000 years ago. In: Gamble, C., O. Soffer, editors. *The World at 18,000 BP.* London: Unwin Hyman.
47. Close, A., F. Wendorf. 1990. North Africa at 18,000 BP. In: Gamble, C., O. Soffer, editors. *The World at 18,000 BP.* London: Unwin Hyman.
48. Hawkes, K. 1991. Showing off: tests of an hypothesis about men's foraging goals. *Ethol Sociobiol* 12:29–54.
49. O'Connell, J. F., K. Hawkes, N. G. Blurton Jones. 1999. Grandmothering and the evolution of *Homo erectus. J Hum Evol* 36:461–485.
50. Gurven, M., W. Allen-Arave, K. Hill, A. M. Hurtado. 2000. It's a wonderful life: signaling generosity among the Ache of Paraguay. *Evol Hum Behav* 21:263–282.
51. Gurven, M., K. Hill, H. Kaplan, M. Hurtado, R. Lyles. 1999. Food transfers among Hiwi foragers of Venezuela: tests of reciprocity. *Hum Ecol* 28:171–218.
52. Hames, R. 1987. Relatedness and garden labor exchange among the Ye'Kwana: a preliminary analysis. *Ethol Sociobiol* 8:259–284.
53. Hurtado, A. M., K. Hill, H. Kaplan, I. Hurtado. 1992. Tradeoffs between female food acquisition and childcare among Hiwi and Ache foragers. *Hum Nat* 3:185–216.
54. Hill, K., A. M. Hurtado. 1991. The evolution of reproductive senescence and menopause in human females. *Hum Nat* 2:315–350.
55. Hurtado, A. M., K. Hill. 1992. Paternal effects on child survivorship among Ache and Hiwi hunter-gatherers: implications for modeling pair-bond stability. In: Hewlett, B., editor. *Father-Child Relations: Cultural and Biosocial Contexts.* Hawthorne, N. Y.: Aldine de Gruyter. p. 31–56.
56. Pennington, R., H. Harpending. 1988. Fitness and fertility among the Kalahari !Kung. *Am J Phys Anthropol* 77:303–319.
57. Blurton Jones, N. G., K. Hawkes, J. F. O'Connell. 1996. The global process, the local ecology: how should we explain differences between the Hadza and the !Kung? In: Kent, S., editor. *Cultural Diversity in Twentieth Century Foragers.* Cambridge: Cambridge University Press. p. 159–187.
58. Kaplan, H., H. Dove. 1987. Infant development among the Ache of Eastern Paraguay. *Dev Psychol* 23:190–198.
59. Kaplan, H. K. 1996. A theory of fertility and parental investment in traditional and modern human societies. *Yearbook Phys Anthropol* 39:91–135.
60. Hawkes, K., J. F. O'Connell, H. G. Blurton Jones. 1995. Hadza children's foraging: juvenile dependency, social arrangements and mobility among hunter-gatherers. *Curr Anthropol* 36:688–700.
61. Hill, K. 1988. Macronutrient modifications of optimal foraging theory: An approach using indifference curves applied to some modern foragers. *Hum Ecol* 16:157–197.
62. Hurtado, A. M., K. Hill, H. Kaplan, J. Lancaster. 1999. *The Origins of the Sexual Division of Labor.* In preparation.
63. Hawkes, K., J. F. O'Connell, N. G. Blurton Jones. 2000. Hadza hunting and the evolution of nuclear families, submitted.
64. Allman, J., E. McGuiness. 1983. Visual cortex in primates. In: *Comparative Primate Biology.* New York: Alan, R. Liss. p. 279–326.
65. Allman, J. 1987. Primates, evolution of the brain. In: Gregory, R. L., editor. *The Oxford Companion to the Mind.* Oxford: Oxford University Press. p. 663–669.
66. Allman, J., T. McLaughlin, A. Hakeem. 1993. Brain weight and life-span in primate species. *Proc Natl Acad Sci* 90:118–122.
67. Hakeem, A., G. R. Sandoval, M. Jones, J. Allman. 1996. Brain and life span in primates. In: Birren, J. E., K. W. Schaie, editors. *Handbook of the Psychology of Aging.* San Diego: Academic Press. p. 78–104.
68. Roff, D. A. 1992. *The Evolution of Life Histories.* London: Chapman and Hall.
69. Altman, J. C. 1987. *Hunter-Gatherers Today: An Aboriginal Economy of North Australia.* Canberra: Australian Institute of Aboriginal Studies.
70. Bahuchet, S. 1990. Food sharing among the pygmies of Central Africa. *Afr Stud Monogr* 11:27–53.
71. Endicott, K. 1988. Property, power and conflict among the Batek of Malaysia. In: Ingold, T., D. Riches, J. Woodburn, editors.

Hunters and Gatherers: Property, Power and Ideology. New York: St. Martin's Press. p. 110–128.
72. Kaplan, H., K. Hill. 1985. Food-sharing among Ache foragers: Tests of explanatory hypotheses. *Curr Anthropol* 26:223–245.
73. Lee, R. 1972. The !Kung bushmen of Botswana. In: Bicchieri, M. G., editor. *Hunters and Gatherers Today*. New York: Holt, Rinehart and Winston. p. 326–368.
74. Marshall, L. 1976. Sharing, talking and giving: relief of social tensions among the !Kung. In: Lee, R., I. DeVore, editors. *Kalahari Hunter-Gatherers*. Cambridge: Harvard University Press. p. 350–371.
75. Winterhalder, B. 1996. Social foraging and the behavioral ecology of intragroup resource transfers. *Evol Anthropol* 5:46–57.
76. Goodall, J. 1986. *The Chimpanzees of the Gombe: Patterns of Behavior*. Cambridge: Cambridge University Press.
77. McGrew, W. C. 1996. Dominance status, food sharing, and reproductive success in chimpanzees. In: Weisner, P., W. Schiefenhovel, editors. *Food and the Status Quest*. Providence, RI: Berghahn Books. p. 39–46.
78. Sugiyama, L., R. Chacon. 2000. Effects of illness and injury on foraging among the Yora and Shiwar: pathology risk as adaptive problem. In: Cronk, L., W. Irons, N. Chagnon, editors. *Human Behavior and Adaptation: An Anthropological Perspective*. New York: Aldine de Gruyter. In press.
79. Bailey, R. C. 1991. *The Behavioral Ecology of Efe Pygmy Men in the Ituri Forest, Zaire*. Ann Arbor: University of Michigan. Museum of Anthropology.
80. Fleagle, J. G. 1999. *Primate Adaptation and Evolution*. New York: Academic Press.
81. Wrangham, R. W. 1979. On the evolution of ape social systems. *Soc Sci Information* 18:335–368.
82. Hiraiwa-Hasegawa, M. 1990. A note on the ontogeny of feeding. In: Nishida, T., editor. *The Chimpanzees of the Mahale Mountains: Sexual and Life History Strategies*. Tokyo: Tokyo University Press. p. 277–283.
83. Howell, N. 1979. *Demography of the Dobe !Kung*. New York: Academic Press.
84. Austad, S., K. E. Fisher. 1991. Mammalian aging, metabolism, and ecology: evidence from the bats and marsupials. *J Gerontol* 46(2):B47–53.
85. Austad, S., K. E. Fischer. 1992. Primate longevity: its place in the mammalian scheme. *Am J Primatol* 28:251–261.
86. Hladik, C. M. 1988. Seasonal variations in food supply for wild primates. In: de Garine, I., G. A. Harrison, editors. *Coping with Uncertainty in Food Supply*. Oxford: Clarendon Press. p. 26–32.
87. Milton, K. 1993. Diet and primate evolution. *Sci Am* 269:70–77.
88. Altmann, S. A. 1998. *Foraging for Survival: Yearling Baboons in Africa*. Chicago: University of Chicago Press.
89. Milton, K. 1988. Foraging behaviour and the evolution of primate intelligence. In: Byrne, R. W., A. Whiten, editors. *Machiavellian Intelligence*. Oxford: Clarendon Press. p. 285–305.
90. Terborgh, J. 1983. *Five New World Primates: A Study in Comparative Ecology*. Princeton: Princeton University Press.
91. Barton, R. A. 1998. Visual specialization and brain evolution in primates. *Proc R Soc London* B 265:1933–1937.
92. Barton, R. A. 1999. The evolutionary ecology of the primate brain. In: Lee, P. C., editor. *Primate Socioecology*. Cambridge: Cambridge University Press.
93. Byrne, R. 1995. *The Thinking Ape: Evolutionary Origins of Intelligence*. Oxford: Oxford University Press.
94. Menzel, C. R. 1997. Primates' knowledge of their natural habitat. In: Whiten, A., R. Byrne, editors. *Machiavellian Intelligence II*. Cambridge: Cambridge University Press. p. 207–239.
95. Byrne, R. 1997. The technical intelligence hypothesis: an additional evolutionary stimulus to intelligence? In: Whiten, A., R. Byrne, editors. *Machiavellian Intelligence II*. Cambridge: Cambridge University Press. p. 289–311.
96. Andrews, P., L. Martin. 1992. Hominoid dietary evolution. In: Whiten, A., E. M. Widdowson, editors. *Foraging Strategies and Natural Diet of Monkeys, Apes and Humans*. Oxford: Clarendon Press. p. 39–50.
97. Wrangham, R. W., N. L. Conklin-Brittain, K. D. Hunt. 1998. Dietary response of chimpanzees and cercopithecines to seasonal variation in fruit abundance. I. antifeedants. *Int J Primatol* 19:949–970.
98. Conklin-Brittain, N. L., R. W. Wrangham, K. D. Hunt. 1998. Dietary response of chimpanzees and cercopithecines to seasonal variation in fruit abundance. II. micronutrients. *Int J Primatol* 26:951–962.
99. Gibson, K. R. 1986. Cognition, brain size and the extraction of embedded food resources. In: Else, J. G., P. C. Lee, editors. *Primate Ontogeny, Cognition, and Social Behavior*. Cambridge: Cambridge University Press. p. 93–105.
100. Gibson, K. R. 1990. New perspectives on instincts and intelligence: brain size and the emergence of hierarchial mental constructional skills. In: Parker, S., K. Gibson, editors. *"Language" and Intelligence in Monkeys and Apes: Comparative Developmental Perspectives*. Cambridge: Cambridge University Press.
101. Whiten, A., J. Goodall, W. C. McGrew, T. Nishida, V. Reynolds, Y. Sugiyama, C. E. G. Tutin, R. Wrangham, C. Boesch. 1999. Cultures in chimpanzees. *Nature* 399:682–685.
102. Lancaster, J., H. Kaplan, K. Hill, A. M. Hurtado. 1999. The evolution of life history, intelligence and diet among chimpanzees and human foragers. In: Tonneau, F., N. S. Thompson, editors. *Evolution, Culture and Behavior*. New York: Plenum Press. p. 47–72.
103. Jerison, H. 1973. *Evolution of the Brain and Intelligence*. New York: Academic Press.
104. Jerison, H. J. 1976. Paleoneurology and the evolution of mind. *Sci Am* 234:90–101.
105. Clutton-Brock, T. H., P. H. Harvey. 1980. Primates, brains and ecology. *J Zool, London* 109:309–323.
106. Milton, K. 1981. Distribution patterns of tropical plant foods as an evolutionary stimulus to primate mental development. *Am Anthropol* 83:534–548.
107. Barton, R. A., R. I. M. Dunbar. 1997. Evolution of the social brain. In: Whiten, A., R. W. Byrne, editors. *Machiavellian Intelligence II*. Cambridge: Cambridge University Press. p. 240–263.
108. Dunbar, R. I. M. 1996. The social brain hypothesis. *Evol Anthropol* 6:178–190.
109. Harcourt, A. H. 1988. Alliances in contests and social intelligence. In: Byrne, R., A. Whiten, editors. *Machiavellian Intelligence*. Oxford: Clarendon Press. p. 132–152.
110. Harcourt, A. H. 1988. Cooperation as a competitive strategy in primates and birds. In: Ito, Y., J. L. Brown, editors. *Animal Societies: Theories and Facts*. Tokyo: Japan Scientific Societies Press. p. 147–157.
111. Walters, J. R., R. M. Seyfarth. 1987. Conflict and cooperation. In: Smuts, B. B., D. L. Cheney, R. M. Seyfarth, R. W. Wrangham, T. T. Struhsaker, editors. *Primate Societies*. Chicago: University of Chicago Press. p. 306–318.

112. de Waal, F. 1996. Conflict as negotiation. In: McGrew, W. C., L. F. Marchant, T. Nishida, editors. *Great Ape Societies*. Cambridge: Cambridge University Press. p. 159–172.
113. Harcourt, A. H. 1987. Dominance and fertility among female primates. *J Zool* 213:471–487.
114. Smith, B. H. 1991. Dental development and the evolution of life history in Hominidae. *Am J Phys Anthropol* 86:157–174.
115. Foley, R. A., P. C. Lee. 1992. Ecology and energetics of encephalization in human evolution. In: Whiten, A., E. M. Widdowson, editors. *Foraging Strategies and Natural Diet of Monkeys, Apes, and Humans*. Oxford: Oxford University Press. p. 63–72.
116. Kaplan, H., S. Gangestad, T. C. Muller, J. B. Lancaster. 2000. The evolution of primate life histories and intelligence. Submitted.
117. Cowlishaw, G., R. I. M. Dunbar. 1991. Dominance rank and mating success in male primates. *Anim Behav* 41:1045–1056.
118. de Waal, F. 1992. Intentional deception in primates. *Evol Anthropol* 1:86–92.
119. de Waal, F. 1994. Overview: culture and cognition. In: Wrangham, R. W., W. C. McGrew, F. De Waal, P. Helte, editors. *Chimpanzee Cultures*. Cambridge: Harvard University Press. p. 263–266.
120. Smith, B. H. 1993. The physiological age of KNM-WT 15000. In: Walker, A., R. Leakey, editors. *The Nariokotome Homo erectus Skeleton*. Cambridge: Harvard University Press. p. 196–220.
121. Leonard, W. R., M. L. Robertson. 1997. Comparative primate energetics and hominid evolution. *Am J Phys Anthropol* 102:265–281.
122. Potts, R. 1998. Variability selection in Hominid evolution. *Evol Anthropol* 7:81–96.
123. Aiello, L., P. Wheeler. 1995. The expensive-tissue hypothesis: the brain and the digestive system in human and primate evolution. *Curr Anthropol* 36:199–221.
124. Galdikas, B. M. F., J. W. Wood. 1990. Birth spacing patterns in humans and apes. *Am J Phys Anthrop* 83:185–192.
125. Piers, L. S., S. N. Diggavi, S. Thangam, J. M. Van Raaij, P. S. Shetty, J. G. Hautvast. 1995. Changes in energy expenditure, anthropometry, and energy intake during the course of pregnancy and lactation in well-nourished Indian women. *Am J Clin Nutr* 61:501–513.
126. Lawrence, M., R. G. Whitehead. 1988. Physical activity and total energy expenditure of childbearing Gambian village women. *Eur J Clin Nutr* 42:145–160.
127. Poppitt, S. D., A. M. Prentice, E. Jequier, Y. Schutz, R. G. Whitehead. 1993. Evidence of energy sparing in Gambian women during pregnancy: a longitudinal study using whole-body calorimetry. *Am J Clin Nutr* 57:353–364.
128. Lancaster, J. B. 1978. Carrying and Sharing in human evolution. *Human Nature Magazine* 1:82–89.
129. Lancaster, J. B. 1997. The evolutionary history of human parental investment in relation to population growth and social stratification. In: Gowaty, P. A., editor. *Feminism and Evolutionary Biology*. New York: Chapman and Hall. p. 466–489.
130. Hawkes, K., J. F. O'Connell, N. G. Blurton Jones, H. Alvarez, E. L. Charnov. 1998. Grandmothering, menopause, and the evolution of human life histories. *Proc Natl Acad Sci* 95: 1336–1339.
131. Bennett, P. M., P. H. Harvey. 1985. Relative brain size and ecology in birds. *J Zool London* (A) 207:151–169.
132. Isaac, G. L. 1978. Food sharing in human evolution: archaeological evidence from the Plio-Pleistocene of East Africa. *J Anthropol Res* 34:311–325.
133. Walker, A., Shipman, P. 1989. The cost of becoming a predator. *J Hum Evolution* 18:373–392.
134. Jones, K. 1984. Hunting and scavenging by early hominids: a study in archeological method and theory. University of Utah.
135. Sillen, A., G. Hall, R. Armstrong. 1995. Strontium calcium ratios (Sr/Ca) and strontium isotopic ratios (87Sr/86Sr) of *Australopithcus robustus* and *Homo* sp. from Swartkrans. *Hum Evol* 28:277–285.
136. Blurton Jones, N. G., F. W. Marlowe. 1999. *The Forager Olympics: Does It Take 20 Years to Become a Competent Hunter-Gatherer?* Salt Lake City: Human Behavior and Evolution Society.
137. Bock, J. 2000. The socioecology of children's activities: a new model of children's work and play. *Am Anthropol*. Submitted.
138. Hill, K., H. Kaplan. 1999. Life history traits in humans: theory and empirical studies. *Ann Rev Anthropol* 28:397–430.
139. Sugiyama, Y. 1989. Population dynamics of chimpanzees at Bossou, Guinea. In: Heltne, P., L. Marquandt, editors. *Understanding Chimpanzees*. Cambridge: Harvard University Press. p. 134–145.
140. Pusey, A. 1990. Behavioral changes at adolescence in chimpanzees. *Behaviour* 115:203–246.
141. Tutin, C. E. G. 1994. Reproductive success story: variability among chimpanzees and comparisons with gorillas. In: Wrangham, R. W., W. C. McGrew, F. DeWaal, P. Heltne, editors. *Chimpanzee Cultures*. Cambridge: Harvard University Press. p. 181–193.
142. Wallis, J. 1997. A survey of reproductive parameters in the free-ranging chimpanzees of Gombe National Park. *J Reprod Fertil* 109:297–307.
143. Nishida, T., H. Takasaki, Y. Takahata, editors. 1991. *The Chimpanzees of the Mahale Mountains: Sexual and Life History Strategies*. Tokyo: University of Tokyo Press.
144. Bose, S. 1964. Economy of the Onge of Little Andaman. *Man in India* 44:298–310.
145. Meehan, B. 1982. *Shell bed to shell midden*. Canberra: Australian Institute of Aboriginal Studies.
146. Silberbauer, G. 1981. *Hunter and Habitat in the Central Kalahari Desert*. Cambridge: Cambridge University Press.
147. Tanaka, J. 1980. *The San, Hunter-Gatherers of the Kalahari: A Study in Ecological Anthropology*. Tokyo: Tokyo University Press.
148. Hurtado, A. M., K. Hill. 1986. Early dry season subsistence ecology of the Cuiva (Hiwi) foragers of Venezuela. *Hum Ecol* 15:163–187.
149. Hill, K. 1983. Adult male subsistence strategies among Ache hunter-gatherers of Eastern Paraguay. Ph.D. dissertation, University of Utah.
150. McArthur, M. 1960. Food consumption and dietary levels of groups of aborigines living on naturally occurring foods. In: Mountford, C. P., editor. *Records of the American-Australian Scientific Expedition to Arnhem Land*. Melbourne: Melbourne University Press. p. 90–135.
151. Politis, G. 1996. *Nukak*. Colombia: Instituto Amazónico de Investigaciones Científicas-SINCHI.
152. Wrangham, R. W., C. A. Chapman, A. P. Clark-Arcadi, G. Isabirye-Basuta. 1996. Social ecology to Kanyawara chimpanzees: implications for understanding the costs of great ape groups. In: McGrew, W., L. Marchant, T. Nishida, editors. *Great Ape Societies*. Cambridge: Cambridge University Press. p. 45–57.
153. Uehara, S. 1997. Predation on mammals by the chimpanzee (*Pan troglodytes*): a review. *Primates* 38:193–214.
154. Stanford, C. B. 1998. *Chimpanzee and Red Colobus: The Ecology of Predator and Prey*. Cambridge: Harvard University Press.

PART II

Pleistocene Foragers and Colonists

While the papers in the previous section of *Evolutionary Ecology and Archaeology* address the biological evolution of our species, the works in this section focus on the adaptations and spread of *Homo sapiens* once derived. During the later part of the Pleistocene (ca. 200,000 to 10,000 years ago) people colonized many new areas including Australia, the higher latitudes of Eurasia, and the Americas. The papers reprinted in this section use EE to address a variety of topics related to population dynamics, dispersal, and human-environment interrelationships during this period. Three of them (Belovsky, Kelly, Byers and Ugan) focus on issues related to New World colonization and have far-reaching implications for long-standing debates related to this topic, while the two others (Stiner et al. and Grayson and Delpech) attend to the European Paleolithic. All of these papers employ models from foraging theory and, in general, are more narrowly focused than those of the preceding section, addressing specific issues pertaining to subsistence, mobility, or population dynamics.

OLD WORLD

M. C. Stiner et al. draw on long-term faunal records from the northern and eastern rims of the Mediterranean Sea to infer abrupt increases in human population densities during the late Middle Paleolithic and again during the Upper and Epi-Paleolithic periods. The demographic signals that the authors describe are derived from the logic of the prey model combined with increases in the relative abundance of more costly, lower-return small-game resources—particularly partridges, hares and rabbits—in archaeological faunal assemblages (see the next section of this volume for further discussion of the relationship between taxonomic relative abundance, the prey model, and human population growth). Also important are size-based indications of increased harvest pressure on lower-cost, higher-return resources, particularly slow-moving tortoises and sessile limpets. The initial increase in human population density that these authors find is consistent with genetic data that also indicate a substantial population increase at this time (e.g., Rogers 1995). Most significantly, Stiner and colleagues (see also Kuhn and Stiner 1998; Stiner and Munro 2002; Stiner et al. 2000) suggest that many of the documented behavioral differences between Middle Paleolithic populations (including Neandertals) and those of the Upper Paleolithic may stem principally from the very different demographic contexts in which those groups found themselves. This argument challenges the idea that a "creative explosion" occurred during the Upper Paleolithic due to a dramatic increase in human intelligence or the appearance of modern human behavioral capabilities (e.g., Binford 1984; Klein 2000; see also Bird and O'Connell 2006; Marean and Assefa 1999; McBrearty and Brooks 2000; Speth 2004).

In the next paper, D. K. Grayson and F. Delpech explore Upper Paleolithic faunal data from France in order to detect possible temporal shifts in diet breadth. In this case, diet breadth is measured using prey model logic and taxonomic richness (numbers of taxa), rather than the relative abundances of taxa with different return rates as in the approach of Stiner et al. Working with the ungulate fauna from the site of Le Flageolet I, Grayson and Delpech conclude that changes in diet breadth did occur through time in the region. Grayson and Delpech attribute these shifts to changes in environmental conditions, an issue that they explore in greater detail elsewhere using the faunal assemblage from the nearby site of Grotte XVI (Grayson and Delpech 2003, 2005; Grayson et al. 2001). Importantly, the conclusion that Grayson and Delpech reach concerning changing diet breadth at Le Flageolet becomes clear only after detailed consideration of the possible effects of such factors as sample size, bone fragmentation, and body part representation on archaeological measures of diet breadth. They also present an important discussion of time-averaging, which creates differences between archaeological and ethnographic measures of this variable (see also Broughton and Grayson 1993; Jones 2004). In the context of this volume, the greatest contribution of this paper lies in its compelling demonstration of steps that can and must be taken in order to produce robust tests of predictions derived from foraging theory using archaeological data.

New World

In the first paper in this section, G. E. Belovsky develops and tests a model of hunter-gatherer population dynamics that employs the optimization technique of linear programming, an approach not widely used in anthropological applications of Optimal Foraging Theory (OFT; for example, see Kaplan and Hill 1992 for discussion; also see Winterhalder et al. 1988 for another model that addresses many of the same issues). The model that he presents assumes a goal of maximizing nutrient intake, rather than maximizing foraging efficiency as is the case in most foraging theory models. Key constraints in the model involve digestive tract capacity, potential harvest yields and rates for different foods, and the amount of each resource type that must be ingested to satisfy physiological requirements. Also incorporated are explicit conversions of nutritional intake into population growth and the effects of hunting and gathering on animal and plant densities. Belovsky uses the model to generate a set of predictions that are tested against the New World Paleoindian archaeological record. The model predicts (1) a rapid colonization of the New World, (2) colonizing populations that had variable and diverse diets (which include plant resources), and (3) prey population depressions and extinctions that were limited to certain highly productive environments and dictated not by the abundance or productivity of those prey but by the productivity of gathered foods.

The issues that Belovsky addresses continue to resonate in Paleoindian archaeology, and many of them are taken up by the following papers in this section. Belovsky's prediction about a rapid colonization of the Americas foreshadows the main prediction of the paper by R. L. Kelly that is reprinted here, and the paper by D. A. Byers and A. Ugan echoes Belovsky's prediction about Paleoindian diet breadth. Elsewhere, Cannon and Meltzer (2004) have argued that the early Paleoindian archaeofaunal record supports Belovsky's prediction concerning variable diets. It is also important to note that Belovsky's model does not predict the range of extinctions that occurred in North America at the end of the Pleistocene, which supports the argument that humans were not responsible for those extinctions (e.g., Grayson and Meltzer 2003). On the other hand, the predictions of Belovsky's model concerning the context of substantial prey depressions have been confirmed in a variety of examples that date to the late Holocene, many of which are discussed in the following section of this volume (e.g., Bayham 1986; Broughton 2002; Cannon 2003; Nagaoka 2005, all reprinted here; also see Butler and Campbell 2004; Byers and Broughton 2004).

The paper by Kelly, like Belovsky's paper, provides a model that predicts a rapid colonization of the western hemisphere. Kelly's model considers the costs of moving from one resource patch to another in relation to the costs of remaining in a patch that is being depleted of high-return resources, and based on this relationship it predicts the time at which foragers should move to the next patch. In Kelly's application of this model to the colonization of the Americas, he assumes—given the arctic, Beringian origins of the colonists (Clovis or otherwise)—that large-bodied vertebrate prey were the primary type of resource that they used. Naïve about human predators, Kelly argues, local populations of large mammals would have quickly become depressed, and higher expected returns in virgin territories would have rapidly pulled colonists from patch-to-patch through the Americas (see Surovell 2000 for further development). It is important to note that this model, unlike Belovsky's formulation, does not rely upon human population growth and subsequent group fissioning to account for rapid colonization; indeed, Kelly suggests that dramatic population growth may not have been required. Consistent with the apparent rapid spread of Clovis-bearing peoples throughout North America, Kelly's model also accounts for other unique features of Paleoindian archaeology. These include (1) the dominance of bifacial tools with long use-lives, (2) strong stylistic similarity in Clovis points from coast to coast, (3) virtually no use of hard-to-find caves, and (4) minimal evidence for caching or food storage.

The notion that early Paleoindians were specialized hunters of megafauna such as mammoth—a key assumption of Kelly's argument about the colonization of the Americas—has a deep history in North American archaeology (e.g., Meltzer 1993). As implied by the previous papers in this section, this issue has important implications for a range of issues from late Pleistocene extinctions to early Paleoindian mobility and the tempo of New World colonization. In the final paper in this section, Byers and Ugan use foraging theory to evaluate how reasonable it is to expect that Paleoindians were big game specialists. Using the logic of the prey model, a variety of actualistic data on resource return rates, and theoretically derived estimates of encounter rates for mammoths and other North American taxa, they conclude that megafaunal specialization would have been profitable only under a very narrow range of circumstances. While mammoths would appear to have yielded extremely high post-encounter return rates, they would not likely have been encountered frequently enough have to have excluded a wide range of other ungulates and even smaller game from the diet. The conclusions that Byers and Ugan derive provide a good illustration of the insights that can result from well-reasoned application of EE logic, and these conclusions also accord well with a recent assessment of the archaeofaunal record of early Paleoindian prey choice in North America (Cannon and Meltzer 2004). For other views about early Paleoindian subsistence that also draw on foraging theory, see Haynes (2002a, 2002b) and Waguespack and Surovell (2003).

In sum, the five EE-based papers in this section have improved our knowledge and understanding of a number of important issues in the prehistory of the Pleistocene, in both the New and Old Worlds. The more substantive propositions or conclusions of these papers include the following:

1. Human diet breadths were inherently fluid during the Paleolithic, with broad spectrum shifts occurring first during the late Middle Paleolithic.

2. Dramatic increases in human population densities occurred during the late Middle Paleolithic and again during the Upper and Epi-Paleolithic periods.
3. The florescence of complexity in the Upper Paleolithic may stem from the above-mentioned demographic shifts, rather than any sudden change in human cognitive abilities.
4. Rapid colonization of the New World is predicted from several EE models and is consistent with available archaeological evidence.
5. Early Paleoindians were likely not megafaunal "specialists"; rather, their diets were likely diverse with smaller game and plant resources being variably important.
6. The depression of prey animals is dependent on the local abundance and productivity of gathered resources.
7. Paleoindian hunting may not have been a primary cause of Pleistocene extinctions.

References

Binford, L. R.
1984 *Faunal Remains from Klasies River Mouth*. Academic Press, Orlando.

Bird, D. W., J. F. O'Connell
2006 Behavioral Ecology and Archaeology. *Journal of Archaeological Research* 14:143–188.

Broughton, J. M., D. K. Grayson
1993 Diet Breadth, Adaptive Change, and the White Mountains Faunas. *Journal of Archaeological Science* 20:331–336.

Butler, V., S. Campbell
2004 Resource Intensification and Resource Depression in the Pacific Northwest of North America: A Zooarchaeological Review. *Journal of World Prehistory* 18:327–405.

Byers, D. A., J. M. Broughton
2004 Holocene Environmental Change, Artiodactyl Abundances, and Human Hunting Strategies in the Great Basin. *American Antiquity* 69:235–256.

Cannon, M. D., D. J. Meltzer
2004 Early Paleoindian Foraging: Evaluating the Faunal Evidence for Large Mammal Specialization and Regional Variability in Prey Choice. *Quaternary Science Reviews* 23:1955–1987.

Grayson, D. K., F. Delpech
2003 Ungulates and the Middle-to-Upper Paleolithic Transition at Grotte XVI (Dordogne, France). *Journal of Archaeological Science* 30:1633–1648.
2005 Pleistocene Reindeer and Global Warming. *Conservation Biology* 19:557–562.

Grayson, D. K., D. J. Meltzer
2003 A Requiem for North American Overkill. *Journal of Archaeological Science* 30:585–593.

Grayson, D. K., F. Delpech, J.-P. Rigaud, J. F. Simek
2001 Explaining the Development of Dietary Dominance by a Single Ungulate Taxon at Grotte XVI, Dordogne, France. *Journal of Archaeological Science* 28:115–125.

Haynes, G.
2002a *The Early Settlement of North America: The Clovis Era*. Cambridge University Press, Cambridge.
2002b The Catastrophic Extinction of North American Mammoths and Mastodons. *World Archaeology* 33:391–416.

Jones, E. L.
2004 Dietary Evenness, Prey Choice, and Human-Environment Interactions. *Journal of Archaeological Science* 31:307–317.

Kaplan, H. S., K. Hill
1992 The Evolutionary Ecology of Food Acquisition. In *Evolutionary Ecology and Human Behavior*, edited by E. A. Smith and B. Winterhalder, pp. 167–202. Aldine de Gruyter, New York.

Klein, R. G.
2000 Archaeology and the Evolution of Human Behavior. *Evolutionary Anthropology* 9:17–36.

Kuhn, S. L., M. C. Stiner
1998 Middle Paleolithic "Creativity": Reflections on an Oxymoron? In *Creativity in Human Evolution and Prehistory*, edited by S. Mithen, pp. 143–164. Routledge, London.

Marean, C. W., Z. Assefa
1999 Zooarchaeological Evidence for the Faunal Exploitation Behavior of Neandertals and Early Modern Humans. *Evolutionary Anthropology* 7:22–37.

McBrearty, S., A. S. Brooks
2000 The Revolution that Wasn't: A New Interpretation of the Origin of Modern Human Behavior. *Journal of Human Evolution* 39:453–563.

Meltzer, D. J.
1993 Is There a Clovis Adaptation? In *From Kostenki to Clovis: Upper Paleolithic–Paleo-Indian Adaptations*, edited by O. Soffer and N. D. Praslov, pp. 293–310. Plenum Press, New York.

Rogers, A. R.
1995 Genetic Evidence for a Pleistocene Population Explosion. *Evolution* 49:608–615.

Speth, J. D.
2004 News Flash: Negative Evidence Convicts Neanderthals of Gross Mental Incompetence. *World Archaeology* 36:519–526.

Stiner, M. C., N. D. Munro
2002 Approaches to Prehistoric Diet Breadth, Demography, and Prey Ranking Systems in Time and Space. *Journal of Archaeological Method and Theory* 9:181–214.

Stiner, M. C., N. D. Munro, T. A. Surovell
2000 The Tortoise and the Hare: Small Game Use, the Broad-Spectrum Revolution, and Paleolithic Demography. *Current Anthropology* 41:39–73.

Surovell, T. A.
2000 Early Paleoindian Women, Children, Mobility, and Fertility. *American Antiquity* 65:493–508.

Waguespack, N. M., T. A. Surovell
2003 Clovis Hunting Strategies, or How to Make out on Plentiful Resources. *American Antiquity* 68:333–352.

Winterhalder, B., W. Baillargeon, F. Cappelletto, I. R. Daniel Jr., C. Prescott
1988 The Population Ecology of Hunter-Gatherers and their Prey. *Journal of Anthropological Archaeology* 7:289–328.

5

An Optimal Foraging-Based Model of Hunter-Gatherer Population Dynamics

Gary E. Belovsky

INTRODUCTION

Optimal foraging models developed in behavioral ecology have been employed to examine diet selection by human hunter-gatherers with varying degrees of predictive success (see Belovsky 1987, and references therein). The underlying assumption of foraging ecology is that either the nutritional status of an individual determines his/her fitness (survival and reproduction) or the time needed to acquire his/her necessary nutritional requirements determines fitness. In the first case, greater nutritional intake leads to higher fitness, and in the latter case greater time foraging leads to lower fitness because of increased exposure to deleterious environmental factors or reduced time for other fitness-increasing behaviors (e.g., care of young). The former goal is called nutrient maximization and the latter is time minimization (Belovsky 1987).

The original rationale in animal ecology for using these models was to gain insights into how animals perceive potential food resources, so that a food-based definition of carrying capacity could be constructed (MacArthur and Pianka 1966; Emlen 1966, Schoener 1969a, 1969b, 1971); i.e., the models could be used to make more realistic population assessments. While the study of optimal foraging models has become a major area of research with implications for neurobiology and psychology, the original intent of using these models to understand population-level processes has been largely overlooked (Schoener 1987). Only a few studies have actually built population models around optimal foraging models (Werner 1977, Belovsky 1984a, 1984b, 1986, Winterhalder et al. 1988; Abrams 1984; Comins and Hassell 1979; Hassell and May 1973).

In this paper, I use an optimal foraging model developed explicitly for human hunter-gatherers which employs the optimization technique of linear programming (Belovsky 1987) to construct a hunter-gatherer population model. The foraging model appears to predict the observed hunter-gatherer ingestion of meat and vegetable foods in environments of different primary productivities, based upon a foraging goal which maximizes nutrient intake (Belovsky 1987). The foraging model is used to construct a population model based upon the assumption that nutritional intake limits population growth. Included in the population model are explicit conversions of nutritional intake into birth and death rates and the depletion of animal and plant populations by exploitation.

A foraging-based population model for hunter-gatherers can be very useful to anthropologists and archaeologists as a means of constructing and testing hypotheses about the role subsistence plays in human population dynamics and the dynamics of their food populations. For example, some demographers argue that human hunter-gatherer populations are maintained at densities well below the "carrying capacity" of the environment, where "carrying capacity" is defined as the total amount of food that the environment can provide divided by per capita human requirements (see Hassan 1981, and references therein). From many of these analyses, the conclusion is that hunter-gatherer densities are maintained below computed maximum densities ("carrying capacities") and cultural means such as infanticide, abortion, and birth control (e.g., Hassan 1981; Glassow 1978) are invoked to explain their observations.

The above definition of "carrying capacity" is an obsolete usage which has largely been abandoned in ecological studies, since it does not include the feedback between consumer populations and their food populations. All food in the environment cannot be consumed without decreasing its recovery rate or decreasing the ability of the consumer to harvest it in the available feeding time (also see Winterhalder et al. 1988). Consequently, a more realistic definition of carrying capacity might indicate that hunter-gatherers are ultimately food limited and cultural means of reducing population size may represent proximate manifestations of an ultimate food limit.

Also, a population model based upon hunter-gatherer foraging models might enable us to assess more fully the "overkill"

*Originally published in *Journal of Anthropological Archaeology* 7(1988):329–372.

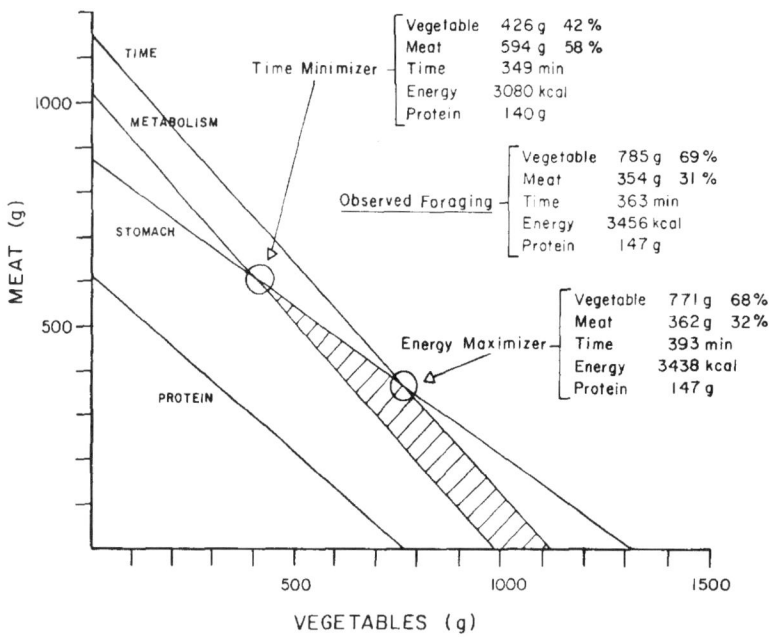

FIGURE 5.1. This example of a linear programming diet model for the !Kung San is taken from Belovsky (1987: Fig. 2). The feeding time (time line), digestive (stomach line), energy (metabolism line), and protein requirement constraints are presented. The cross-hatched region represents the feasible diet combinations of gathered (vegetable) and hunted (meat) foods that satisfy the constraints. The solutions that maximize nutritional intake (energy maximizer) and minimize feeding time (time minimizer) are presented and compared with the observed foraging behavior of these hunter-gatherers. The nutrition-maximizing solution is not statistically different from the observed diet, but the time-minimized solution is (Belovsky 1987). This is the observed pattern for a number of modern hunter-gatherers (Belovsky 1987).

hypothesis for megafaunal extinction at the end of the Pleistocene by more thoroughly examining human population expansion into new environments such as the New World and Australia (Martin 1966, 1967, 1973, 1982, 1984; Mosimann and Martin 1975; Budyko 1967, 1974; Haynes 1982; Whittington and Dyke 1984). In other words, under what conditions might hunter-gatherers deplete their food populations to extinction?

To demonstrate the applicability of the model for hunter-gatherer populations, their observed population dynamics will be shown to be consistent with the model's predictions for environments differing in primary productivity. The model will then be used to examine the severity of exploitation of food resources by hunter-gatherers and the archaeological consequences of their colonization of new areas. Furthermore, the models employed to argue for megafaunal "overkill" will be shown to be biologically unrealistic and structurally unstable in comparison to the model developed here. These preliminary uses of the model, I hope, will indicate its utility for further investigations.

Model Construction

The Diet Model

The hunter-gatherer population model presented here is constructed around a linear programming model for diet choice, where the foraging goal is nutrient maximization (maximum energy and/or protein intake). The foraging model's structure, assumptions, and verification are presented in a previous paper (Belovsky 1987). However, to summarize, the model employs three constraints to hunter-gatherer diet choice:

(1) the amount of each type of food that people can digest given their digestive tract's capacity, the turnover rate of different foods through the digestive tract, and the amount of the digestive tract filled by a unit intake of each food;

(2) the amount of each type of food that people can harvest in some foraging period (e.g., day) which is set by climatic and physiological limits to activity, and the rate at which each food can be harvested given its abundance as well as the time required to prepare it for consumption and make tools to harvest it; and

(3) the amount of each food type that must be ingested by people to satisfy their physiological demands for protein and energy, given the digestible protein and energy content of each food.

Given the above constraints, the combination of food types that best satisfies the nutrient-maximizing or time-minimizing goals can be predicted. From my previous work, hunter-gatherers appear to be nutrient maximizers when two food types are used to define their diets: meat (hunted) and vegetable (gathered) foods. The model is constructed to predict the diet of an average adult in the population (a hypothetical individual doing both male and female work in proportion to each sex's frequency in the population; this accounts for sexual division of labor) and his/her dependents (Belovsky 1987). A graphical analysis of this model is presented in Fig. 5.1; however, the model is fully described elsewhere (Belovsky 1987). This model for human foraging is more appropriate than other optimal foraging models (Belovsky 1987), since it can explicitly include in the constraints:

(1) physiological and behavioral limits to activities,
(2) different environmental conditions (e.g., climate, food abundances),
(3) differences in the spatial distributions of plant and animal foods (Belovsky et al. 1989), and
(4) different methods needed to hunt and gather.

These variations on the model's parameter values can be used to examine the trade-offs that people must confront in making diet choices in different environments.

The Population Model

The solution for the optimal diet can be readily converted into nutritional intake for individuals in a population and, given the hunter-gatherer population size, the diet prediction can be converted into depletion of plant and animal food populations. The human, plant, and animal population densities will then provide feedback to the diet model through changes in the foraging constraint parameters.

The manner in which the diet model changes with changing human, plant, and animal populations is presented graphically in Fig. 5.2. The paper presenting the diet model (Belovsky 1987) provides the necessary constraint parameters and how they change given changes in human, plant, and animal population densities discussed below.

The densities of the animal and/or plant populations impact on only one foraging constraint, the feeding time constraint, by moving it upward (Fig. 5.2a) as more food is available. This occurs because the time needed to crop foods diminishes as more food is available (see below).

Changes in the human population, caused by changes in the number of dependents per foraging adult in the group and/ or total density, act in several ways on the foraging model. As the number of dependents per foraging adult increases, the digestive and energy/protein requirements increase to reflect the added digestive capacity provided by the dependents and their added nutritional requirements (Fig. 5.2b: see Belovsky 1987). In addition, the feeding time constraint moves upward since the dependents aid the foraging adults with some of the food acquisition or preparation, effectively increasing the amount of food cropped per unit time by the adult (Fig. 5.2b: see Belovsky 1987). As the human population density increases only one constraint is modified, feeding time, which moves down-

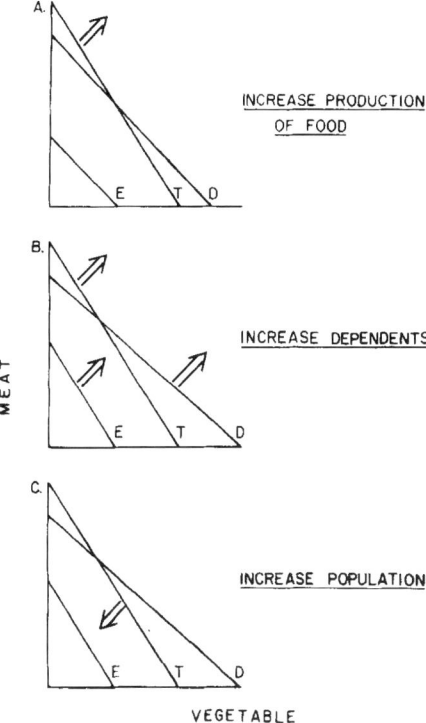

FIGURE 5.2. This figure presents how changes in the population model affect the linear programming diet model's constraint equations. The arrows refer to the expected directional changes in constraint equations with population changes. E is the energy requirement constraint, T is the feeding time constraint, and D is the digestive capacity constraint.

ward because greater human densities lead to depletion of food populations, requiring more time for humans to acquire a unit of food (Fig. 5.2c: see below).

Converting Diet into Nutrition

The optimal foraging model's solution provides an estimate for the intake of meat and vegetable foods measured in mass per unit time (grams dry weight per day) provided by each average foraging adult for consumption by the adult and his/her dependents. It is very easy to convert these values into nutrition per adult forager. The consumption of each food is multiplied by its nutritional value which is already included in the nutritional (protein and energy) constraints (Belovsky 1987) and these products for each food are then summed. The important aspect of nutritional status is not the amount ingested but the amount ingested above the nutritional requirements of the forager and his/her dependents (total intake-requirements). If this value is positive, a surplus is indicated and it can be used for reproduction, while if the value is negative, a shortfall is indicated and it must be eliminated by mortality.

Previous work with the foraging model indicated that the nutritional maximization goals of protein or energy maximization were interchangeable (Belovsky 1987). Therefore, for simplicity the optimal diet model was assumed to include a single nutritional constraint, energy. This makes the population model, including the diet choice model, more tractable.

Converting Nutrition into Births and Deaths

Conversion of nutritional surpluses or shortfalls into births and deaths is not difficult but one must be careful how this is accomplished so population changes are correctly reflected. The correct way to incorporate nutritional intake into a population model requires two parameters (Schoener 1973):

(1) the nutritional requirements of the foragers for maintaining their own body functions plus those of any dependents (survival requirement: M), and
(2) the nutritional requirement to produce a new offspring (reproductive requirement: R), who will be added to the dependents that an individual will be caring for in future time periods.

For humans, M is a measure of the physiological requirements (maintenance metabolism, protein daily allowance, etc.) for the adult plus the requirements for dependents (number of dependents multiplied by R). R represents those nutritional requirements that a parent must supply to each child to enable it to survive from conception to independence; these are averaged over the time period being addressed by the model (e.g., day).

If we assume that children begin to help forage at 13 years of age after 12 years of dependence, during which time they provided only occasional aid in foraging (Hassan 1981; Howell 1976, 1979; Lee 1979), then in a population with a stable age distribution the proportion of children becoming independent (actively foraging) each year would be 1/12. Independence only refers to foraging, not reproductive maturity. If the maximum age for an individual is estimated to be 60 years, where the survivorship curve begins to drop precipitously (Hassan 1981; Howell 1976, 1979), then in a population with a stable age distribution the proportion of adults dying from senescence each year would be $1/(60 - 12)$. The assumption of a stable age distribution is very robust, having little effect upon the population model's predictions.

M is defined as $M_A + R(C_t/A_t)$, where M_A is the adult nutritional requirement and $R(C_t/A_t)$ is the nutritional requirement for dependents. A_t is the number of adults at time t and C_t is the number of children at time t. The value for M_A, adult energy requirement, is 2190 kcal/day; this requirement includes the nutritional needs of the adult (1862 kcal/day: see studies reviewed in Belovsky 1978; Frisancho 1981) plus a sufficient nutritional intake to ensure that an adult living to senescence would leave one surviving offspring (replacement reproduction: 328 kcal/day = 12 years of dependence × R/(60 years to senescence − 12 years of dependence)) (Schoener 1973). The nutritional requirements for dependents (R) is 1312 kcal/day, which is an average for children from conception until 13 years of age (Belovsky 1987; Hassan 1981; Frisancho 1981).

Combining the two measures of nutritional requirements (M and R) with the nutritional intake (I) and the proportions of children becoming adults and adults dying from senescence, the change in the hunter-gatherer population density can be written as:

$$A_{t+1} = C_t/12 + A_t(1 - 1/(60 - 12)), \text{ for } I \geq M,$$
$$C_{t+1} = C_t(1 - 1/12) + A_{t+1}((I - M)/R)$$

and

$$A_{t+1} = A_t(I/M)(1 - 1/(60 - 12)), \text{ for } I < M,$$
$$C_{t+1} = C_t(I/M)(1 - 1/12).$$

Since there is an upper limit to the number of children under 3 years of age (Blurton Jones and Sibley 1978) that a hunter-gatherer woman can carry while she is working, this was accounted for in the model by never permitting more than 0.6 children per adult in the population. This restriction and senescence were the only nonnutritional limits placed on population dynamics, and the limit to carrying young was found to enter into the model's solution only for environments with the highest productivity (see below).

The Hunter-Gatherer's Food Environment

In the original development of the optimal foraging model for hunter-gatherers (Belovsky 1987), the cropping rates for gathered and hunted foods were expressed as functions of environmental primary productivity. These cropping rate relationships are presented in Fig. 5.3. While these relationships were adequate to address how diet choice might change in different global environments (Belovsky 1987), a population model requires inputs on how the food resources are depleted by human utilization, how they recover from exploitation, and what the effects of depletion are on hunter-gatherer cropping rates. Since the diet model examines the consumption of two food types (hunted and gathered foods), how the foraging time constraints might change beyond that discussed in the original paper (Belovsky 1987) will be presented below for each food separately.

Gathered foods (vegetable foods) are the easiest to incorporate in the model since they are a direct function of primary productivity. Most vegetable foods for humans are fruits, tubers, and seeds; these plant parts average about 2% of total primary productivity (Minnis 1985; Harper 1977; McNaughton and Wolfe 1973). Since the original cropping rate function (Belovsky 1987) for gathered foods was based on total primary productivity, the relationship in Fig. 5.3 was modified to include only the production of fruits, seeds, and tubers, 2% of total production. Furthermore, on average, plants are able to replace losses due to consumption at a rate 0.22 times the remaining plants per annum (Harper 1977).

With the above information a logistic equation for fruit, seed, and tuber availability at time $t + 1$ (P_{t+1}) can be written as

$$P_{t+1} = S_t + 0.22S_t((K_P - S_t)/K_P)$$

and

$$S_t = (P_t - G_P - 0.10H_t),$$

where $0.10H_t$ is the consumption of plant parts by animals of H biomass at time t for plant parts that are also used by people

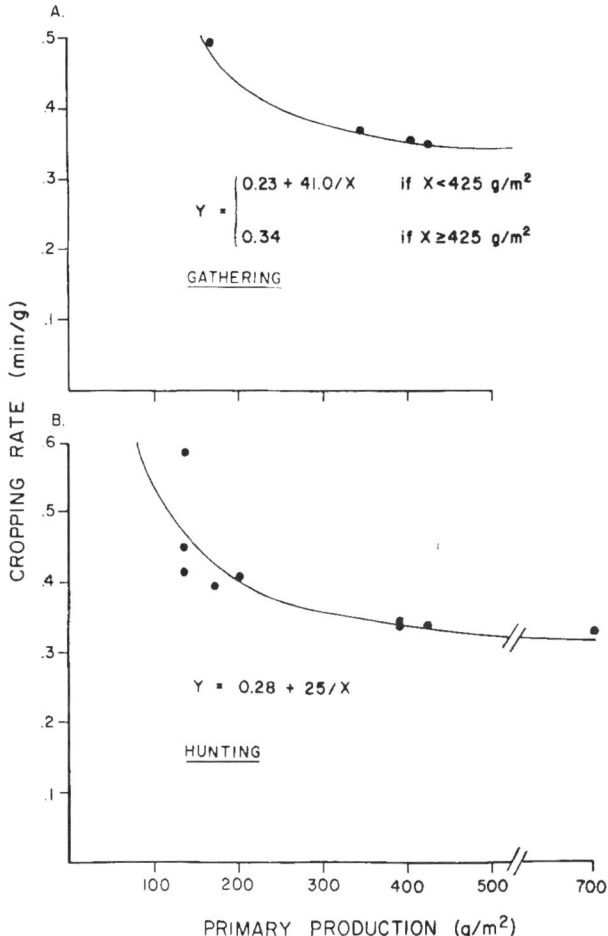
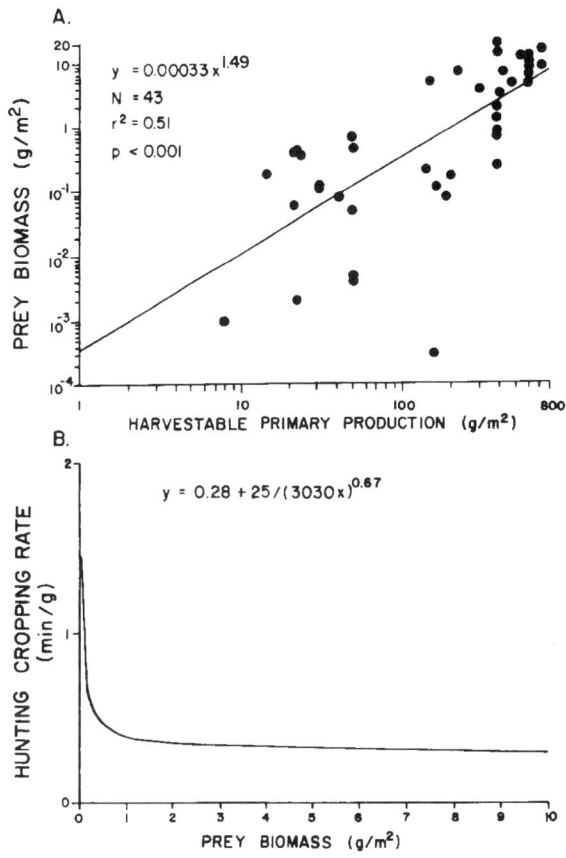

FIGURE 5.3. The observed relationship between the primary productivity in different environments and the cropping rates of hunter-gatherers for gathered (A) and hunted (B) foods is presented (Belovsky 1987). The primary productivity in these plots represents total harvestable production of nonstructural components of the vegetation within 3 m of the ground, so that it would be within reach of humans or their prey animals.

FIGURE 5.4. The relationship between the maximum observed biomass of large prey animals in the environment (>10 kg) and the harvestable primary productivity is presented using data from Farlow (1976) and Belovsky (1986) in Fig. (A). The relationship between the cropping rates for hunted foods in Fig. 5.3b and the biomass of prey animals in the environment is presented in Fig. 5.4b. This is the relationship used in the population model described in the paper.

(Farlow 1976), K_P is the maximum environmental production of seeds, fruit, and tubers, and G_P is the quantity of plant food harvested by humans. K_P is 0.02 times the maximum environmental primary productivity (P). G_P can be written as

$$G_P = 365 A_t (4.2)(I_{P,t})/10^8,$$

where 4.2 is a conversion factor relating the mass of plants harvested to acquire a unit of plant food ingested (Lee 1979; Hawkes et al. 1982), $I_{P,t}$ is the quantity of ingested plant food provided by each forager (adult) at time period t predicted from the foraging model, 365 converts daily consumption to annual consumption, and 10^8 converts human consumption into the appropriate units to compare with primary production (grams dry weight per year per 100 km² into grams dry weight per year per m²).

The seeds, fruits, tubers, etc., eaten by people cannot be totally eliminated from the environment, unlike the hunted animals, because the plant itself is not eliminated by human consumption, guaranteeing recovery (sensu Caughley and Lawton 1981, noninteractive-reactive herbivory). Therefore, the plant biomass used by people was never allowed to fall below 0.0043 times maximum environmental primary productivity (0.02 edible plant parts × 0.22 plant biomass replaced each time period × 0.98 inedible plant parts).

Hunted food (meat) cannot be included in the population model as easily, since the populations of animal prey will be a function of human exploitation and plant primary productivity. If we restrict hunting to larger species of mammals (>10 kg), ecological studies demonstrate that the maximum environmental biomass of these prey animals is a function of plant primary productivity (Fig. 5.4b); however, hunting success will vary with the abundance of prey, not its maximum potential abundance. Since the cropping rates for meat were originally developed as a relationship with plant primary productivity (Fig. 5.3) (Belovsky 1987), it is necessary to express these cropping rates in terms of prey biomass rather than plant primary productivity. Using the relationship in Fig. 5.4a, a simple conversion provides the new function for the cropping rate of meat (Fig. 5.4b).

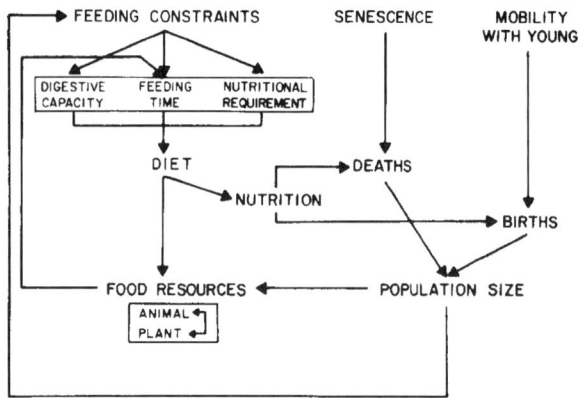

FIGURE 5.5. A flow chart for the population model is presented with the arrows representing the linkages between different aspects of the hunter-gatherer population and food resources. Separate sub-models are represented in the chart as boxes: the linear programming foraging model constraints are enclosed in one box, and the plant and animal population dynamics are enclosed in another box.

With the above information the availability of animals for hunting at time $t + 1$ (biomass: H_{t+1}) can be written as
$$H_{t+1} = 1.19\,(H_t - H_{H,t}),$$
if $H_{t+1} \leq H$, otherwise
$$H_{t+1} = H$$
or
$$H_{t+1} = 0, \text{ if } H_{t+1} \text{ at any time equals or falls below 0.}$$

H is the maximum animal biomass that can be supported by the available primary production after human consumption of plants,
$$H = K_H(50K_P - (K_P - P_{t+1}))/50K_P,$$
where K_H is the maximum animal biomass given maximum environmental primary production ($50K_P$) and the relationship in Fig. 5.4a. $H_{H,t}$ is the harvest of animal biomass by each adult forager at time t and the constant 1.19 is the recovery rate of the animal populations after consumption by predators other than humans (Farlow 1976).

$H_{H,t}$ can be written as
$$H_{H,t} = 365\,A_t(6.7)(I_{H,t})/10^8.$$

$I_{H,t}$ is the edible meat provided by each adult forager in the population based on the foraging model. The constant, 6.7, represents the amount of prey that must be killed to provide a unit of ingested meat, given inedible portions, spoilage, consumption by dogs, etc. (Lee 1979; Budyko 1967, 1974; Martin 1973; Mosimann and Martin 1975); 365 converts daily to annual consumption, and 10^8 converts consumption per 100 km^2 to m^2.

Environmental primary production is assumed to be the maximum plant production that the environment can provide before human consumption; i.e., it is the annual plant carrying capacity or maximum biomass that can be produced each year in the environment and harvested by humans or their prey. The distinction between total and harvestable primary production is discussed in Belovsky (1987). This distinction is critical in distinguishing between grassland and forest environments, since much of the primary production in a forest cannot be harvested because (1) it is above the reach of humans and/or their prey, and/or (2) the plant structural products are inedible (e.g., wood, bark, etc.). The maximum animal biomass is also set by this plant production prior to human exploitation (Fig. 5.4a).

As pointed out above, environmental productivity sets upper limits to the human food base. These are not estimates of human carrying capacity, since hunter-gatherers will deplete the standing crop biomass of foods and reduce the productivity below that which the environment might maximally provide. Therefore, the human carrying capacity, although a function of maximum environmental productivity, will usually be lower than these values imply due to depletion by harvesting.

Solving the Foraging and Population Models

The combined foraging and population models are summarized in Fig. 5.5 as a flowchart, indicating the interplay between variables.

The linear programming foraging model can be solved using the Simplex algorithm (Belovsky 1987). The difficulty of incorporating the foraging behavior model into a population framework involves the large number of times it must be solved, once for each time period (year). Not only must the foraging model be solved, but the changing nutritional effects on population dynamics for each time period (year) must be calculated.

To keep track of the population and foraging parameter changes over time, the model was built into a Lotus 1-2-3 spreadsheet as a large macro. The spreadsheet organizes and keeps track of the parameter changes. The spreadsheet macro requests the input of necessary initial model parameters, accumulates the changes in model parameters over time in a tabular form, and at the conclusion provides the population changes as graphical output. This macro is presented in Appendix 1 in the original publication with a list of all the variables, definitions and equations.

The Lotus spreadsheet macro does not include the complete model since it does not solve the linear programming foraging model. The foraging model is solved for the energy-maximizing diet using a second program, What's Best!. What's Best! is a program that solves linear programming problems using the Simplex algorithm within Lotus spreadsheets. The interface between these two programs was automated using a third program, Superkey.

While the biomass of food in the model is expressed in grams per square meter, the human population is presented as individuals/100 km^2. The model was started with one adult male-female pair entering a previously unexploited environment (i.e., no other humans are present). Finally, the harvestable primary productivity was assigned within the range observed in nature, 50–800 g/m^2. The model was run for more than 300 iterations (years) and no more than 600 iterations, to determine if and when the population trajectory stabilized. A variety of values for M, R, and other model parameters were used to assess the model's sensitivity to changes in parameter values. These parameter changes were restricted to the range reported in the literature.

THE MODEL'S PREDICTIONS

The model's general characteristics can be examined (Fig. 5.6) using a harvestable primary productivity of 200 g/m² as an example, since this productivity is near the average for the environments that hunter-gatherers inhabit today. Starting with two adults the population rapidly reaches a maximum per annum growth rate which depends upon attaining a stable age distribution (children/adult: C_t/A_t). For a maximum harvestable primary productivity of 200 g/m², the maximum growth rate is approximately 1.3% per annum which doubles the population every 53 years. When the population reaches a density of 12.5 individuals/100 km² (including children), the growth rate begins to decline and eventually falls below zero, indicating that the population is declining (Fig. 5.6).

If the population is examined over a longer time period (Fig. 5.7b), it goes through periods of repeated increases and declines which are called stable limit cycles. Stable limit cycles are population trajectories that if perturbed (e.g., artificially increased or decreased), will return to the same trajectory they were following before the perturbation. The period of these cycles (time between successive peaks) is approximately 90 years (Fig. 5.7b).

Even though population growth rates, the time before first population decline, and the period between subsequent population cycles vary with model parameters, a general pattern of hunter-gatherer stable limit cycles emerges (Figs 5.7a–d). Therefore, using a foraging model shown to predict hunter-gatherer diets (Belovsky 1987) and observed nutritional and environmental parameters for these people, a much more restricted range of population dynamics than predicted by Winterhalder et al. (1988) is obtained. The greater specificity of the model presented here also permits comparison of predicted and observed hunter-gatherer demographic parameters.

Before comparing the model's predictions with observed populations, general characteristics of the predicted population dynamics must be examined. Stable limit cycles are found to arise at all primary productivity levels. However, the intensity of the cycles, as measured by the ratio of population peaks to lows and the time between successive peaks in population size, varies with primary productivity (Figs. 5.7a–d). First, the limit cycles increase in severity as harvestable primary productivity increases from the lowest values where a population of hunter-gatherers can persist (50–200 g/m²). However, as productivity continues to increase, limit cycle severity decreases drastically and then once again increases dramatically (200–800 g/m²).

The changes in stable limit cycle severity depend on two factors: (1) the potential growth rate (predicted maximum) of the hunter-gatherer population, and (2) the predicted optimal diet

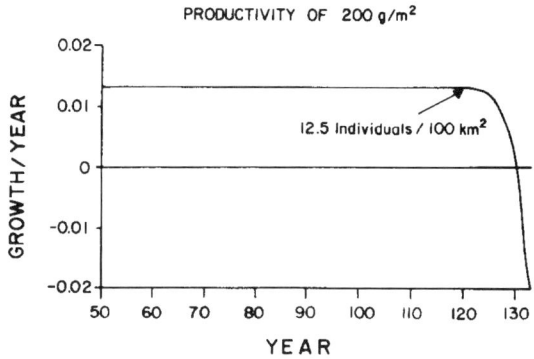

FIGURE 5.6. An example of the model's solution is provided for an environment with a 200 g/m² harvestable primary productivity. Changes in the population's growth rate over time are presented.

for these people. If the hunter-gatherers' potential population growth rate is high, their population can increase rapidly and deplete foods, possibly causing a limit cycle. If the diet can be switched from predominately one food to an alternate when one is depleted, population fluctuations will be reduced even though hunter-gatherer potential population growth rates are high and food depletion periodically occurs.

The least severe limit cycles occur at intermediate primary productivities where hunter-gatherers have a predicted diet more nearly equal in meat and vegetable consumption (Fig. 5.10), even though their population will have a high potential growth rate. The lowest harvestable primary productivities at which hunter-gatherer populations can persist produce stable limit cycles which are not the most severe, even though a diet composed only of hunted foods is predicted. Here the people have little impact on food abundance (see below) and have a low potential population growth rate. However, more severe oscillations occur at slightly higher productivities and at the highest primary productivities that also have a predicted diet composed primarily of one food. In these environments sufficient primary production occurs to maintain a high potential population growth rate.

The parameter changes that have the greatest effect on the model's predictions are the age at which children are considered to be adults, the adult and child nutritional requirements and the harvestable primary productivity. In the model, the age at which a child becomes an adult is a reference to his/her ability to forage independently, not a reference to reproductive ability. Dependents only occasionally help in food preparation and acquisition, while adults are active independent foragers. Changes in this age affect population growth rate and limit cycle periodicity, but have almost no effect on peak and low population densities in the stable limit cycle. By affecting population growth rates, this value determines the minimum harvestable primary productivity that hunter-gatherers require to maintain a viable population (a positive population size over time).

Changes in adult and child nutritional requirements and harvestable primary productivity affect the population growth

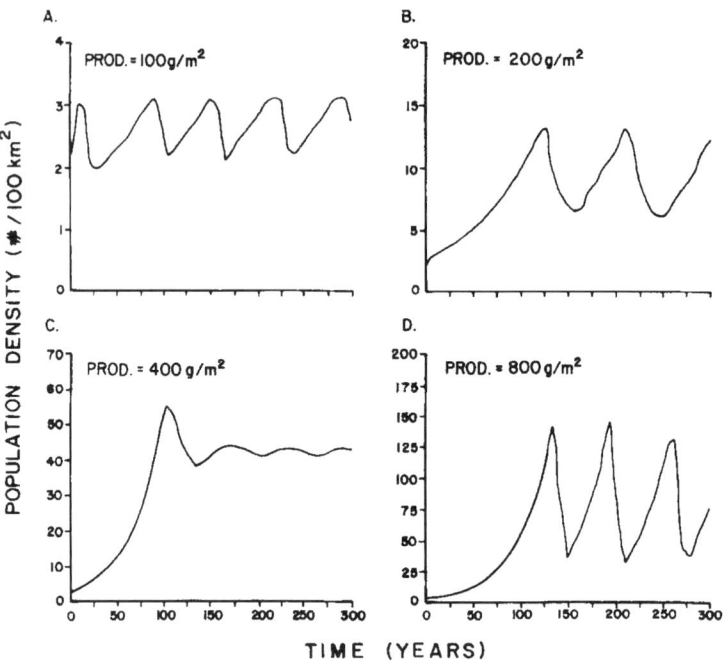

FIGURE 5.7. The population trajectories emerging at four different harvestable primary productivities are presented: 100 g/m² (A), 200 g/m² (B), 400 g/m² (C), and 800 g/m² (D). In all cases a stable limit cycle emerges; however, the severity of this limit cycle, as measured by the ratio of peaks to lows and the length of the period between peaks is greater at low and high harvestable primary productivities.

rates and the peak and low densities in the stable limit cycles. Decreases in nutritional requirements and/or increases in harvestable primary productivity increase population growth rates, up to the limit set by a woman's ability to carry infants and still work (see above). These same changes also result in greater population densities at peaks and lows in the stable limit cycle. Equal relative increases in harvestable primary productivity and declines in nutritional requirements lead to equivalent changes in population parameters over time.

Greater harvestable primary productivity and lower nutritional requirements do not yield a 1:1 change in population density; i.e., doubling primary productivity does not double the peak or low hunter-gatherer density. This pattern was also predicted by Winterhalder et al.'s (1988) model. Therefore, many of the earlier anthropological and archaeological attempts to determine whether or not human populations are near their environmental carrying capacity were conceptually flawed since they assumed a 1:1 relationship between food abundance and human density (Hassan 1981; Glassow 1978).

The model's sensitivity to other cropping rate changes will be discussed below in relation to the ways hunter-gatherers might adapt to different environments.

Comparing the Model with Hunter-Gatherer Demography

The model presented here is based on the idea that hunter-gatherer populations are simply a function of nutrition and nutrition depends upon the availability of food. Food availability is modeled as a function of the environment's maximum harvestable primary productivity and the level of exploitation by people. The only factors in the model that are not related to nutrition are the age at senescence, age at independence, and the maximum number of children a woman can carry and still work. Therefore, a correspondence between observed hunter-gatherer demographic parameters and those predicted by the model leaves nutrition as the most parsimonious explanation. This would indicate that cultural explanations of demography, such as abortion, infanticide, and euthanasia (see below), are proximate regulating factors and hunter-gatherer populations are ultimately limited by their environments, as are most other animals.

Three demographic parameters for hunter-gatherers will be compared with the model's predictions. The first parameter is the average densities of hunter-gatherers observed in environments of different harvestable primary productivities. This is the most important comparison since a failure to find a correspondence would indicate that these human populations are not nutritionally limited. The other two parameters are maximum population growth rates and life expectancy. These two parameters are not as important to the question of what limits hunter-gatherer populations since a population might be limited in its average density by nutrition, while cultural traits might change the population growth rates and life expectancies to reduce population oscillations. For example, the oscillations of the stable limit cycles might be damped leading to a constant density through modified birth and death rates.

Observed hunter-gatherer population densities at different harvestable primary productivities were found using the literature reviews of Hassan (1981) and Hayden (1981) (40 hunter-gatherer groups). The densities for people using marine foods or anadromous fish were not included (14 cases) since the foraging model was not developed using these foods, and estimates of marine productivity are not as easy to derive as productivity estimates for terrestrial environments. The remaining 26 hunter-gatherer peoples were located on a map showing global maximum primary productivities (Leith 1975) and these values were converted into harvestable quantities (Belovsky 1987). The maximum harvestable primary productivity values and the hunter-gatherer densities were compared using linear regression (Fig. 5.8a), and a highly significant correlation was found ($r^2 = .82$, $N = 26$, $p < .001$). Therefore, hunter-gatherer densities are related to food availability.

To examine this relationship more closely, the regression of the model's predicted hunter-gatherer densities versus harvestable primary productivities was compared with the regression line for the observed values. The regression line for the model's predictions (Fig. 5.8a) was not statistically different from the regression line for the observed data (ANOVA: $F = 1.14$, $df = 1$, N.S.; ANCOVA: $F = 1.47$, $df = 1$, N.S.). Furthermore, the model's predictions for average population densities during the stable limit cycles match the observed densities very well ($r^2 = .85$, $N = 26$, $p < .001$), indicating that hunter-gatherers are probably food-limited and the model represents the relationship rather well.

The model's solutions also indicate that the severity of stable limit cycles should increase, then decline precipitously to a minimum, and then increase again as harvestable primary productivity increases (see above). Given the limited population data on hunter-gatherers, this seems to be observed (Fig. 5.8b). Therefore, the population model, although not predicting exactly the observed hunter-gatherer population trajectories, does portray the general pattern very well.

Observed mortality and growth rates for hunter-gatherers can also be compared with the model's predictions. Reports of hunter-gatherer maximum population growth rates vary between 0.3 and 4% per annum (Hassan 1975, 1981; Birdsell 1957, 1968; Howell 1979; Angel 1975; Martin 1973; Mosimann and Martin 1975). The model provides maximum growth rate estimates varying between 0.8 and 2.9%, depending upon harvestable primary productivities increasing as primary productivity increases. Therefore, observed maximum population growth rates and the model's estimates are comparable.

Using data on mortality rates from present day hunter-gatherer populations and skeletal remains from prehistoric *Homo sapiens* hunter-gatherers (Hassan 1975, 1981; Wobst 1974;

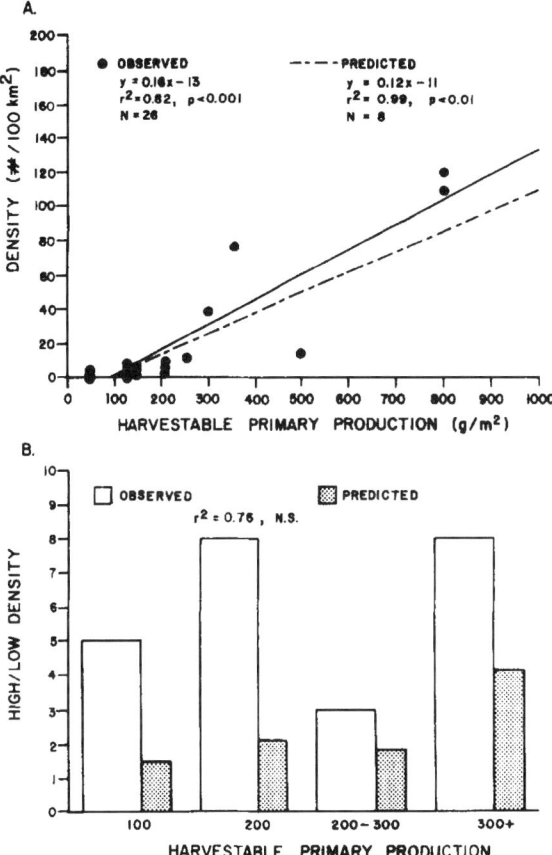

FIGURE 5.8. The observed relationship between the densities of modern hunter-gatherers and their harvestable primary productivity is compared with the model's predicted relationship (A). The observed variation in modern hunter-gatherer densities, ratio of highest to lowest densities reported for environments with comparable primary productivities, but from different studies, are compared with the model's predicted ratio (B). A better comparison could be made if the same hunter-gatherer populations were studied over time, but these data are not available.

Howell 1979; Angel 1975; Deevey 1960), the life expectancy for adults can be estimated to be 41.4 ± 5.5 years ($N = 7$) and 34.1 ± 4.7 years ($N = 16$), respectively. The model at different primary productivities provides an average life expectancy of 38.7 ± 6.6 years ($N = 5$). This estimate is not significantly different from the observed values (Modern: $t = 0.73$, $df = 9$, N.S.; Prehistoric: $t = 1.74$, $df = 19$, N.S.).

The data on growth and mortality rates further support the idea that hunter-gatherer population dynamics are a function of nutrition, rather than cultural restrictions. This means that these people are limited by their environment in much the same way as any other animal population. While not exactly predicting the demographic parameters, the model does very well; perhaps further additions to the model for the degree of nomadic vs. sedentary existence, etc., would explain the observed variations. Nonetheless, the overall agreement between the observed

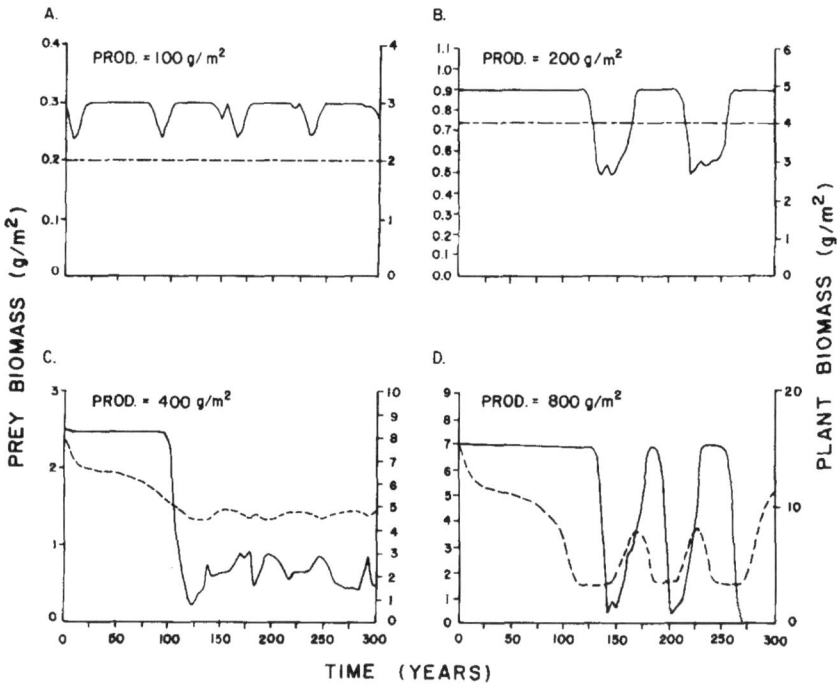

FIGURE 5.9. The model's predictions for changes in hunted and gathered food abundances over time for different harvestable primary productivities are presented: 100 g/m² (A), 200 g/m² (B), 400 g/m² (C), and 800 g/m² (D). The results of these simulations indicate that hunter-gatherers decrease their hunted foods (solid line) to a greater extent (abundance in the stable limit cycle/the maximum initial abundance) than their gathered foods (dashed line) as the harvestable primary productivity increases.

and predicted population densities and parameters raises questions concerning the hunter-gatherers' impacts on food resources.

Hunter-Gatherers' Impacts upon Food Resources

While the impact of the foraging environment on hunter-gatherer demography has been addressed above, there is a reciprocal question: the human impact on the hunted and gathered foods over time. Traditionally anthropologists and archaeologists have posed this question using two very different perspectives: human destruction of the environment (Martin 1973, 1982, 1984; Mosimann and Martin 1975; Budyko 1967, 1974; Whittington and Dyke 1984) or human cultural traits that prevent overexploitation or lead to maximum sustained yield from the environment (Webster 1981; Webster and Webster 1984; Hames 1987; Hayden 1981). What insights are provided by a population model based upon explicit foraging decisions?

First, the models previously built on the assumption of human overexploitation of foods possess some strange characteristics which have been ignored. It has been claimed that these models provide a stable coexistence between the human populations and their hunted foods with certain parameter values (Martin 1973; Mosimann and Martin 1975; Whittington and Dyke 1984), but the characteristics of this coexistence have never been examined. To address this problem, mathematical models must be subjected to a stability analysis which examines the eigenvalues. Eigenvalues provide information on how a population changes when slightly perturbed away from an equilibrium between the interacting populations (May 1973).

A stability analysis of the overexploitation models is presented under Appendix 5.1. Contrary to claims that humans and their prey can achieve stable coexistence in these models with certain parameters, it was found that this equilibrium is unstable and humans will always hunt their prey to extinction. These models are even more unstable than suggested by the eigenvalues. The stability analysis was conducted using differential equations (continuous changes over time) while the original models used difference equations (changes over discrete time) which are inherently less stable (May 1973).

The main reason for this instability is that none of the overexploitation models set the human population's growth rate or equilibrium density (carrying capacity) as functions of prey availability. Therefore, all of these models are constructed without a direct feedback between human demography and food resources. Even by varying hunting success with prey abundance (Mosimann and Martin 1975), these models cannot provide coexistence between humans and their prey.

On the other hand, the model presented here is too complex to allow a stability analysis using eigenvalues. The simulations, however, show that the model can provide stable coexistence or overexploitation depending upon the parameter values (Fig. 5.9). Similar results have been provided by a human population model based upon a different foraging model (Winterhalder et al. 1988). The potential for stable coexistence in these models depends upon the direct linkage between human demography and food availability.

The model presented here will be used to examine under what conditions hunter-gatherers overexploit their gathered and hunted foods. Changes in the abundances of hunted and gathered foods for environments of different primary productivities are presented in Fig. 5.9. Two results are surprising (Fig. 5.9):

(1) at low harvestable primary productivities, hunter-gatherers reduce their hunted food populations but do not cause their extinction, and the hunter-gatherers have no effect upon their gathered foods; and
(2) at high harvestable primary productivities, hunter-gatherers cause the extinction of their hunted foods and reduce the abundance of the gathered foods.

The key to the overexploitation and extinction of hunted foods is the abundance of gathered foods. A greater abundance of gathered foods allows larger human populations to be maintained when hunted foods are rare; this results in the demise of the hunted foods.

The results from the model presented here are counter to the predictions from overexploitation models, where prey extinction is more likely at low primary productivities, not high productivities as predicted here. Also, the key to the demise of hunted foods is not their productivity or abundance, but the productivity of gathered foods. These results depend upon the foraging model's prediction that hunted foods are generally more preferred than gathered foods (Belovsky 1987) and gathered foods increase in the diet as hunted foods are reduced in abundance. If the gathered foods are abundant enough to maintain the humans at high densities, the exploitation of hunted foods will remain high even if the hunted foods are reduced.

Analysis of arguments that hunter-gatherers culturally prevent overexploitation or manage their food resources for maximum sustained yield is more difficult to examine (Hames 1987). Various cultural restrictions to food exploitation are known (Webster 1981; Webster and Webster 1984), but their significance for human and food population dynamics is not known. In a recent study, Vickers (1988) argues that the movement of some South American Indian villages cannot be explained by the depletion of prey populations, since game populations are not depleted. These people, however, are not hunter-gatherers, but agriculturalists who also hunt. To test the idea that hunter-gatherers culturally prevent overexploitation of their foods will require far more detailed data on the dynamics of food utilization (Hames 1987). These data can then be compared with the predictions of a model (like that presented here); a failure to find agreement might indicate cultural restrictions to exploitation.

However, there already is some indication that hunter-gatherers do not prevent overexploitation of their foods by cultural means. First, there are anecdotal accounts of hunter-gatherers reducing the abundances of their foods (Webster 1981; Webster and Webster 1984). Second, the observed densities of hunter-gatherers, as presented above, are comparable to those predicted by the model (Fig. 5.8a). If these people were reducing their exploitation or attempting to manage their food for maximum sustained yield, the observed hunter-gatherer average densities might be higher and their population variation would be much lower than predicted by the model. Therefore, the very limited data on hunter-gatherer demography are not consistent with culturally reduced food exploitation.

Hunter-Gatherer Dietary Changes during Population Growth

Because hunter-gatherers have changing densities and food abundances over time, we should expect their diets also to change over time. These dietary changes never end since the hunter-gatherer populations are predicted to have a stable limit cycle. This means that the diet will also have a stable cycle as long as one food is not overexploited to extinction.

More important are the patterns in diet change that occur during the growth of the human population from an initial low density to its stable limit cycle. The predicted average diets at different stages of population growth in environments with different harvestable primary productivities can be compared. Three stages of population growth can be denoted: the initial colonizing population with few dependents (initial: until a stable age distribution is achieved); the period of population establishment (growth: from the attainment of a stable age distribution until the stable limit cycle is achieved); and the stable limit cycle (established).

The average diets at each stage of population growth in environments of different harvestable primary productivities are presented in Fig. 5.10. At all harvestable primary productivities except 100 g/m², the initial stages of population growth are marked by a diet that is lower in hunted foods than during the next period. The diet in the initial phase is lower in hunted foods because the diet model constraints are based on a low ratio of dependents to adults; while a constant higher ratio occurs in the growth phase. Once the population becomes established within its stable limit cycle, the proportion of hunted foods decreases in the diet, unless the hunted foods have previously been overexploited to extinction. The diet changes for established populations arise with modifications in the foraging model's constraints due to:

(1) a lower dependent to adult ratio as the population stops its continual growth and enters a stable limit cycle, and
(2) the reduction in the abundance of hunted foods to a lower average level within the stable limit cycle.

An Archaeological Example of the Model's Utility

The model provides a set of predictions about changes in hunter-gatherer population densities, food abundances, and diet composition in different environments. These critical

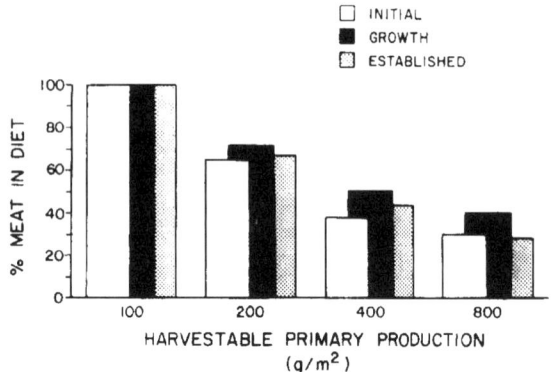

FIGURE 5.10. Dietary changes in environments with different harvestable primary productivities and at different stages of the population growth trajectory are presented. Three stages of the population growth trajectory are denoted: initial (prior to establishment of a stable age distribution), growth (stable age distribution prior to achieving a stable limit cycle), and established (a stable limit cycle). The model predicts that the diets at a given harvestable primary productivity are lower or equal in hunted foods during the initial and established phases than during the growth phase.

aspects affect how people have changed their lifeways over time. As an example, the model was used to examine the colonization of the New World by Paleo-Indians. There are several reasons for choosing this example. It is a topic widely discussed due to the claim that this colonization by humans may have caused the extinction of North American Pleistocene megafauna (Fiedel 1987; Butzer 1971). Also, the "semi-controlled" nature of observations on cultural development in a previously uninhabited land provides an ideal data set to compare with the model's predictions, since complications of past human influences on the environment are eliminated. Finally, it is an example for which other models of hunter-gatherer population dynamics, in particular models of overexploitation of hunted foods, have been developed (Martin 1973, 1982, 1984; Mosimann and Martin 1975; Whittington and Dyke 1984; Budyko 1967, 1974).

The model presented here, or any model, cannot be used to prove or disprove a particular view. The model can be used as a hypothesis against which observations can be compared to see whether or not the hypothesis (model) can be falsified. Therefore, the model may raise more questions than provide answers, which can be as useful.

The New World environment must be described at the time of first human occupation. Although there still is debate over when humans first colonized the New World, there is a growing concensus based upon a variety of lines of evidence that the first colonizations occurred between 15 and 12 thousand years ago (Fiedel 1987; Lewin 1987). At this time, the approximate habitable area of North and South America would have been 9×10^6 and 17×10^6 km², respectively, due to the effects of glaciation (Fiedel 1987; Martin 1973; Mosimann and Martin 1975).

The New World can be dissected into vegetation communities as reconstructed by paleobotanists and paleozoologists (e.g., Jennings 1983; McDonald 1984) and a primary productivity level can be assigned to each vegetation type (e.g., Whittaker 1970). It is important to remember that the primary productivity values must reflect the primary productivity harvestable by humans or their prey animals (Belovsky 1987). This value is much smaller than the total primary productivity in forested regions and these modifications in total primary productivity are accounted for using figures provided in Belovsky (1987).

The hunter-gatherer population model can be solved for each vegetation type. However, rather than tracing the growth of the human population from vegetation type to vegetation type across the two continents, an average primary productivity was used. An average harvestable primary productivity of approximately 200 g/m² in the New World was estimated.

Before examining the spread of hunter-gatherers across the two continents, a simple set of rules must be established on how humans disperse into uninhabited areas. This dispersal can be defined using ideas from animal ecology. "Patch selection" theory (Charnov 1976; Stephens and Krebs 1986) argues that an animal should leave a patch of habitat when its fitness (survival and reproduction) falls below what could be attained on average in the environment. This theory is related to another ecological theory, "free distribution," in which the "patch selection" criteria would lead to individuals in a population being distributed in different patches at densities that would provide equal fitness (Fretwell and Lucas 1970).

Using these ecological theories and the predictions for hunter-gatherer populations with an average productivity of 200 g/m² (Fig. 5.6), human dispersal can be examined. Once a stable age distribution is attained, the human population will grow at a constant high rate until the population density peaks and enters the stable limit cycle phase (Fig. 5.6). Once the population enters the stable limit cycle, the growth rate declines precipitously (Fig. 5.6). Therefore, a population starting from an initial small number of colonists (e.g., a male-female pair) will grow to the density where the population growth begins to decline (approximately 12.5 individuals/100 km²). At this point additional individuals will attain higher fitness by dispersing into uninhabited areas, if available, rather than staying within the original population.

The outcome of the dispersal process is maintenance of the original population at the maximum density where the growth rate also is maximum (approximately 12.5 individuals/100 km²), and establishment of new populations by the surplus individuals produced. The new populations in turn will grow until they reach a density of 12.5 individuals/100 km² and then their surplus individuals will begin to disperse. The time required for a population to reach the dispersal phase, where the only influx of dispersing individuals to each population is the original colonizing pair, is approximately 120 years. When all areas are inhabited by human populations at a density of 12.5/100 km², then the populations will continue to grow and enter the stable limit cycle phase.

Predictions about the colonization of the New World can be made and compared with the predictions of other models developed to examine the overexploitation of prey animals (Fig. 5.11). As in the overexploitation models, the Paleo-Indians are assumed to have arrived in the vicinity of Edmonton, Canada at approximately 11,500 BP. The model presented here and the overexploitation models provide comparable time frames for colonization, a wave-like movement of colonizers (a colonizing front), and comparable densities when the continents are fully populated (Fig. 5.11). Therefore, there is no difference in terms of the general picture of colonization by Paleo-Indians of the New World provided by the model presented here or models for overexploitation of prey (Haynes 1970, 1982; Martin 1973, 1982; Mosimann and Martin 1975).

Whether or not a widespread overexploitation and extinction of the large prey animals occurs with this colonization is a main difference generated by the model presented here and those developed to examine overexploitation. The model presented here does not predict a general demise of large prey animals (Fig. 5.10). However, it is possible that the extinction of prey by overexploitation might occur in some of the environments composing the hypothetical 200 g/m² average productivity.

As presented above, prey are predicted to go extinct by overexploitation in the more productive environments (>500 g/m²); these areas would not include forested areas but could include the most productive savanna-like habitats of the Mexican highlands and tallgrass prairie (Whittaker 1970; Leith 1975). Another habitat where prey extinction might occur by overexploitation would be the very productive areas in river floodplains. Nonetheless, these habitats where megafaunal overexploitation and extinction might occur are not widespread enough to account for the range of observed extinctions at the end of the Pleistocene.

Furthermore, even the highest estimates of plant productivity for areas with the highest observed megafaunal extinctions are not large enough for the model presented here to predict that hunter-gatherers would overexploit their prey. These areas include the tundra bordering the glaciated areas and the Beringian grasslands (Redmann 1982; Schweger 1982; Bliss and Richards 1982; Hibbert 1982). Therefore, the observed megafaunal extinctions cannot be accounted for by the model presented here, suggesting that factors other than overexploitation (e.g., climatic changes) might be the cause (Wesler 1981; Guthrie 1982, 1984; Grayson 1984).

The model's predictions for megafaunal extinction might be criticized since the large prey animals (hunted foods) are treated as a composite group. This means that a single or a few species (e.g., mammoths and mastodons) might be overexploited by hunter-gatherers and this distinction might not be identified

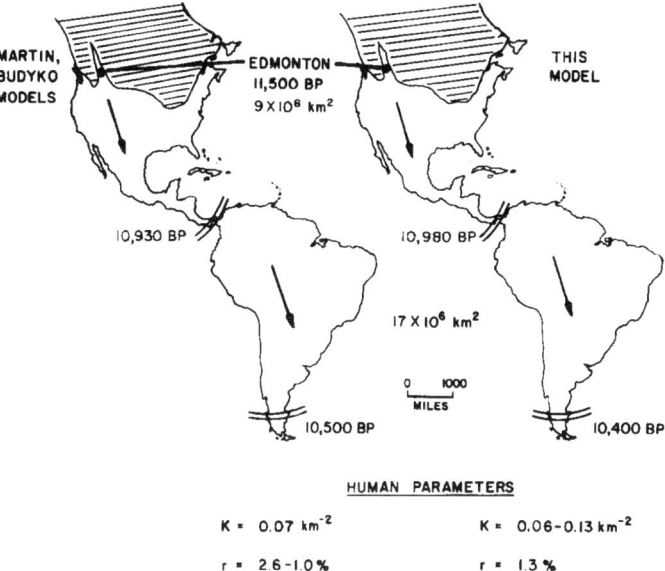

FIGURE 5.11. Predictions for the chronology of colonization in the New World by Paleo-Indians are compared for the model presented here and the models developed to examine megafaunal overkill. (K) refers to human carrying capacity and (r) is the maximum population growth rate.

by the model. The models of overexploitation (see references above) also fail to make this distinction. This, however, is a moot point since the large-scale megafaunal extinction at the end of the Pleistocene in the New World was well beyond one or a few species going extinct at each location (Martin 1984).

One aspect of megafaunal extinction cannot be addressed by the model presented here. If hunter-gatherers overexploit the largest prey (e.g., mammoths, mastodons, etc.) and the prey's demise eliminates their effect on the habitat (disturbance) which in turn reduces food production for other megafauna, then additional extinctions might occur (Owen-Smith 1987). Although I do not place much stock in this hypothesis since the largest megafauna do not seem to have been very abundant in the New World (Guthrie 1968, 1982, 1984) and this hypothesis is not consistent with some observations on modern megafaunal communities (Belovsky in preparation), this is a possibility that cannot be discarded.

While the model of hunter-gatherer populations presented here is not consistent with claims of megafaunal overexploitation, there are other provocative predictions that the model provides. The first deals with the archaeological evidence for diet changes and the onset of the Archaic in the New World, and the second deals with the adoption of new technologies for food acquisition.

Dietary changes for Paleo-Indians as indicated in the archaeological record of the New World pose two problems. Some Paleo-Indian sites of comparable age in a locale indicate a diet based on a more equitable harvesting of hunted and gathered foods and a tool kit reflecting this more diverse diet, while other sites in the same locale and of comparable age indicate a diet dominated by hunted foods and a tool kit for big-game hunting (Fiedel 1987; Butzer 1971; Lynch 1978, 1983; Jennings 1983). Two explanations which have been suggested are: (1) the existence of sympatric but distinct cultures, one specializing in big game hunting and the other exploiting a more diverse diet, or (2) the possibility of a single culture with seasonal differences in food use. The model presented here suggests another intriguing possibility: could the sites reflecting a more diversified diet (lower in hunted foods) represent different population growth phases, either the initial or the established phase (Fig. 5.10)? This would be difficult to ascertain from the archaeological record unless a single stratified site containing both types of tool kits could be found, since the resolution of radiocarbon dating is not accurate enough to distinguish the time frame of the transition from the initial to the growth phase of the population trajectory (less than 25 years), or for the transition from the growth to the established phase (approximately 100 years).

Diet changes also are used, in part, to delineate the end of the Paleo-Indian period and the onset of the Archaic. The model presented here provides the tantalizing suggestion that initiation of the Archaic might reflect the transition from a population's growth phase to its established phase (Fig. 5.10), when it enters a stable limit cycle and the diet includes a lower proportion of hunted foods. Therefore, this would imply that Paleo-Indians sites with a lower intake of hunted foods (see above) are possibly sites from the initial population phase, when people were first colonizing a locale, and that early Archaic sites are the onset of the established population phase.

The population model presented here provides an estimate that the populations in the New World would have reached the established phase (no more areas to colonize) in some parts of the New World by 9500 BP. With the retreat of the glaciers opening new areas to colonization, an established phase may not have been achieved until 7000 BP in other parts of the New World. Furthermore, in areas which had a low harvestable primary productivity at the time of colonization and after Holocene vegetation changes, a diet high in hunted foods would have been perpetuated, showing little dietary change (Fig. 5.10). Therefore, the emergence of the Archaic in different parts of the New World at different times is consistent with this scenario (Fiedel 1987). An earlier emergence is predicted and found in environments of higher harvestable primary productivity (e.g., some Mexican and South American savanna-like habitats) and a late appearance of the Archaic is predicted and found in the plains areas of North America with lower harvestable primary productivities (Flannery 1986; Frison 1974, 1978; Lynch 1978, 1983; Butzer 1971; Fiedel 1987).

Questions dealing with dietary change indicate the need to investigate plant use (gathered foods) by Paleo-Indians more thoroughly, since the model suggests that Paleo-Indians might have relied extensively on plant foods, even when meat intake was great (Fig. 5.10). The abundance of plant foods and their use also influence the severity of exploitation of the prey. Therefore, this investigation of changing lifeways with demography and food resources should be viewed as a suggestion requiring much greater refinement in the model and data collection.

The adoption of new technologies poses another important archaeological question that can be addressed using the model. A new technology reflects new tools or methods for food procurement that decrease the time it takes to harvest a unit of food (decreases the model's cropping rates). The result is that the forager will be able to harvest more food in the same amount of time even though the amount of harvestable food in the environment does not change. For the purposes of this paper, new methods of food procurement that change the amount of harvestable food in the environment (e.g., agriculture) are not examined; these will be addressed in a future paper.

To address the role of technology in the model, the portion of the cropping rates due to food harvesting (not including search or preparation) was increased by 5% (food acquired per unit time). The search for food was not included in the change since this is primarily a function of the food's abundance in the environment; preparation was not included since this was considered to be a constant characteristic for the food type (butchering, cooking, etc.). Averaged over environments of different productivities, a 5% increase in harvesting ability leads to an increase in energy intake of 7–8%, a decrease in meat consumption of 12–13%, and an increase in hunter-gatherer density and growth rates of 17–18%. Therefore, a small change in technology can have a very large effect on population density and growth rate.

Population growth rates and densities reflect the fitness of individuals in small populations, like hunter-gatherer groups. Therefore, we might expect advances in technology to be rapidly adopted, since a small change in cropping rate has a large impact on energy intake and population dynamics. As an example of how important this cultural evolution might be for humans, changes in hominid densities can be examined. Estimates of densities of *Homo habilis* in East Africa 2.5 million years ago average 0.01 individuals/km^2 (Tanner 1981) and modern hunter-gatherers in these areas have an average density of 0.1 individuals/km^2 (Hassan 1981). The 900% greater density observed today, ignoring environmental changes, could have arisen from a mere 25–26% increase in food gathering technology over the last 2.5 million years!

The observed rapid changes in Paleo-Indian lithic technologies could easily be a reflection of the cultural evolution of people to changing demography, its impact upon food resources, and environmental changes. Therefore, we should not be surprised by the rapid appearance and disappearance of technologies in the archaeological record. These changes could easily have occurred by people adopting another group's technology when it was found to be more efficient or by the competitive exclusion of hunter-gatherer groups with different (less efficient) technologies. The importance of technological innovations for hunter-gatherer demography, however, emphasizes the need to provide experimental data on the use of ancient technologies (e.g., Frison 1974, 1978, 1982; Oswalt 1979; Bleed 1986; Browne 1940; Hames 1979; Hill 1948; Kellar 1955; Lahren and Bonnichsen 1974; Peets 1960; Stanford et al. 1981; Huckell 1979; Kleindienst and Keller 1975). This will permit the cropping rate values employed in models like the one presented here to be made more realistic, so model predictions can have greater utility to archaeologists.

Conclusion

The model presented here hopefully illustrates the utility of combining models of hunter-gatherer foraging behavior with their demography and the demography of their food resources. The model's predictions are consistent with the limited available data on present day hunter-gatherers. Furthermore, the model provides predictions that are consistent with archaeological evidence on the colonization of the New World by Paleo-Indians. Like any other hypothesis, however, the model presented here raises many new questions and emphasizes the need for new data to be collected in ways that may not have been previously considered.

Most importantly, the model indicates that, at least for hunter-gatherers, their demography and foraging technologies may be environmentally determined (Buikstra and Mielke 1985). These people must be viewed as adapted to their environments and subject to the same environmental restrictions that apply to other animals. Past claims that these people culturally limit their population sizes or are not subject to short-falls in food that appear in an environment with limited food resources cannot be reconciled with the model presented here or the consistent empirical data (Hassan 1981; Glassow 1978). The key to these past misconceptions is an outdated view of carrying capacity, where carrying capacity was viewed as a static quantity that did not change with or change hunter-gatherer demography.

Refinement of the model presented here or Winterhalder et al.'s (1988) model will depend upon a better understanding of foraging models for hunter-gatherers and their nutritional needs. The model presented here and Winterhalder et al.'s are based upon different foraging models (see Belovsky 1987). While both population models provide similar general results, the detailed quantitative predictions will be very different. Therefore, as these models are used to address more specific questions about hunter-gatherers, distinctions as to which foraging model is more appropriate will be necessary. In this paper a quantitative argument is presented for the utility of the foraging model construct developed in Belovsky (1987).

Appendix 5.1 Stability Analysis of Overexploitation Models

A set of simple differential equations can be used to analyze the stability of models which various authors have employed to determine whether or not Paleolithic peoples might have overexploited game animals and caused their extinction (Martin 1973, 1982, 1984; Mosimann and Martin 1975; Whittington and Dyke 1984; Budyko 1967, 1974). Stability analysis examines whether or not population growth trajectories (changes over time) will generally approach an equilibrium (constant outcome) regardless of the initial conditions (population densities at the start) and whether or not they will return if disturbed from an equilibrium. Differential equations (continuous changes over time) were employed even though many of the original formulations (Martin 1973, 1982, 1984; Mosimann and Martin 1975; Whittington and Dyke 1984) employed difference equations (discrete changes over time) because differential equations are easier to use in a stability analysis and are more stable; i.e., if they are unstable so are the difference equations (May 1973).

The set of differential equations employed is:
$$F_1(P,H) = dP/dt = r(P - gH) \text{ for } P < K;$$
otherwise
$$dP/dt = 0 \text{ and } P = K,$$
and
$$F_2(H) = dH/dt = bH(L - H)/L,$$
where P = the density of game animals; K = the carrying capacity of the game animals, their maximum density; H = the density of humans; r = the intrinsic growth rate of the prey animals; g = the number of game animals killed by each human; b = the intrinsic growth rate of the humans; and L = the carrying capacity of the humans, their maximum density.

Several characteristics of these proposed population models should be noted. First, the number of game animals killed per human (g) does not change with game or human population densities. There is no feedback between prey abundance and human hunting success. This has been argued for on the grounds that humans are so efficient that they can find and kill all large prey animals in the environment (Budyko 1967, 1974; Martin 1973; Mosimann and Martin 1975). Mosimann and Martin (1975) did allow g to be dependent upon prey abundance, but this had no effect on the model's general predictions. Second, the human carrying capacity, L, is not a function of prey abundance, but a constant. Therefore, prey abundance cannot provide feedback to human population density.

$F_2(H)$ is the classic logistic equation from ecology. This equation was not used by Budyko (1967, 1974) who employed the exponential growth equation. Since Budyko's equation does not account for any limit to human population growth, the people increase in number indefinitely until all the prey animals are killed (extinction). Therefore, Budyko's model does not

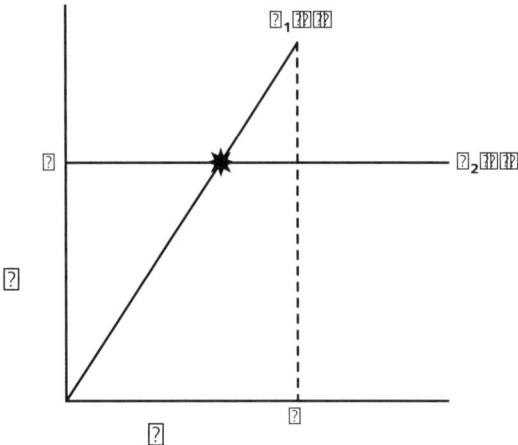

FIGURE 5.A1. Plots of the population isoclines ($F_1 = 0$ and $F_2 = 0$) as functions of prey (P) and human (H) densities. The intersection of the isoclines (*) reflects the equilibrium densities. The dashed line represents the disjunct prey isocline at its carrying capacity.

provide the possibility for a stable coexistence between humans and their prey animals; megafaunal extinction is a foregone conclusion.

The use of the logistic growth equation for humans with a maximum population density, L, provides a potential equilibrium between humans and their prey animals, which has been recognized by the authors of these models. This equilibrium is shown in Fig. 5.A1 and has the following characteristics:

$$P^* = gH, \text{ if and only if } L \leq rK/g(1+r),$$

and

$$H^* = L,$$

where the * denotes the equilibrium solution. If $L > rK/g(1+r)$, then the prey animals are driven to extinction. Even though an equilibrium is possible, it is not necessarily stable, meaning that these constant densities cannot be attained unless the populations are placed at these densities as an initial condition, nor will they return to the equilibrium if perturbed.

To examine the stability of this potential equilibrium, the eigenvalues must be examined for this system of interacting populations. Eigenvalues describe a population's growth trajectory in the vicinity of an equilibrium to determine whether or not the trajectory approaches the equilibrium or moves away from it. Also, the eigenvalues explain whether the trajectory moves in a unidirectional or in an oscillatory fashion. For a system to be stable all eigenvalues, one for each population equation, must have a negative real part; if they do not, the system is not stable (May 1973). If the eigenvalues also have an imaginary part, then the trajectory will be oscillatory (May 1973).

To compute the eigenvalues we need to know the partial derivatives of F_1 and F_2 with respect to H and P in the vicinity of the equilibrium. These partial derivatives are found by taking the derivative of F_1 and F_2 with respect to H and P and then substituting the equilibrium values, H^* and P^*, for any remaining H and P values. These partial derivatives are presented in the matrix

$$A = \begin{vmatrix} r & -rg \\ 0 & -b \end{vmatrix}$$

and the eigenvalues are solved given the relationship

$$\det|A - \lambda I| = 0,$$

where I is the identity matrix and λ is the column matrix of eigenvalues.

For the population models constructed to examine the potential for overexploitation of prey animals, the eigenvalues are $\lambda_1 = r$ and $\lambda_2 = -b$. Because one of the eigenvalues is not negative, this system of population equations is unstable. This means that even though an equilibrium between humans and their prey animals is possible in this set of equations it is unstable. Therefore, humans will always hunt their prey to extinction in these models; there is no other alternative.

Acknowledgments

I thank R. Whallon, J. B. Slade, and an unidentified reviewer for detailed comments on the manuscript. I also acknowledge conversations with R. Whallon, J. Speth, B. Winterhalder. and K. Flannery on aspects of this paper.

References

Abrams, P.
1984 Variability in resource consumption rates and the coexistence of competing species. *Theoretical Population Biology* 25:106–124.

Angel, J. L.
1975 Paleoecology, paleodemography and health. In *Population, Ecology, and Social Evolution*, edited by S. Polgar, pp. 167–190. Mouton, The Hague.

Belovsky, G. E.
1978 Diet optimization in a generalist herbivore: The moose. *Theoretical Population Biology* 14:105–134.

1984a Herbivore optimal foraging: A comparative test of three models. *American Naturalist* 124:97–115.

1984b Summer diet optimization by beaver. *American Midland Naturalist* 111:209–222.

1986 Generalist herbivore foraging and its role in competitive interactions. *American Zoologist* 26:51–69.

1987 Hunter-gatherer foraging: A linear programming approach. *Journal of Anthropological Archaeology* 6:29–76.

Belovsky, G. E., M. E. Ritchie, J. Moorehead
1989 Foraging in complex environments: When prey availability varies over time and space. *Theoretical Population Biology*. 36: 144–160.

Birdsell, J. B.
1957 Some population problems involving Pleistocene man. Population studies: Animal ecology and demography. *Cold Springs Harbor Symposium on Quantitative Biology* 22:47–69.

1968 Some predictions for the Pleistocene based on equilibrium systems among recent hunter-gatherers. In *Man the Hunter*, edited by R. Lee and I. DeVore, pp. 229–240. Aldine, Chicago.

Bleed, P.
1986   The optimal design of hunting weapons: Maintainability or reliability. *American Antiquity* 5:737–747.

Bliss, L. C., J. H. Richards
1982   Present-day arctic vegetation and ecosystems as a predictive tool for the arctic steppe mammoth biome. In *Paleoecology of Beringia*, edited by D. M. Hopkins, J. V. Matthews, Jr., C. E. Schweger, and S. B. Young, pp. 241–257. Academic Press, New York.

Blurton Jones, N. G., R. M. Sibley
1978   Testing adaptiveness of culturally determined behavior: Do Bushmen women maximize their reproductive success by spacing births widely and foraging seldom? In *Human Behaviour and Adaptation*, edited by N. G. Blurton Jones and V. Reynolds, pp. 135–157. Society for the Study of Human Biology Symposium No. 18. Taylor & Francis, London.

Browne, J.
1940   Projectile points. *American Antiquity* 5:209–213.

Budyko, M. I.
1967   On the causes of the extinction of some animals at the end of the Pleistocene. *Soviet Geography: Review and Translation* 8:783–793.
1974   *Climate and Life*. Academic Press, New York.

Buikstra, J. E., J. H. Mielke
1985   Demography, diet, and health. In *The Analysis of Prehistoric Diets*, edited by R. I. Gilbert, Jr., and J. H. Mielke, pp. 359–422. Academic Press, New York.

Butzer, K. W.
1971   *Environment and Archeology: An Ecological Approach to Prehistory*. Aldine-Atherton, Chicago.

Caughley, G., J. H. Lawton
1981   Plant-herbivore systems. In *Theoretical Ecology: Principles and Applications*, edited by R. M. May, pp. 132–166. Sinauer Associates, Sunderland, MA.

Charnov, Eric L.
1976   Optimal foraging, the marginal value theorem. *Theoretical Population Biology* 9:129–136.

Comins, H. N., M. P. Hassell
1979   The dynamics of optimally foraging predators and parasitoids. *Journal of Animal Ecology* 48:335–351.

Deevey, E. S., Jr.
1960   The human population. *Scientific American* 203:194–204.

Emlen, J. M.
1966   The role of time and energy in food preference. *The American Naturalist* 100:611–617.

Farlow, J. O.
1976   A consideration of the trophic dynamics of a late Cretaceous large-dinosaur community (Oldman Formation). *Ecology* 57:841–857.

Fiedel, S. J.
1987   *Prehistory of the Americans*. Cambridge Univ. Press, Cambridge.

Flannery, K. V.
1986   The research problem. In *Guila Naquitz: Archaic Foraging and Early Agriculture in Oaxaca, Mexico*, edited by K. V. Flannery, pp. 3–18. Academic Press, Orlando, FL.

Fretwell, S. D., H. L. Lucas, Jr.
1970   On territorial behavior and other factors influencing habitat distribution in birds. I. Theoretical development. *Acta Biotheoretica* XIX(l):16–36.

Frisancho, A. R.
1981   *Human Adaptation: A Functional Interpretation*. Univ. of Michigan Press, Ann Arbor.

Frison, G. C.
1974   Archaeology of the Casper site. In *The Casper Site: A Hell Gap Bison Kill on the High Plains*, edited by G. C. Frison, pp. 1–112. Academic Press, New York.
1978   *Prehistoric Hunters of the High Plains*. Academic Press, New York.
1982   Bison procurement. In *The Agate Basin Site: A Record of the Paleoindian Occupation of the Northwestern High Plains*, edited by G. C. Frison and D. J. Stanford, pp. 263–269. Academic Press, New York.

Glassow, M. A.
1978   The concept of carrying capacity in the study of culture process. *Advances in Archaeological Method and Theory* 1:31–48.

Grayson, D. K.
1984   Nineteenth-century explanations of Pleistocene extinctions: A review and analysis. In *Quaternary Extinctions: A Prehistoric Revolution*, edited by P. S. Martin and R. C. Klein, pp. 5–39. Univ. of Arizona Press, Tucson.

Guthrie, R. D.
1968   Paleoecology of the large-mammal community in interior Alaska during the Late Pleistocene. *American Midland Naturalist* 79:346–363.
1982   Mammals of the mammoth steppe as paleoenvironmental indicators. In *Paleoecology of Beringia*, edited by D. M. Hopkins, J. V. Matthews, Jr., C. E. Schweger, and S. B. Young, pp. 307–326. Academic Press, New York.
1984   Mosaics, allelochemics and nutrients: An ecological theory of late Pleistocene megafaunal extinctions. In *Quaternary Extinctions: A Prehistoric Revolution*, edited by P. S. Martin and R. C. Klein, pp. 259–298. Univ. of Arizona Press, Tucson.

Hames, R. B.
1979   A comparison of the efficiencies of the shotgun and the bow in neotropical forest hunting. *Human Ecology* 7:219–252.
1987   Game conservation or efficient hunting? In *The Question of the Commons: The Culture and Ecology of Communal resources*, edited by B. J. McCay and J. M. Acheson, pp. 92–107. Univ. of Arizona Press, Tucson.

Harper, J. L.
1977   *Population Biology of Plants*. Academic Press, London.

Hassan, F. A.
1975   Determination of the size, density, and growth rate of hunting-gathering populations. In *Population, Ecology and Social Evolution*, edited by S. Polgar, pp. 27–52. Mouton, The Hague.
1981   *Demographic Archaeology*. Academic Press, New York.

Hassell, M. P., R. M. May
1973   Stability in insect host-parasite models. *Journal of Animal Ecology* 42:693–726.

Hawkes, K., K. Hill, J. F. O'Connell
1982   Why hunters gather: Optimal foraging and the Ache of eastern Paraguay. *American Ethnologist* 9:379–398.

Hayden, B.
1981   Subsistence and ecological adaptations of modern hunter/gatherers. In *Omnivorous Primates: Gathering and Hunting in*

Human Evolution, edited by R. S. O. Harding and G. Teleki, pp. 344–421. Columbia Univ. Press, New York.

Haynes, C. V.
1970 Geochronology of man-mammoth sites and their bearing on the origin of the Llano complex. In *Pleistocene and Recent Environments of the Central Great Plains*, edited by W. W. Dort and A. E. Johnson, pp. 77–92. Special Publication 3. University of Kansas, Lawrence.
1982 Were Clovis progenitors in Beringia? In *Paleoecology of Beringia*, edited by D. M. Hopkins, J. V. Matthews, Jr., C. E. Schweger, and Steven B. Young, pp. 383–398. Academic Press, New York.

Hibbert, D.
1982 History of the steppe-tundra concept. In *Paleoecology of Beringia*, edited by D. M. Hopkins, J. V. Matthews, Jr., C. E. Schweger, and S. B. Young, pp. 153–156. Academic Press, New York.

Hill, M. W.
1948 The atlatl or throwing stick: A recent study of atlatls in use with darts of various sizes. *Tennessee Archaeological Society* 4:37–44.

Howell, N.
1976 The population of the Dobe area !Kung. In *Kalahari Hunter-Gatherers: Studies of the !Kung San and their Neighbors*, edited by R. B. Lee and I. DeVore, pp. 137–151. Harvard Univ. Press, Cambridge.
1979 *Demography of the Dobe !Kung*. Academic Press, New York.

Huckell, B. B.
1979 Of chipped stone tools, elephants, and the Clovis hunters: An experiment. *Plains Anthropologist* 24:177–189.

Jennings, J. D.
1983 Origins. In *Ancient North Americans*, edited by J. D. Jennings, pp. 25–67. Freeman, San Francisco.

Kellar, J. H.
1955 The atlatl in North America. *Indiana Historical Society, Prehistory Research Series*, III, No. 3:281–352.

Kleindienst, M. R., C. M. Keller
1975 Towards a functional analysis of handaxes and cleavers: The evidence from eastern Africa. *Man* New Series, 11:176–187.

Lahren, L., R. Bonnichsen
1974 Bone foreshafts from a Clovis burial in southwestern Montana. *Science* 186:147–150.

Lee, R. B.
1979 *The !Kung San: Men, Women and Work in a Foraging Society*. Cambridge Univ. Press, Cambridge.

Leith, H.
1975 Modeling the primary productivity of the world. In *Primary Productivity of the Biosphere. Ecological Studies 14*, edited by H. Leith and R. H. Whittaker, pp. 237–263. Springer-Verlag, New York.

Lewin, R.
1987 The first Americans are getting younger. *Science* 238:1230–1232.

Lynch, T. F.
1978 The South American Paleo-Indians. In *Ancient Native Americans*, edited by J. D. Jennings, pp. 455–489. Freeman, San Francisco.
1983 The Paleo-Indians. In *Ancient South Americans*, edited by J. D. Jennings, pp. 87–137. Freeman, San Francisco.

MacArthur, R. H., E. R. Pianka
1966 On optimal use of a patchy environment. *The American Naturalist* 100:603–609.

Martin, P. S.
1966 Africa and Pleistocene overkill. *Nature* (London) 212:339–342.
1967 Prehistoric overkill. In *Pleistocene Extinctions: The Search for a Cause*, edited by P. S. Martin and H. E. Wright, Jr., pp. 75–120. Yale Univ. Press, New Haven.
1973 The discovery of America. *Science* 179:969–974.
1982 The pattern and meaning of Holarctic mammoth extinction. In *Paleoecology of Beringia*, edited by D. M. Hopkins, J. V. Matthews, Jr., C. E. Schweger, and S. B. Young, pp. 399–408. Academic Press, New York.
1984 Prehistoric overkill: The global model. In *Quaternary Extinctions: A Prehistoric Revolution*, edited by P. S. Martin and R. C. Klein, pp. 354–403. Univ. of Arizona Press, Tucson.

May, R. M.
1973 Stability and complexity in model ecosystems. *Monographs in Population Biology* No. 6. Princeton Univ. Press, Princeton, NJ.

McDonald, J. N.
1984 The reordered North American selection regime and late Quaternary megafaunal extinctions. In *Quaternary Extinctions: A Prehistoric Revolution*, edited by P. S. Martin, and R. C. Klein, pp. 404–439. Univ. of Arizona Press, Tucson.

McNaughton, S. J., L. L. Wolf
1973 *General Ecology*. Holt, Rinehart and Winston, New York.

Minnis, P. E.
1985 *Social Adaptation to Food Stress: A Prehistoric Southwestern Example*. Univ. of Chicago Press, Chicago.

Mosimann, J. E., P. S. Martin
1975 Simulating overkill by Paleoindians. *American Scientist* 63:304–313.

Oswalt, W. H.
1979 *An Anthropological Analysis of Food Getting Technology*. Willey-Interscience, New York.

Owen-Smith, N.
1987 Pleistocene extinctions: The pivotal role of megaherbivores. *Paleobiology* 13:351–362.

Peets, O. H.
1960 Experiments in the use of atlatl weights. *American Antiquity* 26:108–110.

Redmann, R. E.
1982 Production and diversity in contemporary grasslands. In *Paleoecology of Beringia*, edited by D. M. Hopkins, J. V. Matthews, Jr., C. E. Schweger, and S. B. Young, pp. 223–239. Academic Press, New York.

Schoener, T. W.
1969a Models of optimal size for solitary predators. *American Naturalist* 103:277–313.
1969b Optimal size and specialization in constant and fluctuating environments: An energy-time approach. *Brookhaven Symposium Biology* 22:103–114.
1971 Theory of feeding strategies. *Annual Review of Ecology and Systematics* 2:369–404.
1973 Population growth regulated by intraspecific competition for energy or time: Some simple representations. *Theoretical Population Biology* 4:56–84.
1987 A brief history of optimal foraging ecology. In *Foraging Behavior*, edited by A. C. Kamil, J. R. Krebs, and H. R. Pulliam, pp. 5–67. Plenum, New York.

Schweger, C. E.
1982 Primary production and the Pleistocene ungulates: the productivity paradox. In *Paleoecology of Beringia*, edited by D. M.

Hopkins, J. V. Matthews, Jr., C. E. Schweger, and S. B. Young, pp. 219–221. Academic Press, New York.

Stanford, D., R. Bonnichsen, R. E. Morlan
1981 The Ginsberg experiment: Modern and prehistoric evidence of a bone-flaking technology. *Science* 212:438–440.

Stephens, D. W., J. R. Krebs
1986 Foraging theory. *Monographs in Behavioral Ecology*. Princeton Univ. Press, Princeton, NJ.

Tanner, N. M.
1981 *On Becoming Human*. Cambridge Univ. Press, Cambridge.

Vickers, W. T.
1988 Game depletion hypothesis of Amazonian adaptation: Data from a native community. *Science* 239:1521–1522.

Webster, D.
1981 Late Pleistocene extinction and human predation: A critical overview. In *Omnivorous Primates: Gathering and Hunting in Human Evolution*, edited by R. S. O. Harding and G. Teleki, pp. 556–594. Columbia Univ. Press, New York.

Webster, D., G. Webster
1984 Optimal hunting and Pleistocene extinction. *Human Ecology* 12:275–289.

Werner, E. E.
1977 Species packing and niche complementarity in three sunfishes. *American Naturalist* 111:553–578.

Wesler, K. W.
1981 Models for Pleistocene extinction. *North American Archaeologist* 2:85–100.

Whittaker, R.
1970 *Communities and Ecosystems*. MacMillan, New York.

Whittington, S. L., B. Dyke
1984 Simulating overkill: Experiments with the Mosimann and Martin model. In *Quaternary Extinctions: A Prehistoric Revolution*, edited by P. S. Martin and R. C. Klein, pp. 451–465. Univ. of Arizona Press, Tucson.

Winterhalder, B., W. Baillargeon, F. Cappelletto, I. R. Daniel, Jr., C. Prescott
1988 The population ecology of hunter-gatherers and their prey. *Journal of Anthropological Archaeology* 7:289–328.

Wobst, H. M.
1974 Boundary conditions for Paleolithic social systems: A simulation approach. *American Antiquity* 39:147–178.

Changing Diet Breadth in the Early Upper Palaeolithic of Southwestern France

Donald K. Grayson and Françoise Delpech

INTRODUCTION

During the past few decades, evolutionary ecologists have developed powerful quantitative models for understanding the decisions predators make in acquiring food (Stephens & Krebs, 1986; Kaplan & Hill, 1992; Smith, 1991; Kelly, 1995). Some of these models predict how long a predator will remain in a given part of its habitat, while others, termed prey-choice models, predict which prey types will be taken, and which ignored, on encounter. Foraging theorists have had a great deal of success in applying these models to contemporary human contexts, both in the sense that the models have been shown to have significant predictive power (e.g. O'Connell & Hawkes, 1981; Hawkes, Hill & O'Connell, 1982; Hill et al., 1987), and in the sense that they have led to the precise formulation of novel questions about human behaviour (e.g. Hawkes, 1993).

The archaeological potential of foraging theory has been apparent for some time (Bayham, 1979; O'Connell, Jones & Simms, 1982; Simms, 1985, 1987; O'Connell, Hawkes & Blurton-Jones, 1988; Szuter & Bayham, 1989), but detailed and compelling applications have only recently begun to appear (e.g. Broughton, 1994a, b, 1995). In these applications, concepts that are meant to apply in ecological time must be translated to archaeological time, and variables that are readily measured when they can be observed directly must now be estimated from very different kinds of information (O'Connell, 1995; Grayson & Cannon, 1998). For instance, while energy, measured in calories, is used to calibrate foraging theory models applied to living peoples, past energy returns have, up to this point, either been estimated from prey sizes (e.g. Bayham, 1979; Broughton, 1994a, b), or from experiments directed towards measuring return rates directly, under the assumption that those rates applied in the past as well (e.g. Simms, 1987; Madsen & Kirkman, 1988; Jones & Madsen, 1989, 1991; see the discussion in Grayson & Cannon, 1998).

Since prey-choice models are designed to predict that set of resources which will be included in a predator's diet, they are often referred to as "diet breadth" models, with diet breadth defined as "the total number of resources in the diet" (Kaplan & Hill, 1992:171). In this formulation, resources, or prey types, are in theory defined according to their expected energetic return rates, and are thus not necessarily equivalent to biological species (see the discussion in Smith, 1991). In practice, however, resources are often equated with species (e.g. Winterhalder & Goland, 1997), and we follow this equation here.

Some aspects of foraging theory models can be applied archaeologically without measuring diet breadth. However, many applications will require that this variable be quantified in such a way as to allow predictions derived from contemporary ecological considerations to be transferred to predictions that are archaeologically appropriate. Here, we discuss the obvious archaeological measure of diet breadth: the number of taxa incorporated in an archaeological assemblage, using an example drawn from the early Upper Palaeolithic faunal assemblages provided by the site of Le Flageolet I.

LE FLAGEOLET I

Le Flageolet I is a small, well-stratified rockshelter overlooking the north side of the Dordogne River near the small town of Bézenac, southwestern France. Excavated under the direction of Jean-Philippe Rigaud, the site has provided substantial samples of both lithic and faunal material (Rigaud, 1982, 1993; Delpech, 1983; Simek, 1984, 1987; see also Enloe, 1992). These materials are distributed across a series of eight cultural strata deposited between about 34,000 and 20,000 ^{14}C years ago (Rigaud, 1993). Analysis of the stone tools by Rigaud (1982) has allowed assignment of the Le Flageolet I lithic assemblages to both Aurignacian and Perigordian industries (Table 6.1).

Of these eight depositional units, one, composite Stratum 0-III, provided only 27 identifiable ungulate specimens. Because the analyses that we present below require samples larger than this, Stratum 0-III is not considered here. Numbers of identi-

*Originally published in *Journal of Archaeological Science* 25(1998):1119–1129

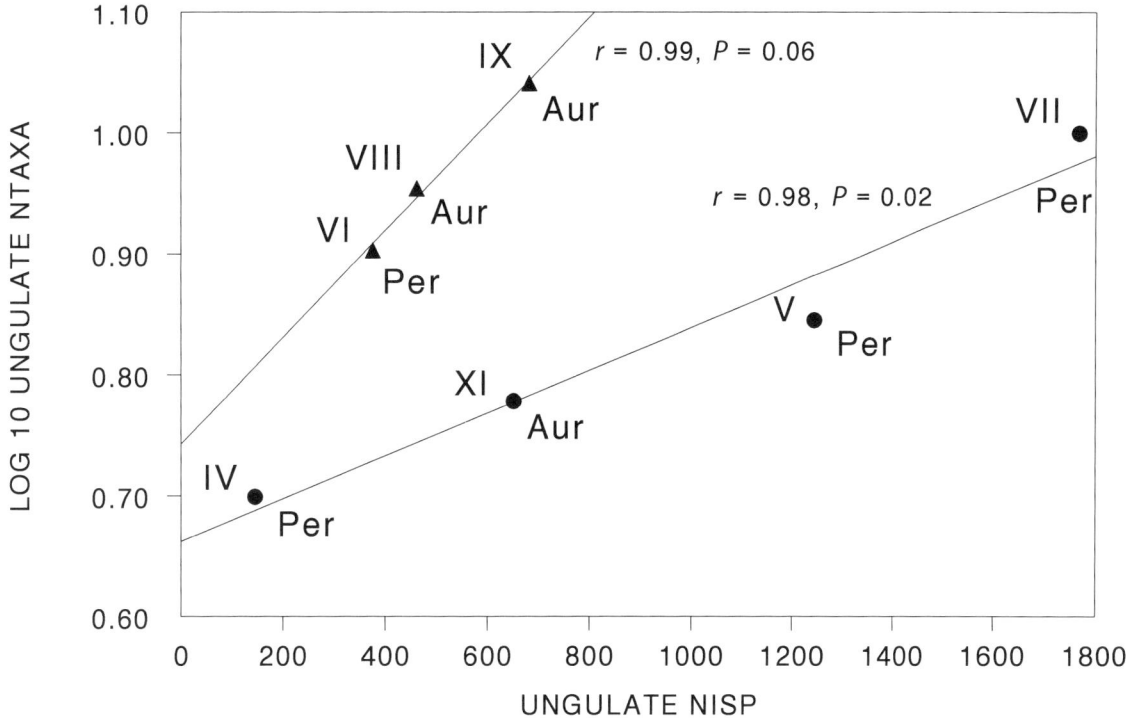

FIGURE 6.1. The relationship between ungulate NISP and (Log10) ungulate NTAXA in the Le Flageolet I faunal assemblages (Aur = Aurignacian; Per = Perigordian).

fied specimens per ungulate and carnivore taxon for strata IV through XI are provided in Tables 6.2 and 6.3. Because the focus of our discussion is on the Le Flageolet I ungulates, and because there are many pathways by which ungulate remains can be introduced into an "archaeological" site, we note that these ungulate assemblages are clearly of human origin. This issue will be treated in greater detail in the forthcoming Le Flageolet I monograph (Rigaud, 1997), but we note that while 219 (4.1%) of the ungulate specimens from this site show cut marks, only 25 (0.5%) show evidence of having been altered by carnivores. These and other indicators lead us to conclude that the Le Flageolet I ungulate assemblage has been overwhelmingly, and perhaps entirely, introduced by people.

Numbers of Ungulate Taxa at Le Flageolet I

A total of 12 ungulate taxa are represented in the Le Flageolet I faunal assemblages, but not all taxa are present in all assemblages. Not surprisingly, the number of taxa (NTAXA) varies across assemblages, and the number present in any given assemblage scales to the number of identified specimens (NISP) in that assemblage (see Grayson, 1991 for the protocol used to count non-overlapping taxa).

However, unlike many other faunal assemblages that have been analysed (e.g. Grayson, 1984), there are two distinct, and statistically significant, NISP-NTAXA relationships represented at Le Flageolet I (see Figure 6.1 and Table 6.4), with the slopes of these relationships significantly different (at $P<0.05$) from one another. Importantly, "cultural" affiliation is unrelated

TABLE 6.1. Summary of the contents and chronology of Le Flageolet I (from Rigaud, 1982, 1993).

Stratum	Cultural Assignment	^{14}C Dates
O–III	Late Perigordian	18,610 ± 440 (Ly-2185)
		22,240 ± 680 (Ly-1606)
		24,600 ± 700 (OxA-448)
IV	Perigordian	21,190 ± 920 (Ly-1607)
		23,250 ± 500 (OxA-596)
V	Perigordian	22,520 ± 500 (Ly-2721)
		25,700 ± 700 (OxA-447)
VI	Perigordian	24,280 ± 500 (Ly-2722)
		26,500 ± 900 (OxA-579)
VII	Perigordian	25,720 ± 610 (Ly-1748)
		26,150 ± 600 (Ly-2723)
VIII	Late Aurignacian	23,280 ± 670 (Ly-1608)
		24,800 ± 600 (OxA-597)
		26,800 ± 1000 (Ly-2724)
		27,350 ± 1400 (Ly-2725)
IX	Aurignacian	20,070 ± 1760 (Ly-1749)
XI	Early Aurignacian	33,800 ± 1800 (OxA-598)

to position in these relationships: Aurignacian and Perigordian faunal assemblages appear on both high-slope (upper) and low-slope (lower) curves.

Measuring the diet breadth of modern human populations may be time-consuming, but it is at least conceptually straightforward: one counts the resources that are included in the diet. Accordingly, given our equation of resources with species, the

TABLE 6.2. The Le Flageolet I ungulates: NISP by taxon.[a]

	Couche							
Taxon	IV	V	VI	VII	VIII	IX	XI	Totals
Bos/Bison	4	4	12	124	41	34	38	257
Capra spp.	2	1	22	10	10	12	15	72
Cervus elaphus		29	115	1223	126	79	18	1590
Rupicapra sp.	3	16	6	14	15	10	3	67
Capreolus sp.		1	3	71	2	9		86
Equus spp.		2	3			1		6
E. caballus	4	21	45	28	22	50	66	236
E. hydruntinus				8		11		19
Mammuthus sp.						1		1
Megaceros sp.				1	1	1		3
Rangifer tarandus	132	1170	169	283	240	468	511	2973
Rhinoceros			1					1
Sus sp.				6	4	5		15
Totals	145	1244	376	1768	461	681	651	5326
NTAXA	5	7	8	10	9	11	6	

[a] Includes specimens identified as "cf.".

TABLE 6.3. The Le Flageolet I carnivores: NISP by taxon.[a]

	Couche							
Taxon	IV	V	VI	VII	VIII	IX	XI	Totals
Canis lupus		1	3	4	4	3		15
Felis sylvestris					1			1
Lynx sp.			1	8		1		10
Lynx spelaea				1				1
Mustela erminea					1			1
Mustela cf. putorius					1			1
Panthera spelaea				1				1
Ursus sp.						1		1
Vulpes vulpes				1				1
Vulpes/Alopex		3	4	10	8	18	5	48
Totals	0	4	8	25	15	23	5	80

[a] Includes specimens identified as "cf.".

TABLE 6.4. Regression coefficients for the relationship between ungulate NISP and (Log10) ungulate NTAXA across the Le Flageolet I ungulate assemblages.

Relationship	Intercept (S.E.)	Slope (S.E.)	Correlation
High slope	0.743 (0.023)	0.00044 (0.00004)	0.995 ($P = 0.06$)
Low slope	0.662 (0.030)	0.00018 (0.00003)	0.978 ($P = 0.02$)

S.E. = standard error; P = probability.

simplest interpretation of the two NISP-NTAXA relationships at Le Flageolet I is that these relationships reflect two different diet breadths. In this interpretation, the high-slope relationship resulted from broader diets than did the low-slope one. This is the case because, at any given sample size, the high-slope assemblages contain more taxa than the low-slope assemblages. For instance, at an NISP of 500, the low slope regression equation predicts the presence of 5.7 taxa, while the high-slope equation predicts the presence of 9.2 taxa.

Unfortunately, things are not this simple. First, diet breadths measured archaeologically are not comparable to those measured ethnographically. Second, it is possible that these curves do not measure diet breadth at all. We treat each of these issues in turn.

What Do Archaeological Numbers of Taxa Mean?

Ethnographic applications of foraging theory are built from detailed observations of the results of activities that are generally archaeologically invisible: single hunting or gathering events, for instance. Although there are rare exceptions, archaeological faunal assemblages reflect the results of an uncontrolled number of indistinguishable collecting events distributed over an uncontrolled, but often long, period of time.

As a result, the number of taxa present in an archaeological faunal assemblage is not directly comparable to diet breadths measured ethnographically. Instead, and as has been argued elsewhere (Broughton & Grayson, 1993), the archaeological measure reflects the maximum diet breadth (in terms of the taxa being monitored—in our case, ungulates) of the human population whose activities accumulated that assemblage across the time period involved.

Effects of Time-averaging

Consider, for instance, a human group living in an environment in the absence of significant climatic change. The potential diet of this group includes 10 species, each of which has a distinct return rate (and hence is a distinct "resource type"). Five of these species provide return rates sufficiently high that they are always taken when encountered, while the remaining five may or may not be taken depending on the rates at which the five highest-ranked taxa are encountered.

Every spring, the group returns to the same site, which it then uses as a base for collecting, and to which it returns to process all materials collected. Only once a century does the abundance of the five highest-ranked taxa decline to a point that all 10 potential food resources are actually included in the diet. In all other years, only the five highest-ranked resources are taken. The archaeological NTAXA provided by the resultant assemblage will be "10".

Consider, on the other hand, a human group existing in an environment that provides the exact same set of resources, but in which the five highest-ranking resources have become rare, perhaps due to continued predation (see the discussion of resource depression by Charnov, Orians & Hyatt, 1976). In this context, all 10 resources are always in the diet. The archaeological NTAXA will again be "10", although the resource structures, and resultant adaptive responses, that have led to these identical measures are quite different.

Finally, consider a situation in which these two contexts follow one another in time at the same site and produce two stratigraphically distinct faunal assemblages with identical NISP values. In both cases, NTAXA will be 10, and one might be tempted to conclude that average diet breadths were identical. This, of course, would be incorrect. Only maximum diet breadths are the same.

Two things follow from these considerations. First, similarities in NTAXA in archaeological faunal assemblages do not necessarily reflect similarities in average diet breadth. However, significant differences in these values might well be meaningful in terms of human adaptation (see Broughton & Grayson, 1993). Second, even if two faunal assemblages provide identical NTAXA values at a given sample size, differences in fine-scaled diet breadth might nevertheless be reflected in the distribution of specimens across taxa. Since high-ranked taxa will always be taken on encounter, their abundances should reflect encounter rates in the surrounding environment. Low-ranked taxa, on the other hand, will be taken only when encounter rates with higher-ranked taxa decline. Accordingly, analyses of the distribution of specimens across taxa may help determine how frequently lower-ranked taxa were incorporated into the diet. There are a number of measures, both simple (e.g. relative abundances) and complex (e.g. evenness and diversity) that can be used to make such assessments. Of course, any such assessment assumes that dietary ranking can somehow be determined, as for instance, from the sizes of the resources involved (e.g. Broughton, 1994a).

Effects of Differential Time-sampling

While time-averaging can cause similar NTAXA values to result from different dietary usages, differential time-sampling can cause NTAXA values for the same population to suggest greater dietary differences than actually exist. Take, for instance, the situation described above, in which only five taxa are included in the local diet in 99 of 100 years. A faunal assemblage that incorporates only the first 99 years will provide an NTAXA of five. An assemblage that incorporates all 100 years will provide an NTAXA of 10. That is, the longer an assemblage takes to accumulate, the greater the chances that it will incorporate a low-probability dietary event. If that event incorporates taxa not otherwise represented in the assemblage, NTAXA will increase.

Comparing assemblages that differentially sample time in this way might, as a result, produce contrasts in NTAXA values that simply reflect these sampling differences. Indeed, even if all archaeological assemblages to be compared cover exactly the same amount of time—50 seasons, say—it is still possible that differences in NTAXA reflect nothing more than the incorporation of rare broad-diet events in one or more of these assemblages. Again, if the reasons for similarities and differences in NTAXA values are to be understood in dietary terms, comparisons of NTAXA must be combined with an analysis of the kinds of taxa represented in the assemblages and with an analysis of the distribution of specimens across those taxa.

These considerations by no means exhaust the potential causes of differences in NTAXA among archaeological faunal assemblages. Clearly, alterations in seasonal use, or even in the duration of use, of a site can also cause changes in NTAXA that do not reflect changing diets, but instead reflect changing site use (see the related discussion in Broughton, 1994a, 1995). More broadly, changes in climate and technology can also cause the numbers and kinds of taxa that enter the diet to change as well (Grayson & Cannon, 1998). We return to these matters below.

Mechanical Effects
The simple fact of a correlation between sample size and numbers of classes across assemblages does not necessarily mean that changing sample sizes have caused the correlation (Grayson, 1984). In fact, it takes no great insight to conceive of situations in which the numbers of specimens is the dependent variable in this relationship.

Consider a situation in which continued predation by hunter-gatherers on a set of high-ranked resources causes a severe decline in the abundance of, and hence encounter rates with, those taxa. In response, the human group involved broadens its diet to include a wider range of lower-ranked taxa, and takes more individuals of those taxa than it took of higher-ranked ones prior to the decline. In addition, it continues to take high-ranked taxa on encounter. The archaeological assemblages that result from this process will contain greater numbers of faunal specimens and contain greater numbers of taxa, but it is the dietary shift that has caused the increase in NISP. A stratified sequence containing the shift may reveal two NISP-NTAXA relationships, one high and one low in slope, with the differences in slope reflecting differences in diet breadth.

Unfortunately, purely mechanical factors can also cause different relationships between NISP and NTAXA. Consider a stratified set of faunal assemblages composed of identical suites of taxa based on initially identical numbers of specimens. For whatever reason, some of these assemblages have undergone greater fragmentation than the rest, but the specimens remain identifiable.

If these assemblages are sampled and the NISP-NTAXA relationships analysed, two relationships will result, with the slope of the relationship for the highly fragmented group of assemblages lower than that for the more intact group. This is the case for the simple reason that the highly fragmented set will contain larger numbers of specimens for a given NTAXA value than the less fragmented set. Below, we refer to this possibility as the "NISP Increase Model."

This situation may change, however, if fragmentation is differentially distributed across assemblages and fragmented specimens of selected taxa do not remain identifiable. If fragmentation proceeds to the point that certain taxa can no longer be identified at all (e.g. Marshall & Pilgrim, 1991), then those assemblages may appear to have fewer taxa for a given NISP value than was, in fact, the case. We will refer to this possibility as the "NTAXA Decrease Model".

Accordingly, before NISP-NTAXA relationships can be analysed for their potential meaning in terms of diet breadth, it must be established that the differences involved have not been caused by differential fragmentation.

Differential Bone Transport and Skeletal Part Representation
Ethnoarchaeological research has established that a wide variety of factors determine how many, and which, skeletal elements will be transported away from a kill site. O'Connell, Hawkes & Blurton Jones (1988, 1990), for instance, documented that bone transport among the Hadza of northern Tanzania is determined by a complex combination of variables, including animal size, the amount of meat removed at a kill or scavenging site, distance from the base camp, and the number of people involved in the transport episode (see the review in O'Connell, 1995). Bartram (1993) found no significant relationship between skeletal part utility and relative skeletal abundance in gemsbok (*Oryx gazella*) kill sites produced by the Kua of the eastern Kalahari. The reason, he observed, was simple: the Kua often stripped much of the meat from their prey and left the bones behind. Indeed, Bartram (1993) found a strong positive correlation between the amount of time spent processing animals at kill sites and the number of bones left at those sites.

Differential bone transport produced by such factors can readily produce different relationships between NISP and NTAXA in archaeological faunas. Consider, for instance, two groups of people preying on an identical set of taxa. Whenever any of these taxa is encountered, not only is it taken, but the same number of individuals is taken. If one of these groups always retrieves the entire skeleton, while the second group only retrieves a subset of that skeleton, the bone assemblages produced by these groups will show two distinct relationships between NISP and NTAXA. The relationship displayed by the assemblages produced by the group that always retrieves the entire skeleton will have a lower slope than the one retrieved by the second group. This is the case even though the number of species taken by the two groups is identical.

Indeed, it is not just differential bone transport that can cause this effect. Any process that causes the skeletons in a given set of assemblages to be better represented than skeletons in a second set of assemblages will result in differences in slope in the resultant NISP-NTAXA relationships. This possibility must also be eliminated if those slopes are to be interpreted in terms of diet breadth. Fortunately, this effect should be readily detected as long as differences in skeletal representation are not randomly distributed across skeletal elements.

THE NISP-NTAXA RELATIONSHIP AT LE FLAGEOLET I

Differential Fragmentation
We first consider whether the two different NISP-NTAXA relationships at Le Flageolet I could be caused by differential fragmentation. To investigate this possibility, we follow Todd &

Rapson (1988; see also Lyman, 1994 and the important discussion in Marean, 1991) and use a very simple fragmentation index: the ratio of proximal and distal ends to shafts for all long bones and ribs. We presume, with Todd & Rapson (1988), that greater degrees of fragmentation will differentially increase the number of shaft fragments in this ratio, and hence trace degree of breakage.

If differential fragmentation has increased NISP counts for the assemblages that define the low-slope curve (the NISP Increase Model), then these assemblages should have higher fragmentation ratios than do those assemblages that define the high-slope relationship. The NTAXA Decrease Model is more difficult to deal with, since we currently lack ways to detect taxa that have been so heavily fragmented as to be unidentifiable (but see Hardy, Raff & Raman, 1997). Nonetheless, if this has occurred, then it is again the low-slope relationship that should have higher fragmentation ratios, since it is the low-slope relationship that has undergone sufficient fragmentation to cause certain taxa to become completely unidentifiable. Table 6.5 provides the relevant raw data and ratios.

We note two aspects of the relationship between degree of fragmentation and position on the high- and low-slope relationships at Le Flageolet I. First, taken as a composite, the four assemblages that form the low-slope relationship have a lower composite fragmentation index (2.59) than do the four assemblages that form the high-slope relationship (3.53), opposite to the prediction of both NISP Increase and NTAXA Decrease Models. Second, chi-square analysis shows that the low-slope relationship has significantly fewer shafts, and significantly more proximal and distal specimens, than does the high-slope curve (chi-square = 12.23, $P < 0.01$). These relationships suggest that differential fragmentation has not produced the NISP-NTAXA relationships that mark the Le Flageolet I assemblages.

While this is the case, it is also true that there is no fully consistent relationship between degree of fragmentation and position on the two curves. For instance, Stratum VII has the lowest fragmentation index of all eight assemblages, but is on the low-slope curve. Similarly, Stratum IX, with a high fragmentation index, is on the high-slope curve. These positions are also opposite those predicted by both of the fragmentation models considered here.

A more precise view of the relationship between degree of fragmentation and location on the Le Flageolet I curves can be gained by examining the adjusted residuals for numbers of shafts across all eight assemblages (Table 6.5). Adjusted residuals are read as standard normal deviates (Everitt, 1977), and thus provide the probability that the cell values in question could have occurred by chance; negative values indicate instances in which the value in question occurs less often than chance allows. According to the fragmentation models we have discussed, shafts should be over-represented in all low-slope assemblages and under-represented in all high-sloped ones. However, as Table 6.5 shows, they are not.

TABLE 6.5. Fragmentation index (shaft/proximal and distal [PD] specimens) and adjusted residual values for shaft specimens for the Le Flageolet I ungulate assemblages.

Stratum	Shaft	PD	S/PD	Shaft Adjusted Residual
High-slope relationship				
VI	158	66	2.39	−1.15 ($P > 0.10$)
VIII	220	70	3.14	0.84 ($P > 0.10$)
IX	362	72	5.03	4.87 ($P < 0.001$)
Totals	749	212	3.53	
Low-slope relationship				
IV	92	18	5.11	2.39 ($P < 0.02$)
V	863	119	7.25	11.83 ($P < 0.001$)
VII	563	489	1.15	−17.88 ($P < 0.001$)
XI	327	84	3.89	2.83 ($P < 0.01$)
Totals	1839	710	2.59	

TABLE 6.6. NISP of major skeletal elements by high- and low-slope assemblages.

Skeletal Part	High-slope	Low-slope
Phalanges	95[a]	351[b]
Metapodials	358[a]	1051[b]
Podials	49	117
Tibia	102	299
Femur	84[b]	158[a]
Innominate/Sacrum	20	45
Vertebrae	67	157
Scapula	12	48
Ribs	201[b]	326[a]
Humerus	57	146
Radius/Ulna	96[a]	314[b]
Skull/Mandible	120	289

[a] Element significantly underpresented ($P < 0.05$)
[b] Element significantly overrepresented ($P < 0.05$)

We conclude that while differential fragmentation may account for some of the patterning in the Le Flageolet I NISP-NTAXA relationships, it cannot account for all of it. We are, as a result, led to seek other causes.

Differential Skeletal Part Representation

Table 6.6 presents NISP values for major skeletal parts by high-slope and low-slope assemblages for the Le Flageolet I ungulates. Chi-square and adjusted residual analyses show that there are, in fact, significant differences in skeletal part representation across these two groups of assemblages (chi-square = 55.76, $P < 0.001$). In particular, phalanges, metapodials, and radioulnae are significantly under-represented in the high-slope assemblages.

Why this has occurred—for instance, whether this pattern represents differential bone transport or some other

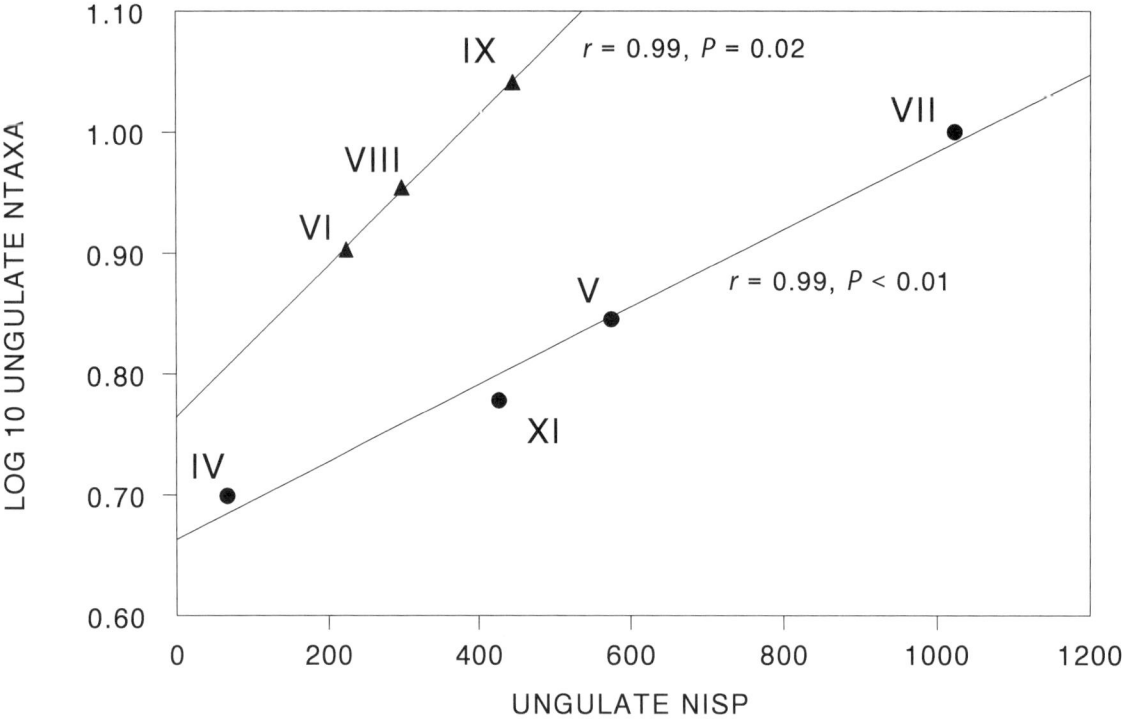

FIGURE 6.2. The relationship between ungulate NISP and (Log10) ungulate NTAXA in the Le Flageolet I faunal assemblages with phalanges, metapodials, and radioulnae excluded from the analysis.

mechanism—is not important here. What is important is that it is precisely this kind of under-representation that can drive NISP-NTAXA slopes up or down and give the impression of differences in diet breadth when none exists. If this has occurred, however, then the effect should disappear when the under-represented body parts are dropped from the analysis. Figure 6.2 replots the data presented in Figure 6.1, with phalanges, metapodials, and radioulnae removed. Clearly, the dual relationships between NISP and NTAXA represented at Le Flageolet I remain. The same results occur if over-represented elements are also excluded from the analysis. We conclude that differential body part representation, whether caused by differential bone transport or by some other mechanism, cannot account for these patterns. Note, however, that our analysis assumes that any such differential representation would not be randomly scattered across skeletal elements. Were it so distributed, our approach could not detect it.

Diet Breadth

It is, in fact, fairly evident that these curves are primarily reflecting the degree to which the Le Flageolet I assemblages are dominated by two taxa: red deer (*Cervus elaphus*) and reindeer (*Rangifer tarandus*). Figure 6.3 (see also Table 6.7) plots the fraction of each assemblage that is accounted for by these two taxa. Those assemblages most dominated by red deer and reindeer form the low-slope relationship.

TABLE 6.7. Red deer–reindeer dominance and evenness values for the Le Flageolet I ungulate assemblages.

STRATUM	RED DEER–REINDEER DOMINANCE	EVENNESS
High-slope relationship		
VI	0.755	0.661
VIII	0.794	0.604
IX	0.803	0.485
Low-slope relationship		
IV	0.910	0.262
V	0.964	0.154
VII	0.852	0.454
XI	0.813	0.446

In addition, the assemblages that form this curve strongly tend to be dominated by either red deer or reindeer, but not both (Figure 6.4). Because the low-slope assemblages are dominated by a single taxon, they are also less even than the high-slope assemblages (Table 6.7; evenness has been calculated as the Shannon Index/ln(NTAXA), and varies from 0 to 1; when evenness = 1, all taxa are equally common: see Magurran, 1988). While there are significant correlations between NISP and both evenness and red deer-reindeer dominance in the high-slope relationship, there are no such correlations in the low-slope relationship. Not surprisingly, there is a very high correlation

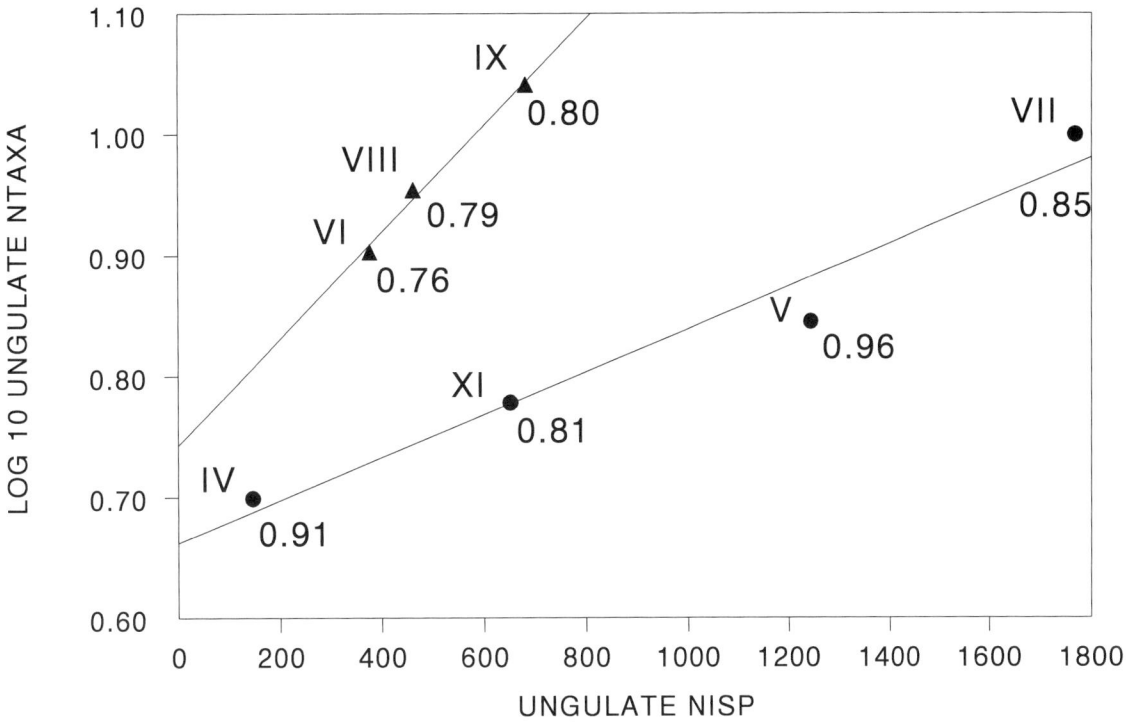

FIGURE 6.3. Red deer–reindeer dominance in Le Flageolet I ungulate assemblages.

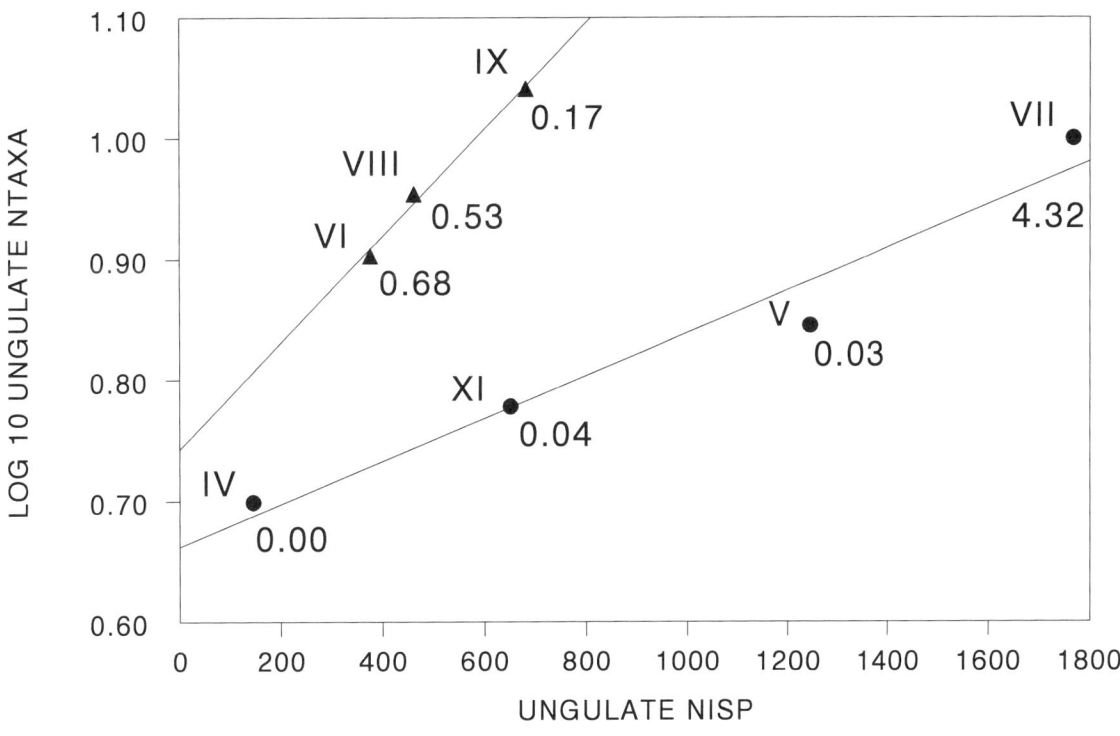

FIGURE 6.4. The ratio of red deer to reindeer in the Le Flageolet I ungulate assemblages.

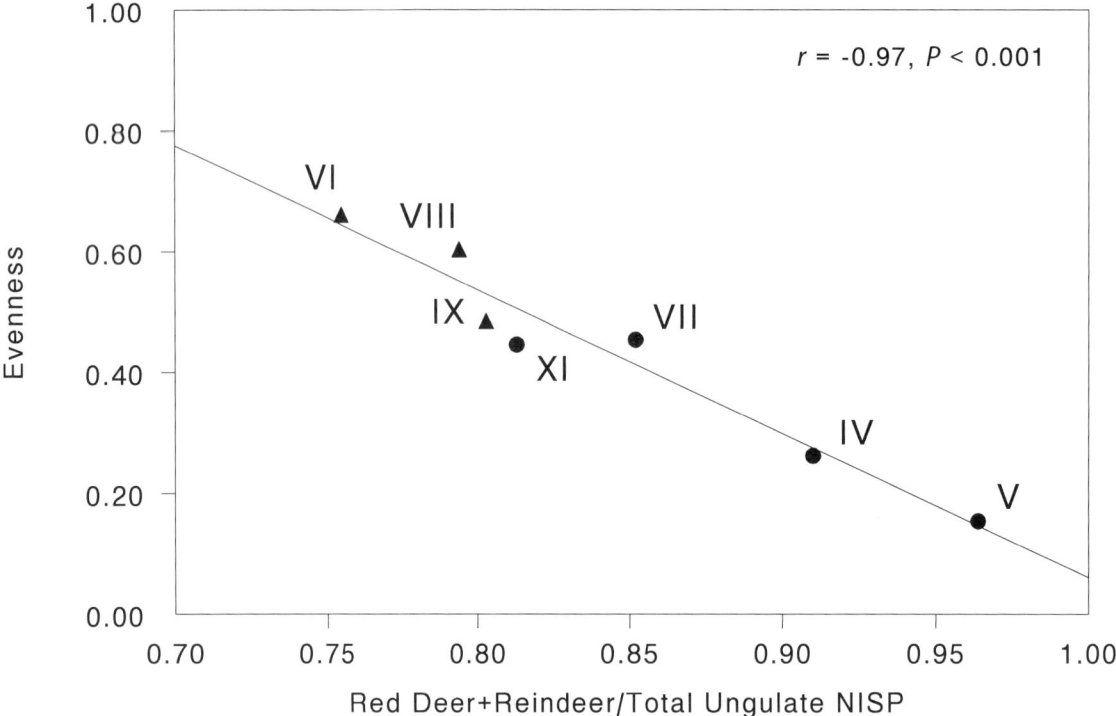

FIGURE 6.5. The relationship between evenness and red deer–reindeer dominance across the Le Flageolet I ungulate assemblages; high-slope assemblages are indicated by triangles, low-slope assemblages by closed circles.

between evenness and red deer-reindeer dominance across all assemblages: the greater the dominance, the less even the assemblage (Figure 6.5).

We thus conclude that the high-slope and low-slope relationships at Le Flageolet do, in fact, reflect diet breadth, even if some of the variability in the NISP-NTAXA relationships can be accounted for by differential fragmentation. The low-slope relationship describes assemblages that are dominated by red deer or reindeer, and hence are less even than those that form the high-slope relationship. Accordingly, the high-slope relationship contains greater numbers of taxa at a given NISP value than does the low-slope one. The low-slope assemblages reflect maximum diet breadths that were lower during the periods sampled by these assemblages than were the maximum diet breadths during the periods that the high-slope assemblages accumulated.

The question remains, of course, as to why did diet breadth change the way it did through time at Le Flageolet I? This is an issue we do not address here. We do, however, note that the reasons for these changes do not appear to reflect technological innovations. This possibility is made extremely unlikely by the fact that the Perigordian and Aurignacian assemblages lie on both curves. On the other hand, we are currently unable to control for different seasonal uses of the site through time. The information that is available suggests that the red deer of Stratum VII were taken between fall and late spring, but comparable information is not available for other taxa and other strata (Pike-Tay, 1991, 1993). As we argue elsewhere, however, the changes in richness that we have detected appear to have been driven by changes in the nature of the environments that surrounded Le Flageolet I (Delpech & Grayson, 1997).

Conclusions

Archaeological data can be used to test hypotheses drawn from foraging theory models, and thus to aid in the development of those models, or the validity of current foraging models can be assumed and then used to further our understanding of prehistoric land use and diet. We are aware of no instances of the former approach, even though the archaeological record would seem to provide a powerful source of information on the predictive strength of foraging theory models. Indeed, uses of foraging theory models to understand past settlement and subsistence change are themselves very much in their infancy (Grayson & Cannon, 1998).

In either approach, variables that are critical to the models need to be translated into archaeological terms. In this paper, we have examined one of these key variables, diet breadth, and have discussed a series of issues that can complicate the simple use of numbers of taxa in faunal assemblages as a diet breadth measure. However, even though numbers of taxa per assemblage may be problematic in this context, combined analyses of numbers of taxa, numbers of specimens, fragmentation and the distribution of specimens across body parts and taxa can provide strong evidence that diet breadth has, or has not, changed through time.

At Le Flageolet I, the two very different relationships that exist between ungulate NISP and ungulate NTAXA cannot be fully explained as a simple function of differential bone fragmentation or of differential skeletal representation. Instead, we argue that these relationships reflect distinctly different maximum diet breadths at distinctly different times as monitored at this site. These differences, we suggest, do not reflect changing technologies, but instead represent dietary responses to changes in the environments that surrounded Le Flageolet I during the early Upper Palaeolithic.

Acknowledgments

We thank M. D. Cannon, J. F. O'Connell, and E. A. Smith for very helpful comments on an earlier version of this manuscript, and J.-Ph. Rigaud and J. F. Simek for assistance throughout the Le Flageolet I project. The research reported here was supported by the National Science Foundation (BNS88-03333) and the Centre National de la Recherche Scientifique.

References

Bartram, L. E., Jr.
1993   Perspectives on skeletal part profiles and utility curves from eastern Kalahari ethnoarchaeology. In (J. Hudson, Ed.) *From Bones to Behavior.* Center for Archaeological Investigations Occasional Papers 21, 115–137.

Bayham, F. E.
1979   Factors influencing the Archaic pattern of animal utilization. *Kiva* 44, 219–235.

Broughton, J. M.
1994a   Declines in mammalian foraging efficiency during the Late Holocene, San Francisco Bay, California. *Journal of Anthropological Archaeology* 13, 371–401.
1994b   Late Holocene resource intensification in the Sacramento Valley: the archaeological vertebrate evidence. *Journal of Archaeological Science* 21, 501–514.
1995   Resource Depression and Intensification during the Late Holocene, San Francisco Bay: Evidence from the Emeryville Shellmound Vertebrate Fauna. Ph.D. Dissertation, Department of Anthropology, University of Washington, Seattle.

Broughton, J. M., D. K. Grayson
1993   Diet breadth, Numic expansion, and the White Mountains faunas. *Journal of Archaeological Science* 20, 331–336.

Charnov, E. L., G. H. Orians, K. Hyatt
1976   Ecological implications of resource depression. *American Naturalist* 110, 247–259.

Delpech, F.
1983   Les faunes du Paléolithique supérieur dans le sud-ouest de la France. *Cahiers du Quaternaire* 6.

Delpech, F., D. K. Grayson
1997   Biostratigraphie et paleoenvironnements animaux. In (J.-P. Rigaud, Ed.) *Le Flageolet I: An Early Upper Paleolithic Site in Southwestern France.* Prehistory Press, in preparation.

Edwards, D., J. F. O'Connell
1995   Broad spectrum diets in arid Australia. *Antiquity* 69, 769–783.

Enloe, J. G.
1992   Subsistence organization in the early Upper Paleolithic: reindeer hunters of the Abri de Flageolet, Couche V. In (Knecht, H., A. Pike-Tay, R. White, Eds.) *Before Lascaux: The Complex Record of the Early Upper Paleolithic.* Boca Raton: CRC Press, pp. 101–116.

Everitt, B. S.
1977   *The Analysis of Contingency Tables.* London: Chapman & Hall.

Grayson, D. K.
1984   *Quantitative Zooarchaeology.* New York: Academic Press.
1991   Alpine faunas from the White Mountains, California: adaptive change in the late prehistoric Great Basin? *Journal of Archaeological Science* 18, 483–506.

Grayson, D. K., M. D. Cannon
1999   Human paleoecology and foraging theory in the Great Basin. In (C. Beck, Ed.) *Models for the Millennium: Great Basin Anthropology Today.* Salt Lake City: University of Utah Press, pp. 141–151.

Hardy, B. L., R. A. Raff, V. Raman
1997   Recovery of mammalian DNA from Middle Paleolithic stone tools. *Journal of Archaeological Science* 24, 601–612.

Hawkes, K.
1993   Why hunter-gatherers work. *Current Anthropology* 34, 341–361.

Hawkes, K., K. Hill, J. F. O'Connell
1982   Why hunters gather: optimal foraging and the Ache of eastern Paraguay. *American Ethnologist* 9, 379–398.

Hill, K., H. Kaplan, K. Hawkes, A. Hurtado
1987   Foraging decisions among Ache hunter-gatherers: new data and implications for optimal foraging models. *Ethnology and Sociobiology* 8, 1–36.

Jones, K. T., D. B. Madsen
1989   Calculating the cost of resource transportation: a Great Basin example. *Current Anthropology* 30, 529–534.
1991   Further experiments in native food processing. *Utah Archaeology* 4, 68–77.

Kaplan, H., K. Hill
1992   The evolutionary ecology of food acquisition. In (Smith, E. A., B. Winterhalder, Eds.) *Evolutionary Ecology and Human Behavior.* New York: Aldine de Gruyter, pp. 167–202.

Kelly, R. L.
1995   *The Foraging Spectrum: Diversity in Hunter-Gatherer Lifeways.* Washington, D.C.: Smithsonian Institution Press.

Lyman, R. L.
1994   *Vertebrate Taphonomy.* Cambridge: Cambridge University Press.

Madsen, D. B., J. E. Kirkman
1988   Hunting hoppers. *American Antiquity* 53, 593–604.

Magurran, A. E.
1988   *Ecological Diversity and its Measurement.* Princeton: Princeton University Press.

Marean, C. W.
1991   Measuring the post-depositional destruction of bone in archaeological assemblages. *Journal of Archaeological Science* 18, 677–694.

Marshall, F., T. Pilgrim
1991   Meat versus within-bone nutrients: another look at the meaning of body part representation in archaeological sites. *Journal of Archaeological Science* 18, 149–164.

O'Connell, J. F.
1995   Ethnoarchaeology needs a general theory of behavior. *Journal of Archaeological Research* 3, 205–255.

O'Connell, J. F., K. Hawkes
1981   Alyawara plant use and optimal foraging theory. In (Winter-

halder, B., E. A. Smith, Eds.) *Hunter-Gatherer Foraging Strategies: Ethnographic and Archaeological Analyses.* Chicago: University of Chicago, pp. 99–125.

O'Connell, J. F., K. Hawkes, N. G. Blurton-Jones
1988 Hadza hunting, butchering, and bone transport and their archaeological implications. *Journal of Anthropological Research* 44, 113–161.
1990 Reanalysis of large mammal body part transport among the Hadza. *Journal of Archaeological Science* 17, 301–316.

O'Connell, J. F., K. T. Jones, S. R. Simms
1982 Some thoughts on prehistoric archaeology in the Great Basin. In (Madsen, D. B., J. F. O'Connell, Eds.) *Man and Environment in the Great Basin.* Society for American Archaeology Papers 2, 227–240.

Pike-Tay, A.
1991 *Red Deer Hunting in the Upper Paleolithic of Southwest France: a Study in Seasonality.* British Archaeological Reports, International Series 569.
1993 Hunting in the Upper Perigordian: a matter of strategy or expedience? In (Knecht, H., A. Pike-Tay, R. White, Eds.) *Before Lascaux: the Complex Record of the Early Upper Paleolithic.* Boca Raton: CRC Press, pp. 85–99.

Rigaud, J.-Ph.
1982 *Le Paléolithique en Périgord: les Données du Sud-ouest Sarladais et Leurs Implications.* Thèse de Doctorat d'Etat ès Sciences, Université de Bordeaux I, Talence.
1993 L'Aurignacien dans le sud-ouest de la France: bilan et perspectives. *Actes de XIIᵉ Congrès International des Sciences Prehistoriques et Protohistoriques, 2: Aurignacien en Europe et au Proche Orient.* Bratislava: Union International des Sciences Prehistoriques et Protohistoriques, pp. 181–186.

Rigaud, J.-Ph. (Ed.)
1997 *Le Flageolet I: an Early Upper Paleolithic Site in Southwestern France.* Prehistory Press, in preparation.

Simek, J.
1984 *A K-Means Approach to the Analysis of Spatial Structure in Upper Paleolithic Habitation Sites: Le Flageolet I and Pincevent.* British Archaeological Reports, International Series 205.

Simek, J. F.
1987 Spatial order and behavioural change in the French Palaeolithic. *Antiquity* 61, 25–40.

Simms, S. R.
1985 Pine nut use in three Great Basin cases: data, theory, and a fragmentary material record. *Journal of California and Great Basin Anthropology* 7, 166–175.
1987 *Behavioral Ecology and Hunter-Gatherer Foraging: An Example from the Great Basin.* British Archaeological Reports, International Series 381.

Smith, E. A.
1991 *Inujjuamiut Foraging Strategies: Evolutionary Ecology of an Arctic Hunting Economy.* New York: Aldine de Gruyter.

Stephens, D. W., J. R. Krebs
1986 *Foraging Theory.* Princeton: Princeton University Press.

Szuter, C. R., F. E. Bayham
1989 Sedentism and prehistoric animal procurement among desert horticulturalists of the North American Southwest. In (S. Kent, Ed.) *Farmers as Hunters.* Cambridge: Cambridge University Press, pp. 80–95.

Todd, L. C., D. J. Rapson
1988 Long bone fragmentation and interpretation of faunal assemblages: approaches to comparative analysis. *Journal of Archaeological Science* 15, 307–325.

Winterhalder, B., C. Goland
1997 An evolutionary ecology perspective on diet choice, risk, and plant domestication. In (K. J. Gremillion, Ed.) *People, Plants, and Landscapes: Studies in Paleoethnobotany.* Tuscaloosa: University of Alabama Press, pp. 123–160.

7

Paleolithic Population Growth Pulses Evidenced by Small Animal Exploitation

Mary C. Stiner, Natalie D. Munro, Todd A. Surovell, Eitan Tchernov, Ofer Bar-Yosef

The size of a population has much to do with its long-term prospects for survival and the potential impact of random events on its evolutionary history. It is difficult to appreciate or quantify just how small early Paleolithic populations were, or how thinly they were spread during much of the Pleistocene. Thus, to assess when human populations grew during the Late Pleistocene, we analyzed trends in the small animal species most commonly eaten by Paleolithic foragers.

Paleolithic humans relied on both small animals and ungulates for meat,[1] but predator-prey relations between humans and small animals are more sensitive indicators of changes in human population density[2] because small prey species vary much more than ungulate species with respect to life history and predator avoidance characteristics. In the Mediterranean Basin (Israel and Italy) small animals were important to human diets throughout the Middle, Upper, and Epi-Paleolithic periods (Tables 7.1 and 7.2).[3] Most common in the Paleolithic refuse middens are the remains of tortoises (mainly *Testudo graeca*), shellfish (in Italy, *Patella* spp., *Mytilus galloprovincialis*, *Ostrea edulis*, *Callista chione*, *Glycymeris* spp., *Monodonta turbinata*, and others), partridges (*Alectoris chukar*, *Perdix perdix*, *Coturnix coturnix*), hares (*Lepus capensis*), and rabbits (*Oryctolagus cuniculus*). Other small-bodied species were also consumed on occasion, including doves, waterfowl, hedgehogs, marmots, squirrels, legless lizards, and large snakes, as well as ostrich eggs. Damage from fire (Fig. 7.1), tool marks, and percussion fractures from stone hammers link the skeletal remains of small vertebrates and invertebrates to the feeding activities of prehistoric humans.[4] Littoral mollusks and tortoises, the most easily caught small prey species, are also the most sensitive to hunting pressure from humans.

We examined two faunal series for changes in small prey composition, one compiled from archaeological sites on the western coast of Italy (Liguria and Lazio) and another from two sites lying 1 km apart in the Nahal Meged, an inland valley in the Galilee of Israel (Fig. 7.2). Each series spans the Middle through Epi-Paleolithic periods, beginning in Israel about 200,000 years ago (200 ka) at Hayonim Cave,[5] and about 110 ka in Italy at Grotta dei Moscerini.[6] Each series ends around the Pleistocene-Holocene boundary 10 ka. The Middle to Upper Paleolithic cultural transition occurred about 44 ka in Israel and 36 ka in Italy. The Upper Paleolithic to Epi-Paleolithic transition in each region coincided roughly with the Last Glacial Maximum, 20

FIGURE 7.1. Remains of a Middle Paleolithic meal: fractured tortoise (*Testudo graeca*) carapace preserved in cemented wood ash, from 200,000 year old stratigraphic layer in Hayonim Cave, Israel.

*Originally published in *Science* 283(1999):190–194

TABLE 7.1. Relative abundances of small prey animals in the Paleolithic faunal series from Italy (Ligurian and Lazio sites combined) time-ordered from most recent to oldest.

Culture Period (and Site)	Age Range (ka)	Small to Large Game (index)[a]	Total Small Game[b]	Tortoise (%)	Hares and Rabbits (%)	Other Small Mammals (%)	Birds (%)	Shellfish (%)
EP Late Epigravettian (RM)[c]	9–12	0.83	802	0	«1	0	0	100
EP Late Epigravettian (GPo)[d]	10–11	0.02	889	0	41	1	58	0
EP Evolved Epigravettian (GPa)[d]	15–16	0.01	30	0	17	0	83	0
EP Early Epigravettian (RM)	17–19	0.40	1176	0	45	6	34	14
UP Gravettian (RM)	24–28	0.21	767	0	23	15	43	19
UP Middle Aurignacian (RM)	27–32	0.46	416	0	2	9	12	76
UP Early Aurignacian (RM)	32–36	0.46	710	0	4	6	18	71
MP Middle Paleolithic (GM)	70–110	0.47	660	6	1	0	0	93

Note: Shellfish are limpets and common Mediterranean bivalves such as mussels. Birds are mostly partridges, with lower frequencies of other game birds such as Columbiforms and Anseriforms. Other small mammals are marmots and hedgehogs. Major culture periods are Epi-Paleolithic (EP), 9 to 19 ka; Upper Paleolithic (UP), 19 to 36 ka; and Middle Paleolithic (MP), 36 to 200 ka; successive phases are indicated therein. Sites are Riparo Mochi (RM), Grotta Polesini (GPo), Grotta Palidoro (GPa), and Grotta dei Moscerini (GM).
[a] Number of small game remains divided by sum of small game and ungulate remains.
[b] Counting units are number of identified skeletal specimens (NISP) for vertebrate remains, but minimum number of individuals (MNI) for marine mollusks; the latter to correct for differences in average specimen sizes.
[c] Special purpose occupation, based on the paucity of terrestrial resources.
[d] Inland site for comparison with Late Epigravettian assemblage from Riparo Mochi.

TABLE 7.2. Relative abundances of small prey animals in the Paleolithic faunal series from Israel (Nahal Meged, Galilee) time-ordered from most recent to oldest.

Culture Period (and Site)	Age Range (ka)	Small to Large Game (index)[a]	Total Small Game[b]	Tortoise (%)	Hares (%)	Ostrich Eggshell (%)	Birds (%)
EP Natufian (Hay)	11 to 13	0.57	1154	35	30	0	35
EP Kebaran (Hay)	14 to 17	0.17	532	77	9	2	13
EP early Kebaran (Meg)	18 to 19	0.37	730	64	12	0	23
UP Pre-Kebaran (Meg)	19 to 22	0.27	160	77	6	0	16
UP Aurignacian (Hay)	26 to 28	0.28	2950	60	5	0	34
MP (Hay 350-419 bd)	~150[c]	0.29	437	89	5	«1	6
MP (Hay 420-469 bd)	~170[c]	0.39	2625	95	<1	2	2
MP (Hay 470-539 bd)	~200[c]	0.52	2371	97	<1	1	1

Note: Birds are mostly partridges, with lower frequencies of other bird taxa; tortoise column includes scant remains of large lizards; hare column includes scant remains of Persian squirrels and hedgehogs. Culture periods are Epi-Paleolithic (EP), 10 to 19 ka; Upper Paleolithic (UP), 19 to 44 ka; and Middle Paleolithic (MP), 44 to 200 ka. Sites are Hayonim Cave (Hay) and Meged Rockshelter (Meg).bd, centimeters below datum. The Hayonim E 200- to 349-bd assemblage is omitted, because analysis is not complete except for the tortoise measurements in Figure 7.5.
[a] Number of small game remains divided by sum of small game and ungulate remains.
[b] Counting unit is always number of identified skeletal specimens.
[c] Approximate age.

to 18 ka. Neolithic cultures, which represent the transition from forager (Epi-Paleolithic hunter-gatherer) to food producing economies began <10.5 ka in Israel, and <8 ka in Italy.

The proportion of small to large game animals taken by Paleolithic humans in the two study areas shows no consistent trend over the past 200,000 years (Tables 7.1 and 7.2). However, the types of small animals most often consumed by prehistoric foragers—tortoises, shellfish, partridges, hares, and rabbits—shifted dramatically in both areas during the Upper and Epi-Paleolithic, even though the compositions of the biotic communities remained relatively stable.[6,7]

The small game species form two simple categories on the basis of how easy they are to catch and intrinsic potentials for population growth. We term these prey types "slow" and "quick."

Tortoises and marine shellfish (in shoreline sites) dominate the Middle Paleolithic record of small game use (Fig. 7.3). Tortoises and most littoral shellfish are immobile or sluggish (essentially sessile) and, once discovered, are easy to collect. Because of their low metabolisms and long lifespans, these animals can exist at high densities in some habitats, where they become attractive prey. Tortoises and shellfish mature relatively slowly, however, requiring several years to reach reproductive age,[2] and individuals continue to grow well into their adult years. Agile, warm-blooded animals that mature rapidly (<1 yr)—mostly partridges and hares in our archaeofaunal series—became important in human diets only later in each series:[8] birds in the early Upper Paleolithic, followed by lagomorphs in the Epi-Paleolithic (Tables 7.1 and 7.2).

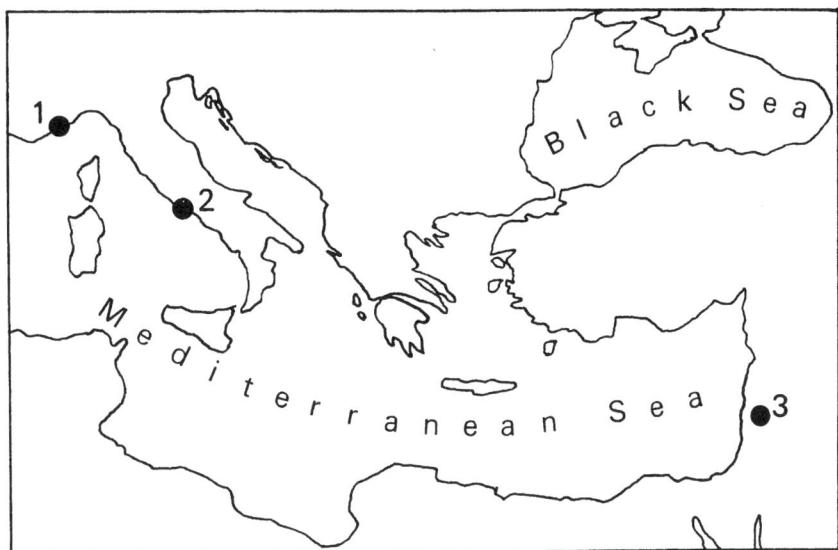

FIGURE 7.2. Sources of archaeological sites in the Mediterranean Basin: In Italy (1) Riparo Mochi on northwestern coast, Province of Liguria, (2) Grotta dei Moscerini on west-central coast and Grotta Polesini and Grotta Palidoro inland, Province of Lazio; in Israel (3) Hayonim Cave and Meged Rockshelter 1 km apart in an inland valley (Nahal Meged) in the western Galilee.

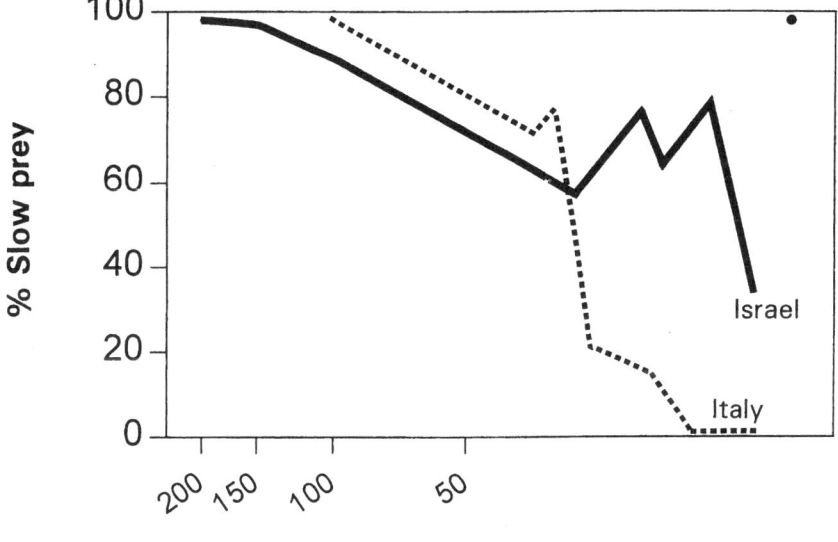

FIGURE 7.3. Trend in the relative contribution (%) of "slow" prey types (relative to "quick" prey) to the Pleistocene small game assemblages from Italy and Israel. These prey are immobile or slow-moving, and have slow population turnover rates. In Italy they are primarily edible shellfish and low frequencies of tortoises; in Israel they are predominantly tortoises accompanied by scant remains of other large reptiles (legless lizards) and ostrich eggs. A mean time (ka) is used for each assemblage in this graph, as opposed to the time range listed in Tables 7.1 & 7.2. The isolated point represents the anomalous Late Epigravettian assemblage in the Italian series; see explanation in text.

The youngest (Late Epigravettian) assemblage from Riparo Mochi in coastal Italy presents an exception to the trend described above in that shellfish were practically the only prey consumed (isolated point in Fig. 7.3), but it is clear that birds and lagomorphs were important at contemporaneous inland sites (Table 7.1) such as at Grotta Polesini[6,9] and Grotta Palidoro.[6,10]

By contrast, where small game are present in Middle Paleolithic sites at all, the focus invariably was on the easily caught types.

Tortoises and shellfish continued to be consumed during the Upper and Epi-Paleolithic periods, but they figured less prominently with respect to total small game intake. The range of species targeted by human foragers changed little with time

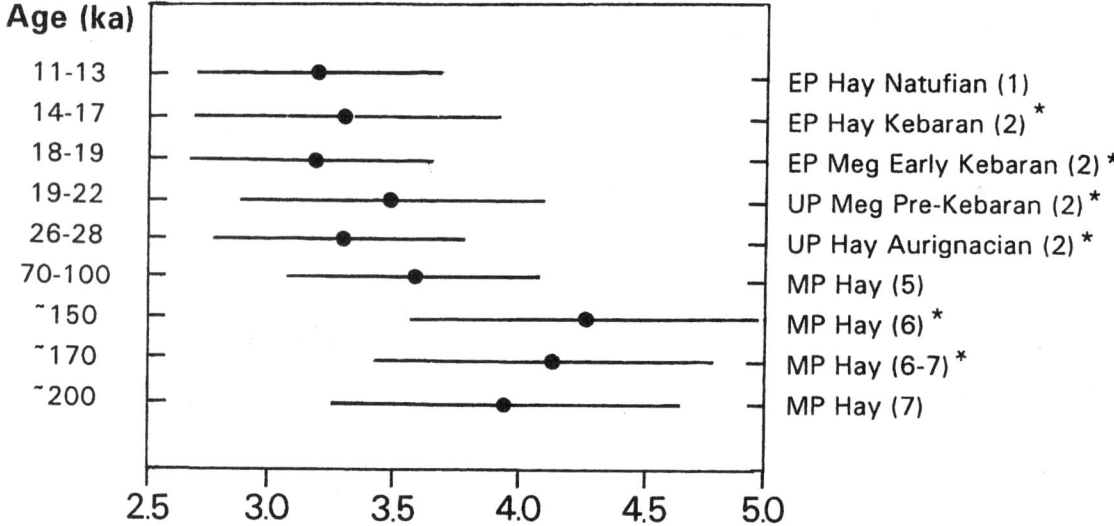

FIGURE 7.4. Size reduction trend in spur-thighed tortoises (*Testudo graeca*) in the time-ordered assemblages from the Nahal Meged, Israel, based on mean values for the mediolateral dimension of the humeral diaphysis (mm) and standard deviations. The size decline does not coincide with changes in global climate as defined by Shackleton and Opdyke's[13] oxygen isotope stages (shown in parentheses). A hiatus in human occupations lasting roughly 40 thousand years separates these size groups in the Nahal Meged series. At Kebara Cave on Mount Carmel (not shown), a marked size decline in tortoises occurs ca. 44 ka, filling the gap in the Nahal Meged sequence. (Hay) Hayonim Cave; (Meg) Meged Rockshelter; (*) generally colder/drier climate; (ka) thousand years ago; (~) ages are preliminary, based on biostratigraphy, and on TL dating by E. Valladas and N. Mercier, and ESR dating by J. Rink and H. P. Schwarcz.

except as conditioned by the categorical increase in the exploitation of birds, an inherently diverse taxonomic group. Overall, however, few of the many bird and small mammal species known to have inhabited the Mediterranean Basin at the time were routinely taken.[6,7] More significant is increasingly even use of high-ranked ("slow") and low-ranked ("quick") small prey types with time.

Further evidence that the shifts in prey species resulted from hunting pressure and not simply climate or environmental change is that the mean sizes of slow-growing prey types decreased. Diminution was sudden for tortoises (*T. graeca*) in Israel (Fig. 7.4), beginning either in the late Middle Paleolithic or in the earliest Upper Paleolithic. Size suppression in the tortoises was sustained through the later periods. A hiatus in human occupations separates the end of the Middle Paleolithic and the Upper Paleolithic in the Nahal Meged series from Israel [Hayonim Cave and Meged Rockshelter[11] combined], however, which makes it difficult to pinpoint the onset of tortoise diminution in this faunal series. The late Middle Paleolithic (about 65 to 50 ka) and early Upper Paleolithic (Ahmarian phase, about 44 ka) levels from Kebara Cave[12] fill this gap, and, while this site is situated in a slightly richer vegetation zone where tortoises were always correspondingly larger, a significant decline in mean size is associated with the earliest Ahmarian (Upper Paleolithic) phase there. Mean size reduction in tortoises therefore took place at least 44 ka and probably earlier. Limpet (always dominated by *Patella caerulea*) diminution at Riparo Mochi in Italy began later, around 23 ka (Fig. 7.5). In each region an abrupt, significant decline in the average size of individual prey collected by foragers was sustained across major oscillations in world climate, as inferred on the basis of oxygen isotope stages and pollen and rodent chronologies.[13] The diminution trends in tortoises and limpets thus are not explained by climate change or by local species replacement.

Predator-prey simulations[2] show that humans' decreasing emphasis on and coincident size suppression in slow-growing taxa could easily have resulted from overharvesting, especially if the intervals between exploitation events were shortened as human populations grew. The predator-prey simulations use life history parameters for modern Mediterranean tortoises (*T. graeca* and *T. hermanni*), partridges (*P. perdix* and *A. chukar*), and hares (*Lepus europeaus*, *L. capensis*, and certain North American species). Hunting offtake was increased incrementally as a percentage of the whole population until that population crashed. Population growth and the effects of hunting pressure were modeled under favorable and subaverage conditions, which produced a tolerance range for each type of prey rather than a single, averaged measure of resilience. Differing maturation rates among the three kinds of prey proved crucial and, because of this, hare and partridge populations were minimally 7 to 10 times as resilient as tortoise populations to sustained hunting pressure (Fig. 7.6). Ease of capture accounts for the appar-

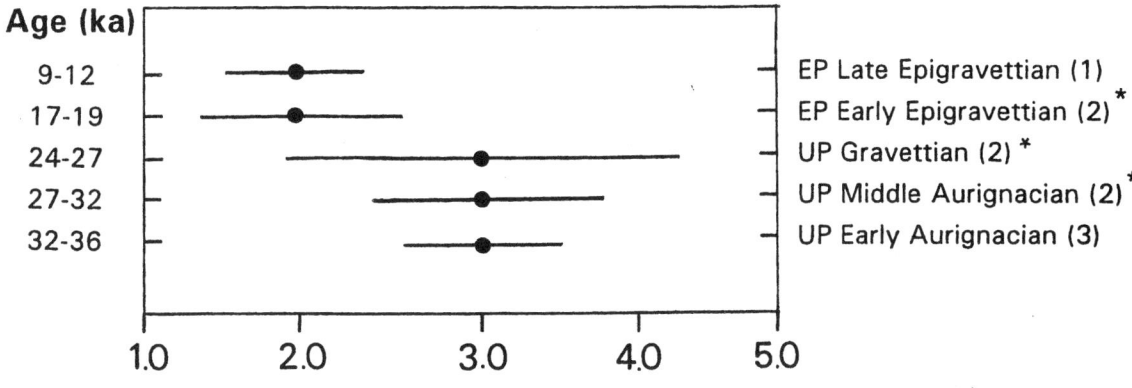

FIGURE 7.5. Size reduction trend in limpets (mainly *Patella caerulea*) in time-ordered Upper Paleolithic assemblages from Riparo Mochi (Liguria) in Italy, based on mean shell diameter (cm) and standard deviations. Size reduction occurred abruptly between the Gravettian and Early Epigravettian phases (ca. 23 ka). Although more than one limpet species is represented, there are no significant differences in species proportions among the cultural layers. Time ranges are in uncalibrated radiocarbon years, based on direct and indirect dating. Symbols as in Fig. 7.4.

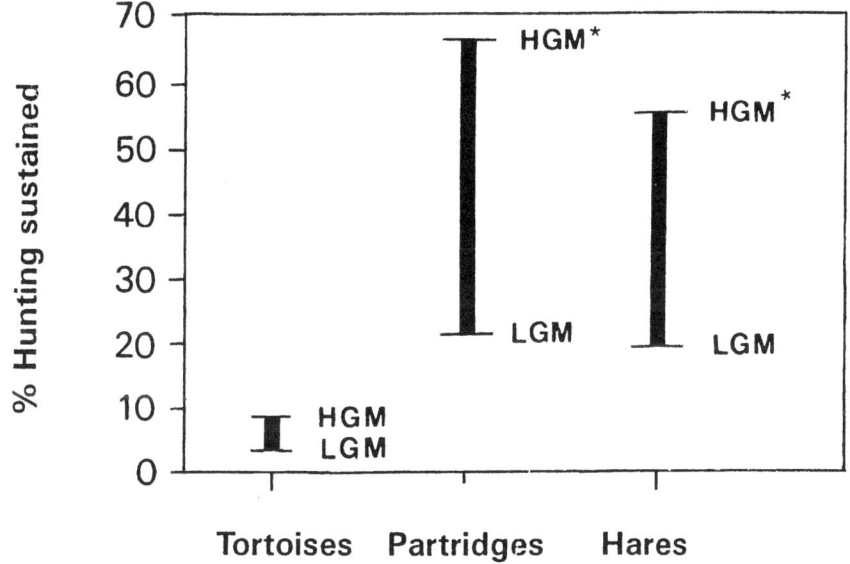

FIGURE 7.6. Comparison of hunting tolerance thresholds for tortoise (4–7%), partridge (22–65%), and hare (18–53%) populations in high (HGM) and low (LGM) growth models. Tortoise populations are far less resilient to predator effects than are partridges or hares. Upper horizontal bars are crash thresholds, above which predators' dependence on that prey type is unsustainable; vertical bar is the range of variation in population resilience between the LGM and HGM; (*) HGM threshold is conservative and could be higher in reality.

ently high ranking of tortoises and littoral shellfish by Middle Paleolithic foragers. We did not model shellfish populations, but earlier studies suggest that, under conditions of intense predation, the marine mollusks normally eaten by humans are only somewhat more resilient than tortoises.[14]

The prevalence of tortoise and shellfish remains in some Middle Paleolithic sites is remarkable in light of the simulation results. Shellfish and tortoises constitute at least 46% of the total number of large- and small-bodied animal remains in the early Middle Paleolithic site of Grotta dei Moscerini, Italy. Tortoises, along with a few large lizards and ostrich eggs, constitute as much as 52% of specimens identified in the early Middle Paleolithic layers of Hayonim Cave, Israel, and the tortoises collected by early Middle Paleolithic foragers were quite large on average. A high and sustained dependence on these slow-maturing animals for complete dietary protein and other nutrients implies that early human populations were exceptionally small, even by later Paleolithic standards, and that early Middle Paleolithic humans did not spend much time foraging in any one vicinity.

The trends in Paleolithic small game use can serve as a demographic barometer.[2,15] Human populations in the northern and eastern Mediterranean appear to have grown in a series of pulses, which began earlier in the east. Paleolithic foragers of the Mediterranean Rim could not dispense with small game when easily collected types were in short supply. What were once occasional shortages of highly-ranked (slow) types in the Middle Paleolithic, evidenced by the rare presence of bird and lagomorph remains in the archaeofaunas, became chronic shortages 44 to 35 ka because of hunting pressure, which forced people's attention to fleeter (lower ranked) prey types. By contrast, early humans were uniformly interested in the large packages of meat that ungulates represent: there is remarkably little evidence for choosiness on the part of Paleolithic hunters when procuring ungulate species in our study areas,[6,16] apart from the elevations at which people were willing to travel to find them,[17] and there are no trends in the relative contribution of ungulates to total game consumed. There appears to have been more room for adjustment in the exploitation of small animals.

Hunter-gatherers of the historic era normally obtain game birds and lagomorphs in quantity only with the help of special tools that take time to make and maintain (nets, snares, or barbed weapon tips[18]) or through communal game drives.[19] We conclude that the key to ancient humans' lower ranking of birds and lagomorphs was the ability of these animals to escape quickly, which translates to a high work of capture that could be reduced only with trap-related technology. In western Asia, the timing of tortoise diminution shows that human populations increased substantially before the remarkable and rapid technologic innovations (radiations) that mark the Upper and Epi-Paleolithic periods.[20]

Our results show that the breadth of human diet increased during the Upper and Epi-Paleolithic[21] and support the notion that population pressure played a significant role in the evolution of Late Pleistocene human cultures.[22] Mobility was the preferred solution to local resource scarcity throughout much of prehistory. Any loss of mobility options is a grave matter for people who live by hunting and gathering. The changes in prey species during the Mediterranean Paleolithic nonetheless indicate demographic packing and associated reductions in mobility.[22] The diversity of prey species consumed by humans did not substantially increase with time as supposed;[23] rather, humans responded to differences in the defense mechanisms of a few key prey species as human population densities grew. Quite by accident, the differences in prey defense mechanisms to human predation correspond closely to differences in prey population resiliencies. These characteristics in turn affected the relative availability of slow and quick types as predator pressure increased. Intensified use of just a few species, a tendency also diagnostic of food-producing economies of the Neolithic, has its origins in the hunting and gathering lifeways of the later Paleolithic.

Anatomically modern humans are thought to have evolved within the Middle Paleolithic period and overlapped for some time with human populations possessing archaic features (for example, the Neandertals). With so few people on earth, replacement of one human population with another could have involved minor differences in the intrinsic rate of increase, such as from a modest improvement in child survivorship. Low human population densities during most of the Middle Paleolithic imply that group sizes and social networks were small, which certainly limited the numeric scope of individual interactions. Under these conditions the possibilities for evolution of complex sharing and exchange behavior as ways to counter the effects of unpredictable resource supplies would also have been quite limited.[20] The value of more diverse and efficient foraging equipment may have been lower as well.

Our data imply that Middle Paleolithic foragers simply did not experience the types of demographic and economic stresses that would tend to accelerate technological and social evolution of the sort typical of the Upper and Epi-Paleolithic. The zooarchaeological data, which can be dated and tied to region, clarify the timing and geographic centers of rapid population growth suggested by research on human molecular phylogenetics.[24]

Acknowledgments

Supported by grants from the National Science Foundation (SBR-9511894 to M. C. S.; and SBR-9409281 to O. B-Y.) and the Levi-Sala (Care) Archaeological Foundation (to N. D. M.). We thank S. L. Kuhn for advice on simulation model development, unpublished data on Meged Rockshelter, and assistance in manuscript preparation. Thanks also to A. Belfer-Cohen, A. Segre, E. Segre-Naldini, A. Bietti, and P. Cassoli for access and information to the faunal collections from Israel and Italy; to A. Recchi for unpublished data on bird remains from Riparo Mochi, and S. Heppel, Y. Werner, and J. Behler for information on Mediterranean tortoise ecology; and to H. Harpending and the *Science* reviewers for many helpful comments on this manuscript.

References and Notes

1. Tortoises require 8 to 12 years to reach adulthood, and most edible shellfish require 2 to 5 years; see M. C. Stiner, N. D. Munro, T. A. Surovell, *Curr. Anthropol.*, in press.
2. H. Harpending and J. Bertram, in *Population Studies in Archaeology and Biological Anthropology*, A. C. Swedlund, Ed. (Society for American Archaeology Memoir 30, Washington DC, 1975), pp. 82–91.
3. R. G. Klein and K. Scott, *J. Archaeol. Science* 13, 515–542 (1986); A. Palma di Cesnola, *Riv. Sci. Preistoriche* 25, 3–87 (1965); *Estratto dagli Scritti sul Quaternario in Onore di Angelo Pasa* (Museo Civico di Storia Naturale, Verona, 1969), 95–135; H. Valladas et al.,

Nature 331, 614–616 (1988); O. Bar-Yosef, in *Humans at the End of the Ice Age: the Archaeology of the Pleistocene-Holocene Transition*, L. G. Straus, B. V. Eriksen, J. M. Erlandson, D. Yesner, Eds. (Plenum, New York, 1996), pp. 61–76.

4. M. C. Stiner, in *Hunting and Animal Exploitation in the Later Palaeolithic and Mesolithic of Eurasia*, G. L. Peterkin, H. Bricker, P. Mellars, Eds. (Archaeological Papers of the American Anthropological Association, Washington DC, 1993), vol. 4, pp. 101–119; M. C. Stiner and E. Tchernov, in *Neanderthals and Modern Humans in West Asia*, T. Akazawa, K. Aoki, O. Bar-Yosef, Eds. (Plenum, New York, 1998), pp. 241–262; S. L. Kuhn and M. C. Stiner, *Curr. Anthropol.* 39, S175 (1998).

5. O. Bar-Yosef, in *Prehistoire du Levant*, J. Cauvin and P. Sanlaville, Eds. (Editions du CNRS., Paris, 1981), pp. 389–408.

6. M. C. Stiner, *Honor Among Thieves: A Zooarchaeological Study of Neandertal Ecology* (Princeton Univ. Press, Princeton, 1994).

7. E. Tchernov, in *The Evolution and Dispersal of Modern Humans in Asia*, T. Akazawa, K. Aoki, T. Kimura, Eds. (Hokusen-Sha, Tokyo, 1992), pp. 149–188; E. Tchernov, in *Late Quaternary Chronology and Paleoclimates of the Eastern Mediterranean*, O. Bar-Yosef and R. S. Kra, Eds. (Radiocarbon, Tucson, AZ, 1994), pp. 333–350.

8. See also Pichon [J. Pichon, in *La Faune du Gisement Natufien de Mallaha (Eynan) Israel*, J. Bouchud, Ed. (Memoires et Travaux du Centre de Recherche Français de Jerusalem, Association Paléorient, Paris), pp. 115–150.

9. A. M. Radmilli. *Gli Scavi Nella Grotta Polesini a Ponte Lucano di Tivoli e la Piu Antica Arte nel Lazio* (Sansoni Editore, Firenze, 1974).

10. P. Cassoli, *Quaternaria* 19, 187 (1976/77).

11. S. L. Kuhn et al., *Report on the 1997 Excavation Season at Meged Rockshelter (Upper Galilee, Israel)*, Permit G-47/97 (Report to Israel Antiquities Authority, Jerusalem, 1998).

12. O. Bar-Yosef et al., *Curr. Anthropol.* 33, 497 (1992); O. Bar-Yosef et al., *J. Archaeol. Sci.* 23, 297 (1996).

13. N. J. Shackleton and N. D. Opdyke, *Quat. Res.* 3, 39 (1973); S. Bottema, in *Chronologies in the Near East*, G. Aurenche, J. Ervin, and P. Hours, Eds. (Bar International Series, Oxford, 1987), pp. 295–310; E. Tchernov, *Paléorient* 23, 209 (1998).

14. S. Botkin, in *Modeling Change in Prehistoric Subsistence Economies*, T. K. Earle and A. L. Christenson, Eds. (Academic Press, New York, 1980), pp. 121–139.

15. G. A. Clark and L. G. Straus, in *Hunter-Gatherer Economy in Prehistory*, G. Bailey, Ed. (Cambridge Univ. Press, 1983), pp. 131–148; R. G. Klein, *The Human Career: Human Biological and Cultural Origins* (Univ. Chicago Press, Chicago, 1989).

16. M. C. Stiner, *Curr. Anthropol.* 33, 433 (1992).

17. C. Gamble, *The Palaeolithic Settlement of Europe* (Cambridge Univ. Press, Cambridge, 1986); M. C. Stiner, *Quat. Nova* 1, 333 (1990/91).

18. W. H. Oswalt, *An Anthropological Analysis of Food-Getting Technology* (Wiley, New York, 1976).

19. J. H. Steward, *Basin-Plateau Aboriginal Sociopolitical Groups* (Bulletin No. 120, Bureau of American Ethnology, Washington, DC, 1938).

20. S. L. Kuhn and M. C. Stiner, in *Creativity in Human Evolution and Prehistory*, S. Mithen, Ed. (Routledge, London, 1998), pp. 143–164.

21. Following Stephens and Krebs [D. W. Stephens and J. R. Krebs, *Foraging Theory* (Princeton University Press, Princeton, 1986)].

22. K. V. Flannery, in *The Domestication and Exploitation of Plants and Animals*, P. J. Ucko and G. W. Dimbleby, Eds. (Aldine, Chicago, 1969), pp. 73–100; L. R. Binford, in *New Perspectives in Archaeology*, S. R. Binford and L. R. Binford, Eds. (Aldine, Chicago, 1968), pp. 313–341; M. N. Cohen, *The Food Crisis in Prehistory: Overpopulation and the Origins of Agriculture* (Yale Univ. Press, New Haven, CT, 1977); O. Bar-Yosef and A. Belfer-Cohen, *J. World Prehistory* 3, 447 (1989); L. H. Keeley, in *Last Hunters—First Farmers*, T. D. Price and A. B. Gebauer, Eds. (School of American Research, Santa Fe, NM, 1995), pp. 243–272; P. J. Watson, in ibid, pp. 21–37.

23. Compare recent studies by D. O. Henry, [*From Foraging to Agriculture: The Levant at the End of the Ice Age* (Univ. Pennsylvania Press, Philadelphia, 1989)]; P. C. Edwards, [*Antiquity* 63, 225 (1989); M. P. Neeley and G. A. Clark, [in *Hunting and Animal Exploitation in the Later Palaeolithic and Mesolithic of Eurasia*, G. L. Peterkin, H. Bricker, P. Mellars, Eds. (Archaeological Papers of the American Anthropological Association, Washington DC, 1993), vol. 4, pp. 221–240.]

24. See, for example, J. C. Long, *Ann. Rev. Anthropol.* 22, 251 (1993); Sherry et al., *Hum. Biol.* 66, 761 (1994); D. E. Reich and D. B. Goldstein, *Proc. Natl. Acad. Sci. U.S.A.* 95, 8119 (1998).

Hunter-Gatherer Foraging and Colonization of the Western Hemisphere

Robert L. Kelly

INTRODUCTION

The human tide that washed over Europe, Asia, and the New World beginning around 1.5 million years ago consisted initially of pre-sapiens hominids and later of *H. sapiens sapiens*. There can be little doubt that pre-sapiens hominids were not human in the way that we know humanity. They were different from us. It is not clear what those differences are, but understanding them is necessary to understanding the process of human evolution.

Understanding the biological difference between pre-sapiens hominids and modern humans comes from an examination of skeletal remains; understanding the cultural difference comes from examining archaeological evidence of behaviour. One potential research area is to examine on broad spatial and temporal scales the differences in how pre-sapiens and *H. sapiens sapiens* colonized large, unpopulated land masses. I suggest broad temporal and spatial scales because the nature of archaeological data makes it difficult to specify dates of colonization very accurately and because the colonization process is one that occurs on a broad spatial scale. Examining how pre-sapiens and modern humans coped with a similar problem may point to ways in which they were similar or different. The three largest land masses that were colonized initially by *H. sapiens sapiens* are Australia, north-eastern Asia, and, by far the largest, the Western Hemisphere. We will examine the last of these.

The colonization of the Western Hemisphere was a historical event (or events) that was part of demographic, economic, and ecological processes. We know that Native Americans came from north-eastern Asia across Beringia, entering lower North America through the ice-free corridor or by traversing the western coast. We also know that people were in North America by at least 11,200 BP. Unsettled, however, is the question of exactly when people first arrived in the New World. The answer to this question bears not only on the culture history of Native Americans, but on theoretical concerns as well because the timing of the arrival tells us something about the way in which people occupied the Americas. Therefore, the question of New World colonization needs to be couched in theoretical terms.

Discussions and critiques of the colonization of the New World often assume that the colonization process was one of demic expansion, population growth and subsequent group fissioning (Belovsky 1988, Martin 1973). There is considerable and not completely unwarranted skepticism of the growth and migration rates used in these models (e.g. Whitley, Dorn 1993). However, Lawrence Todd and I argued (1988) that the colonization of North America could be understood in terms of a model that did not assume population growth and subsequent fissioning as the driving force behind the colonization of the Western Hemisphere. Our major point all along has been that the archaeology of the early Paleo-Indian period (11,200 to 10,500 BP) suggests that Paleo-Indian lifeway was strikingly different than that of ethnographically known foragers, although both must, we assume, represent manifestations of the same decision making processes and evolutionary principles. Elsewhere I have considered criticisms of the model's use of archaeological data, and alternative reconstructions of Paleo-Indian lifeway that employ ethnographic analogy (Kelly 1996). Here, I wish to consider the relationship between foraging and migration. My purpose is to suggest that while population growth undoubtedly occurred during the initial colonization of the Western Hemisphere and helped it along, it may not have been the primary driving force. Consequently, refutations of density-dependent migration rates for Paleo-Indians to account for the apparently rapid occupation of the Americas may be irrelevant (e.g. Whitley, Dorn 1993).

FORAGING AND MODELS OF MIGRATION

Ethnographic data show that there is a relationship between individual foraging and residential movement such that people rarely use all the food they could from a given camp (see Kelly 1995). In the central Kalahari, for example, Kade G/wi women "begin to gather food near the campsite [and] they can com-

---

*Originally published in *Anthropologie* 37 (1999):143–153

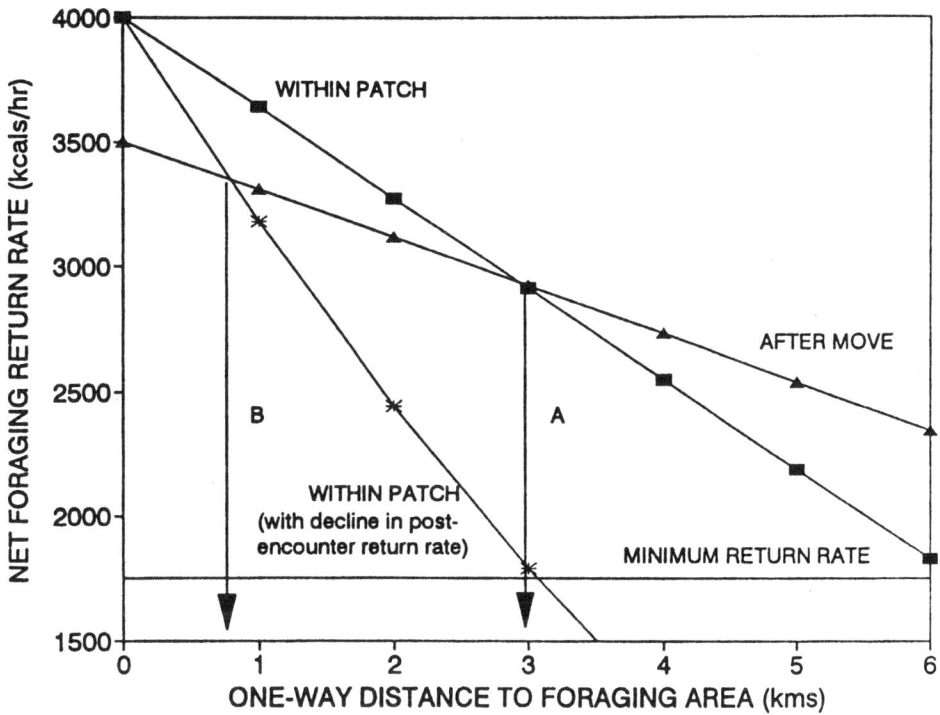

FIGURE 8.1. The relationship between the return rate experienced within a foraging area relative to that which could be expected if the foragers moved to a new area. The model predicts camp movement at the point at which foragers are traveling about 3 km from camp in order to find food (arrow B). If post-encounter return rate declines (at a rate of 500 kcal/km radius) as a function of the presence of humans, then movement is predicted after resources decrease within only 1 km of camp (arrow B).

plete their work in a trip of 1 to 2 km during the first few days of their stay. Then, gradually, as they consume the plants near camp, they must go farther. If the round trip for gathering food plants exceeds 10 km or so, convenience dictates that they move themselves with all their belongings to virgin territory." (Tanaka 1980:66) Optimal foraging theory's marginal value theorem also suggests that if foragers wish to forage efficiently, and maintain as high a rate of food intake as possible, then they should leave a foraging area before depleting it of food. But the marginal value theorem is designed for a forager who eats food at the point of capture, and one who encounters food randomly. For the most part, this does not describe living foragers who transport food back to camp and who, for the most part, know where they are going and what they are seeking before leaving camp.

Imagine, therefore, a foraging family living in an environment where a resource with a post-encounter return rate of 4000 kcal/hr is homogeneously distributed. This food is collected by a forager who walks at a leisurely pace of three km/hr at a caloric cost of 300 kcal/hr that increases by 30 percent when returning home with food. For the sake of simplicity, let's assume this forager must collect 14,000 kcal/day to feed his or her family. Assuming an eight-hour workday, this means that the forager must gather the resource at a minimum net daily return rate of 1750 kcal/hr. The net return rate (RR) declines as the forager spends more time (T) and energy traveling out to and back from the foraging area (Figure 8.1):

$$RR = ((8\ hrs - (2 \times T)) \times 4000) - ((300\ kcal/hr \times T) + (390\ cal/hr \times T))/8$$

This gives an effective foraging radius of 6 km. Food cannot be gathered beyond this point.

We can also compute the return rate if the family were to move to a new foraging area after exploiting the resources within a given radius of the site. Since food is homogeneously distributed, we will assume that they move the minimum distance to position themselves in a pristine foraging area, that is, twice the current foraging radius. The post-move return rate of the individual forager, allowing an hour for camp breakdown and setup is figured as:

$$RR = \{\{[7\ hrs - (2 \times T)] \times 4000\} - [300 \times (2 \times T)]\}/8$$

The after-move line in Figure 8.1 shows the daily return rate if the forager were to move camp after foraging within 1, 2, 3… km of camp. Note that at a return rate of just under 3000 kcal/hour (achieved at a foraging distance of about 3 km) the net after-move return rate is equal to the within patch return rate. After foraging within about 3 km of camp (arrow A), the family would do better to move to the centre of a new foraging area (6 km away). Many other variables alter this model's predictions (see Kelly 1995), but it suggests that even if foragers could remain in their current camp or territory, they will move to maximize their foraging efficiency. We may be justified in assuming, therefore, that if Clovis hunters moved into an environment rich in large game, as late Pleistocene North America apparently was, with no human competitors, and no territorial boundaries, that they would still have been a mobile people. But they might have been even more mobile than this simulation suggests.

The simple model used above assumes that hunter-gatherers have no immediate effect on the resources they exploit. What if this were different? What if a resource responded in such a way that its return rate decreased as a function of the presence of humans? The first inhabitants of North America found a game population that had never encountered people. It is hard to say how these animals would have responded to humans, but perhaps the arctic archaeologist Moreau Maxwell can give us a clue. Conducting an archaeological survey in 1958 on northern Ellsmere Island, Maxwell set foot where no human had been since 1881. He found game there to be downright placid: ptarmigan did not fly away; he could come within 30 m of musk ox; seals only a meter away from his tent ignored him. A hunter in this landscape could have achieved a very high return rate.

But game quickly learn the dangers of humanity. Had Maxwell hunted these animals, they would have figured out soon that he was something to be avoided, and initially high return rates would have decreased rapidly. We try to model this situation also in Figure 8.1. Using the same model as above, we decreased the post-encounter return rate by 500 kcal/hr with each km radius, that is, the return rate does not stay the same but decreases the longer the camp is occupied to reflect the increased time it would take to pursue an increasingly more wary prey (I have no idea what the correct reduction factor should be; the example serves only to show the predicted direction of the effect of a change in prey knowledge). Note that now a higher return rate could be achieved by moving after eating everything within only 1 km from camp (arrow B) before moving to virgin territory where prey have not learned to avoid humans. An initially naïve prey that rapidly learns to avoid a new predator would cause the predator to become more residentially mobile.

Dramatic population growth may not be needed for migration to occur, for the critical variable is the difference between the return rate experienced in the current area and the expected return rate of a neighbouring area, allowing for the costs of movement. By even slightly increasing the rate of local resource depletion, even modest population growth would soon make the cost of moving to virgin territory worthwhile.

Two Models of Migration

As Table 8.1 shows, migration can occur under a number of circumstances. In all of these, migration entails a cost—the cost of giving up a known environment for an unknown environment. Migration occurs when the cost of remaining in a current territory is higher than the potential cost of migration, or when the local net foraging return rate falls below that which could be ob-

TABLE 8.1. Potential causes of migration.

1. When population reaches local carrying capacity.
2. When the local habitat has low stability relative to species generational length.
3. When suitable foraging patches are relatively close together, or when the "hostility" of transition habitat (the land to be crossed between foraging patches) decreases.
4. When mortality creates vacant patches (assumes a previous population was inhabiting the land to be colonized).
5. When social hierarchy reduces rate of food intake or reproductive possibility for some individuals.
6. When patch size decreases relative to population size.

TABLE 8.2. Characteristics of transient explorers and estate settlers (modified from Beaton, 1991).

Characteristic	Transient Explorers	Estate Settlers
Demography		
Threshold to fission	Low	High
Group composition	Stable	Fluid
Genetic factors	Founder effects, Drift, local natural selection	Gene flow, Clinal distributions
Fecundity	Low	High
Population growth	Low	High
Local group extinction probability	High	Low
Parent population	May not be left behind	Will be left behind
Economy		
Territory	Unconstrained	Constrained
Niche differentiation between groups	Low	High
Diet diversity	Low	High
Seasonal mobility	High	Possibly high
Territorial mobility	High	Low
Geographic pattern of colonization	Lineal advances	Wave front

tained by moving to new territory after taking the cost of moving into the new territory into account.

Under different circumstances, however, migration will take different forms. John Beaton (1991) has proposed two potential models to describe colonization. The chief characteristics of each model are listed in Table 8.2. In the Estate Settlers Model (ESM), colonization occurs as a result of population growth and group fissioning as local carrying capacity is reached. (Group fissioning could occur as a function of poorly adapted individuals—the old, infirm, or weak—moving away from areas of high competition. These "colonists" would probably not survive.) In this model, colonization occurs at a rate that is controlled by the local rate of population growth (this is in turn controlled by a number of variables, e.g. maternal nutrition and labour patterns, breast-feeding; see Kelly 1995). Regardless of the speed, however, migration would be incremental, with daughter groups

TABLE 8.3. Rates of population movement, north-eastern Asia versus the Western Hemisphere for different possible migration dates.

From	At	To	At	Distance (km)	kms/year
Across Beringia:					
Lake Baikal	33,000	Tanana Valley	11,800	5,000	0.24
			11,300		0.24
		Bluefish Cave	15,500	5,600	0.32
			13,000		0.28
Aldan River	18,000	Tanana Valley	11,800	3,500	0.56
			11,300		0.52
		Bluefish Cave	15,500	3,730	1.49
			13,000		0.75
Clovis-First:					
Tanana Valley	11,800	Blackwater Draw	11,170	4,500	7.14
Bluefish Cave	15,500	Blackwater Draw	11,170	4,500	1.04
	13,000		11,170	4,500	2.46
Tanana Valley	11,800	Tierra del Fuego	10,700	16,600	15.09
	11,300		10,700		27.67
Bluefish Cave	15,500	Tierra del Fuego	10,700	16,200	3.38
	13,000		10,700		7.04
Blackwater Draw	11,170	Tierra del Fuego	10,700	12,100	25.74
Pre-Clovis:					
Bluefish Cave	15,500	Meadowcroft	14,500	5,040	5.04
			14,000		3.36
Bluefish Cave	15,500	Monte Verde	13,900	15,050	9.41
			12,400		4.85
	13,000	Monte Verde	12,400	15,050	25.08
Meadowcroft	14,500	Monte Verde	13,900	12,800	21.33
			12,400		6.10
	14,000	Monte Verde	12,400	12,800	8.00
Meadowcroft	14,500	Tierra del Fuego	10,700	13,660	3.59
	14,000		10,700	13,660	4.14

budding off from the parent group and moving to an "edge" of the currently inhabited region. This form of colonization probably occurs when new territory is close and the new environment is not radically different from the old one.

However, migration occurs in many species long before carrying capacity is reached. In the Transient Explorers Model (TEM), migration occurs when fluctuations in resources increase the cost of remaining in place for all members, and/or when the distance between foraging patches is great. Migration and colonization in this case would probably be rapid, either as a function of moving long distances between patches or as a function of rapid declines in initially high return rates—as when foragers exploit high-ranked, but naïve game resources.

One major difference in the two models is that in the ESM, a colonizing population leaves a parent population behind; in the TEM, a colonizing population may or may not leave a parent population behind. In the TEM, populations could be separated by long distances, whereas in the ESM populations would be more or less equally spaced and maintain contact with one another, permitting both the continual flow of genes and cultural ideas. The ESM would be susceptible to the effects of gene flow, and might result in clinal gene distributions, whereas in the TEM, migration could result in discrete populations subject to founder effects, genetic drift, and the effects of localized natural selection.

In the ESM it is unlikely that local population extinction could occur (at least, on a scale that is archaeologically visible), although there could be retractions in response to climatic change. In the TSM, local population extinctions could occur regularly as a function of a combination of small population size, isolation and chance events (e.g. winter storms, disease).

In the ESM, some areas could be occupied, then abandoned, then reoccupied as the population along the "front" of the colonizing wave goes through pulses generated by demographic

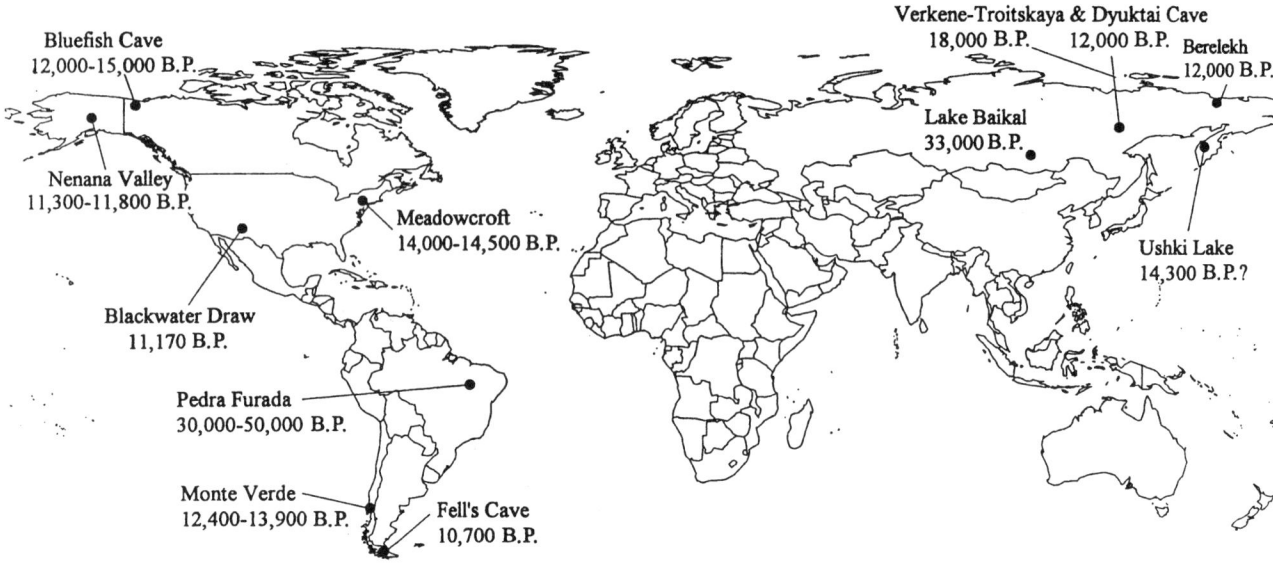

FIGURE 8.2. Locations of the sites used in the text to calculate potential migration rates.

and/or environmental fluctuations. Land eventually occupied permanently by a daughter group would probably not be "unknown" land. It could, therefore, be used in a fashion different from that of the parent population as the daughter population altered its strategy to suit the particular environmental features of the new territory. This could result in clinal distributions of foraging strategies that parallel clinal distributions of resources.

In the TEM, colonizers could move long distances into unknown terrain. Under these circumstances, they might continue to use familiar resources and tried and true strategies, resulting in behavioural homogeneity rather than cultural differentiation across large regions. Models of cultural transmission (Boyd, Richerson 1985) suggest that in cases where an individual confronts circumstances where experimentation is costly, that individual should mimic whatever behaviour or trait appears to be successful for others in that environment. If Clovis peoples moved into an unoccupied environment, then they should have continued doing whatever they had been doing before moving into the new environment since they themselves were the only models available. In this model, we can expect that the original behavioural repertoire of the colonizers would exert strong influence over the character of the colonizers' adaptation and a similar behavioural strategy could be practiced across very different kinds of environments.

Colonization of the Western Hemisphere

Which of these models describes the colonization of the Western Hemisphere? Let us first ask what we know about the colonization of the Western Hemisphere. We know that people came to North America from Siberia via the land bridge. The land bridge was open, most recently, from 60,000 to 10,000 BP, reaching its maximum extent between 20,000 and 18,000 BP. (The bridge was first breached about 14,300 BP, but would have remained passable at times until 10,000 BP. While it probably provided a diversity of large fauna, it was apparently not until 12,000 BP that trees were present in river valleys to provide a source of firewood [see Guthrie 1990, Hoffecker et al. 1993]; this would not, however have prevented migration along the coast, where driftwood would have been available.) To move south from Alaska, people had to pass through the ice-free corridor when it was open—before 30,000 or after 11,500 BP (the ice-free corridor was open earlier at its southern reach, by about 14,500 BP, but was apparently still closed north of 60° until later). Alternatively, people could have traveled along the western coast. The Cordilleran ice sheet did not form until as late as 20,000 BP; before this date, people could have moved along the interior valleys of the Canadian Rockies (Jackson, Duk-Rodkin 1996). By about 13,000 BP an ice-free coast was available to entering populations. Coming along the south coast of Beringia humans could have entered North America south of the ice sheets quite early.

However, archaeological data from eastern Siberia do not provide much support for this possibility. The earliest evidence of human occupation there, in the Aldan River Valley, is at the site of Verkhene-Troitskaya, which dates conservatively to 18,000 BP (Figure 8.2).[1] At Dyuktai Cave, the earliest well-dated levels are between 14 and 12,000 years old (Yi, Clark 1985). On the north coast, Berelekh dates to only about 12,000 BP and Ushki Lake to less than 14,300 BP. Thus, present data suggest that people were not in Beringia before about 13,000 BP, or, if they were, they traveled along a now submerged coast. More research is obviously needed in far eastern Siberia.

From the Tanana Valley sites in Alaska we know that people had crossed the Bering Strait possibly by 11,800 years ago. And, by at least 11,200 years ago, people had moved south into the present day United States, where the distinctive stone tools that

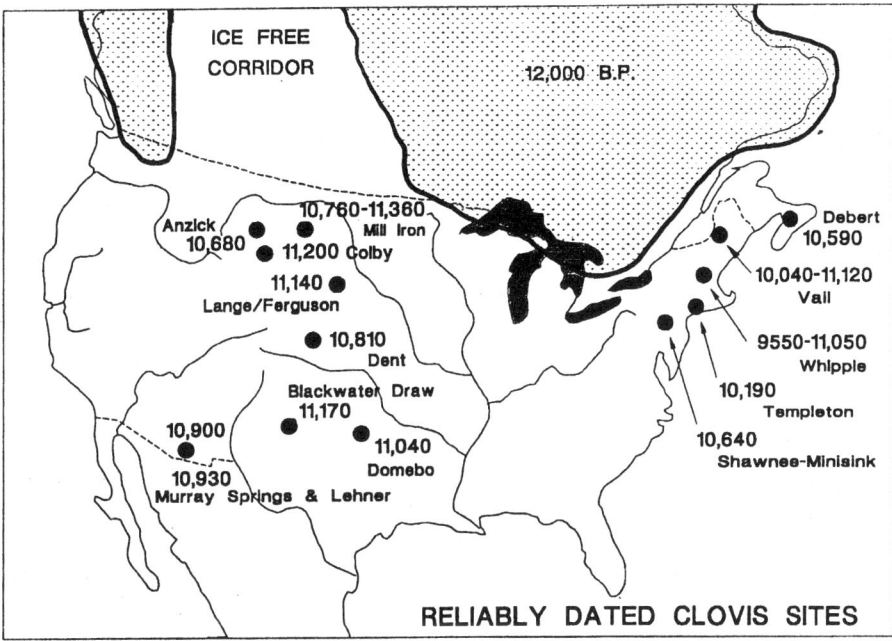

FIGURE 8.3. Location of reliably-dated Clovis sites, based on Haynes (1993) and Frison (1991a).

comprise the Clovis tool kit are well-dated between 11,200 and 10,900 BP (Haynes 1993, Figure 8.3). By at least 10,700 years ago, people had pushed south to Tierra del Fuego.

But were the people who fashioned Clovis points the first in North America? Recently, geneticists and linguists both have answered in the negative. Patterns in the mtDNA of Native Americans suggest that Amerindians separated from Siberian populations 17,000 to 34,000 (Torroni et al. 1993a, b) or possibly as many as 78,000 years ago (Ward et al. 1991), suggesting an occupation of the Americas significantly before Clovis. Other researchers, however, suggest that the divergence occurred between 12,100 and 13,200 BP, more in line with Clovis dates (Shields et al. 1993; these researchers only examined northern populations, however). There are some difficulties with mtDNA as a record of population movements because mtDNA dates lineage divergence and not necessarily population divergence; thus it provides only a maximum age for the latter. Additionally, many of the assumptions of the models have also not yet been tested for the Western Hemisphere (Szathmary 1993). Linguists make claims similar to those made by geneticists. By making some assumptions about the rate of linguistic change, some argue that the diversity of Native American languages cannot be accounted for within the Clovis time frame. Nichols (1990), for example, suggests that different lineages must have entered North America at the rate of one every 3500 years, ascribing considerable antiquity to the earliest migration, or that a population from a single language family entered the New World some 35,000 years ago, and then diversified. Yet there are questions here, too, about the rate of linguistic change in small populations, and the effects of adaptive processes. For me, therefore, the bottom line is the archaeology. If people were here in sizeable numbers before 11,200 BP, then we should be able to find evidence of them.

There is some archaeological evidence of a pre-Clovis occupation, but it is not altogether convincing. Stone tools and their associated dates from Bluefish Cave in the Canadian Yukon (Cinq-Mars 1979) may suggest that people were in northwestern North America between 12,000 and 15,000 years ago (Figure 8.2). From Meadowcroft Rock-shelter, there is perhaps evidence of the presence of humans in the continental U.S. 14,000 or 14,500 years ago (Adovasio et al. 1990) although the site still has some unresolved questions. From South America, Monte Verde suggests an occupation by at least 12,300 and possibly as early as 13,900 years ago (Dillehay 1989, Dillehay et al. 1992). Still greater antiquity is ascribed to the Brazilian sites of Pedra Furada and Toca da Esperanca, but there are significant problems with each of these (Lynch 1990). Most recently, Whitley and Dorn (1993) propose that rock varnish dating of artifacts in the Mohave Desert of southern California indicate that the tools were manufactured in excess of 20,000 years ago, but the accuracy of cation ratio dating and AMS dating of organics recovered from rock varnish is still debated (see Bierman, Gillespie 1994, Reneau et al. 1991).

Given the best evidence at hand, therefore, we have four major scenarios of colonization:

1. Clovis is the first migration and there are currently unknown problems with the dating or analysis of the purported pre-Clovis sites. The linguistic and genetic arguments would also have to be incorrect.

2. There was at least one pre-Clovis occupation that moved only along the western coast of North America, but which turned east into the interior of South America (or which

jumped over the isthmus of Panama and moved along the northern and eastern coasts of South America, to eventually turn inward and leave artifacts behind at Pedra Furada). A migrating population, those who would develop the Clovis technology, then occupied (the still empty interior of) North America at a later date by coming through the ice-free corridor soon after it opened.

3. A pre-Clovis population in North America became extinct relatively soon after its arrival (but perhaps not in South America, although whether it did is not critical to the argument here); Clovis-bearing peoples then later migrated into the empty North American continent.

4. A pre-Clovis population in North America was initially small, but grew in size, not becoming large enough to be recognizable archaeologically until about 11,200 years ago (Meltzer 1993:162). This could be complicated by geomorphic processes at the end of the Pleistocene in North (but not South) America that may have prevented site burial and/or eroded earlier deposits (Butzer 1991).

As others have suggested, it seems likely that if there were a pre-Clovis population, that we would have found substantial evidence of it by now—especially if that population entered North America some 30 or 40,000 years ago. The Pleistocene population of eastern Siberia was undoubtedly sparse, and yet evidence of that population has still been recovered (in an area that has seen far less intensive archaeological research than the United States). Compare this to the eastern United States, where there is only one good pre-Clovis candidate (Meadowcroft) but where there are some 50 sites containing fluted points (Meltzer 1988, 1991; although only a few have good radiocarbon dates) and over 9000 isolated fluted points (Anderson 1990).

The systematic removal or non-formation of pre-Clovis archaeological sites is a real possibility. The earliest evidence of the use of caves and rock-shelters on the High Plains of the United States is about 10,000 BP (Frison 1991a:69), but earlier deposits may have been scoured out of these, as may have been the case at Medicine Lodge Creek Cave (Frison, personal communication, 1992). The same may have happened at some rock-shelters in Tennessee (Daniel Amick, personal communication, 1992). Pre-Clovis rock-shelters may also have completely eroded away, hiding the presence of pre-Clovis deposits to rock-shelter-fixated archaeologists (Collins 1991). Nonetheless, there are many rock-shelters that contain late Pleistocene deposits, but no cultural materials. And some archaeologists have intentionally sought out deposits of the appropriate age, but have come home empty-handed. While we would continue to look for pre-Clovis deposits, at present it is hard to draw conclusions from negative evidence.

In sum, in three of these four scenarios, we may have two or more instances of colonization of an empty landscape in the Western Hemisphere; Meadowcroft and Monte Verde, along with perhaps some other, less secure localities may record one and Clovis may record another. But in these three, Clovis-bearing peoples may have still been the successful occupants of an empty continent. Perhaps they cannot claim the distinction of being first, but they may still have been an instance of *H. sapiens sapiens* inhabiting an empty continent.

Rates of Migration

Whether we accept Clovis, or those who left tools behind at Meadowcroft and Monte Verde as the first occupants, migration through the Western Hemisphere appears to have been rapid when compared to the rate of colonization of Beringia. Using available radiocarbon dates from sites such as Dyuktai Cave, the Nenana Valley and Bluefish Caves (Figure 8.2), and including scenarios in which the Nenana Valley or Bluefish Caves date the earliest occupation, people migrated across Beringia at an average rate of 0.5 km/yr (range 0.24–1.49 km/yr). From Alaska southward, and looking only at pre-Clovis sites, the migration rate is 10.3 km/yr (range 3.6–25 km/yr); using only terminal Pleistocene sites in Alaska, those containing fluted points in North America and post-11,200 BP sites in South America, the rate is 11.2 km/yr (range 3.4–27.7 km/yr). No matter when the migration occurred, it appears to have occurred faster than that in Beringia.

Is the postulated rate of Clovis migration too fast? Certainly, it is not physically impossible: even the most unathletic of us could migrate on foot 11 km/yr. But is it too fast for a population? We have almost no other instances of colonization of unpopulated land for comparison. From archaeological data, however, we can look to the initial occupation of the eastern Arctic of North America. Here, bearers of the Arctic Small Tool Tradition migrated from Alaska across the northern reaches of Canada to Labrador and northern Greenland, a distance of 5500 km. Maxwell (1985) suggests the migration probably took about 300–500 years for a migration rate of 11–18 km/yr. However, the 300–500 year range is only a guess, and available radiocarbon dates indicate an instantaneous occupation of the far eastern and western arctic (and the same happened again during the Thule occupation around 900 BP)—in much the same way that radiocarbon dates on fluted point sites suggest a nearly instantaneous occupation of the continental U.S. Thus, a migration rate of 11–18 km/yr for the Arctic Small Tool Tradition is only a minimum and it makes the postulated Clovis rate of migration appear reasonable. However, we must be cautious in offering this as an analogy, for Arctic Small Tool Tradition people moved along a strip of land that was similar environmentally, rather that across a continental land mass and through radically different biomes. It is not clear how long it would take Paleo-Indians to adapt to these new environments.

Having calculated a migration rate for Clovis from archaeological data (rather than from historical analogies or from population-density driven models), we must hasten to point out that this does not imply a continual year by year migration. As Todd and I argued (Kelly, Todd 1988), Paleo-Indians probably used large annual ranges that shifted long distances every few years. Thus, migration may have been, on the scale of an individual living at that time, more jerky, rather than continuous.

Colonization of the Western Hemisphere

Lawrence Todd and I proposed a model to account for what we saw as unique features of early Paleo-Indian archaeology and that we thought was expectable given what is known about the ecology of foraging peoples. That is, we arrived at this model both inductively and deductively.

Coming from an arctic climate, colonizers would have been pre-adapted to hunting. But the environment they entered in North America was undergoing rapid change. Game was initially more abundant and more homogeneously distributed than today, but rapidly diminishing in abundance and becoming more localized as Late Pleistocene climate changed from non-seasonal to seasonal. Local fluctuations in game density dramatically lowered return rates periodically. In such instances, the colonizers would either have turned to new resources or moved to a new location. It is difficult to predict which option they would have selected. However, since experimenting with new resources is costly in the short term, Paleo-Indians may have chosen to move, following familiar game resources, rather than stay in place and shift to new foods (Kelly, Todd 1988). Modern foragers frequently move in preference to using secondary or less familiar resources because this gives them a higher mean return rate from foraging. Given the foraging model described above, we can expect a people who (a) move into an unknown region, (b) exploit a faunal base whose post-encounter return rates decline with exploitation, and (c) experience declines in foraging returns, relative to the cost of moving to a new region, from even slight increases in local population density to shift their ranges long distances after occupying them for fairly short periods of time.

We thought that Paleo-Indians faced a unique difficulty in moving into a new hunting area in response to local resource failure or diminishing returns. When modern foragers move, they move to be with relatives or friends who know their home region well. But the first occupants of North America had few or no neighbors—they moved into unknown terrain. Under these circumstances Paleo-Indians would have continued to rely upon game as their primary food source, because knowledge of game is more easily transferred from one region to another than knowledge of plants. This strategy would have been possible in the late Pleistocene given that game was more homogeneously distributed than during the Holocene. To a carnivore, the different environments Paleo-Indians crossed would have looked more similar than they would to a herbivore. This strategy, however, also put Paleo-Indians in a bind: to survive in a new place using familiar skills, they had to rely on animals that were rapidly disappearing or responding to humans by making themselves harder to catch (obviously, this would not be a problem if plants were the focus of diet). This required Paleo-Indians to move constantly into new territory and thus they were pulled south, throughout the Americas.

We proposed that Paleo-Indians would move into a region, learn a few features, and use them redundantly as they continued to use familiar resources and strategies. Paleo-Indians had no recourse to "resource geography" to fall back upon, that is, knowing where to find food in a particular region, during a particular season, under a range of conditions. Resource areas used initially could have been used repetitively because they were locations that, although they may not have been the best places to use, could nonetheless be counted on to meet minimal needs.

Todd and I concluded that early Paleo-Indians were like Arctic hunters in some ways, but like tropical foragers in others. Just as there are no modern analogues for Pleistocene environments, there are no modern analogues for Pleistocene society in North America.

Patterns in Paleo-Indian Archaeology

The unique features of early Paleo-Indian archaeology, other than the apparently rapid spread of Clovis-bearing peoples throughout North America, can be accounted for within the proposed model. These are discussed elsewhere in more detail (Kelly, Todd 1988, Kelly 1995, 1996), and are only briefly reviewed here.

First, there is a reliance upon a bifacial technology and the use of high-quality stone for manufacturing tools. These tools are often transported long distances from their sources, several hundred or even over a thousand kilometers (e.g. Tankersley 1991). Made from high quality material, bifacial implements provided Paleo-Indians with long use-life tools. These would have been essential to a lifeway that was both residentially and logistically mobile, that is, where people moved frequently and long distances individually and in residential groups.

Second, the tools found in fluted point sites, beginning with the fluted points themselves, are more similar to one another, even from opposite coasts of North America, than tools from later time periods. While there are some stylistic differences among fluted points, the flute itself, which is difficult to learn to make and which frequently results in the breakage of the point, testifies to strong cultural continuity across the continent. The similarity of the stone tools (see e.g. Goebel et al. 1991), despite the fact that they were deposited in many different kinds of paleoenvironments, suggests that a similar niche was occupied across North America. Todd and I argued that Paleo-Indians were generalists with regard to fauna (not just large game) and opportunists in regard to everything else in a diversity of environments.

Third, relative to later time periods, there is virtually no fluted point material found in caves or rock-shelters. In the entire eastern U.S., there are no fluted points in good context in a cave or rock-shelter (Meltzer 1988). In the state of Wyoming, a state thoroughly searched for Paleo-Indian sites by George Frison, there is not a single fluted point in a cave or rock-shelter.

This could be a function of preservation of the deposits or the shelters themselves, as noted above. On the other hand, it is not an unexpected consequence of the model we proposed since shelters and caves are unique attributes of a region, and may not have been found and used by a people who remain in a region for relatively short periods of time. Some may point out that Paleo-Indians managed to locate high quality stone sources, which are also unique features of a landscape. But these sources occur over large exposures (geologically and through secondary deposits), increasing the probability that they will be encountered (Kelly 1996).

Fourth, relative to later time periods, there are very few cases of tool caching or food storage. While there are a few instances of Clovis tool caching (none for Folsom), most of these are associated with burials or cannot be accurately dated (Kelly 1996). For fluted point sites, there is only one demonstrated case of food storage, the Colby mammoth cache (Frison, Todd 1986) a location to which no one apparently returned. In sum, storage and caching were unimportant strategies in Paleo-Indian life. This, too, is expectable when a people cannot always anticipate how much of a resource will be available before a lean period, or where they will be or what will be available after the lean period. Storage implies a commitment to a particular region that is predicated upon an intimate knowledge of the resource potential of that region, the location and timing of resources.

Conclusions

As noted above, the nature of migration is partially contingent upon the nature of the host population. In this regard, three things are important here to the colonization of the Western Hemisphere. First, the colonizers came from an arctic environment where they were well adapted to a hunting way of life, one in which plant collection did not figure prominently.

Second, they were adapted to a way of life in which technology was undoubtedly important. Data from ethnographically known hunting and gathering societies shows that technology is far more complex and diversified for arctic groups than for tropical groups. The initial inhabitants of the Western Hemisphere may have seen the maintenance of technology to be of great importance.

Third, the initial inhabitants of the continental U.S., regardless of when they entered, came either along the coast, or through the ice-free corridor. Neither route would have been pleasant, and I guess that foraging groups would have moved through them rapidly. A slow demic expansion through the ice-free corridor may not even have been possible (see Aoki 1993, Mandryk 1993); a rapid intentional trek may have been the only way through it. Since both routes are linear paths, migrating groups would probably have been aware of the fact that they were moving in a single direction, southward, and that they were moving into new, unexplored terrain, someplace that even their grandparents knew nothing about. Exploration might very well have been part of their cultural ethic, but this was an ethic that was grounded in the unique economy of hunting in the rapidly changing, unpopulated environment of late Pleistocene North America. Knowing they were in unknown terrain, Paleo-Indians may have channeled their movements along major geographic features, e.g. the rivers feeding into the Mississippi (Anderson 1990) or the Rocky Mountains. In so doing, they may have been able to move across North America but not by a process of regional population saturation and group fissioning that would result if population growth were the primary driving factor in the colonization process.

In sum, the first colonists would have been hunting-adapted people who were accustomed to movement, exploration, and to a life dependent on technology. Entering North America just after 12,000 years ago, Clovis hunters entered a world undergoing a rapid climatic change, from a non-seasonal to a seasonal environment and in which many species of fauna were going extinct. Local environments were probably very unstable, giving these hunters a reason to move, and the rapid decrease in return rates that would come from the hunting of a naïve fauna would make other areas look all the more rewarding. Paleoenvironmental data indicate that there was less environmental differentiation between regions than exists today, thus lowering the cost of migration and encouraging movement. A dependence on game is probably what permitted these first hunters to move across North America, although it is also probably what forced them to move frequently as well, since they could not turn to other resources as a response to declines in the return rate from hunting. In sum, Paleo-Indians have every reason to move, and no reason not to, but little time in which to learn much about each new local environment.

These facts suggest an initial Paleo-Indian lifeway that was highly mobile and that fits a model of Transient Explorers rather than Estate Settlers, and points to a rapid occupation of North America (at least) but not one that necessarily was driven by high population growth and density-dependent migration. I expect that the variation that exists within Paleo-Indian archaeology can be accounted for by variation in the variables described here: density of fauna, fluctuations in game, responses by different species to humans, etc. Looking further afield, I would not expect the colonization of other land masses by *H. sapiens sapiens* to have been the same as that reconstructed here for North America, given differences in environment and the nature of the colonizing population, but I would expect it to have operated in terms of the same economic and evolutionary principles.

NOTE

1. On the Chuckchee Peninsula there is a site that is a moraine deposit containing stone flakes some 32–33 meters below the surface (Laukhin, Drozdov 1991). This location may predate 30,000 BP, but the dating of the site is not certain and the stone tools, based on the published illustrations, may only be fractured cobbles.

REFERENCES

Adovasio, J. M., J. Donahue, R. Stuckenrath
1990 The Meadowcroft rockshelter radiocarbon chronology 1975–1990. *American Antiquity* 55:348–354.

Anderson, D. G.
1990 The Paleoindian colonization of eastern North America: a view from the southeastern United States. In: K. B. Tankersley, B. L. Isaac (Eds.): *Early Paleoindian Economies of Eastern North America, Research in Economic Anthropology. Supplement 5.* pp. 163–216. JAI Press, Greenwich, CT.

Aoki, K.
1993 Modelling the dispersal of the first Americans through an inhospitable ice-free corridor. *Anthropol. Sci.* 101:77–87.

Beaton, J.
1991 Colonizing continents: some problems from Australia and the Americas. In: T. D. Dillehay, D. J. Meltzer (Eds.): *The First Americans: Search and Research.* Pp. 209–230. CRC Press, Boca Raton.

Belovsky, G.
1988 An optimal foraging-based model of hunter-gatherer population dynamics. *J. Anthr. Archaeo.* 7:329–372.

Bierman, P., A. Gillespie
1994 Evidence suggesting that methods of rock varnish cation ratio dating are neither comparable nor consistently reliable. *Quaternary Research* 41:82–90.

Boyd, R., P. Richerson
1985 *Culture and the evolutionary process.* University of Chicago Press, Chicago.

Butzer, K. W.
1991 An Old World perspective on potential mid-Wisconsinian Settlement of the Americas. In: T. D. Dillehay, D. J. Meltzer (Eds.): *The First Americans: Search and Research,* p. 137–156. CRC Press, Boca Raton, FL.

Cinq-Mars, J.
1979 Bluefish Cave I: A Late Pleistocene eastern Beringian cave deposit in the northern Yukon. *Can. J. Archaeo.* 3:1–32.

Collins, M. B.
1991 Rockshelters and the early archaeological record in the Americas. In: T. D. Dillehay, D. J. Meltzer (Eds.): *The First Americans: Search and Research.* Pp. 157–182. CRC Press, Boca Raton, FL.

Dillehay, T. D.
1989 *Monte Verde: A Late Pleistocene Settlement in Chile. Volume I.* Smithsonian Institution Press, Washington, D.C.

Dillehay, T. D., G. I. Ardila Calderón, G. Politis, M. C. de M. C. Beltrão
1992 Earliest hunters and gatherers of South America. *J. of World Prehistory* 6:145–204.

Frison, G.
1991a The Goshen Paleoindian complex: New data for Paleoindian research. In: R. Bonnichsen, K. Turnmire (Eds.): *Clovis: Origins and Adaptations.* Pp. 133–152. Center for the Study of the First Americans, Corvallis, Oregon.
1991b *Prehistoric Hunters of the High Plains.* 2nd ed. Academic Press, New York.

Frison, G., L. C. Todd
1986 *The Colby Mammoth Site: Taphonomy and Archaeology of a Clovis Kill in Northern Wyoming.* University of New Mexico Press, Albuquerque.

Goebel, T., R. Powers, N. Bigelow
1991 The Nenana complex of Alaska and Clovis origins. In: R. Bonnichsen, K. L. Turnmire (Eds.): *Clovis: Origins and Adaptations.* Pp. 49–80. Center for the Study of the First Americans, Corvallis, OR.

Goddard, I., L. Campbell
1994 The History and classification of American Indian languages: What are the implications for the peopling of the Americas? In: R. Bonnichsen, D. G. Steele (Eds.): *Method and Theory for Investigating the Peopling of the Americas.* Pp. 189–208. Center for the Study of the First Americans, University of Oregon.

Guthrie, R. D.
1990 *Frozen Fauna of the Mammoth Steppe: The Story of Blue Babe.* University of Chicago Press, Chicago.

Haynes, C. V., Jr.
1993 Clovis-Folsom geochronology and climatic change. In: O. Soffer, N. D. Praslov (Eds.): *From Kostenki to Clovis: Upper Paleolithic-Paleo-Indian Adaptations.* Pp. 219–236. Plenum Publishing Corp., New York.

Hoffecker, J. F., W. R. Powers, T. Goebel
1993 The Colonization of Beringia and the peopling of the New World. *Science* 259:46–53.

Jackson, L. E., A. Duk-Rodkin
1996 Quaternary geology of the ice-free corridor: Glacial controls on the peopling of the New World. In: T. Akazawa, E. Szathmary (Eds.): *Prehistoric Dispersals of Mongoloid Peoples.* Pp. 214–227. Oxford University Press, Oxford.

Kelly, R. L.
1995 *The Foraging Spectrum.* Smithsonian Institution Press, Washington.
1996 Ethnographic analogy and migration to the Western Hemisphere. In: T. Akazawa, E. J. E. Szathmary (Eds.): *Prehistoric Mongoloid Dispersals.* Pp. 228–240. Oxford University Press, Oxford.

Kelly, R. L., L. C. Todd
1988 Coming into the country: Early Paleoindian hunting and mobility. *Amer. Antiquity.* 53:231–244.

Laukhin, S. A., N. I. Drozdov
1991 Discovery of artifacts in the north of Eastern Chukotka and migration of Paleolithic Man from Asia to Northern America. *Ancient Prehistory.* 1:175–186.

Lynch, T. F.
1990 Glacial-Age Man in South America? A Critical review. *Amer. Antiquity.* 55:12–36.

Mandryk, C. A. S.
1993 Hunter-gatherer social costs and the nonviability of submarginal environments. *J. of Anthrop Research* 49:39–71.

Martin, P. S.
1973 The Discovery of America. *Science* 179:969–974.

Meltzer, D. J.
1988 Late Pleistocene human adaptations in eastern North America. *J. of World Prehistory* 2:1–52.
1993 Pleistocene peopling of the Americas. *Evol. Anthrop.* 1:157–169.

Maxwell, M. S.
1985 *Prehistory of the Eastern Arctic*. Academic Press, Orlando, FL.

Nichols, J.
1990 Linguistic diversity and the first settlement of the New World. *Language* 66:475–521.

Reneau, S. L., T. M. Oberlander, C. D. Harrington
1991 Accelerator mass spectrometry radiocarbon dating of rock varnish: Discussion and reply. *Geol. Soc. Am. Bull.* 103:310–314.

Schurr, T. G., S. W. Ballinger, Y.-Y. Gan, J. A. Hodge, D. A. Merriwether, D. N. Lawrence, D. C. Wallace
1990 Amerindian mitochondrial DNAs have rare Asian mutations at high frequencies, suggesting they derived from four primary maternal lineages. *Am. J. Hum. Genet.* 46:613–623.

Shields, G. F., A. M. Schmiechen, B. L. Frazier, A. Redd, M. I. Voevoda, J. K. Reed, R. Ward
1993 mtDNA sequences suggest a recent evolutionary divergence for Beringian and northern North American populations. *Am. J. Hum. Genet.* 53:549–562.

Szathmary, E.
1993 mtDNA and the peopling of the Americas. *Am. J. Hum. Genet.* 53:793–799.

Tanaka, J.
1980 *The San Hunter-Gatherers of the Kalahari: A Study in Ecological Anthropology*. University of Tokyo Press, Tokyo.

Tankersley, K. B.
1991 A geoarchaeological investigation of distribution and exchange in the raw material economies of Clovis groups in eastern North America. In: A. Montet-White, S. Holen (Eds.): *Raw Material Economies Among Prehistoric Hunter-Gatherers*. University of Kansas Publications in Anthropology 19. Pp. 285–304. University of Kansas, Lawrence, KS.

Torroni, A., T. G. Schurr, M. F. Cabell, M. D. Brown, J. V. Neet, M. Larsen, D. G. Smith, C. M. Vullo, D. C. Wallace
1993a Asian affinities and continental radiation of the four founding Native American mtDNAs. *Am. J. Hum. Genet.* 53:563–590.
1993b mtDNA variation of aboriginal Siberians reveals distinct genetic affinities with Native Americans. *Am. J. Hum. Genet.* 53:591–608.

Ward, R. H., B. L. Frazier, K. Drew, S. Pääbo
1991 Extensive mitochondrial diversity within a single Amerindian tribe. *Proc. Natl. Acad. Sci.* (USA) 88:8720–8724.

Whitley, D. S., R. I. Dorn
1993 New perspectives on the Clovis vs. Pre-Clovis controversy. *Amer. Antiquity.* 58:626–647.

Yi, S., G. Clark
1985 The "Dyuktai Culture" and New World origins. *Curr. Anthrop.* 26:1–20.

9

Should We Expect Large Game Specialization in the Late Pleistocene?

An Optimal Foraging Perspective on Early Paleoindian Prey Choice

David A. Byers and Andrew Ugan

INTRODUCTION

Several recent papers appeal to foraging theory in support of the view that early Paleoindians were big game specialists whose subsistence focused on hunting large-bodied, high-return animals such as mammoths (*Mammuthus columbi*), mastodonts (*Mammut americanum*) and bison (*Bison antiquus*) (e.g. Refs. [44, 45, 106]). The importance of fully understanding the place of proboscideans and other large game in the late Pleistocene diet cannot be overstated. If mammoths and other Pleistocene megafauna played a dominant role in shaping early Paleoindian prey choice, then subsistence patterns focused on these extinct animals may have important implications for questions ranging from late Pleistocene extinctions to early Paleoindian mobility and the pace and tempo of the colonization of the New World [27, 28, 44, 45, 60, 61, 69, 70, 98]. To better understand the interplay between late Pleistocene fauna and human foragers, we employ the prey choice model [68, 95] as a framework to evaluate the potential for these extinct taxa to have shaped early Paleoindian subsistence patterns.

Here we address two questions crucial to any discussion of optimal foraging models and megafaunal specialization: what were the likely on-encounter return rates for prey such as mammoths, and how often must they have been taken to exclude smaller animals from the diet? Most discussions assume that the large size and caloric value of mammoths and other megafauna would have made them highly attractive, yet numerous studies of both human and non-human foragers have demonstrated that prey choice decisions are structured not only by the energetic value of prey, but by their worth relative to acquisition costs once encountered (e.g. Refs. [35, 68, 90, 91]). Unfortunately, current models of early Paleoindian foraging lack the energetic and time investment data critical to evaluating the assumption that proboscideans were the highest ranked prey on the Late Pleistocene landscape. Furthermore, arguments in favor of big game specialization also identify the importance of being able to acquire large prey regularly [59, 74, 106]. From an optimal foraging perspective, overall return rates and diet breadth depend on encounter rates with higher ranked resources, with returns declining and additional prey being added to the diet as high ranked resources are found less frequently. Thus the issue of specialization depends on demonstrating that 1) animals such as mammoths are higher ranked than other potential prey taxa and, 2) that they could be found frequently enough to make searching for them more attractive than taking smaller animals when encountered.

To answer these important questions, we use anatomical data from African elephants (*Loxodonta africana*) to estimate the caloric yield from a mammoth. These data provide a framework for identifying the range of handling times and encounter rates within which mammoth specialization might occur. We continue by estimating encounter rates for mammoths and other North American species using allometric relationships between body size and population density in mammals. By combining these two pieces of information, we construct an optimal diet curve and evaluate just how narrow the late Pleistocene diet might have been. Our results suggest that a specialized diet focused on large game such as mammoth or extinct bison would be profitable within a very narrow range of circumstances where such prey were extremely abundant, easily procured, and could be processed at minimal time cost. Given the data presented below, we suspect such instances were uncommon at best. Instead, our results challenge the view of early Paleoindians as megafaunal specialists and suggest that these foragers would have pursued a wide array of taxa including megafauna, the full range of ungulates and smaller game as well.

*Originally published in *Journal of Archaeological Science* 32(2005):1624–1640

TABLE 9.1. Weight of various body portions from four adult male *Loxodonta africana*.

Carcass Portion	Mass (kg)[a]	Mass (kg)[b]	Mass (kg)[b]	Mass (kg)[b]	Mean (kg)
Bodily fluids	534.50	596.20	533.60	565.40	557.43
Brain	n/a	n/a	n/a	4.00	4.00
Cranium/mandible (excluding brain)	180.50	188.10	179.90	158.10	176.65
Ears	42.50	50.20	42.10	43.90	44.68
Heart	25.00	n/a	24.80	n/a	24.90
Innominate	91.00	98.30	90.70	71.90	87.98
Kidneys	8.00	8.50	8.10	9.00	8.40
Leg elements (all legs, includes feet)	303.50	413.60	302.90	358.20	344.55
Liver	77.50	83.30	77.40	73.40	77.90
Lungs	29.50	32.10	29.40	27.60	29.65
Meat	1936.00	2327.20	1931.60	1945.80	2035.15
Penis	51.00	57.40	50.70	53.40	53.13
Ribs, vertebrae and scapulae	383.50	460.70	378.20	381.30	400.93
Skin	414.00	500.60	413.10	412.20	434.98
Spleen	18.00	19.90	18.10	18.50	18.63
Stomach and intestines (empty)	302.00	304.20	301.50	n/a	302.57
Stomach and intestines (with contents)	n/a	n/a	n/a	886.50	n/a
Stomach contents	539.50	625.20	538.10	n/a	567.60
Tail	11.50	7.60	11.20	9.00	9.83
Testicles	5.50	6.30	5.40	6.30	5.88
Tongue	12.50	14.00	12.60	13.00	13.03
Trunk	113.50	127.30	117.90	95.60	113.58
Tusk nerves	12.00	9.90	12.10	n/a	11.33
Tusks	69.50	72.90	69.20	41.20	63.20
Total	5160.50	6003.50	5148.60	5174.30	5385.93
Percent meat	0.38	0.39	0.38	0.38	0.38
Total edible tissue[c]	2182.00	2602.00	2181.60	2171.70	2284.33
Percent edible tissue	0.42	0.43	0.42	0.41	0.42

[a] Data from Wilson (in Ref.) [30].
[b] Data from Robertson-Bullock [80].
[c] Includes brain, heart, liver, meat, tongue, trunk.

The Meat Utility of a Mammoth

Estimating return rates for mammoths requires a basic knowledge of the edible components of proboscidean carcasses. Table 9.1 presents the weight of various organs, bones, and muscle tissue from the necropsies of four adult male *L. africana*. Marrow is not included here because there are no data on the amount of marrow in an elephant carcass and because evidence for marrow extraction at sites showing unambiguous early Paleoindian associations is rare [43]. Carcasses average 38 percent muscle tissue by weight with little variation. Including other portions such as the brain, heart, liver, tongue, and trunk raises the edible mass to 42 percent of total weight, exceeding the 25–35 percent edible portion suggested by Frison and Todd [22]. Because our estimate is more generous, we use this value in subsequent calculations to approximate the edible tissue in a mammoth.

Although African elephants and late Pleistocene mammoths overlap in stature [43], they likely differed in average body mass. Mammoth long bone diameters are often greater relative to their length when compared to those of *L. africana*, which suggests mammoths were more massive for their height [13, 88]. Based on an allometric relationship between longbone size and body mass, Christiansen [13] estimates *M. columbi* averaged 6560 kg. This number is the product of measurements from two specimens, one a humerus and the other a tibia. In a more comprehensive analysis focused on North American mammoths, Shipman [88] uses data collected from 28 sets of remains in combination with an allometric relationship between body mass and the circumference of femora and humeri to suggest that *M. columbi* were in some cases 2–2.5 times as large as *L. africana*. In this sample, body mass estimates range from 3838 kg to 14,930 kg with an average of 7368 kg. Here we use the round figure of 7500 kg in our model and given the sexual dimorphism of proboscideans [43, 64], suggest this value would be consistent with a female mammoth. Combining this mass estimate with the edible portion data, the average mammoth would have yielded 3150 kg of edible tissue, or slightly more than 1.5 metric tons. We are by no means married to this estimate and simply note that smaller size estimates will broaden diet breadth as developed below.

While the weights of elephant organs and meat are well documented, the nutritional values of the tissues are not. In fact, we were unable to locate any food composition data for elephant. To compensate for the lack of nutritional information, we consider an array of values shared by other game species (Table 9.2). At the low end, we calculate the caloric yield for raw edible tissue at 1000 kcal/kg, comparable to lean ungulate species such as water buffalo [103]. In this case, a mammoth would have offered ~3.15 million calories (7500 kg * 42% edible * 1000 kcal/kg). An estimate assuming 1250 kcal/kg, similar to fattier ungulates such as deer and caribou, would provide ~3.94 million kilocalories. Finally, we calculate an upper boundary at 1670 kcal/kg. This number represents the value for beluga whale, a fat-rich arctic marine mammal [91]. A mammoth with similar body composition would provide ~5.26 million kilocalories. Given that these values encompass game species varying from very lean to fat-rich, we see no reason to expect that the caloric yield of mammoth meat would not fall somewhere within this range.

Handling Time and On-Encounter Return rates

While a mammoth would have offered a bonanza of calories, elephant meat comes at a cost in pursuit and handling. We emphasize that in order for a group of early Paleoindians to have specialized in mammoth hunting, procuring these animals must have provided on-encounter return rates greater than other potential prey available in the late Pleistocene environment. This observation leads us to evaluate the handling time thresholds for mammoth specialization relative to the return rates from other game species.

Handling Time and Return Rates

Table 9.3 identifies the minimum handling times required to provide return rates of 10,000, 20,000, and 30,000 kcals/hr given a 7500 kg mammoth of varying caloric value. It is important to note that unless post-encounter returns for a mammoth exceed ~30,000 kcals/hr, early Paleoindians would have found proboscideans no more attractive than a variety of medium sized artiodactyls such as deer (*Odocoileus* sp.), mountain sheep (*Ovis canadensis*) and pronghorn (*Antilocapra americana*) (Table 9.4). Prehistoric hunters would have been indifferent to which of these taxa they exploited *regardless of how many mammoths were encountered*, and diets would have been correspondingly broad as a result. Returns of 10,000 kcals/hr carry similar implications for the incorporation of lagomorphs, beaver, and other small to medium-sized mammals. With this in mind, fat-rich mammoths (1670 kcals/kg) need to be pursued, killed, and processed in less than 175 hours in order for mammoth hunting to provide on-encounter returns exceeding those for medium artiodactyls. Comparatively leaner mammoths would require even shorter handling times to maintain the required threshold, ranging from 131 total hours at 1250 kcal/kg to 105 hours for fat-poor animals in the 1000 kcal/kg range.

TABLE 9.2. Kcals/1000 grams of edible portion for selected prey species.

Species	Common Name	Kcals/kg
Bubalus bubalis	Water buffalo	990
Alces alces	Moose	1020
Cervus elaphus	Elk	1110
Antilocapra americana	Pronghorn	1140
Sylvilagus sp.	Cottontail	1140
Odocoileus sp.	Deer	1200
Sciuridae sp.	Squirrel	1200
Bison bison	Bison	1220
Sus scrofa	Wild boar	1220
Rangifer sp.	Caribou	1270
Equus caballus	Horse	1330
Castor canadensis	Beaver	1460
Ursus arctos	Bear	1610
Ondatra zibethicus	Muskrat	1620
Delphinapterus leucas[a]	Beluga whale	1670

All values from USDA [103] unless otherwise noted.
[a] Value from Smith [91].

TABLE 9.3. Handling time thresholds.

Kcal/hr	1000 Kcal/kg	1250 Kcal/kg	1670 Kcal/kg
10,000	315.0	393.8	527.6
20,000	157.5	196.9	263.8
30,000	105.0	131.3	175.9

Ethnographic Estimates of Handling Time

Although specific time allocation data on elephant hunting and processing are unavailable, ethnographic accounts suggest that proboscidean hunting comes at a substantial time cost. Several ethnographic records of recent elephant butchery events illustrate this point. While conducting ethnographic research with the Efe and Lese in the Ituri Forest of Zaire, Bailey [3, personal communication 2004] observed two elephant butchery events. In these instances, parties ranging from 40 to 120 people moved from the local village to the kill site and made camp. Approximately ten men defleshed an elephant carcass in about 2.5 hours, while subsequent processing of the meat, cutting it into strips, building drying racks and fires and tending the meat while it dried, occupied up to 30 people for an additional 24 hours. The Efe and Lese also processed the carcasses with metal knives, machetes, and axes, all implements that likely reduced handling times relative to the use of stone tools. If everyone in the group performed some task related to processing the carcass for just the time spent on butchery, then handling time would vary from 100 to 300 hours, depending on group size. If we include the effort spent drying the meat, then the butchery and processing of an animal the size of an African elephant could take as many as 745 person-hours, not including the time to get the processing party to the kill site.

TABLE 9.4. Return rates for various prey species.

Animal	Taxa	Hunting Method	Return Rate		Reference
			Low	High	
Bison	*Bison bison*	Encounter hunting	32,400	32,400	[47]
Pronghorn	*Antilocapra americana*	Encounter hunting	15,725	31,450	[90]
Mtn. sheep	*Ovis canadensis*	Encounter hunting	17,971	31,450	[90]
Mule deer	*Odocoileus hemionus*	Encounter hunting	17,971	31,450	[90]
Bearded seal	*Erignathus barbatus*	Encounter hunting	15,000	25,680	[91]
Caribou	*Rangifer tarandus*	Encounter hunting	25,370	25,370	[91]
Jackrabbit	*Lepus californicus*	Encounter hunting	13,475	15,400	[90]
Snowshoe hare	*Lepus americanus*	Trapping	8260	15,220	[108]

Fisher [21] provides complementary evidence from three additional Efe and Lese butchery events. Here, groups of 25 to 35 adults spent 3 to 5 hours defleshing the elephants, setting up temporary habitation shelters, and building meat drying racks. No estimates were given for the time spent processing the meat, and we assume that metal tools were again the norm. Given that the smaller group took longer, this amounts to 105–125 hours of handling time excluding costs associated with drying the meat.

Actualistic Data and Handling Time

Actualistic experiments involving elephant carcasses also point to the substantial numbers of people and subsequent investments in man-hours inherent in elephant processing. Huckell [52] recounts that simply processing the intestines of a female Asian elephant required almost 3½ hours. Moreover, a group of seven people was unable to flip the carcass to access the meat on its downside, even after it had been gutted, limbs and ribs removed, and head disarticulated from the body. In this instance, Huckell finally resorted to a pick-up truck and tow chain to roll the carcass. Laub [63] also notes the difficulty of flipping an elephant carcass and his crew of eight ultimately resorted to using a back-hoe to access the meat on the downside of the body. Finally, in their accounts of elephant culls, Frison [22] documents that it took 15 individuals to flip the carcass of a large male African elephant once the upside had been butchered and Haynes [43] observed that this task took 10 to 12 individuals.

In the only instance that speaks directly to total butchery time, Laub [63] observes that his crew of eight took 10.5 hours to deflesh an already eviscerated carcass, including the examination of anatomy and note taking. He suggests that if his team had systematically butchered the animal without stopping they could have accomplished the task in ½ to ⅔ the time. This translates into 42 to 56 person-hours of butchering time, omitting the cost of removing the internal organs. Finally, we note that although these experiments were in part designed to evaluate the use of stone tools, both Laub [63] and Huckell[52] used modern tools including knives, crowbars, and chainsaws while processing their carcasses.

While none of these examples can be used as direct analogs for early Paleoindians, who may have been much more adept at elephant processing than anyone today, they do suggest that the costs associated with elephant hunting are far from trivial. Indeed, mammoth processing would have required a substantial number of people, especially considering that the animals studied in both the ethnographic examples and the actualistic experiments were likely much smaller than their late Pleistocene counterparts. Given the data available, 50 to 125 hours appears to be a reasonable range for elephant handling times, excluding costs associated with pursuit time, moving camps, and drying meat. We note that pursuit time might add significantly to the total cost of exploiting proboscideans and that by excluding it we increase the returns from mammoth hunting. We also note that spoilage would have been an issue throughout most of the year even in the late Pleistocene and that ignoring partial processing also inflates return rates; partial processing of the carcasses to avoid the problem comes at the cost of lower overall returns as discussed later. Assuming that the ethnographic times are for 5000 kg animals and that we use a mass for mammoths that is 50% larger, handling times for mammoths would range from 75 to 187.5 hours.

Using this range of handling times translates into post encounter return rates of 16,800 to 70,140 kcal/hr (minimum energy, maximum time to maximum energy, minimum time). Since actualistic butchering experiments suggest elephants lack large pockets of subcutaneous fat [52, 63], we are skeptical of the proposition that the caloric value of mammoths approached that of marine mammals like beluga whale. We are equally skeptical that they were as lean as water buffalo. If we take 1250 kcal/kg as a reasonable middle ground, then on-encounter return rates for mammoths would vary from 21,028–52,500 kcal/hr with a midpoint return of 30,029 kcal/hr at 131.125 hours of handling (Fig. 9.1). A return rate of 30,000 kcal/hr is greater than that for almost all other small and medium bodied animals common to North America [90]. This implies that big-game specialization in the terminal Pleistocene may have been sustainable *provided that* 1) we are not being overly generous in our estimates of handling time and, 2) that mammoths and other large animals were encountered frequently enough. If encoun-

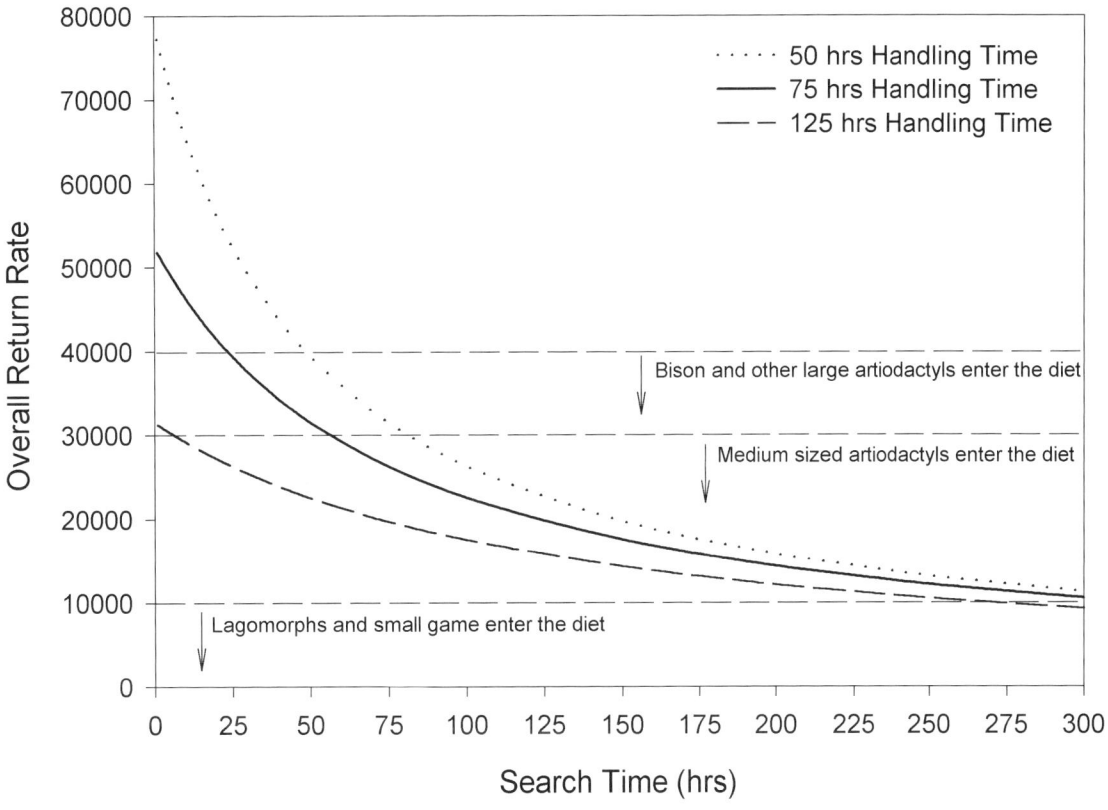

FIGURE 9.1. Relationship between overall return rate and search time for a mammoth only diet. Overall returns are plotted for three different handling times.

ter rates with megafauna were low, then the addition of search time will cause overall return rates to drop and make smaller animals attractive on encounter.

Prey densities and search time

Whether a predator pursues only one or a range of prey types depends not only on energetic returns for the animals once encountered, but also on how often they are found. Table 9.5 presents data for encounter rates and overall returns for both moderate (1250 kcal/kg) and fat-rich (1670 kcal/kg) mammoths. Areas marked in bold indicate where overall return rates are high enough to favor mammoth specialization and the exclusion of most small mammals and medium-sized artiodactyls from the diet. In both cases, overall returns decline as time invested in either search or handling increase. Because we are interested in "best case" scenarios, we omit the 1000 kcal/kg example discussed above. Lowering the caloric value into this range simply requires much lower handling times and much higher encounter rates in order to maintain high overall returns.

While it might appear that mammoth specialization was sustainable, several important observations merit further discussion. First, the 50 hour handling time estimate is provided for completeness and represents best case handling times associated with modern elephants weighing less than 5000 kg. This estimate is almost certainly too low when modeling the 7500 kg mammoth used here. Second, the more favorable case for specialization makes the problematic assumption that mammoths are the caloric equivalent of beluga whales. Focusing on the more probable middle ground estimates of 1250 kcal/kg and 125 hours handling time, early Paleoindians would have needed to take mammoth once per hour of search to exclude medium artiodactyls (>30,000 kcal/hr) from the diet and once every 150 hours to exclude jackrabbits (>10,000 kcal/hr; Table 9.5, upper panel; Fig. 9.1).

This encounter rate threshold points directly to a third issue, namely the need to understand large game abundances on the late Pleistocene landscape. Recent ethnographic data underscore this point. Modern hunter-gatherers such as the Hadza regularly pursue a variety of large-bodied animals, yet successful encounters with *any* of them occur only once every 30 hunter-days [77]. If early Paleoindians enjoyed comparable success rates, then mammoth specialization would be indefensible regardless of their caloric value or associated handling times. Even if one argues that early Paleoindians enjoyed higher success rates than the Hadza because Hadza territory had been depleted of most of its big game, some estimate of how much higher is still in order. A mammoth every 150 hours of hunting, for example, implies an encounter rate of 21 kg of meat per hour. By way of comparison, Simms' [90] estimates suggest that prehistoric hunters in the North American Great Basin would have acquired 1.4 kg of meat per hour from deer, mountain sheep, and pronghorn combined, *assuming all taxa were encountered at the maximum rate.*

TABLE 9.5. Overall return rates for mammoth hunting.

Hours Search	Encounter Rate		Handling Time in Hours					
	Ind/Hr	Kg/Hr	50	75	100	125	150	175
1250 kcal/kg								
1	1.0000	3150	**77,206**	**51,809**	**38,985**	**31,250**	26,076	22,372
10	0.1000	315	**65,625**	**46,324**	**35,795**	29,167	24,609	21,284
20	0.0500	158	**56,250**	**41,447**	**32,813**	27,155	23,162	20,192
30	0.0333	105	**49,219**	**37,500**	**30,288**	25,403	21,875	19,207
40	0.0250	79	**43,750**	**34,239**	28,125	23,864	20,724	18,314
50	0.0200	63	**39,375**	**31,500**	26,250	22,500	19,688	17,500
70	0.0143	45	**32,813**	27,155	23,162	20,192	17,898	16,071
100	0.0100	32	26,250	22,500	19,688	17,500	15,750	14,318
150	0.0067	21	19,688	17,500	15,750	14,318	13,125	12,115
200	0.0050	16	15,750	14,318	13,125	12,115	11,250	10,500
250	0.0040	13	13,125	12,115	11,250	10,500	9844	9265
300	0.0033	11	11,250	10,500	9844	9265	8750	8289
1670 kcal/kg								
1	1.0000	3150	**103,147**	**69,217**	**52,084**	**41,750**	**34,838**	29,889
10	0.1000	315	**87,675**	**61,888**	**47,823**	**38,967**	**32,878**	28,435
20	0.0500	158	**75,150**	**55,374**	**43,838**	**36,279**	**30,944**	26,977
30	0.0333	105	**65,756**	**50,100**	**40,465**	**33,939**	29,225	25,661
40	0.0250	79	**58,450**	**45,743**	**37,575**	**31,882**	27,687	24,467
50	0.0200	63	**52,605**	**42,084**	**35,070**	**30,060**	26,303	23,380
70	0.0143	45	**43,838**	**36,279**	**30,944**	26,977	23,911	21,471
100	0.0100	32	**35,070**	**30,060**	26,303	23,380	21,042	19,129
150	0.0067	21	26,303	23,380	21,042	19,129	17,535	16,186
200	0.0050	16	21,042	19,129	17,535	16,186	15,030	14,028
250	0.0040	13	17,535	16,186	15,030	14,028	13,151	12,378
300	0.0033	11	15,030	14,028	13,151	12,378	11,690	11,075

Data reflect a 7500 kg mammoth and a 42 percent edible portion (3150 kg). Here returns rates are "overall" returns rather than those achieved on encounter and incorporate the cost of searching for as well as handling the animal. Search time is presented in terms of raw hours. Encounter rates with mammoth are expressed as both individuals per hour and kilograms of meat per hour. Regions highlighted in bold provide more than 30,000 kcal/hr and would be sufficient to exclude most small and medium sized animals from the diet.

This striking contrast raises questions about the kinds of prey densities which might have been sustained during the late Pleistocene. There is a temptation to view late Pleistocene North America as a landscape teeming with large game prior to the arrival of Paleoindian hunters (e.g. Refs. 61, 106), but quantitative estimates of actual prey densities are generally lacking.

Some arguments in favor of megafaunal specialization appeal to paleoenvironmental reconstructions which suggest patchy resource distributions associated with drought conditions during the early Paleoindian period (11,200 to 10,900 ^{14}C yr. BP) (e.g. Refs. 44, 45, 93). C.V. Haynes [39–42] for example, interprets the presence of wells at sites such as Murray Springs and Blackwater Draw, in combination with stratigraphic evidence, as indicating a period of drought coincident with the timing of the initial Paleoindian occupation of the American Southwest. If important water sources were widely distributed across the terminal Pleistocene landscape, then megafauna would have been concentrated at these oasis water-holes. Early Paleoindians would have been aware of the patchy distribution of mammoth and such knowledge in combination with well-worn and highly visible elephant trails would have resulted in decreased search time and locally elevated encounter rates such that the diet narrowed to the exclusion of medium artiodactyls and other smaller taxa.

Three points need to be made, however. First, even if true this scenario would only apply around the proposed late Pleistocene oases. Densities of mammoth and other megafauna across much of the rest of the continent would be lower and diets broader. Even within such oases, the fundamental questions continue to be how often do we think mammoth were encountered and how quickly must they have been pursued and processed in order to exclude smaller animals. Second, more recent geoarchaeological treatments of the early Paleoindian period find little evidence for the severe and wide-spread drought necessary to restructure the environment in this way [48, 49]. In fact, Holliday [48, 49] argues that this time was more likely cool and wet, especially when compared to the environmental conditions that would immediately follow. Finally, there is no strong evidence to date for intensive megafauna hunting at oasis water holes as predicted by the oasis hunting model [76]. Here we focus exclusively on this first issue: how often must mammoth and other megafauna have been encountered to exclude smaller game.

Modeling Early Paleoindian Prey Choice

Model Assumptions

We turn to building a diet breadth model for early Paleoindian hunters using two pieces of data to reconstruct late Pleistocene encounter rates. The first is a body of biological literature concerning the allometric relationship between population density and body size in plants and animals [7, 8, 18, 20, 58, 82, 109]. For this study, we use data from Damuth [15, 16], who demonstrates that population densities among mammalian primary consumers ($n = 368$) scale at $-¾$ the power of body mass.[1] Absolute densities vary geographically and by habitat type [16, Table 3], and are generally higher in non-tropical environments. We use the intercept for all North American taxa ($b_0 = 4.33$) as the baseline for our study such that $\log(\text{density}) = 4.33 - 0.75 * \log(\text{mass})$. This allows us to reconstruct mean population densities for prey that prehistoric foragers were likely to have encountered, including now-extinct taxa such as mammoths (for a similar approach to reconstructing fish communities, see Ref. 56).

The second data set derives from several studies describing the relationship between population densities and artiodactyl encounter rates cited in Simms [90] (Table 9.6). Here encounter rates and population density for a sample of deer and bighorn sheep hunts are tightly and linearly correlated ($b = 0.007$; $r = 0.98$; $P < 0.01$). We use the slope of this regression line as a "scale factor" in order to translate density estimates into encounter rates. By extending the linear relationship to larger and smaller taxa and combining it with the allometric data from Damuth [16], we can estimate encounter rates for a range of potential resources and model diet breadth accordingly. Body size dictates prey density, the scale factor translates density into prey encounter rates, and prey encounter rates coupled with our knowledge of post-encounter return rates predicts diet breadth.

Potential Sources of Error

Whether either of these two relationships provide a valid basis for comparison is certainly a reasonable question. In their defense, we make the following observations. In terms of prey abundances during the late Pleistocene, the use of modern North American density data will only be inappropriate if late Pleistocene landscapes supported substantially higher numbers of primary consumers than the modern densities on which Damuth's estimates are based. Several considerations make this unlikely. First, late Pleistocene herbivores supported a large predator guild. These predators would have helped keep prey densities low even in the absence of human hunters. Second, several studies suggest that many of the large herbivores such as horse (*Equus caballus/ferrus*, Fig. 9.2) and bison (*Bison* sp.)

TABLE 9.6. Population density and encounter rates.

DENSITY (Ind/km²)	ENCOUNTER RATE (Ind/hr)	REFERENCE
0.000	0.000	No animals, no encounters
0.502	0.010	[57]
1.930	0.012	[104]
3.860	0.018	[104]
7.722	0.060	[104]
11.583	0.096	[51]

were in decline before Clovis hunters made an appearance [29, supplemental; 86]. These paleontological studies provide reason to expect that prehistoric prey densities may not have been substantially greater than we would expect today for animals of their size. Finally, our model results remain consistent even using the largest intercept identified by Damuth ($b_0 = 4.62$). That intercept implies carrying capacities and prey densities twice those of modern N. American herbivore taxa ($n = 84$), a value that is almost surely too high.

The relationship between prey density and encounter rates also poses problems. First, whether trends identified for medium artiodactyls can be reliably extended to very large animals such as mammoth or smaller game such as rabbits, hares, and gophers is unclear. Doing so implies that the larger and smaller animals are distributed in the same fashion as the artiodactyls, which is unlikely. Jetz et al. [58, supplemental] note a significant relationship between body size and group size, suggesting that larger animals not only occur at lower densities but are more patchily distributed. If female and juvenile mammoths tended to form herds in the same way as modern elephants, their effective density would be even lower than predicted. Scaling encounters in a linear fashion will overestimate encounter rates unless increased group size makes mammoth easier to locate, thereby offsetting the additional search associated with a more patchy distribution. Encounters with small animals, which are likely to be more evenly distributed, will show the reverse pattern. Scaling up and down as we do may therefore favor rather than diminish encounters with the larger animals, though this is by no means proven.

A second problem revolves around the appropriateness of the linear model itself. The estimate of 0.007 encounters/animal/km² derives from a very small sample. While it appears linear, a larger sample might exhibit a sigmoidal response (c.f. Ref. [50]), with low encounter rates at low densities, higher than expected encounters at intermediate densities, and a leveling off as density and encounter rates become saturated. We are unaware of any study which addresses this issue and cannot evaluate the possibility directly. If we consider encounters with all organisms, however, then the effect of such a response would be higher encounters with intermediate-sized animals. Megafauna densities would remain low and the potential impacts on megafaunal specialization would be small.

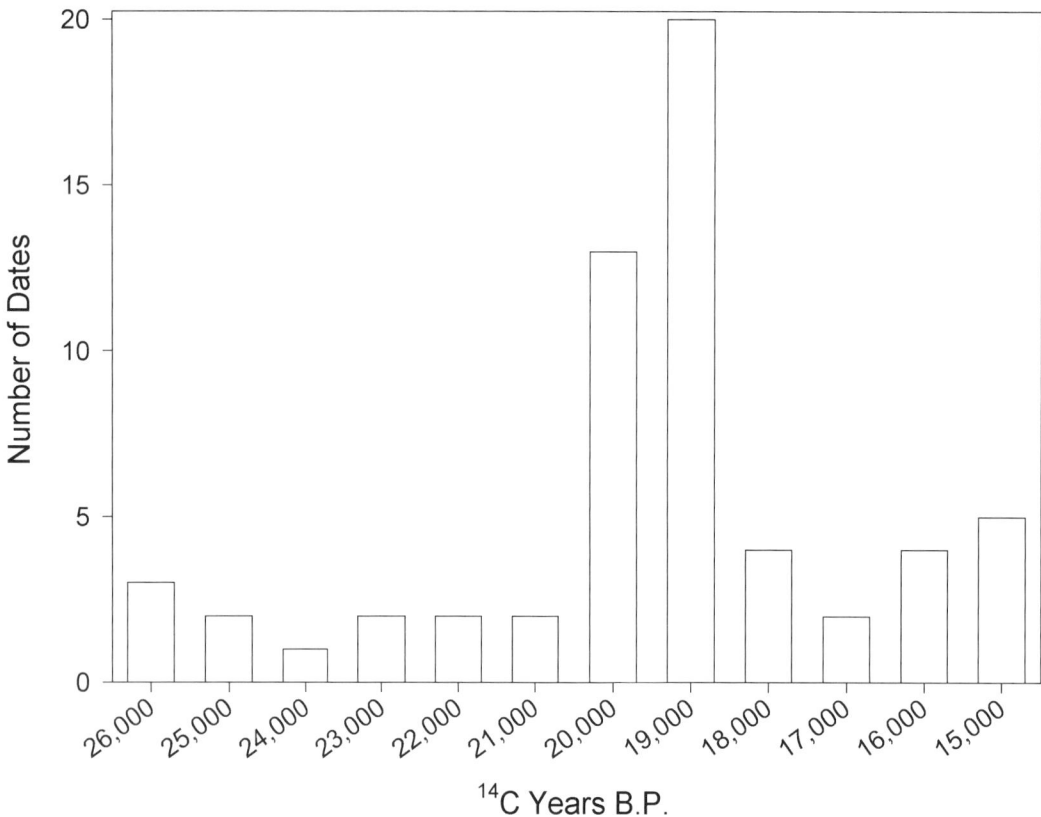

FIGURE 9.2. Frequency of radiocarbon dates for Alaskan *Equus caballus/ferrus*. *Equus* is most common during the last glacial maximum but declines in numbers substantially thereafter. Raw data from Guthrie.[29, supplemental]

Model Results

While empirical measurements of encounters with large game at varying densities would be preferable to our indirect estimates, quantification of any sort for late Pleistocene environments is lacking. We recognize the potential problems inherent in our approach and encourage due caution, but we feel it is still sufficiently robust to be constructive. Using this relationship between body size, population density, and encounter rates allows us to reconstruct overall returns associated with taking any of a hypothetical suite of taxa. Table 9.7 shows the results of such a model. All fauna are ranked from high to low on the basis of their post-encounter returns. The taxa chosen loosely reflect the kinds of animals one might encounter in prehistoric western North America. They include a range of small mammals, birds, and medium artiodactyls that would have been typical of many late Pleistocene environments, plus bison and mammoth. Late Pleistocene bison (*B. antiquus*) are modeled by increasing size and handling time estimates for modern bison (after Ref. [47]) by 50%. This scales extant bison into a size range generally consistent with that demonstrated by the mean difference in horn core measurements between early and late Holocene specimens [107]. The data for mammoths assume 7500 kg per mammoth, a 42 percent edible portion, 1250 kcal/kg, and 75 hours of handling time. By using the low-end estimate for handling time, mammoths become the largest and highest-ranked resource available and provide 52,500 kcal/hr on encounter. Bison rank only slightly lower and encounter rates with the two megafauna in terms of kilograms of meat per hour are 37 percent greater than the next three taxa combined.

The rightmost three columns in Table 9.7 show the overall return rates enjoyed by adding successive, lower-ranked, prey taxa to the diet. The first of these columns computes returns assuming that encounters with mammalian prey scale at 0.007 times prey density (scale factor (SF) = 0.007, Table 9.6). This is the density/encounter rate relationship derived from Simms data and given this value, the optimal diet yields 6503 kcal/hr and includes megafauna, medium artiodactyls, and lagomorphs (Fig. 9.3). A single forager enjoying these encounter rates could acquire a little over half a metric ton of edible meat per month working just three hours per day (100 hrs total). About 60 percent of that would come from megafauna, and the workload would include searching for, pursuing, and processing the animal. As a comparison, these rates are four to five times greater than those reported for the Hadza of Tanzania, who are specialized big-game hunters with access to a wide variety of species including elephant, giraffe, and other megaherbivores [37, 38].

The second to last column in Table 9.7 assumes encounters scale at one-half the derived value (0.0035) and is provided as an additional basis for comparison. Here the optimal diet provides 3986 kcal/hr and includes not only the megafauna, artiodactyls,

TABLE 9.7. Optimal diet breadth.

Resource	Scientific Name	Energy (kcal/kg)[a]	Live Weight (kg/ind)	Edible Fraction	Edible Weight (kg/ind)	Total (kcal)	Handling Time (hr)[c]	Handling Time (hr/kg)	Encounter Rate (ind/hr)[d,e]	Encounter Rate (kg/hr)	Return Rate On-Encounter (kcal/hr)	Overall Return Rate @1*SF (kcal/hr)[f]	Overall Return Rate @½*SF (kcal/hr)	Overall Return Rate @2*SF (kcal/hr)
Mammoth	*Mammuthus columbi*	1250	7500.00	0.42	3150.00	3937500	75.000	0.0238	0.00066	2.0778	52500	2475	1267	4727
Bison[b]	*Bison antiquus*	1090	900.00	0.60	540.00	588600	14.609	0.0271	0.00324	1.7470	40291	4105	2147	7544
Deer	*Odocoileus hemionus*	1200	85.00	0.60	51.00	61200	2.517	0.0493	0.01899	0.9685	24318	4949	2641	8787
Mtn sheep	*Ovis canadensis*	1200	75.00	0.60	45.00	54000	2.517	0.0559	0.02086	0.9387	21457	5673	3091	9742
Pronghorn	*Antilocapra americana*	1140	56.50	0.60	33.90	38646	2.017	0.0595	0.02580	0.8745	19163	6234	3462	10,396
Hare	*Lepus* sp.	1140	2.42	0.60	1.45	1655	0.125	0.0861	0.27399	0.3978	13242	6421	3609	10,520
Cottontail	*Sylvilagus* sp.	1140	1.10	0.60	0.66	752	0.083	0.1263	0.49494	0.3267	9029	6503	3705	10,446
Gopher	*Thomomys* sp.	1200	0.25	0.85	0.21	255	0.042	0.1961	1.50362	0.3195	6120	6485	3769	10,140
Lg Squirrel	*Spermophilus* sp.	1200	0.35	0.85	0.30	357	0.058	0.1961	1.16827	0.3476	6120	6468	3834	9854
Sage grouse	*Centrocercus urophasianus*	1340	1.50	0.70	1.05	1407	0.258	0.2460	0.81250	0.8531	5446	6339	3961	9060
Sm Squirrel	*S. tridecemlineatus*	1200	0.20	0.85	0.17	204	0.042	0.2451	1.77755	0.3022	4896	6278	3986	8811
Ducks	*Anas* sp.	1230	1.00	0.70	0.70	861	0.258	0.3690	0.81250	0.5688	3333	5961	3940	8017

[a] Caloric values are taken from USDA Poultry Products; Lamb, Veal, & Game; Ethnic Foods. [103]
[b] Bison weights and handling times are from Henrikson [47] and scaled up 50% to reflect the larger size of *B. antiquus*.
[c] All other handling times were computed using median times from Simms [90]; sage grouse use estimated times for ducks.
[d] Encounter rates for rodents, rabbits, hares, and ungulates use data from Damuth [16] and Simms [90]. Log density in individuals per square kilometer is 4.33−.75 * log(mass in kg). Encounter rates are computed as (density * ScaleFactor).
[e] Encounter rates for birds use the data from Simms [90]. Because they cannot use the same allometric equations as mammals, these values are adjusted up or down to reflect changes in encounter rate relative to a base scale factor of .0035.
[f] Overall return rates are computed using encounter rate scale factors of .007, .0035, and .014. The first estimate uses the raw scale factor computed from the regression in Table 9.6. Given the second value, encounter rates with all medium artiodactyls (in kg/hr) are equivalent to the maximum levels estimated by Simms [90]. The third value is twice that estimated by the regression equation. Although encounters with mammoth and bison are much higher, the optimal diet continues to include medium artiodactyls and lagomorphs.

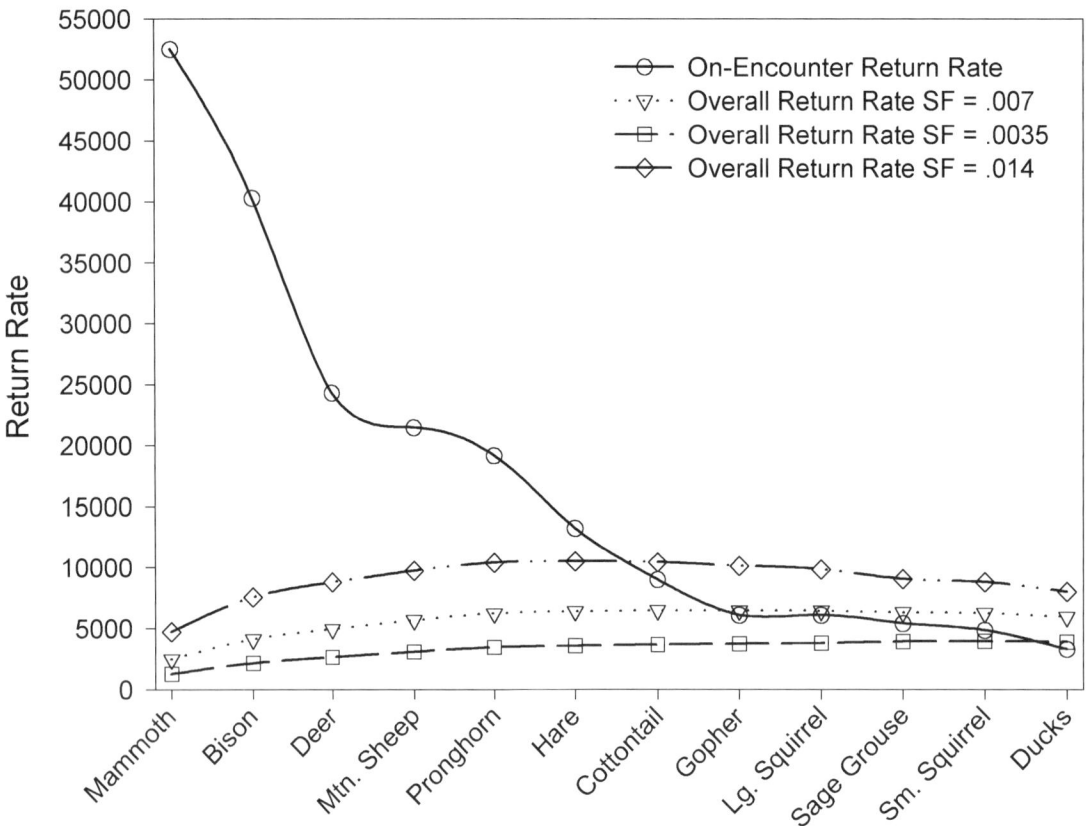

FIGURE 9.3. A diet breadth model for early Paleoindians. Overall returns are plotted for the encounter rate scaling factors of SF=0.007, SF=0.0035, and SF=0.014.

and lagomorphs, but also all small animals except the ducks. At this level, encounters with the deer, bighorn sheep, and antelope total 1.4 kg/hr, corresponding to the maximum estimate provided by Simms [90].[2] The appropriateness of using this measure depends on how accurately Simms' maximum estimated encounter rates for Holocene hunters captures prey availability in the Pleistocene. Unless early Paleoindian hunters regularly encountered game at densities higher than Holocene maxima, diets would have been quite broad. This is true even after including megafauna in the list of potential prey. Even if encounters with all taxa were twice as high as those predicted by the derived scaling factor (SF = 0.014), hares remain firmly in the diet with cottontails on the cusp (last column, Table 9.7). Such observations provide reason to question whether Paleoindian subsistence would ever have been narrow and highly specialized as some have contended [44, 45, 60, 106].

Modeling Returns from Partial Carcass Utilization

Up to this point, we have only considered a scenario where an entire mammoth was procured and then completely processed. As a result, we have presented one extreme of a continuum of processing intensity that would have varied not only in the proportion of the edible products harvested but also in the handling times needed to do so. In fact, the early Paleoindian faunal record suggests that partial carcass utilization may indeed have been the norm [43]. If so, our model overestimates both total caloric yields and handling times. Accordingly, we evaluate the impact of partial carcass utilization on both on-encounter returns and the average return rate earned by a group of early Paleoindians.

Partially processing likely offered much higher return rates than would be gained by using an entire animal. When a resource provides diminishing returns with handling, a hunter should leave off processing as soon as the marginal returns for the resource at hand fall below the average returns for the environment as a whole [95:31–33]. If completely processing a mammoth provides substantially higher returns than the overall environment, as we suggest here, then partial processing is a sub-optimal choice. Even though the chosen part of the carcass provides extremely high returns, lucrative pieces still go to waste and the animal is underutilized. Large mammals contribute less to the overall diet than they should given their size and diets actually broaden. While energetically sub-optimal, such partial processing decisions may have been common where spoilage, transport constraints, band size or mobility decisions limited the amount of flesh a group of early Paleoindians could effectively use.

To illustrate this point consider a case where a forager takes one third of a mammoth's edible tissue (0.14 total weight) in $\frac{1}{15}$ of our 75 hr handling time estimate (5 hrs). Also assume encounters scale at the higher rate (SF = 0.007). In this instance, on-encounter returns increase from 52,500 kcal/hr to 262,500 kcal/hr. Because only part of the carcass was used, however, the maximum overall return rate drops from 6503 kcal/hr to 5451 kcal/hr and a range of small animals including gophers, large squirrels and sage grouse enter the diet. Scavenging partial carcasses would result in similar outcomes. On-encounter returns increase dramatically while maximum overall returns fall and diets widen. Because we are interested in showing that a wider range of animals may have been favored even under very auspicious circumstances, we chose to focus on complete rather than partial carcass utilization.

To summarize, megafauna may well have been the highest-ranked prey on the late Pleistocene landscape. In terms of kg/hr of edible tissue, they were probably encountered more frequently than other prey. Based on these results however, we simply question whether early Paleoindians encountered mammoth and other megafauna taxa often enough to support a specialized hunting economy.

The Importance of Encounter Rates

Prey encounter rates are the fundamental reason why diets are broad in this model. As a matter of empirical fact, one could reduce handling times for completely processing a mammoth from 75 to 50 hours (making them worth 78,750 kcal/hr on-encounter) or increase them to 150 hours (reducing returns to 26,250 kcal/hr). Neither would alter the optimal diet. Likewise, if we assume partial butchering of a carcass in a relatively short time, the model still predicts a wide diet. This is because overall returns from exploiting high-ranked resources tend to be much more sensitive to search time than processing [36]; high-ranked animals require relatively little time to process by definition. Moreover, whether or not a forager ignores small game should be a function of encounters with all of the larger, higher-ranked taxa, to include the medium artiodactyls.

We recognize that attention to Paleoindian diets has focused on whether they were specialized big-game hunters, not mammoth hunters per se (e.g. Ref. [106]), which is part of the reason bison were added to the list of potential prey. Adding multiple, high-ranked taxa serves to increase overall returns and substantially narrow the diet in much the same way as increasing encounter rates with mammoth or bison. If we were to drop bison entirely, overall returns would fall to the point where sage grouse become attractive, even while using the higher encounter rate scaling. Conversely, adding additional large taxa such as giant ground sloth (*Megatherium* sp.), mastodont and horse to the diet would cause it to narrow further. Consequently, one might ask whether our results are simply a function of failing to consider the full range of megafauna available to early Paleoindian hunters.

While we acknowledge this latter possibility, there are four issues to consider. The first is niche overlap, how many different large taxa can be supported within a particular area, and the degree to which there is partial competitive exclusion [2, 6:247–290, 670–699, 67, 72, 84] (for specific examples for herbivores similar to those used here see Refs. [23, 54, 73, 85]). Here we included mammoth and bison, both grazers, a suite of smaller ungulate browsers, and a variety of birds and small mammals representative of similar animals found in almost any habitat. Making a case for big game specialization in the Pleistocene must start by identifying a larger set of megafauna sufficiently differentiated to coexist and that early Paleoindians might have encountered frequently enough to drive small animals from the diet. While such subsets may exist for particular ecosystems, we are skeptical about their ubiquity on a continental scale. Even if we wish to concede that such high-density patches existed, it seems premature to talk about "megafuanal specialization" if it only occurs in certain favorable but potentially rare circumstances. The late Pleistocene Great Basin provides a case in point, since there is clear evidence of Paleoindian occupation along pluvial lakes throughout the region yet no evidence for the use of extinct megafauna [5, 25, 46].

Second, it is highly unlikely that any subset of taxa would ever lead foragers to exclude the medium artiodactyls (i.e. pure megafaunal specialization). Doing so would require either mammoth encounter rates 15 times greater than those predicted by the body mass allometry data, encounters with 30 different taxa similar in size and post-encounter returns to *Bison antiquus*, or overall encounter rates with bison 30 times higher than suggested here. Although any of these options would raise overall returns enough to exclude "average" artiodactyls (providing 21,000 kcal/hr or so, Table 9.7), none seems likely. Given that the upper limit for medium artiodactyl returns is ca. 30,000 kcal/hr [90], excluding them entirely would require even greater megafauna densities.

Third, there is the empirical observation that bison and mammoth are the only megafauna taxa regularly identified in firm association with early Paleoindian artifacts [12, 27, 28]. We include them in our list of potential prey for this reason. If Paleoindians were capable of regularly taking animals the size of mammoths and mastodonts, smaller animals such as horse, capybara (*Hydrochaeris hydrochaeris*), giant ground sloth, camel (*Camelops hesternus*) and the many other now-extinct late Pleistocene herbivores would seem obvious targets. That neither these nor a host of other large mammals show signs of being actively hunted suggests they may have been rare or extinct by the early Paleoindian period and as a result are omitted from this discussion. Any evidence of their use would obviously deserve reconsideration, though doing so will likely have little effect on the overall results for the reasons just discussed.

Fourth, and perhaps most importantly, our discussion so far has been decidedly androcentric. There is substantial evidence that men, women, and children differ in foraging ability and foraging goals [9, 19, 31, 32, 34, 53, 66, and others]. Women or children may have participated in the production and maintenance of hunting technologies and household goods and also in processing animals once acquired. At the same time, each of these groups could also be expected to provision themselves, especially to the degree that large game exploitation is risky and highly variable in terms of success rates [31–33]. Such variability is often at odds with the needs of mothers and young children, requiring both groups to exploit lower-ranked, more readily and reliably acquired resources including both plants and smaller game animals. Since the Paleoindian subsistence record should reflect the contributions of all participants, the presence of a variety of small game should come as no surprise even where the overall encounter rate with megafauna was quite high and men were able to focus exclusively on their exploitation.

Proboscideans and Prey Choice: Prehistoric and Recent Subsistence Records

Our model predicts that megafaunal specialization would have rarely provided an optimal foraging solution for early Paleoindians. An optimal prey set would almost certainly have included both megafauna and medium artiodactyls, while rabbits would have been pursued often and we would not be surprised to see larger rodents in some contexts as well. We arrive at this conclusion even though the model parameters we chose were biased in favor of the specialist model whenever possible. The overall returns earned by early Paleoindians would surely have been better than many Holocene foragers would later experience (e.g. Ref. [11]), but never sufficient to exclude prey such as deer and pronghorn at any set of megafauna encounter rates. We proceed by comparing these expectations against not only the early Paleoindian faunal record, but also data from a range of Old World prehistoric and recent ethnographic contexts offering encounters with proboscideans.

Variability in the Early Paleoindian Subsistence Record

The expectation that early Paleoindians exploited a broad range of taxa does not imply that every faunal assemblage should contain every prey item in the diet. In fact, we expect the faunal record to be highly variable in the numbers and types of taxa identified in any given assemblage, particularly if sites represent short term, non-redundant occupations across a range of different environments. In this case, it is important to recognize a central prediction of the prey model: foragers always take high-ranked animals upon encounter. Consequently large game kill/processing sites simply demonstrate that, as expected, early Paleoindians took large animals such as mammoth whenever possible.

To underscore this point, we use information compiled by Cannon and Meltzer [12, references therein, Tables 3 and 4, pp. 7–10], to illustrate the variability in early Paleoindian prey choice. Fig. 9.4 shows the proportion of the total number of taxa (NTAXA) represented by large game (>500 kg) at 17 early Paleoindian sites. This sample provides a conservative estimate that only employs taxa demonstrating strong evidence for subsistence use. While mammoths dominate at eleven of the sites, the faunal remains at Lewisville and Aubrey include artiodactyls, lagomorphs and small rodents, not to mention birds, snakes, turtles, fishes, snails and fresh water mussel. Importantly, many of these species represent small-bodied prey not considered in our analysis and document diets even wider than our simulation might suggest. Likewise, the assemblages from Whipple and Udora are limited to medium- and small-bodied species such as caribou (*Rangifer* sp.), beaver (*Castor* sp.) and hares (*Lepus* sp.).

Because the conservative approach potentially obscures the variability in early Paleoindian diet breadth and because our model predicts a wide range of prey choice, we construct a second, inclusive index (Fig. 9.5) using all taxa identified at these 17 sites (c.f. Ref. [106]). Thirteen of the assemblages include a wide variety of large-, medium- and small-bodied prey. We note that in both the conservative and inclusive examples, several of the assemblages consistently suggest diets focused on prey other than megafauna, while a number of others contain only the remains of proboscideans. The range of species identified in the remaining assemblages varies depending on whether or not one opts for a conservative or inclusive estimate. In these cases, we suspect that the actual number of prey species lies in between the two sets of values. Given the ambiguity of much of the available data, however, we simply note that the record appears more variable than a big game specialist model would predict.

The Old World Record and Megafauna Hunting

Human foragers and megafauna also shared much of the Old World throughout the Pleistocene [43, 79] and one might reasonably expect Old World foragers to have specialized on their exploitation under favorable circumstances. Yet faunal assemblages from archaeologically well-known areas of the Old World contrast sharply with the view of early Paleoindians as megafaunal specialists and instead suggest the routine exploitation of a broad spectrum of prey dominated by artiodactyls and perrisodactyls. Throughout Eastern Europe, for example, faunal assemblages include red deer (*Cervus elaphus*), reindeer (*Rangifer tarandus*), musk ox (*Ovibos moschatus*), bison (*Bison priscus*), horse (*Equus* sp.), fox (*Vulpes* sp.), wolf (*Canis lupus*), boar (*Sus scrofa*), hares (*Lepus* sp.), fish and birds (e.g. Ref. [10, 92]). Across 14 occupational layers at Kosoutsy on the Dnestr River [10], reindeer account for 40 percent ($n = 80$) of the total MNI, followed by horses (21%, MNI = 38), bison (17%, MNI = 31) and hares (7%, MNI = 13,). In contrast to the abundances of artiodactyls and horses, only two individuals (3%) represent mammoth (*M. primigenius*). Similarly, sites on the Central Russian Plain have produced the remains of substantial numbers of marmots (*Marmota bobac*) as well as the bones of numerous artiodactyls and mammoths, and there is good evidence for the systematic procurement of hares as well [92].

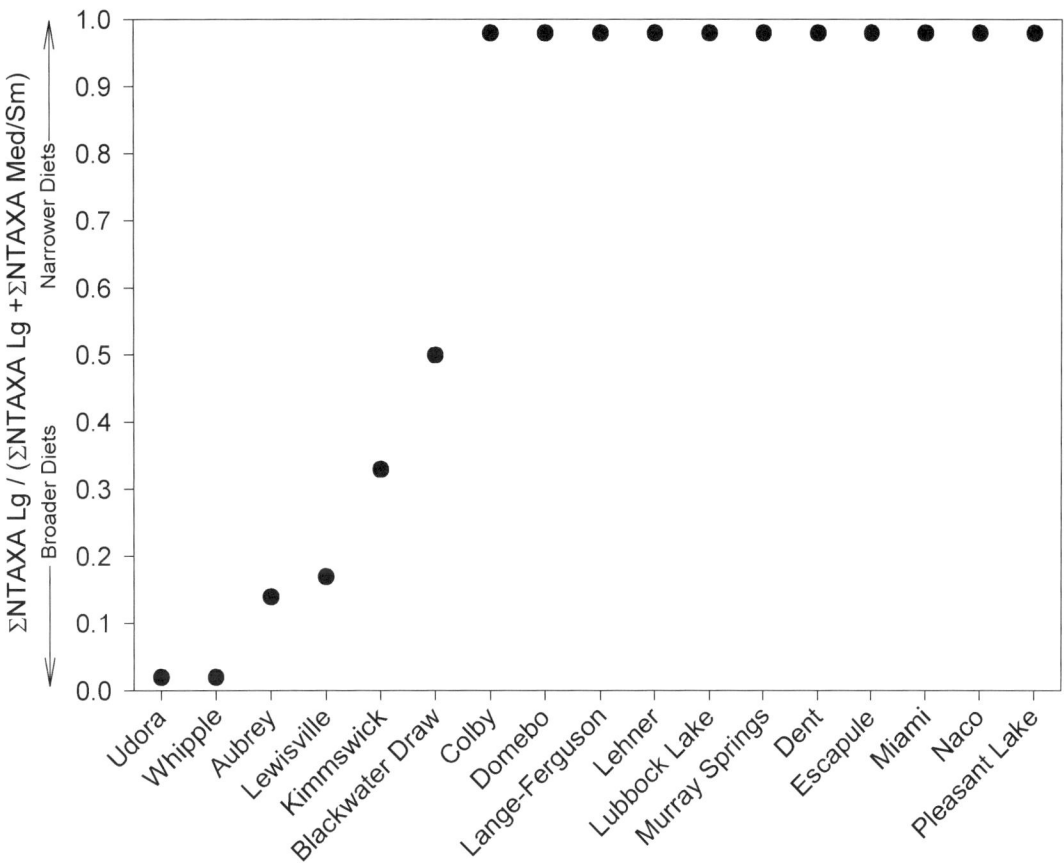

FIGURE 9.4. Large game index, conservative estimate, plotted as point values. Values closer to 1 indicate assemblages limited mainly to megafauna. Values closer to 0 document assemblages containing a wide range of prey. Raw data from Cannon and Meltzer.[12]

Upper Paleolithic faunal records from Western Europe also document a wide range of species frequently dominated by medium artiodactyls. In 33 Aurignacian assemblages from southern France reviewed by Grayson and Delpech [26], *Rangifer tarandus* occurred in 76 percent of the collections, bison in 13 percent, and mammoth in none. The total number of taxa at these sites ranged from two species to ten, and more than five taxa are recovered in over 80 percent of the cases. Upper Paleolithic sites in Spain demonstrate not only large numbers of medium-sized artiodactyl remains, but in several cases lagomorphs appeared to have played a substantial role [101, 105]. Within the Aurignacian/Gravettian components of Cova Beneito and Cueva de Malladetes, lagomorph remains are by far the most common taxa with NISPs of 5544 (86%) and 83 (55%) respectively [17, 71], and artiodactyls such as ibex (*Capra pyrenacia*) and red deer comprise the bulk of the remaining faunas. We note that in each of the cited instances, all of the assemblages predate the youngest radiocarbon date on mammoth bone for each region [97]. Similarly wide diets from contexts where megafauna are known to have been present include the Levant (e.g. Ref. [96]), Africa (e.g. Ref. [62]), and Australia (e.g. Ref. [1]).

The Prey Choice of Recent African Elephant Hunters

The wide range of animal prey taken by recent hunting and gathering groups further underscores our conclusions about the importance of proboscideans to early Paleoindian subsistence. While recent African hunter-gatherers cannot provide direct analogs for late Pleistocene foragers living in North America, they do provide a valuable ethnographic context. Insofar as mammoths can be argued to have provided returns substantial enough to narrow the diet to the exclusion of all but a few other large-bodied prey, then there is little reason to view elephants otherwise today. If elephants are highly ranked, then they should regularly be taken on encounter; if they are frequently enough encountered, overall diets should be narrow. This is simply not the case.

While numerous accounts of African hunter-gatherers describe elephant hunting (e.g. Refs. [3, 4, 14, 21, 55, 81, 83, 87, 94, 100, 102]), a close look at prey choice reveals that these animals were rarely taken. This observation is nothing new. Roosevelt [81:299] for instance, noted that "very few of the native tribes in Africa hunt the elephant systematically." Indeed, Bailey's [3, 4] ethnographic accounts of Efe subsistence mention elephant

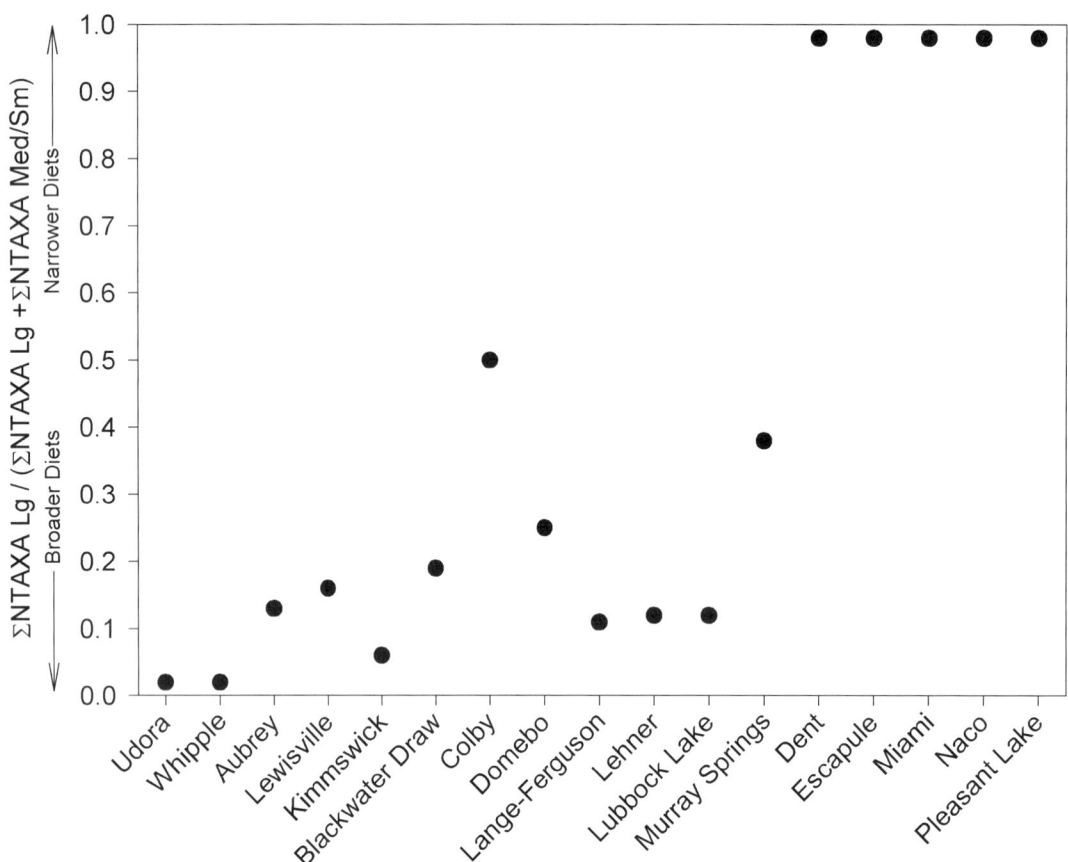

FIGURE 9.5. Large game index, inclusive estimate, plotted as point values. Values closer to 1 indicate assemblages limited mainly to megafauna. Values closer to 0 document assemblages containing a wide range of prey. Raw data from Cannon and Meltzer.[12]

hunting only briefly and mainly within the context of photo captions. Instead of elephants, more than 45 species of smaller prey constitute the bulk of the diet. South African Bantu-speaking groups took a similar range of prey: including not only elephant, but also artiodactyls such as buffalo and eland in addition to monkeys, small game and birds [83, 100]. Moreover, elephants are notably absent from subsistence accounts of many other recent African foraging groups, including the !Kung [65] of southern Africa and the Hadza of east Africa [77]. While both groups share environments with elephants, accounts of their pursuit are largely absent and hunters from both groups focus on a wide range of smaller artiodactyls.

In sum, these records appear to support our expectations. In areas with elephants, African foragers pursue these animals but regularly take a variety of other prey ranging from large game to medium- and small-bodied taxa. Given the lack of ethnographic accounts documenting systematic elephant hunting and the paucity of evidence for the widespread use of similar-sized animals in either the New World, Old World (see Ref. [99]), or even Australia, one might question whether hunter-gatherers focused on megafauna anywhere during the Pleistocene.

While both the Old World Pleistocene and ethnographic records are generally consistent with our predictions, proboscidean-only faunas do represent a comparatively large portion of known Paleoindian sites. This empirical observation begs the question of why these animals are so common. One suggestion is that the prevalence of mammoths in the early Paleoindian record may simply be a function of discovery bias [12, 24, 74, 75]. Not only are sites with mammoth remains more likely to be discovered, but excavations often focus spatially on bone beds derived from kill/processing events [12]. As a result, investigations limited to megafauna remains may fail to recover the full range of prey present at a site. Hadza intercept hunting locations, for instance, often document multiple kills representing a range of taxa spread over several thousand square meters [78]. Likewise, transport decisions and the selective deletion of the remains of smaller taxa by scavengers can also bias kill/processing locations in favor of larger-bodied animals [78]. This might be especially so in the context of a mammoth kill where the ethnographic record predicts a residential move to the carcass as opposed to the transport of meat back to a residential base. Whatever the case, published excavations appear to have left us with a record skewed towards easily recognizable cultural deposits containing the highly visible bones of mammoths.

Conclusions

Our study suggests that early Paleoindians should have pursued a broad range of mammalian prey. Moreover, these results appear very robust. One can vary the model parameters across a wide range of values without changing the outcome to any great degree. Consequently, we expect that megafaunal specialization would have been profitable within a very narrow range of circumstances where large game species were extremely abundant, easily procured, and could be processed at minimal time cost. Given the data presented here, we suspect such instances were uncommon at best.

While we will never know how often early Paleoindians actually encountered megafauna, we suspect that encounter rates never exceeded the levels necessary to exclude medium artiodactyls from the diet and would rarely have been sufficient to exclude lagomorphs. In this respect, the late Pleistocene of North America appears to have been little different from most other terrestrial hunter-gatherer contexts through time and across space. The highest-ranked taxa (in this case, megafauna) were taken whenever possible, but were likely never plentiful enough to result in the sort of narrow hunting specialization often argued to characterize early Paleoindian subsistence. We note that as the faunal record receives more and more attention, the evidence points to a broader spectrum of prey choice than has often been suggested. However favored megafaunal prey might have been, a wide range of game was still necessary to keep these hunting and gathering groups fed.

It is also worth pointing out that there is a potential semantic argument lurking in our discussion of "specialization". Throughout this paper, we take specialization to mean exploiting megafauna to the exclusion of other, smaller prey and go to some lengths to suggest this was unlikely. At the same time, we also assume that megafauna were regularly and successfully hunted. Consequently, the actual proportion of the overall diet made up by megafauna in Table 9.7 is on the order of 60 percent. While we would expect early Paleoindians to take a wide range of taxa, which we consider "generalist" in this context, megafauna would still make up the bulk of men's contributions to the diet and we might expect them to exert a strong influence on material culture and behavior as a result.

That being said, the operative word is "might." Optimal diet models such as this one are static constructs applied to particular places and points in time. We do not consider how prehistoric hunters may have depleted particular patches. We also do not address the issue of how often they may have been able to exploit "megafauna-rich" patches. If megafauna were quickly depleted, scarce across much of the landscape, or subject to high hunting failure rates, diets would have been correspondingly broad and the contribution of these rarer megafauna more subdued. The influence of mammoth and other large game on human subsistence decisions, material culture, mobility, and other aspects of behavior and social organization would be reduced as a result. How these factors played out prehistorically is a different question which deserves separate consideration.

Finally, it should be obvious from this study that evaluating early Paleoindian prey choice requires that we take the time to fully understand the environments these foragers lived in, the types and abundances of food items they hunted and gathered, and the costs and benefits involved in procuring and processing both plants (which we have admittedly shorted here) and animals. In this case, much of the strength of our conclusions rests on whether or not we have adequately modeled late Pleistocene prey population densities. Indeed, if this paper points to any one issue, it is that estimates of encounter rates with ancient faunas are critical to interpreting any record of prehistoric hunting, especially within an optimal foraging context. Whatever the true breadth of the early Paleoindian diet, a great deal of research still lies before us.

Acknowledgments

We would like to thank Jack Broughton, Mike Cannon, Judson Finley, David Meltzer, James O'Connell, Craig Smith, Todd Surovell, and two anonymous reviewers for their helpful comments on earlier versions of this paper and Gary Haynes and Bob Kelly for their discussion at the 2005 SAAs. We also give special thanks to Robert Bailey, who was kind enough to share his observations on elephant handling time.

Notes

1. There is substantial debate about the nature of the relationship between animal density and body mass, including the appropriateness of −0.75 as a slope. We evaluated Silva et al.'s [89] alternative derivation for terrestrial herbivores, which is log(density) = 1.43−.68 ∗ log(mass). Proboscidean densities are comparable to those predicted using Damuth's [16] derivation, but encounters with smaller animals drop off sharply. Using this formula would substantially depress overall return rates and increase diet breadth compared to Damuth. We feel Damuth's relationship provides more reasonable estimates, especially for small and medium sized animals, and use it instead.
2. The specific encounter rates used in Table 9.7 differ slightly from Simms [90] because here they vary with body weight. Encounters with deer and antelope are more frequent than reconstructed by Simms, bighorn sheep a bit less frequent.

References

1. H. Allen, Reinterpreting the 1969–1972 Willandra Lakes archaeological surveys, in: H. Johnston, P. Clark, J. P. White (Eds.), *Willandra Lakes: People and Palaeoenvironments*, 33, Archaeology in Oceania, 1998, pp. 207–220.
2. P. Amarasekare, Competitive coexistence in spatially structured environments: a synthesis, *Ecology Letters* 6 (2003) 1109–1122.
3. R. C. Bailey, The Efe: archers of the African rain forest, *National Geographic* 176 (1989) 664–686.

4. R. C. Bailey, *The Behavioral Ecology of Efe Pygmy Men in the Ituri Forest, Zaire*, Anthropological Papers, Museum of Anthropology, University of Michigan, Ann Arbor, 1991.
5. C. Beck, G. Jones, The terminal Pleistocene/Early Holocene archaeology of the Great Basin, *Journal of World Prehistory* 11 (1997) 161–236.
6. M. Begon, J. L. Harper, C. R. Townsend, *Ecology: Individuals, Populations, and Communities*, Sinauer Associates, Inc., Sunderland, 1986.
7. A. Belgrano, A. P. Allen, B. J. Enquist, J. F. Gillooly, Allometric scaling of maximum population density: a common rule for marine phytoplankton and terrestrial plants, *Ecology Letters* 5 (2002) 611–613.
8. T. M. Blackburn, K. J. Gaston, A critical assessment of the form of the interspecific relationship between abundance and body size in animals, *The Journal of Animal Ecology* 66 (1997) 233–249.
9. N. Blurton-Jones, K. Hawkes, P. Draper, Differences between Hadza and !Kung children's work: original affluence or practical reason? in: E. S. Burch Jr., L. J. Ellanna (Eds.), *Key Issues in Hunter Gatherer Research*, Berg, Oxford, 1994, pp. 189–215.
10. I. A. Borziyak, Subsistence practices of Late Pleistocene groups along the Dnestr River and its tributaries, in: O. Soffer, N. D. Praslov (Eds.), *From Kostenki to Clovis: Upper Paleolithic–Paleoindian Adaptations*, Plenum Press, New York, 1993, pp. 67–84.
11. D. A. Byers, C. S. Smith, J. M. Broughton, Artiodactyl population histories and large game hunting in the Wyoming Basin, USA, *Journal of Archaeological Science* 32 (2005) 125–142.
12. M. D. Cannon, D. J. Meltzer, Early Paleoindian foraging: examining the faunal evidence for large mammal specialization and regional variability in prey choice, *Quaternary Science Reviews* 23 (2004) 1955–1987.
13. P. Christiansen, Body size in proboscideans, with notes on elephant metabolism, *Zoological Journal of the Linnean Society* 140 (2004) 523–549.
14. D. C. Crader, Recent single-carcass bone scatters and the problem of "butchery" sites in the archaeological record, in: J. Clutton-Brock, C. Grigson (Eds.), *Animals and Archaeology: Hunters and Their Prey*, BAR International Series 163, British Archaeological Reports, Oxford, 1983, pp. 107–141.
15. J. Damuth, Population density and body size in mammals, *Nature* 290 (1981) 699–700.
16. J. Damuth, Interspecific allometry of population density in mammals and other animals: the independence of body mass and population energy use, *Biological Journal of the Linnean Society* 31 (1987) 193–246.
17. I. Davidson, La economía del final del paleolítico en la España oriental, *Serie de Trabajos Varios del SIP* 85, 1989.
18. F. S. Dobson, B. Zinner, M. Silva, Testing models of biological scaling with mammalian population densities, *Canadian Journal of Zoology* 81 (2003) 844–851.
19. R. G. Elston, D. W. Zeanah, Thinking outside the box: a new perspective on diet breadth and sexual division of labor in the Prearchaic Great Basin, *Journal of World Prehistory* 34 (2002) 103–130.
20. B. J. Enquist, J. H. Brown, G. B. West, Allometric scaling of plant energetics and population density, *Nature* 395 (1998) 163–165.
21. J. W. Fisher Jr., Observations on the Late Pleistocene bone assemblage from Lamb Spring, Colorado, in: D. J. Stanford, J. S. Day (Eds.), *Ice Age Hunters of the Rockies*, Denver Museum of Natural History and University Press of Colorado, Niwot, 1992, pp. 51–82.
22. G. C. Frison, L. C. Todd, *The Colby Mammoth Site: Taphonomy and Archaeology of a Clovis Kill in Northern Wyoming*, University of New Mexico Press, Albuquerque, 1986.
23. H. Fritz, P. Duncan, I. J. Gordon, A. W. Illius, Megaherbivores influence trophic guild structure in African ungulate communities, *Oecologia* 131 (2002) 620–625.
24. D. K. Grayson, Perspectives on the archaeology of the first Americans, in: R. C. Carlisle (Ed.), *Americans before Columbus: Ice-Age Origins*, Ethnology Monographs 12 (1988) 107–123.
25. D. K. Grayson, *The Desert's Past, a Natural Prehistory of the Great Basin*, Smithsonian Institution Press, Washington D.C., 1993.
26. D. K. Grayson, F. Delpech, Specialized early Paleolithic hunters in France? *Journal of Archaeological Science* 29 (2002) 1439–1449.
27. D. K. Grayson, D. J. Meltzer, Clovis hunting and large mammal extinction: a critical review of the evidence, *Journal of World Prehistory* 14 (2002) 313–359.
28. D. K. Grayson, D. J. Meltzer, A requiem for North American overkill, *Journal of Archaeological Science* 28 (2003) 585–593.
29. R. D. Gutherie, Rapid body size decline in Alaskan Pleistocene horses before extinction, *Nature* 426 (2003) 169–171.
30. J. Hanks, *The Struggle for Survival: The Elephant Problem*, Mayflower Books, New York, 1979.
31. K. Hawkes, Showing off: tests of a hypothesis about men's hunting goals, *Ethology and Sociobiology* 12 (1991) 29–54.
32. K. Hawkes, Why hunter-gatherers work: an ancient version of the problem of public goods, *Current Anthropology* 34 (1993) 341–361.
33. K. Hawkes, Foraging differences between men and women, in: J. Steele, S. Shannen (Eds.), *The Archaeology of Human Ancestry: Power, Sex, and Tradition*, Routeledge, London, 1996, pp. 283–305.
34. K. Hawkes, R. Bliege Bird, Showing off, handicap signaling, and the evolution of men's work, *Evolutionary Anthropology* 11 (2002) 58–67.
35. K. Hawkes, J. F. O'Connell, Optimal foraging models and the case of the !Kung, *American Anthropologist* 87 (1985) 401–405.
36. K. Hawkes, J. F. O'Connell, On optimal foraging models and subsistence transitions, *Current Anthropology* 33 (1992) 63–66.
37. K. Hawkes, J. F. O'Connell, N. G. Blurton-Jones, Hunting income patterns among the Hadza: big game, common goods, foraging goals, and the evolution of the human diet, *Philosophical Transactions of the Royal Society* B 334 (1991) 243–251.
38. K. Hawkes, J. F. O'Connell, N. G. Blurton-Jones, Hadza meat sharing, *Evolution and Human Behavior* 22 (2001) 113–142.
39. C. V. Haynes Jr., Geoarchaeological and Paleohydrological evidence for a Clovis-age drought in North America and its bearing on extinction, *Quaternary Research* 35 (1991) 438–450.
40. C. V. Haynes Jr., Clovis-Folsom geochronology and climate change, in: O. Soffer, N. D. Praslov (Eds.), *From Kostenki to Clovis: Upper Paleolithic–Paleoindian Adaptations*, Plenum Press, New York, 1993, pp. 219–236.
41. C. V. Haynes, Jr., Geochronology of paleoenvironmental change, Clovis type site, Blackwater Draw, New Mexico, *Geoarchaeology: An International Journal* 10 (1995) 317–388.
42. C. V. Haynes Jr., D. J. Stanford, M. Jodry, J. Dickenson, J. L. Montgomery, P. H. Shelly, I. Rovner, G. A. Agogino, A Clovis well at the type site 11,500 BC: the oldest prehistoric well in America, *Geoarchaeology: An International Journal* 14 (1999) 455–470.
43. G. Haynes, *Mammoths, Mastodonts, and Elephants: Biology, Behavior and the Fossil Record*, Cambridge University Press, Cambridge, 1991.

44. G. Haynes, The catastrophic extinction of North American mammoths and mastodons, *World Archaeology* 33 (2002) 391–416.
45. G. Haynes, *The Early Settlement of the Americas: The Clovis Era*, Cambridge University Press, Cambridge, 2002.
46. R. F. Heizer, M. A. Baumhoff, Big game hunters in the Great Basin: a critical review of the evidence, *University of California Archaeological Research Facility Reports* 7 (1970) 1–12.
47. L. S. Henrickson, Frozen bison and fur trapper's journals: building a prey choice model for Idaho's Snake River Plain, *Journal of Archaeological Science* 31 (2004) 903–916.
48. V. T. Holliday, *Paleoindian Geoarchaeology of the Southern High Plains*, University of Texas Press, Austin, 1997.
49. V. T. Holliday, Folsom drought and episodic drying on the Southern High Plains from 10,900–10,200, *Quaternary Research* 53 (2000) 1–12.
50. C. S. Holling, Some characteristics of simple types of predation and parasitism, *Canadian Entomologist* 91 (1959) 385–398.
51. W. N. Hollsworth, Hunting efficiency and white-tailed deer density, *Journal of Wildlife Management* 37 (1973) 336–342.
52. B. B. Huckell, Of chipped stone tools, elephants and the Clovis hunters: an experiment, *Plains Anthropologist* 24 (1979) 177–189.
53. C. Hudecek-Cuffe, Engendering Northern Plains Paleoindian Archaeology: Decision Making and Gender/Sex Roles in Subsistence and Settlement Strategies. Unpublished Ph.D. dissertation, Department of Anthropology, University of Alberta, Edmonton, 1996.
54. I. A. R. Hulbert, R. Andersen, Food competition between a large ruminant and a small hindgut fermenter: the case of the roe deer and mountain hare, *Oecologia* 128 (2001) 499–508.
55. J. Janmart, Elephant hunting as practiced by the Congo Pygmies, *American Anthropologist* 54 (1952) 146–147.
56. S. Jennings, J. L. Blanchard, Fish abundance with no fishing: predictions based on macroecological theory, *Journal of Animal Ecology* 73 (2004) 632–642.
57. G. K. Jense, J. S. Burruss, *Big Game Harvest Report 1978*, Utah State Division of Wildlife Resources Publication 76-6, 1978.
58. W. Jetz, C. Carbone, J. Fulford, J. H. Brown, The scaling of animal space use, *Nature* 306 (2004) 266–268.
59. M. A. Jochim, *Strategies for Survival: Cultural Behavior in an Ecological Context*, Academic Press, New York, 1981.
60. R. L. Kelly, Hunter-gatherer foraging and colonization of the Western Hemisphere, *Anthropologie* 37 (1999) 143–153.
61. R. L. Kelly, L. C. Todd, Coming into the country: early Paleoindian hunting and mobility, *American Antiquity* 53 (1988) 231–244.
62. R. G. Klein, The Mammalian fauna of the Klasies River Mouth sites, southern Cape Province, South Africa, *South African Archaeological Bulletin* 31 (1976) 75–98.
63. R. S. Laub, On disassembling an elephant: anatomical observations bearing on Paleoindian exploitation of Proboscidea, in: J. W. Fox, C. B. Smith, K. T. Wilkins (Eds.), *Proboscidean and Paleoindian Interactions*, Baylor University Press, Waco, 1992, pp. 99–110.
64. R. M. Laws, I. S. C. Parker, R. C. B. Johnstone, *Elephants and Their Habitats: The Ecology of Elephants in Northern Bunyoror, Uganda*, Clarion Press, Oxford, 1975.
65. R. B. Lee, *The !Kung San: Men, Women and Work in a Foraging Society*, Cambridge University Press, Cambridge, 1979.
66. K. D. Lupo, D. N. Schmitt, Upper Paleolithic net-hunting, small prey exploitation, and women's work effort: a view from the ethnographic and ethnoarchaeological record of the Congo Basin, *Journal of Archaeological Method and Theory* 9 (2002) 147–179.
67. R. H. MacArthur, R. Levins, Competition, habitat selection, and character displacement in a patchy environment, *Proceedings of the National Academy of Sciences* 51 (1964) 1207–1210.
68. R. H. MacArthur, E. Pianka, On optimal use of a patchy environment, *The American Naturalist* 100 (1966) 603–609.
69. P. S. Martin, Prehistoric overkill, in: P. S. Martin, H. E. Wright (Eds.), *Pleistocene Extinctions: The Search for a Cause*, Yale University Press, New Haven, 1967, pp. 75–120.
70. P. S. Martin, Prehistoric overkill: the global model, in: P. S. Martin, R. G. Klein (Eds.), *Quaternary Extinctions: a Prehistoric Revolution*, University of Arizona Press, Tucson, 1984, pp. 354–403.
71. R. Martínez Valle, Fauna del Pleistoceno Superior del País Valenciano: Aspectos Económicos, Huellas de Manipulación y Valoración Paleoambiental, Tesis Doctoral, Universitat de Valencia, Valencia, 1996.
72. R. M. May, R. H. MacArthur, Niche overlap as a function of environmental variability, *Proceedings of the National Academy of Sciences* 69 (1972) 1109–1113.
73. M. L. McInnis, M. Vavra, Dietary Relationships among Feral Horses, Cattle, and Pronghorn in Southeastern Oregon, *Journal of Range Management* 40 (1987) 60–66.
74. D. J. Meltzer, Late Pleistocene human adaptations in eastern North America, *Journal of World Prehistory* 2 (1988) 1–52.
75. D. J. Meltzer, Is there a Clovis adaptation? in: O. Soffer, N. D. Praslov (Eds.), *From Kostenki to Clovis: Upper Paleolithic–Paleoindian Adaptations*, Plenum Press, New York, 1993, pp. 293–310.
76. D. J. Meltzer, Peopling of North America, *Developments in Quaternary Science* 1 (2004) 539–563.
77. J. F. O'Connell, K. Hawkes, N. Blurton-Jones, Hadza hunting, butchering and bone transport and their archaeological implications, *Journal of Anthropological Research* 44 (1988) 113–161.
78. J. F. O'Connell, K. Hawkes, N. Blurton-Jones, Patterns in the distribution, site structure, and assemblage composition of Hadza kill-butchering sites, *Journal of Archaeological Science* 19 (1992) 319–345.
79. H. F. Osborn, *Proboscidea: A Monograph of the Discovery, Evolution, Migration and Extinction of the Mastodonts and Elephants of the World*, Vol II: Stegontoisea, Elephantoidea. American Museum of Natural History, New York, 1942.
80. W. Robertson-Bullock, The weight of the African elephant *Loxodonta africana*, *Proceedings of the Zoological Society of London* 138 (1962) 133–135.
81. T. Roosevelt, *African Game Trails: An Account of the African Wanderings of an American Hunter-Naturalist*, Volume 1, Charles Scribner's Sons, New York, 1920.
82. V. M. Savage, J. F. Gillooly, J. H. Brown, G. B. West, E. L. Charnov, Effects of body size and temperature on population growth, *The American Naturalist* 163 (2004) 429–441.
83. I. Schapera, A. J. H. Goodwin, Chapter VII: Work and Wealth, in: I. Schapera (Ed.), *The Bantu-Speaking Tribes of South Africa: An Ethnographic Survey*, George Routledge and Sons, London, 1937, pp. 131–170.
84. T. W. Schoener, Resource partitioning in ecological communities, *Science* 185 (1974) 27–39.
85. C. C. Schwartz, J. E. Ellis, Feeding ecology and niche separation in some native and domestic ungulates on the shortgrass prairie, *Journal of Applied Ecology* 18 (1981) 343–353.
86. B. Shapiro, A. J. Drummond, A. Rambaut, M. C. Wilson, P. E. Matheus, A. V. Sher, O. G. Pybus, M. Thomas, P. Gilbert, I. Barnes, J. Binladen, E. Willerslev, A. J. Hansen, G. F. Baryshnikov,

J. A. Burns, S. Davydov, J. C. Driver, D. G. Froese, C. R. Harrington, G. Keddie, P. Kosintsev, M. Kunz, L. Martin, R. O. Stephenson, J. Storer, R. Tedford, S. Zimov, A. Cooper, Rise and fall of the Beringian Steppe Bison, *Science* 306 (2004) 1561–1565.

87. M. Shaw, Material culture, in: W. D. Hammond Tooke (Ed.), *Bantu-Speaking Tribes of South Africa*, Routledge and Kegan Paul Ltd, London, 1974, pp. 85–131.

88. P. Shipman, Body size and broken bones: preliminary interpretations of proboscidean remains, in: J. W. Fox, C. B. Smith, K. T. Wilkins (Eds.), *Proboscidean and Paleoindian Interactions*, Baylor University Press, Waco, 1992, pp. 75–98.

89. M. Silva, M. Brimacombe, J. A. Downing, Effects of body mass, climate, geography, and census area on population density of terrestrial animals, *Global Ecology and Biogeography* 10 (2001) 469–485.

90. S. R. Simms, *Behavioral ecology and hunter-gatherer foraging: an example from the Great Basin*, BAR International Series 381, Oxford, 1987.

91. E. A. Smith, *Inujjuamiut Foraging Strategies*, Aldine De Gruyter, Hawthorn, 1991.

92. O. Soffer, *The Upper Paleolithic of the Central Russian Plain*, Academic Press, San Diego, 1985.

93. D. Stanford, Paleoindian archaeology and late Pleistocene environments in the Plains and southwestern United States, in: R. Bonnichsen, K. L. Turnmire (Eds.), *Ice Age Peoples of North America: Environments, Origins and Adaptations*, Center for the Study of the First Americans, Oregon State University Press, Corvallis, 1999, pp. 281–339.

94. H. S. Stannus, Notes on some tribes of British Central Africa, *The Journal of the Royal Institute of Great Britain and Ireland* 40 (1910) 285–335.

95. D. W. Stephens, J. R. Krebs, *Foraging Theory*, Princeton University Press, Princeton, 1986.

96. M. Stiner, N. D. Munro, T. A. Surovell, The tortoise and the hare: Small-game use, the broad spectrum revolution and Paleolithic demography, *Current Anthropology* 41 (2000) 39–79.

97. A. J. Stuart, D. L. D. Sulerzhitsky, L. A. Orlova, Y. V. Kuzmin, A. M. Lister, The latest woolly mammoths (*Mammuthus primigenius* Blumenbach) in Europe and Asia: a review of the current evidence, *Quaternary Science Reviews* 21 (2002) 1559–1569.

98. T. A. Surovell, Early Paleoindian women, children, mobility and fertility, *American Antiquity* 65 (2000) 495–509.

99. T. Surovell, N. Waguespack, P. J. Brantingham, Global archaeological evidence for proboscidean overkill, *Proceedings of the National Academy of Sciences* 102 (2005) 6231–6236.

100. B. K. Taylor, *The Western Lacustrine Bantu (Nyoro, Toro, Nyankore, Kiga, Haya, and Zinza, with sections on the Amba and Konjo)*, International African Institute, 1962.

101. J. E. A. Tortosa, V. Villaverde Bonilla, M. Perez Ripoll, R. Martínez Valle, P. Guillem Calatayud, Big game and small prey: Paleolithic and Epipaleolithic economy from Valencia (Spain), *Journal of Archaeological Method and Theory* 9 (2002) 215–268.

102. C. M. Turnbull, *The Forest People*, Simon and Schuster, New York, 1961.

103. U. S. Department of Agriculture, Agricultural Research Service, USDA National Nutrient Database for Standard Reference, Release 17, Nutrient Data Laboratory Home Page, <http://www.nal.usda.gov/fnic/foodcomp>, 2004.

104. R. C. Van Etten, D. F. Switzenberg, L. Eberhardt, Controlled deer hunting in a square mile enclosure, *Journal of Wildlife Management* 29 (1965) 59–73.

105. V. Villaverde, J. E. Aura, C. M. Barton, The Upper Paleolithic in Mediterranean Spain: a review of the current evidence, *Journal of World Prehistory* 12 (1998) 121–198.

106. N. M. Waguespack, T. A. Surovell, Clovis hunting strategies or how to make out on plentiful resources, *American Antiquity* 68 (2003) 333–352.

107. M. Wilson, Archaeological kill site populations and the Holocene evolution of the genus *Bison*, in: L. B. Davis, M. Wilson (Eds.), *Bison Procurement and Utilization: a Symposium*, 14, Plains Anthropologist Memoir, 1978, pp. 9–22.

108. B. Winterhalder, Foraging strategies in the boreal forest: an analysis of Cree hunting and gathering, in: B. Winterhalder, E. A. Smith (Eds.), *Hunter-Gatherer Foraging Strategies*, University of Chicago Press, Chicago, 1981, pp. 66–98.

109. L. Witting, The body mass allometries as evolutionarily determined by the foraging of mobile organisms, *Journal of Theoretical Biology* 177 (1995) 129–137.

PART III

Post-Glacial Adaptations

After the Ice Age, human populations the world over expanded and adapted to a variety of ever-changing Holocene environments. Archaeologists have long recognized that this period witnessed a shift in human foraging behavior to incorporate a broader array of resources than was typically used during the Pleistocene, although, as noted in the introduction to the preceding section, we now also appreciate greater variability in Pleistocene economies than was once the case. Terms such as "Archaic," "Mesolithic," "resource intensification," and "broad spectrum revolution" are all used to refer to the changes in subsistence, settlement, and technology that occurred following the Last Glacial Maximum (ca. 20,000 years ago).

The papers collected here in part three of *Evolutionary Ecology and Archaeology* use the EE approach to address the processes underlying such changes. These papers explore human foraging behavior, resource intensification and depression, resource extraction and transport, settlement patterns, and technology during the Holocene. More EE-based archaeological analyses have been conducted in this context than in any other, and we have therefore divided the works in this part of the volume into two sections. Papers in the first section focus on issues relating to foraging, predator-prey interactions and the sexual division of labor, while papers in the second section examine issues involving tools and technology.

PART III, SECTION I

Foraging, Predator-Prey Interactions, and the Sexual Division of Labor

The papers here address settlement location, resource choice, the transport of resources once selected, the impacts that prehistoric human foragers had on hunted resources, and the consequences of those impacts for other aspects of ancient human lifeways. The patterns of subsistence, mobility, and settlement that are documented in these papers provide a basis for the exploration of trends in tools and technology that occurs in the following section of the volume. All of the papers in this first section illustrate that the connection between foraging behavior and its archaeological consequences is complex, making it difficult to derive straightforward predictions even from the simplest of models. The hard-fought successes represented here are thus encouraging; the examples are geographically diverse but are centered on western North America, the region in which archaeological applications of EE are exceptionally well developed. Two of the papers (Bayham, Cannon) focus on hunting at the transition to agriculture in the American Southwest, two pertain to Great Basin foragers (Barlow and Metcalfe, Zeanah), and two focus on coastal hunter-gatherers: one in California (Broughton) and the other in New Zealand (Nagaoka).

The foraging theory logic and basic methodology developed by F. E. Bayham in the first paper provides a foundation not only for most of the subsequent papers in this section, but also for applications of foraging theory to zooarchaeological contexts in general (see Broughton 2002, reprinted here, for further discussion). The empirical heart of Bayham's research is faunal data from south-central Arizona, and especially from Ventana Cave, an upland site located in the Castle Mountains that was occupied by Archaic and Hohokam peoples for much of the Holocene. Building upon earlier arguments that body size is the most tangible archaeological measure of post-encounter return rate for vertebrate prey (Bayham 1979, 1982), Bayham documents a significant increase through time at Ventana Cave in the abundance of artiodactyls relative to lagomorphs (rabbits), a pattern that indicates an increase in hunting efficiency at this locality.

This trend, Bayham argues, was driven by a settlement system shift that altered the function of Ventana Cave from initial use as a residential base to later use as a logistic hunting camp. He argues in turn that this change in site function altered transport costs for hunters using the cave such that they could economically narrow their diet (at least while foraging from the cave) to include only high-return artiodactyl prey.

This site-specific pattern is then used to derive a more general hypothesis regarding trends in the regional faunal record. Insofar as the earlier Archaic period was characterized by greater residential mobility and lower logistic mobility, Bayham suggests, faunal assemblages from sites dating to this period should exhibit greater uniformity and vary less in their proportions of high- and low-ranked prey. By contrast, he argues, with the shift to lowland sedentism, associated resource depression, and higher logistic mobility, intersite variability in faunal composition should increase. Bayham tests these predictions, as well as others that involve the abundances of stone tool types and vertebrate body part representation, and shows that patterns in a large regional database are largely consistent with them (see also Szuter and Bayham 1989).

Although this particular paper is from an obscure venue, it represents a succinct summary of Bayham's seminal work in the late 1970s and early 1980s (e.g., Bayham 1979, 1982). This body of work also represents the first detailed archaeological application of Orians's and Pearson's (1979) influential central place foraging models, which recognize that costs of travel and transport between residential bases and points of acquisition have important implications for prey choice. Central place foraging models are further adapted and developed by many other papers in this and the succeeding section of this volume. Bayham's work also establishes that diachronic trends in hunting behavior at a specific locale can be reflective of broader changes in the socioeconomic context of foraging, and that these can be understood using a hierarchy of appropriate foraging models.

153

The next paper, by K. R. Barlow and D. Metcalfe, continues with the theme of central place foraging. In an extremely influential earlier paper, Metcalfe and Barlow (1992) developed a model to address the tradeoff that central place foragers face between transport costs and field processing costs. Assuming a goal of maximizing the net rate of calories delivered to a residential base, their model predicts the amount of time that foragers should devote to field processing a resource given round-trip travel time and the "utility function" of the resource, or the manner in which the utility of a load of that resource is increased with processing time. In the paper reprinted here, Barlow and Metcalfe apply this model to the archaeological record of plant use in the eastern Great Basin. They first experimentally derive utility functions for two regionally important plant foods, pinyon pine nuts and pickleweed seeds. They then use the model to derive expectations about the types and quantities of pinyon and pickleweed remains in regional archaeological sites. They predict, among other things, that pinyon should be underrepresented relative to pickleweed in residential sites—despite a greater importance of pinyon in the diet—because of the benefits that accrue from processing pinyon nuts in the field. Archaeobotanical records from the region are consistent with this prediction.

Most broadly, this analysis demonstrates that the energetics of processing and transport must be understood before interpretations can be derived from archaeological remains about the relative importance of different plant foods—or any resource, for that matter—in the diet (also see Bird et al. 2002). Applications of the field processing model, or derivations thereof, have now been carried out for resources as diverse as big game, toolstone, shellfish, and acorns (e.g., Beck et al. 2002, this volume; Bettinger et al. 1997; Bird et al. 2002; Cannon 2003, this volume; Elston and Raven 1992; Lupo 2001; Lupo and Schmitt 1997; O'Connell and Marshall 1989; Zeanah 2000; see also Bird and O'Connell 2006).

Context for the next paper, by J. M. Broughton, is provided by traditional anthropological and archaeological descriptions of aboriginal lifeways in California, which emphasized the unparalleled richness and abundance of resources available to human consumers here. These traditional descriptions contrast with more recently obtained late Holocene archaeological data, which have been read to indicate both increasing human population densities and an increasing use of low-return plant resources, especially acorns (e.g., Bettinger and Baumhoff 1981; Basgall 1987). The declining foraging efficiency implied by these trends has far-reaching implications for our understanding of California prehistory, touching on issues ranging from increases in violence, warfare, and territoriality, to reduced stature and declining health, to the emergence of hereditary inequalities (see Broughton and O'Connell 1999; Raab and Jones 2004).

In the late 1980s, Broughton initiated a series of analyses to systematically test the hypothesis that foraging efficiency declined in California during the late Holocene by focusing on the faunal evidence for resource depression and intensification (e.g., Broughton 1988, 1994a, 1994b, 1995, 1997, 1999, 2004; Broughton et al. 2007). The paper reprinted here summarizes one such test conducted with the rich archaeological vertebrate record from San Francisco Bay's Emeryville Shellmound, an enormous, deeply stratified residential site that was occupied by native Californians between about 2,600 and 700 BP.

Following Bayham, Broughton uses prey body mass as a proxy measure of return rate for taxa that arguably would not have exhibited significant differences in prey handling costs, and he reasons that resource depression should be signaled by declining relative abundances of large-bodied vertebrate taxa in archaeofaunal assemblages. Adhering to the prey model's important, but often overlooked, fine-grained search assumption (see the introductory chapter to this volume), separate analyses were conducted for faunal resources that were likely derived from different local patches or hunt types (such as terrestrial mammals, waterfowl, and estuarine fishes). Temporal declines in the relative abundances of large-bodied taxa were detected in almost every case, but one exception involved a major increase in the relative abundance of deer late in the occupational sequence. Drawing on patch-use models, this was hypothesized to reflect an increase in the use of less-depleted, more distant patches. This hypothesis was tested by using Metcalfe and Barlow's (1992) field processing model to derive the prediction that the mean utility of skeletal elements returned to Emeryville should increase if greater use were being made of distant patches. The taxonomic relative abundance and body part data that Broughton presents are well in accord with this prediction (but see Rogers and Broughton 2001), and a variety of prey-specific demographic signals of harvest pressure corroborate this evidence for resource depression.

These results suggest late Holocene foragers had significant impacts on populations of fishes, birds, and mammals in the area around Emeryville. The declining foraging efficiency that is implied by such patterns, documented both here and elsewhere, not only has important implications for a variety of issues in the late Holocene prehistory of California, as noted above, but it also undercuts a long-held utopian perception of aboriginal lifeways that has been the cornerstone of traditional California ethnography and archaeology. This evidence for prehistoric resource depression is directly relevant to the modern management of wildlife and wilderness in the region because it reveals past conditions and processes in a way that can inform on the creation of ecological benchmarks (for further discussion, see Broughton 2002, 2004; Broughton et al. 2007; Hildebrandt and Jones 2004; Porcasi et al. 2000; Raab and Jones 2004).

Building upon work exemplified by the previous papers in this section, the next paper, by M. D. Cannon, develops a unique model that integrates Metcalfe and Barlow's (1992) field processing model with a model of prey choice for central place foragers. This model addresses the decision of what kinds of prey and what parts of those prey central place foragers should bring home from foraging trips. Patch choice, prey choice, field processing and transport are thus addressed with a single model, rather than with nested series of separate models as has been the case in previous works (e.g., Bayham 1986; Broughton 2002, above). Compared to previous approaches, Cannon argues that this approach makes assumptions that better match the ways in which humans often forage because central place foraging violates the fine-grained search assumption of the standard prey model (see also Cannon 2000). In addition, it enables transport costs—which are very relevant to decisions about resource choice, as other papers in this volume also discuss—to be factored into resource return rates. This is something the standard prey model does not do, but which is very important for situations involving central place foragers. Finally, the model allows simultaneous evaluation of the effects of prey choice, field processing, and transport on foraging efficiency, which cannot be done when these issues are addressed using separate models.

Cannon uses the central place forager prey choice model to address the issue of resource depression with archaeofaunal data from the Mimbres Valley of southwestern New Mexico. There, declines in the relative abundance of deer and pronghorn indicate depression of these resources and significant declines in foraging efficiency between about AD 400 and 800. This is roughly synchronous with evidence for increasing investment in farming in the Mimbres Valley, and we note that models described in detail in the next section predict that such a shift to more intensive agriculture should occur as returns from wild resource foraging decline. In addition, while changes in taxonomic relative abundance in Mimbres faunal assemblages are consistent with predictions of the central place forager prey choice model concerning resource depression, changes in body part representation are not. Cannon suggests that the predictive failure in this case may be a result of changes over time in the social context of hunting in the Mimbres valley, manifested in an increase in the size of hunting parties that altered the constraints on carcass transport.

The aim of the works in this section that have been discussed so far has been to understand and explain patterns in resource selection as documented by subsistence remains recovered from residences or sites of resource acquisition. In the next paper, D. W. Zeanah develops a model to predict where residences should be established in the first place. His work attends specifically to the different reproductive goals and strategies that might apply to males and females and to how optimal residential site locations will vary depending on whether women's or men's resources are most productive. In Zeanah's formulation, residential sites are predicted to be located in places that enable foragers to maximize the caloric delivery rate of men's and women's resources combined. Applying this model to the late Holocene archaeological record of the Carson Desert of western Nevada, and adjusting for climate-induced variation in the return rates of different resources and resource patches, residential sites are predicted in most contexts to be located in the best habitats for women's resources. In conjunction with this, men are predicted to have adjusted to those settlement locations and to have exhibited greater logistic mobility in their own foraging. Patterns in prehistoric site distributions and other data are generally consistent with these predictions.

This paper engages the issue of men's hunting goals, raised also in the papers by O'Connell et al. and Kaplan et al. that are reprinted in the first part of this volume, and it points out the need for further modeling work on this issue as discussed earlier (see also Smith 2004 and the debate involving Hildebrandt and McGuire 2002; Broughton and Bayham 2003; Byers and Broughton 2004). In addition, whereas previous papers in this section have focused on factors intrinsic to human populations as catalysts for diachronic changes in resource densities and foraging efficiencies—for instance, changes in site function or growing human populations—the paper by Zeanah considers the influence of extrinsic factors—especially climate change through its effect on habitat productivities—on settlement locations (see also Byers and Broughton 2004; Wolverton 2005). Noteworthy as well in this context is the observation that the same climate shift—increased moisture during the Neoglacial (ca. 4,500 to 1,250 BP)—can have very different effects on the returns from hunted versus gathered resources: estimated men's hunting returns skyrocket at this time, while those for women increase only modestly. This reflects the greater sensitivity of resources with high search and low handling costs (such as large game) to increased encounter rates, and it underlines the point that temporal trends in men's and women's returns need not be congruent.

In the final chapter in this section, L. Nagaoka takes us to southern New Zealand during the late Holocene (AD 1250 to 1450) to explore trends in the butchery and processing of moas—large, flightless birds that ranged in size from 20 to 250 kg. Prior to human colonization, New Zealand was inhabited by about ten species of these birds, but all were unfortunately lost shortly after people arrived—in a pattern typical of island settings—due not only to increased predation, but also to biotic introductions (particularly of rats and dogs) and to deforestation (e.g. Steadman 1995; Grayson 2001; Worthy and Holdaway 2003).

Working with faunal remains from the Shag River Mouth site, Nagaoka has shown in previous studies that human foragers responded to the decline and eventual loss of moas by pursuing higher proportional frequencies of smaller prey, by increasing the number of prey they incorporated into their diets, and by using previously ignored, more distant patches—all indicators of declining foraging efficiency (Nagaoka 2001, 2002a, 2002b). With this foundation, Nagaoka's paper uses Cannon's central place forager prey choice model, described above, to address moa carcass field butchery and transport, as well as the Marginal

Value Theorem (MVT) to address the extraction of marrow and grease from moa bones returned to camp.

With respect to the transport issue, Nagaoka follows both Broughton and Cannon (see above), and predicts that the greater use of distant patches should be signaled by an increase in the mean utility of moa skeletal parts returned to the residential base. The MVT, originally designed to model patch residence time (Charnov 1976), is adapted here to address the decision of how much time consumers should devote to the processing of particular body parts once returned to camp (see also Burger et al. 2005; Sih 1980). The scale of analysis is thus adjusted so that each skeletal element is treated as a discrete patch. The prediction of the MVT given declining foraging efficiency is that more time should be spent extracting marrow and grease from moa elements and that this should be reflected archaeologically by evidence for an increase in bone fragmentation. After a careful evaluation of potentially confounding taphonomic issues, Nagaoka shows that patterns in both skeletal part abundances and bone breakage are consistent with her deductively derived predictions. This work nicely illustrates the integrative nature of EE, as Nagaoka shows how prey choice, patch use, field butchery, transport, and in-camp processing are inextricably linked, each responding predictably to declining returns brought on by foragers' own activities.

The papers in this section have improved our knowledge and understanding of Holocene subsistence issues in a variety of ways. The more substantive conclusions that can be derived from them are as follows:

1. Foraging theory models gain purchase on the causes and consequences of broad spectrum dietary shifts (variously reflected in the labels Archaic, Mesolithic, or resource intensification), which are of long-standing interest to archaeologists.
2. Foraging theory models can be used to develop archaeologically operational measures of foraging efficiency involving taxonomic abundances, skeletal part representation, and bone processing intensity.
3. Holocene declines in foraging efficiency and resource depression occur in numerous contexts characterized by expanding human population densities and either high-abundances of gathered resources or economies fueled by agriculture.
4. The energetics of processing and transport must be understood before secure interpretations can be made about the relative importance of subsistence resources derived from archaeological remains.
5. Declines in foraging efficiency may be causally related to, and theoretically integrated with, many other aspects of human behavior (e.g., warfare, territoriality, violence, hereditary inequalities).
6. Differences in male and female foraging goals and the sexual division of labor are essential considerations in determining residential site location.
7. Residential sites are predicted in many contexts to be located in the best habitats for resources that are typically harvested by women.
8. The same environmental change can disproportionately affect returns for resources with different search and handling characteristics and produce different trajectories in gathered versus hunted resources.

References

Basgall, M. E.
1987 Resource Intensification among Hunter-Gatherers: Acorn Economies in Prehistoric California. *Research in Economic Anthropology* 9:21–52.

Bayham, F. E.
1979 Factors Influencing the Archaic Pattern of Animal Utilization. *Kiva* 44:219–235.
1982 A Diachronic Analysis of Prehistoric Animal Exploitation at Ventana Cave. Ph.D. dissertation, Department of Anthropology, Arizona State University, Tempe.

Bettinger, R. L., M. A. Baumhoff
1981 The Numic Spread: Great Basin Cultures in Contact. *American Antiquity* 47:485–503.

Bettinger, R. L., R. Mahli, H. McCarthy
1997 Central Place Models of Acorn and Mussel Processing. *Journal of Archaeological Science* 24:887–899.

Bird, D. W., J. F. O'Connell
2006 Behavioral Ecology and Archaeology. *Journal of Archaeological Research* 14:143–188.

Bird, D. W., J. L. Richardson, P. M. Veth, A. J. Barham
2002 Explaining Shellfish Variability in Middens on the Meriam Islands, Torres Strait, Australia. *Journal of Archaeological Science* 29:457–469.

Broughton, J. M.
1988 Archaeological Patterns of Prehistoric Fish Exploitation in the Sacramento Valley. Master's thesis, Department of Anthropology, California State University, Chico.
1994a Declines in Mammalian Foraging Efficiency During the Late Holocene, San Francisco Bay, California. *Journal of Anthropological Archaeology* 13:371–401.
1994b Late Holocene Resource Intensification in the Sacramento Valley, California: The Vertebrate Evidence. *Journal of Archaeological Science* 21:501–514.
1995 Resource depression and intensification during the late Holocene, San Francisco Bay: evidence from the Emeryville Shellmound vertebrate fauna. Doctoral dissertation, Department of Anthropology, University of Washington, Seattle.
1997 Widening diet breadth, declining foraging efficiency, and prehistoric harvest pressure: ichthyofaunal evidence from the Emeryville Shellmound, California. *Antiquity* 71:845–862.
1999 *Resource Depression and Intensification during the Late Holocene, San Francisco Bay: Evidence from the Emeryville Shellmound Vertebrate Fauna*. University of California Anthropological Records 32. University of California Press, Berkeley.
2002 Pre-Columbian Human Impact on California Vertebrates: Evidence from Old Bones and Implications for Wilderness Policy. In *Wilderness and Political Ecology: Aboriginal Influences and*

the Original State of Nature, edited by C. E. Kay and R. T. Simmons, pp. 44–71. University of Utah Press, Salt Lake City.

2004 *Prehistoric Human Impacts on California Birds: Evidence from the Emeryville Shellmound Avifauna*. Ornithological Monographs No. 56. The American Ornithologists' Union, Washington, DC.

Broughton, J. M., F. E. Bayham
2003 Showing Off, Foraging Models, and the Ascendance of Large-Game Hunting in the California Middle Archaic. *American Antiquity* 68:783–789.

Broughton, J. M., J. F. O'Connell
1999 On Evolutionary Ecology, Selectionist Archaeology, and Behavioral Archaeology. *American Antiquity* 64:153–165.

Broughton, J. M., D. Mullins, T. Ekker
2007 Avian Resource Depression or Intertaxonomic Variation in Bone Density?: A Test with San Francisco Bay Avifaunas. *Journal of Archaeological Science* 34:374–391.

Burger, O., M. J. Hamilton, R. Walker
2005 The Prey as Patch Model: Optimal Handling of Resources with Diminishing Returns. *Journal of Archaeological Science* 32:1147–1158.

Byers, D. A., J. M. Broughton
2004 Holocene Environmental Change, Artiodactyl Abundances, and Human Hunting Strategies in the Great Basin. *American Antiquity* 69:235–256.

Cannon, M. D.
2000 Large Mammal Relative Abundance in Pithouse and Pueblo Period Archaeofaunas from Southwestern New Mexico: Resource Depression Among the Mimbres-Mogollon? *Journal of Anthropological Archaeology* 19:317–347.

Charnov, E. L.
1976 Optimal Foraging: The Marginal Value Theorem. *Theoretical Population Biology* 9:129–136.

Elston, R. G., C. Raven
1992 *Archaeological Investigations at Tosawihi, A Great Basin Quarry*. Intermountain Research, Silver City, Nv.

Grayson, D. K.
2001 The Archaeological Record of Human Impacts on Animal Populations. *Journal of World Prehistory* 15:1–68.

Hildebrandt, W. R., T. L. Jones
2004 Depletion of Prehistoric Pinniped Populations along the California and Oregon Coasts: Were Humans the Cause? In *Wilderness and Political Ecology: Aboriginal Influences and the Original State of Nature*, edited by C. E. Kay and R. T. Simmons, pp. 72–110. University of Utah Press, Salt Lake City.

Hildebrandt, W. R., K. R. McGuire
2002 The Ascendance of Hunting during the California Middle Archaic: An Evolutionary Perspective. *American Antiquity* 67:231–256.

Lupo, K. D.
2001 On the Archaeological Resolution of Body Part Transport Patterns: An Ethnoarchaeological Example from East African Hunter-Gatherers. *Journal of Anthropological Archaeology* 20:361–378.

Lupo, K. D., D. Schmitt
1997 Experiments in Bone Boiling: Nutritional Yields and Archaeological Reflections. *Anthropozoologica* 25-26:137–144.

Metcalfe, D., K. R. Barlow
1992 A Model for Exploring the Optimal Tradeoff between Field Processing and Transport. *American Anthropologist* 94:340–356.

Nagaoka, L.
2001 Using Diversity Indices to Measure Changes in Prey Choice at the Shag River Mouth Site, Southern New Zealand. *International Journal of Osteoarchaeology* 11:101–111.

2002a Explaining Subsistence Change in Southern New Zealand using Foraging Theory Models. *World Archaeology* 34:84–102.

2002b The Effects of Resource Depression on Foraging Efficiency, Diet Breadth, and Patch Use in Southern New Zealand. *Journal of Anthropological Archaeology* 21:419–442.

O'Connell, J. F., B. Marshall
1989 Analysis of Kangaroo Body Part Transport among the Alyawara of Central Australia. *Journal of Archaeological Science* 16:393–405.

Orians, G. H., N. E. Pearson
1979 On the Theory of Central Place Foraging. In *Analysis of Ecological Systems*, edited by D. J. Horn, R. D. Mitchell, and G. R. Stairs, pp. 154–177. The Ohio State University Press, Columbus.

Porcasi, J. F., T. L. Jones, L. M. Raab
2000 Trans-Holocene Marine Mammal Exploitation on San Clemente Island, California: A Tragedy of the Commons Revisited. *Journal of Anthropological Archaeology* 19:200–220.

Raab, L. M., T. L. Jones
2004 *Prehistoric California: Archaeology and the Myth of Paradise*. University of Utah Press, Salt Lake City.

Rogers, A. R., J. M. Broughton
2001 Selective Transport of Animal Parts by Ancient Hunters: A New Statistical Method and an Application to the Emeryville Shellmound Fauna. *Journal of Archaeological Science* 28:763–773.

Sih, A.
1980 Optimal Foraging: Partial Consumption of Prey. *The American Naturalist* 116:281–290.

Smith, E. A.
2004 Why do Good Hunters have Higher Reproductive Success? *Human Nature* 15:343–364.

Steadman, D. W.
1995 Prehistoric Extinctions of Pacific Island Birds: Biodiversity Meets Zooarchaeology. *Science* 267:1123–1131.

Szuter, C. R., F. E. Bayham
1989 Sedentism and Prehistoric Animal Procurement among Desert Horticulturalists of the North American Southwest. In *Farmers as Hunters: the Implications of Sedentism*, edited by S. Kent, pp. 80–95. Cambridge University Press, Cambridge.

Wolverton, S.
2005 The Effects of the Hypsithermal on Prehistoric Foraging in Missouri. *American Antiquity* 70:91–106.

Worthy, T. H., R. N. Holdaway
2003 *The Lost World of the Moa*. Indiana University Press, Bloomington, IN.

Zeanah, D.
2000 Transport Costs, Central Place Foraging, and Hunter-Gatherer Alpine Land-Use Strategies. In *Intermountain Archaeology*, edited by D. B. Madsen and M. D. Metcalf, pp. 1–14. University of Utah Anthropological Papers No. 122. University of Utah, Salt Lake City.

Effects of a Sedentary Lifestyle on the Utilization of Animals in the Prehistoric Southwest

Frank E. Bayham

Introduction

The prehistoric occupation of the desert Southwest spans the Archaic and Hohokam cultural periods, an interval of time extending minimally from 3000 BC to AD 1300. During this period, it is generally acknowledged that a mobile hunter-gatherer subsistence strategy was supplanted by an agricultural one (Cordell 1984:182–211; Haury 1976). Considerable attention has been devoted to the botanical aspect of this economic and cultural tradition (e.g. Ford 1981, Gasser 1980), yet relatively little is known about the impact of the adoption of cultigens on the utilization of animals. Hunter-gatherers of the prehistoric Southwest, like most non-agricultural peoples of the world, relied on animals as an important source of physical, and perhaps spiritual sustenance. The transition to a more sedentary, agricultural lifestyle affected the nature of animal use, but did not diminish their significance. To slight the zoologic aspect of this important transition is to not fully appreciate the social and economic impact of agriculture.

The analysis of archaeofaunal data is the primary avenue by which patterns of animal utilization are reconstructed. Faunal studies undertaken in the desert Southwest from the 1940s through the 1960s documented which taxa were being exploited, and established that Archaic, as well as many Hohokam, faunal assemblages were dominated by various species of rabbit with other desert-dwelling species showing up consistently but in lesser proportions (e.g. Haury 1950, Sayles and Antevs 1941). During the Archaic Period animal exploitation is commonly thought to have been relegated to a subsidiary role in comparison to the utilization of vegetal resources. Agricultural developments were presumed to diminish the importance of animal resources even further; hunting is perceived as declining in importance from the Archaic through the Hohokam Periods (Haury 1950:543–548). The characterization of Archaic hunters and gatherers and later Hohokam agriculturalists desperately searching for rodents, lizards, and rabbits to supplement a protein deficient-diet developed out of this early data base. As summarized by Emil Haury (1976:114):

> ...the impoverished nature of animal utilization [among the Hohokam] is not readily explained. If a scant meat supply was available and if the yield of vegetal protein was small, for which there is some indication, then the Hohokam diet may have suffered seriously from protein deficiency.

Archaeofaunal research in the past decade has shown that the impact of cultigens in the Southwest did not result in a simple linear or progressive decline in the importance of animal use. Perhaps as a consequence of the accelerated pace of contract archaeology in the Southwest during the middle and late 1970s, the number and geographic distribution of faunal reports increased dramatically. These studies were from small as well as large sites, the sites themselves were situated across the physiographic range of variation, and they provided a perspective over time that had previously been inaccessible. It eventually became apparent that a wide range of intersite variation existed in the character of animal use. As documented in Ventana Cave, for instance, not all later period occupations exhibited a greater proportion of low-yield smaller species such as rodents (Bayham 1982). Several socioeconomic factors which may have influenced both temporal and spatial variation in the archaeofaunal record may be related to agricultural intensification and a decrease in residential mobility.

In this paper I discuss how factors related to increased sedentism may have influenced the logistics of hunting behavior from the Archaic through the Hohokam Periods in the Southwest. I assume that some sort of selection among available prey species was operative prehistorically, and that it may significantly contribute to the composition of and variability

*Originally published in *Agriculture: Origins and Impacts of a Technological Revolution*, Occasional Papers of the Archaeological Research Facility, California State University, Fullerton No 5(1986):54–78.

among different faunal assemblages. Two socioeconomic factors which appear to have an important influence on the process of selective utilization include costs associated with travel and transport, and resource depression. Employing only these two factors, which are related to the economic structure of a cultural system, I illustrate how they may condition the process of selection among prey species at different loci. I will first review some pertinent findings generated from analysis of the faunal material from Ventana Cave.

Observations on Animal Use at Ventana Cave

Ventana Cave is located in the Castle Mountains of south-central Arizona and overlooks a broad alluvial valley (Figure 10.1). Emil Haury (1950) excavated the site in 1941 and 1942, and it has since anchored nearly all discussions of the prehistoric occupation of south-central Arizona. The occupational sequence at Ventana Cave spanned a major portion of the Archaic and Hohokam Periods. The upper portion of the midden deposits in Ventana Cave (level 1 through the middle of level 3) are associated with the Hohokam Period which dates roughly from AD 500 to about AD 1350. The lower levels of the midden (from the middle of level 3 to level 8) were designated Chiricahua-Amargosa II and San Pedro by Haury (1950:522–525) who estimated this occupation to span the period from 2500 BC to the beginning of the Hohokam Period (Figure 10.2). In 1982 I completed the analysis of a large quantity of animal bone from Ventana Cave which had lain in storage for about 35 years (Bayham 1982). The results of this faunal study indicated that a wide variety of taxa were exploited, and that some variation existed in the proportionate representation of those taxa over time.

Vertebrate remains from Ventana Cave are dominated by mammals, many of which show evidence of having been used as food (Table 10.1). The most common species represented in the assemblage are jackrabbits (*Lepus* spp.) and deer (*Odocoileus* spp.), but a wide variety of species were recovered. Other taxa of economic significance included the pronghorn antelope (*Antilocapra americana*), bighorn sheep (*Ovis canadensis*), coyotes (*Canis latrans*), and some ground squirrels (e.g. *Spermophilus* spp., *Ammospermophilus* sp.). An interesting temporal trend is evident when we compare the relative frequencies of the dominant species through time. The middle Archaic levels (8–6) document a much stronger utilization of rabbits, particularly the black-tailed jackrabbit (*L. californicus*) and to a lesser degree the desert cottontail (*Sylvilagus audubonii*), in comparison to the artiodactyls. The later Archaic horizons (5–3) provide evidence of a much greater exploitation of deer and other endemic artiodactyls. This trend continues through the Hohokam Periods.

This empirically based observation at Ventana Cave is not wholly consistent with the prevailing conception of subsistence change during the prehistoric occupation of southern Arizona. The Archaic is frequently characterized as the "long span of time during which hunter-gatherers made increasing use of the edible resources of the environment, particularly plants" (Schiffer and McGuire 1982:235). The role of hunting is generally thought to have been on the decline slowly giving way to a sedentary agricultural lifestyle (Haury 1950:544). As suggested by the pattern at Ventana Cave, the regional scenario may be considerably more complex. The cultural, ecological, and perhaps evolutionary, significance of this trend can more readily be ascertained by considering the energetic contribution of different species over time.

It is possible to comparatively assess the degree of efficiency represented in an archaeofaunal assemblage using potential food value and the relative abundance of different prey items in the diet. In short, because larger prey provide more energy per individual, they can be ranked higher on a preference scale than smaller animals. An index to measure selective efficiency (SEI) has been developed that weighs the relative contribution of different-sized animals in a given assemblage (Bayham 1982:241):

$$\text{SEI} = \sum M_i E_i / (1 + \sum M_i) 1000,$$

where M_i is an estimate of the minimum number of individuals (MNI) for a given species i, and E_i is the caloric value per individual prey item i (in situations were MNIs cannot be determined with confidence, a standardized estimate of relative abundance such as the NISP [number of individual specimens] can be substituted). This value may be construed as an aggregate output/input efficiency estimate over some specified time interval.

Prey items at Ventana Cave were ranked in terms of preference by their energetic value, and a proportionate estimate of their relative abundance (both MNIs and NISPs) was used to determine the SEI by stratum. A significant pattern emerged: there is a strong linear trend towards the utilization of larger game items throughout the middle/late Archaic and Hohokam sequence. This translates to an increase in the selective efficiency of animal utilization in the vicinity of the cave accounting for 91% of the variation over time in the faunal assemblage (Figure 10.3; $r^2 = .91$; $p < .001$; see Bayham 1982 for a more thorough discussion).

This dramatic shift over time is intriguing for several reasons. First, it does not support the conventional view that gathering activities increased in importance from the early Archaic through the later Archaic Periods culminating in agriculture, and that hunting decreased to a level of lesser importance. Secondly, this gross change in proportionate use was rather unexpected since no animals in the Sonoran Desert are excessively abundant. Lastly, the strength of the linear trend over time in probabilistic terms is far less than would be expected by chance implicating causal processes that were not precipitous, but sequentially gradual or punctuated over intervals not detectable in the archaeological record.

Patterning in animal utilization similar to that at Ventana Cave has been explained with reference to changes in the environment. If, for instance, environmental changes increased the abundance of artiodactyls over time, the pattern could be ac-

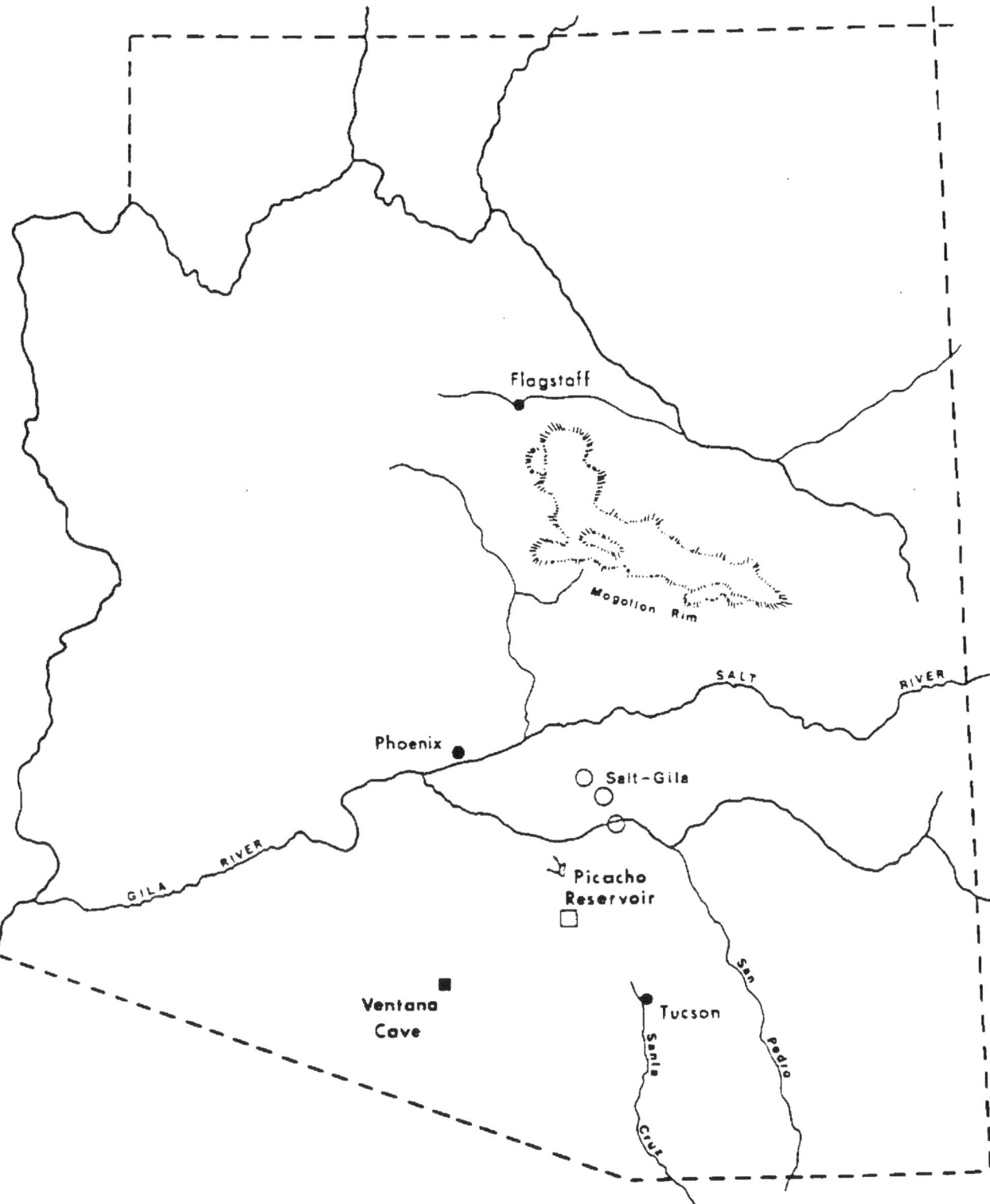

FIGURE 10.1. Map of Study Area.

counted for. This, however, does not appear to be the case. Packrat midden studies document a trend in effective moisture that is declining over time: from a woodland in the early Holocene, to a wet summer desertscrub, to the dryer arid regime of today (Van Devender 1977; Van Devender and Spaulding 1979). While fluctuations no doubt occurred since 4500 BP, no evidence suggests an environmental transformation which would have increased the relative or absolute availability of mule or whitetail deer, bighorn sheep, or pronghorn antelope through time. Natural changes in the local environment around Ventana Cave cannot by themselves account for the diachronic trend in animal procurement. Also, no technological innovations in the

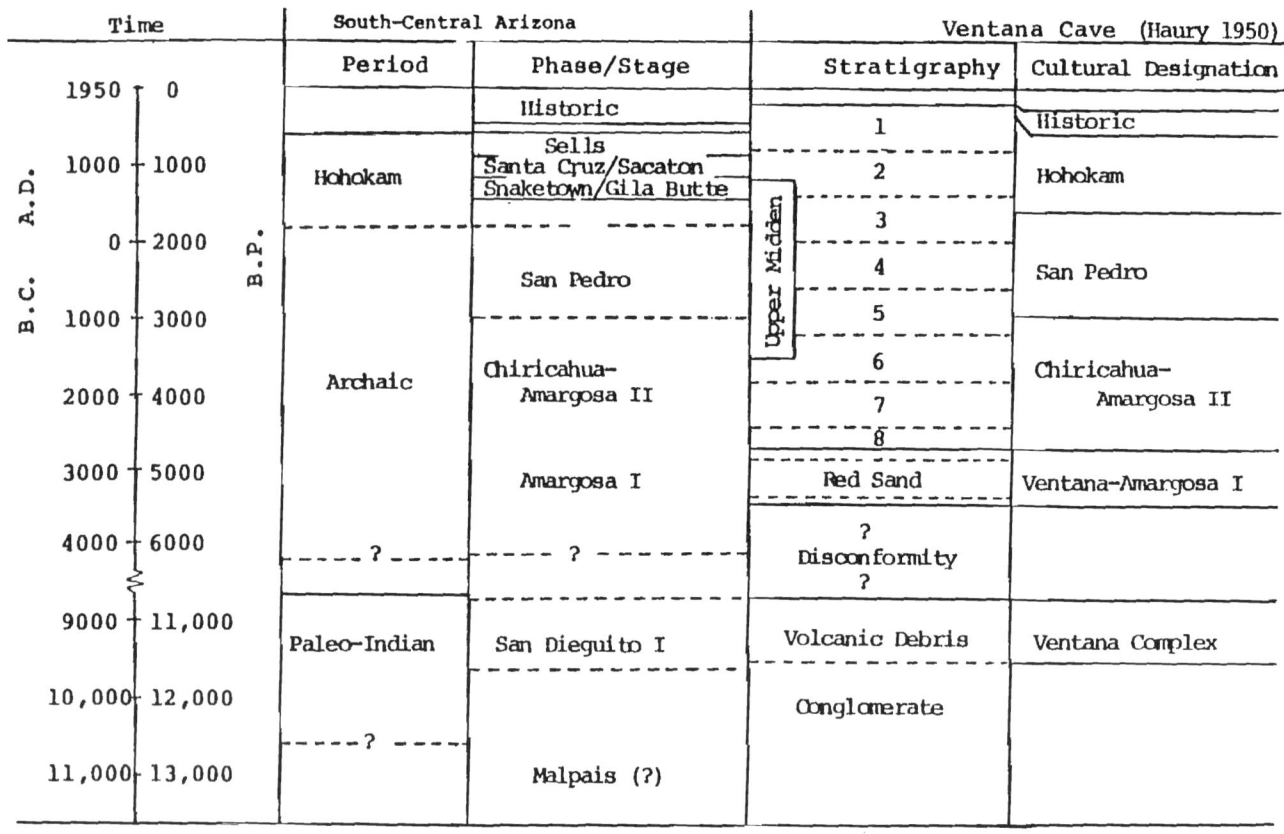

FIGURE 10.2. Correlation of prehistoric sequence of south-central Arizona with Ventana Cave.

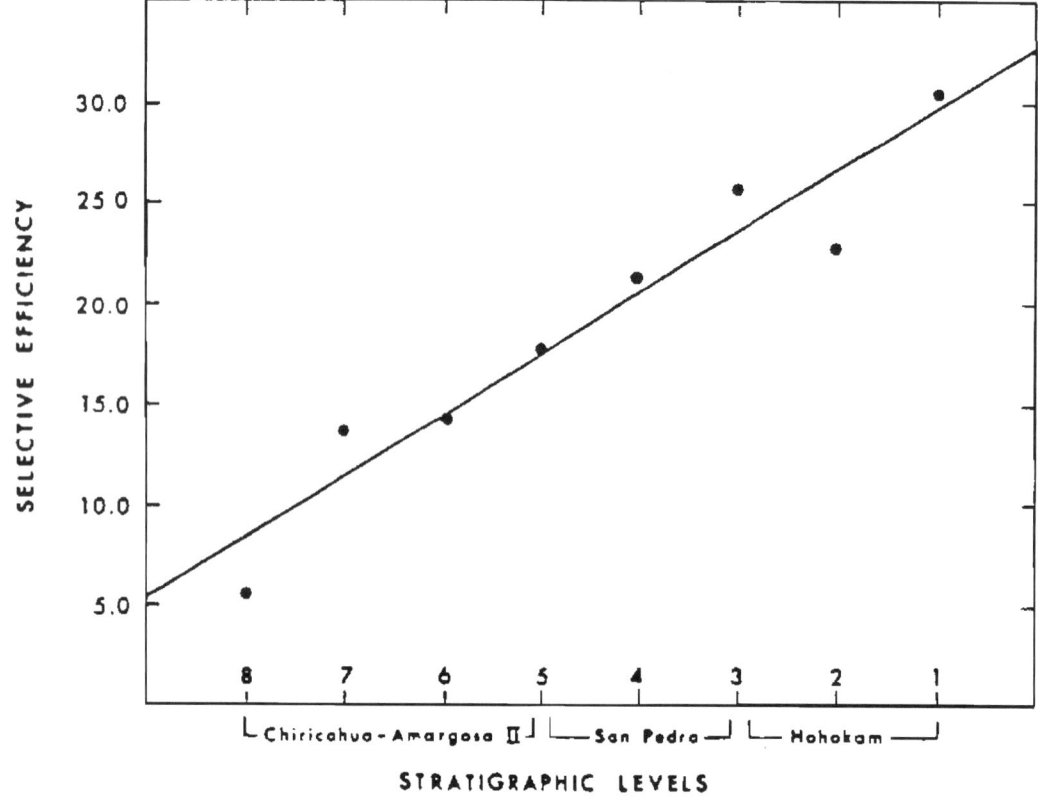

FIGURE 10.3. Change in the selective efficiency index (SEI) over time at Ventana Cave illustrating increasing specialization on larger game from Middle to Late Archaic (y = 3.02x + 5.46; r^2 = .909; Bayham 1982:243).

TABLE 10.1. Distribution of identified mammals and reptiles by stratigraphic level, Ventana Cave.*

	STRATIGRAPHIC LEVELS																
	1		2		3		4		5		6		7		8		TOTALS	
TAXA	N	M	N	M	N	M	N	M	N	M	N	M	N	M	N	M	N	M
Mammalia																		
Leporidae																		
Lepus spp.	6																6	
L. alleni	46	5	55	4	28	6	11	2	19	2	1	1	2	1	2	1	164	22
L. californicus	107	11	184	17	81	6	224	14	114	9	167	14	108	8	42	5	1027	84
Sylvilagus sp.	13		9		1		15		10				2				50	
S. audubonii	7	3	11	3	15	4	14	3	2	4	9	2	6	2	4	1	68	22
Sciuridae															1		1	
Cynomys ludovicianus									1	1			1	1	1	1	3	3
Spermophilus variegatus	4	1	4	3	3	2	7	1	1	1					1	1	20	9
Cricetidae																		
Neotoma spp.			1				1	1									2	1
Neotoma albigula			2	1													2	1
Erethizontidae																		
Erethizon dorsatum	10	3	1	1													11	4
Carnivora	3		11		6		6				7						33	
Canidae	1						1				5						7	
Canis spp.	18		10		1		7		2								38	
Canis latrans	91	5	88	5	40	3	76	6	23	3	36	3	15	2	7	1	376	28
Canis familiaris	2	1															2	1
Vulpes macrotis	2	1	3	1	1	1											6	3
Urocyon cinereoargenteus	6	1	5	2	2	1	6	1	2	1	2	1			1	1	24	8
Mustelidae/Procyonidae																		
Bassariscus astutus			1	1													1	1
Taxidea taxus	22	3	33	4	22	4	61	6	47	7	41	5	29	3	10	3	265	35
Spilogale putorius			1	1													1	1
Mephitis cf. *M. mephitis*			1	1			1	1									2	2
Felidae																		
Felis concolor	1	1	2	1													3	2
Lynx rufus	9	1	10	1	3	1	15	2	10	1	2	1	3	1			52	8
Artiodactyla	56		96		17		43		13		22		10		3		260	
Odocoileus sp.	77	3	67		16		40		15		14		6		2		237	3
Odocoileus hemionus	94	5	84	3	22	2	18	2	9	1	1	1	1	1	1	1	230	16
Odocoileus virginianus	17	1	20	2	3	1	14	2	11	1	3	1			1	1	69	9
Antilocapra americana	35	2	45	2	14	2	31	2	13	1	10	1	4	1	1	1	153	12
Ovis canadensis	42	2	59	2	19	2	9	1	1	1	1	1			2	1	133	10
Reptilia																		
Gopherus agassizii	141	5	62	4	20	3	4	1	2	1			2	1	3	1	234	16
small reptiles	1	1	5	1									2	1			8	3
Domesticates																		
Equus sp.	1	1	1	1														
Bos taurus	2	1																
Ovis/Capra	3	1																
Totals	811	55	870	60	314	38	604	45	295	34	321	31	191	22	82	19	3488	304

*N=Number of identified elements. M=Minimum number of individuals.

hunting repertoire appear to have been developed which would have increased their capturability.

What then could have happened to bring about the demonstrable patterned change in resource use at Ventana Cave? How might this be related to regional changes in subsistence strategies? One hypothesis which can effectively account for the pattern suggests that the panregional socioeconomic system underwent significant changes from the Archaic to the Hohokam Period.

Resource Imbalance and Socioeconomic Change: A Hypothesis

Irwin-Williams and Haynes (1970) present an argument for the depopulation of south-central Arizona during the height of the proposed Altithermal (based on the absence of dated contexts in this time period). They further contend that after 2500 BC, population increases appear to characterize the desert regions of the Southwest. Although the methodological aspects of this population reconstruction can be questioned, it appears to adequately crystallize the collective knowledge that has been accumulated. Let us assume, for the sake of hypothesis testing, that this notion has some merit. No doubt, increasing population size would have altered the arrangement of human-environmental interactions. Increased population density among modern hunter-gatherers, in general, intensifies competition and creates the following effects:

(1) decrease in the per capita availability of food resources;
(2) increased use of less preferred (using energetic or nutritional criteria) food resources, resulting in dietary diversification and/or intensification;
(3) decreased mobility of local bands;
(4) greater amount of expressed territoriality; and
(5) perhaps, increased cooperation among groups.

In seasonal environments that characterized south-central Arizona during the occupational sequence (Van Devender 1977, 1980), an increase in population density would exacerbate the problem of survival during the less productive winter months. Avoidance of the late winter/early spring starvation period necessitates the implementation of storage strategies to assure adequate provisioning (Binford 1980). The gradual trend towards specialization in animal exploitation at Ventana Cave may be related to population increases and an increased need for food storage. Yet, the relationship is not direct. It is likely that the directional change at Ventana was affected by an organizational shift in the socioeconomic system. Specifically, the costs of hunting at Ventana Cave may have been changed due to reduced residential mobility.

During the middle Archaic, Ventana Cave may have operated as a base camp like many other localities in the regional procurement system. At that time, decisions concerning which resources to take likely were made from the cave functioning as a base camp. As plant and animal resources in the vicinity of the cave declined, the local band would have moved on. Foraging theorists suggest that groups should move on when the energetic return rate at a specific location falls below the average return rate for that habitat; patches of lower quality would be left sooner (Winterhalder 1981; Charnov 1976).

Population increases, even at an extremely slow rate, could have altered the basis of an effective, stable socioeconomic system. The likely consequence of regional population increase (or in-migration) during the late Archaic/Hohokam Period would have been an increase in the need for storage facilities and a decrease in residential mobility. Storage, by its very nature, is incompatible with a high degree of mobility. It is likely that during the late Archaic, and continuing into the Hohokam Period, the major alluvial basins of south-central Arizona were occupied for longer periods of time in the annual cycle. Rather than moving base camps to upland localities, as may have been done in the middle Archaic, it probably became more efficient to engineer expeditions from base camps to acquire those resources not immediately available (see idealized diagram in Figure 10.4).

This hypothesis of regional socioeconomic reorganization focuses on the consequences of increased sedentism and does not purport to account for reduced mobility. Two important simultaneous effects of increased sedentism, independent of its cause, which may have influenced the change toward large game utilization at Ventana Cave, are resource depression and increased travel costs.

Resource Depression and Central Place Foraging

The very existence of a human population in an area may create non-natural changes in prey density and distribution. One important consequence of decreased residential mobility, or increased sedentism, is resource depression. Resource depression refers to an actual or perceived reduction in the availability of prey species due either to shifts in their natural distribution, or learned avoidance responses. Animal ecologists have observed this patterned response among animals such as ants or various nesting species which forage from a stable location for a period of time. It has been suggested that these prey responses should evolve to relocate prey items where predators find it more difficult to encounter and capture them (Charnov, Orians, and Hyatt 1976:248).

Hamilton and Watt (1970) have explored how populations may modify the spatial relationships of non-domesticated resources. They define three concentric broadly overlapping zones which surround a core area (Figure 10.5). Whether the central habitation area is seasonally occupied or permanent, the core is generally devoid of resources during the period of occupation. They also note that as the intensity of occupation, measured in the amount of disturbed space, increases, the distance from the core area to the beginning of resource producing areas should also increase. A second zone with virtually no resource production "may result from trampling or roadway effects near the core

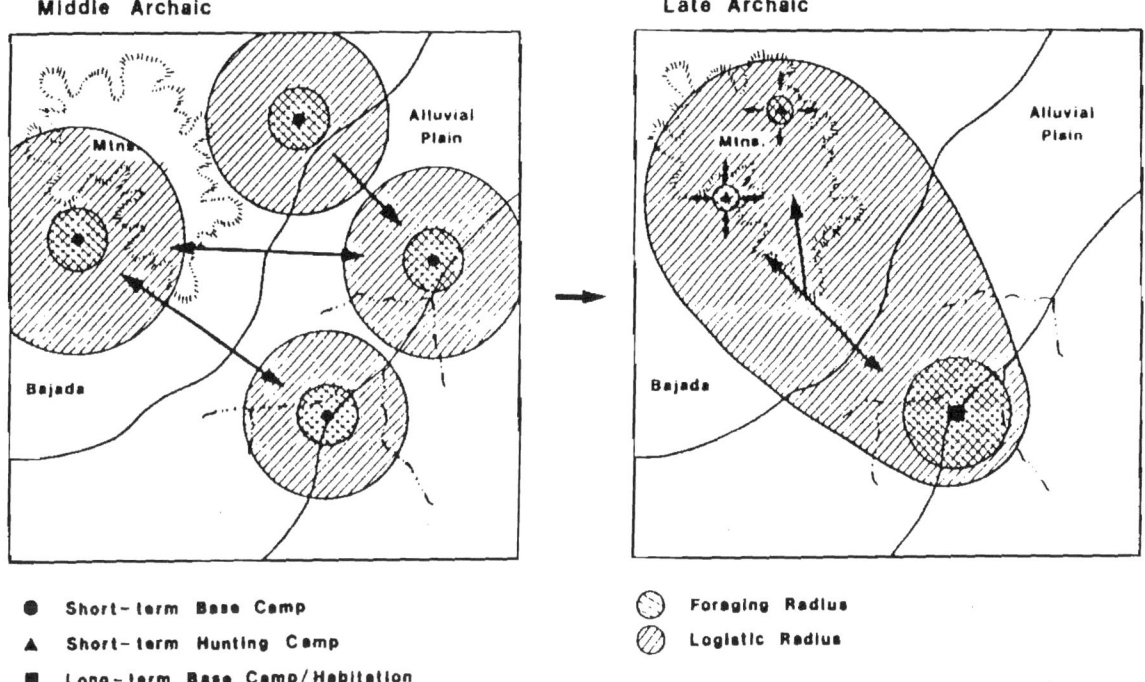

FIGURE 10.4. Schematic representation of hypothetical regional socioeconomic changes from the middle to the late Archaic in south-central Arizona.

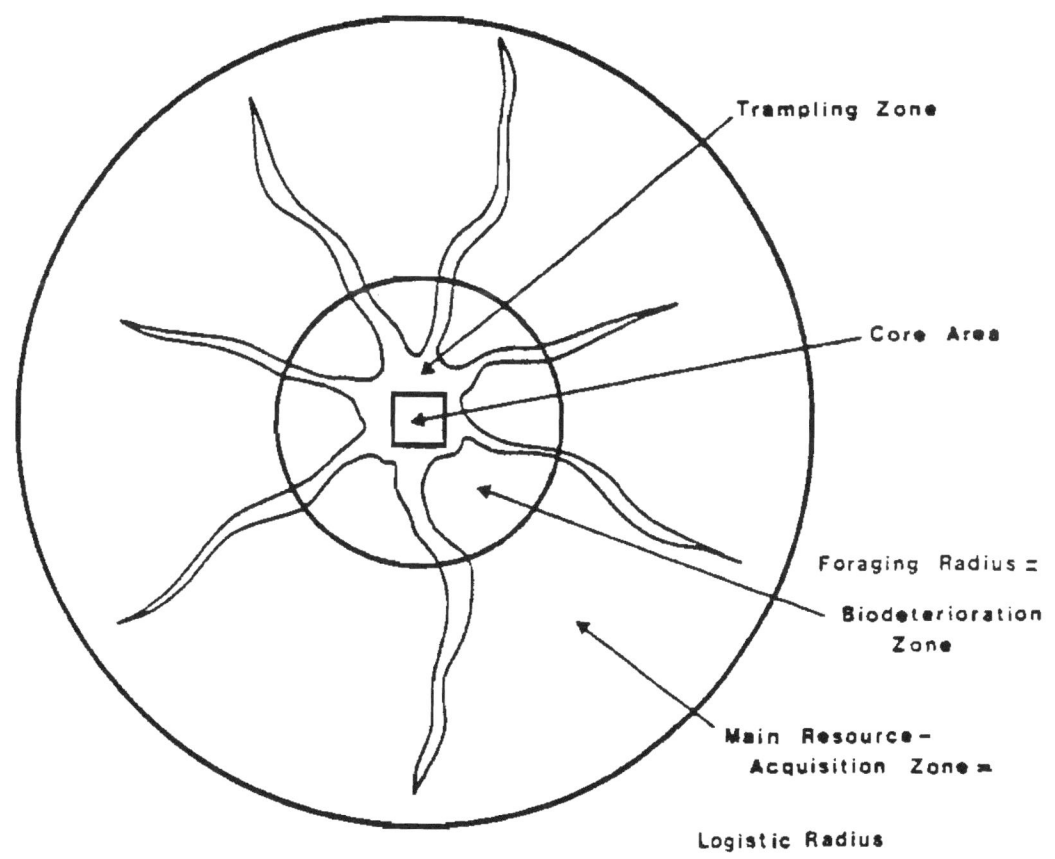

FIGURE 10.5. Patterns of economic zonation resulting from resource depression (Hamilton and Watt 1970).

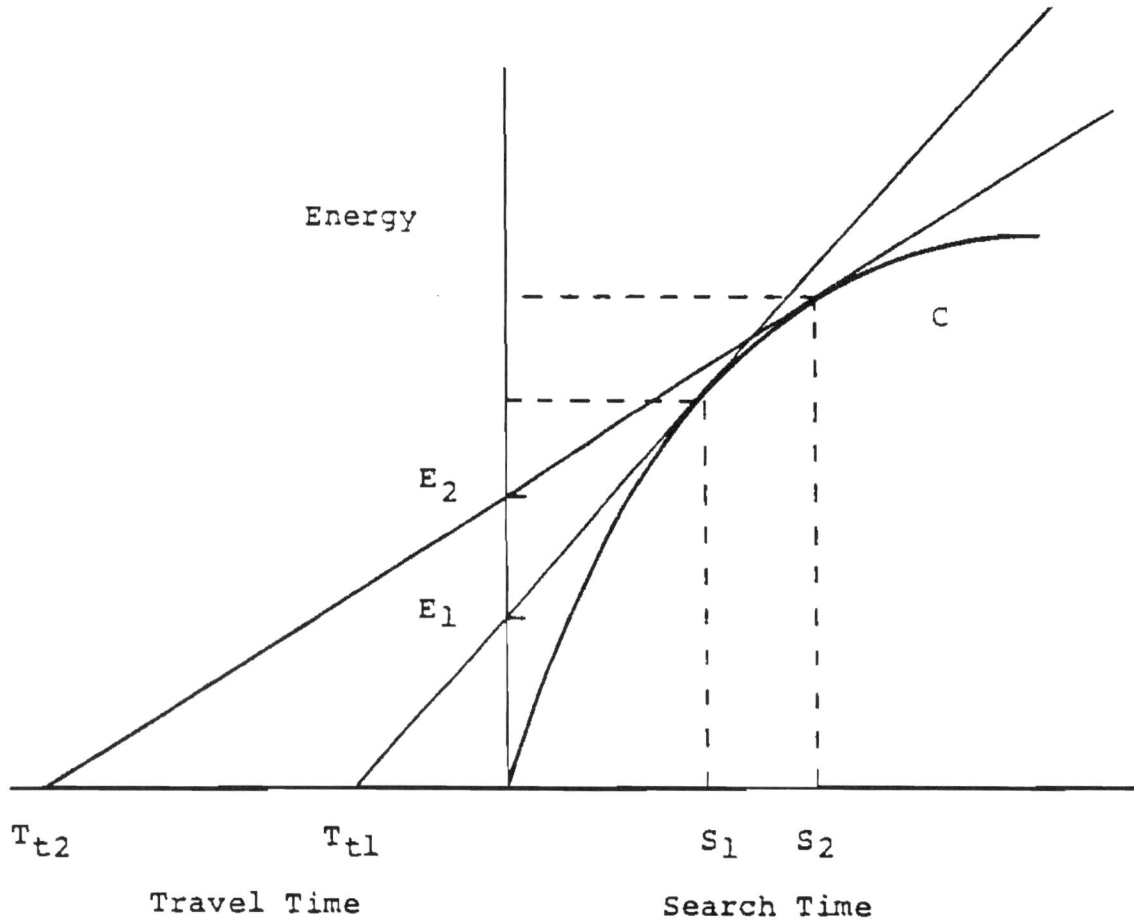

FIGURE 10.6. Model of relationship between travel time and prey selection; a forager may efficiently exploit prey of low energy content when $T_t < 1$, but when $T_t > 2$ items with higher energy value should be taken (from Orians and Pearson 1979).

area" (1970:265). A third area, which Hamilton and Watt term the biodeterioration zone, is one of low availability due to the habitually intensive use of resources within that zone. Beyond this zone lies the main resource-acquisition zone where prey species have mostly been unaffected by human occupation; this zone may or may not blend gradually into the other zones (see Binford 1982 for a discussion of economic zonation among the Nunamuit which concurs with this scheme).

Human ecologists studying patterns of animal use among sedentary horticulturalists have documented prey responses to environmental modification and hunting pressure which lend support to the refuging model. Linares (1976) observed that certain habitat changes associated with horticulture increase the availability of small game; she terms the human behavioral response to this change "garden hunting." It is relevant that Linares did not observe an increase in the availability of large game. Vickers (1980) examined the effect of settlement age on hunting efficiency, and demonstrated a decline in hunting yield that is a function of the length of settlement occupation. This observed relationship is the crux of resource depression.

Another related factor that may alter prey selection at an outlying locality is the increased cost of traveling to and returning from a resource patch. Models dealing with this general problem are structured around the concept of central place foraging. These models assume that prey are not consumed where they are captured, but returned to some fixed locality where they are eaten, stored, or distributed. In general, as the distance and hence traveling time from a central place to a foraging locality is increased, selection should proceed in favor of those items providing more energy per load (Orians and Pearson 1979:161; Killeen, Smith, and Hanson 1981).

How simple variation in travel time influences prey selection among items of different value is illustrated in Figure 10.6. This model is derived from Orians and Pearson and assumes no constraints on loading different combinations of prey items. This assumption seems reasonable in the case of humans. The x-axis represents costs in units of time: to the left of the y-axis is travel time, and to the right of the y-axis is foraging time, or search time within a patch. The curve (C) plots the expected energy (E') of a given prey type versus its expected encounter time (S).

The exact shape of this curve depends on prey densities in the patch. The concavity results from decreasing marginal effectiveness of increased foraging time in a patch. Maximum efficiency is obtained if the predator chooses prey with a minimum energy content fixed by the point at which the line from the x-axis corresponding to the travel time (T_t) becomes tangent to the foraging curve.

If the round-trip traveling time is short ($<T_{t1}$) the predator may effectively exploit prey of lower energy content, but if traveling time is long ($>T_{t2}$), the predator should only take prey items with higher energy values ($>E_2$). Thus, as distance, measured in travel time, from a patch to the central place is increased, the greater must be the prey food value selected by the predator. For intermediate travel times mixed combinations should occur.

The interaction of these two related, but independently acting, processes from the middle Archaic through the Hohokam Period may account for the observable changes through time in the faunal assemblage at Ventana Cave. First, increased transport and travel costs associated with expedition hunts in the uplands would have shifted selection towards large game. Simply, it would not have been efficient to travel a great distance to upland localities and settle for a mere cottontail (although this certainly may have occurred). Second, unintentional habitat destruction in the vicinity of more permanent base camps would have created unnatural, mosaic distributions in the availability of certain resources, necessitating the increased travel to productive hunting areas. Both of these factors would engender greater amounts of selectivity and specialization at upland localities, even if the same basic constellation of resources were being utilized.

Regional Patterns of Animal Use: Predictions and Data

Several predictions can be generated from the previously outlined model in order to assess the accuracy of this explanation, only one of which I will elaborate on here. Two predictions not discussed concern the composition of lithic and ground stone assemblages from Hohokam sites in the region, and the spatial distribution of artiodactyl body parts [these predictions and the quantified results are presented elsewhere in some depth (Bayham 1982)]. The results of these tests were mixed. It was demonstrated that hunting equipment such as projectile points, bifaces, and flake knives increase through time at Ventana Cave, while vegetal processing tools, such as manos and metates, proportionally decline. This pattern supports the main thesis and suggests that the function of Ventana Cave shifted from a base camp to a more specialized hunting camp. However, the distributional analysis of artiodactyl body parts yielded inconclusive results. Carnivores had impacted the faunal assemblage at Ventana to the extent that no clear human behavioral use pattern could be recognized.

The main hypothesis suggests that the position of sites within the regional procurement system should influence the pattern of animal food selection, and consequently, the character of the faunal assemblage from those sites. If we hypothesize that a greater amount of residential mobility characterized the middle Archaic Period, then middle Archaic faunal assemblages from residential camps should reflect the potential energetic return rate for that particular segment of the habitat at a particular point in time. High-ranking prey items should be represented in proportion to their local abundance. Due to the greater abundance of large game in upland habitats, lowland or basin residential campsites during the middle Archaic should contain a lesser amount of large game than do residential sites in upland contexts such as the middle Archaic levels at Ventana Cave.

Faunal assemblages from more permanently occupied sites, i.e. residential camps or habitation sites, during the late Archaic/Hohokam Periods should be characterized by resource intensification measurable by the greater prevalence of low-ranking taxa in the diet. Contemporaneous, intermittently occupied localities used more for logistic animal procurement should exhibit more variability in the pattern of selective utilization. The additional travel and transport costs of logistical resource procurement suggests that a greater proportion of high-ranking taxa should be found in sites of this type.

An attempt to test these predictions several years ago was limited by the absence of Archaic Period faunal data from open sites in south-central Arizona. No data from lowland valley contexts were available for comparison (a situation which still remains). At that time, I proceeded with a comparison of Sedentary Hohokam faunal assemblages in valley settings to the later Hohokam Period horizons at Ventana Cave. The results were positive, and tended to confirm the hypothesis. Lowland Hohokam faunal assemblages were dominated by the lower-ranking taxa, and markedly contrasted with the later occupational levels at Ventana Cave (Bayham 1982, 1985). The temporal bias of this comparison to the later, bona fide agricultural communities of the Hohokam, however, restricted any generalizations regarding mobility during the Archaic Period.

Recently, two open Archaic sites in south-central Arizona yielded faunal data which supplements the regional data set. Both of these sites were buried up to a meter deep, and based on artifactual diversity, evidence of hearth features, and considerable trash accumulation, they appear to be remnants of residential campsites occupied on an intermittent basis. These sites were investigated as part of the mitigation effort on the Tucson Aqueduct branch of the Central Arizona Project (Bayham et. al. 1983), and lie in the drainage basin of the Santa Cruz River at an elevation of approximately 1500 feet (Figure 10.1). The region is punctuated by a series of north-south trending mountain ranges the closest of which are the Picacho Mountains.

One of the two sites, the Gate Site (AZ AA:3:8 (ASU)), is situated just below the foothills of the Picacho Mountains; this site consisted of dispersed sheet trash and yielded radiocarbon dates of 4350±90 BP (A-2964; Czaplicki, ed. 1984:42–45)

TABLE 10.2. Frequency of identified mammals and reptiles at the Arroyo and the Gate sites.

Taxa	AZ:AA:3:28					AZ:AA:3:8
	F.12	F.14	F.15	TP 1	TP 2	F.15
Reptilia						
Gopherus agassizii		1				
Colubridae						2
Iguanidae	1					
Phrynosoma solare	1					
Mammalia						
Lagomorpha	1			1		
Lepus spp.	1			1	2	122
L. alleni	1			3	1	2
L. californicus	5	3	1	4	23	55
Sylvilagus spp.						11
S. audubonii	2	1		1	4	12
Rodentia					1	1
Spermophilus tereticaudus						6
Spermophilus/Ammospermophilus						3
Dipodomys deserti/spectabilis						1
Neotoma albigula						1
Carnivora					1	1
Canis latrans			1		1	1
Taxidea taxus	1			2		
Vulpes macrotis/velox	1				1	
Lynx rufus						3
Artiodactyla					2	
Odocoileus sp.					1	
Ovis canadensis					1	
Unidentifiable	81	13	4	67	184	120
Totals	95	18	6	79	222	342

and 4910±95 BP (Beta-12271). The other Archaic site is known as the Arroyo Site (AZ AA:3:28 (ASU)) and is a more consolidated trash deposit adjacent to an arroyo in a stabilized dune field. Radiocarbon dates from the extensive trash area ranged from 3910±290 BP to 4840±100 BP (Beta-12270 and Beta-12268, respectively). Both of these sites suggest an occupation between 4000 and 4500 BP, placing them roughly contemporaneous with the lower midden levels of Ventana Cave. The faunal material from these two sites, summarized in Table 10.2, temporally augments the Archaic faunal data from Ventana Cave situated in the uplands and fills a geographic lacuna in the regional settlement data.

In order to evaluate the prediction concerning residential mobility, a comparison of these lowland sites with the various levels at Ventana Cave and later Hohokam habitation sites is warranted. However, comparability among these various faunal assemblages is confounded by such problems as divergent excavation procedures affecting the proportionate representation of small species, the impact of intrusives upon the assemblage, and drastically different sample sizes affecting determinations of species abundance. Ventana Cave, for instance, was excavated in 1941–1942, and although the excavation procedures were exemplary for the time, there is an underrepresentation of small vertebrate species in the assemblage. Excavation units from the two open sites, on the other hand, were screened through ¼" mesh and flotation sampled; these procedures insured the recovery of small vertebrate remains. These different recovery techniques would obviate a test weighing the relative contribution of small species in the diet. Even if the above disparity could be reconciled, the problem of differential preservation between open and protected sites still remains. Discarded bone at open sites is subject to intense abiotic deterioration unlike a protected cave site, and certain species might have been adversely affected.

In an effort to circumvent these problems and offer a preliminary evaluation of the hypothesis, I decided to focus on the relative contribution of lagomorphs and artiodactyls in the various assemblages. These taxonomic groups are the most economically important ones in southwestern faunal assemblages, and

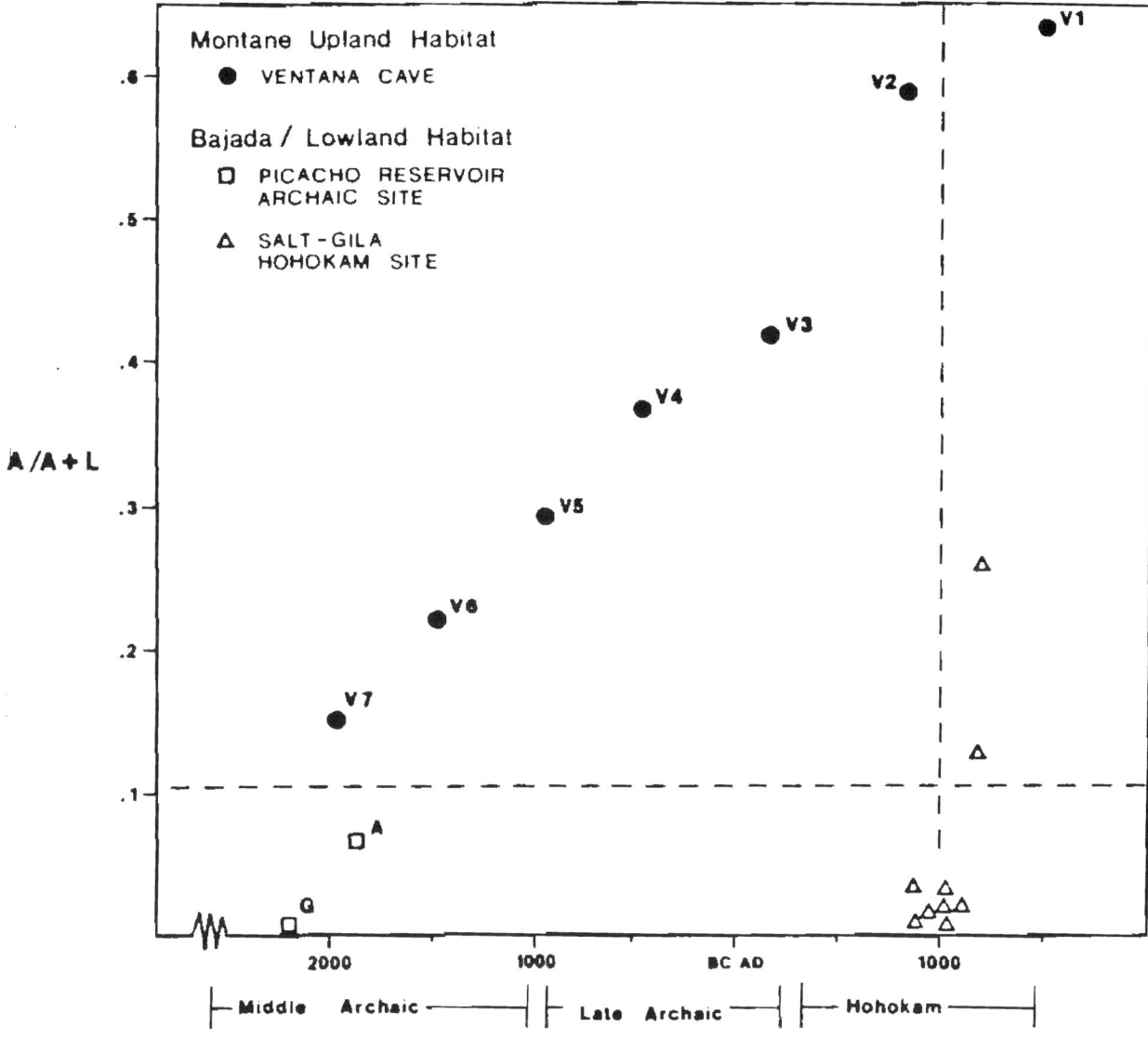

FIGURE 10.7. Proportionate representation of artiodactyls at Ventana Cave and various sites from the Picacho Reservoir and Salt-Gila study areas.

much of the measurable variation among assemblages is contained in the proportionate representation of these two groups. Lagomorphs are low-ranking in terms of their food value relative to any of the artiodactyls, and their proportionate abundance in an assemblage may be a good index to the degree of exploitative efficiency contained in that assemblage.

In order to insure comparability among species abundance estimates from the various assemblages, the MNI/NISP value was calculated for artiodactyls and lagomorphs from the two sites mentioned above, and each level within Ventana Cave. Employing Grayson's (1978) criteria, it was established that MNI values from the two open sites were primarily a function of small sample sizes (i.e. the MNI/NISP value exceeded .15 in most cases). This observation suggested that comparison of MNI values would be a tenuous unnecessary extrapolation from the data.

For this reason, I relied on the NISP value as a relative measure of species abundance instead of the MNI.

The proportionate representation of artiodactyls relative to the lagomorph-artiodactyl (A/A+L) total was determined for each of the open sites and the levels within Ventana Cave. This value, hereafter referred to as the artiodactyl index, summarizes the contribution of high-ranking artiodactyls in an assemblage and is analogous to the previously described index of selective efficiency: low values indicate intensive utilization of the species which provide less return per item, and vice versa. This index is plotted through time for a number of sites in south-central Arizona in Figure 10.7. The situation in the uplands is depicted by dots. Due to a lack of faunal material from other montane upland contexts, only strata from Ventana Cave are included in this category. The two lowland Archaics and some lower-elevation

Hohokam sites from the Salt-Gila Project are plotted as open squares and triangles.

The results of this comparison are interesting and tend to support the major hypothesis of regional socioeconomic reorganization. First, in the upper portion of Figure 10.7, the chronologic sequence of stratigraphic levels from Ventana Cave (V7 through V1) indicates essentially the same basic trend that was noted earlier: from the middle through the late Archaic, and continuing to the Hohokam Period, the sequence shows an increasing dependence on larger game in what typifies upland habitat utilization through time.

In the lowland basin sites, a very different situation obtains: low-ranking species dominate the assemblage and elicit no marked change over time until the latter part of the Hohokam sequence. The middle Archaic Gate Site (G) contains no artiodactyls and the Arroyo Site (A) contains only a few artiodactyl element fragments. In terms of the proportionate representation of artiodactyls, these two sites are similar and typify the lowland basin pattern of animal use during the Archaic Period. Interestingly, these two sites more closely resemble the earliest middle Archaic levels at Ventana Cave than they do any other known upland campsite. The Hohokam habitation sites are geographically close to the Archaic sites and show a similarity to the lowland Archaic sites in the proportion of artiodactyls represented (Szuter 1984:147); it is noteworthy, however, that greater variation in this pattern characterizes the later end of the continuum.

The increasing disparity over time between upland and lowland habitat utilization is arguably related to decreased residential mobility and an increase in the use of Ventana Cave as an outlying hunting camp. The overall similarity among lowland valley residence sites in terms of their low degree of selective efficiency is predicted by the proposed diachronic model. A reduction in hunter-gatherer mobility over time would encourage intensification on the local available resources resulting in the low degree of selective efficiency. It does appear, however, that even from the middle Archaic, the degree of selective efficiency at residence camps is sufficiently low that intensification on the animal resource base through time was either not feasible or imperceptible using only the species selected for this index. In no lowland residence site or later habitation site prior to AD 1000 does the proportion of artiodactyls relative to the lagomorph-artiodactyl total increase over time. It is possible that finer discrimination of efficiency patterns at residence sites would be obtainable if other small taxa such as rodents and reptiles, demonstrably used as food, were incorporated. Yet, I think the results of this comparison imply a pattern of changing settlement and resource use which supports the main thesis.

Summary

In this paper I have documented several axes of variability in archaeofaunal assemblages from Archaic and Hohokam Period sites in south-central Arizona. First, during the middle Archaic the differences between upland and lowland residence sites (probably base camps) are minimal, and can be explained with reference to variation in environmental productivity. The slightly greater abundance of artiodactyls in upland middle Archaic sites can be explained with reference to their greater availability in those montane habitats.

Variation over time from the middle Archaic through the Hohokam Period within a single upland site, Ventana Cave, appears not to be a function of environmental change over time, but related more to the macroscale changes in the socioeconomic structure of the cultural system. These changes are presumably tethered to the adoption of cultigens into the subsistence regime. Specifically, the phenomena of resource depression and central place foraging in the vicinity of lowland residence camps appear to be influencing the character of selection at upland sites. This situation concomitantly increases variability between upland and lowland sites through time and is evident by the late Archaic. The disparity is pronounced by the time Hohokam villages have become established. Lastly, the similarity or lack of variation in lowland residence sites over time is more likely due to the functional similarity of the site use pattern than to any form of environmental stability.

These observations imply that patterns of animal utilization may be conditioned by complex socioeconomic factors as well as environmental ones. The regional scenario of animal exploitation that emerges in south-central Arizona is one of constant emphasis on low-ranking prey through time. It appears as though the basic strategy of animal exploitation in terms of predator-prey dynamics changed very little from the middle to the late Archaic, and from the Archaic to the Hohokam Period. Not only were essentially the same animals being exploited but the net proportions of differently valued prey items in the aggregate diet remained rather constant. Within this basic subsistence strategy, however, there exists a surprising degree of heterogeneity characterized by both temporal and spatial variation that appears to be influenced by the establishment of an agricultural subsistence base.

Acknowledgments

The author extends his thanks to the members of the Archaeological Research Facility and the Department of Anthropology at CSU Fullerton for the opportunity to participate in this symposium. I would especially like to extend my gratitude to Susan Mershon-Spraker and Constance Cameron who graciously and subtly spurred me on to write this paper.

The research issues and the ideas addressed in this paper grew out of my dissertation research, and have benefited from conversations with numerous colleagues at Arizona State University and the University of Arizona. My debt to many individuals over the past ten years is acknowledged: S.J. Olsen, Emil Haury, John Olsen, Ms. Chris Szuter, Sylvia Gaines, Don Morris, Glen Rice, and Neal and Steve Shackerly. Data from the Archaic sites was obtained through funding from the Bureau of Reclamation ably managed by Gene Rogge.

References

Bayham, F. E.
1982 A Diachronic Analysis of Prehistoric Animal Exploitation at Ventana Cave. Ph.D. dissertation, Arizona State University. University Microfilms. Ann Arbor.
1985 Socioeconomic Inferences from Regional Patterns of Hohokam Animal Utilization. In *Proceedings of the 1983 Hohokam Symposium*, edited by D. Dove and A. E. Dittert, Occasional Paper 12 of the Phoenix Chapter, Arizona Archaeological Society.

Bayham, F. E., D. H. Morris, R. Most, G. Rice, M. Waters
1983 *The Picacho Reservoir Archaic Complex: A Research Design*. Anthropological Field Studies No. 5, Arizona State University, Tempe.

Binford, L. R.
1980 Willow Smoke and Dog's Tails: Hunter-Gatherer Settlement Systems and Archaeological Site Formation. *American Antiquity* 45(1):4–20.
1982 The Archaeology of Place. *Journal of Anthropological Archaeology*. 1(1):5–31.

Charnov, E. L.
1976 Ecological Foraging: The Marginal Value Theorem. *Theoretical Population Biology* 9:129–136.

Charnov, E. L., G. H. Orians, K. Hyatt
1976 Ecological Implications of Resource Depression. *American Naturalist* 110:247–249.

Cordell, L. S.
1984 *Prehistory of the Southwest*. Academic Press, Inc.

Czaplicki, J. (editor)
1984 *A Class III Survey of the Tucson Aqueduct Phase A Corridor, Central Arizona Project*. Arizona State Museum Archaeological Series No. 165. Tucson.

Ford, R. I.
1981 Gardening and Farming Before AD 1000: Patterns of Prehistoric Cultivation North of Mexico. *Journal of Ethnobiology* 1(1):6–27.

Gasser, R. E.
1980 Gu Achi: Seeds, Seasons, and Ecosystems. In *Excavations at Gu Achi, A Reappraisal of Hohokam Subsistence in the Arizona Papagueria*, edited by W. Bruce Masse, pp.314–342. Western Archaeological Center, Tucson.

Grayson, D. K.
1978 Minimum Numbers and Sample Size in Vertebrate Faunal Analysis. *American Antiquity* 43(1):53–64.

Hamilton, W. J., K. E. F. Watt.
1970 Refuging. *Annual Review of Ecology and Systematics* 1:263–286.

Haury, E. W.
1950 *The Stratigraphy and Archaeology of Ventana Cave*. University of Arizona Press, Tucson.
1976 *The Hohokam, Desert Farmers and Craftsmen: Excavations at Snaketown, 1964–1965*. University of Arizona Press, Tucson.

Irwin-Williams, C., C. V. Haynes
1970 Climatic Change and Early Population Dynamics in the Southwest United States. *Quaternary Research* 1:59–71.

Killeen, P. R., J. P. Smith, S. J. Hanson
1981 Central Place Foraging in *Rattus norvegicus*. *Animal Behavior* 29(1):64–70.

Linares, O. F.
1976 Garden Hunting in the American Tropics. *Human Ecology* 4(4):331–349.

Orians, G. H., N. E. Pearson
1979 On the Theory of Central Place Foraging. In *Analysis of Ecological Systems*, edited by D. J. Horn, G. R. Stairs, R. D. Mitchell, pp.155–177. Ohio State University Press, Columbus.

Sayles, E. B., E. Antevs
1941 The Cochise Culture. *Medallion Papers* 29. Gila Pueblo, Arizona.

Schiffer, M. B., R. H. McGuire
1982 The Study of Cultural Adaptations. In *Hohokam and Patayan*, edited by R. H. McGuire, M. B. Schiffer, pp.223–274. Academic Press, New York.

Szuter, C. R.
1984 Faunal Exploitation and the Reliance on Small Animals among the Hohokam. In, *Hohokam Archaeology along the Salt-Gila Aqueduct, Central Arizona Project, Volume VII: Environment and Subsistence*, edited by Lynn S. Teague, Patricia L. Crown, pp.139–169. Cultural Resource Management Section, Arizona State Museum, University of Arizona, Archaeological Series No.150.

Van Devender, T. R.
1977 Holocene Woodlands in the Southwestern Deserts. *Science* 198(4313):189–192.
1980 The Earliest Record of Sonoran Desert Vegetation: Castle Mountains and Picacho Peak Packrat Middens. Unpublished manuscript., Arizona-Sonoran Desert Museum, Tucson.

Van Devender, T. R., W. G. Spaulding
1979 Development of Vegetation and Climate in the Southwestern United States. *Science* 204:701–710.

Vickers, W. T.
1980 An Analysis of Amazonian Hunting Yields as a Function of Settlement Age. In *Working Papers on South American Indians*, Volume 2, edited by R. B. Hames, pp.7–29. Bennington College, Vermont.

Winterhalder, B.
1981 Optimal Foraging Strategies and Hunter-Gatherer Research in Anthropology: Theory and Models. In *Hunter-Gatherer Foraging Strategies*, edited by Bruce Winterhalder, E. A. Smith, pp.13–35. University of Chicago Press, Chicago.

Plant Utility Indices: Two Great Basin Examples

K. Renee Barlow and Duncan Metcalfe

INTRODUCTION

In the last 20 years the quantity of plant remains recovered from archaeological sites has increased dramatically. Paleoethnobotanists have developed more efficient and less destructive methods to retrieve plant specimens from archaeological sites and are employing more sophisticated statistical techniques to identify temporal and spatial patterning in macrofossil assemblages (Ford, 1988; Hastorf & Popper, 1988; Pearsall, 1989). Although the measures of taxonomic variation employed in the analysis of macrofossil assemblages vary (e.g. frequency, ubiquity or proportion of taxa represented), differences in the types and quantities of these remains are routinely investigated to recover information about the role of plant resources in prehistoric economies.

Researchers have identified a number of behavioural and post-depositional processes that likely contribute to inter- and intra-site patterning in the distribution of plant remains in archaeological sites. Studies of behaviours that structure the deposition of plant remains include investigations of the process of selecting plant resources for consumption (O'Connell & Hawkes, 1981; Hawkes, Hill & O'Connell, 1982; Hawkes, 1987); the types of plant waste associated with harvesting and processing cereal grains (Hillman, 1981; Harlan, 1989; Sikkink, 1989; Hillman, Colledge & Harris, 1989); and the use of different areas of residential sites for storage, processing, consumption and waste disposal (Hillman, 1984; Jones, 1984; Hastorf, 1988; Sikkink, 1988, 1989). In addition, some researchers have discussed the potential effects of natural agents of deposition, differential preservation, and cleaning by site residents on macrofossil assemblages (e.g. Dennell, 1972; Hally, 1981; Pearsall, 1988; Metcalfe & Heath, 1990).

This study investigates another aspect of plant processing activities likely to introduce patterning in the types and quantities of plant waste deposited in archaeological sites. Field processing is the removal of waste or low utility parts (e.g. hulls, chaff or shells) from resources at the location of procurement. If prehistoric foragers or farmers sometimes processed all or some subset of these parts from collected plant foods before transporting them, the potential effects of these behaviours on the subsequent removal and deposition of plant remains at residential sites may be significant.

We present the methods and results of processing experiments with two plant resources often recovered from archaeological sites in the Great Basin of western North America. Pinyon pine (*Pinus monophylla*) and pickleweed (*Allenrolfea occidentalis*) were collected from wild stands of plants growing in the vicinity of the Great Salt Lake (Figure 11.1) and experimentally processed using techniques and equipment chosen to approximate those used by prehistoric foragers.

The experiments were conducted to estimate increases in the economic utility of pinyon and pickleweed with time spent field processing, and associated changes in the types and quantities of inedible waste remaining with pine nuts and pickleweed seeds. We develop expectations about the circumstances in which prehistoric foragers operating from base camps should have removed different components of these plants at locations of procurement, and describe the types and relative quantities of waste associated with pinyon and pickleweed at different processing stages.

We also explore relationships between the costs and benefits of collecting, field processing and transporting loads of these resources to base camps and the implications for overall efficiency while foraging.

The results of these experiments indicate that archaeologists should expect patterned variation in macrofossil assemblages simply as a function of differences in the processing characteristics and spatial distribution of plant resources. We predict fairly dramatic differences in the types and quantities of pinyon remains recovered from sites within pinyon groves versus sites located within several days travel from the grove, with very little

---

*Originally published in *Journal of Archaeological Science* 23(1996):351–371

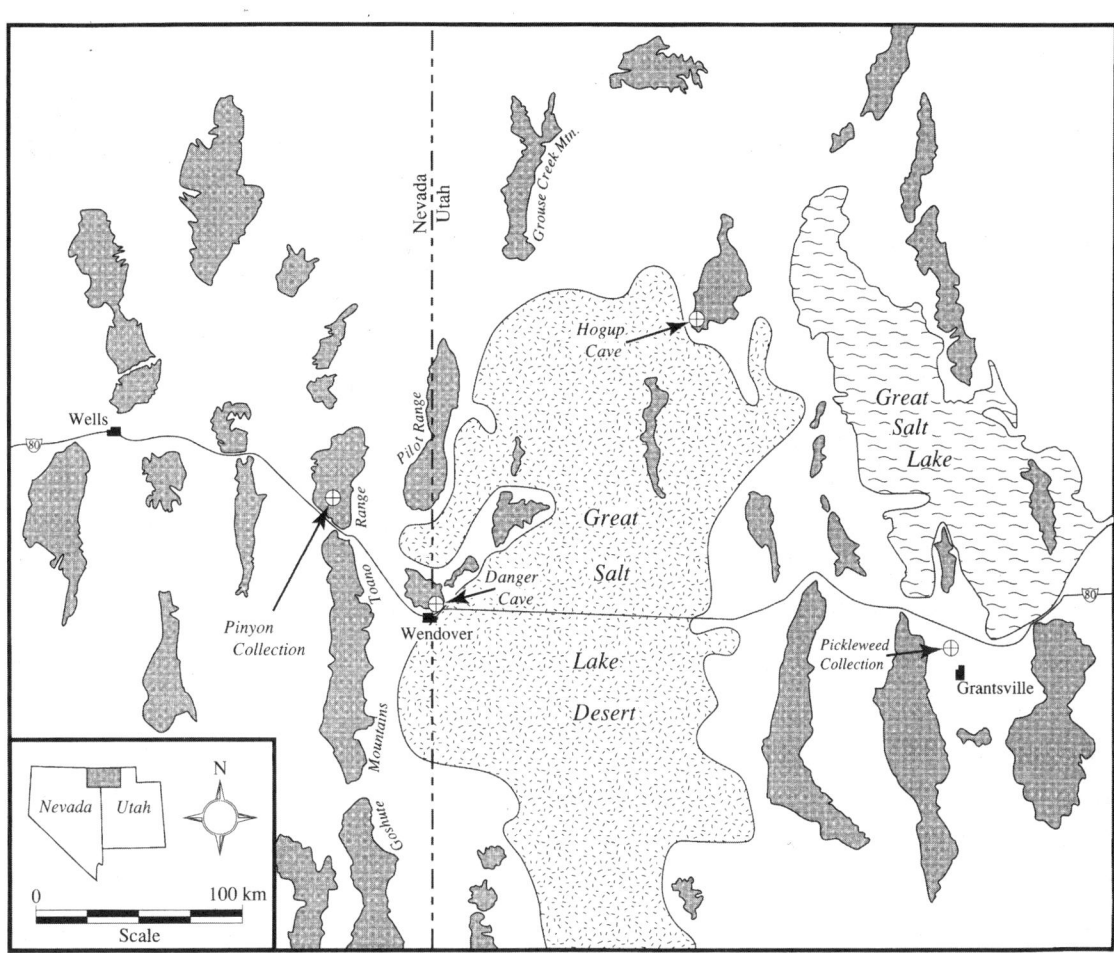

FIGURE 11.1. Map of the Great Salt Lake Desert and surrounding area. Macrofossil assemblages from Danger and Hogup Caves yielded large quantities of pickleweed plant parts, but very little pinyon waste. Pinyon cones with nuts were collected for the processing experiments from a grove in the Toana Range northwest of Wendover, Nevada. Pickleweed plants were collected from a patch north of Grantsville, Utah.

waste expected in more distant sites. In contrast, we expect large quantities of waste to have been carried with pickleweed seed even if the resource was transported great distances.

Our analyses also suggest that the caloric return rates associated with exploiting plant foods vary dramatically with different stages of field processing and transport distances. Foragers would always have gained greater amounts of food energy per time spent in acquisition and processing if they moved residences to pinyon and pickleweed patches while exploiting them. On the other hand, foragers exploiting these resources simultaneously would have increased caloric return rates by locating residential sites on pickleweed patches and field processing and transporting pine nuts up to large distances, even if pinyon was an important component of the diet. Where differences in the costs and benefits of field processing and transport vary greatly between resources, plant utility indices may provide the basis for developing inferences about which resources should have the greatest influence on the locations of residential sites.

THE PROBLEM: GREAT BASIN SUBSISTENCE

Reconstructions of prehistoric subsistence in the eastern Great Basin, especially during the Archaic period from approximately 9000 to 1500 years ago, are largely based on assemblages recovered from deeply stratified cave sites located near springs at the edges of lake basins and salt flats (Jennings, 1978; Madsen, 1982; Aikens & Madsen, 1986). Sites such as Danger (Jennings, 1957, 1978; Fry, 1976; Hall, 1988) and Hogup Caves (Aikens, 1970; Fry, 1976; Aikens & Madsen, 1986) yielded large collections of plant and animal remains associated with cultural deposits, including pinyon and pickleweed (Figure 11.1). Archaeologists routinely develop inferences about the importance of local plant resources in prehistoric diets based on patterned variation in the types and quantities of plant parts recovered from these sites. When pinyon cones and hulls are rare, either absolutely or relative to the remains of other species, this is generally interpreted as evidence that pine nuts did not play a significant role in the diet of site occupants (e.g. Harper & Alder, 1970:218; Madsen, 1982:216). Similarly, large quantities of pickleweed chaff and

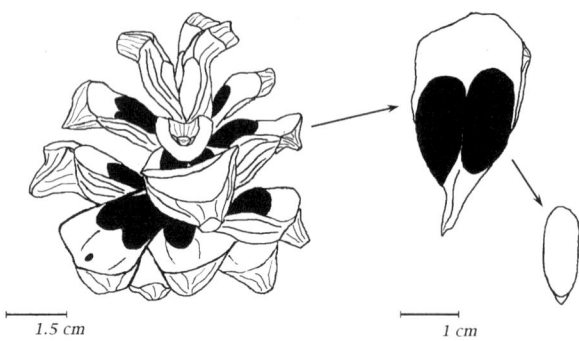

FIGURE 11.2. Open pinyon cone with nuts (*Pinus monophylla*), enlarged cone scale with nuts, and hulled nutmeat. The woody cone, cone scales, and hulls are waste; the soft, white nutmeat is the edible component of pinyon.

plant parts in cultural deposits, and seeds and plant parts in coprolites, have led researchers to hypothesize that pickleweed was an important component of prehistoric diets (Jennings, 1957: 64, 1978:30; Harper & Alder, 1970; Fry, 1970, 1976; Madsen, 1982:214; Bettinger, 1993).

These interpretations present a striking contrast with historic accounts of aboriginal plant use. Pine nuts are consistently identified as an important food source, and ethnographic studies and observations made by explorers and early settlers throughout the Great Basin indicate that local pinyon crops often provided the primary source of winter food (e.g. Steward, 1938; Wheat, 1967; Fowler, 1989). In contrast, references to the use of pickleweed are rare. Several older Northern Paiute informants identified pickleweed seeds as a food that had been collected in the past (Stewart, 1941), and a Gosiute story entitled "The Pickleweed Winter" suggests that pickleweed was an important winter food "long ago" (Miller, 1972:44–46). The story describes an apparently unusual event when pickleweed seeds were concentrated in windrows along the salt flats by melting snow, allowing it to be gathered in large quantities. "During the winter, one ate all he wanted. It was over there at Big Springs… they called it the pickleweed winter. They ate it with pine nuts, they say. They ate it with jack rabbits. Times were good, they say" (Miller, 1972:45). On the other hand, pickleweed was excluded in the most detailed list and description of plant foods used by Gosiutes (Chamberlin, 1911), and Julian Steward asserted that the seed bearing shrubs of the salt flats, including pickleweed, "are worthless for food" (Steward, 1938:17).[1]

These accounts suggest that the importance of pickleweed varied through time even in locations where the plant is locally abundant. It appears to have been dropped from the diets of indigenous people completely around the time that Euro-Americans settled in the region, while pinyon continued to comprise a large portion of aboriginal diets in the early 1900s and is still collected commercially and for private consumption (Lanner, 1981).

One possible explanation of the apparent difference between these observations and the representation of pinyon and pickleweed in macrofossil assemblages is that pinyon was locally less abundant during the time the sites were occupied. Alternatively, patches of these resources are often separated by substantial distances, and they may have been differentially processed near procurement locations. If so, differences in the types and quantity of pinyon and pickleweed remains in archaeological sites may reflect variation in the amount of time prehistoric foragers spent removing low-utility but archaeologically visible plant parts from these resources prior to transporting them to base camps, regardless of their relative importance in past diets.

Pinyon

Single-leaf pinyon are relatively small, scrubby pines found interspersed with juniper trees or in sparse groves in many of the foothills and small mountain ranges of the Great Basin. Beginning in the early fall, these pines are sometimes laden with green, sappy cones bearing pine nuts. As autumn progresses, the cones dry and turn brown, opening and releasing the nuts (Lanner, 1981). An early, hard freeze will hasten this process.

Pinyon cones may grow singly or in clusters, range in size from about 4 to 7 cm in length, and usually contain between 6 and 16 nuts (Figure 11.2). The nuts occur either singly or in pairs on the cone scales, and are about 2 cm long and 1 cm wide. Each nut consists of a nutmeat surrounded by a soft leathery covering (nucellus) and a hard, brittle shell (integument or seed coat) (Sporne, 1965; Farjon, 1984). The plant parts relevant to this study are the cones, shells, and nutmeats. Nutmeat is the only edible component with energetic value to humans; the other components are waste.

There are several detailed accounts of aboriginal processing techniques (Palmer, 1878; Coville, 1892; Dutcher, 1893; Chamberlin, 1911; Wheat, 1967; Fowler, 1989), from which the following was synthesized.

Pinyon nuts were often gathered when the cones were still green, requiring that the cones be knocked or picked from the trees. They were then either placed in piles or storage pits and left to open by drying, or were opened by exposure to fire. After the cones were open, the nuts were removed either by shaking the cones, beating the cones with a stick, or beating a container filled with cones with a stick. Pinyon nuts were also gathered when the cones were brown and open. The pine nuts were collected by picking them off the ground after they had fallen, and/or placing mats under the tree and beating and shaking the nuts out of the cones on the trees. Once the nuts were separated from the cones, they were either dried and stored, or parched in a tray with hot coals. After parching, the shells were separated from the nutmeats with a mano and metate, and removed by winnowing. The nutmeats were again parched, ground and eaten.

Pickleweed

Pickleweed is a perennial shrub of the Goosefoot family that occupies a fairly exclusive niche in the poorly drained sediments of Great Basin salt flats (Figure 11.3). Pickleweed plants grow in cir-

FIGURE 11.3. Caprielle and Aaron Barlow in pickleweed patch near Grantsville, Utah.

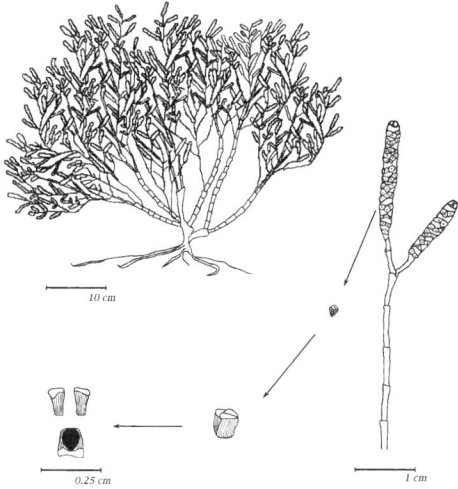

FIGURE 11.4. Dried pickleweed plant (*Allenrolfea occidentalis*), enlarged branchlet with flower spikes, and papery bracts encasing the pickleweed seed. The jointed plant parts and chaff created from crushing the bracts are waste; the tiny, flat seed is the edible component.

cular, sparsely distributed clumps, trapping sediments and creating low mounds on the desert floor. Rarely more than 50 cm tall, but often spreading to a diameter of greater than a meter, clumps of bright green, succulent pickleweed plants are often conspicuous across otherwise barren stretches of salt or alkali flats in late summer to early fall. Each plant has numerous jointed, vertical branches supported by a shallow woody root system. In late fall pickleweed turns reddish and the seed-bearing spikes at the terminal ends of branches become dry and brittle.

Each branch of the plant has c. 90 to 120 inconspicuous flower spikes (Figure 11.4). Each spike has c. 30 to 60 seeds arranged in a linear spiral. The tiny, black seeds range from about 0.4 to 0.9 mm in diameter. Each seed is encased in a pericarp, which in turn is encased within three specialized leaves called bracts. When dried, the pericarps and bracts are thin papery shells, or chaff, that can be separated from seeds by hand-rubbing and removed by winnowing. The components of pickleweed relevant to this study are branches, spikes, bracts, pericarps, and seeds. Seeds are assumed to be the only component having nutritional value for humans; the other components were assigned a caloric value of zero.

We failed to locate any description of how pickleweed was processed by natives of the Great Basin. However, ethnographic descriptions of the aboriginal procurement and processing of small seed plants in the Great Basin contain a number of striking similarities in technique. Perhaps the most variable is the method of procurement, which ranges from beating or threshing the seeds from plants to collecting whole plants by pulling them out of the ground, roots and all. When entire plants were collected, they were generally processed in the following manner (from Fowler, 1989:46–49). First, the plants were allowed to dry for several days to a week. Once dry, the seeds were separated from other plant parts by beating, pounding, grinding, or hand "crushing" and the chaff was removed by winnowing in a tray. When seeds were encased in a hard hull, the seeds were next parched in a tray, the hulls cracked using a metate, and the hulls removed by winnowing. The seeds were then either stored for later use, or parched again, ground and eaten. These general steps guided the experimental processing procedures developed for pickleweed.

THE FIELD PROCESSING/TRANSPORT MODEL

Recently, we cast the problem of the differential utility and transport of resource components in very general terms: when should central-place foragers remove low-utility parts at procurement locations, and what are the archaeological implications of processing resources in the field? Based on a series of explicit assumptions, we developed an optimality model that predicts variation in field processing behaviour with differences in roundtrip travel time between the place of procurement and consumption, and increases in the utility of a resource with processing time, as the critical factors structuring the type and amount of waste components returned to residential camps (Metcalfe, 1989; Metcalfe & Barlow, 1992).

A resource's utility function is simply the relationship between time spent field processing and the resulting increase in

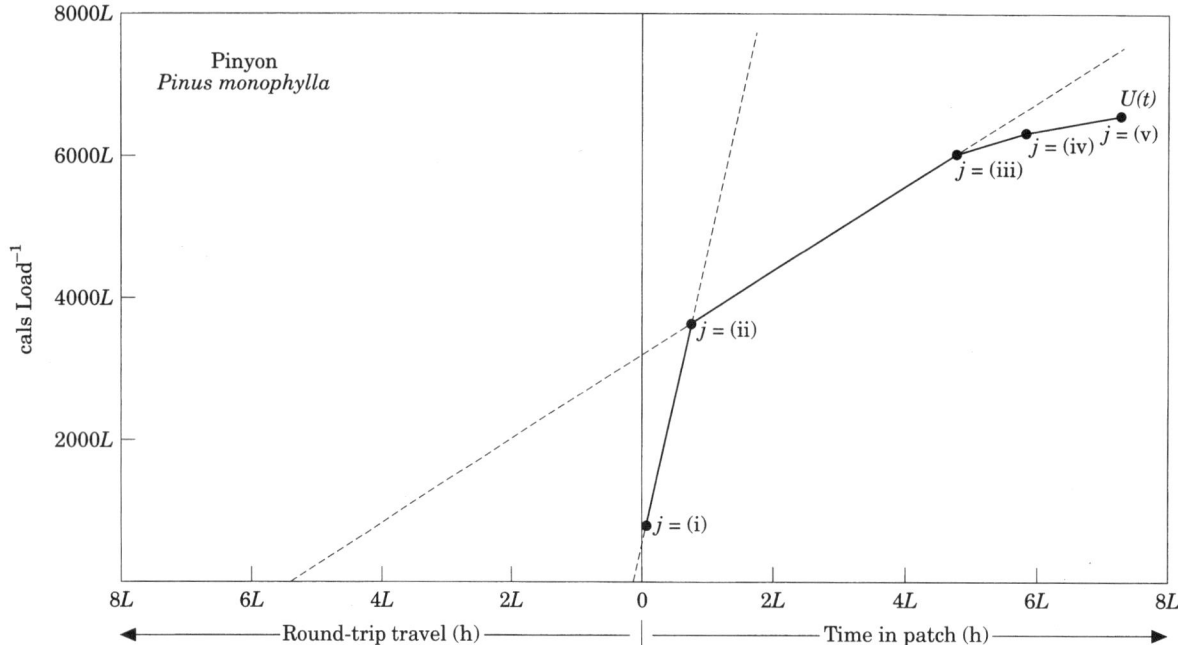

FIGURE 11.5. Changes in the utility of pinyon with field processing time. The right side of the graph displays increases in the calories per load of pinyon with field processing time. The two dashed lines drawn through the x-axis on the left side of the graph indicate the hours of round-trip travel time between an archaeological site and a pinyon grove and when removing pine nuts from cones [$j = $ (ii)] and hulling pine nuts [$j = $ (iii)] are expected at the grove.

the utility of the transported load. When the collected resource consists of a number of different parts of varying utility, the field processing/transport tradeoff may determine the types and quantities of parts returned to a central-place forager's camp, and therefore the composition of assemblages that enter the archaeological record. For a load of food resource, utility can be measured in calories. In this case, *the model predicts that collectors will spend the amount of time field processing that yields the maximum calories per hour spent travelling to the resource patch, collecting and field processing the resource, and transporting it back to camp.*

To develop expectations about when different parts of pinyon or pickleweed should be discarded at procurement locations, we measured changes in the utilities of these resources associated with processing time (Metcalfe & Barlow, 1992). We constructed utility functions by calculating the caloric values of these resources at different processing stages and plotting caloric value against the amount of time required to accomplish each processing stage (Figures 11.5 & 11.6).

Calculating Utility

For resources with only one edible component (e.g. pinyon nutmeats and pickleweed seeds), changes in utility can be calculated by determining the proportional weight of nutmeats or seeds in the resource after each processing stage and the caloric value of the nutmeats or seeds. The general equation presented in Metcalfe and Barlow (1992:347) for calculating the utility of a load at any processing stage was simplified to:

$$y_j = a\beta_j L \qquad (11.1)$$

where y_j is the utility of the load at processing stage j, a is the caloric value of nuts or seeds measured in calories per kilogram, β_j is the proportion of the resource at processing stage j made up of nuts or seeds, and L is the weight of the transported load.[2] This equation monitors the benefit associated with field processing: increases in the utility of resource in terms of the total number of calories transported per load.

Calculating Cost

The cost of collecting and field processing a load of resource can be calculated by recording the weight of the resource after collection and each processing stage, and the time spent during each of these procedures. This equation was also simplified from the general case presented in Metcalfe and Barlow (1992:348). Here:

$$x_j = \frac{t_j}{w_j} L \qquad (11.2)$$

where x_j is the time spent collecting and field processing a load of resource to processing stage j, t_j is the time required to collect and field process a sample of the resource to stage j and w_j is the weight of that sample after stage j of processing.

Calculating the benefits and costs of field processing resources for transport using Equations 11.1 and 11.2 requires an estimate of load size (L): the amount of resource foragers carry from the resource patch to the residential site. For this study, weights of 3 and 15 kg were selected for the minimum and maxi-

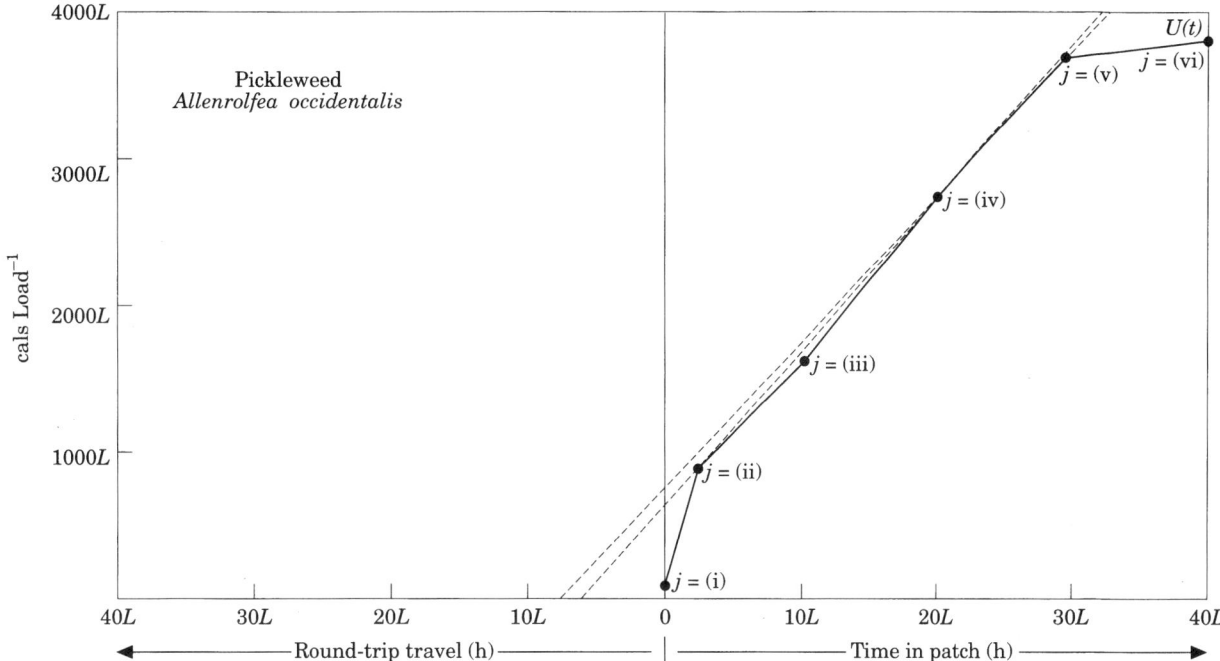

FIGURE 11.6. Changes in the utility of pickleweed with field processing time. Stripping the plants and removing large branch parts to produce loads that consist of 23% seed by weight, or 10% by volume [$j = (ii)$], is expected for very small round-trip travel times between sites and pickleweed patches. The dashed lines drawn through the x-axis indicate the minimum travel times when winnowing to produce loads of approximately 70% [$j = (iv)$] or 90% [$j = (v)$] seed is expected at the patch.

mum weights of potential load sizes. There are several reasons for this. First, weight was chosen rather than volume because it was easy to measure the entire range of package types encountered during the processing experiments (i.e. pine cones to nutmeats and whole plants to pickleweed seeds) with a consistent level of precision. Comparable measures of changes in the weight of resource samples, and the proportions of edible components in those samples with processing time, allowed us to construct comparable utility functions for pinyon and pickleweed. Second, historical descriptions indicate Great Basin women typically carried plant foods in large, conical burden baskets (Steward, 1938; Wheat, 1967; Fowler, 1989; Jones & Madsen, 1989). Three to 15 kg appears to be a reasonable estimate of the weight of burden baskets filled with pinyon and pickleweed (Barlow et al., 1993),[3] and encompasses the range of load sizes typically carried by modern female hunter–gatherers who transport plant resources on a regular basis (Lee, 1969; Blurton Jones & Sibley, 1978; O'Connell & Hawkes, 1981; Hawkes, O'Connell & Blurton Jones, 1989).

Predicting Field Processing Stages

Once the benefits (y_j) and costs (x_j) associated with field processing pinyon and pickleweed were calculated, they were used to develop expectations about when the various processing stages should occur prior to transport. By determining the minimum amount of travel and transport time when field processing would increase the amount of calories returned to camp per hour spent in travel, collection, field processing and transport, relationships between round-trip travel time and expected field processing stages were modelled (Figures 11.5 & 11.6). The expected minimum round-trip travel times (z_j) when each processing stage (j) will increase the caloric return rate for time spent away from camp can be calculated by modifying the equation in Metcalfe & Barlow (1992:344) to the more general form:

$$z_j = \frac{y_{j-1} x_j - y_j x_{j-1}}{y_j - y_{j-1}} \quad (11.3)$$

Equation 11.3 was used to calculate the minimum travel and transport times when different stages of field processing pinyon and pickleweed were expected to occur at procurement locations. By identifying the types and quantities of plant parts remaining with pine nuts and pickleweed seeds at each processing stage, we also developed expectations about relationships between transport distance and the range and relative quantities of waste in macrofossil assemblages.

PROCESSING EXPERIMENTS

Pinyon

The pinyon processing experiments were conducted on closed to partially open green cones collected in the Toana Range northwest of Wendover, Nevada (Figure 11.1). Cones were picked from trees in September, before the nuts began dropping. The cones were laid out to dry indoors. After opening, the cones were placed in a cardboard box and beaten with a stick. The nuts were collected from the cone debris and put in a basketry parching tray. The small quantities of pine needles and cone

TABLE 11.1. Results of nutritional analysis of pinyon nutmeats.

Sample	% Moisture	% Protein	% Lipid	% Ash	% Carbohydrate	Kcal/100 gm
Clean nutmeat sample						
1	10.5	8.6	56.4	2.4	22.1	630
2	10.1	8.0	57.6	2.4	21.9	638
Mean	10.3	8.3	57.0	2.4	22.0	634
Parched nutmeat sample						
3	7.1	7.6	59.0	2.7	23.6	656

fragments remaining with the nuts were removed, and the nuts were parched using live coals. The nuts were then hulled on a flat milling stone. A loaf-shaped stone was used to break the shells and separate them from the nutmeats. The nutmeats were winnowed to remove remaining hull fragments, further cleaned by hand, and parched with coals in a basketry tray. Following this procedure resulted in five processing stages: (i) collect whole cones; (ii) beat cones to remove nuts; (iii) clean, parch, hull, and winnow nuts; (iv) clean nutmeats; and (v) parch nutmeats.

These collection and processing procedures were timed. Separating the nuts from the cones, Stage (ii), was conducted until Barlow felt she had reached the point of diminishing returns. Careful inspection of the cone debris after the experiments demonstrated that over 96% of the available nuts were recovered during this stage. Nutmeats "lost" during a particular stage of processing (usually less than 10 g) were not used in calculating β for that stage, but were included in the calculation of β for all preceding stages. Two samples of clean nutmeats and one sample of parched nutmeats were sent to the Department of Nutrition and Food Science at Utah State University for nutritional analyses. The clean nutmeat samples averaged 6340 cal kg^{-1}; the sample of parched nutmeats had a caloric value of 6560 cal kg^{-1} (Table 11.1). Previous published estimates of the energetic value of *Pinus monophylla* nutmeats include 4880 (Farris, 1980) and 7335 (Little, 1938) cal kg^{-1}.

Table 11.2 presents a brief summary of the field processing procedures, waste products, cost, and various measures of the utility associated with each stage. Table 11.3 presents the basic data for each of the experiments reported here and the calculated values for x_j and y_j.

Figure 11.5 illustrates the relationship between time spent field processing at the pinyon grove and the utility of a load of pinyon transported to a residential camp. The utility function ($U(t)$) for pinyon is based on the average collection and processing time (x_j) and utility (y_j) of the samples for each processing stage. The y-axis measures the utility of the resource in calories per kilogram. Time spent procuring and field processing is scaled on the x-axis to the right of the y-axis, round-trip travel time on the x-axis to the left of the y-axis. Round-trip travel times where each subsequent processing stage is expected to occur prior to transport (z_j) were calculated using Equation (11.3). These times can also be determined graphically by identifying the x-intercept of a line passing through j and $j-1$ of the utility function. Round-trip travel time is expressed in terms of L to indicate that this is a relative measure that varies proportionally with expected load size.

Table 11.4 presents the time and distance expectations for the relationship between minimum round-trip length and the expected field processing stage. Figures are expressed as ranges because they were calculated for loads of 3 kg and 15 kg, the range reported for modern hunter–gatherers transporting vegetal resources. Two points are important. First, for this size range of transported loads, expectations about the amount of field processing are unambiguous; there is no overlap in the round-trip travel times/distances for the three processing stages that result in the transport and deposition of plant waste in residential sites. For loads of 3 to 15 kg, relatively large quantities of cone debris and hull fragments and a few pine needles are expected in residential sites within approximately 0.5 to 2.5 km (c. 1–5 km round-trip) of pinyon groves. Large quantities of hull fragments but few needles and cone fragments are expected in residential sites up to approximately 25–120 km away (c. 49–243 km round-trip). Small quantities of pinyon waste are expected in residential sites farther away, and these should consist only of a few hull fragments. Second, the estimated minimum distances for the processing stages that remove all plant waste at the pinyon grove are enormous—for stage (iv), the minimum round-trip distance for 3–15 kg loads is 138 to 692 km between residential base and the pinyon patch, distances at or exceeding the far end of the range noted for ethnographic groups (Rhode, 1990).

Pickleweed

The pickleweed processing experiments were conducted on plants recovered from a large, dense patch near Grantsville, Utah, south of the Great Salt Lake (Figure 11.1). The plants were collected when the seeds were mature but the plants were still green and succulent. A total of 11.9 kg (51.5 l) of the whole plants were collected in just under 8 min. After the plants had dried indoors for several days, the branches and spikes were stripped by hand, allowing the seeds, pericarps and bracts to fall into a basket. Because the branches are jointed and the plant parts are brittle when dry, many branch and spike fragments also fell into the basket. Fortunately, these fragments are very light, and shaking the basket sorts the larger fragments to the top, where they can be quickly retrieved and discarded. At this point, the basket is filled with hundreds of thousands of pickleweed seeds, each encased in a pericarp and bracts, as well as numerous very small branch and spike fragments.

The next processing step consisted of rubbing handfuls of the contents of the basket vigorously between the palms in order to separate the bracts and pericarps from the seeds. Hand-rubbing is an essential step since winnowing is ineffective without it. Samples were then winnowed for 20 min, divided into 5 min intervals. Hand-rubbing was included as part of the first

TABLE 11.2. Pinyon processing stages and proportion of nutmeats to inedible components.

Processing Stage	Remaining Waste Components	Proportion Nutmeat by Weight	Proportion Nutmeat by Volume
(i) Collect whole cones	Mostly cones and cone bracts; hulls, some pine needles	0.13	0.03
(ii) Beat cones and separate nuts	Mostly hulls; some needles and a few cone bracts	0.57	0.43
(iii) Clean, parch, hull, and winnow nuts	Few hulls	0.95	0.94
(iv) Clean nutmeats	None	1.00	1.00
(v) Parch nutmeats	None	1.00	1.00

TABLE 11.3. Results of pinyon processing.[a]

Field Processing Stage	Sample	Time (min)	Weight (kg)	t_j (min)	x_j (min kg^{-1})	Mean x_j (min kg^{-1})	β_j	y_j (Cal kg^{-1})	Mean y_j (Cal kg^{-1})	z (h)
(i) Cone collection	1	10.15	2.2282	10.15	4.56	4.36	0.1026	650.73	800.75	0.00
	2	16.25	2.4136	16.25	6.73		0.1051	666.41		
	3	9.00	2.5088	9.00	3.59		0.1566	993.15		
	4	10.12	3.9366	10.12	2.57		0.1408	892.72		
(ii) Separate nuts from cones	1	12.90	0.4289	23.05	53.74	45.04	0.5164	3274.21	3644.38	0.12
	2	17.18	0.4532	33.43	73.76		0.5410	3430.20		
	3	7.25	0.6100	16.25	26.64		0.6302	3995.24		
	4	12.98	0.8876	23.10	26.03		0.6116	3877.86		
(iii) Clean, parch, hull, and winnow nuts	1	58.28	0.2245	81.33	362.27	286.60	0.9550	6054.77	6028.95	5.40
	2	55.17	0.2487	88.60	356.25		0.9429	5978.01		
	3	69.35	0.4014	85.60	213.25		0.9544	6050.96		
	4	97.97	0.5641	121.07	214.63		0.9514	6032.05		
(iv) Clean nutmeats	1	8.07	0.2144	89.40	416.98	348.98	1.0000	6340.00	6340.00	15.37
	2	12.30	0.2345	100.90	430.28		1.0000	6340.00		
	3	20.63	0.3831	106.23	277.29		1.0000	6340.00		
	4	24.33	0.5358	145.40	271.37		1.0000	6340.00		
(v) Parch nutmeats	1	11.52	0.1846	100.92	546.70	435.74	1.0000	6560.00	6560.00	35.85
	2	10.62	0.2067	111.52	539.53		1.0000	6560.00		
	3	13.70	0.3587	119.93	334.35		1.0000	6560.00		
	4	19.50	0.5115	164.90	322.39		1.0000	6560.00		

[a] All derived variables (t_j, x_j, mean x_j) were calculated from the original times and weights using a computer spreadsheet that maintains 15 significant figures. Consequently, calculating some values (e.g. mean x_j) from other derived values (e.g. t_j and x_j) as listed in this table may give slightly different results because of rounding error.

TABLE 11.4. Results of pinyon processing experiments for 3 and 15 kg loads.

Processing Stage	Minimum Round-Trip Travel Times (h)	Minimum Round-Trip Travel Distances (est. 3 km h^{-1})	Time Spent Procuring & Field Processing (h)	Maximum Net Energetic Gain/Loss (Cal h^{-1})
(i) Cones	0	0	0.22–1.09	10891–10891
(ii) Nuts	0.35–1.77	1.06–5.32	2.25–11.26	4069–4067
(iii) Nutmeats	16.21–81.03	48.62–243.10	14.33–71.65	466–461
(iv) Clean nutmeats	46.12–230.61	138.36–691.82	17.45–87.24	173–165
(v) Parched nutmeats	107.56–537.82	322.69–1613.45	21.79–108.93	26–17

TABLE 11.5. Results of nutritional analysis of pickleweed seed.

Sample	% Moisture	% Protein	% Lipid	% Ash	% Carbohydrate	Kcal 100 g^{-1}
1	6.18	27.18	10.72	5.87	50.05	405
2	6.18	27.40	9.75	6.18	50.49	399
3	6.21	27.27	9.92	6.02	50.58	401
Mean	6.19	27.28	10.13	6.02	50.37	402

winnowing stage because it does not, by itself, change the proportion of seed to inedible components. Consequently, the experiments were divided into six processing stages: (i) plant collection; (ii) drying, hand-stripping, and removing larger branch fragments; (iii) hand-rubbing for 3 min and winnowing for 5 min; (iv) second winnowing; (v) third winnowing; and (vi) fourth winnowing.

These experiments were timed, and the cost of field processing pickleweed was calculated using Equation (11.2). Calculating utility for pickleweed required an additional step in the analysis. Measuring the proportional weight of seeds (β) in whole and partially processed pickleweed was problematic because of the extremely small size of seeds. A regression formula was used to transform measures of sample density to estimates of β_j.[4] The resulting β_j values were then employed to calculate utility using Equation (11.1).

Three samples of pickleweed seed were sent to the Department of Nutrition and Food Science at Utah State University for nutritional analyses and proved to have an average caloric value of approximately 4020 cal kg^{-1} (Table 11.5).

Table 11.6 presents a brief summary of the field processing procedures, waste products, cost, and various measures of the utility of the resource after each processing stage. Table 11.7 presents the basic data for each of the pickleweed experiments, and the values for x_j and y_j.

Figure 11.6 illustrates the relationship between time spent field processing at the pickleweed patch and the utility of a load of pickleweed transported to a residential camp. The utility function ($U(t)$) for pickleweed is based on the average collection and processing time (x_j) and the average estimated utility (y_j) of the samples for each processing stage. Again, the y-axis measures the utility of the resource in cal/kg, and time spent procuring and field processing is scaled on the x-axis to the right of the y-axis, round-trip travel time on the x-axis to the left of the y-axis. These values are again expressed in terms of load size (L) to indicate that they vary with the size of the transported load. Round-trip travel times where each subsequent processing stage is expected to occur prior to transport (z_j) were calculated using Equation (11.3).

Table 11.8 presents the range of minimum times and distances when each stage of field processing is expected for loads of 3–15 kg of pickleweed. Base camps located within several kilometers of a pickleweed patch would be expected to contain large quantities of plant parts if the plant was exploited by site occupants. For sites located up to relatively large distances from patches, the resource should have been partially winnowed. The major waste component transported to these sites should have been large quantities of chaff. Extensive winnowing at the patch is not expected unless the camp site is hundreds of kilometers away. It is probably not economically profitable to exploit and transport pickleweed these distances, which are well outside the range reported for Great Basin foragers (Rhode, 1990). Round-trip travel times were not calculated for stage (iii) of pickleweed processing because it produces a lower increase in utility than the next 5 min of winnowing. Under these circumstances, the model predicts the forager will transport the resource after Stage (ii), or continue winnowing to remove more chaff.

Drying Time and Opportunity Costs at Procurement Locations

Although drying both fresh pickleweed plants and sappy pinyon cones was necessary to facilitate processing, we did not include drying time in our calculations of field processing costs because it required no work effort. We initially assumed foragers could spend drying time collecting more pinyon or pickleweed; processing previously collected, dried pinyon or pickleweed; collecting or processing other resources; or in other activities less directly related to subsistence. It is difficult to imagine a forager waiting for a resource to dry. However, field processing behaviours are embedded in a larger set of foraging decisions that likely affect the costs associated with time spent at procurement locations and base camps. These include the range of resources to exploit; the set of patches to utilize; the degree to which collection trips from a base camp or residential moves are employed to collect those resources; and where to locate base camps and auxiliary camps and facilities.

The field processing/transport model was developed to predict the behaviour of foragers making collection trips to resource patches from a base camp, and assumes time spent processing at procurement locations is costly because the forager loses opportunities to search for, handle and transport other resources, or engage in activities at a base camp that would benefit the forager or the forager's offspring. On the other hand, time spent at procurement locations may range from relatively costly when foragers employ logistic collection trips from long-term residential base camps, to equal to time spent at home when foragers employ residential moves to resource patches. To illustrate, we imagine several situations in which the costs associated with drying pickleweed and pinyon at procurement locations may vary, and would result in different expectations about the removal and deposition of plant parts than those presented earlier.

1. *The forager travels from a base camp to a pinyon grove or pickleweed patch, collects enough resource for one load only, waits for it to dry, processes and transports it.* This situation may be most likely when a small group of individuals travels from

TABLE 11.6. Pickleweed processing stages and proportion of seed to inedible components.

Processing Stage	Remaining Waste Components	Proportion Seed by Weight	Proportion Seed by Volume
(i) Collect green plants	As represented in the plant, including branches, spikes, bracts and pericarps	0.02	0.01
(ii) Strip and remove large branches	Spikes, small spike and branch fragments (most <3 cm), bracts and pericarps	0.23	0.10
(iii) Hand-rub and 1st winnow	Mostly chaff (crushed bracts and pericarps) with very small branch and spike fragments (most <0.5 cm)	0.47	0.31
(iv) 2nd winnow	Mostly chaff and pericarps attached to seeds	0.71	0.58
(v) 3rd winnow	Same as above	0.91	0.85
(vi) 4th winnow	Same as above	0.95	0.92

TABLE 11.7. Results of pickleweed processing.[a]

Field Processing Stage	Sample	Time (min)	Weight (kg)	t_j[b] (min)	x_j (min kg^{-1})	Mean x_j	Density (w/v)	β_j	y_j	Mean y_j
(i) Fresh plant collection		7.75	11.9	7.75	0.65	0.65	—	—	94.00[c]	94.00
(ii) Strip and remove branches (Total for all samples, including those not included in this analysis, is 0.734 kg)	a	30.05	0.200	32.16	160.81	147.07	0.21	0.20	796.32	890.29
	b	27.20	0.231	29.64	128.31		0.22	0.21	838.20	
	c	14.72	0.104	15.82	152.10		0.26	0.26	1036.34	
Subsample for later stages of processing	1(a)	—	0.0537	8.63	—	—				
	2(a)	—	0.0558	8.97	—					
	3(c)	—	0.0589	8.96	—					
(iii) Hand-rub and 1st winnow	1	8.00	0.0251	16.64	662.77	615.73	0.36	0.43	1746.16	1621.22
	2	8.00	0.0260	16.97	652.81		0.37	0.46	1834.00	
	3	8.00	0.0319	16.96	531.61		0.30	0.32	1283.51	
(iv) 2nd winnow	1	5.00	0.0159	21.64	1360.72	1218.98	0.45	0.66	2646.96	2733.09
	2	5.00	0.0173	21.97	1270.12		0.49	0.78	3126.97	
	3	5.00	0.0214	21.96	1026.10		0.43	0.60	2425.33	
(v) 3rd winnow	1	5.00	0.0141	26.64	1889.04	1787.23	0.47	0.72	2880.84	3687.66
	2	5.00	0.0150	26.97	1798.21		0.58	1.09	4386.23	
	3	5.00	0.0161	26.96	1674.44		0.54	0.94	3795.92	
(vi) 4th winnow	1	5.00	0.0127	31.64	2490.98	2397.37	0.51	0.84	3385.36	3803.07
	2	5.00	0.0135	31.97	2368.38		0.56	1.02	4084.95	
	3	5.00	0.0137	31.96	2332.74		0.55	0.98	3938.90	

[a] All derived variables (t_j, x_j, mean x_j, β_j, y_j, mean y_j) were calculated from the original times, weights and density using a computer spreadsheet that maintains 15 significant figures. Consequently, calculating some values (e.g. mean x_j) from other derived values (e.g. t_j and x_j) as listed in this table may give slightly different results because of rounding error.

[b] For calculating t_j for Stage (ii), the sample weight was divided by 0.734 kg, the quotion multiplied by 7.75 min (time for total fresh plant collection) to which the sample time to strip and remove branches was added. For calculating t_j for the subsamples, the sample weight was multiplied by the x_j of the sample from which it was recovered, indicated by letter in parentheses.

[c] To estimate the utility of whole, wet plants, the y_j for dry plants was calculated using the density of the entire sample of dry plants. The resulting value (Cal kg^{-1} dry plants) was multiplied by the total weight of the dry plants, and divided by the total weight of the wet plants: (Cal kg^{-1} dry plants × kg dry plants) ÷ kg wet plants.

TABLE 11.8. Results of pickleweed processing experiments for 3 and 15 kg loads.

Processing Stage	Minimum Round-Trip Travel Times (h)	Minimum Round-Trip Travel Distances (est. 3 km h^{-1})	Time Spent Procuring & Field Processing (h)	Maximum Net Energetic Gain/Loss (Cal h^{-1})
(i)	0	0	0.03–0.16	8535
(ii)	0.83–4.16	2.49–12.47	7.35–36.77	201–200
(iii)	—	—	—	—
(iv)	18.54–92.70	55.62–278.09	60.95–304.74	(−22)–(−25)
(v)	20.40–102.00	61.20–306.00	89.36–446.81	(−25)–(−27)
(vi)	855.44–4,427.18	2656.31–13,281.55	119.87–599.34	(−115)–(−124)

a residential base to a procurement patch for the collection of a specific resource (*sensu* Binford, 1980:9). In this case, drying time should be included in the cost of subsequent processing stages. In our experiments (indoors in a dry laboratory) green pine cones required a minimum of approximately 7 days to open and pickleweed plants were dry enough to strip in about 13 days. Including this time in field processing costs results in substantial increases in minimum round-trip travel times before any field processing is expected (approximately 60 km for pinyon and 50 km for pickleweed). However, the estimated caloric return rates for collection trips become so low that we do not expect foragers to employ this type of land use/resource procurement strategy for exploiting resources that require long periods of drying or curing. When drying time cannot be spent in other activities, foragers might be expected to decrease drying costs by varying collection strategies (e.g. collect brown open pinyon cones or nuts from the ground or strip seeds directly from ripe plants) or processing techniques (e.g. roasting the cones or parching plants).

2. *The forager makes two round-trips from a base camp per load of resource, first to collect the cones or plants and leave them to dry, and second to retrieve the dry resource, field process as appropriate and transport back to camp.* If two trips are made per load, the forager is free to engage in any number of activities during the days or weeks required for drying, but the cost of travel/transport doubles. The forager is expected to maximize caloric return rates over two round-trips to the resource patch. Although some additional time may be required to construct or move resources to a drying facility (e.g. a cache or a cave), the minimum round-trip travel times when processing increases caloric return rates during a trip (Z_j) may be roughly half those predicted in Figures 11.5 and 11.6.

3. *The forager collects and stores multiple loads of pinyon cones or pickleweed plants at one or a series of procurement locations during a single trip from a residential site or seasonal base camp, and later returns to field process and transport loads.* If the collection of numerous loads of pinyon and pickleweed were "embedded" in one or a sequence of residential moves, drying time would again be free to the forager, and the cost of the extra trip could be amortized over the number of loads collected. The cost of the collection trip likely affects overall foraging efficiency and should influence the forager's decision about whether to employ this type of foraging strategy. However, we still expect the forager to employ field processing strategies that maximize caloric return rates during trips to retrieve, field process and transport loads of pinyon or pickleweed. During these trips, collection is no longer part of handling time at the patch. With a reduction in field processing costs, the minimum travel time when field processing increases caloric return rates (Z_j), would be smaller than predicted in Figures 11.5 and 11.6. Differences are likely to be most dramatic for early to moderate stages of field processing, with less effect on later stages where collection time comprises only a small proportion of overall handling in the field.

4. *The forager employs residential moves to collect and process resources for consumption and/or transport to the next resource patch.* The field processing/transport model was not developed to address processing strategies employed during residential moves. In this situation, the procurement location is also the base camp. The opportunity costs associated with time spent in the field and at home are equal. If the forager processes resources for consumption at the procurement location/base camp, waste components should be removed and discarded in the same relative proportions in which they occur in the collected resource. If resources are carried to the next resource procurement location/base camp, *when processing costs are equal at both locations* we expect the forager to always process resources into a high-utility load before transport. We do not expect the removal and deposition of waste at base camps to be structured by distance from resource patches.

These few hypothetical situations suggest archaeologists should evaluate the strengths and limitations of the field processing/transport model in predicting variation in processing behaviour among modern people employing different foraging strategies. To the extent that these studies yield alternate sets of predictions about spatial patterning in the deposition of material remains, they may provide a useful tool for identifying spatial and

TABLE 11.9. Estimated energetic return rates for transporting 3 and 15 kg loads of pinyon.

DISTANCE TO PATCH (km)	TIME TO PATCH (h)	EXPECTED PROCESSING STAGE	TIME SPENT PROCURING & FIELD PROCESSING (h)	ESTIMATED NET ENERGETIC GAIN/LOSS (Cal h^{-1})
0.5	0.17	(i) Cones	0.22–1.09	4231–8311
5	1.67	(ii) Nuts	2.25–11.26	1831–3618
25	8.33	(ii) Nuts	2.25–11.26	451–1825
50	16.67	(iii) (3 kg)–(ii) (15 kg)	11.26–14.33	253–1092
100	33.33	(iii) Nutmeats	14.33–87.24	97–523

temporal variation in the land-use strategies employed by prehistoric foragers.

FIELD PROCESSING/TRANSPORT STRATEGIES AND ENERGETIC EFFICIENCY

Once the round-trip travel times were calculated for the various processing stages and load sizes, we investigated how field processing pinyon and pickleweed to various stages would affect net caloric gains per hour associated with exploiting these resources from a base camp.

Calculating Net Caloric Return Rates

Equation (11.4) was used to estimate net caloric return rates for exploiting pinyon and pickleweed at various distances. The net return rates for 3 and 15 kg loads were calculated using the following formula:

$$n = \frac{a\beta L - (\frac{1}{2}Te_o + \frac{1}{2}Te_b + te_t)}{T + t} \quad (11.4)$$

where n is the net gain (or loss) in calories, T is the round-trip travel time, t is the time spent procuring and processing the load to be transported, e_o is the rate of energy expenditure travelling to the patch, e_b is the rate of energy expenditure travelling to camp with the load, and e_t is the rate of energy expenditure while in the patch. The utility decay functions developed from these changes in net rate provide insights into the transportability of various resources.

To calculate net energetic return rates, it was necessary to obtain estimates of the energetic costs of travel and processing. Based on the results of experimental studies with human subjects, Pandolf, Givoni & Goldman (1977) provide a nondimensional equation for such estimates:

$$M = 1.5W + 2.0(W+L)\left(\frac{L}{W}\right)^2 + \eta(W+L)(1.5V^2 + 0.35VG) \quad (11.5)$$

where M is the metabolic cost expressed in watts; W is the weight of the forager in kilograms; L is the weight of the transported load, also in kilograms; V is the speed of walking measured in meters per second; G is the grade measured as percent slope; and η is the terrain coefficient. Because we are interested in determining a conservative estimate of the relationship between net return rate and round-trip travel time, in estimating the parameters in Equation (11.5) we tried to err on the side of making the costs of transport unrealistically low. We assumed the forager weighed 55 kg, walked at a speed of 3 km h^{-1} (V = 0.833 m s^{-1}), was always on perfectly level terrain (G = 0), and traveled along a dirt trail (η = 1.1). With these parameters, the cost of travelling without a load is 125 cal h^{-1}, with a 3 kg load about 128 cal h^{-1}, and a 15 kg load costs about 149 cal h^{-1} (50 cal km^{-1}).

Using Equation (11.4) we calculated net caloric return rates for foragers exploiting pinyon and pickleweed from base camps (Tables 11.4 & 11.8). While calculating these we noticed several general trends. Larger load sizes increased estimated net caloric return rates at every processing stage, and although caloric return rates decreased rapidly with field processing and transport time, the rate of this decrease varied between resources.

Effects of Load Size on Energetic Efficiency

To illustrate the effects of load size on energetic efficiency, Table 11.9 displays the estimated net energetic return rates for travelling to a grove and collecting, field processing and transporting 3 and 15 kg loads of pinyon. In this example, the expected processing stage, time spent procuring and field processing in the patch, and net return rates were calculated for hypothetical camps located varying distances from the grove. At every distance and processing stage, caloric return rates are highest for 15 kg loads. This difference in return rates increases with travel and transport time. For pinyon groves near base camps, foragers double net caloric return rates during collection trips by carrying 15 kg loads. For pinyon groves 100 km away, transporting a 15 kg load yields a greater than five-fold increase in energetic return rates. The forager will always be more efficient transporting larger, rather than smaller loads.

Transport Decay Curves

To compare the decreases in energetic efficiency associated with field processing and transporting loads of pinyon and pickleweed, we calculated changes in net caloric return rates associated with transporting loads of each over distances from zero to 100 km. This relationship is modelled as a utility decay function that relates variation in the expected net return rate for the central place forager as a function of (1) the gross caloric value of the transported load, (2) the energetic expenditure of travelling to the patch, procuring and field processing a load of resources, and returning with the added weight of the load, (3) the time required to complete the trip.

The relationship between transport distance and the net rate of energetic return for collecting, field processing and transporting loads of pinyon is illustrated in Figure 11.7. Three functions are plotted: one for a 15 kg load of cones and nuts, one for the same load of unhulled nuts, and one for nutmeats. Where the functions intersect is the travel time at which increased field processing will increase the net rate of return. For instance, the intersection of the transport decay function for transporting nuts still in their cones and the function for transporting the same size load of just nuts and hulls indicates the distance at which it becomes profitable to cull the cones in the field. This graphic technique also calculates the minimum round-trip travel time when a particular field processing stage increases caloric return rates (z_j), but can include different energetic costs associated with time spent travelling, collecting, processing and transporting resources.[5]

Figure 11.8 illustrates the net rate of energetic return for transporting 15 kg loads of pickleweed various distances, calculated using Equation (11.5). Again, where the functions intersect is the round-trip travel time at which increased field processing will increase the net rate of return. In this example, the loads consist of whole plants, threshed pickleweed, and partially winnowed seed. The net energetic return rates for transporting pickleweed decrease even more rapidly with distance from the patch than those associated with transporting pinyon.

The transport decay curves clarify three relationships between field processing/transport strategies and overall foraging efficiency. First, it is always more efficient to move the consumer to the resource patch than it is to move the resource to the consumer. Even relatively short transport times dramatically lower the net return rate for exploiting pinyon and pickleweed. The procurement of either of these plants from a base camp outside the resource patch is only expected when the choice of the camp location is a tradeoff between other competing demands, such as access to water, fuel, other foods, or perhaps previously stored foods. Second, when the forager stages collection trips from a base camp, the transport decay curves illustrate large differences in caloric return rates associated with transporting pinyon or pickleweed at different field processing stages. Foragers can increase net energetic efficiency substantially if they vary field processing strategies with transport distance. For example, at 20 h round-trip travel time, a forager gains approximately 500 cal h^{-1} transporting unprocessed pine cones, nearly 2000 cal h^{-1} transporting pine nuts, or about 1000 cal h^{-1} transporting hulled nutmeats. Finally, our estimated net caloric return rates indicate that pinyon may be efficiently collected, field processed and transported while operating from a base camp, but return rates associated with trips to pickleweed patches drop rapidly with increases in travel/transport time. This suggests intertaxonomic differences in processing characteristics should structure variation in which resources are likely to be exploited through logistic collection trips, and which through residential moves. In general, when both types of resources are taken simultaneously, the spatial distribution of resources with processing characteristics like pickleweed should strongly affect where foragers locate base camps or residential sites.

Pickleweed Return Rates

Several researchers have noted that the estimated caloric return rates associated with the exploitation of pickleweed are extremely low, both absolutely and relative to other Great Basin resources (Simms, 1987; Jones & Madsen, 1989). This observation has been enigmatic to Great Basin archaeologists, mostly because large quantities of remains in archaeological sites indicate it was a food resource for prehistoric inhabitants. Some archaeologists have discounted the utility of experimental estimates, arguing that modern investigators are not likely to achieve aboriginal levels of efficiency in plant processing, and variation in the way resources were collected and processed results in substantial variation in return rates (Bettinger, 1993; Bettinger & Baumhoff, 1983). Our calculations also suggest that the caloric return rates associated with exploiting pickleweed are low, and large amounts of time are required to collect and process loads of pickleweed even to intermediate processing stages. Consequently, we attempted to identify aspects of the processing experiments that likely influenced our estimated energetic return rates, and may limit energetic efficiency during the processing of small seed resources in general.

Loss of pickleweed seed during each stage of the processing experiments may have lowered our estimated return rates. The most significant loss occurred during the stripping/threshing of whole plants, at which point an estimated 47% of the seed occurring in whole plants was lost. The maximum seed loss occurring during any other procedure was approximately 6%. Because of their small size, it is likely that pickleweed seeds were also lost by aboriginal foragers threshing, crushing or winnowing this resource (cf. Sikkink, 1989, threshing quinoa seeds). If they lost significantly less, however, their caloric returns rates may have been nearly twice those estimated for processing stages (ii) through (vi).

Our energetic return rates for pickleweed processing were also constrained by the amount of resource winnowed during the experiments. A standard sample size of 0.25 l was used for each winnowing batch; partly to facilitate monitoring increases in efficiency during winnowing, and also because we used a small winnowing tray. We did not anticipate the potential effects of batch size on calculating changes in the utility of pickleweed with winnowing time. However, the small batch size resulted in large estimates of the time required to collect and process a kilogram of pickleweed at each processing stage (x_j) that included winnowing (Table 11.7). To assess the effect of batch size on our estimated caloric return rates we recalculated our estimates of

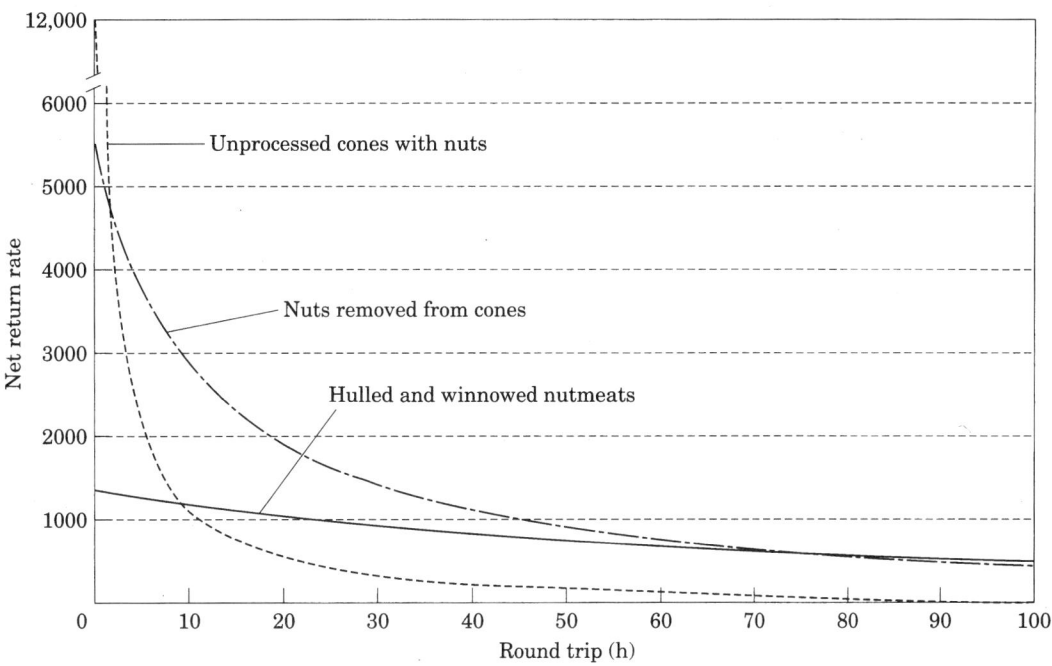

FIGURE 11.7. Pinyon transport decay functions. The three decreasing curves show estimated net caloric return rates for collecting, field processing and transporting 15 kg loads of cones, nuts or nutmeats to camps at increasing distances from pinyon groves. Energetic estimates based on Pandolf et al. (1977).

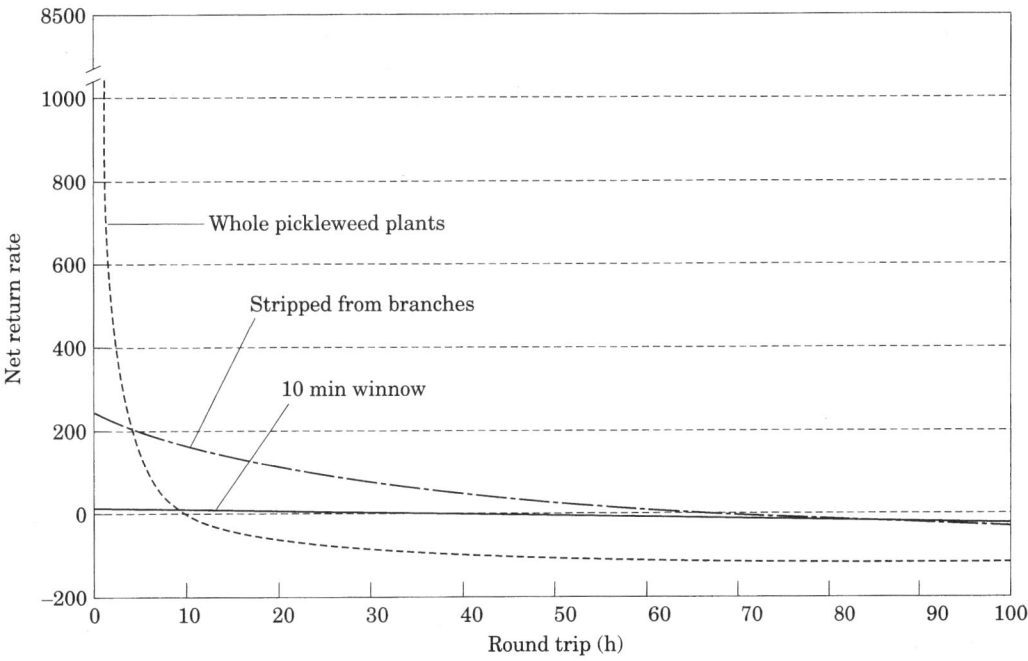

FIGURE 11.8. Pickleweed transport decay functions. The three decreasing curves show estimated net caloric return rates for collecting, field processing and transporting 15 kg loads of whole plants, material stripped from plants, or moderately winnowed pickleweed seed to camps at increasing distances from pickleweed patches. Energetic estimates based on Pandolf et al. (1977).

TABLE 11.10. Results of pickleweed processing experiments recalculated for samples winnowed in 2 l batches.

Processing Stage	Minimum Round-Trip Travel Times (h)	Minimum Round-Trip Travel Distances (est. 3 km h^{-1})	Time Spent Procuring & Field Processing (h)	Maximum Net Energetic Gain/Loss (Cal hr^{-1})
(i)	0	0	0.03–0.16	8552–8552
(ii)	0.83–4.16	2.49–12.47	7.35–36.77	201–200
(iii)	—	—	—	—
(iv)	—	—	—	—
(v)	1.42–7.08	4.25–21.25	34.91–174.55	179–179
(vi)	186.98–934.90	560.94–2804.69	41.85–209.27	(–77)–(–85)

TABLE 11.11. Values relating β to density for pickleweed samples.

β	Density (g ml^{-1})
0.34	0.23
0.27	0.23
0.31	0.33
0.24	0.31
0.33	0.34
0.64	0.45
0.76	0.44
0.11	0.04
1.00	0.56

the costs and benefits associated with field processing pickleweed using a winnowing batch size of 2 l.[6]

The results of these changes are shown in Table 11.10. Increasing the amount of resource winnowed in each batch substantially decreased the minimum round-trip distance to camp before moderate amounts of time spent winnowing at the pickleweed patch would be expected. The estimated time in patch for collecting and processing 3 and 15 kg loads of winnowed pickleweed is less than half that calculated for smaller batches, and the estimated net energetic return rates associated with collecting, processing and transporting pickleweed increased dramatically. Even so, our estimated net caloric return rates for threshed and winnowed pickleweed are at the low end of caloric return rates estimated for plant foods, and similar to the post-encounter return rates calculated for other small seed resources (O'Connell & Hawkes, 1981; Simms, 1987; Cane, 1989; Wright, 1994).

Implications for Great Basin Plant Macrofossil Assemblages

As noted earlier, the analyses and interpretation of plant macrofossils from Danger and Hogup Caves have structured interpretations of Archaic subsistence in the Great Basin. Both sites are located within a kilometer of pickleweed communities today, and cultural deposits are reported to have been dominated by "organic chaff" and "stems and twigs" (Jennings, 1957; Aikens, 1970). Although we were not able to determine the relative quantities of the different parts of pickleweed plants recovered from these sites, these descriptions are not inconsistent with the kinds of waste that would be expected at base camps or processing locations adjacent to pickleweed communities.

Our results do not support inferences about the relative importance of pickleweed in past diets based solely on large quantities of chaff and plant parts in sites near procurement locations. The processing experiments indicate that 98% of wet pickleweed plants brought to a site would consist of chaff and inedible plant parts (Table 11.6). Less than half a kilogram of each 15 kg load consists of edible seed, so large quantities of chaff and plant parts may not indicate a large dietary contribution. Even if pickleweed was only stripped from plants at the sites for transport elsewhere, waste would accumulate rapidly. Each 15 kg load transported away from the site would include more than 11 kg of chaff and small branch fragments, but approximately seven 15 kg loads of whole plants are needed for one 15 kg load of pickleweed at this processing stage. Again, the deposition of large quantities of pickleweed waste (in this case c. 350 l per 3.5 kg of transported pickleweed seed) may not be associated with large quantities of the resource in the diet of the forager. If pickleweed was also winnowed at the site, even greater quantities of waste would have been deposited relative to the amount of seed field processed or consumed at the site.

In contrast, pinyon remains were relatively rare in these sites. Early reports indicate that Danger and Hogup Caves yielded "a few seed coats" to "one or two pine cones and a few hulls" (Jennings, 1957; Aikens, 1970). However, the recent meticulous removal and laboratory excavation of a 2 by 2 m column of deposits from Danger Cave suggests that pinyon remains may have been more common than previously thought. Pinyon remains were recovered in varying quantities from 27 levels radiocarbon dated to between 7900 and 330 years BP. Fourteen of the strata yielded 0 to 15 pinyon hull fragments, while greater quantities of pinyon remains were recovered from 13 levels, ranging from 20 to 130 hull and cone fragments (Madsen & Rhode, 1990).

Today, the nearest pinyon groves are located in the Toana, Pilot and Grouse Creek Mountain Ranges, approximately 20–40 km from the caves (Figure 11.1). Our experiments suggest that if prehistoric foragers staged collection trips from these sites to pinyon groves located between approximately 3 and 120 km away, they should have transported 15 kg loads of nuts still in hulls (Table 11.4). The waste remaining with our experimental samples in this field processing stage consisted mainly of hulls, with some needles and a few fragments of cone bracts (Table 11.2). The remains from several levels are not inconsistent with our expectations about waste components returned to base camps. No pine needles were recovered, but five of nine cultural levels dating from approximately 7400 to 6000 years BP yielded relatively large numbers of hull fragments, and three of these also contained a few cone fragments. Higher in the stratigraphic profile, two levels dating to approximately 5300 and 5000 years BP, and one level deposited before 880 years BP, yielded similar assemblages of pinyon remains. Conversely, our predictions about the deposition of large quantities of pinyon hulls appear to be falsified for at least half of the occupational units if foragers were staging collection trips to pinyon groves during these times.

It is interesting to note that a sharp decrease in the frequency of cultural levels with large quantities of pinyon remains at approximately 6000 years BP appears to coincide with proposed ecological changes and increases in the frequency of sites located in upland environments in the region (Aikens & Madsen, 1986). Although the period from approximately 7500 to 4500 years ago is generally thought to have been a time of increased aridity throughout the Great Basin (Antevs, 1948; O'Connell, 1975; Mehringer, 1986; Grayson, 1993), some local paleoenvironmental records from the central and eastern Basin suggest that most post-Pleistocene changes in vegetation communities occurred prior to approximately 6000 years ago. Plant and macrofossil and pollen records indicate rapid elevational and latitudinal changes in the distribution of plant species between approximately 7000 and 6000 years BP, possibly associated with intense aridity or a shift towards summer-dominated precipitation. With the exception of two species (Ephedra and Rocky Mountain Juniper), plant communities appear to have been essentially modern in the range and relative proportions of species by or shortly after 6000 BP in the eastern Great Basin (Mehringer, 1986; Thompson, 1990). Around 6000 BP a trend towards increased precipitation is suggested by pollen records from a number of Great Basin locations, including the Great Salt Lake area (Madsen & Currey, 1979; Currey & James, 1982; Grayson, 1993).

If the stratigraphic distribution of pinyon remains in the Danger Cave sediment column accurately reflects temporal variation in processing activities, this may indicate differences in the use of Danger Cave within the larger, regional strategy of land use employed by prehistoric foragers. Estimates of length of occupation range from several weeks (Aikens, 1970) to the hypothesis that Danger Cave was a winter base camp where inhabitants stored local resources and foods collected during trips to other resource patches in the area (e.g. Madsen, 1982:214–215; Hall, 1988). It may be that when the site was used as a base camp foragers returned relatively large quantities of hulls to the site during trips to nearby pinyon groves, while at other times it was occupied primarily for the collection and processing of local resources such as pickleweed (e.g. Jennings, 1957; Aikens, 1970; Fry, 1976). If so, we might also expect temporal variation in the types and quantities of plant components representing other local and transported plant foods. An analysis of the relative proportions of pickleweed components from the Danger Cave column, and pinyon and pickleweed remains recently recovered from the nearby Floating Island rockshelter site (K. Jones, D. Madsen, personal communication) may provide a preliminary assessment of these predictions. A more informative test requires the analyses of macrofossil assemblages from habitation sites located varying distances from both pickleweed and pinyon patches.

Discussion

Plant resources in the Great Basin are distributed in widely scattered, environmentally segregated patches. Given this situation, the costs associated with plant procurement and processing strategies have important implications for expecting differential field processing, and subsequent patterning in the deposition of plant waste in archaeological sites. If prehistoric collectors varied field processing strategies as a consequence of differences in the costs and benefits of transporting plant foods, the results of these experiments suggest that reconstructions of dietary importance based on relative quantities of remains are suspect. The importance of plant foods represented by waste that can be quickly removed for large increases in utility may be underestimated, while the contribution of plants that require extensive field processing for similar gains in utility may be exaggerated.

In addition to making potentially testable predictions about field processing behaviour and its material consequences, identifying the costs and benefits of field processing may contribute to our understanding of resource exploitation systems. Field processing decisions are embedded in a larger resource procurement strategy which includes resource choice, patch choice and settlement pattern decisions. In this paper transport distance, or the distance between the resource patch and the residential camp, was not modelled as a decision variable. From the perspective of the ethnographer or archaeologist who monitors field processing behaviour, transport distance has already been determined by the location of the residential camp. Within the larger framework of logistic organization, however, field processing resources for transport (especially those with predictable locations) occurs near the end of the exploitation continuum. It is possible that the processing characteristics of important resources in the diet influence settlement patterns in general, and specifically the locations of base camps or residential sites.

The estimated net energetic return rates associated with collecting, field processing and transporting loads of pinyon and pickleweed to base camps strongly suggest that foragers would always increase their energetic efficiency by carrying larger loads of these resources and locating camps close to collection patches. If transporting resources is generally characterized by rapid decreases in energetic return rates (Figures 11.7 & 11.8), such strategies should only be employed when the locations of other critical resources [e.g. water, firewood or other collected foods, sensu Binford (1980)] require a forager to consider the costs and benefits of transporting several resources to the same place. Given an array of resources in the diet, the processing characteristics of each may influence which resources are most effectively exploited through strategies of logistic procurement and long distance transport, and which are more likely to affect the locations of residential camps.

The transport decay functions calculated for pinyon and pickleweed illustrate how differences in processing characteristics might affect camp location. A forager's expected return rate is high for a camp located at either patch. But if camp is located 15 km away (a round-trip distance of 10 h or 30 km), the estimated net return rates for collecting, field processing and transporting loads of pinyon and pickleweed are approximately 3000 and 190 cal h^{-1}, respectively (Figures 11.7 & 11.8). A resource like pinyon, considered a "high-ranked" plant resource, yields relatively high caloric returns, even with extensive field processing and transport time. Alternatively, small seed resources and possibly many "low-ranked" resources, may not be very attractive for transportation (Jones & Madsen, 1989). By reducing the transport time of this kind of resource when it becomes an important component of the diet, the expected costs of field processing and transporting it could be minimized, and may increase the caloric return rate for the entire set of resources exploited.

Studies which detail the field processing and transport of plant resources among modern hunter–gatherer populations are needed to test this model. A number of ethnographic and actualistic studies have already employed the methodology needed to document these relationships. Researchers collected cost/benefit data, recorded procurement and processing times and calculated the caloric benefits of resources after processing (e.g. O'Connell & Hawkes, 1981; Hawkes et al., 1982, 1989; Simms, 1987; Bleige Bird et al., 1995; Jones & Madsen, 1991). If resources were also weighed and sampled at timed intervals during processing to document changes in utility, expectations about variation in field processing strategies could easily be developed. Ethnographic situations that also include observations of foragers or farmers travelling to resource patches and transporting loads of resources back to camp would provide the opportunity to test those expectations (e.g. Hawkes et al., 1989; O'Connell & Hawkes, 1981; Sikkink, 1989).

One research project has recently been undertaken that includes calculating the costs and benefits of field processing several kinds of shellfish for transport among the Meriam of the Eastern Torres Strait Islands, Australia (Bird & Bird, in press). This study will determine if some individuals vary field processing strategies as expected with different transport distances and processing characteristics. Initial observations indicate that Meriam children are efficient foragers, and routinely vary the locations where shellfish are processed with differences in the size of shells and the relative ease or difficulty of removing them (Bleige Bird et al., 1995). More studies of this kind, documenting relationships between field processing and transport in a variety of environments and foraging contexts, would provide insights about the types of resources most likely to be processed prior to transport and the situations in which archaeologists should expect field processing behaviour to strongly influence assemblage composition.

Acknowledgements

This research was supported in part by a University of Utah graduate research fellowship. An earlier version of this paper was presented in 1990 at the 22nd Great Basin Anthropological Conference. We wish to thank Robert Bettinger, Doug Edwards, Robert Elston, Kristen Hawkes, Penny Henriksen, Steve Josephson, Ken Juell, David Madsen, Sarah Mason, James O'Connell, John Parkington, Alan Rogers, Andrew Ugan and David Zeanah for comments on earlier drafts. We extend a special thanks to Robert Bettinger, Kristen Hawkes and David Madsen for comments that forced us to consider the likely costs associated with drying resources at procurement locations. We thank Professor Deloy G. Hendricks, Utah State University Department of Nutrition and Food Sciences, for the nutritional analyses of pickleweed seeds and pine nuts, and Ronald M. Lanner for an enlightening discussion of pinyon morphology. We also thank Lauri Travis for assisting in several pickleweed collection experiments, Margaret Trepl for assistance in weighing, measuring and recording pinyon data, and Jennifer Graves for editorial assistance. We are especially grateful to James O'Connell for numerous discussions, critiques and commentaries. Finally, we thank Aaron and Caprielle Barlow for their assistance and remarkable patience during numerous "lost in the desert tours" to these and other resource patches.

Notes

1. Robert Bettinger (personal communication) has suggested Chamberlin or his informants incorrectly identified pickleweed (*Allenrolfea occidentalis*) as "brittlewort or samphire (*Salicornia herbacea*)", or perhaps included it in descriptions of aboriginal use of the latter plant (Chamberlin, 1911:340, 380). Although the plants are morphologically distinct, *Salicornia* is found today in very wet locations on salt flats, sometimes within pickleweed communities.

2. Equations 11.1 and 11.2 are less complex than those presented in Metcalfe & Barlow (1992) because it is not necessary to deal with multiple components with varying utility; only one component in each plant was modelled as having utility. However, for those interested in using plant utility indices to better understand prehistoric behaviour, we caution that these simpler equations focus attention on the utility of a load of the resource rather than the individual components of plants. As with animals, some plant resources have several edible components which may vary in utility to the consumer. Archaeologists must use these equations thoughtfully because it is variation in the types and frequencies of the waste components that is likely to be most informative for

interpreting prehistoric land use patterns (Metcalfe & Barlow, 1992:341, 352–353). In this study, we qualitatively characterized the types and frequencies of waste components after each stage of processing. Quantitative estimates would have allowed a more precise description of the proportions of waste components associated with each processing stage.

3. We calculated the volumes of 18 used burden baskets from museum collections and catalog descriptions, and employed weight per unit volume measures to estimate the weights of basketloads of pinyon and pickleweed (Barlow, Henriksen & Metcalfe, 1993). The estimated weights of most basketloads range from 3 to 20 kg, and the average basket could hold approximately 6 kg of unprocessed cones, or 18 kg of processed nuts or seeds if filled to the rim.

4. Originally we planned to take a sample of the material resulting from each stage of each experiment and manually separate the seed from the inedible components to determine the resulting increase in utility. This procedure was extremely time consuming; each sample required between 10 and 20 h to process. Consequently, it was necessary to find a proximate, simpler measure of changes in β. Increases in the w/v density of the resource with successive processing stages provided such a measure. Nine samples, collected at various points in the processing experiments, were manually separated to directly determine the proportion of seed to inedible parts, and their weight and volume were measured (Table 11.11). The relationship between β and density is well described by a second order polynomial regression (β = 0.15539 − 0.59691 (density) + 3.8107 (density)^2: r^2 = 0.92). Density was measured for each sample of each experiment and converted to $β_j$ using this regression.

5. The graphic technique illustrated in Figures 11.5 and 11.6 is only precise when the costs associated with (a) travelling to the patch, (b) transporting the load home, and (c) procurement and field processing are equal. It also requires that further processing for consumption at the central place has no cost. The graphic technique illustrated in Figures 11.7 and 11.8 provides estimates when those costs vary, although we have not attempted to include variable costs for processing at home.

6. The reported sizes of winnowing trays in several ethnographic collections suggest women often winnowed large batches of small seed resources. Generally, trays used to winnow seeds were shallow, closely woven circular or fan-shaped baskets measuring approximately 35 to 50 cm across by 8 to 12 cm deep (Bowers Museum, 1977; Cahodas, 1979; Fowler & Matley, 1979; Bedford, 1980; Fowler, 1989). Two tightly woven Paiute winnowing trays were located in the Utah Museum of Natural History collections. These were fan-shaped and measured 41 × 38 × 7 cm, and 51 × 48.5 × 11.5 cm. The volumes were approximately 4 and 9.5 l, respectively, calculated by filling them to the rim with Styrofoam "peanuts" and measuring the volume of the "peanuts." Even the smaller tray would allow a forager to winnow more than the 0.25 l samples we employed in the experiments. To assess how an increase in the size of winnowing samples would affect estimated return rates, we recalculated these values using half the volume of the smallest tray measured (2 l). The weight of each 0.25 l subsample and the time required to collect and process it (Table 11.7) were multiplied by 8. The weights of samples at subsequent winnowing intervals were also multiplied by 8. These changes increased the sample weights (w_j) and times (t_j) for each winnowing stage. We made the assumption that the volume increase would not decrease processing efficiency, and did not change the w/v density or utility (y_j) of samples.

References

Aikens, C. M.
1970   *Hogup Cave.* University of Utah Anthropological Papers 93, Utah.

Aikens, C. M., D. B. Madsen
1986   Prehistory of the eastern area. In (W. L. D'Azevedo, Ed.) *Handbook of North American Indians Vol 11: The Great Basin,* Washington: Smithsonian Institution, pp. 149–160.

Antevs, E.
1948   Climate changes and pre-white man. In *Bulletin of the University of Utah Vol 38, No 20: The Great Basin with Emphasis on Glacial and Postglacial Times,* Salt Lake City: University of Utah, pp. 168–191.

Barlow, K. R., P. R. Henriksen, D. Metcalfe
1993   Estimating load size in the Great Basin: data from conical burden baskets. *Utah Archaeology* 1993, 27–37.

Bedford, C. P.
1980   *Western North American Indian baskets from the collection of Clay P. Bedford.* San Francisco, CA: California Academy of Sciences.

Bettinger, R. L.
1993   Doing Great Basin archaeology recently: coping with variability. *Journal of Archaeological Research* 1, 43–66.

Bettinger, R. L., M. A. Baumhoff
1983   Return rates and intensity of resource use in Numic and Prenumic adaptive strategies. *American Antiquity* 48, 830–834.

Binford, L. R.
1980   Willow smoke and dogs' tails: hunter-gatherer settlement systems and archaeological site formation. *American Antiquity* 45, 4–20.

Bird, D. W., R. L. Bird
In press   Meriam intertidal gathering strategies: shellfish processing and transport. *Journal of Archaeological Science.*

Bleige Bird, R. B., D. W. Bird, J. M. Beaton
1995   Children and traditional subsistence on Mer (Murray Island), Torres Strait. Canberra: Australian Aboriginal Studies, pp. 2–7.

Blurton Jones, N. G., R. M. Sibley
1978   Testing adaptiveness of culturally determined behavior: do Bushman women maximize their reproductive success by spacing births widely and foraging seldom? In (N. G. Blurton Jones & V. Reynolds, Eds.) *Society for the Study of Human Biology Symposium 18: Human Behavior and Adaptation,* London: Taylor & Francis, pp. 135–157.

Bowers Museum
1977   *Indian Basketry of Western North America.* Los Angeles: Brooke House.

Cahodas, M.
1979   *Degikup: Washoe Fancy Basketry 1890–1935.* Vancouver: University of British Columbia Fine Arts Gallery.

Cane, S.
1989   Australian Aboriginal seed grinding and its archaeological record: a case study from the Western Desert. In (D. R. Harris & G. C. Hillman, Eds.) *Foraging and Farming: The Evolution of Plant Exploitation.* London: Unwin Hyman, pp. 99–119.

Chamberlin, R. V.
1911 The ethno-botany of the Gosiute Indians of Utah. *American Anthropological Association Memoirs* 2, 331–405.

Coville, F. V.
1892 The Panamint Indians of California. *American Anthropologist* 5, 351–361.

Currey, D. R., S. R. James
1982 Paleoenvironments of the northeastern Great Basin and Northeastern rim region: a review of geological and biological evidence. In (D. B. Madsen, J. F. O'Connell, Eds.) *Man and Environment in the Great Basin.* SAA Papers No. 2, pp. 27–52.

Dennell, R. W.
1972 The interpretation of plant remains: Bulgaria. In (E. S. Higgs, Ed.) *Papers in Economic Prehistory.* Cambridge: Cambridge University Press, pp. 149–159.

Dutcher, B. H.
1893 Pinon gathering among the Panamint Indians. *American Anthropologist* 6, 376–380.

Farjon, A.
1984 *Pines: Drawings and Descriptions of the Genus Pinus.* Leiden: E. J. Brill.

Farris, G. J.
1980 A reassessment of the nutritional value of *Pinus monophylla*. *Journal of California and Great Basin Anthropology* 2, 132–136.

Ford, R. I.
1988 Commentary: little things mean a lot—quantification and qualification in paleoethnobotany. In (C. A. Hastorf, V. S. Popper, Eds.) *Current Paleoethnobotany: Analytical Methods and Cultural Interpretations of Archaeological Plant Remains.* Chicago, IL: University of Chicago Press, pp. 215–222.

Fowler, C. S., Ed.
1989 *Willard Z. Park's Ethnographic Notes on the Northern Paiute of Western Nevada, 1933–1944.* University of Utah Anthropological Papers 114, Utah.

Fowler, D. D., J. F. Matley
1979 *Material Culture of the Numa: The John Wesley Powell Collection 1867–1880.* Washington, D.C.: Smithsonian Institution.

Fry, G. F.
1970 *Prehistoric Human Ecology in Utah: Based on the Analysis of Coprolites.* Ph.D. dissertation (unpublished), Department of Anthropology, University of Utah, Salt Lake City, Utah.
1976 *Analysis of Prehistoric Coprolites from Utah.* University of Utah Anthropological Papers 97, Utah.

Grayson, D. K.
1993 *The Desert's Past.* Washington, D.C.: Smithsonian Institution Press.

Hall, H. J.
1988 Preliminary analysis of human paleofeces from Danger and Floating Island Caves. Paper presented at the 21st Great Basin Archaeological Conference, Park City.

Hally, D. J.
1981 Plant preservation and the content of paleobotanical samples: a case study. *American Antiquity* 46, 732–742.

Harlan, J. R.
1989 Wild-grass seed harvesting in the Sahara and Sub-Sahara of Africa. In (D. R. Harris, G. C. Hillman, Eds.) *Foraging and Farming: The Evolution of Plant Exploitation.* London: Unwin Hyman Ltd., pp. 79–98.

Harper, K. T., G. M. Alder
1970 The macroscopic plant remains of the deposits of Hogup Cave, Utah, and their paleoclimatic implications. In (C. M. Aikens) *Hogup Cave,* University of Utah Anthropological Papers 93, Utah, pp. 215–240.

Hastorf, C. A.
1988 The use of paleoethnobotanical data in prehistoric studies of crop production, processing, and consumption. In (C. A. Hastorf, V. S. Popper, Eds.) *Current Paleoethnobotany: Analytical Methods and Cultural Interpretations of Archaeological Plant Remains.* Chicago, IL: University of Chicago Press. pp. 119–144.

Hastorf, C. A., V. S. Popper, Eds.
1988 *Current Paleoethnobotany: Analytical Methods and Cultural Interpretations of Archaeological Plant Remains.* Chicago, IL: University of Chicago Press.

Hawkes, K.
1987 How much food do foragers need? In (M. Harris, E. B. Ross, Eds.) *Food and Evolution: Toward a Theory of Human Food Habits.* Philadelphia, PA: Temple University Press, pp. 341–355.

Hawkes, K., K. Hill, J. F. O'Connell
1982 Why hunters gather: optimal foraging and the Ache of eastern Paraguay. *American Ethnologist* 9, 379–398.

Hawkes, K., J. F. O'Connell, N. G. Blurton Jones
1989 Hardworking Hadza Grandmothers. In (V. Standen, R. A. Foley, Eds.) *Comparative Socioecology: The Behavioral Ecology of Humans and Other Mammals.* Oxford: Blackwell Scientific Publications, pp. 341–366.

Hillman, G.
1981 Reconstructing crop husbandry practices from charred remains of crops. In (R. Mercer, Ed.) *Farming Practice in British Prehistory.* Edinburgh: Edinburgh University Press, pp. 123–162.
1984 Interpretation of achaeological plant remains: the application of ethnographic models from Turkey. In (W. Van Zeist, W. A. Casparie, Eds.) *Plants and Ancient Man: Studies in Palaeoethnobotany.* Netherlands: A. A. Balkema Publishers, pp. 1–41.

Hillman, G. C., S. M. Colledge, D. R. Harris
1989 Plant-food economy during the Epipalaeolithic period at Tell Abu Hureyra, Syria: diet, diversity, seasonality and modes of exploitation. In (D. R. Harris, G. C. Hillman, Eds.) *Foraging and Farming: The Evolution of Plant Exploitation.* Oxford: Unwin Hyman Ltd., pp. 79–98.

Jennings, J. D.
1957 *Danger Cave.* University of Utah Anthropological Papers 27, Utah.
1978 *Prehistory of Utah and the Eastern Great Basin.* University of Utah Anthropological Papers 98, Utah.

Jones, G. E. M.
1984 Interpretation of archaeological plant remains: ethnographic models from Greece. In (W. Van Zeist, W. A. Casparie, Eds.) *Plants and Ancient Man: Studies in Palaeoethnobotany.* Netherlands: A. A. Balkema Publishers, pp. 43–61.

Jones, K. T., D. B. Madsen
1991 Further experiments in native food procurement. *Utah Archaeology* 1991, 68–77.
1989 Calculating the cost of resource transportation: A Great Basin example. *Current Anthropology* 30, 529–534.

Lanner, R. M.
1981 *The Piñon Pine: A Natural and Cultural History.* Reno: University of Nevada Press.

Lee, R.
1969 !Kung bushman subsistence: an input-output analysis. In (A. P. Vayda, Ed.) *Environment and Cultural Behavior: Ecological Studies in Anthropology.* Garden City: the Natural History Press, pp. 47–79.

Little, E. L., Jr.
1938 *Food Analyses of Pinon Nuts (A Compilation of Existing Data).* Southwestern Forest and Range Experiment Station Research Notes 48, Tucson.

Madsen, D. B.
1982 Get it where the gettin's good: a variable model of Great Basin susbistence and settlement based on data from the Eastern Great Basin. In (D. B. Madsen, J. F. O'Connell, Eds.) *Man and Environment in the Great Basin.* Washington D.C.: SAA Papers No. 2, pp. 207–226.

Madsen, D. B., D. R. Currey
1979 Late Quaternary glacial and vegetation changes, Little Cottonwood Canyon Area, Wasatch Mountains, Utah. *Quaternary Research* 12, 254–270.

Madsen, D. B., D. R. Rhode
1990 Early Holocene Pinyon (*Pinus monophylla*) in the Northeastern Great Basin. *Quaternary Research* 33, 94–102.

Mehringer, P. J., Jr.
1986 Prehistoric Environments. In (W. L. D'Azevedo, Ed.) *Handbook of North American Indians Vol 11: The Great Basin.* Washington, D.C.: Smithsonian Institution, pp. 31–50.

Metcalfe, D.
1989 *A General Cost/Benefit Model of the Tradeoff Between Transport and Field Processing.* Paper presented at the 54th annual meeting of the Society for American Archaeology, Atlanta, GA.

Metcalfe, D., K. R. Barlow
1992 A Model for Exploring the Optimal Tradeoff Between Field Processing and Transport. *American Anthropologist* 94, 340–356.

Metcalfe, D., K. M. Heath
1990 Microrefuse and site structure: the hearths and floors of the Heartbreak Hotel. *American Antiquity* 55, 781–796.

Miller, W. R.
1972 *Newe Natekwinappeh: Shoshoni Stories and Dictionary.* University of Utah Anthropological Papers 94, Utah.

O'Connell, J. F.
1975 *The Prehistory of Surprise Valley.* Ramona: Ballena Press.

O'Connell, J. F., K. Hawkes
1981 Alyawara plant use and optimal foraging theory. In (E. Smith, B. Winterhalder, Eds.) *Hunter-Gatherer Foraging Strategies.* Chicago, IL: University of Chicago Press, pp. 99–125.

Palmer, E.
1878 Plants used by the Indians of the United States. *American Naturalist* 12, 592–606, 646–655.

Pandolf, K. B., B. Givoni, R. F. Goldman
1977 Predicting energy expenditure with loads while standing or walking very slowly. *Journal of Applied Physiology* 43, 577–581.

Pearsall, D.
1988 Interpreting the meaning of macroremain abundance: the impact of source and context. In (C. A. Hastorf, V. S. Popper, Eds.) *Current Paleoethnobotany: Analytical Methods and Cultural Interpretations of Archaeological Plant Remains.* Chicago, IL: University of Chicago Press, pp. 97–118.
1989 *Paleoethnobotany: A Handbook of Procedures.* San Diego: Academic Press.

Rhode, D.
1990 On transportation costs of Great Basin resources: an assessment of the Jones-Madsen Model. *Current Anthropology* 31, 413–419.

Sikkink, L.
1988 From field to house: ethnoarchaeology and ethnobotany of harvest and crop-processing in Andean peasant households. MSc thesis, Department of Anthropology, University of Minnesota, MN.
1989 Traditional crop-processing in central Andean households: an ethnoarchaeological perspective. In (V. Vitzthum, Ed.) *Multidisciplinary Studies in Andean Anthropology, Michigan Discussions in Anthropology Fall 1988 Volume 8.* Ann Arbor, MI: University of Michigan, pp. 65–86.

Simms, S. R.
1987 *Behavioral Ecology and Hunter-Gatherer Foraging.* Oxford: BAR International Series 381.

Sporne, K. R.
1965 *The Morphology of Gymnosperms: the Structure and Evolution of Primitive Seed-plants.* London: Hutchinson University Library.

Steward, J. H.
1938 *Basin-Plateau Aboriginal Sociopolitical Groups.* Bureau of American Ethnology Bulletin 120.

Stewart, O. C.
1941 *Culture Element Distributions: XIV Northern Paiute.* University of California Anthropological Records 4, pp. 360–446.

Thompson, R. S.
1990 Late Quaternary vegetation and climate in the Great Basin. In (J. L. Betancourt, T. R. Van Devender, P. S. Martin, Eds.) *Packrat Middens: The Last 40,000 Years of Biotic Change.* Tuscon, AZ: University of Arizona Press, pp. 200–239.

Wheat, M. M.
1967 *Survival Arts of the Primitive Paiutes.* Reno, NV: University of Nevada Press.

Wright, K. I.
1994 Ground-stone tools and hunter-gatherer subsistence in southwest Asia: implications for the transition to farming. *American Antiquity* 59, 238–263.

Prey Spatial Structure and Behavior Affect Archaeological Tests of Optimal Foraging Models

Examples from the Emeryville Shellmound Vertebrate Fauna

Jack M. Broughton

In the last decade, optimal foraging theory has increasingly been used as a framework to analyze prehistoric *resource depression*, or declines in the capture rates of prey species that result from the activities of foragers (Charnov et al. 1976). These studies, based on analyses of archaeological vertebrate data, reflect a growing interest in understanding the details of how non-industrial peoples structured the environments in which they lived and many far-reaching implications for human prehistory have emerged from them (Alvard 1994; Steadman 1995; Redman 1999; Cannon 2000; Smith and Wishnie 2000; Grayson 2001; Kay and Simmons 2001).

In the late Holocene of California, for example, faunal evidence for resource depression and the associated declines in foraging efficiency have been causally linked to technological changes, increases in human violence and warfare, reduced stature, declining health and the emergence of social hierarchies (e.g. Hildebrandt and Jones 1992; Raab et al. 1995; Broughton and O'Connell 1999). Implications for modern wilderness conservation have also been derived from the evidence suggesting that late prehistoric faunal landscapes in this setting were fundamentally anthropogenic (Broughton 2001a; Grayson 2001).

Faunal resource depression and the resulting declines in foraging returns may be linked to many other important behavioral transitions in human prehistory. These range from late Pleistocene broad-spectrum transitions and the origins and adoption of agriculture (Hawkes and O'Connell 1992; Madsen and Simms 1998; Stiner et al. 2000) to changes in foraging behavior and material culture that occurred during hominid evolution (e.g. Marean and Assefa 1999).

So far, quantitative trends in relative taxonomic abundances and the age composition of exploited prey have been the primary measures of prehistoric resource depression. Based on implications of the fine-grained prey model of foraging theory (hereafter, simply the "prey model"), quantitative indices summarizing the relative abundance of "high-return" or large species in faunal samples have been used to measure hunting "efficiency", or the overall caloric return rate derived from hunting. All other things being equal, such indices are commonly expected to *decline through time* in the context of increasing harvest rates and prey depression and in many cases they appear to do just that (e.g. Grayson 2001; Hildebrandt and Jones 1992; Broughton 1994a, 1994b, 1997, 1999; Janetski 1997; Butler 2000, 2001; Cannon 2000; Porcasi et al. 2000; Nagaoka 2000, 2001, this volume). But, in some settings, such trends have not been revealed where they are expected, leading some to suggest that the underlying foraging models may be flawed (McGuire and Hildebrandt 1994; Hildebrant and McGuire 2000).

The ontogenetic age structure of harvested prey has also been used as a measure of human predation pressure to corroborate evidence for resource depression based on relative abundance indices. The standard prediction is that resource depression will drive a decline through time in the average age of a prey species and such trends have been empirically documented in many different prehistoric contexts as well (e.g. Anderson 1981; Klein and Cruz-Uribe 1984; Broughton 1997; Stiner et al. 2000; Butler 2001). But, again, increases in the average age of prey species, or no detectable changes at all, have appeared in several recent studies where all other data suggest the species were substantially depressed by intensive human harvesting (e.g. Broughton 1999; Butler 2001).

In this paper, I demonstrate with the Emeryville Shellmound faunal sequence how resource depression can produce different trends in large game relative abundances and prey age structure. This analysis suggests that variation in the spatial distribution and behavior of prey species are the important factors that influence the different trends and hence can be critical considerations in applying optimal foraging models to the archaeological record of ancient hunting behavior.

*Originally published in *World Archaeology* 34(2002):60–83

Archaeological Vertebrate Measures of Resource Depression and Foraging Efficiency

Relative Abundance Indices

In the late 1970s, Bayham (1979, 1982) pioneered a unique approach that used foraging theory to suggest how patterns in the relative abundances of different archaeological prey species could be used as measures of ancient hunting efficiency. The foundation for this approach came from two important papers in the ecological literature. The first was Wilson's (1976) application of the prey model to interpret patterns in insect prey derived from the stomach contents of birds. Wilson argued that the abundance of high- relative to low-ranked insect prey found in the stomach contents of birds could be used to measure the quality of their foraging environments. Since the prey types that yield the highest post-encounter return rates should always be pursued, their relative abundances measure both how frequently they were encountered and the amount of energy "available in the environment". Second, Griffiths (1975) contributed a detailed rationale for using prey body size as a proxy, positive measure of the "post-encounter return rates" or ranks of prey species, an argument that is now supported by many empirical studies conducted among living foragers, both human and non-human (Broughton 1999; but see Madsen and Schmitt 1998; Lindstrom 1996 for interesting exceptions).

Adapting these arguments to the archaeological context, Bayham suggested that "the relative abundance of small species in the diet should not only be a function of, but an indirect index to the availability of larger species or the higher ranking ones" (1979:229). Ratios incorporating the numbers of archaeologically identified large and small prey species could thus be used as a measure of foraging efficiency: the more large, high-ranked prey in an assemblage, the higher the foraging efficiency.

Bayham discussed a variety of factors that could cause the abundances of high-ranking prey and, hence, foraging efficiency to decline and suggested that resource depression was likely one of the more ubiquitous among them. He also reasoned that the higher-ranked prey would exhibit depression prior to lower-ranking prey because the former should always be pursued upon encounter. Large prey should also be more sensitive to increased mortality because they typically have slower reproductive rates. Declines in large prey abundances could thus signal resource depression, if other causes, such as environmental change, could be ruled out (Bayham 1982:63). This approach was applied to track variation in foraging efficiency through the Archaic and Hohokam periods in the American Southwest (Bayham 1979, 1982; Szuter and Bayham 1989).

It is important to note that this research strategy, as applied in either an ecological or archaeological context, does not actually test predictions of the prey model; it assumes them to be qualitatively true so as to "use the behavior of the consumer" (Wilson 1976:960) to measure inductively relative differences in encounter rates and foraging returns of foragers operating in different environments.

Recent archaeological applications, however, have adapted the Wilson-Bayham logic to a deductive research strategy by using regional archaeological data to leverage *predictions* about how foraging efficiency and relative abundance indices should have varied in the past. Quantitative analyses of archaeological vertebrate data are then conducted to decipher these trends. Insofar as the critical assumptions of the prey model apply, these analyses thus constitute tests of its implications as they apply to prehistoric human foraging behavior.

One of those critical assumptions involves the spatial distribution of prey types and may be unrealistic in many archaeological contexts. This *fine-grained search assumption* stipulates that different prey types are searched for simultaneously and that the chance of encountering any prey type is independent of previous encounters with it or any other type. In other words, the model assumes prey types are encountered in a fine-grained manner. This assumption allows search time to be detached from individual prey types and assigned to the set of resources as a whole; it is also required for the relative abundances of high-ranked prey to be a valid measure of foraging efficiency. If, however, different prey types are spatially clumped across the environment surrounding a site, variation in the overall net caloric returns from those clumps or patches should dictate the extent to which they are used; patch use decisions, not just prey ranks, can thus determine prey choice (Smith 1991:206).

An obvious strategy to deal with this constraint is to examine changes in differently ranked prey types within single "patches" or "hunt types". These are defined as groups of prey taxa that were likely encountered in a fine-grained manner; that is, they are found in the same habitat types and were pursued and captured with similar technologies. Resource depression and declining foraging efficiency should still be signaled by declines in the relative abundances of high-ranked prey *within each patch*. In this approach, the fine-grained search assumption is analytically maintained.

The depression of prey within resource patches directly adjacent to a residential base has implications for changes in patch-use strategies in the wider environment surrounding a locality. Researchers investigating the effects of prehistoric resource depression on differential time allocation to resource patches have drawn on two separate models: Charnov's (1976) marginal value theorem and Orians and Pearson's (1979) central-place forager patch choice model (Broughton 1999; Cannon 2000; Nagaoka 2000). Both predict that more foraging time will be devoted to higher return patches but only the Orians and Pearson model explicitly considers the round-trip travel time between the resource patch and the central place. It is thus a more realistic model for the hunting behavior most commonly addressed in resource depression studies (Cannon 2000). Both models, however, can and have been used to derive the following prediction: as once high-return patches located closer to home become depleted, more use should be made of distant, less-depleted patches located farther away if such patches are available. Since the more distant patches were previously underutilized, they should contain higher densities of high-ranked prey. Resource depression and overall reduced foraging efficiencies can thus be signaled by temporal *increases* in the abundances of high-ranked prey. Indeed, this phenomenon has been empirically documented in several ethnographic settings (see Hames 1989; Hames and Vickers 1982; Vickers 1980, 1988). Evaluating archaeologically whether or not increasing abundances of large prey are reflecting distant patch use will, however, require analyses of additional faunal data, as I discuss below.

Considered together, the prey and patch models suggest that the depression of high-ranked prey within local resource patches should lead to the selection of more abundant but lower-ranked prey species in those patches and/or increased foraging effort devoted to less-depleted patches located farther away from the central place. Both options entail substantial declines in foraging efficiency. The transport models developed by Metcalfe and Barlow (Metcalfe and Barlow 1992; Barlow and Metcalfe 1996), for instance, indicate that even modest increases in round-trip travel times to resource patches for central-place foragers rapidly diminish the return rates of resource types exploited in those patches.

Prey Age Composition

Patterns in the age composition of archaeological prey species are one of the most frequently used (Broughton 1995, 1997, 1999; Stiner et al. 2000; Butler 2001) or suggested (Grayson and Cannon 1999; Cannon 2000; Wolverton 2001) means of corroborating evidence of resource depression derived from relative abundance indices. And, for several reasons, resource depression is commonly argued to cause declines in the mean and maximum ages of individuals in exploited populations. However, resource depression can also lead to *increases* in the mean age of harvested prey in certain contexts as well. Fortunately, these very different trends follow predictably from differences in the behavior and spatial characteristics of the particular prey species involved. This allows prey-specific predictions to be made regarding trends in age composition in analyses of resource depression.

A variety of factors can cause increasing harvest rates to produce declines in the mean and maximum ages of individuals in the harvested populations. First, for species that grow throughout their lives, older individuals can be substantially larger than younger ones and would likely represent higher-return prey items. In such cases, human predation should be biased towards older individuals, and increased harvest rates on species with indeterminate growth may thus be reflected by reductions in the proportion of adults in a population. Second, increasing mortality rates, even if they are unbiased with respect to age, can also produce this effect by reducing intra-specific competition and increasing recruitment rates (Caughley 1966, 1977). Third, increasing mortality rates may reduce the probability that individuals survive to enter older age classes so that average lifespans decrease and survivorship curves become steeper (Taber et al. 1982; Lyman 1987; Wolverton 2001). Finally, increased adult mortality rates may lead to the alteration of life-history schedules and the selection of phenotypes that mature early at the expense of later somatic growth (Charnov 1993).

While the reasons that suggest predation pressure should cause declines in mean age are sound in an abstract sense—making this the default prediction of resource depression—there are several factors that may overshadow their importance in certain zooarchaeological applications. In particular, these arguments do not account for the reality that archaeological faunas often result from foragers that encounter different-aged individuals within species in a coarse-grained manner. That is, some prey species occur in clumps or patches that are disproportionately comprised of juvenile or adult individuals. Differential use of such patches would thus have an important effect on the age composition of exploited prey.

The standard prediction also takes no accounting of the potential for prey "behavioral" and "microhabitat" depression. These phenomena refer collectively to behavioral responses prey species can make in response to increasing predation risk. Such responses may involve flocking behavior, greater alertness or shifting the time and duration of particular activities (e.g. foraging, courtship) that place prey at high risks of predation. Prey species may also shift positions spatially as a response to an increased presence of predators such as, for example, by moving into denser vegetation cover, up steeper slopes or higher into the forest canopy (Charnov et al. 1976). Since individuals of different age and sex experience different trade-offs involving the use of habitat types that vary in predation risk and forage quality, there can be substantial variation in the degree to which individuals are sensitive to behavioral depression (e.g. Main and Coblentz 1996).

Pinnipeds and colonial waterbirds provide the best examples of taxa that are characterized both by patchy age-related spatial structure and the propensity to respond behaviorally to human hunting pressure. High-density, seasonally based, breeding colonies are typical components of the reproductive strategies of these animals. Such colonies are often quite sensitive to predation and even to the mere presence of predators. Sustained persecution will cause survivors to abandon colonies and form new ones in areas offering higher security—out of reach to human foragers (Boekelheide et al. 1990; Carney and Sydeman 1999; Gonzalez 1999). Regions that lose breeding colonies lose not only the highest-return patches for the species but the major local source of sub-adult animals. In such cases, resource depression would cause relative increases in the encounter rates with adults and an increase in the mean age of exploited individuals.

A similar phenomenon may even occur for species that do not form spatially concentrated breeding colonies. For example, in many polygynous ungulates, the sexes are spatially segregated with female-young groups occupying habitats that provide greater security from predators (Bowyer 1984; Miquelle

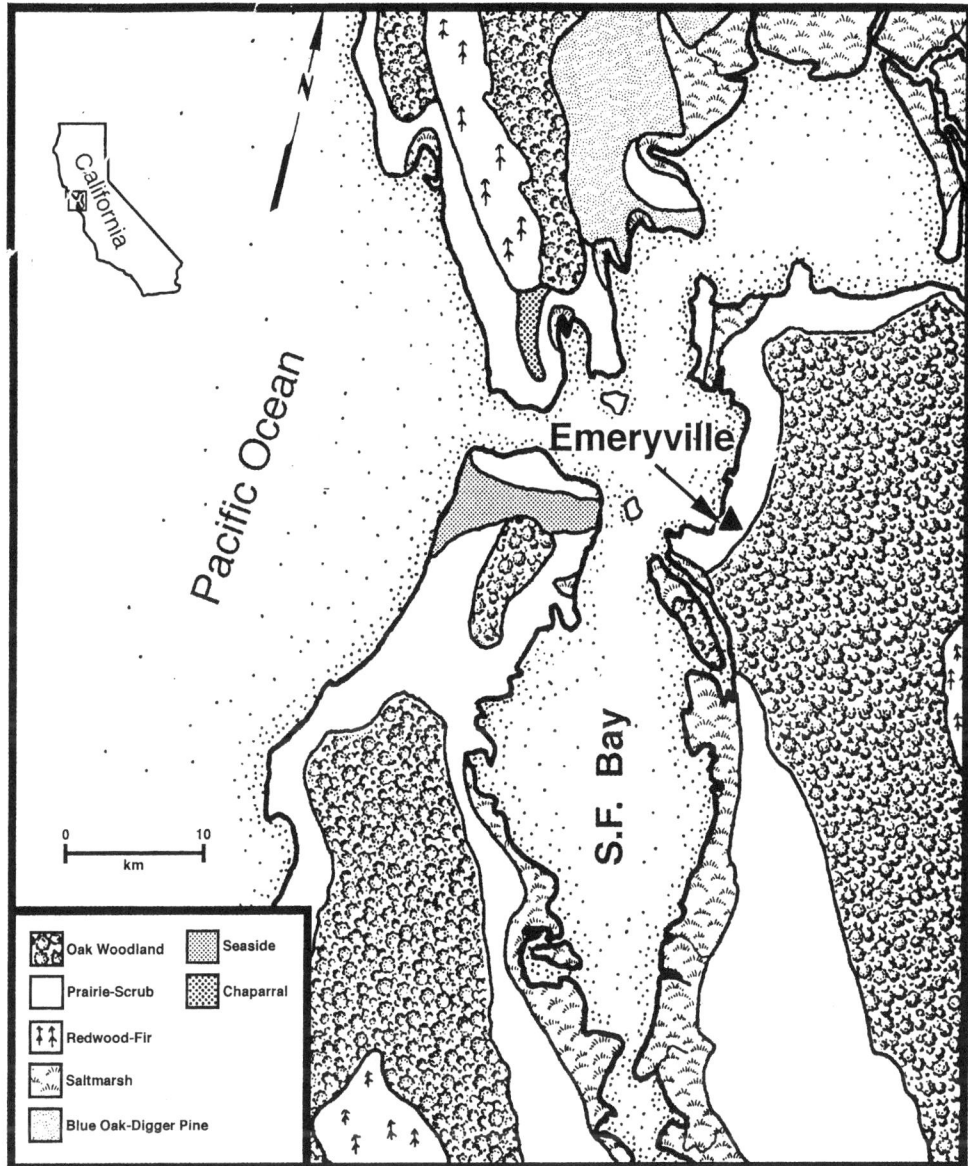

FIGURE 12.1. Map of the San Francisco Bay area indicating the location of the Emeryville Shellmound and historic period vegetation.

et al. 1992; Bowyer et al. 1996; Main and Coblentz 1996). Insofar as females with attendant young are more sensitive to predation risk or behavioral/microhabitat depression, areas subjected to elevated predator densities should be characterized by reductions in the relative frequencies of both females and sub-adult animals. For species with these characteristics, resource depression should again entail increases in the mean age of exploited animals.

Late Holocene Resource Depression in California: Evidence from the Emeryville Shellmound Fauna

In the California setting, regional late Holocene archaeological data document both expanding human population densities and increasing use of lower-return plant resources, namely acorns (e.g. Bettinger and Baumhoff 1981:499; Basgall 1987). These patterns have been read to suggest that overall foraging efficiencies were declining over this period of time. If so, evidence for resource depression and declining foraging efficiency, as described above, should be found among the vertebrate resources used by these foragers. The rich vertebrate record from the stratified Emeryville Shellmound provides an excellent test of this hypothesis since it reflects the activities of central-place human foragers using a single spot on the landscape for nearly 2000 years (Broughton 1999).

The Emeryville Shellmound was situated in a complex mosaic of terrestrial and aquatic habitat types on the east shore of San Francisco Bay before it was demolished in 1924 (Fig. 12.1 and Fig. 12.2). The mound itself was huge, measuring 100 × 300 m in area and extending to a depth of over 10 m. Prior to its

FIGURE 12.2. Demolition of the Emeryville Shellmound by steam shovel, 1924. (Photo courtesy of the Phoebe A. Hearst Museum of Anthropology.)

destruction, the mound was excavated on three occasions, each time in a different location. Most of the sediments were excavated stratigraphically and sieved with coarse-mesh screens; ten primary strata were revealed during this work. Fourteen radiocarbon dates range from ~2600 to 700 ^{14}C year BP and exhibit no stratigraphic inconsistencies. On average, each stratum took about 200 years to accumulate.

During this time, the eastern San Francisco Bay region appears to have witnessed significant human population growth, at least to judge from the increase through time in the number of dated site components (Broughton 1999). Thus the Emeryville region follows the more general California trend of expanding human population densities during the late Holocene.

The excavations at Emeryville provided seventeen faunal samples that could be placed into stratigraphic context. Collectively, about 24,000 fish, bird and mammal specimens, representing 102 different species, have been identified from these samples. The Emeryville deposits thus allow a fairly high-resolution, ordinal-scale analysis of change in human foraging behavior over much of the late Holocene. (For a full reporting of the mammal and fish data summarized below see Broughton 1995, 1999; for the birds see Broughton 2001b).

Relative Abundance Indices from the Emeryville Vertebrates
Trends from local patches
As indicated above, the Emeryville site is surrounded by a number of distinct habitats so different that vertebrate prey species would have been spatially distributed in a patchy or coarse-grained manner. In order to satisfy the prey model's fine-grained search assumption, relative abundance indices should be constructed so as to include only taxa that were derived from the same patches or hunt types. Several of the more obvious resource patches surrounding the Emeryville locality are the terrestrial mammals patch, the estuarine fishes patch, and the waterfowl (Anatidae) patch. These different patches each contain large and small-sized vertebrate prey types from which to derive relative abundance indices that can be analyzed separately for evidence of resource depression.

The largest taxa, derived from the terrestrial mammals, estuarine fishes and waterfowl (Anatidae) patches, are tule elk (*Cervus elaphus*), white sturgeon (*Acipenser transmontanus*), and geese (*Anser, Branta, Chen*), respectively. If resource depression occurred across the occupational history of Emeryville, the abundances of these large-sized prey should decline through time relative to the smaller prey types that occurred in their respective patches. Figures 12.3–12.5 show that this is precisely what occurred at Emeryville. None of these changes appear to be caused by changes in the estuarine or terrestrial paleoenvironments (see Broughton 1999 for a detailed discussion of this issue). These patterns are also fully consistent with the data Gifford (1916) collected in his early twentieth-century study of the Emeryville molluscs. Specifically, the surface-dwelling molluscs of largest size (California oyster, *Ostrea lurida*, and bay mussel, *Mytilus edulis*) decline significantly ($r_s = -.54$; $p < .05$) across the Emeryville strata relative to smaller species (Broughton 1999:71).

It is also important to emphasize that these resources occupied habitats that were, for the most part, confined to the region immediately surrounding the site. That is, sturgeon could have been found only within the San Francisco Bay estuary, geese would have been concentrated on the Bay itself or wetland areas on its flanks and tule elk were largely restricted to the marsh-

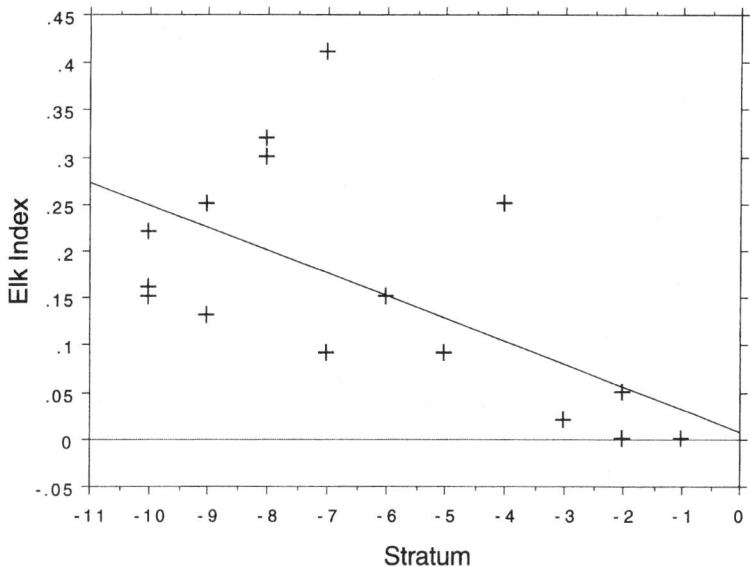

FIGURE 12.3. The distribution of the Elk Index ($\sum$NISP Elk/[$\sum$NISP Terrestrial Mammals]) by stratum at the Emeryville Shellmound ($r_s = -0.658$, $p < 0.01$). Does not include the potentially intrusive rodents and lagomorphs. A least-squares regression line indicates the direction of the trend.

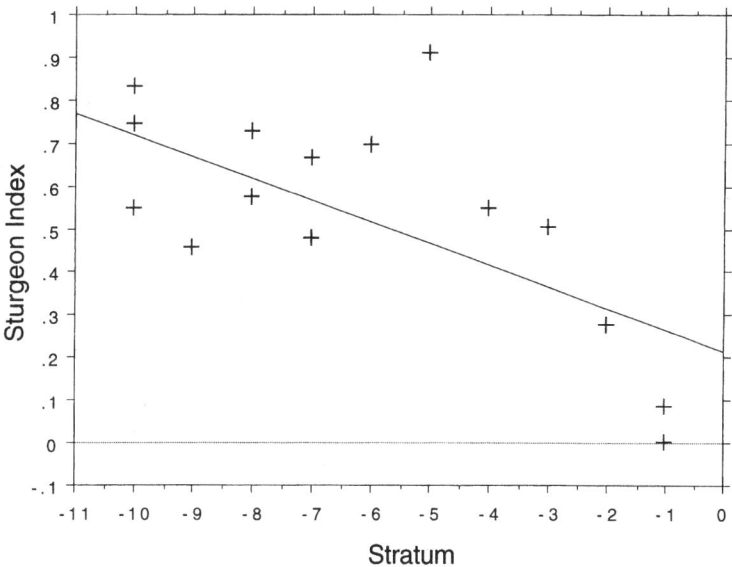

FIGURE 12.4. The distribution of the Sturgeon Index ($\sum$NISP Sturgeon/[$\sum$NISP Estuarine Fishes]) by stratum at the Emeryville Shellmound ($r_s = -0.49$, $p = 0.05$). A least-squares regression line indicates the direction of the trend.

lands and a narrow swath of grassland that surrounded the margin of the Bay (McCullough 1969; Broughton 1999). The molluscs, of course, are found only on the mudflats and the littoral zone of the estuary, just adjacent to the site. Travel times to and from these patches are thus not likely to have been important foraging costs. Collectively, these trends provide compelling evidence for resource depression in local patches over the period that Emeryville was occupied.

Increasing Use of Distant Patches?
But these are not the only changes in vertebrate utilization that occurred across the Emeryville sequence. Against this background of steady declines in large prey from local patches, the relative abundance of black-tailed deer (*Odocoileus hemionus*), the second largest prey type from the terrestrial mammals patch, first declines over the four lower strata of the site but then increases across the upper six strata (Fig. 12.6). This resurgence in

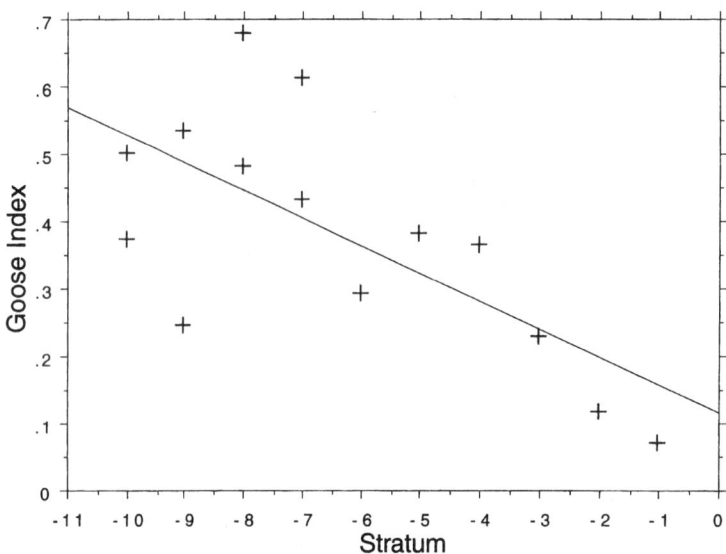

FIGURE 12.5. The distribution of the Goose Index ($\sum$NISP Medium + Large Geese/[$\sum$NISP Anatids]) by stratum at the Emeryville Shellmound ($r_s = -0.62$, $p = 0.02$). A least-squares regression line indicates the direction of the trend.

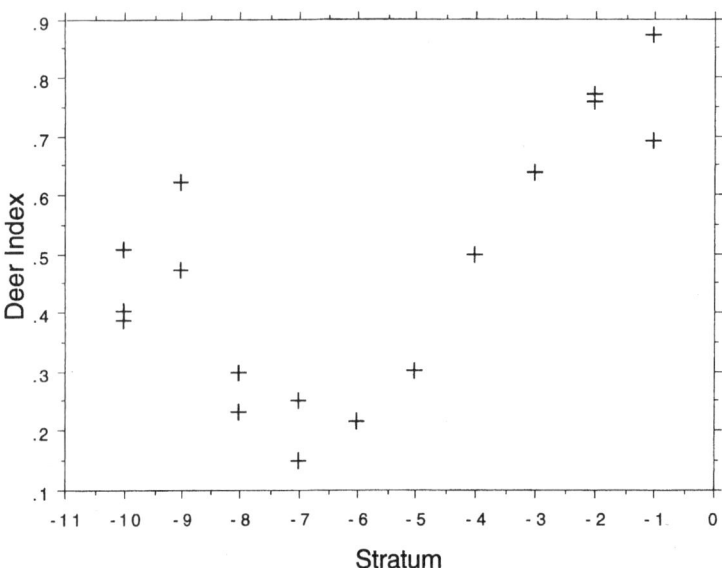

FIGURE 12.6. The distribution of the Deer Index ($\sum$NISP Deer/[$\sum$NISP Mammals]) by stratum at the Emeryville Shellmound. Does not include the potentially intrusive rodents and lagomorphs. The Deer Index declines across the lower four strata (strata ten through seven: $r_s = -0.721$, $p = 0.04$) but increases over the upper six strata (strata six through one: $r_s = 0.68$, $p = 0.01$).

deer abundance is striking: no matter what set of mammalian taxa deer are arrayed against, deer overwhelmingly dominates the mammalian fauna of the upper-most strata of the mound. Given that all signals from local patches suggest foraging returns were declining across the entire occupational history of the mound, could the resurgence in deer use be reflecting the increasing use of less-depleted deer patches located in the oak woodlands that extend far to the east of the site? Such distant patches might have existed for a variety of reasons, including, for instance, buffer zones (Hickerson 1965; Mech 1977; Martin and Szuter 1999).

That the upper-strata deer were derived from distant patches has received support from patterns in skeletal part representation in relation to the economic utility of elements (Broughton 1999). Recent theoretical and ethnoarchaeological research (O'Connell et al. 1990; Metcalfe and Barlow 1992; Bird and

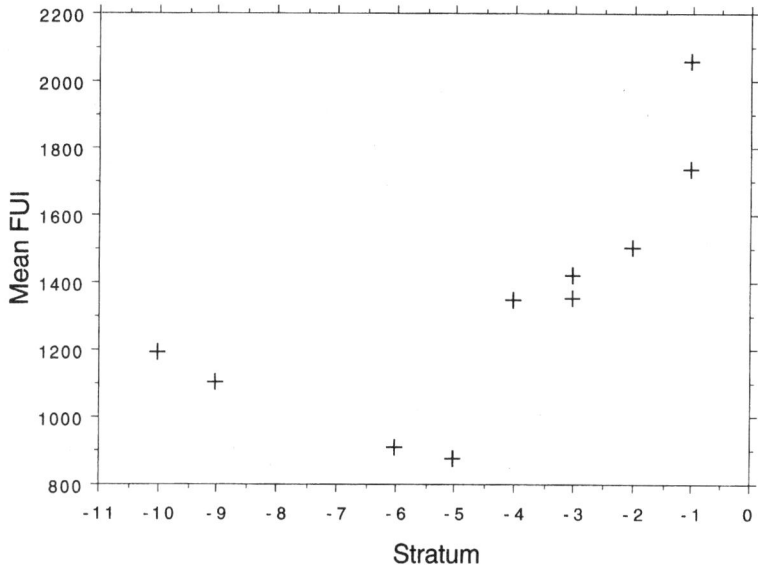

FIGURE 12.7. Mean utility of artiodactyl skeletal parts across the Emeryville strata. Samples showing a significant correlation between part density and part representation have been excluded (see Broughton 1999 for details). The Food Utility Index (FUI) is defined by Metcalfe and Jones (1998).

Bliege Bird 2000) suggests that transport cost is one of the important variables that determine the degree to which low-utility body parts (those parts with minimal amounts of associated nutrition) are returned from kill sites to base camps, and transport distance is one of the most important components of transport costs. Other things being equal, as the distance from kill site to home base increases, so should the degree of field processing to remove low-utility parts from the transported carcass. It follows that, if the resurgence in deer abundance documented across the upper strata at Emeryville is reflecting an ever-increasing use of distant, less-depleted deer patches, it should be associated with increasing relative frequencies of high-utility body parts, or an increase in the "mean utility" of represented parts. Figure 12.7 shows that this is the pattern evident in the Emeryville fauna: mean utility increases significantly across the upper six strata ($r_s = .96, p < .01$). Such a pattern was also revealed in an analysis of the anatomical part data based on the statistical method of maximum likelihood, although the trend could not be maintained when samples were pooled in order to produce acceptably narrow confidence intervals (Rogers and Broughton 2001).

Conclusions
The increase in black-tailed deer use has yet to be revealed at any other site or sub-region in the San Francisco Bay area—settings that all show steady declines through time in the relative abundances of both elk and deer (Broughton 1994b)—and may have been conditioned by the availability of, and distance to, less-depleted deer patches. Emeryville sits just west of an extensive oak woodland zone—prime deer habitat—that reaches uninterrupted far to the east. This feature does not characterize other described sites in the region or other large taxa represented at Emeryville: elk, sturgeon and geese are, again, largely confined to the immediate area of the site locality. Once they are gone, there is no other place to get them. Thus, the complex set of signals in relative abundance indices documented at Emeryville—all large prey from local patches steadily decline through time, but large prey with a more extensive regional distribution first decline and then increase—seems to be linked to the unique spatial structure of the available vertebrate species. Similar factors could possibly underlie the documented increases in large game in several other late Holocene contexts in California (McGuire and Hildebrandt 1994; Hildebrandt and McGuire 2000).

Trends in Age Composition for the Emeryville Vertebrates

As discussed in detail above, for prey species not susceptible to behavioral depression and not characterized by spatially discrete breeding colonies or rookeries, resource depression should lead to declines through time in both their relative abundances and the average age of exploited individuals. *Positive* correlations should thus be found between the relative abundance of a taxon and its mean age. Conversely, for species sensitive to predator disturbance and/or characterized by discrete breeding areas, resource depression should be reflected by *negative* correlations between relative abundance indices and mean age. That is, declining abundances of a species due to depression should correlate with increases in mean age. Several prey species well-represented in the Emeryville fauna exhibit these fundamental differences and clearly exemplify the distinctive demographic trends.

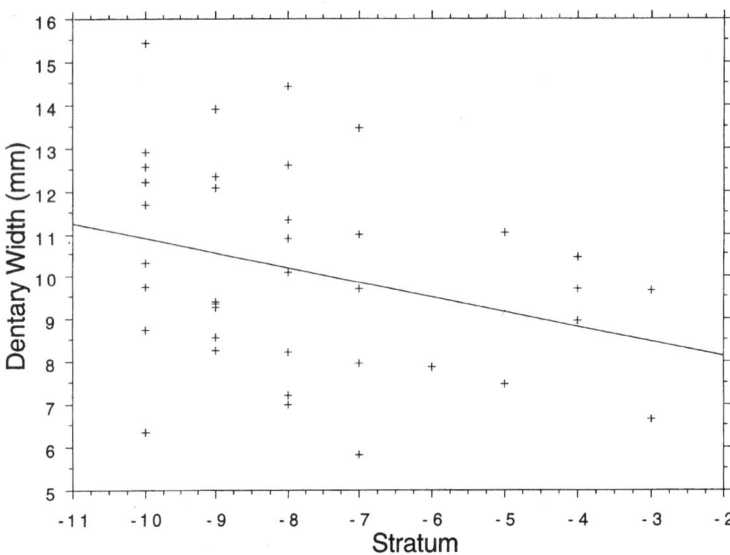

FIGURE 12.8. Dentary widths by stratum for the Emeryville Shellmound sturgeon (*Acipenser*).

Fish
The largest species derived from the estuarine fishes patch type is white sturgeon and, again, the relative abundance of this species declines steadily over the period that Emeryville was occupied. This species does not form seasonal high-density breeding congregations and exhibits no detectable spatial structure by age/size within San Francisco Bay (David Kohlhurst pers. comm. 2001). It is also unlikely that sturgeon could perceive a human fishing threat in such a way that might lead them to abandon particular areas of the bay. White sturgeon then, probably like most fishes and molluscs, is a prey species for which resource depression should cause declines in mean age. As Figure 12.8 shows, both the mean and maximum size of measured sturgeon dentaries decline significantly with the relative abundance of this high-ranked fish across the occupational history of the site (maximum width by stratum: $r_s = -0.833, p < 0.03$; mean width by stratum: $r_s = -0.76, p < 0.05$).

Birds and Mammals
Other well-represented vertebrate species in the Emeryville assemblage clearly form seasonal breeding congregations and are susceptible to behavioral/microhabitat depression. These include the cormorants, sea otters, tule elk, and black-tailed deer. For these taxa, I assigned the identified specimens to three broad ontogenetic age classes—newborns (or chicks), sub-adults, and adults—based primarily on bone size and development. More details on the criteria I used to make those assignments are described in Broughton (1999, 2001b).

Double-crested cormorants (*Phalacrocorax auritus*) and Brandt's cormorants (*P. penicillatus*) are the most abundant avian taxa in the Emeryville fauna, next to the winter visitant anatids. These birds form large rookeries during the spring and summer months in central California, typically on small islands within estuaries or off the coast. In no other contexts do chicks and sub-adult birds attain such high proportional abundances. Adult cormorants do not defend flightless young from large predators but cormorant colonies are well known for their sensitivity to disturbance: vandalized or disturbed colonies are routinely abandoned (Boekelheide et al. 1990:165–6). Treasure Island, located 9 km west of Emeryville, is the nearest location that may have supported cormorant rookeries in the past. Such colonies would have thus presented the highest-return patches for the cormorant resource.

The abundance of cormorants, relative to all other birds, increases steadily over the initial 800 years of site occupation (strata ten through seven; Figure 12.9), as elk, geese, sturgeon and deer decline. It is clear that cormorant rookeries were exploited here as the remains of baby chicks and subadults are extremely abundant in the collection (Broughton 2001b). Cormorant exploitation, however, then falls off dramatically to judge from the significant decline in the relative abundance of cormorant bones across the upper six strata of the mound. If this decline reflects resource depression and the loss of local breeding colonies it should be reflected by an increase in the proportion of adult birds. This is, in fact, the pattern revealed in the data: the relative abundance of cormorants is negatively correlated with the proportion of adults in the collection (Cormorant Index versus percentage adult cormorants: $r_s = -0.72, p = 0.01$; Figure 12.10). In other words, the average age of the harvested cormorants *increases* as the local population appears to decline through resource depression (Broughton 2001b).

Sea otters represent another taxon, well represented in the Emeryville fauna, that are characterized by "rookery-like" breeding colonies and are very sensitive to human persecution (Hall

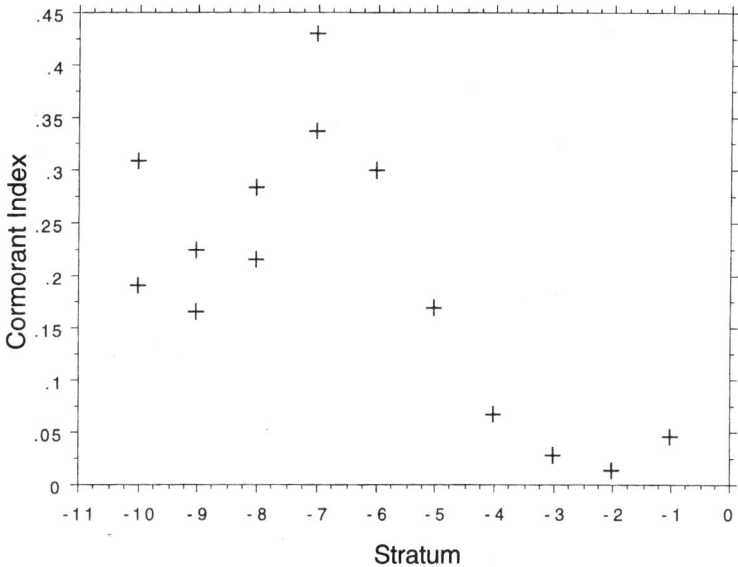

FIGURE 12.9. The distribution of the Cormorant Index ($\sum$NISP *Phalacrocorax*/[$\sum$NISP Birds]) by stratum at the Emeryville Shellmound. The Cormorant Index increases across the lower four strata (strata ten through seven: $r_s = 0.595$, $p = 0.11$) but declines over the upper six strata (strata six through one: $r_s = -0.77$, $p < 0.02$).

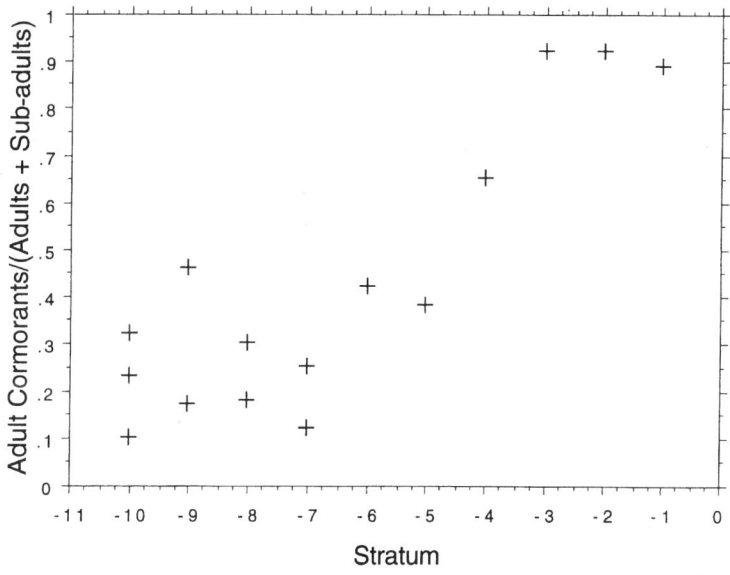

FIGURE 12.10. Changes in the proportional contribution of adult cormorants (Adult Cormorants/[Adults + Sub-adults]) across the Emeryville Shellmound strata. Sub-adults includes chicks. (Does not include stratum U1, that was represented by fewer than ten cormorant specimens for which age could be determined.)

1945; Riedman and Estes 1990). Sea-otter mother and pup pairs are known to congregate in extremely large nursery groups in sheltered bays (Riedman and Estes 1990:68) and appear to have done so on land prior to their near-extinction at the hands of nineteenth-century fur trade hunters (Hall 1945; Broughton 1999). Depression of sea otters should thus be signaled by the local abandonment of these high-density nursery congregations and an increase in the relative encounter rate with adult animals.

As in the case of the cormorants, sea-otter use intensifies across the lower four Emeryville strata, but then declines dramatically over the upper strata. If this latter decline resulted from harvest pressure, it should thus be associated with the loss of breeding colonies, and an increase in the proportion of adult

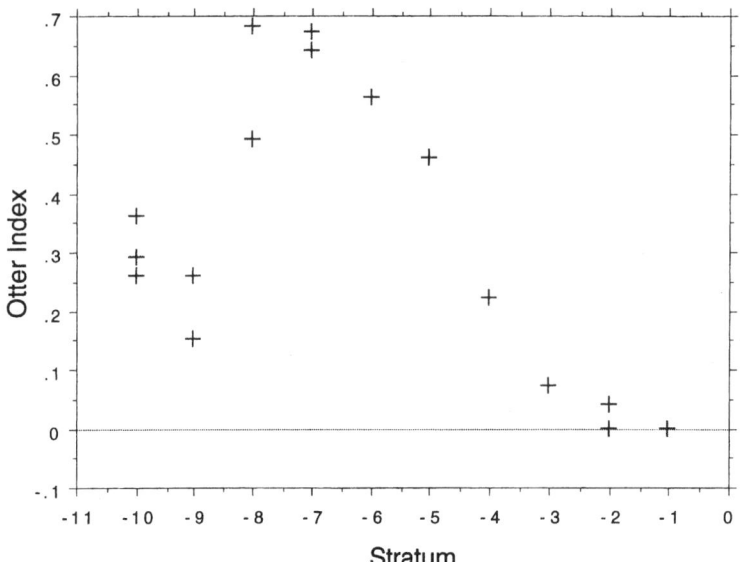

FIGURE 12.11. The distribution of the Otter Index ($\sum$NISP *Enhydra lutris*/[$\sum$NISP Mammals]) by stratum at the Emeryville Shellmound. Does not include the potentially intrusive rodents and lagomorphs. The Otter Index increases across the lower four strata (strata ten through seven: $r_s = 0.668$, $p = 0.05$) but declines over the upper six strata (strata six through one: $r_s = -0.73$, $p < 0.01$).

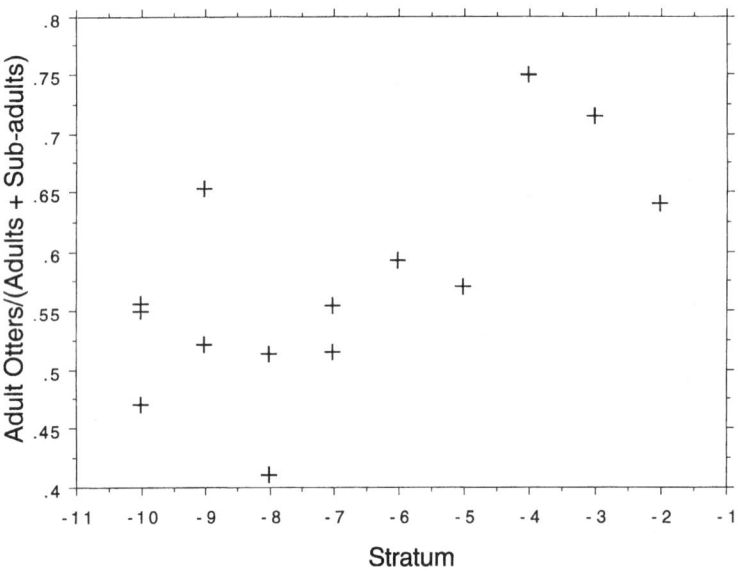

FIGURE 12.12. Changes in the proportional contribution of adult sea otters (Adult Sea Otters/[Adults + Sub-adults]) across the Emeryville Shellmound strata. Sub-adults includes newborns. (Does not include stratum U1, that was represented by fewer than ten sea-otter specimens for which age could be determined.)

animals. Figures 12.11 and 12.12 show that pattern at Emeryville. As sea otters decrease in abundance, adults comprise a greater proportion of individuals in the samples; mean age thus *increases* significantly with declines in their abundances (Otter Index versus percentage adult otters: $r_s = -0.68$, $p = 0.01$).

As for many ungulates, spatial separation of the sexes outside the mating season is well documented for both of the cervids represented in the Emeryville fauna (Clutton-Brock et al. 1987; Bowyer et al. 1996; Main and Coblentz 1996). Although there are several competing hypotheses about why this occurs, evidence from black-tailed deer suggests it is driven at least in part by differential sensitivity to predation. In particular, females with young tend to occupy areas that offer greater protective cover and security for their offspring. Insofar as female-

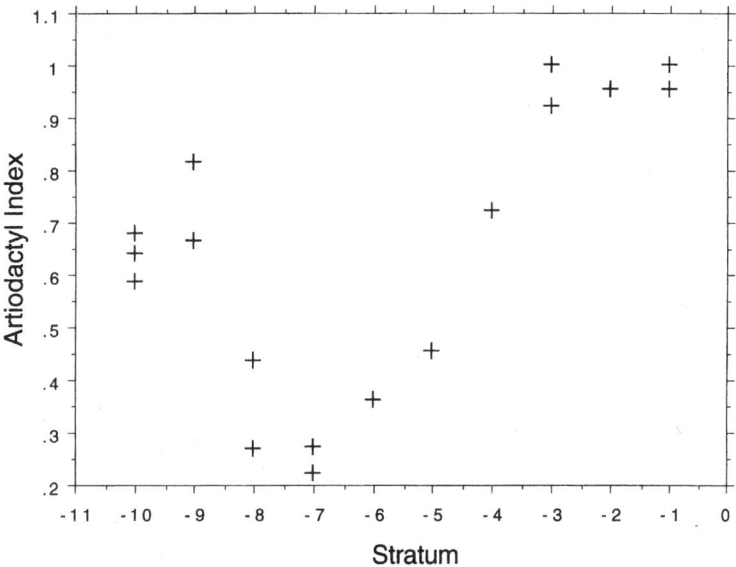

FIGURE 12.13. The distribution of the Artiodactyl Index ($\sum$NISP Artiodactyls/[$\sum$NISP Artiodactyls + Sea otters]) by stratum at the Emeryville Shellmound.

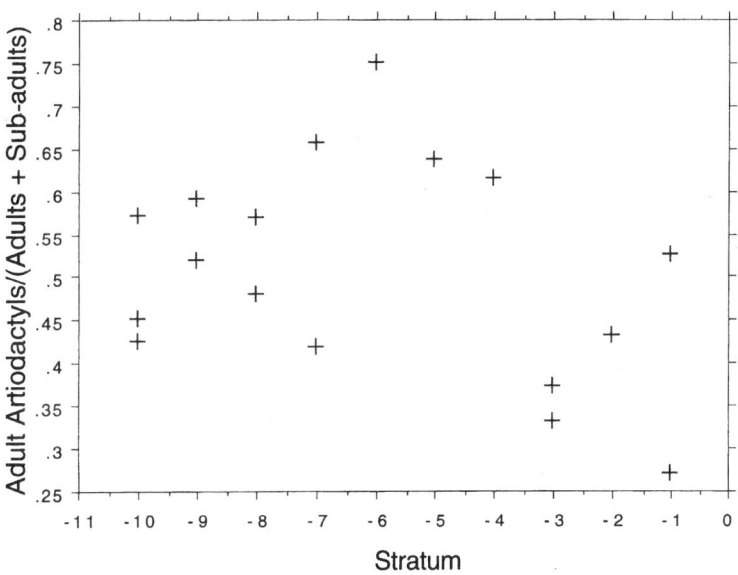

FIGURE 12.14. Changes in the proportional contribution of adult artiodactyls (Adult Artiodactyls/[Adults + Sub-adults]) across the Emeryville Shellmound strata. (The subadult category does not include foetal and newborn specimens; see Broughton 1999.)

young groups are more sensitive to the presence of predators than males, resource depression may cause disproportionate local losses of such groups and ultimately lead to an increase in the mean age of animals encountered in areas depressed from hunting.

As illustrated in Figures 12.3 and 12.6 above, both elk and deer decline significantly in abundance across strata ten through six or the initial 800 years of site occupation. And, while elk continue to decline in abundance, deer numbers steadily increase after that. Figure 12.13 shows the relative abundance of artiodactyls—the conjoined sample of elk and deer—compared to sea otter, the other most abundant mammal species in the Emeryville fauna. As a group, the relative abundance of artiodactyls follows the striking V-shaped pattern exhibited for deer.

As expected, changes in the proportions of adult artiodactyls mirror this trend precisely (Fig. 12.14). The initial decline in artiodactyl abundance is associated with an *increase* in adult animals and average age, while the upper strata increase in their abundance is associated with decreasing proportions of adults. Indeed, the abundance of artiodactyls is negatively correlated with the proportion of bones assigned as adults ($r_s = -.50$; $p < .05$). It appears that, as local artiodactyl populations first decline due to hunting pressure, fewer young are encountered. Later in the sequence, however, the proportion of young animals increases as more use is made of the less-depleted herds located far away. This demographic signal is thus fully consistent with the proposed distant patch origin of the upper strata deer.

Conclusions

Two of the most commonly used indicators of prehistoric resource depression are temporal declines in the relative abundances of high-ranked species and declines in the mean age of harvested prey. However, resource depression may also be reflected by *increases* in both high-ranked prey abundances and mean age. The vertebrate sequence from Emeryville uniquely displays each of these different patterns that seem to reflect systematic differences in the behavior and spatial distributions of the represented prey species.

At Emeryville, the relative abundances of large-sized prey species more-or-less restricted to habitat types located near the site invariably exhibit significant declines through time. However, black-tailed deer, a large species whose preferred habitat extends far to the east of the mound, first declines in abundance but then steadily increases across the later period of site occupation. This resurgence in deer hunting seems to reflect continued resource depression from local patches and an ever-increasing use of distant, less-depleted patches located far away. That this resurgence in deer exploitation resulted from increasing distant patch use is supported by patterns in skeletal part abundances. An increase in large game use is thus one of the options rate-maximizing foragers can potentially take in response to a deteriorating local environment.

Temporal patterns in the age composition of the Emeryville vertebrates strongly support the evidence for prey depression based on relative abundance indices, but in ways that depart from standard expectations. White sturgeon show a significant decline in mean age/size through time in a pattern consistent with their demographically fine-grained spatial distribution and low susceptibility to behavioral depression. While this follows the commonly cited expectation, one that has been revealed in several other archaeological mollusc, fish, and turtle faunas, the well-represented resident bird and mammal taxa at Emeryville show just the opposite pattern.

Sea otters, cormorants, tule elk and black-tailed deer are taxa that exhibit pronounced age-related spatial structure and are responsive behaviorally to predation risk and human hunting activities. For these species, breeding colonies or microhabitats favored by females with young, may have been abandoned with increasing predation pressure. As the relative abundances of these prey species decrease, so does the proportion of specimens that represent sub-adult animals. For these species, resource depression thus appears to cause increases in the mean age of exploited animals.

Since increases or decreases in the mean age of harvested prey can follow from resource depression depending on the spatial structure and behavioral characteristics of the species involved, these features must be carefully evaluated *prior to* deriving expectations for prey response in any particular application. This is essential, of course, to maintain the deductive element of the enterprise. It will not be possible in most cases, however, to anticipate the contexts in which declining foraging returns will result in increasing distant patch use and increases in large game abundances. Detecting this phenomenon will require further tests involving other faunal data, such as, for example, anatomical part representation.

The documentation of resource depression and the associated declines in foraging efficiency and average fitness may entail predictions for variation in many other aspects of human behavior and morphology. Such features include, but are not limited to, technological changes associated with intensive processing of lower-ranked resources, increasing disease loads and declines in body size and stature, greater attention to resource defense with the development of more tightly circumscribed territories and higher levels of interpersonal violence (Broughton and O'Connell 1999; Bright et al. this volume). These trends, in fact, characterize many regional sequences in world prehistory.

Perhaps of greater significance, we are starting to realize how knowledge of prehistoric human impacts on animal populations may have implications for the future conservation of species currently threatened by human influences (Broughton 2001a; Grayson 2001). For example, one of the fundamental goals of modern conservation biology has been to establish benchmarks for what the "natural" or "pristine" conditions of a region are so that management can be directed toward restoring and maintaining those conditions (Hunter 1996:695). In the Americas, pristine is often equated with pre-European times, but archaeofaunal analyses of resource depression clearly show that this approach is aimed at a moving target (cf. Martin and Szuter 1999).

In California, for instance, many early historic records described a land of "inexpressible fertility" (Margolin 1978) teeming with a wide variety of large, tame vertebrates. While these vast herds of elk and deer may have darkened the California landscape for miles during the early historic period, detailed records like Emeryville suggest that those taxa were anything but abundant during late prehistoric times. Prior to the disease-based near decimation of California Indians, beginning perhaps as early as the sixteenth century (Preston 1996), those peoples had reached extremely high densities and appear to have nearly eaten their way through many populations of large fishes, birds and mammals. The early European explorers, it now seems, had "traversed an Edenic blip" on the California landscape (Grayson 2001:8). The latest prehistoric and early historic benchmarks for California ecosystems, although separated in time by mere decades, would appear to be worlds apart. As I have discussed in detail elsewhere (Broughton 1997, 1999, 2001a), whether or not these very different benchmarks and the processes that created them are recognized has implications for how the restoration and maintenance of wilderness areas will be most effectively carried out.

These far-reaching implications suggest that there will be more riding on the accurate documentation of resource depression from archaeological faunas in the years to come. And it is clear from the Emeryville sequence that this task can be far more complex than many of us once thought. By giving more detailed attention to the spatial structure and behavioral characteristics of the prey resources involved, more robust applications of foraging theory to analyses of prehistoric resource depression can be made.

Acknowledgments

I am grateful to Frank Bayham, Virginia Butler, Michael Cannon, Donald Grayson and Andrew Ugan for extremely helpful comments on the manuscript. Analysis of the Emeryville Shellmound vertebrate fauna was funded by the National Science Foundation (DBS-9218417 and SBR-9707997).

References

Alvard, M. S.
1994   Conservation by native peoples: prey choice in a depleted habitat. *Human Nature*, 5:127–54.

Anderson, A.
1981   A model of prehistoric collecting on the rocky shore. *Journal of Archaeological Science*, 8:109–20.

Barlow, K. R., D. Metcalfe
1996   Plant utility indices: two Great Basin examples. *Journal of Archaeological Science*, 23:351–71.

Basgall, M.
1987   Resource intensification among hunter-gatherers: acorn economies in prehistoric California. *Research in Economic Anthropology*, 9:21–52.

Bayham, F. E.
1979   Factors influencing the Archaic pattern of animal utilization. *Kiva*, 44:219–35.
1982   A diachronic analysis of prehistoric animal exploitation at Ventana Cave. Doctoral dissertation, Department of Anthropology, Arizona State University, Tempe.

Bettinger, R. L., M. A. Baumhoff
1981   The Numic spread: Great Basin cultures in contact. *American Antiquity*, 47:485–503.

Bird, D. W., R. Bliege Bird
2000   The ethnoarchaeology of juvenile foragers: shellfishing strategies among Meriam children. *Journal of Anthropological Archaeology*, 19:461–76.

Boekelheide, R. J., D. G. Ainley, S. H. Morrell, T. J. Lewis
1990   Brandt's cormorant. In *Seabirds of the Farallon Islands* (eds. D. G. Ainley, R. J. Boekelheide). Stanford, CA: Stanford University Press, pp. 163–94.

Bowyer, R. T.
1984   Sexual segregation in southern Mule Deer. *Journal of Mammalogy*, 65:410–17.

Bowyer, R. T., J. G. Kie, V. Van Ballenberghe
1996   Sexual segregation in black-tailed deer: effects of scale. *Journal of Wildlife Management*, 60:10–17.

Broughton, J. M.
1994a   Declines in mammalian foraging efficiency during the late Holocene, San Francisco Bay, California. *Journal of Anthropological Archaeology*, 13:371–401.
1994b   Late Holocene resource intensification in the Sacramento Valley, California: the vertebrate evidence. *Journal of Archaeological Science*, 21:501–14.
1995   Resource depression and intensification during the late Holocene, San Francisco Bay: evidence from the Emeryville Shellmound vertebrate fauna. Doctoral dissertation, Department of Anthropology, University of Washington, Seattle.
1997   Widening diet breadth, declining foraging efficiency, and prehistoric harvest pressure: ichthyofaunal evidence from the Emeryville Shellmound, California. *Antiquity*, 71:845–62.
1999   Resource depression and intensification during the late Holocene, San Francisco Bay: Evidence from the Emeryville Shellmound vertebrate fauna. *University of California Anthropological Records*, 32.
2001a   Pre-Columbian human impact on California vertebrates: evidence from old bones and implications for wilderness policy. In *Wilderness and Political Ecology: Aboriginal Influences and the Original State of Nature* (eds. C. Kay, R. Simmons). Salt Lake City: University of Utah Press, in press.
2001b   A stratigraphic reanalysis of the Emeryville Shellmound avifauna. Manuscript in preparation for Contributions in Science, Natural History Museum of Los Angeles County.

Broughton, J. M., J. F. O'Connell
1999   On evolutionary ecology, selectionist archaeology, and behavioral archaeology. *American Antiquity*, 64:153–65.

Butler, V. L.
2000   Resource depression on the northwest coast of North America. *Antiquity*, 74:649–61.
2001   Changing fish use on Mangaia, southern Cook Islands: resource depression and the prey choice model. *International Journal of Osteoarchaeology*, 11:88–100.

Cannon, M. D.
2000   Large mammal relative abundance in Pithouse and Pueblo period archaeofaunas from southwestern New Mexico: resource depression among the Mimbres-Mogollon? *Journal of Anthropological Archaeology*, 19:317–47.

Carney, K. M., W. J. Sydeman
1999   A review of human disturbance effects on nesting colonial waterbirds. *Waterbirds*, 22:68–79.

Caughley, G.
1966   Mortality patterns in mammals. *Ecology*, 47:906–17.
1977   *The Analysis of Vertebrate Populations*. London: Wiley.

Charnov, E. L.
1976   Optimal foraging: the marginal value theorem. *Theoretical Population Biology*, 9:129–36.
1993   *Life History Invariants: Some Explorations of Symmetry in Evolutionary Ecology*. Oxford: Oxford University Press.

Charnov, E. L., G. H. Orians, K. Hyatt
1976   Ecological implications of resource depression. *American Naturalist*, 110:247–59.

Clutton-Brock, T. H., G. R. Iason, F. E. Guinness
1987   Sexual segregation and density-related changes in habitat use in male and female red deer (*Cervus elaphus*). *Journal of Zoology* (London) 211:275–89.

Gifford, E. W.
1916 Composition of California shellmounds. *University of California Publications in American Archaeology and Ethnology*, 12:1–29.

Gonzalez, J. A.
1999 Effects of harvesting of waterbirds and their eggs by native people in the northeastern Peruvian Amazon. *Waterbirds*, 22:217–24.

Grayson, D. K.
2001 The archaeological record of human impacts on animal populations. *Journal of World Prehistory*, 15:1–68.

Grayson, D. K., M. D. Cannon
1999 Human paleoecology and foraging theory in the Great Basin. In *Models for the Millennium: Great Basin Anthropology Today* (ed. C. Beck). Salt Lake City: University of Utah Press, pp. 141–51.

Griffiths, D.
1975 Prey availability and the food of predators. *Ecology*, 56:1209–14.

Hall, E. R.
1945 Chase Littlejohn, 1854–1943: observations by Littlejohn on hunting sea otters. *Journal of Mammalogy*, 26:89–91.

Hames, R. B.
1989 Time, efficiency, and fitness in the Amazonian food quest. *Research in Economic Anthropology*, 11:43–85.

Hames, R. B., W. T. Vickers
1982 Optimal foraging theory as a model to explain variability in Amazonian hunting. *American Ethnologist*, 9:358–78.

Hawkes, K., J. F. O'Connell
1992 On optimal foraging models and subsistence transitions. *Current Anthropology*, 33:63–6.

Hickerson, H.
1965 The Virginia deer and intertribal buffer zones in the upper Mississippi Valley. In *Man, Culture, and Animals* (eds. A. Leeds, A. P. Vayda). American Association for the Advancement of Science Publication, No. 78. Washington, DC: AAAS, pp. 43–66.

Hildebrandt, W. R., T. L. Jones
1992 Evolution of marine mammal hunting: a view from the California and Oregon coasts. *Journal of Anthropological Archaeology*, 11:360–401.

Hildebrandt, W. R., K. McGuire
2000 The ascendance of hunting during the California Middle Archaic: an evolutionary perspective. Paper presented at the 34th Annual Meeting of the Society for California Archaeology. 19 April, Riverside, California.

Hunter, M.
1996 Benchmarks for managing ecosystems: are human activities natural? *Conservation Biology*, 10:695–7.

Janetski, J. C.
1997 Fremont hunting and resource intensification in the eastern Great Basin. *Journal of Archaeological Science*, 24:1075–89.

Kay, C., R. Simmons
2001 *Wilderness and Political Ecology: Aboriginal Influences and the Original State of Nature* (eds. C. Kay, R. Simmons). Salt Lake City: University of Utah Press, in press.

Klein, R. G., K. Cruz-Uribe
1984 *The Analysis of Animal Bones from Archaeological Sites*. Chicago: University of Chicago Press.

Lindstrom, S.
1996 Great Basin fisherfolk: optimal diet breadth modeling the Truckee River aboriginal subsistence fishery. In *Prehistoric Hunter-gatherer Fishing Strategies* (ed. M. Plew). Boise, ID: Boise State University, pp. 114–79.

Lyman, R. L.
1987 On the analysis of vertebrate mortality profiles: sample size, mortality type, and hunting pressure. *American Antiquity*, 52:125–42.

McCullough, D.
1969 The tule elk: its history, behavior, and ecology. *University of California Publications in Zoology*, 88:1–209.

McGuire, K. R., W. R. Hildebrandt
1994 The possibilities of women and men: gender and the California Millingstone Horizon. *Journal of California and Great Basin Anthropology* 16:41–59.

Madsen, D. B., D. Schmitt
1998 Mass collecting and the diet-breadth model: a Great Basin example. *Journal of Archaeological Science*, 25:445–55.

Madsen, D. B., S. R. Simms
1998 The Fremont Complex: a behavioral perspective. *Journal of World Prehistory*, 12:255–336.

Main, M., B. Coblentz
1996 Sexual segregation in Rocky Mountain mule deer. *Journal of Wildlife Management*, 60:497–507.

Marean, C. W., Z. Assefa
1999 Zooarchaeological evidence for the faunal exploitation behavior of Neandertals and Early Modern Humans. *Evolutionary Anthropology*, 8:22–37.

Margolin, M.
1978 *The Ohlone Way: Indian Life in the San Francisco–Monterey Bay Area*. Berkeley, CA: Heyday Books.

Martin, P. S., C. Szuter
1999 War zones and game sinks in Lewis and Clark's west. *Conservation Biology*, 13:36–45.

Mech, L. D.
1977 Wolf-pack buffer zones as prey reservoirs. *Science*, 198:320–1.

Metcalfe, D., K. R. Barlow
1992 A model for exploring the optimal trade-off between field processing and transport. *American Anthropologist*, 94:340–56.

Metcalfe, D., K. T. Jones
1988 A reconsideration of animal body-power utility indices. *American Antiquity*, 53:486–504.

Miquelle, D. G., J. M. Peek, V. Van Ballenberghe
1992 Sexual segregation in Alaskan moose. *Wildlife Monographs*, 122:1–57.

Nagaoka, L. A.
2000 Resource depression, extinction, and subsistence change in prehistoric southern New Zealand. Doctoral dissertation, Department of Anthropology, University of Washington, Seattle.

2001 Using diversity indices to measure changes in prey choice at the Shag River Mouth site, southern New Zealand. *International Journal of Osteoarchaeology*, 11:101–11.

O'Connell, J., K. Hawkes, N. Blurton Jones
1990 Reanalysis of large animal body part transport among the Hadza. *Journal of Archaeological Science*, 17:301–16.

Orians, G. H., N. E. Pearson
1979 On the theory of central place foraging. In *Analysis of Ecological Systems* (eds. D. J. Horn, R. D. Mitchell, G. R. Stairs). Columbus, OH: Ohio State University, pp. 154–77.

Porcasi, J., T. Jones, L. M. Raab
2000 Trans-Holocene marine mammal exploitation on San

Clemente Island, California: a tragedy of the commons revisited. *Journal of Anthropological Archaeology*, 19:200–20.

Preston, W.
1996 Serpent in Eden: dispersal of foreign diseases into pre-mission California. *Journal of California and Great Basin Anthropology*, 18:2–37.

Raab, L. M., J. F. Porcasi, K. Bradford, A. Yatsko
1995 Debating cultural evolution: regional implications of fishing intensification at Eel Point, San Clemente Island. *Pacific Coast Archaeological Society Quarterly*, 31:3–27.

Redman, C. L.
1999 *Human Impacts on Ancient Environments*. Tucson, AZ: University of Arizona Press.

Riedman, M. L., J. A. Estes
1990 The sea otter (*Enhydra lutris*): behavior, ecology, and natural history. *United States Fish and Wildlife Service Biological Report*, 90:1–126.

Rogers, A. R., J. M. Broughton
2001 Selective transport by ancient hunters: a new statistical method and an application to the Emeryville Shellmound fauna. *Journal of Archaeological Science*, 28:763–73.

Smith, E. A.
1991 *Inujjuamiut Foraging Strategies: Evolutionary Ecology of an Arctic Hunting Economy*. New York: Aldine De Gruyter.

Smith, E. A., M. Wishnie
2000 Conservation and subsistence in small scale societies. *Annual Review of Anthropology*, 29:493–524.

Steadman, D. W.
1995 Prehistoric extinctions of Pacific Island birds: biodiversity meets zooarchaeology. *Science*, 267:1123–31.

Stiner, M. C., N. D. Munro, T. A. Surovell
2000 The tortoise and the hare: small game use, the broad spectrum revolution, and Paleolithic demography. *Current Anthropology*, 41:39–73.

Szuter, C., F. E. Bayham
1989 Sedentism and prehistoric animal procurement among desert horticulturalists of the North American Southwest. In *Farmers as Hunters: The Implications of Sedentism* (ed. S. Kent). Cambridge: Cambridge University Press, pp. 80–95.

Taber, R. D., K. J. Raedeke, D. K. Paige
1982 Population characteristics. In *Elk of North America: Ecology and Management* (eds. J. W. Thomas, D. E. Toweill). Harrisburg, PA: Stackpole Books, pp. 279–98.

Vickers, W. T.
1980 An analysis of Amazonian hunting yields as a function of settlement age. In *Working Papers on South American Indians*, Vol. 2. (ed. R. B. Hames). Vermont: Bennington College, pp. 7–29.
1988 Game depletion hypothesis of Amazonian adaptation: data from a native community. *Science*, 239:1521–2.

Wilson, D. S.
1976 Deducing the energy available in the environment: an application of optimal foraging theory. *Biotropica*, 8:96–103.

Wolverton, S.
2001 Environmental implications of zooarchaeological measures of resource depression. Doctoral dissertation, Department of Anthropology, University of Missouri, Columbia.

13

A Model of Central Place Forager Prey Choice and an Application to Faunal Remains from the Mimbres Valley, New Mexico

Michael D. Cannon

Many researchers have used assemblages of animal bones from archaeological sites to document cases of *resource depression* (Charnov et al., 1976), or reductions in the prey capture rates of human foragers that were the result of their own harvesting of those prey (e.g., Bayham, 1979; Broughton, 1994a,b, 1997, 1999; Cannon, 2000; Grayson, 1991; Hildebrandt and Jones, 1992; Janetski, 1997; Nagaoka, 2000, 2002; Szuter and Bayham, 1989; also see Broughton, 2002a,b; Grayson and Cannon, 1999; Hildebrandt and McGuire, 2002; Jones and Hildebrandt, 1995; Lyman, 1995; Lyman and Wolverton, 2002; Ugan and Bright, 2001). These studies have important implications for our understanding of the ways in which people have structured our planet's ecosystems, particularly through their effects on populations of large-bodied vertebrates (e.g., Broughton, 2002b; Grayson, 2001; Kay, 1994, 1998).

Also important is that these studies provide an archaeologically operational measure of *foraging efficiency*, or the efficiency with which people obtained calories from their environments through the harvest of wild resources. Reductions in foraging efficiency have been suggested to have led to several changes that are evident in the archaeological record of various parts of the world, ranging from increases in the importance of agriculture (e.g., Barlow, 1997, 2001; Cannon, 2000, 2001a; Winterhalder and Goland, 1997) to increases in the degree of violent conflict (e.g., Broughton and O'Connell, 1999). By providing an empirical measure of foraging efficiency, resource depression studies offer a means of testing hypotheses about the causes of major changes that occurred in the human past.

In this paper I present a model of central place forager prey choice that unifies several issues that have previously been addressed in archaeological resource depression studies through the use of separate models. In comparison to the models invoked in most resource depression analyses, the model that I present here makes assumptions that more closely match the ways in which human hunting is often carried out. Perhaps more important, however, is that, in comparison to the approaches taken in previous studies, the model that I present makes it easier to determine how decisions about the processing of prey at their point of capture will combine with decisions about prey choice to influence overall foraging efficiency for central place foragers. I illustrate the benefits that arise from the use of such a model by applying it to archaeofaunal data from the Mimbres Valley, southwestern New Mexico, where it appears that people experienced depression of large mammal resources, and declining hunting efficiency, during the period between about AD 400 and AD 1200.

Resource Depression and Central Place Foraging

Most previous archaeological resource depression studies have drawn explicitly on the prey model—also known as the prey choice or diet breadth model—from foraging theory (e.g., Kaplan and Hill, 1992; Stephens and Krebs, 1986). This model shows that, given certain assumptions (see Broughton, 1994a, 1997; Smith, 1991; Stephens and Krebs, 1986), the most energy-efficient foraging strategy is to pursue resources with higher post-encounter caloric return rates whenever they are encountered, and to pursue lower-return resources only when higher-return resources are encountered relatively infrequently.

Since post-encounter return rate is positively correlated with body size for most vertebrate prey, it is argued that, if energetic returns were important as a decision-making currency among a group of foragers, larger-bodied taxa should have been pursued by those foragers whenever they were encountered (e.g., Bayham, 1979; Broughton, 1994a, 1997, 1999; Szuter and Bayham, 1989). From this it follows that a decline over time in the relative abundance of a larger-bodied taxon in an archaeological assemblage would indicate a reduction in the rates at which

---

*Originally published in *Journal of Anthropological Archaeology* 22(2003):1–25

individuals encountered that taxon (Broughton and Grayson, 1993). Assuming that other causes of the reduction in encounter rates such as climate change can be ruled out (see Grayson and Cannon, 1999), the decline in relative abundance would indicate that people experienced resource depression, or a reduction in per capita prey capture rates resulting from the effects of their own predation on the taxon involved. In addition, because the resource whose availability was reduced was one that provided a high caloric return rate, the decline in its relative abundance would indicate a decline in overall foraging efficiency.

Foraging theory models, however, make very specific assumptions about the real-world cases to which they are applicable, and their use in situations in which their assumptions are not met may not provide a valid test of any hypotheses derived from them (e.g., Stephens and Krebs, 1986). Some researchers have recognized that most archaeological situations likely violate a key assumption of the prey model (e.g., Broughton, 1994a, 1999; Nagaoka, 2000), and they have dealt with this by employing other foraging models in conjunction with it. As I describe next, though, this approach is less than ideal when research goals include documenting changes over time in foraging efficiency, particularly in cases involving people who hunt from a central place to which they return with captured prey.

The aspect of the prey model that has been acknowledged to be problematic is its assumption that the probability of encountering any prey type is independent of previous encounters either with it or with any other prey type. This "fine-grained search assumption" will be violated, for example, when individuals of a given prey type have a better than random chance of being found close to one another within certain areas of a habitat; that is, when prey exhibit "patchy", or heterogeneous, distributions (e.g., Broughton, 1994a; Smith, 1991:206–207, 228). In such instances the prey model applies to a forager while foraging within a homogeneous resource patch, but another model is required to address the issue of which patches to exploit in the first place (e.g., Smith, 1991).

Most archaeofaunal assemblages are likely to have been deposited over long periods of time by multiple individuals who hunted in a variety of resource patches (e.g., Broughton, 1994a; Broughton and Grayson, 1993), and for this reason the fine-grained search assumption of the prey model will be violated unless the model is applied only to sets of resources that could have been collected within a single patch. Archaeologists who have recognized this have appropriately dealt with the nested decisions of patch choice and prey choice by using models designed to address each decision independently (e.g., Broughton, 1999; Nagaoka, 2000). These researchers have first divided their study areas into patches: for example, a coastal setting might be divided into a marine resource patch and a terrestrial resource patch.[1] They have then used the prey model to address the decision of which prey to pursue within a patch, the solution to which depends on the abundances of high-return prey within that patch, and a patch choice model (based on the marginal value theorem; see Charnov, 1976) to address the decision of how much time to spend foraging in each patch, the solution to which depends both on the marginal rate of energetic gain obtained from each patch per unit time spent foraging within it and on the average amount of time that it takes to travel between patches.

Because this approach uses separate models to address what are conceptualized as separate decisions, it is difficult to determine what effects prey choice and patch choice will combine to have on overall foraging efficiency. In addition, because the patch choice model used in these studies assumes that patches are encountered in the same manner in which prey are assumed to be encountered by the prey model, it is not appropriate for cases involving central place foragers who begin each foraging trip from a single point on the landscape. For central place foragers, patch choice will affect overall foraging efficiency because different patches will be located at varying distances from the central place and will thus entail varying costs in terms of travel and resource transport time. Using the approach taken previously, however, it is difficult to integrate travel and transport costs, prey encounter rates, and prey post-encounter return rates into a single theoretical measure of foraging efficiency for central place foragers.

As a potential solution to these problems, I have argued that a more appropriate foraging theory model for use in most archaeological situations is Orians and Pearson's (1979) model of central place forager patch choice (Cannon, 2000). This model assumes a habitat that consists of some number of internally homogeneous resource patches located at varying distances from a central place, and it asks which of the available patches will maximize the rate of energy delivery to the central place. The decision variable in this model is which patch to exploit, but the solution to this decision depends both on the time that it takes to travel to each patch and on the energetic gain function provided by each patch, which depends in turn on the abundances and post-encounter return rates of the various prey types found within each patch. Thus, this model allows direct analysis of the combined effects on foraging efficiency of both transport costs and the availability of different prey types within patches.

I do not go into the details of the Orians and Pearson patch choice model here because I have done so elsewhere (Cannon, 2000). In this paper, I am primarily concerned with one factor that may be extremely important in cases of central place foraging that the Orians and Pearson model does not explicitly take into account. Foragers may often be able to increase their rate of energy delivery to a central place by processing resources at their location of acquisition in such a way that parts of low caloric "utility" are left behind (e.g., Binford, 1978; Perkins and Daly, 1968; Thomas and Mayer, 1983; White, 1954). Metcalfe and Barlow (1992) (see also Bettinger et al., 1997) have developed a theoretical model of the tradeoff between transport costs and field processing costs that is central to this issue, and implications of this model have been explored using both ethnographic and

TABLE 13.1. Definitions of the variables included in the central place forager prey choice model.

Variable	Definition
R_i	The delivery rate provided by a unit of the *i*th prey type, or the average amount of nutritional energy taken home per unit of foraging time when an item of this prey type is harvested
E_i	The energy of a load, or the average amount of energy contained in a load of prey type *i* that is transported home from the place where it is acquired and processed; I use this term interchangeably with the term utility[a]
e_i	The average amount of energy contained in one complete unit of prey type *i*
m_i	The average weight of one complete unit of prey type *i*
s_i	Search and transport time, or the average amount of time that it takes to find a unit of prey type *i* once search has begun plus the average amount of time that it takes to transport a load of that prey type home
h_i	Handling time, or the average amount of time that it takes to pursue and obtain a unit of prey type *i* following encounter plus any additional time that is required to initially prepare the prey item for transport
c_i	Capture time, which equals $s_i + h_i$
p_i	Processing time, or the amount of time spent processing a load of prey type *i* in order to increase its utility *after any initial processing necessary to prepare it for transport (i.e., "handling") has been completed*
T_i	Total foraging time, which equals $c_i + p_i$

[a] As used here, "energy" is similar to the concept of "utility" employed in previous field processing models (e.g., Bettinger et al., 1997; Metcalfe and Barlow, 1992), but it is expressed in units of energy (e.g., kilocalories) rather than in units of energy per unit of weight or volume (e.g., kcal/kg or kcal/L).

archaeological data (e.g., Barlow and Metcalfe, 1996; Beck et al., 2002; Bettinger et al., 1997; Bird and Bliege Bird, 1997; Bird et al., 2002; see also O'Connell et al., 1988, 1990).

One of the main points to be taken from this work is that an efficiency-minded forager should spend more time processing prey at the point of capture as the distance back to the central place increases. In turn, this relation between transport distance and optimal field processing time has been used in archaeological resource depression analyses to draw inferences about changes in patch use based on the abundances of different parts of vertebrate prey carcasses recovered at residential sites (e.g., Broughton, 1999; Nagaoka, 2000; Rogers and Broughton, 2001; also see Rogers, 2000). The logic behind doing so is simple. If a hunter spends more time processing a carcass in the field, then more parts of low food value should be removed from the load that is taken home so that the utility of that load, measured in calories per unit weight, is increased. If more time is spent field-processing carcasses as transport distance increases, then a smaller proportion of low-utility parts should be taken home when more distant patches are used.

It is possible to combine a model of field processing like the one developed by Metcalfe and Barlow (1992) with a model of central place forager prey choice that is similar to the patch choice model presented by Orians and Pearson (1979), and I do so next. The model that I present facilitates exploration of the combined effects of prey selection, transport distance, and field processing on overall foraging efficiency. This model can also be used to develop predictions about the patterns that should be observed in archaeofaunal assemblages deposited at residential sites over spans of time during which central place foragers were experiencing resource depression, and I apply these predictions to the Mimbres Valley case below.

THE CENTRAL PLACE FORAGER PREY CHOICE MODEL

As with the Orians and Pearson (1979) patch choice model, the model that I present here assumes that the goal of a central place forager is to maximize his or her "delivery rate," or the amount of energy carried back to the central place per unit foraging time. The decision that I model here, however, is different from the one that Orians and Pearson model: rather than "what patch should I forage in on my next foraging trip?" it is, "what kind of prey, and what parts of that prey, should I bring home from my next foraging trip?" Central place foraging will often require foragers to choose between prey with different post-encounter return rates that they can expect to encounter at different distances from home, and what follows is an attempt to model decisions of this sort.

The delivery rate in this model is essentially a measure of foraging efficiency, or the amount of energy obtained per unit of time spent foraging. As I illustrate below, this rate is affected by the abundances of prey, by transport time, by the post-encounter return rates of prey, and by the amount of time spent processing prey in the field. As a result, all of these variables are incorporated into the measure of foraging efficiency that the model provides.

I first present the basic elements of the model and then discuss them in greater detail. My presentation is geared towards the mammalian prey that I consider in my analysis of Mimbres Valley faunal assemblages: large-bodied artiodactyls like deer (*Odocoileus* spp.) and pronghorn (*Antilocapra americana*) and small-bodied leporids like jackrabbits (*Lepus* spp.) and cottontails (*Sylvilagus* spp.). It should require little modification, however, for the model to be applicable to other kinds of resources.

Central to this model is the assumption that there is a maximum load size, measured in units of weight, that a forager will transport home (e.g., Jones and Madsen, 1989; Zeanah, 2000), and I designate this parameter "L_{max}." In addition, there are four characteristics of resources that are important in the model: (1) the number of calories provided by a prey item, (2) the weight of a prey item, (3) the length of time that it takes to encounter a prey item, "handle" it, and transport it back home, and (4) the manner in which the utility of a load of a prey type can be increased through field processing.[2] The variables that reflect these characteristics are defined in Table 13.1, and the relations between these variables are depicted graphically in Figure 13.1.

Regarding the assumption of a maximum load size (L_{max}),

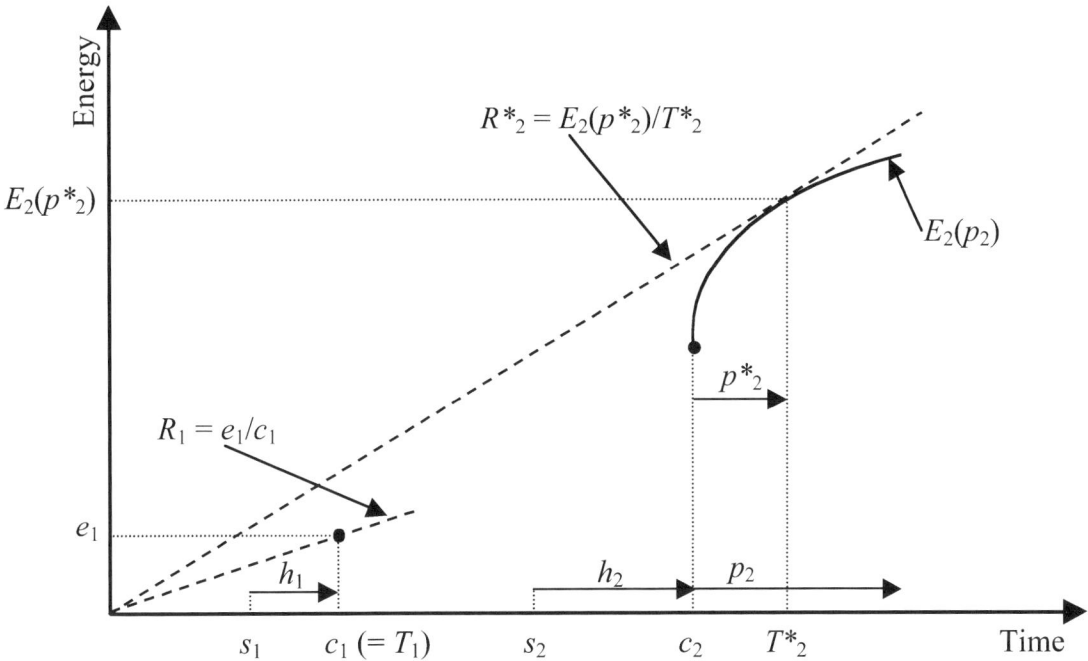

FIGURE 13.1. The central place forager prey choice model. Prey type 1 is a "small" prey type ($m_1 \leq L_{max}$ and prey type 2 is a "large" prey type ($m_2 > L_{max}$). See text for details.

it is obvious that some resources, such as large mammals, come in packages that are too heavy to be carried by a single person or perhaps even by several people. A central place forager who harvests such a prey item will have to make a decision about which parts of it to take home and which to leave behind, and the model that I present here integrates this decision into the decision about prey choice. It is not necessary to know the precise value of the maximum size of the loads transported by foragers in order to apply this model: archaeological predictions derived from it can be tested qualitatively—that is, by considering the directions of changes in various aspects of resource acquisition (see Kaplan and Hill, 1992)—if certain assumptions can be made. First, it must be assumed that L_{max} remained relatively constant over time, or, more exactly, since different individual foragers may well have had different L_{max} values, and since different numbers of people may have been involved in different foraging trips, it must be assumed that the distribution of L_{max} values across foraging trips did not change substantially over time. It must also be possible to make assumptions about whether the package sizes of important resources fell above or below the maximum load size, and I note that the weight of the large mammal resources that I consider below certainly must have fallen above this size for hunters in the Mimbres Valley.[3]

Because this model assumes that there is a maximum load size that a forager will transport home, prey types can be divided into two classes that I will call "small" and "large": small prey types are those for which the weight of a complete prey item is less than or equal to the maximum load size (i.e., $m_i \leq L_{max}$), and large prey types are those for which the weight of a complete prey item is greater than the maximum load size (i.e., $m_i > L_{max}$).

I discuss small prey types first, and in order to develop the model I make the unrealistic assumption that only one unit of a small prey type will be harvested per foraging trip. In other words, I assume that a forager will return home as soon as the first prey item is captured and "handled", even if that item is small enough that additional items might easily be collected and transported home on the same trip. I present this model in greater detail elsewhere (Cannon, 2001a), and I show there that relaxing this assumption has no effect on the qualitative predictions that can be derived from it.

Prey type 1 in Figure 13.1 is an example of a small prey type that is harvested singly. When a central place forager leaves home to forage, a certain amount of time will pass before a unit of prey type 1 is encountered. The average amount of this "search" time is included in the variable s_i, and I envision this to be search time averaged over a relatively short period, perhaps on the order of days or weeks. The average amount of time that it takes to encounter a unit of a given prey type might vary for many reasons, but the abundance of that prey type on the landscape, or at least its abundance in areas close to a forager's central place, should be especially important.

This model makes no specific assumptions about such issues as whether search time consists of time spent traveling to a distant location to harvest prey or whether it consists of time spent searching or waiting for prey close to home. However, in addition to search time, the average amount of return trip "transport" time for a prey type is also included in the variable s_i. Thus, if a forager does travel a long distance, on average, to harvest prey type i, s_i will be greater than it would be if the resource could be obtained closer to the central place.

This model also makes no specific assumptions about the distribution of prey on the landscape. It does not necessarily assume that a forager searches for all prey types simultaneously, as the standard prey model does, and it entails no "fine-grained search assumption": search time is treated as a distinctive characteristic of each individual prey type and is not distributed equally among all prey types. Likewise, the model does not assume the existence of discrete resource patches, although it is compatible with them.

If a forager decides to pursue a prey item once it is encountered, "handling" time (h_i) begins. For now, I use the term handling time in a manner similar to its usual usage in foraging theory (e.g., Stephens and Krebs, 1986:13–24), and I consider it to include time spent pursuing, capturing, and preparing a prey item for transport. If a prey item is smaller than the maximum load size, and if only one such prey item is acquired per foraging trip, then it is relatively straightforward to calculate the delivery rate because processing time, as defined in this model, can be disregarded. It makes no sense for a delivery rate-maximizing forager to "process" such a prey item after "handling" it because any time spent doing so cannot increase the amount of energy taken home: it can only lower the delivery rate by increasing the amount of time spent in the field. As I discuss below, however, this is not the case for large vertebrate prey types.

For a small prey item that is acquired singly, the delivery rate is simply the amount of transported energy provided by that prey item (e_i) per total amount of time spent foraging. In this case total foraging time is just "capture" time (c_i, which equals $s_i + h_i$), and the delivery rate for a unit of prey type i is given by:

$$R_i = e_i/c_i \quad (13.1)$$

This rate can also be represented by a line that begins at the origin and then passes through the point defined by c_i on the time axis and e_i on the energy axis: the steeper the slope of this line, the higher the delivery rate.

I now turn to large prey types, an example of which is provided by prey type 2 in Figure 13.1. The concepts of search and transport time and handling time apply to these prey types just as they do to small prey types. With large prey types, however, the issue of processing time becomes important.

If the weight of a unit of a prey type is greater than the heaviest load that a forager will transport (i.e., if $m_i > L_{max}$), then that prey item must be processed in the field in order to reduce its weight before it can be taken home. Adult male mule deer, for example, can weigh over 150 kg, which is far too heavy for a single person to carry. Thus, when one of these animals is killed, its carcass will have to be prepared to some degree before any of it can be transported, and even if several people are available to carry the carcass, some butchery may be required to divide up the load.[4] In terms of the model presented here, the time that it takes to perform such processing is considered to be "handling time" rather than "processing time."

However, it may be possible to increase the utility of a load by spending additional time processing the prey item beyond the minimum amount of time required initially to prepare it for transport. Consider, for example, two loads of equal weight: one consists of two complete deer hind limbs, and the other consists of a section of ribs plus two hind limbs with phalanges, metatarsals and tarsals removed. Because the lower limbs provide fewer calories than ribs (e.g., Metcalfe and Jones, 1988:table 3), less of the weight in the second load will be taken up by low-calorie body parts, and the utility of this load will be higher than the utility of the first load. On the other hand, it will also take longer to prepare the second load because time must be spent removing the section of ribs from the carcass and disarticulating the tarsals from the tibiae. In the model presented here, any such "extra" time spent processing a load in order to increase its utility constitutes "processing time" rather than "handling time."

More generally, for any prey type for which a prey item consists of parts that vary in caloric content per unit weight, the way in which the utility of a load of that prey type changes with processing time can be described by a "processing function."[5] I denote this function as $E_i(p_i)$ because a processing function will be specific to the ith prey type. In Figure 13.1 the processing function for prey type 2 is represented by the curve labeled $E_2(p_2)$, and Figure 13.2 is a more detailed depiction of such a function. If a prey item weighs more than the maximum load size, then it will provide no transportable utility to a forager until handling is complete because it is only at this point—when a load of size L_{max} is produced—that any of the item can be carried home. In Figure 13.2, the utility of a prey item after handling is denoted as E_{0i}. If the utility of a load can be increased further through processing, then this can be described by a processing function, $E_i(p_i)$. Eventually, however, a point should be reached at which processing no longer increases the utility of a load (utility may even decline with further processing), and the utility of the load at this point is designated $E_{max\,i}$.

Many prey types should exhibit processing functions that take the shape of a diminishing returns curve, as shown in Figure 13.2. Barlow and Metcalfe (1996) and Bettinger et al. (1997) have shown that the processing functions for at least some plant resources approximate diminishing returns curves, but data are not currently available to determine empirically whether this is the case for vertebrate resources.[6] However, Metcalfe and Barlow (1992:350–351) present a logical argument as to why the processing functions for certain resources will resemble diminishing returns curves, and their argument probably applies to vertebrate prey.

These authors first characterize some kinds of resources as "structured" and others as "unstructured." A structured resource is one for which the morphology of the prey item determines the order in which processing steps must be undertaken: for example, a pinyon nut must first be removed from the cone before the hull can be removed from the nut (e.g., Barlow and Metcalfe, 1996). For an unstructured resource, on the other hand, a forager can choose the order in which to undertake processing steps. When processing a resource of this sort, Metcalfe and Bar-

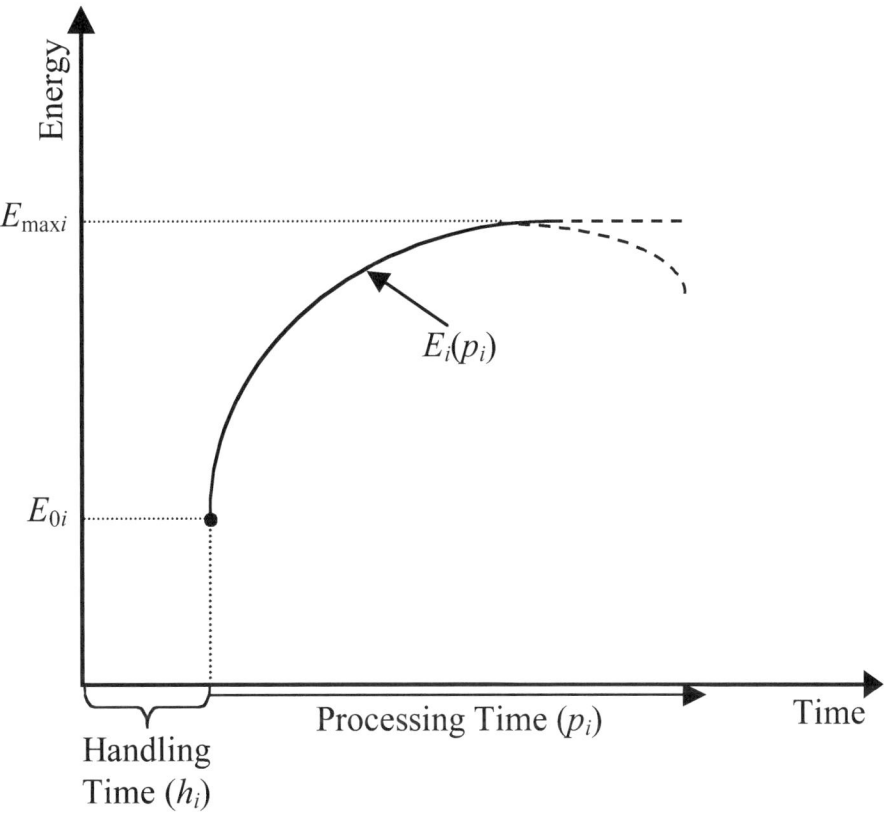

FIGURE 13.2. Hypothetical processing function for a vertebrate prey type for which $m_i > L_{max}$. Handling time is the time that it takes to reduce load size to the maximum load size (L_{max}); a prey item provides no transportable utility until this point is reached, and the utility of a load at this point is designated as E_{0i}. Processing time is any additional time that is spent increasing the utility of the load, and the way that utility changes with processing time is described by the function $E_i(p_i)$. A point should eventually be reached at which the utility of a load can no longer be increased, and the utility of a load at this point is designated as E_{maxi}.

low argue, a forager should first take whatever processing step results in the greatest increase in load utility per unit of processing time, then take the step that results in the next greatest increase in load utility per unit of processing time, and so on. This will be the strategy that produces the load of the highest utility for any given amount of processing time, and the processing function that results from such a strategy will exhibit diminishing marginal returns.

Vertebrate resources are partially structured—for example, the scapula and fore limbs of a large mammal must be removed before all of the ribs can easily be removed from the vertebrae—but they are also partially unstructured—a fore limb can be removed before a hind limb, for example, and vice versa. In fact, there is probably a nearly infinite number of combinations of body parts that a forager could remove from a vertebrate carcass and transport home (Rogers, 2000, calls such combinations "configurations"), and there are certainly many different ways in which the removal of various body parts might be ordered. Thus, there should be sufficient flexibility in the way in which a vertebrate carcass can be processed to allow processing functions for this kind of resource to approximate diminishing returns curves.

I assume here that this is the case. Regarding the handling time-processing time distinction that the central place forager model makes, I assume that the initial handling stages of vertebrate carcass preparation (h_i) will be carried out in such a way that a load of size L_{max} will be produced in the most efficient manner possible, meaning that whichever processing steps produce a load of the highest utility per unit of processing time will be carried out first during these initial stages. For many vertebrate resources, however, it should be possible to further increase the utility of a load with additional processing time (p_i). I assume that this additional processing will be performed in the manner in which Metcalfe and Barlow (1992) argue that it should be carried out when an unstructured resource is being processed. This will result in processing functions for vertebrate prey that take the shape of diminishing returns curves.

If the processing function for a large prey item does approximate a diminishing returns curve, then the highest delivery rate that is obtainable from that prey item can be calculated as shown in Figure 13.1. In solving for this highest delivery rate, the optimal amount of time to spend processing the item after handling is also necessarily determined. It can be seen graphically that the highest delivery rate that can be obtained for prey

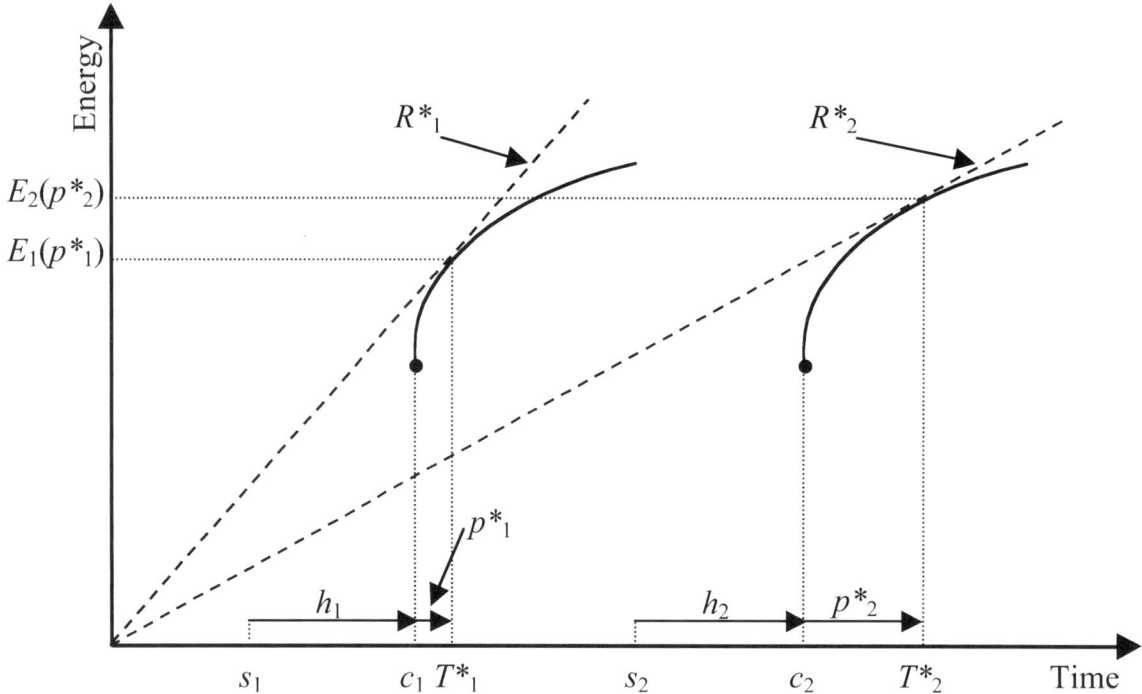

FIGURE 13.3. Two prey types that illustrate the effect of search and transport time on maximum delivery rate and optimal processing time.

type 2 in Figure 13.1 is defined by a line beginning at the origin that is tangential to the processing function, and I demonstrate this mathematically elsewhere (Cannon, 2001a:Appendix A). This line also gives the optimal amount of processing time, which I denote as p^*_2 in Figure 13.1 or more generally as p^*_i: for any amount of processing time greater than or less than p^*_i, the delivery rate will be lower than the delivery rate obtained when p^*_i units of time are spent processing. The maximum delivery rate provided by a large resource can be described by the equation:

$$R^*_i = \frac{E_i(p^*_i)}{T^*_i} \tag{13.2}$$

where

$$T^*_i = c_i + p^*_i \tag{13.3}$$

To explore the effects that transport distance will have on optimal processing time and the overall delivery rate, consider Figure 13.3. Two prey types with identical handling times and processing functions are depicted here, and the only way in which these prey types differ is in average search and transport time: s_1 is considerably lower than s_2. This would be the case, for example, if prey type 2 could only be found in areas more distant from the central place, so that more time was required to transport units of this prey type home. It is apparent that the highest delivery rate that prey type 1 can provide is much greater than the highest delivery rate that could be obtained from prey type 2. In addition, however, note that the optimal amount of time to spend processing prey type 1—that is, the amount of processing time that results in the highest delivery rate—is much lower than the optimal amount of processing time for prey type 2.

This latter result is consistent with the prediction that can be drawn from the model presented by Metcalfe and Barlow (1992) that more time should be spent processing a resource in the field (i.e., removing low utility parts from the load that is carried home) the greater the distance over which that resource must be transported. What the model presented here makes easier to see, however, is that, all else being equal, spending less time processing a resource acquired close to home will always result in higher overall foraging efficiency than can be obtained by spending more time processing that resource when it is acquired far from home, even if additional processing increases the utility of the load that is carried back (assuming that actual processing time approximates p^*_i).

In addition, since the variable s_i is the sum of average transport time plus average search time, this model can also be used to predict that more time should be spent processing a resource in the field the greater the average amount of time it takes to encounter a unit of that resource. This prediction cannot be derived from the Metcalfe and Barlow model because that model does not take search time into account. This prediction can be understood intuitively by considering that, when search time is high, a little extra processing time will increase total foraging time by a proportion that is small relative to the proportion by which it will increase the utility of the load that is taken home. When search time is low, however, that same amount of extra processing time would increase total foraging time by a proportion that would be larger in relation to the increase in load utility that could be obtained.

It is now possible to address the issue of prey choice. Figure

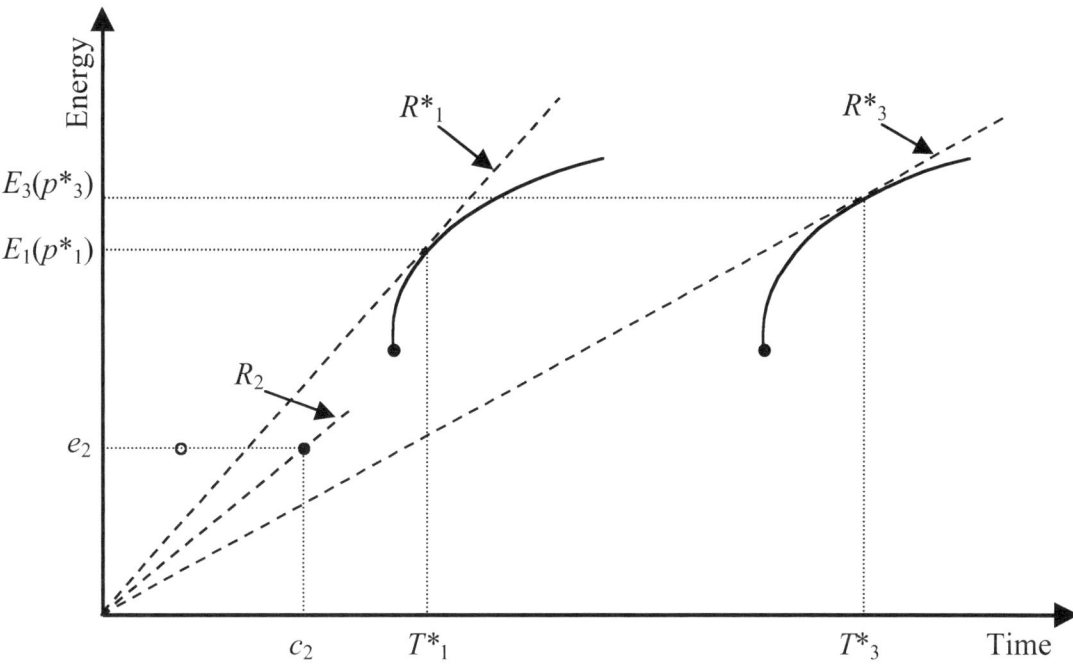

FIGURE 13.4. Three prey types that provide different average delivery rates. Prey types 1 and 3 are "large" prey types, and prey type 2 is a "small" prey type.

13.4 depicts three prey types that provide different average delivery rates. If the goal of a forager is to maximize the rate of energy delivery to the central place, then that forager should necessarily choose to harvest the prey type that provides the highest delivery rate. Of the prey types shown in Figure 13.4, prey type 1 will do so on average, and it can thus be predicted that a delivery rate-maximizing forager will most often harvest this prey type.

I point out that this model does not predict that a forager will harvest only the resource with the highest average delivery rate during a given span of time. The variable s_i represents short-term *average* search and transport time, and search and transport time will vary around this mean for most resources: on some foraging trips it will take less than the average amount of time to encounter a unit of prey type i and carry it home, while on other trips an above-average amount of search and transport time will be required. In terms of Figure 13.4, for example, a forager may occasionally encounter a unit of prey type 2 soon enough that it could provide a delivery rate higher than the average delivery rate provided by prey type 1, as represented by the open dot to the left of the solid dot for prey type 2. In this case the forager should, of course, take the unit of prey type 2. Because the average time-to-encounter for prey type 2 is much longer than this, however, occurrences of this sort will necessarily be rare. Likewise, it may sometimes take longer than usual for a forager to encounter a unit of prey type 1, or one might not be found at all. In such a case the forager's best option may be to pursue a resource with a lower average delivery rate that is encountered late in the foraging trip, so that at least something is taken home.

In essence, then, this model assumes that a forager makes an encounter-contingent decision about whether or not to pursue a resource based on reasonably accurate knowledge of such factors as the distribution of search and transport times for each potential prey type and the post-encounter return rate of each prey type. In other words, the model assumes that a forager decides whether to pursue a resource upon encounter based on an evaluation of the probability that he or she will encounter an alternative resource on that same foraging trip that can provide a higher delivery rate.

Unlike the standard prey model, which assumes that a forager searches for all prey types simultaneously, the central place forager model is consistent with a variety of search strategies because it does not treat search time as being distributed equally among all prey types. For this reason, it is better suited to the range of search strategies used by human foragers. As Kaplan and Hill (1992:184–185) discuss, some ethnographically observed central place foragers do search for all potential resources simultaneously, but others set out on foraging trips with specific resources in mind, while yet others follow strategies that fall somewhere between the two extremes (also see Smith, 1991). The model that I present here should be applicable to all such strategies. Moreover, as Kaplan and Hill also note, central place foragers who have a specific resource in mind when they leave on foraging trips sometimes end up harvesting resources other than the one that they had initially targeted. The central place forager prey choice model may often be helpful for explaining why such "target-switching" occurs because it can be used to explore the consequences of encountering resources more quickly or more slowly than a forager might expect at the outset of a foraging trip.

Regarding the assumption made by this model that a forager knows the relevant characteristics of potential prey types, I point out that most foraging theory models assume that foragers possess accurate information about the environments in which they live (e.g. Stephens and Krebs, 1986). For human foragers, this assumption is justified since it has been well documented that people in small-scale societies usually possess a remarkable degree of ecological knowledge (e.g., Felger and Moser, 1985; Hill et al., 1997; Nelson, 1983; Rea, 1998).

Returning to Figure 13.4, prey type 1, which provides the highest average delivery rate, is a "large" prey type ($m_1 > L_{max}$), while prey type 2, which provides a lower average delivery rate, is a "small" prey type ($m_2 \leq L_{max}$). As I mentioned above, previous archaeological resource depression studies that have employed the standard prey model have used vertebrate prey body size as a proxy measure of post-encounter return rate because there is a strong correlation between the two variables (e.g., Bayham, 1979; Broughton, 1994a,b, 1997, 1999; Szuter and Bayham, 1989). The concept of "the post-encounter return rate of a resource" takes on a slightly different meaning in the central place forager prey choice model because this model recognizes that post-encounter returns will vary with processing time, which should vary, in turn, with search and transport time. I show elsewhere, however, that it remains appropriate to use body size as a proxy for post-encounter return rate when applying this model to such vertebrate resources as artiodactyls and leporids that are encountered sequentially (Cannon, 2001a).

Since this is the case, if a small prey type and a large prey type have equivalent average search and transport times, then the large prey type will provide the higher average delivery rate, and this should be the resource that is most often pursued by a delivery-rate maximizing forager. In fact, the large prey type would provide the higher average delivery rate even if it took somewhat longer on average to encounter and/or transport this prey type; only when the average search and transport time of the large prey item is sufficiently higher than that of the small prey item will the small prey item provide the higher average delivery rate.[7]

This brings us finally to the issue of long-term change over time. As I discussed above, the variable s_i is the average amount of search and transport time for a resource calculated over some relatively short period of time. Over longer periods, however, the average search and transport time for a resource might change. If this were to happen, then the resource would, in effect, become a new prey type from the point of view of the central place forager model.

If the average amount of time that it took to find a unit of a particular resource increased over the long term (as would occur if the resource became less abundant on the landscape) and/ or if foragers had to travel further to harvest that resource (perhaps due to declines in its abundance in nearby areas), then the average delivery rate that the resource provided would decline. This is illustrated by prey type 3 in Figure 13.4: this prey type has a processing function identical to that of prey type 1, and the difference between these two prey types can be accounted for entirely by an increase in average search and transport time. Search and transport time for prey type 3 is so much greater, in fact, that the average delivery rate that it provides is lower than the average delivery rate provided by prey type 2, a small prey type. In a case like this a delivery rate-maximizing forager should pursue the small prey type (prey type 2) more often than the large one (prey type 3).

Given this, the same key predictions about the effects of resource depression can be derived from the central place forager model as have been derived from the separate models used in previous archaeofaunal resource depression studies. One set of these predictions involves the kinds of prey that foragers harvest and another involves the parts of prey that foragers transport home. For each of these issues, predictions can be made both about the changes in foraging behavior that should occur due to long-term resource depression and about the patterns that should be observed in archaeological assemblages as a result of such changes in behavior.

Regarding the kinds of prey harvested, larger-bodied vertebrates will usually provide higher post-encounter caloric return rates than will smaller-bodied vertebrates. As a result, if a large-bodied prey type initially has a relatively low average search and transport time, it will likely provide a higher average delivery rate than would any smaller-bodied prey type, and it should be the resource that foragers pursue most often. If the average search and transport time for the larger-bodied resource increases, however, there will come a point at which some smaller prey type begins to provide the highest average delivery rate, and at this point foragers should begin to pursue the smaller resource more often. Since the average search and transport time of a prey type will be highly dependent on the abundance of that prey type on the landscape, at least in areas close to the central place, such a change in prey selection will often reflect a decline in the abundance of the large-bodied resource. It will also reflect a decline in foraging efficiency.

As I discussed above, variability around short-term mean search and transport times should lead foragers occasionally to take resources other than the one that provides the highest average delivery rate. In addition, for any given resource, one or more of the key factors e_i, m_i, s_i, h_i, and $E_i(p_i^*)$ may well have changed in a non-directional manner during some span of time. These factors could have varied seasonally, for example, or yearly if climate fluctuated from year to year, and such variation would likely have led foragers to pursue different resources in different seasons or in different years.

However, if a long-term directional trend occurred in any factor such as the average search and transport time of a resource, then a long-term directional trend should also have occurred in the proportions of different resources that were harvested. This would cause, in turn, a directional trend in the relative abundances of various taxa in archaeological assemblages (see Broughton and Grayson, 1993 for a similar point in the context of diet breadth analysis). In the specific case of depression of a large-bodied vertebrate resource, we would see a decline over time in the archaeological abundance of the large-bodied taxon relative to smaller-bodied taxa.

As for the kinds of vertebrate prey body parts that hunters transport home, the central place forager model predicts that foragers should spend increasing amounts of time in the field removing low utility parts from the loads that they carry home as average search and transport time increases. Archaeologically, this leads to the expectation that increases in search time and transport distance, such as might occur due to resource depression, will result in the deposition of lower proportions of low-utility body parts at residential sites. As noted above, the prediction that processing time should increase with transport distance can also be derived from the model presented by Metcalfe and Barlow (1992). However, by integrating field processing into a unified theoretical measure of the rate of energy delivery to a central place, the model presented here makes it easier to see that such increases in transport distance and field processing time will generally reflect declines in overall foraging efficiency.

Faunal Assemblages from the Mimbres Valley

To show how the model that I have presented can be used archaeologically to test hypotheses about resource depression and reduced foraging efficiency, I apply it to faunal data from the Mimbres Valley of southwestern New Mexico. These data were collected as part of a larger study of the relation between hunting efficiency and the importance of agriculture during the Mimbres pithouse and pueblo periods (Cannon, 2001a), and they come from the Galaz, Mattocks and McAnally sites, which are located in the central portion of the valley (Figure 13.5). Because these were all sizable residential settlements, and because people surely would have had to leave these settlements to capture the animals whose bones they deposited at them, a model designed specifically for central place foragers is preferable to the standard prey model in this case for reasons that I have discussed above.

Background

Based primarily on apparent changes over time in the relative abundances of artiodactyl specimens in archaeofaunal assemblages from the Mimbres Valley, several researchers have suggested that Mimbres hunters experienced depression of large mammal resources as the human population of the valley grew. However, the data necessary to test this hypothesis fully have not previously been collected (see overviews in Cannon, 2000, 2001a; also see Sanchez, 1996). In particular, factors that might interfere with the use of taxonomic relative abundance as a measure of resource depression—for example, taphonomic effects or climate change—have not been considered in any detail, nor has artiodactyl body part representation been analyzed with the subject of resource depression in mind. The results of my analysis of data pertaining to these issues, a portion of which is summarized here, are largely consistent with the hypothesis that Mimbres hunters reduced local abundances of artiodactyls, as I illustrate below.

Galaz (LA 635) was located on the west side of the Mimbres River on the first alluvial terrace above the river's floodplain. The site effectively no longer exists, having been bulldozed by pothunters, but workers associated with the Mimbres Foundation research group were able to excavate pithouse deposits here during the 1970s while the site was in the process of being dismantled (Anyon and LeBlanc, 1984). Mattocks (LA 676) is located approximately 6 km north of Galaz, and it also sits on the first terrace above the floodplain on the west side of the Mimbres River; Mimbres Foundation researchers conducted excavations in and around four pueblo room blocks at this site (Gilman and LeBlanc, in preparation; LeBlanc, 1975, 1976a,b). McAnally (LA 12110) lies on top of a small hill directly across the Mimbres River from Mattocks; this site consists of several pithouse depressions, two of which were excavated by the Mimbres Foundation (Arthur, 1994; Diehl and LeBlanc, 2001; LeBlanc, 1975, 1976b). The faunal assemblages from all three of these sites have been analyzed previously (e.g., Diehl and LeBlanc, 2001; Powell, 1977), but the data that I present here come from my reanalysis of them. This reanalysis necessitates some modification of the conclusions that have been drawn from earlier analyses, particularly regarding the timing of important changes.

Variability in the use of faunal resources by people living at settlements in the Mimbres region appears to have been strongly influenced by differences in the local biotic communities surrounding those settlements (e.g., Cannon, 2000, 2001a; Sanchez, 1996; Shaffer and Schick, 1995). I have shown elsewhere, however, that such habitat effects are unlikely to be a problem for the relative abundance analyses that I present in this paper, which is understandable because the sites that I consider here are located fairly close together within the same type of vegetational community (Cannon, 2001a).

The culture history sequence that has been developed for the Mimbres region is presented in Table 13.2; this sequence is divided into periods and phases, which are distinguished mainly on the basis of ceramic and architectural attributes (see Anyon et al., 1981; Diehl, 1994; Diehl and LeBlanc, 2001; Hegmon et al., 1999; Nelson and Anyon, 1996; Wills, 1996). The Early Pithouse period (ca. AD 400–600) apparently represents the earliest occupation of the Mimbres region by people who grew crops. McAnally is one of only a few sites with structures dating to the Early Pithouse period that have been excavated in the Mimbres Valley, and LeBlanc (1976a) argues that the ceramic assemblage from this site indicates that these structures were occupied only during this period (see also Diehl and LeBlanc, 2001).

For the purposes of this study I have assigned individual deposits from Galaz and Mattocks to time periods based primarily

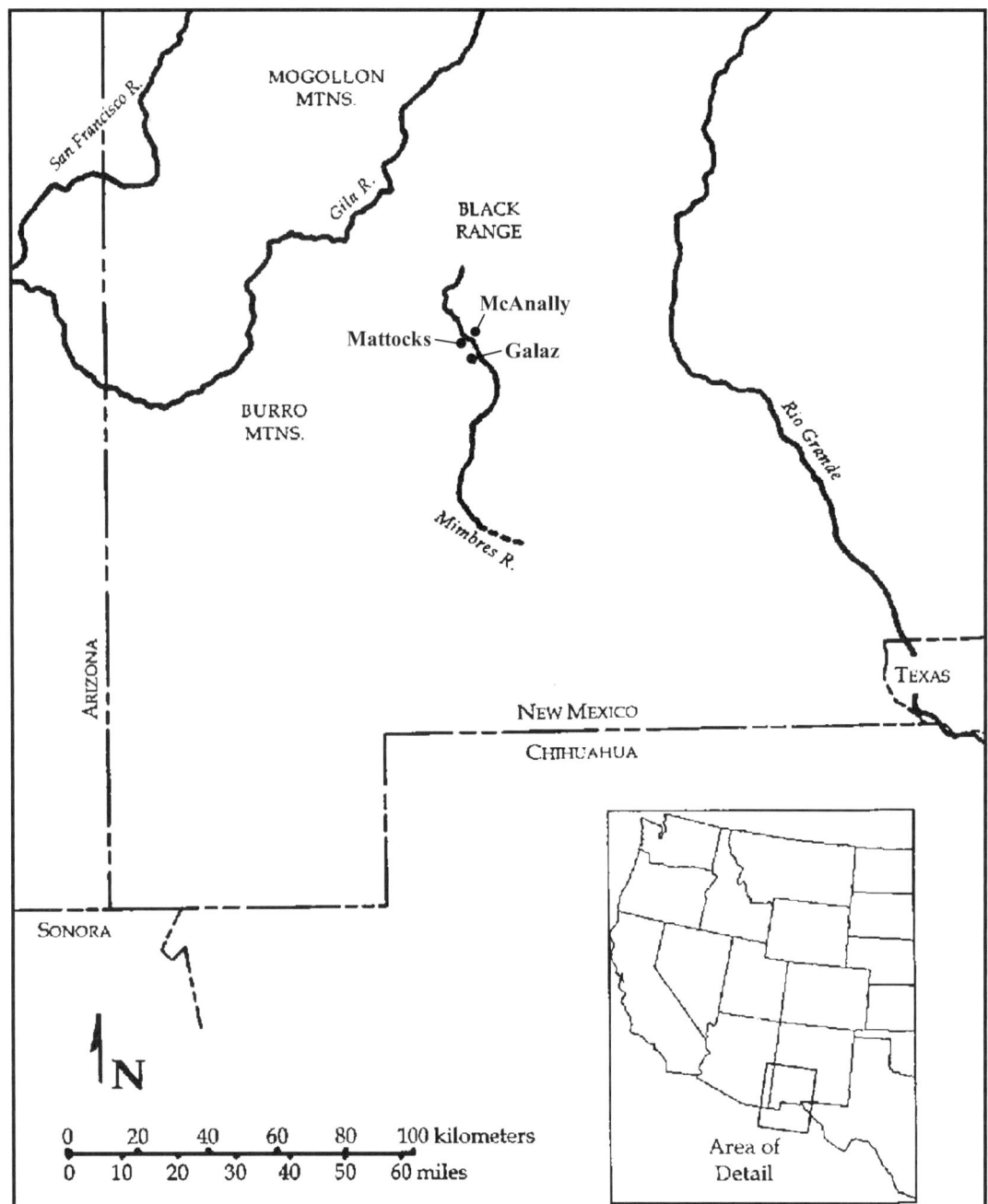

FIGURE 13.5. Map of the Mimbres Valley and surrounding region showing the location of the sites included in this analysis.

on analyses of ceramic content, the results of which are entirely consistent with the stratigraphic relationships that exist among these deposits and with the chronometric dates that are available for them (Cannon, 2001a). Deposits at Galaz with useful faunal samples contain material dating to the Georgetown (AD 600–700) and/or San Francisco (AD 700–early 800s) phases, the Three Circle phase (early AD 800s–AD 1000), and the Classic Mimbres phase (AD 1000–1130), which appears to be the phase during which the human population of the Mimbres Valley was the largest (e.g., Blake et al., 1986). The majority of the useful deposits at Mattocks contain material dating to the Classic Mimbres phase, though some of the material at this site dates to the preceding Three Circle phase and some dates to the subsequent Terminal Classic (which lasts until the late AD 1100s).

Table 13.3 presents numbers of identified faunal specimens, aggregated by time period, for the combined assemblages from Galaz, Mattocks and McAnally.[8] To ensure that the most accurate and precise information possible is obtained concerning changes over time in these assemblages, I include in this analysis only samples from undisturbed, well-dated deposits that were screened through ¼" mesh during excavation (see Cannon 2001a, for further details about analytical procedures). I also include here only those taxa that are most useful for answering the research questions that I am addressing: deer (*Odocoileus* sp.), pronghorn (*Antilocapra americana*), jackrabbits (*Lepus* sp.) and cottontails (*Sylvilagus* sp.).

Deer and pronghorn are among the largest-bodied vertebrates that occur in the Mimbres Valley today, and the extreme rarity of such larger-bodied taxa as elk (*Cervus elaphus*) and bison (*Bison bison*) in archaeological assemblages from the valley suggests that people encountered these larger animals very infrequently during the span of time considered here. Deer and pronghorn thus certainly provided the highest post-encounter return rates of any of the wild resources that were common in the area, and for this reason any decline in the abundances of these taxa on the landscape would likely have caused a substantial reduction in overall wild resource foraging efficiency. To determine how search and transport times for these artiodactyls changed over time in the Mimbres Valley, I examine their abundance in faunal samples relative to the abundance of smaller-bodied leporids. Bones of jackrabbits and cottontails dominate most faunal assemblages from the valley, and these two taxa are the only small-bodied vertebrates that are common in the assemblages included in this study that are also likely to have been deposited by people after having been captured and eaten by them.

Artiodactyl Relative Abundance

The model that I outlined above can be used to predict that reductions over time in the archaeofaunal abundance of a large-bodied taxon relative to the abundances of smaller-bodied taxa will be observed in cases in which central place foragers experienced depression of the large-bodied taxon. Table 13.4 presents total numbers of identified artiodactyl and leporid specimens per time period from Galaz, Mattocks and McAnally, and it also provides "Artiodactyl Index" values, calculated as the ratio of all artiodactyl specimens relative to all artiodactyl specimens plus all leporid specimens (e.g., Broughton, 1994a,b; Janetski, 1997; Szuter and Bayham, 1989): if Mimbres Valley hunters experienced long-term depression of artiodactyl resources, this measure of taxonomic relative abundance should decline over time.

This does occur here. A χ^2 test indicates that there are highly significant differences in artiodactyl relative abundance among the samples in Table 13.4 ($\chi^2 = 53.23$, $p < 0.001$), and examination of the adjusted standardized residuals from this test (e.g., Everitt, 1977:46–48) shows that leporids are substantially under-represented in the two earliest samples. In addition, Cochran's test of linear trend among proportions (see Cannon, 2000, 2001b; Zar, 1999:565–568) indicates that the temporal reduction in artiodactyl relative abundance that is apparent across these samples is highly significant ($\chi^2_{trend} = 19.28$, $p < 0.001$).[9] The combined samples from these sites thus show a clear decline in the relative abundance of artiodactyls between the Early Pithouse period and the Three Circle phase.

As I describe elsewhere (Cannon, 2001a), the patterns that occur at other sites in the Mimbres region for which large faunal assemblages are available are consistent with the pattern that is evident among the sites that I consider here: artiodactyl relative abundance declines during the period of time leading up to the Three Circle phase, and there are no major changes

TABLE 13.2. Mimbres–Mogollon culture historical time periods (after Anyon et al., 1981; Diehl, 1994; Diehl and LeBlanc, 2001; Hegmon et al., 1999; Nelson and Anyon, 1996; Wills, 1996).

PERIOD	PHASE	ABBR.[a]	DATES (AD)
Late Pueblo	Cliff (Salado)	—	1300–1450
	Black Mountain (Animas, El Paso)	—	1200–1300
Early Pueblo	Terminal Classic	TCM	1130–1200
	Classic Mimbres	CM	1000–1130
Late Pithouse	Three Circle	TC	825/850–1000
	San Francisco	SF	700–825/850
	Georgetown	GT	600–700
Early Pithouse	Cumbre	EP	400–600[b]

[a] Abbreviations for time periods used in subsequent tables.
[b] The beginning of the Early Pithouse period has traditionally been placed at AD 200. However, Wills (1996) notes that all of the pre-AD 450 dates that are available for early ceramic sites in the upland Mogollon region are radiocarbon dates that may be subject to the "old wood problem," and he suggests that AD 400 is a more appropriate date to use for the start of the Early Pithouse period in the Mimbres Valley.

TABLE 13.3. Numbers of identified specimens of artiodactyl and leporid taxa.

PHASE	ARTIODACTYLA[a]	*ODOCOILEUS* SP. (DEER)	*ANTILOCAPRA AMERICANA* (PRONGHORN)	LEPORIDAE[b]	*SYLVILAGUS* SP. (COTTONTAILS)	*LEPUS* SP. (JACKRABBITS)
CM/TCM	11	4	1	5	23	24
CM	56	16	3	20	113	135
TC/CM	26	15	4	16	62	83
GT/SF	3	2	0	2	3	0
EP	10	6	1	0	1	0
Total	106	43	9	43	202	242

[a] These specimens are from deer- or pronghorn-sized artiodactyls but were not identifiable below the level of order. Four specimens from bison and from bison- or elk-sized artiodactyls are present in the deposits from Mattocks that are incorporated into this table (which date to the Classic Mimbres phase), but these specimens are not included in the analyses presented here.
[b] Identification of leporid specimens to genus was based on size; these specimens are those that are too fragmentary to determine the size of the animal from which they came.

TABLE 13.4. Total numbers of identified artiodactyl and leporid specimens in the combined samples from the sites included in this analysis; also provided are "Artiodactyl Index" values for these samples (see text).

PHASE	ARTIODACTYLS	LEPORIDS	TOTAL	ARTIODACTYL INDEX
CM/TCM	16 (−0.20)	52 (0.20)	68	0.235
CM	75 (−1.66)	268 (1.66)	343	0.219
TC/CM	45 (−1.07)	161 (1.07)	206	0.218
GT/SF	5 (1.89)	5 (−1.89)	10	0.500
EP	17 (7.00)	1 (−7.00)	18	0.944
Total	158	487	645	0.245

Values in parentheses are the adjusted standardized residuals from a χ^2 test performed on the bone counts in this table ($\chi^2 = 53.23$, $p < 0.001$; $\chi^2_{trend} = 19.28$, $p < 0.001$).

in artiodactyl relative abundance at any site within the valley from the beginning of the Three Circle phase through the Terminal Classic. An analysis of artiodactyl age profiles in Mimbres Valley faunal assemblages, which should be sensitive to changes in harvest rates (e.g., Broughton, 1997, 1999; Koike and Ohtaishi, 1985, 1987), is also consistent with this pattern (Cannon, 2001a).

Of course, a variety of factors other than predation by people might also affect the taxonomic relative abundance values observed in archaeofaunal assemblages, and efforts must be made to control for such factors before it can be concluded that a decline in artiodactyl relative abundance truly indicates that hunters experienced depression of these resources (e.g., Grayson and Cannon, 1999). I present detailed analyses of potential confounding factors elsewhere (Cannon, 2001a), and it does not appear that the changes in artiodactyl relative abundance that I have described can be explained by variability in the taphonomic processes to which these faunal samples have been subjected, by changes over time in the technologies or the tactics that people used to harvest prey, or by changes in climate or vegetation that might have affected the abundances of prey on the landscape independently of human hunting. The relative abundance data that I have discussed here thus seem to indicate that Mimbres hunters experienced depression of large mammal resources, and a decline in hunting efficiency, sometime during the period between about AD 400 and AD 800 or 850.

Artiodactyl Body Part Representation

The second set of predictions that can be derived from the model that I presented above concerns the effects of resource depression on the kinds of large mammal body parts that central place foragers transport back to their residential sites. As I described, if the average amount of search time required before an artiodactyl was encountered increased, and/or if the average distance over which such prey had to be transported increased, then people should have begun to spend more time processing these prey in the field, and this should be reflected by a decline in the proportion of low-utility artiodactyl body parts that they carried home.

To determine whether this occurred at the Mimbres Valley sites that I consider here, I explore whether there are differences among time periods in the mean utility of the artiodactyl elements that are present at these sites. I employ Metcalfe and Jones' (1988) standardized, whole-bone Food Utility Index (FUI), derived for caribou (*Rangifer tarandus*), which provides a measure of the amount of calories contained in each part of the body of an artiodactyl.[10] A reduction in the proportion of low-utility body parts transported to a residential site, such as might occur due to resource depression, should result in an increase in mean FUI.

To evaluate whether mean FUI changes across the samples that I use, I first assigned to each faunal specimen in these samples the FUI value of the element from which it comes. I then calculated the mean FUI values for each phase, which can be compared statistically using analysis of variance. I present the results of such analyses below.

This analysis can show whether there are significant differences in mean utility among a set of samples, but it does not show whether the mean utility of any individual sample is particularly high or low relative to a value that might occur if people were transporting large mammal body parts without any consideration for their utility. In other words, before it can be concluded that people were especially "selective" about the body parts that they carried home, it must be shown that the mean utility of those body parts is significantly higher than the mean utility that would be observed if people simply brought home a random sample of body parts.

To determine whether this is the case, I also present the results of a second type of analysis in which I compare the distribution of FUI values within each archaeofaunal sample to a null distribution composed of all of the elements within the body of a single artiodactyl. There are 118 bones in the body of an artiodactyl to which FUI values can be assigned: the mean FUI value for these bones is 38.2, and the standard deviation of their FUI values is 20.3. If a t-test shows that the mean FUI of an archaeological sample is significantly higher than the mean FUI of such a null sample, then there is reason to conclude that people selectively excluded low-utility artiodactyl body parts from the loads that they transported home.

Of course, analyses of changes in body part transport practices of the sort that I present here are of little value unless it can be demonstrated that observed patterns are not simply the result of density-mediated taphonomic processes (e.g., Grayson, 1989; Lyman, 1985; Rogers, 2000). As Lyman (1985) has shown, there is a weak negative correlation between the caloric utility of artiodactyl body parts and the volume density of the bones within those body parts as measured by photon densitometry. It is thus possible that a pattern in mean utility among a set of faunal samples might be the result not of differences in the kinds of body parts that people transported, but rather of variability in the intensity of taphonomic processes that differentially affect bones of different densities. Specifically, because skeletal elements associated with higher-utility parts tend to be lower in density, intense density-mediated attrition of bones within a sample might reduce the mean utility value observed for that sample.

To control for such effects, I recorded all of Lyman's (1984) densitometer "scan sites" that are present on each artiodactyl specimen in the assemblages that I use, which allows calculation of the mean of the volume density (VD) values for the scan sites present on the specimens within a given sample.[11] The standard deviation of the distribution of density values within a sample can also be calculated, allowing the mean density values of samples to be compared using analysis of variance. If mean volume density is found to differ significantly among a set of samples, then there is cause for concern that any pattern in mean utility values observed among those samples might be the result of taphonomic factors rather than the result of variability in body part transport practices.

Just as with my analysis of differences in mean utility, though, such an analysis of differences in mean density does not show whether the mean density value observed in any individual sample is particularly high or low. In other words, before it can be concluded that density-mediated attrition has appreciably affected the bones within a given sample, it must be shown that the mean density of the scan sites present in that sample is significantly higher than might be observed in a sample of bones that was selected randomly. To determine whether this is the case, I use t-tests to compare the distribution of density values present in each of my archaeological samples to a null distribution composed of all of the scan sites present in the body of a single artiodactyl. Such a null distribution consists of 316 scan sites, the mean volume density of which is 0.353 and the standard deviation of which is 0.146.

Artiodactyl body part utility data for the combined samples from the three sites that I consider here are presented in Table 13.5.[12] An analysis of variance shows that the observed differences in mean utility among these samples are significant at the $p < 0.10$ level ($F = 2.20$, $p = 0.071$), which suggests that the proportion of low-utility elements deposited at these sites did vary over time. It appears that this result is due primarily to the differences between the three intermediate samples, all of which have mean utility values between 38 and 41, and the Early Pithouse and Classic Mimbres/Terminal Classic samples, both of which have mean utility values greater than 50.

This conclusion is strengthened by an examination of the results of t-tests comparing the mean utility of each archaeological sample to the mean utility of a null distribution composed of all of the elements in the body of a single artiodactyl (Table 13.5). The observed mean utility values for the three intermediate archaeological samples all approximate the mean utility value of the null sample, and none of them are significantly different from it. However, both the Early Pithouse sample and the Classic Mimbres/Terminal Classic sample have mean utility values that are much higher than the mean of the null sample, and these differences are significant at the $p < 0.05$ level.

These results do not appear merely to be reflecting variability among samples in the degree to which they have been affected by density-mediated taphonomic processes, as shown in Table 13.6. All of these samples do seem to have experienced some degree of density-mediated attrition: they all have mean density values that are higher than that of a null distribution composed of all of the scan sites within the skeleton of an artiodactyl, and the differences for three of them are highly significant, while those for the other two, which are smaller, are nearly significant at the $p < 0.10$ level. However, an analysis of variance gives an insignificant result for a test of the hypothesis that mean scan site density differs among these samples ($F = 0.60$, $p = 0.662$). Thus, although each sample has likely experienced some density-mediated attrition, the effects of such attrition have evidently not varied among them in a manner that could have produced the pattern observed among their mean utility values.

TABLE 13.5. Mean artiodactyl body part utility for the combined samples from Galaz, Mattocks, and McAnally, aggregated by time period (result of ANOVA, testing for differences in mean utility among phases: $F = 2.20$, $p = 0.071$).

PHASE	MEAN FUI	NISP[a]	SD	RESULTS OF t-TESTS FOR DIFFERENCE FROM MEAN OF NULL SAMPLE[b]
CM/TCM	56.5	13	30.7	$t = 2.10$, $p = 0.038$
CM	40.4	97	25.2	$t = 0.69$, $p = 0.488$
TC/CM	39.6	44	22.2	$t = 0.37$, $p = 0.715$
GT/SF	38.2	5	20.0	$t = 0.00$, $p = 1.000$
EP	53.7	15	22.2	$t = 2.57$, $p = 0.011$

[a] Number of identified adult and sub-adult specimens to which FUI values could be assigned. The Classic Mimbres phase sample is larger than the sample of artiodactyl specimens for this phase that is included in Table 13.4 because I include specimens from a larger number of depositional contexts in this analysis (see Cannon, 2001a for details).
[b] Mean = 38.2, $n = 118$, SD = 20.3. All p values are for two-tailed tests.

TABLE 13.6. Mean artiodactyl scan site volume density for the combined samples from Galaz, Mattocks, and McAnally, aggregated by time period (result of ANOVA, testing for differences in mean density among phases: $F = 0.60$, $p = 0.662$).

PHASE	MEAN VD	n[a]	SD	RESULTS OF t-TESTS FOR DIFFERENCE FROM MEAN OF NULL SAMPLE[b]
CM/TCM	0.421	14	0.170	$t = 1.47$, $p = 0.142$
CM	0.429	88	0.144	$t = 4.37$, $p < 0.001$
TC/CM	0.459	50	0.140	$t = 4.95$, $p < 0.001$
GT/SF	0.483	5	0.188	$t = 1.54$, $p = 0.125$
EP	0.464	21	0.174	$t = 2.86$, $p = 0.005$

[a] Sum of the number of scan sites present on all of the specimens within each sample.
[b] Mean = 0.353, $n = 316$, SD = 0.146. All p values are for two-tailed tests.

I also note that there is no significant correlation between the mean utility of the body parts represented in these samples and the mean density of the scan sites that are present in them ($r = -0.44$, one-tailed $p = 0.230$). If higher degrees of density-mediated attrition have systematically lowered mean utility values among a set of samples, then a negative correlation should be apparent between their mean utility values and their mean density values. The absence of a significant correlation between mean density and mean utility among the Mimbres Valley samples that I use here suggests that density-mediated taphonomic processes have not had a substantial systematic effect on the mean utility values observed in them.

It thus seems that the proportion of low food value artiodactyl body parts that hunters transported to sites in the central Mimbres Valley increased between the Early Pithouse period and the Georgetown or San Francisco phases and then declined again either late in the Classic Mimbres phase or during the Terminal Classic (Table 13.5). It can also be concluded that hunters in the valley selectively excluded low-utility body parts from the loads that they carried home during the earliest and the latest time periods represented here, but there is no reason to think that they were particularly selective in this regard during the intervening time periods.

Considering the implications of the central place forager prey choice model that I presented above, the decline in mean utility that occurs between the Early Pithouse sample and the Georgetown/San Francisco sample would seem to indicate that search and transport times for artiodactyl prey declined in the central Mimbres Valley during this span of time. The taxonomic relative abundance data from these samples, on the other hand, suggest that artiodactyl search and transport times increased over this period. A possible reason for this apparent discrepancy between the results of my taxonomic relative abundance analysis and those of my body part analysis may be that this case violates a key assumption that I made in order to apply the model.

Specifically, I noted that archaeological application of the central place forager model requires assuming that the distribution of maximum load sizes across foraging trips remained constant over time. However, there is reason to think that effective load sizes may have increased in the Mimbres Valley during the period of time in question. Settlements in the valley appear to have grown substantially during the Early and Late Pithouse periods (e.g., Anyon et al. 1981), and it is possible that the average size of hunting parties grew as communities became larger. If this occurred, then the size of the total load of meat that could have been carried home by an average-sized hunting party would have increased, and if the number of animals captured per party increased by a lesser proportion, then the benefits that could be obtained from spending additional time removing low-utility body parts prior to transport would have been reduced. The mean utility of the body parts that hunters carried home might therefore have declined over this period due to an increase in the average size of hunting parties.

I point out that if such an increase in average hunting party size did occur, and if the average number of animals captured per party did not increase by a proportionately equivalent amount, thereby lowering the benefits to be gained by removing low-utility body parts from the loads that were carried home, then the energetic delivery rates obtained by individual hunters would necessarily have declined. In conjunction with the reduction in the abundance of large mammals on the Mimbres Valley landscape that is implied by the artiodactyl relative abundance data that I have presented, this suggests that foraging efficiency declined substantially for Mimbres hunters between the Early Pithouse period and the beginning of the Three Circle phase.

The other major change in mean artiodactyl body part utility that occurs among these samples is an increase between the Classic Mimbres sample and the Classic Mimbres/Terminal Classic sample. Taking the logic of the central place forager prey choice model into account, the reduction in the transport of low-utility body parts that is entailed by this increase in mean utility suggests that average search and transport times for artiodactyl prey increased over this span of time. Such an increase in artiodactyl search and transport times is not evident in the taxonomic relative abundance data from these samples, but when the possible cause of this increase in artiodactyl search and transport time is considered, an explanation also appears for the absence of a coincident decline in the taxonomic abundance of artiodactyls relative to leporids.

The most striking feature of the dendroclimatic records that are available for southwestern New Mexico is a period of several severe drought years that occurred during the AD 1130s, and a reduction in the size of the human population in the Mimbres Valley that apparently began at this time has long been considered to have been a result of this drought (e.g., Creel 1999; Minnis 1985). The latest faunal sample that I use in this study comes from deposits that contain substantial quantities of pottery types that date to the Terminal Classic (i.e., post-AD 1130), and it is therefore possible that the increase in artiodactyl search and transport times that is suggested by the increase in mean body part utility observed in this sample was a response to this period of drought.

Droughts can reduce the population densities of artiodactyls such as deer and pronghorn by lowering fawn recruitment rates (e.g., Bradybaugh and Howard 1982; Kitchen and O'Gara 1982; Mackie et al. 1982). Thus, the drought of the AD 1130s may have reduced the density of artiodactyls on the Mimbres Valley landscape, thereby raising the average amount of time required to locate these prey and transport them home. The model that I have presented here predicts that such an increase in average search and transport times for artiodactyl prey will be reflected by a reduction in the proportion of low food value body parts transported to residential sites. However, such an increase in artiodactyl search and transport times would not result in a decline in the numbers of artiodactyls captured relative to the numbers of leporids captured if the density of leporids on the landscape were reduced to approximately the same degree as was the density of artiodactyls. Annual precipitation can affect the size of leporid populations in arid western North America (e.g., Davis et al. 1975), and it is thus possible that abundances of both leporids and artiodactyls declined in the Mimbres Valley at the beginning of the Terminal Classic. This would account for the patterns observed in archaeofaunal samples from the valley, and it would certainly also have resulted in a considerable decline in hunting efficiency.

Conclusions

My application of the central place forager prey choice model to faunal data from the Mimbres Valley has provided a more complete test than has yet been attempted of a longstanding hypothesis about the effects of human predation on populations of large mammals in the region. It has also led to a novel hypothesis about certain social aspects of hunting in the Mimbres area that would not have been generated had such a model not been employed.

By integrating several issues that have previously been approached through separate theoretical models, the model that I have presented allows easier determination of the effects that factors such as prey choice, transport distance, and field processing will combine to have on overall foraging efficiency. Using this model to track changes over time in foraging efficiency in the Mimbres Valley, some important substantive conclusions can be drawn. First, the dramatic early decline in artiodactyl relative abundance that occurs in faunal assemblages from the central part of the valley suggests that average search and transport times for artiodactyl prey increased here between the initial occupation of the Mimbres region by farmers and the beginning of the Three Circle phase. In other words, there is good reason to believe that deer and pronghorn were more abundant on the landscape of the Mimbres Valley during the few centuries before AD 800 or 850 than they were during the few centuries after this. This reduction in the abundance of artiodactyls in the valley, which is best explained as being the result of increases in absolute rates of harvest by human hunters (Cannon 2001a), likely caused a substantial decline in the per capita efficiency with which those hunters were able to capture prey and transport them home.

From the beginning of the Three Circle phase through the Terminal Classic, however, there are no appreciable changes in the relative abundances of artiodactyls in faunal samples from Mimbres Valley sites. The absence of a continued decline in artiodactyl relative abundance during this span of time suggests that the density of these animals on the landscape remained steady, perhaps because they were now so rare that people set out to hunt them only at occasional times of the year when search and transport times for them were likely to be low. The increase in the mean utility of the artiodactyl body parts deposited at residential sites that occurred either late in the Classic Mimbres phase or during the Terminal Classic suggests that the situation changed at this time as people began to travel farther from home to hunt large mammals. The model that I have presented shows that such an increase in travel and transport distance, which may have been necessitated by the drought of the AD 1130s, would have entailed further declines in hunting efficiency beyond those that occurred early in the Mimbres sequence.

These results are consistent with those of recent work on other aspects of the Mimbres-Mogollon archaeological record, which suggest that important economic changes occurred much earlier in the region than has previously been thought. Archaeologists working here have traditionally devoted most of their attention to the Classic Mimbres phase, and they have usually argued that, aside from the changes that occurred at the end of this phase (i.e., ca. AD 1130), the most important changes—including the depression of large mammal resources and major increases in the importance of agriculture—occurred at the beginning of it (i.e., ca. AD 1000), coincident with the transition from pithouse to pueblo architecture (see, for example, the overviews in Anyon et al. 1981; Anyon and LeBlanc 1984; LeBlanc 1989; Minnis 1985; Sanchez 1996). However, Diehl (1996, 1997; see also Diehl and LeBlanc 2001) has argued that residential mobility began to decline throughout the region, and that agriculture began to increase in importance, well before AD 1000, during the same span of time in which the archaeofaunal taxonomic relative abundance data that I discussed above suggest that people in the Mimbres Valley experienced depression of large mammal resources. I have argued elsewhere that the available evidence from the valley is consistent with the hypothesis that people began to devote more time to farming in response to the decline in foraging efficiency that would have resulted from this early occurrence of resource depression. (Cannon 2001a).

In addition to providing insights into changes in foraging efficiency in the Mimbres Valley, my application of the central place forager prey choice model illustrates how the use of explicit theoretical constructions provides benefits regardless of whether the specific predictions that can be derived from them are met (e.g., Stephens and Krebs 1986). The central place forager model points out an aspect of the Mimbres Valley empirical record that cannot be explained straightforwardly in terms of the declining availability of high-return prey on the landscape: the reduction in mean artiodactyl body part utility that occurs between the Early Pithouse period and the Georgetown or San Francisco phase, which is the same period of time over which the taxonomic relative abundance of artiodactyls declines dramatically. Without a model that can be used to link both prey choice and body part transport practices to changes in the abundances of prey around a central place, it would be far more difficult to see that the taxonomic relative abundance data and the body part data from the Mimbres Valley may be telling us different things, and a potential avenue for future research would likely have gone unnoticed. Though the hypothesis that the average size of hunting parties increased over time requires further evaluation, if hunting parties in the Mimbres Valley did become larger, this would likely reflect an increase in the degree to which hunting was carried out communally that would have important implications in its own right (e.g., Giraldeau and Caraco 2000).

Acknowledgments

Jack Broughton, Ben Fitzhugh, Donald Grayson and Eric Smith provided very helpful insights and suggestions concerning earlier incarnations of the ideas contained here, and their input improved those ideas immensely. I would also like to thank Lee Lyman and an anonymous reviewer for helpful comments on this paper. I would not have been able to use the faunal material discussed here without the generous assistance of Roger Anyon, Patricia Gilman and Steven LeBlanc of the Mimbres Foundation, nor could I have done so without the help of the staff of the Maxwell Museum of Anthropology at the University of New Mexico, which loaned this material to me and provided me with access to Mimbres Foundation records; I would particularly like to thank Bruce Huckell, Mary Anne Jaffe, Dorothy Larson and Michael Lewis of this institution. The research summarized in this paper was supported by a National Science Foundation Dissertation Improvement Grant (BCS-9909339).

Notes

1. In archaeology, of course, it is possible to do this only in situations in which each prey type can be found exclusively in a single kind of patch (Broughton 1994a; Nagaoka 2000). In an island setting, for example, pelagic fishes will not be found in the forest and large flightless birds will not be found out at sea. In other settings, however, a single prey type might be present in several patches. In western North America, for example, jackrabbits and deer can be found in both woodland and grassland habitats, but they will occur in different frequencies in each type of habitat, and it will not usually be possible to determine from which of these patches any individual archaeological specimen was obtained.

2. I use the terms "prey type" and "prey item" as they are used in Smith (1991:205), though the model that I present requires slight modification of the definition of a prey type: a prey item is a single unit of harvest, and a prey type is defined by the variables e_i, m_i, and c_i and by the processing function $E_i(p_i)$.

3. In his model of optimal site location for central place foragers, Zeanah (2000) assumes a maximum load weight of 30 kg. O'Connell et al. (1988; see also O'Connell et al. 1990) found that Hadza foragers in East Africa usually carry loads of meat that average 10 to 20 kg per person, though when they kill or scavenge the carcasses of giraffes (which weigh more than twice as much as the next largest prey species regularly taken), loads might weigh as much as 45 kg per person. For plant resources, Bettinger et al. (1997) have analyzed ethnographically collected baskets from California, and they suggest that the average load size transported on daily "residential" foraging trips was approximately 15 kg, while the average load size carried on longer distance "logistical" foraging trips was approximately 36 kg (see also Barlow et al. 1993). Even if the largest value of 45 kg were used as an estimate of maximum load size for Mimbres Valley hunters, most individuals of the artiodactyl species that are the subject of this study would fall above this value (see weights in Chapman and Feldhamer 1982, for example).

4. Large-bodied vertebrates, of course, are not the only kinds of resources that might fall into this class of prey types. For example, if individuals of a smaller-bodied vertebrate taxon are encountered and harvested simultaneously, then the entire group of harvested individuals is the relevant prey item (e.g., Madsen and Schmitt 1998), and such aggregate prey items might also exceed L_{max}. In cases like this foragers will have to do one or both of two things: leave some of the individuals that are collected behind, or process at least some of them so that a load of the maximum size is obtained (see Cannon 2001a for further details on this and related points).

5. This is similar to the "utility function" of Metcalfe and Barlow (1992), but I have changed the name to distinguish this function from the overall energetic delivery rate in the model that I present here (R_i), which could also be considered to be a "utility function."

6. Presently available measures of vertebrate body part utility (e.g., Binford 1978; Lyman et al. 1992; Metcalfe and Jones 1988) include no information about the amount of time required to remove various parts, or various combinations of parts, from a carcass. Thus, though useful for some purposes, these measures cannot be used to determine processing functions for vertebrate resources (cf. Metcalfe and Jones 1988).

7. Specifically, this will occur when $R_2 > R_1$, which, if prey type 1 is a "large" resource and prey type 2 is a "small" resource, will occur when $s_1 > [(c_2/e_2)E_1(p_1^*)] - (h_1 + p_1^*)$; see equations 13.1 and 13.2.

8. The Georgetown/San Francisco sample comes from Galaz, and this material cannot be dated any more precisely than "Georgetown phase and/or San Francisco phase." The Three Circle/Classic Mimbres and Classic Mimbres samples include material from both Galaz and Mattocks, and the former of these two samples contains material that likely dates to both the Three Circle and the Classic Mimbres phases. The Classic Mimbres/Terminal Classic sample comes from Mattocks and contains material that dates to both the Classic Mimbres phase and the Terminal Classic. There is also a faunal sample from Mattocks that contains material dating to perhaps as late as the AD 1300s or 1400s, but I do not include this sample here because it has apparently been subjected to a taphonomic history that is quite different from those experienced by the other faunal samples from this site (see Cannon 2001a).

9. The statistical methods that I use here all take sample size into account; in other words, the differences in artiodactyl relative abundance among these samples are large enough to be statistically significant despite the small size of the two earliest samples.

10. An objection might be raised to my application of utility values derived for caribou to the remains of deer and pronghorn. However, because I do not use the absolute caloric content of caribou body parts, but instead use Metcalfe and Jones's (1988) standardized FUI values, in which the caloric content of a body part is expressed as a percentage of the caloric content of the highest utility body part, all that must be assumed is that the utility of various body parts is proportionately the same for all of the taxa involved. Such an assumption would be unwarranted if these taxa varied greatly in anatomical structure, as do deer and bison, for example (e.g., Kruetzer 1992), but given the overall similarity in body form among caribou, deer and pronghorn, this assumption is not likely to be problematic for my analysis.

11. I include in my analysis all of the scan sites that are present on each individual bone specimen, rather than simply a "typical" scan site for each portion of each element as has been done in previous studies (e.g., Lyman 1985, Grayson 1989), because this should provide a more accurate representation of the degree to which density-mediated taphonomic processes have affected these samples. For example, if both high- and low-density scan sites are present on specimens, and if only the highest-density scan sites on those specimens are included in a density analysis, then

the effects of density-mediated attrition will be exaggerated. The density value that I use for each scan site is the mean of all of the volume density measurements for that scan site published in Lyman (1984) for both deer and pronghorn.

12. I exclude specimens from very young individuals from my body part analyses; see Cannon (2001a) for further details about the procedures followed here.

References Cited

Anyon, R., P. A. Gilman, S. A. LeBlanc
1981   A reevaluation of the Mogollon-Mimbres archaeological sequence. *Kiva* 46:209–225.

Anyon, R., S. A. LeBlanc
1984   *The Galaz Ruin: A prehistoric Mimbres Village in Southwestern New Mexico.* University of New Mexico Press, Albuquerque.

Arthur, J. W.
1994   Ceramic function during the Early Pithouse period in the Mimbres River Valley. Masters thesis, University of Texas at San Antonio.

Barlow, K. R.
1997   Foragers that farm: A behavioral ecology approach to the economics of corn farming for the Fremont case. Ph.D. dissertation, University of Utah, Salt Lake City.
2002   Predicting maize agriculture among the Fremont: An economic comparison of farming and foraging in the American southwest. *American Antiquity* 67:65–88.

Barlow, K. R., P. R. Henriksen, D. Metcalfe
1993   Estimating load size in the Great Basin: Data from conical burden baskets. *Utah Archaeology* 6:27–37.

Barlow, K. R., D. Metcalfe
1996   Plant utility indices: Two Great Basin examples. *Journal of Archaeological Science* 23:351–371.

Bayham, F. E.
1979   Factors influencing the Archaic pattern of animal utilization. *Kiva* 44:219–235.

Beck, C., A. Taylor, G. T. Jones, C. M. Fadem, C. R. Cook, S. A. Milward
2002   Rocks are heavy: Transport costs and Paleoarchaic quarry behavior in the Great Basin. *Journal of Anthropological Archaeology* 21:481–507.

Bettinger, R. L., R. Malhi, H. McCarthy
1997   Central place models of acorn and mussel processing. *Journal of Archaeological Science* 24:887–899.

Binford, L. R.
1978   *Nunamiut Ethnoarchaeology.* Academic Press, New York.

Bird, D. W., R. L. Bliege Bird
1997   Contemporary shellfish gathering strategies among the Merriam of the Torres Straits Islands, Australia: Testing predictions of a central place foraging model. *Journal of Archaeological Science* 24:39–63.

Bird, D. W., J. L. Richardson, P. M. Veth, A. J. Barham
2002   Explaining shellfish variability in middens on the Merriam Islands, Torres Strait, Australia. *Journal of Archaeological Science* 29:457–469.

Blake, M., S. A. LeBlanc, P. E. Minnis
1986   Changing settlement and population in the Mimbres Valley, SW New Mexico. *Journal of Field Archaeology* 13:439–464.

Bradybaugh, J. S., V. W. Howard, Jr.
1982   Effects of severe weather on pronghorn reproduction and fawn survival in southeastern New Mexico. *Proceedings of the Tenth Pronghorn Antelope Workshop, Dickinson, North Dakota, April 5–7, 1982*, pp. 143–155.

Broughton, J. M.
1994a   Declines in mammalian foraging efficiency during the late Holocene, San Francisco Bay, California. *Journal of Anthropological Archaeology* 13:371–401.
1994b   Late Holocene resource intensification in the Sacramento Valley, California: The vertebrate evidence. *Journal of Archaeological Science* 21:501–514.
1997   Widening diet breadth, declining foraging efficiency, and prehistoric harvest pressure: Icthyofaunal evidence from the Emeryville Shellmound, California. *Antiquity* 71:845–862.
1999   *Resource Depression and Intensification During the late Holocene, San Francisco Bay: Evidence from the Emeryville Shellmound Vertebrate Fauna.* University of California Publications: Anthropological Records 32. University of California Press, Berkeley.
2002a   Prey spatial structure and behavior affect archaeological tests of optimal foraging models: Examples from the Emeryville Shellmound vertebrate fauna. *World Archaeology* 34:60–83.
2002b   Pre-Columbian human impacts on California vertebrates: Evidence from old bones and implications for wilderness policy. In *Wilderness and Political Ecology: Aboriginal Influences and the Original State of Nature*, edited by C. Kay and R. Simmons. University of Utah Press, Salt Lake City.

Broughton, J. M., D. K. Grayson
1993   Diet breadth, adaptive change, and the White Mountains faunas. *Journal of Archaeological Science* 20:331–336.

Broughton, J. M., J. F. O'Connell
1999   On evolutionary ecology, selectionist archaeology, and behavioral archaeology. *American Antiquity* 64:153–165.

Cannon, M. D.
2000   Large mammal relative abundance in pithouse and pueblo period archaeofaunas from southwestern New Mexico: Resource depression among the Mimbres-Mogollon? *Journal of Anthropological Archaeology* 19:317–347.
2001a   Large mammal resource depression and agricultural intensification: An empirical test in the Mimbres Valley, New Mexico. Ph.D. dissertation, University of Washington, Seattle.
2001b   Archaeofaunal relative abundance, sample size, and statistical methods. *Journal of Archaeological Science* 28:185–195.

Chapman, J. A., G. A. Feldhamer (Eds.)
1982   *Wild Mammals of North America: Biology, Management and Economics.* Johns Hopkins University Press, Baltimore.

Charnov, E. L.
1976   Optimal foraging, the marginal value theorem. *Theoretical Population Biology* 9:129–136.

Charnov, E. L., G. H. Orians, K. Hyatt
1976   Ecological implications of resource depression. *The American Naturalist* 110:247–259.

Creel, D. G.
1999   The Black Mountain phase in the Mimbres area. In *The Casas Grandes World*, edited by Curtis F. Schaafsma and Carroll L. Riley, pp. 107–120. University of Utah Press, Salt Lake City.

Davis, C. A., J. A. Medlin, J. P. Griffing
1975   *Abundance of Black-Tailed Jackrabbits, Desert Cottontail Rabbits, and Coyotes in Southeastern New Mexico.* Research Report 293. Agricultural Experiment Station, New Mexico State University, Las Cruces.

Diehl, M. W.
1994 Subsistence economies and emergent social differences: A case study from the prehistoric North American Southwest. Ph.D. dissertation, State University of New York at Buffalo.
1996 The intensity of maize processing and production in upland Mogollon pithouse villages AD 200–1000. *American Antiquity* 61:102–115.
1997 Changes in architecture and land use strategies in the American southwest: Upland Mogollon pithouse dwellers, AC 200–1000. *Journal of Field Archaeology* 24:179–194.

Diehl, M. W., S. A. LeBlanc
2001 *Early Pithouse Villages of the Mimbres Valley and Beyond: The McAnally and Thompson Sites in their Cultural and Ecological Contexts*. Papers of the Peabody Museum of Archaeology and Ethnology 83. Harvard University, Cambridge.

Everitt, B. S.
1977 *The Analysis of Contingency Tables*. John Wiley and Sons, New York.

Felger, R. S., M. B. Moser
1985 *People of the Desert and Sea: Ethnobotany of the Seri Indians*. University of Arizona Press, Tucson.

Gilman, P. A., S. A. LeBlanc
In prep. *Early Aggregation in the Prehistoric Southwest: The Mattocks Site in the Mimbres Valley, New Mexico*.

Giraldeau, L.-A., T. Caraco
2000 *Social Foraging Theory*. Princeton University Press, Princeton.

Grayson, D. K.
1989 Bone transport, bone destruction, and reverse utility curves. *Journal of Archaeological Science* 16:643–652.
1991 Alpine faunas from the White Mountains, California: Adaptive change in the prehistoric Great Basin? *Journal of Archaeological Science* 18:483–506.
2001 The archaeological record of human impacts on animal populations. *Journal of World Prehistory* 15:1–68.

Grayson, D. K., M. D. Cannon
1999 Human paleoecology and foraging theory in the Great Basin. In *Models for the Millennium: Great Basin Anthropology Today*, edited by Charlotte Beck, pp. 141–151. University of Utah Press, Salt Lake City.

Hegmon, M., M. C. Nelson, R. Anyon, D. Creel, S. A. LeBlanc, H. J. Shafer
1999 Scale and time-space systematics in the post-AD 1100 Mimbres region of the North American southwest. *Kiva* 65:143–166.

Hildebrandt, W. R., T. L. Jones
1992 Evolution of marine mammal hunting: A view from the California and Oregon coasts. *Journal of Anthropological Archaeology* 11:360–401.

Hildebrandt, W. R., K. R. McGuire
2002 The ascendance of hunting during the California Middle Archaic: An evolutionary perspective. *American Antiquity* 67:231–256.

Hill, K., J. Padwe, C. Bejyvagi, A. Bepurangi, F. Jakugi, R. Tykuarangi, T. Tykuarangi
1997 Impact of hunting on large vertebrates in the Mbaracayu Reserve, Paraguay. *Conservation Biology* 11:1339–1353.

Janetski, J. C.
1997 Fremont hunting and resource intensification in the eastern Great Basin. *Journal of Archaeological Science* 24:1075–1088.

Jones, K. T., D. B. Madsen
1989 Calculating the cost of resource transportation: A Great Basin example. *Current Anthropology* 30:529–534.

Jones, T. L., W. R. Hildebrandt
1995 Reasserting a prehistoric tragedy of the commons: A reply to Lyman. *Journal of Anthropological Archaeology* 14:78–98.

Kaplan, H., K. Hill
1992 The evolutionary ecology of food acquisition. In *Evolutionary Ecology and Human Behavior*, edited by Eric A. Smith and Bruce Winterhalder, pp. 167–201. Aldine de Gruyter, New York.

Kay, C. E.
1994 Aboriginal overkill: The role of Native Americans in structuring western ecosystems. *Human Nature* 5:359–398.
1998 Are ecosystems structured from the top-down or bottom-up: A new look at an old debate. *Wildlife Society Bulletin* 26:484–498.

Kitchen, D. W., B. W. O'Gara
1982 Pronghorn: *Antilocapra americana*. In *Wild Mammals of North America*, edited by Joseph A. Chapman and George A. Feldhamer, pp. 960–971. Johns Hopkins University Press, Baltimore.

Koike, H., N. Ohtaishi
1985 Prehistoric hunting pressure estimated by the age composition of excavated sika deer (*Cervus nippon*) using the annual layer of tooth cement. *Journal of Archaeological Science* 12:443–456.
1987 Estimation of prehistoric hunting rates based on age composition of sika deer (*Cervus nippon*). *Journal of Archaeological Science* 14:251–269.

Kreutzer, L. A.
1992 Bison and deer bone mineral densities: Comparisons and implications for the interpretation of archaeological faunas. *Journal of Archaeological Science* 19:271–294.

LeBlanc, S. A.
1975 *Mimbres Archaeological Center: Preliminary report of the first season of excavation, 1974*. Institute of Archaeology, University of California, Los Angeles.
1976a Mimbres Archaeological Center: Preliminary report of the second season of excavation 1975. *Journal of New World Archaeology* 1:1–23.
1976b The 1976 field season of the Mimbres Foundation in southwestern New Mexico. *Journal of New World Archaeology* 2:1–24.
1989 Cultural dynamics in the southern Mogollon area. In *Dynamics of Southwest prehistory*, edited by Linda S. Cordell and George J. Gumerman, pp. 179–207. Smithsonian Institution Press, Washington, DC.

Lyman, R. L.
1984 Bone density and differential survivorship of fossil classes. *Journal of Anthropological Archaeology* 3:259–299.
1985 Bone frequencies: Differential transport, in situ destruction, and the MGUI. *Journal of Archaeological Science* 12:221–236.
1995 On the evolution of marine mammal hunting on the west coast of North America. *Journal of Anthropological Archaeology* 14:45–77.

Lyman, R. L., J. M. Savelle, P. Whitridge
1992 Derivation and application of a meat utility index for phocid seals. *Journal of Archaeological Science* 19:531–555.

Lyman, R. L., S. Wolverton
2002 The late prehistoric-early historic game sink in the northwestern United States. *Conservation Biology* 16:73–85.

Mackie, R. J., K. L. Hamlin, D. F. Pac
1982 Mule deer: *Odocoileus hemionus*. In *Wild Mammals of North America*, edited by Joseph A. Chapman and George A. Feldhamer, pp. 862–877. Johns Hopkins University Press, Baltimore.

Madsen, D. B., D. N. Schmitt
1998 Mass collecting and the diet breadth model: A Great Basin example. *Journal of Archaeological Science* 25:445–455.

Metcalfe, D., K. R. Barlow
1992 A model for exploring the optimal tradeoff between field processing and transport. *American Anthropologist* 94:340–356.

Metcalfe, D., K. T. Jones
1988 A reconsideration of animal body-part utility indices. *American Antiquity* 53:486–504.

Minnis, P. E.
1985 *Social Adaptation to Food Stress: A Prehistoric Southwestern Example*. University of Chicago Press, Chicago.

Nagaoka, L. A.
2000 Resource depression, extinction, and subsistence change in prehistoric southern New Zealand. Ph.D. dissertation, University of Washington, Seattle.
2002 Explaining subsistence change in southern New Zealand using foraging theory models. *World Archaeology* 34:84–102.

Nelson, B. A., R. Anyon
1996 Fallow valleys: Asynchronous occupations in southwestern New Mexico. *Kiva* 61:275–294.

Nelson, R. K.
1983 *Make Prayers to the Raven: A Koyukon View of the Northern Forest*. University of Chicago Press, Chicago.

O'Connell, J. F., K. Hawkes, N. Blurton Jones
1988 Hadza hunting, butchering, and bone transport and their archaeological implications. *Journal of Anthropological Research* 44:113–161.
1990 Reanalysis of large mammal body part transport among the Hadza. *Journal of Archaeological Science* 17:301–316.

Orians, G. H., N. E. Pearson
1979 On the theory of central place foraging. In *Analysis of Ecological Systems*, edited by David J. Horn, Rodger D. Mitchell, and Gordon R. Stairs, pp. 154–177. The Ohio State University Press, Columbus.

Perkins Jr., D., P. Daly
1968 A hunters' village in Neolithic Turkey. *Scientific American* 219:97–106.

Powell, S. J.
1977 Changes in prehistoric hunting patterns resulting from agricultural alteration of the environment: A case study from the Mimbres River area, New Mexico. Master's thesis, University of California, Los Angeles.

Rea, A. M.
1998 *Folk Mammalogy of the Northern Pimans*. University of Arizona Press, Tucson.

Rogers, A. R.
2000 Analysis of bone counts by maximum likelihood. *Journal of Archaeological Science* 27:111–125.

Rogers, A. R., J. M. Broughton
2001 Selective transport of animal parts by ancient hunters: A new statistical method and an application to the Emeryville Shellmound fauna. *Journal of Archaeological Science* 28:763–773.

Sanchez, J. L. J.
1996 A re-evaluation of Mimbres faunal subsistence. *Kiva* 61:295–307.

Shaffer, B. S., C. P. Schick
1995 Environment and animal procurement by the Mogollon of the southwest. *North American Archaeologist* 16:117–132.

Smith, E. A.
1991 *Inujjuamiut Foraging Strategies: Evolutionary Ecology of an Arctic Hunting Economy*. Aldine de Gruyter, New York.

Stephens, D. W., J. R. Krebs
1986 *Foraging Theory*. Princeton University Press, Princeton.

Szuter, C. R., F. E. Bayham
1989 Sedentism and prehistoric animal procurement among desert horticulturalists of the North American southwest. In *Farmers as Hunters: The Implications of Sedentism*, edited by Susan Kent, pp. 80–95. Cambridge University Press, Cambridge.

Thomas, D. H., D. Mayer
1983 Behavioral faunal analysis of selected horizons. In *The Archaeology of Monitor Valley. 2. Gatecliff Shelter*, by David H. Thomas, pp. 353–391. Anthropological Papers of the American Museum of Natural History 59(pt. 1). American Museum of Natural History, New York.

Ugan, A., J. Bright
2001 Measuring foraging efficiency with archaeological faunas: The relationship between relative abundance indices and foraging returns. *Journal of Archaeological Science* 28:1309–1321.

White, T. E.
1954 Observations on the butchering technique of some aboriginal peoples: Nos. 3, 4, 5, and 6. *American Antiquity* 19:254–264.

Wills, W. H.
1996 The transition from the preceramic to ceramic period in the Mogollon Highlands of western New Mexico. *Journal of Field Archaeology* 23:335–359.

Winterhalder, B., C. Goland
1997 An evolutionary ecology perspective on diet choice, risk, and plant domestication. In *People, Plants, and Landscapes: Studies in Paleoethnobotany*, edited by Kristen J. Gremillion, pp. 123–160. University of Alabama Press, Tuscaloosa.

Zar, J. H.
1999 *Biostatistical Analysis (Fourth Edition)*. Prentice Hall, Upper Saddle River, New Jersey.

Zeanah, D. W.
2000 Transport costs, central-place foraging, and hunter-gatherer alpine land-use strategies. In *Intermountain Archaeology*, edited by David B. Madsen and Michael D. Metcalf, pp. 1–14. University of Utah Anthropological Papers 122. University of Utah Press, Salt Lake City.

14

Sexual Division of Labor and Central Place Foraging

A Model for the Carson Desert of Western Nevada

David W. Zeanah

Ethnographers emphasized the contribution of women's labor to the diet of Great Basin hunter-gatherers, but whether or not the ethnographic record serves as a valid analogy for prehistoric economies is a longstanding debate. Given the importance of gathered foods among ethnographic foragers, it seems no surprise that archaeologists seeking prehistoric variability that diverged from ethnographic accounts often envision prehistoric subsistence-settlement systems that emphasized hunting.[1] Many archaeologists suspect that Early Holocene (10,000 to 6000 BP) foragers stressed hunting (Butler 1986; Elston 1982, 1986; O'Connell et al. 1982; Tuohy 1974; Warren 1967),[2] and some see evidence that hunting-oriented adaptations persisted into the Late Holocene (Bettinger 1999; Bettinger and Baumhoff 1982; Hildebrandt and McGuire 2002; Kelly 1985).

Such interpretations imply that the relative influence of men's and women's labor on subsistence and settlement decisions changed over time. However, archaeologists usually have paid little attention to behavioral implications of such variability, perhaps because a cultural ecological perspective assumes that sexual division of labor served common household needs (Steward 1938:44, 230). In recent decades, archaeologists who employ evolutionary theory have realized that sexual division of labor has profound implications for models of subsistence change (Bettinger and Baumhoff 1982; Hildebrandt and McGuire 2002; Kelly 1997; Leach 1999; Simms 1987). This paper draws an evolutionary perspective from human behavioral ecology that recognizes men and women may forage to achieve different goals and sometimes their foraging interests conflict. It develops a central place foraging model for the Carson Desert of western Nevada (Fig. 14.1) by evaluating men's and women's foraging strategies separately and considering implications of conflicting foraging interest for subsistence-settlement variability.

A Behavioral Ecology Perspective of the Sexual Division of Labor Among Hunter-Gatherers

Many anthropologists perceive sexual division of labor as fundamental to hunter-gatherer subsistence organization (Ember 1975; Hiatt 1978), and many suspect provisioning of offspring by hunting fathers was a significant force in hominid evolution (Washburn and Lancaster 1967). However, they often assume that male and female foragers equally benefit from and participate in child provisioning. Recent investigations of modern hunter-gatherers based in behavioral ecology are persuasive that this assumption is unwarranted (Bliege Bird 1999; Hawkes 1996; Hawkes and Bliege Bird 2002).

Both men and women have the option of investing resources either to provision children or have additional offspring. They must monitor costs and benefits of each alternative to maximize reproductive fitness. However, tradeoffs differ between sexes. Females are likely to benefit most from parental effort because they are certain which offspring are theirs and have relatively few reproductive opportunities, each of which is relatively costly and risky. In contrast, males have no absolute certainty of paternity, but may have many more mating opportunities bearing relatively low costs and risks. Therefore, natural selection is more likely to favor male reproductive strategies that stress mating effort and female strategies that emphasize parental investment (cf. Maynard Smith 1978; Williams 1975). This does not mean that selection never favors men who provision offspring or women who pursue additional mates, but the point at which these options become most beneficial for individual fitness differs between the two sexes. Differences in fitness goals create conflicts of interest between men and women that are expressed in sexual division of subsistence labor.

*Originally published in *Journal of Anthropological Archaeology* 23(2004):1–32

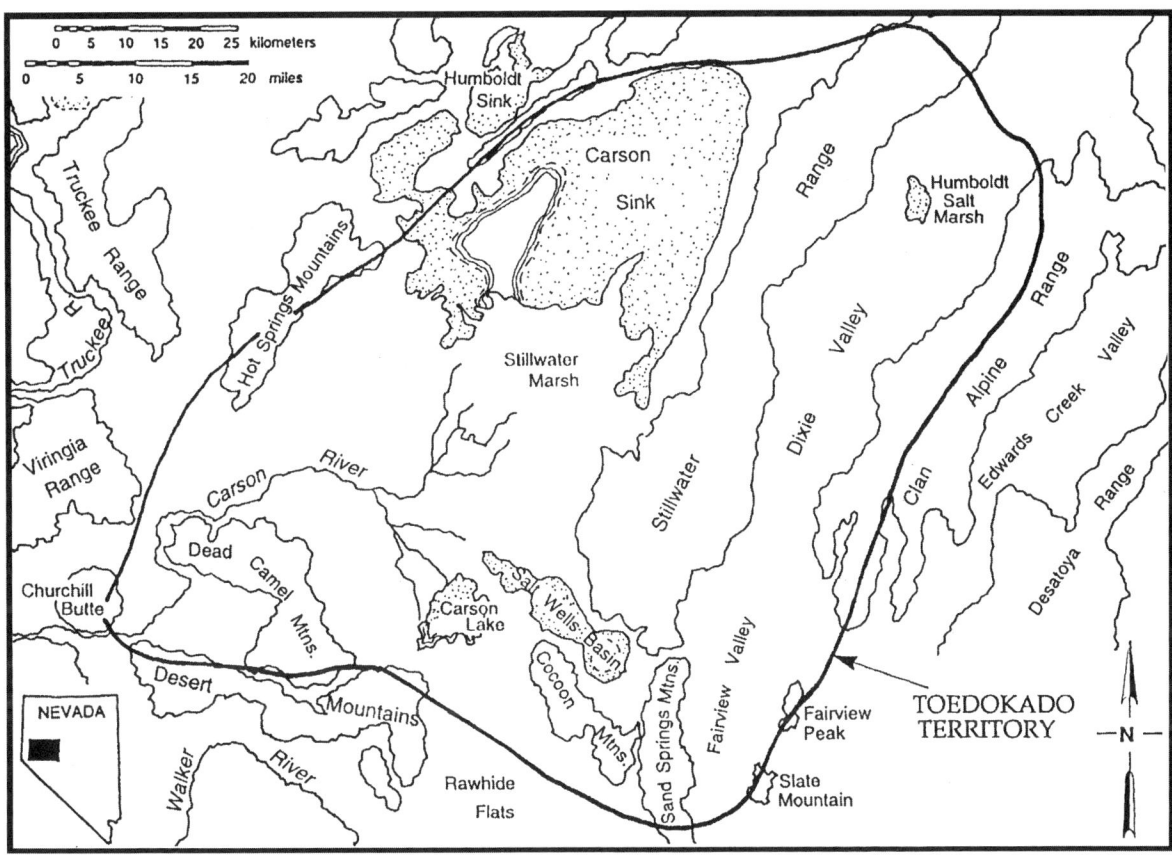

FIGURE 14.1. Map of the Carson Desert region showing boundary of ethnographic Toedokado Territory and physiographic landmarks (after Shimkin and Reid, 1970; Thomas, 1985).

Among hunter-gatherers, men and women often procure different arrays of prey; although women hunt large game in some ethnographic cases (Estioko-Griffen and Griffen 1981), men usually hunt larger game whereas women gather plants and smaller animals (Bliege Bird 1999). To some extent, child rearing may constrain women's mobility (Hurtado et al. 1985) and limit their opportunities to learn hunting skills (Kaplan et al. 2000), preventing them from pursuing larger, elusive quarry while still overseeing their children's safety. However, the crucial difference between male and female subsistence effort may be variability of foraging returns. Women must provision children daily, so the uncertainty of procuring large game may encourage them to pursue more predictable, but smaller, prey. Also, women benefit whenever another adult supplies food to children, possibly providing a selective advantage for women who mate with hunters that feed offspring (Hawkes 1990).

However, this does not necessarily favor fathers who provision their own children. Men's fitness may profit most by widely sharing occasional large packages of meat in hopes of gaining mating opportunities from several women, rather than by reliably provisioning a single mate and her offspring (Hawkes 1990). Indeed, the fitness benefits offered by hunting may be in the information it reveals about innate qualities of the hunter to prospective mates and competitors, rather than as a means of provisioning mates and offspring (Hawkes and Bliege Bird 2002; Smith and Bliege Bird 2000). Such circumstances would select males who specialize in hunting larger game even if overall hunting returns are low, rare, and unpredictable compared to those to be gained by gathering. Evolutionary ecologists disagree about whether large packages of meat offer nutritional benefits to women and children that are unavailable from the plants and small animals they normally gather (Bliege Bird 1999; Hawkes 1991; Hill 1988; Kaplan et al. 2000). However, it is clear that among some foragers, women may receive no more calories from the fathers of their children than do other prospective mates of the hunter. Also, women may garner fewer calories from hunting fathers than they would have from fathers who gather (Hawkes 1990; Hawkes et al. 2001).

Human behavioral ecologists warn that sexual division of labor cannot be overlooked when applying optimal foraging models to humans because men and women often have different motives for seeking different prey under different constraints (Hawkes 1996; Hill et al. 1987; Simms 1987). However, their intent is not to replace an old stereotype of cooperating parents with a new generalization about deadbeat dads. Instead they advocate an approach that models how foraging strategies might reflect conflicting interests between men and women in different environments.

A Reconsideration of Sexual Division of Labor Among Ethnohistoric Great Basin Foragers

Ethnographers observed a division of subsistence labor by gender among Great Basin foragers (Kelly 1932:79; Steward 1938: 44; 1941:312–313; Stewart 1941:406), which they interpreted as cooperative effort directed toward ensuring household self-sufficiency (Steward 1938:44, 230). However, they also observed aspects of men's and women's foraging effort that do not necessarily accord with a model of inter-gender cooperation in provisioning families. Today, these observations suggest that a human behavioral ecology perspective of sexual division of labor among Great Basin foragers is warranted.

First, men's and women's prey overlapped considerably. Women participated in antelope, rabbit, and waterfowl drives (Fowler 1992:78; Kelly 1932:79) and fished and hunted small game on their own (Fowler 1989:23; 1992:62, 79; Kelly 1932:79). In turn, men harvested pinyon (Fowler 1989:51; Steward 1941: 253; Wheat 1967:29), as well as helped women cultivate (Steward 1938:119; 1941:232), gather (Fowler 1989:46; Steward 1938: 96), and process (Fowler 1989:11; Wheat 1967:29) smaller seeds. Clearly food procurement and processing labor among Great Basin hunter-gatherers were not so sharply divided by sex as sometimes portrayed.

Second, Great Basin subsistence strategies were notable for the contribution of plant food to the diet (Steward 1938:230–231), the importance of women as food providers (Steward 1938:242), the role of seed gathering in annual rounds (Steward 1938:245), and the significance of seed storage in ensuring survival through periods of scarcity (Steward 1938:19–20). This presumably reflected the poverty of large-game hunting opportunities in the Great Basin (Steward 1938:33, 232, 254).

Third, the ethnographic record implies that men and women shared food differently. Women foraged to feed their own families (Steward 1938:20); although they might share gathered foods, they seem not to have been obligated to do so. Productive seed patches were often claimed as private property (Steward 1938:254–256; 1941:254), whereas women owned their seed caches (Steward 1938:20) and occasionally hid them from needy neighbors (Steward 1938:73, 128; 1941:254). In contrast, men widely shared large game outside family households (Steward 1938:231). Although relatives might receive preferential shares, a hunter's family had limited rights to his own kills (Steward 1938: 74, 253). Nevertheless, men's hunting skills were critical for their prospects to marry one or more wives (Steward 1938:240).

If sexual specialization of subsistence effort is a shared strategy for provisioning families, why did the prey of men and women overlap? Why should men hunt at all in an environment impoverished of large game, would they not have fed their families more efficiently by gathering? Also, why should hunters readily share food with non-family members, whereas gatherers only reluctantly share outside the family? These questions arouse suspicion that Great Basin men and women foraged to achieve different ends; men hunted because of the mating opportunities they gained by sharing meat, whereas women gathered to provision their children.

Sexual Division of Labor and Subsistence-Settlement Systems in the Carson Desert

Before the late 1980s, Carson Desert archaeologists rarely considered gender (but see Tuohy 1974:103–104) even though their models implied the relative contribution of men's and women's foraging effort influenced prehistoric subsistence and settlement. Instead, their interest lay in the role of wetlands in forager subsistence-settlement systems. To many it seemed obvious that rich wetlands encouraged intensive use of lacustrine resources and prolonged occupation of marsh-edge base camps, an adaptation Heizer (1967:7) described as "limnosedentary." Although Early Holocene assemblages hinted that the Carson Desert once hosted a more mobile, hunting-oriented adaptation (Tuohy 1974), the prevailing view was that wetlands were the focus of hunting and gathering throughout the Holocene (Heizer and Baumhoff 1970).

Thomas (1985) derived the alternative "limnomobile" model by reinterpreting local ethnohistoric accounts in light of the forager-collector model (cf. Binford 1980). In this scenario, Carson Desert hunter-gatherers maintained high residential and logistic mobility to procure resources from a variety of habitats, including wetlands. Thomas' interpretation of Hidden Cave as one of several logistic cache caves bolstered the limnomobile model by suggesting that subsistence routines took Carson Desert hunter-gatherers away from marshes for prolonged periods.

Kelly (1985) introduced an optimal foraging perspective to the limnosedentary/limnomobile debate by observing that upland game yield higher post-encounter caloric return rates than many wetland resources (cf., Simms 1987), and inferring that the dispersion of large game encouraged hunter-gatherers to remain mobile. Kelly predicted intensive use of marsh resources by sedentary foragers was unlikely except when upland game was scarce. Consequently, marsh resources served as a reliable backup to more profitable upland game.

However, Kelly's survey data suggested that subsistence-settlement patterns in the Carson Desert had shifted emphasis from terrestrial hunting (limnomobile) to marsh-edge gathering (limnosedentary) over time. Kelly reasoned that Neoglacial climatic conditions of the early Late Holocene (4500–1250 BP) fostered higher populations of large game than the Carson Desert uplands were capable of supporting in ethnohistoric times. This allowed foragers to pursue large game by maintaining residential mobility and camping in wetlands only when game was hard to come by. More severe winters after 1250 BP depleted game populations, causing foragers to expand diet breadth and rely on the harvest and storage of lower-ranked seeds. This later strategy emphasized residence near marshes, may have been sedentary at times, and employed logistic mobility to exploit distant upland patches (Kelly 1985:303–308).

Kelly's formulation of limnosedentary and limnomobile models carried profound implications for the contribution of men's and women's foraging effort to subsistence and settlement patterns. Limnomobile strategies emphasized men's resources, whereas limnosedentary strategies stressed women's resources. A shift from camping in men's best hunting patches to residence in women's best gathering patches signaled the transition between strategies. Yet Kelly's optimal foraging perspective treated men's and women's foraging strategies collectively. In a sense, Kelly modeled men and women as the same forager; changes in abundance of high-ranked men's prey caused women's diet breadth to contract or expand accordingly.

Investigators at Intermountain Research undertook a different approach in studies elsewhere in the Carson Desert by modeling men's and women's prey items separately (Raven and Elston 1989; Zeanah et al. 1995). They based their analysis on an *a priori* assumption that childcare tethers women to the vicinity of residential bases; men accommodate this constraint by hunting logistically. Therefore, they reasoned that foragers most efficiently camp in locations that are adjacent to women's best foraging patches but central to men's best patches (Raven and Elston 1989:153). Like Kelly, Intermountain Research investigators ranked resources to evaluate costs and benefits, but employed a fine-grained characterization of resource distributions in the Carson Desert. Their model evaluated wetlands as the best foraging patches for women. Although the fish, waterfowl, and small mammals of wetlands also made marshes among the highest-ranked of men's foraging patches, men frequently found more profitable opportunities hunting upland large game, particularly during the Neoglacial. Nevertheless, women's childcare constraints and foraging opportunities firmly anchored base camps to wetlands, with men logistically hunting in uplands. Intermountain Research investigators concluded marshes were the focus of subsistence in the Carson Desert "for more than five millennia" (Raven 1990:135), contrary to Kelly's (1985) view of variability over that time span.

Independently, Kelly (1990 1995, 1999, 2001), reconsidered Carson Desert foraging strategies using a model based on the costs of transporting resources from marsh and upland patches to lowland residential camps. Kelly also concluded that wetlands were the focus of women's foraging effort, with men logistically hunting in mountains. Thus, predictions of Kelly's revised model duplicated those of the Intermountain Research model.

These predictions received empirical support from bioarchaeological analyses of skeletal populations from Stillwater Marsh (Larsen and Kelly 1995). Stable carbon and nitrogen isotopes collected from bone collagen are consistent with diet emphasizing marsh resources and notable for the lack of evidence for pinyon (an upland resource) consumption (Schoeninger 1995, 1999).[3] Further, osteoarthritis is common in the Stillwater Marsh skeletal sample, particularly in the hips, shoulders, and ankles of men. Long bone cross-sections are sexually dimorphic with the cross-sections of men's long bones being sufficiently more robust than those of women to suggest greater male mobility (Larsen et al. 1995; Larsen and Hutchinson 1999). These differences probably result from mobility over severe terrain, supporting the premise that men were more logistically mobile and traveled over rugged uplands more often than women.

Analysis of 19 coprolites from Hidden Cave provides empirical evidence that women's gathering strategies emphasized wetland resource procurement and hints that women foraged separately from men. Most plant and animal resources in the coprolites originate from wetland habitats, with resources of adjacent dry deserts only secondarily present. A single pinyon hull was the only montane resource identified. Low ratios of progesterone and testosterone to estrogen in the fecal material are evidence that women, rather than men, probably deposited the Hidden Cave coprolites. The absence of men's coprolites suggests that women foraged separately from men in the Carson Desert (Rhode 2003; Rhode et al. 2000:55–56).

Survey and excavation data also support Intermountain Research's and Kelly's predictions about foraging in the Carson Desert. Catastrophic flooding of Stillwater Marsh in the 1980s exposed large, complex archaeological sites containing pit structures, storage facilities, refuse middens, human burials, and diverse artifact assemblages (traits suggestive of prolonged residential occupation). Moreover, these sites contain abundant evidence of intensive marsh exploitation with only traces of upland resource use (Brooks et al. 1988; Raven 1990; Raven and Elston 1988; Raymond and Parks 1989, 1990; Tuohy et al. 1987).

However, Kelly (1995:21–23; 2001:289–294) notes that chronometric evidence from excavated occupation sites and human burials in Stillwater Marsh fall within the period from 1250 BP to 650 BP. This contrasts with cache caves, such as Hidden Cave, that tend to predate 1250 BP, suggesting that they relate to an earlier adaptation. According to Kelly, caching specialized lacustrine subsistence gear implies a residentially mobile strategy that took pre-1250 BP foragers out of marshes for extended periods. Such mobility may have reflected the increased effective moisture of the Neoglacial, which enhanced overall productivity. Earlier foragers used Stillwater Marsh as one stop in a seasonal round that included neighboring wetlands and uplands (including large game hunting opportunities). Pre-1250 BP mobility strategies may also have been driven by periodic crashes of wetland productivity that forced foragers out of marshes.

Although Kelly's later explanations de-emphasize the role of large-game hunting in driving the earlier subsistence-settlement system, the effects of climatic change on hunting opportunities still play a role in determining settlement pattern change over time. In another context,[4] Kelly (1997:26; 2001:299) conjectures that earlier hunters may have been reluctant to adopt subsistence-settlement strategies geared toward women's foraging decisions because of the prestige that they gained by sharing meat.

Sexual division of labor has grown pivotal to understanding hunter-gatherer ecology in the Carson Desert. The prevailing

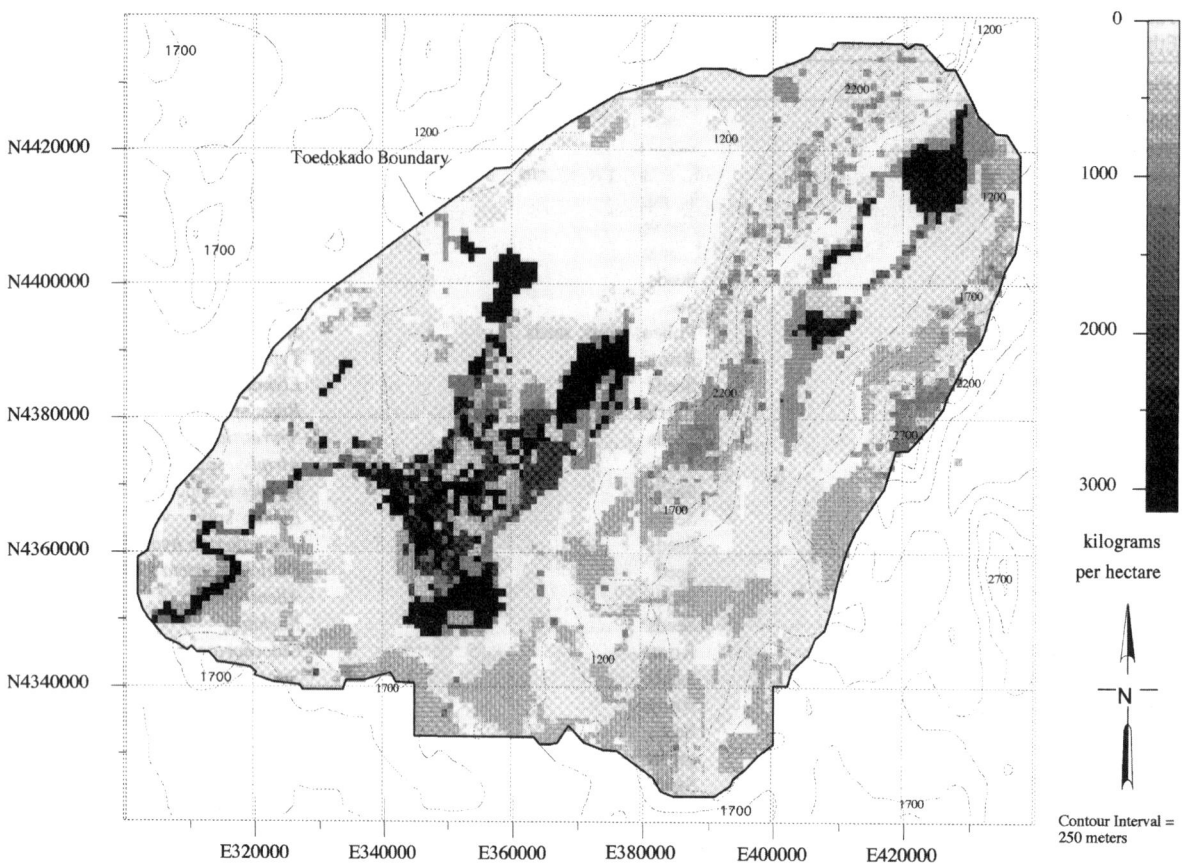

FIGURE 14.2. Normal year annual air-dry production (kg/ha) of biomass by square kilometer habitat in Carson Desert.

limnosedentary model implies that women's foraging determined base camp locations and both men's and women's mobility strategies. Since marshes offered women's best foraging opportunities, wetlands were residential hubs of the subsistence-settlement system. However, supporting archaeological evidence is biased toward a restricted span of the Late Holocene, and a hunting-oriented, limnomobile strategy may have operated earlier. Moreover, Kelly's observation of the prestige gained by men from sharing meat suggests that something other than common foraging efficiency might determine men's and women's foraging strategies.

Clearly, we must understand how men and women resolve conflicts of foraging interest if we are to understand whether a hunting-oriented adaptation was ever feasible in the Carson Desert and model circumstances when it might have occurred. Yet how can this be done? Kelly (1997:26) rightly points out that men and women face different tradeoffs that "are by no means simple and straightforward and…require sophisticated modeling" by archaeologists. Men's and women's subsistence strategies cannot be modeled (as Kelly did in 1985) as if both sexes share a common diet breadth. However, it is equally inappropriate to assert that women's foraging choices always determine residential locations as proposed by Intermountain Research; such an assumption simply begs the issue at hand. This paper models the resolution of conflicts of interest between men and women via a central place foraging strategy.

Modeling Men's and Women's Foraging Behavior in the Carson Desert

This study simulates the foraging behavior of hunter-gatherers using a model of the Carson Desert environment developed by Intermountain Research (Raven and Elston 1989; Zeanah 1996; Zeanah et al. 1995). This model defines 41 soil-based "habitats" from descriptions of climax vegetation associated with soils of the Carson Desert. These habitats are assumed representative of the biotic landscape at the time of ethnohistoric observation (ca. 1850 AD). Quantitative estimates of plant biomass accompany each habitat, allowing mapping of the plant productivity of the Carson Desert as it may have been for a normal year in the early 19th century (Fig. 14.2). This database also monitors the percentage composition of various shrubs, grasses, and forbs in each community.

These include 41 plants harvested for food by ethnographic Great Basin foragers (Table 14.1). The distribution of each resource can be mapped as illustrated for Indian ricegrass in Fig. 14.3.

Quantity and quality of forage plants, and physiographic characteristics such as proximity to water and slope rank the

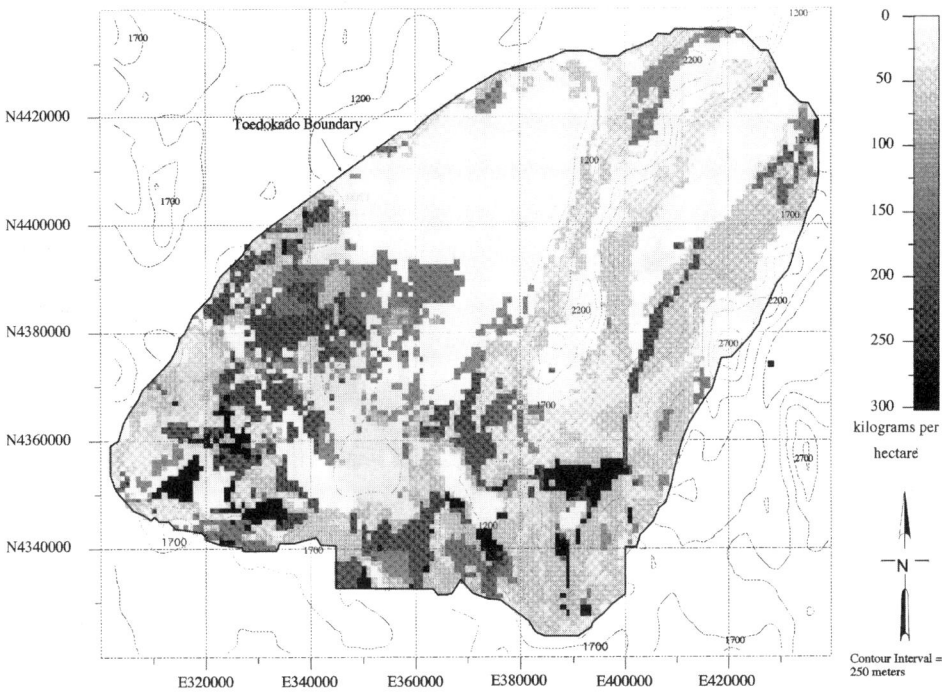

FIGURE 14.3. Normal year productivity (kg/ha) of Indian ricegrass in the Carson Desert.

suitability of each habitat for wildlife. The carrying capacities of these habitats for 14 categories of game are estimated using modern wildlife management guides and inventories (Table 14.1). For example, Fig. 14.4 shows the ranked distribution of suitable bighorn sheep habitats; wildlife studies estimate that this habitat is capable of supporting 475 sheep (USDI Bureau of Land Management 1991:37).

The habitat landscape serves as a baseline for estimating Late Holocene communities when compared with paleoenvironmental evidence. Fig. 14.5 modifies the habitat landscape to reflect mesic conditions before 1250 years ago. The distribution of marshes is expanded to reflect higher lake levels, the productivity of terrestrial habitats increased to that typical of modern favorable years, and pinyon removed from mountain habitats to reflect its probable absence from the region during the early Late Holocene. The model assumes that greater precipitation enhanced the productivity of all plant resources during the Neoglacial with the exception of pinyon.

The effects of such climate change on animal populations can also be estimated with the model. Fig. 14.6 shows the productivity of sheep forage in modern dry, normal, and wet years in the five most productive habitats. Production of large-game forage increases by as much as 75% in wet years. Assuming that modern wet-year productivity was typical of the Neoglacial, the model agrees with Kelly that the Carson Desert was capable of supporting higher densities of large game than it is today. If a 75% increase in forage would produce a comparable increase in sheep population, then about 900 sheep roamed Carson Desert highlands during the early Late Holocene. In a similar manner, the model estimates the increased abundance of all 14 game species during the early Late Holocene based on the enhanced productivity of forage plants and water sources that would have resulted from the greater rainfall of the period.

In Table 14.2, plant and animal resources are ranked according to post-encounter caloric return rates (kilocalories per hour). Several experimental studies estimate return rates of many resources (Barlow and Metcalfe 1996; Bullock 1994; Couture et al. 1986; Hooper 1994; Jones and Madsen 1991; Larralde and Chandler 1980:102–108; Madsen et al. 1997; Raymond and Sobel 1990; Simms 1987; Todt and Hannon 1998). However, other resources must be ranked based on the similarity of harvest techniques, package size, and caloric content with resources of known return. Because of the generality of this approach, resources are grouped into a scale of rank classes defined by ranges of similar return rates.

Simulating Men's and Women's Overall Foraging Return Rates in Carson Desert Habitats

Resources are divided into men's and women's prey sets (Table 14.2); sheep, deer, antelope, fish, rabbits, marmots, woodrats, and waterfowl are classified as exclusively men's resources, ground squirrels, sage grouse and small rodents are listed as both men's and women's resources, and mollusks, bird eggs, cattail pollen, and all remaining plant foods are exclusively women's resources. This classification purposefully oversimplifies the foraging behavior of ethnographic Carson Desert foragers in order to distinguish typical prey sets of men and women from exceptional cooperative or opportunistic efforts. For example,

TABLE 14.1. Ethnographic plant and animal foods monitored in the Carson Desert model.

	Edible Category	Latin Name
Plant Foods		
Alkali sacaton	Seeds	*Sporobolus airoides*
Anderson peachbrush	Berries	*Prunus andersonii*
Annual forbs	Roots, seeds, greens	Various
Balsamroot	Roots	*Balsamorhiza* spp.
Biscuitroot	Roots	*Lomatium* spp.
Bitterroot	Roots	*Lewisia rediviva*
Bluegrass	Seeds	*Poa* sp.
Bottlebrush squirretail	Seeds	*Sitanion hystrix*
Bulrush	Pollen, roots, seeds	*Scirpus* spp.
Cattail	Roots, seeds	*Typha* spp.
Creeping wildrye	Seeds	*Elymus triticoides*
Currant	Berries	*Ribes* sp.
Dropseed	Seeds	*Sporobolus* spp.
Saltbush	Seeds	*Atriplex* spp.
Wild barley	Seeds	*Hordeum* spp.
Wildrye	Seeds	*Elymus* spp.
Indian ricegrass	Seeds	*Achnatherum hymenoides*
Inland saltgrass	Seeds	*Distichlis stricta*
Iodine bush	Seeds	*Allenrolfea* sp.
Kochia	Seeds	*Kochia*
Mariposa lily	Roots	*Calochortus* sp.
Needlegrass	Seeds	*Stipa* sp.
Onion	Roots	*Alium* spp.
Pinyon pine	Seeds	*Pinus monophylla*
Prince's plume	greens	*Stanlea pinnata*
Sago pondweed	Seeds	*Potamogeton pectinatus*
Scratchgrass	Seeds	*Sporobolus asperifolius*
Sedge	Seeds	*Carex* sp.
Seepweed	Seeds	*Suaeda* spp.
Serviceberry	Berries	*Amelanchier* sp.
Shadscale	Seeds	*Atriplex confertifolia*
Silver buffaloberry	Berries	*Shepherdia argentea*
Spikerush seed	Seeds	*Eleocharis palustris*
Sunflower	Seeds	*Helianthus* spp.
Tansymustard	Seeds	*Descurainia pinnata*
Dock	Seeds	*Rumex* spp.
Wheatgrass	Seeds	*Agropyron* spp.
Wild rose	Berries	*Rosa* sp.
Wolfberry	Berries	*Lycium* sp.
Animal Foods		
Tui chub	Game	*Gila bicolor obesus*
Mule deer	Game	*Odocoileus hemionus*
Bighorn sheep	Game	*Ovis canadensis*
Pronghorn antelope	Game	*Antelocapra americana*
Jackrabbit	Game	*Lepus* spp.
Cottontail rabbit	Game	*Sylvilagus* spp.
Large ground squirrel	Game	*Spermophilus* spp.
Ducks	Game, eggs	various
Woodrat	Game	*Neotoma* spp.
Marmot	Game	*Marmota flaviventris*
Muskrat	Game	*Ondatra zibethicus*
Small mammals	Game	various
Sage grouse	Game	*Centrocercus urophasianus*
Shellfish	Game	various

Sources: Chamberlin (1911), Fowler (1986, 1989, 1992), Kelly (1932), Steward (1933, 1938, 1941), Stewart (1941), Wheat (1967).

Fowler (1992:62) reports women using seed baskets to seine fish but I know of no record of women using specialized fishing gear without the participation of men, suggesting that the anecdotal account may refer to an opportunistic event. Women seem to have cooperated with men in communal antelope, rabbit, and waterfowl drives rather than drive animals on their own (Fowler 1992:78; Kelly 1932:79). Similarly, men appear to have harvested ricegrass and pinyon only in cooperation with women (Fowler 1989:46). In contrast, ethnographic accounts are explicit that both men and women trapped small mammals and sage grouse without participation of the other gender (Fowler 1989:23; 1992: 79; Kelly 1932:79).

As shown in Table 14.2, men obtain resources with highest post-encounter caloric return rates; men's and women's resources overlap only in Ranks 5 and 6. Women take all resources in Ranks 1, 2, 3, and 4. However, keep in mind that these rankings ignore search costs. Men's prey is mobile and rare, requiring considerable search costs to procure, whereas women's resources are stationary and abundant, thus relatively cheap to find.

Using the diet breadth model and forager search simulation techniques (Simms 1987:48–53; Winterhalder et al. 1989), an optimal diet and overall foraging return rate were calculated for each habitat, by season and gender (Zeanah 1996).[5] Tables 14.3 and 14.4 present the resulting overall returns rates for ethnohistoric women and men, ranking habitats in sequence from highest to lowest.[6]

Tables 14.5 and 14.6 present the same data adjusted to reflect men's and women's feasible foraging returns between 1250 and 4000 years ago. Obviously care must be taken in interpreting these results, but they suggest three insights into the interactions of Carson Desert ecology and sexual division of labor. These concern the relative return rates of men and women, the return rates of women relative to Rank 1 and 2 resources, and the degree to which men and women achieve their best foraging returns in similar or divergent habitats.

Relative Return Rates of Men and Women

The simulations suggest that ethnohistoric women could earn higher overall foraging returns than men in all seasons even though men pursued higher-ranked prey. This reflects the relative search costs of women's versus men's prey. Indeed, men attain higher overall foraging returns than women only in the late summer simulation after cattail pollen becomes unavailable in marshes (Fig. 14.7). Also, the simulation predicts that in many cases men's optimal prey breadth should include higher-ranked women's resources (such as cattail pollen, winter cattail roots, and large bulrush seed), while lower-ranked men's resources (for example ducks and fish) should occasionally fall out of women's optimal prey set. If correct, men frequently foraged less efficiently by hunting large game to the exclusion of gathering higher-ranked plants.

Both men's and women's foraging benefit from wetter conditions of the Neoglacial, but the foraging returns of men improve most. The overall return rates for early Late Holocene men

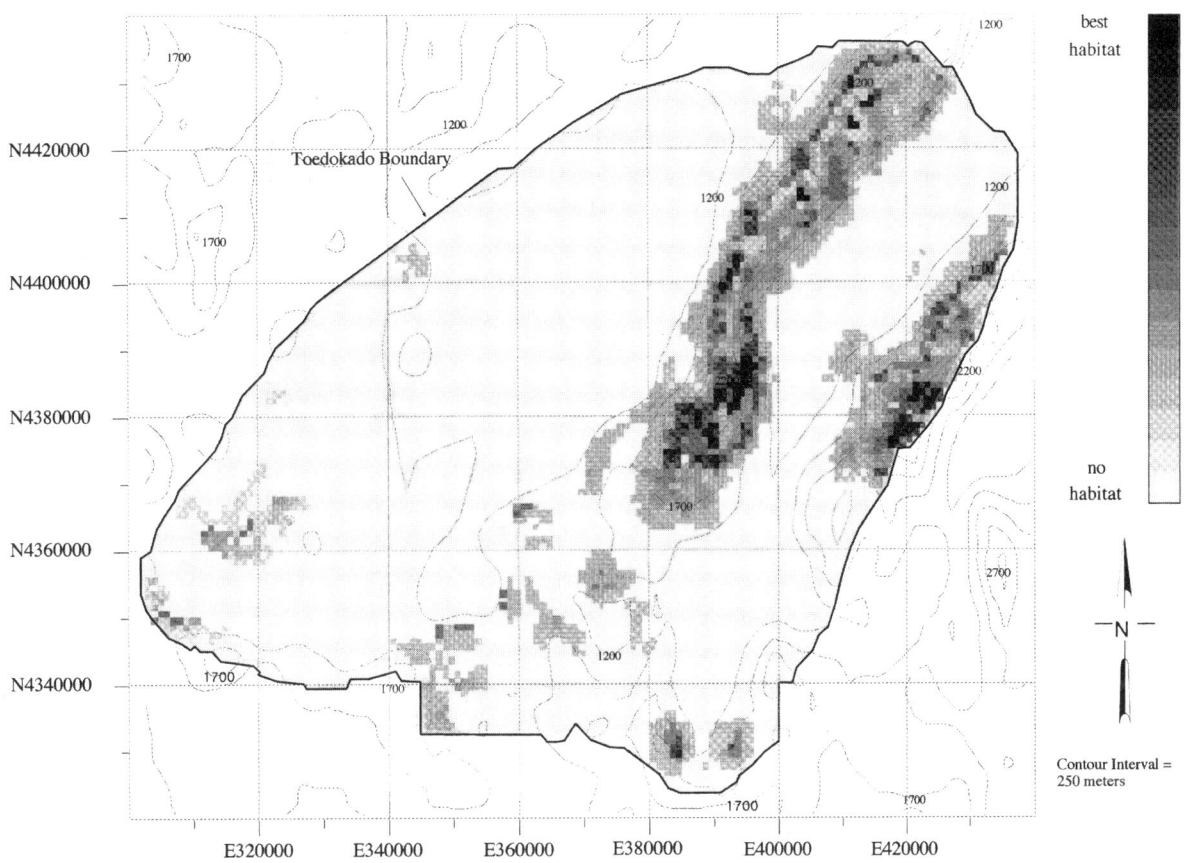

FIGURE 14.4. Bighorn sheep habitat quality rankings in the Carson Desert.

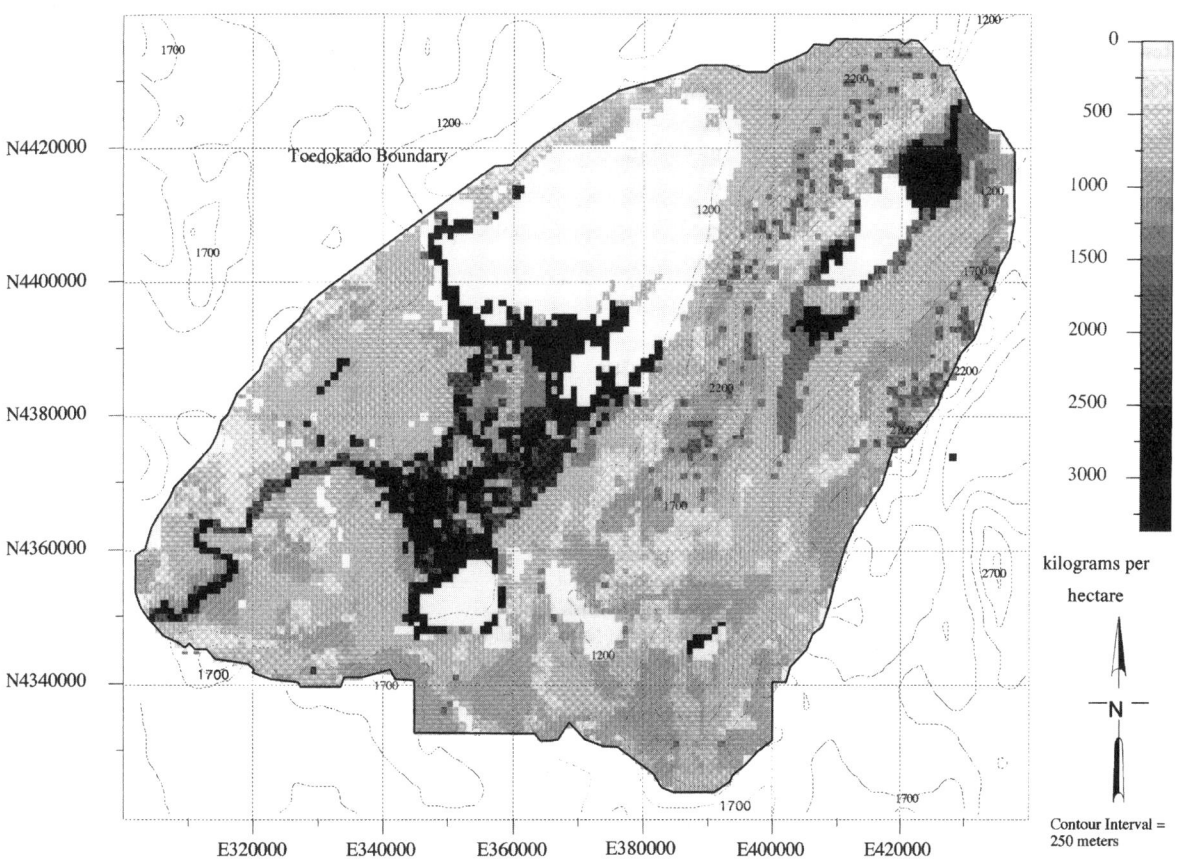

FIGURE 14.5. Wet regime habitat productivity of the early Late Holocene.

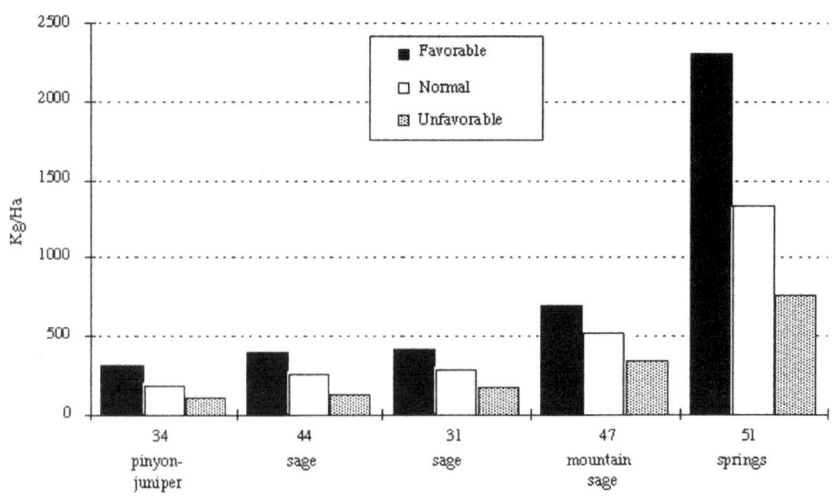

FIGURE 14.6. Forage productivity of top five bighorn sheep habitats in the Carson Desert.

TABLE 14.2. Resource classes ranked according to caloric returns and classified as men's and women's prey sets for the Carson Desert model.

Class Rank	Range of Caloric Return	Resource	Women's Prey[a]	Men's Prey[a]
8	20000+	Deer		X
		Bighorn sheep		X
		Antelope		X
7	9000+	Jackrabbit		X
		Gopher		X
		Cottontail rabbit		X
6	5000+	Ground squirrel (large)	X	X
		Woodrat		X
		Marmot		X
		Cattail Pollen	X	
		Duck eggs	X	
		Muskrat		X
5	1500–3499	Ground squirrel (small)	X	X
		Biscuitroot	x	
		Tui chub (lake)	X	X
		Duck		X
		Sage grouse		X
		Tui chub (pond)	X	X
		Hardstem bulrush seed	X	
		Cattail root (autumn and winter)	X	
4	900–1499	Tansymustard seed	X	
		Bitterroot root	X	
		Pinyon pine nut	X	
		Shellfish	X	
		Arrowleaf balsamroot	X	
		Shadscale seed	X	
		Four-wing saltbush seed	X	
		Onion root	X	
3	600–899	Peppergrass seed	X	
		Torrey quailbush	X	
		Seepweed	X	
		Serviceberry	X	
2	300–599	Sunflower seed	X	
		Salt marsh bulrush seed	X	
		Bluegrass	X	
		Great Basin wildrye seed	X	
		Indian ricegrass seed	X	
		Small bulrush seed	X	
		Arrowhead (wapato)	X	
		Creeping wildrye	X	
		Western wheatgrass	X	
		Slender wheatgrass	X	
1	0–299	Cattail seed	X	
		Currant	X	
		Cooper wolfberry	X	
		Anderson peachbrush	X	
		Silver buffaloberry	X	
		Wild rose	X	
		Serviceberry	X	
		Scratchgrass seed	X	
		Dropseed seed	X	
		Alkali sacaton seed	X	
		Sago pondweed seed	X	
		Other forbs	X	
		Foxtail barley seed	X	
		Meadow barley seed	X	
		Sedge seed	X	
		Western dock seed	X	
		Spikerush seed	X	
		Bulrush root	X	
		Pickleweed seed	X	
		Kochia seed	X	
		Mariposa lily root	X	
		Cattail root (spring and summer)	X	
		Needlegrass seed	X	
		Prince's plume greens	X	
		Saltgrass seed	X	
		Squirreltail grass seed	X	

[a] Sources: Fowler (1989, 1992), Kelly (1932, p. 79), Steward (1938, 1941, pp. 312–313), Stewart (1941, p. 406), Wheat (1967)

TABLE 14.3. Simulated ethnohistoric women's overall foraging returns (kcal/h) and ranks by habitat and season.

Habitat	Spring Return Rate	Spring Rank	Summer Return Rate	Summer Rank	Autumn Return Rate	Autumn Rank	Winter Return Rate	Winter Rank
A2	0	38.5	0	40.5	0	40.5	0	34.5
A54	0	38.5	0	40.5	0	40.5	0	34.5
G10	108	30	321	30	1130	10	0	34.5
G11	807	6	345	15	1068	17	690	10
G13	111	29	319	32	1096	13	637	15
G14	114	28	311	36	1152	6	631	16
G15	584	7.5	342	20	739	29	709	5.5
G16	958	3	342	19	714	30	714	4
G18	0	38.5	280	39	1135	8	449	22
G19	454	11	321	29	1078	16	612	17
G20	584	7.5	344	17	1017	20	709	5.5
G21	118	27	322	28	1088	15	551	21
G26	78	34	315	35	1129	11	285	23
G29	190	24.5	375	10	1006	21	738	3
G3	190	24.5	333	22	1031	19	695	8
G36	419	15.5	352	14	1160	4	0	34.5
G38	265	22	317	34	971	23	653	13
G40	103	32	286	38	1127	12	0	34.5
G42	275	21	317	33	1136	7	0	34.5
G46	0	38.5	326	26	1152	5	0	34.5
G49	105	31	320	31	1130	9	0	34.5
G56	73	35	324	27	1094	14	0	34.5
G7	323	19	360	11	1061	18	651	14
G9	441	12	331	23	926	25	579	18
M34	502	9	401	7	1278	3	190	25
M47	480	10	354	12	575	36	0	34.5
M51	342	18	435	4	378	37	190	25
S27	419	15.5	353	13	948	24	568	20
S31	422	13.5	334	21	924	26	578	19
S35	415	17	344	18	588	35	0	34.5
S37	422	13.5	345	16	273	38	0	34.5
S44	0	38.5	301	37	849	28	0	34.5
S48	96	33	327	25	600	33	0	34.5
S52	0	38.5	329	24	600	34	0	34.5
W1	1901	1	5613	1	1659	1	3134	1
W28	190	24.5	399	8	994	22	707	7
W4	190	24.5	383	9	882	27	692	9
W5	831	4.5	422	6	664	32	664	12
W53	1148	2	5464	2	1633	2	3037	2
W55	290	20	452	3	230	39	190	25
W6	831	4.5	433	5	684	31	684	11

A, abiotic; G, greasewood/saltbush; M, montane; S, sagebrush; W, wetland.

increase on average 41%, with improvements as high as 425% in some habitats. In contrast, women's foraging returns increase an average of only 5%, with 50% maximum improvement. The greater success of early Late Holocene men reflects the sensitivity of resources with high search and low handling costs to increased abundance as compared to the insensitivity of resources with low search and high handling costs (Simms 1987:53–55). For example, the autumn returns available for men in the wetland meadows associated with montane springs and floodplains (Habitat M51) increases from 1230 kcal/h in the ethnohistoric simulation to 2070 kcal/h in the early Late Holocene simulation primarily because of the enhanced carrying capacity of game in this habitat. In contrast, enhanced productivity of bulrush in wetland marshes (Habitat W1) raises women's autumn foraging returns from 1660 kcal/h in the ethnohistoric simulation only to 1665 kcal/h in the Neoglacial simulation. Despite the

TABLE 14.4. Simulated ethnohistoric men's overall foraging returns (kcal/h) and ranks by habitat and season.

Habitat	Spring Return Rate	Spring Rank	Summer Return Rate	Summer Rank	Autumn Return Rate	Autumn Rank	Winter Return Rate	Winter Rank
A2	0	33.5	0	32	0	32.5	0	31.5
A54	0	33.5	0	32	0	32.5	0	31.5
G10	0	33.5	0	32	0	32.5	0	31.5
G11	631	11	573	9.5	253	18.5	253	16.5
G13	317	19	317	14	317	15.5	0	31.5
G14	0	33.5	0	32	0	32.5	0	31.5
G15	190	21.5	190	19.5	190	20.5	190	18.5
G16	887	7	573	9.5	534	12	534	11
G18	0	33.5	0	32	0	32.5	0	31.5
G19	0	33.5	0	32	0	32.5	0	31.5
G20	190	21.5	190	19.5	190	20.5	190	18.5
G21	0	33.5	0	32	0	32.5	0	31.5
G26	0	33.5	0	32	0	32.5	0	31.5
G29	534	15	190	19.5	534	12	534	11
G3	253	20	190	19.5	253	18.5	253	16.5
G36	694	10	419	12.5	304	17	304	15
G38	0	33.5	0	32	0	32.5	0	31.5
G40	0	33.5	0	32	0	32.5	0	31.5
G42	0	33.5	0	32	0	32.5	0	31.5
G46	0	33.5	0	32	0	32.5	0	31.5
G49	0	33.5	0	32	0	32.5	0	31.5
G56	0	33.5	0	32	0	32.5	0	31.5
G7	68	23.5	0	32	68	22.5	68	20.5
G9	68	23.5	0	32	68	22.5	68	20.5
M34	813	8	648	7	977	4	717	4
M47	948	6	496	11	1115	2	851	2
M51	1162	2	736	2	1228	1	985	1
S27	419	17	419	12.5	0	32.5	0	31.5
S31	1272	1	695	3.5	946	5	672	5
S35	5	25	0	32	0	32.5	0	31.5
S37	698	9	695	3.5	317	15.5	0	31.5
S44	379	18	191	16	564	9	564	8
S48	0	33.5	0	32	0	32.5	0	31.5
S52	0	33.5	0	32	0	32.5	0	31.5
W1	1003	3	1003	1	1003	3	808	3
W28	534	15	190	19.5	534	12	534	11
W4	534	15	190	19.5	534	12	534	11
W5	968	4.5	666	5.5	633	6.5	595	6.5
W53	610	13	610	8	610	8	465	14
W55	627	12	290	15	534	12	534	11
W6	968	4.5	666	5.5	633	6.5	595	6.5

A, abiotic; G, greasewood/saltbush; M, montane; S, sagebrush; W, wetland.

disproportionate improvement of men's hunting returns, men enjoy higher returns than women only in spring, autumn, and late summer during the Neoglacial (Fig. 14.8); women obtain higher returns from cattail pollen in early summer and from cattail roots in winter. Although the Neoglacial climate benefits men's foraging more than women's, women continue to garner higher overall returns in some seasons.

Women's Return Rates Relative to Low-Ranked Plant Foods

Simulated ethnohistoric women's foraging returns (Table 14.3) exceed 450 kcal/h in 11 spring habitats, 3 summer habitats, 36 autumn habitats, and 21 winter habitats. In these habitats, Rank 1 and many Rank 2 resources listed in Table 14.2 are too low ranked to fall into women's optimal diet. The presence of so

TABLE 14.5. Simulated early Late Holocene women's overall foraging returns (kcal/h) and ranks by habitat and season.

Habitat	Spring Return Rate	Spring Rank	Summer Return Rate	Summer Rank	Autumn Return Rate	Autumn Rank	Winter Return Rate	Winter Rank
A2	0	38.5	0	40.5	0	40.5	0	34.5
A54	0	38.5	0	40.5	0	40.5	0	34.5
G10	116	30	346	24	1146	9	0	34.5
G11	955	6	372	11	1103	15	708	10
G13	118	29	330	32	1117	12	659	15
G14	120	28	319	38	1161	5	653	16
G15	685	7.5	353	20	761	28	721	5.5
G16	1080	3	348	22	726	29	726	4
G18	0	38.5	329	33	1150	6	496	22
G19	525	13.5	326	35	1101	16	636	17
G20	685	7.5	366	14	1061	19	721	5.5
G21	125	27	368	12	1117	13	595	21
G26	88	34	358	17	1145	10	335	23
G29	253	23	377	10	1042	21	742	3
G3	236	26	348	23	1066	18	708	9
G36	504	17	367	13	1167	3	0	34.5
G38	328	21.5	326	34	1018	22	674	13
G40	111	32	302	39	1143	11	0	34.5
G42	328	21.5	323	36	1149	7	0	34.5
G46	0	38.5	332	29	1163	4	0	34.5
G49	114	31	332	30	1146	8	0	34.5
G56	82	35	351	21	1117	14	0	34.5
G7	397	19	364	15	1092	17	673	14
G9	525	13.5	335	28	980	24	613	18
M34	564	9	423	6	254	38	254	25
M47	540	10	331	31	330	36	0	34.5
M51	521	16	448	4	379	35	299	25
S27	523	15	360	16	1004	23	599	20
S31	528	11.5	340	25	973	25	608	19
S35	439	18	356	19	672	33	0	34.5
S37	528	11.5	357	18	291	37	0	34.5
S44	0	38.5	320	37	909	27	0	34.5
S48	104	33	337	27	672	33	0	34.5
S52	0	38.5	338	26	672	33	0	34.5
W1	1901	1	5613	1	1663	1	3194	1
W28	245	24.5	419	7	1048	20	719	7
W4	245	24.5	409	8	950	26	709	8
W5	1000	4	390	9	684	31	684	12
W53	1148	2	5464	2	1639	2	3093	2
W55	339	20	453	3	241	39	222	25
W6	999	5	441	5	700	30	700	11

A, abiotic; G, greasewood/saltbush; M, montane; S, sagebrush; W, wetland

many foraging patches offering higher foraging returns suggests that ethnohistoric gatherers should have harvested low-ranked seeds, such as basin wildrye and Indian ricegrass, only if they foraged in relatively barren and impoverished settings. Both ethnographic documentation (Fowler 1989, 1992; Wheat 1967), and archaeobotanical remains (Budy 1988:349–350; Napton and Heizer 1970:107; Roust 1967:57–59; Rhode 2003:913–917) directly contradict this inference. If the model is correct, Carson Desert women lowered their overall foraging return rates by harvesting such low-ranked resources.

This discrepancy between model results and empirical evidence may point to a key feature of women's foraging ecology in the Carson Desert. It is possible that the inconsistency reflects temporal variation in foraging opportunities not captured

TABLE 14.6. Simulated early Late Holocene men's overall foraging returns (kcal/h) and ranks by habitat and season.

Habitat	Spring Return Rate	Spring Rank	Summer Return Rate	Summer Rank	Autumn Return Rate	Autumn Rank	Winter Return Rate	Winter Rank
A2	0	33.5	0	32	0	32.5	0	31.5
A54	0	33.5	0	32	0	32.5	0	31.5
G10	0	33.5	0	32	0	32.5	0	31.5
G11	803	11	741	12	581	15	581	13
G13	317	20	402	15	402	17	0	31.5
G14	0	33.5	0	32	0	32.5	0	31.5
G15	190	23.5	246	18.5	246	22.5	246	20.5
G16	1055	9	751	11	952	8	952	5
G18	0	33.5	0	32	0	32.5	0	31.5
G19	0	33.5	0	32	0	32.5	0	31.5
G20	190	23.5	246	18.5	246	22.5	246	20.5
G21	0	33.5	0	32	0	32.5	0	31.5
G26	0	33.5	0	32	0	32.5	0	31.5
G29	716	14	253	17	947	9	947	6.5
G3	441	18	236	22	546	16	546	14
G36	694	17	504	14	371	19	371	17
G38	0	33.5	0	32	0	32.5	0	31.5
G40	0	33.5	0	32	0	32.5	0	31.5
G42	0	33.5	0	32	0	32.5	0	31.5
G46	0	33.5	0	32	0	32.5	0	31.5
G49	0	33.5	0	32	0	32.5	0	31.5
G56	0	33.5	0	32	0	32.5	0	31.5
G7	269	21.5	0	32	357	20	357	18
G9	269	21.5	0	32	354	21	354	19
M34	1763	4	1490	2	1490	3	1180	3
M47	1926	3	1237	3	1550	2	1257	2
M51	2054	2	1823	1	2074	1	1765	1
S27	419	19	523	13	0	32.5	0	31.5
S31	2179	1	861	6.5	753	14	386	16
S35	5	25	0	32	0	32.5	0	31.5
S37	698	16	861	6.5	399	18	0	31.5
S44	1451	5	953	5	953	7	953	4
S48	0	33.5	0	32	0	32.5	0	31.5
S52	0	33.5	0	32	0	32.5	0	31.5
W1	1219	6	1219	4	1219	4	808	12
W28	716	14	245	20.5	916	11	916	10
W4	716	14	245	20.5	918	10	918	9
W5	1166	7.5	860	8	1036	5	947	6.5
W53	790	12	790	10	790	13	465	15
W55	805	10	339	16	835	12	835	11
W6	1166	7.5	859	9	1035	6	946	8

A, abiotic; G, greasewood/saltbush; M, montane; S, sagebrush; W, wetland.

by the seasonal resolution of the simulations. For example, Figs. 14.7 and 14.8 show that women achieved foraging returns of more than 5000 kcal/h by harvesting pollen in early summer. After the pollen harvest, the highest women's overall return rate dropped to 450 kcal/h for harvesting Rank 2 and 3 seeds. Women garner winter returns exceeding 3000 kcal/h by harvesting cattail roots, but alternative resources may be hard to find in late winter and early spring. If inclement winter weather denied women access to marshes, women's foraging returns may have fallen sufficiently low to allow women to economically harvest Rank 1 and 2 resources on a daily or weekly basis.

Alternatively, women may have harvested low-ranked seeds to bank them for more prolonged future food shortages. Ethnographic accounts note that food storage was essential for overwinter survival (Fowler 1976:3, Wheat 1967:16). Simms (1987:83, 95) suggests that women harvest seeds that fell out of their diet

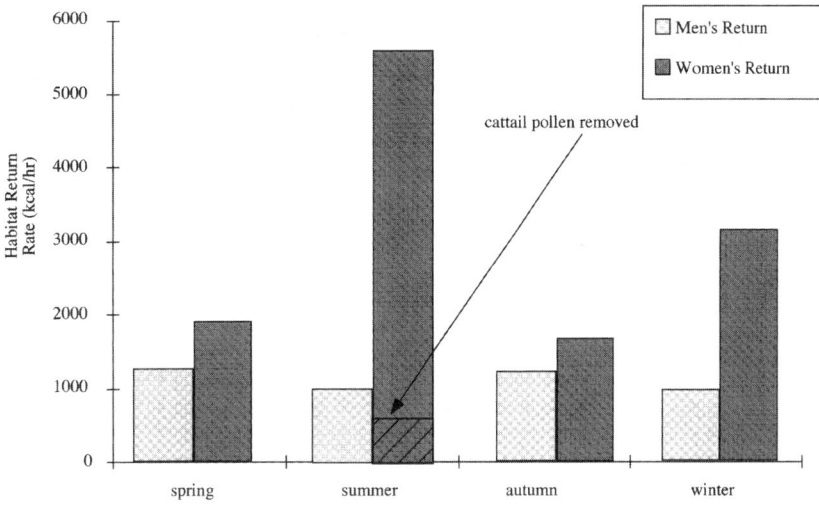

FIGURE 14.7. Highest habitat return rates by season for ethnohistoric men and women in the Carson Desert.

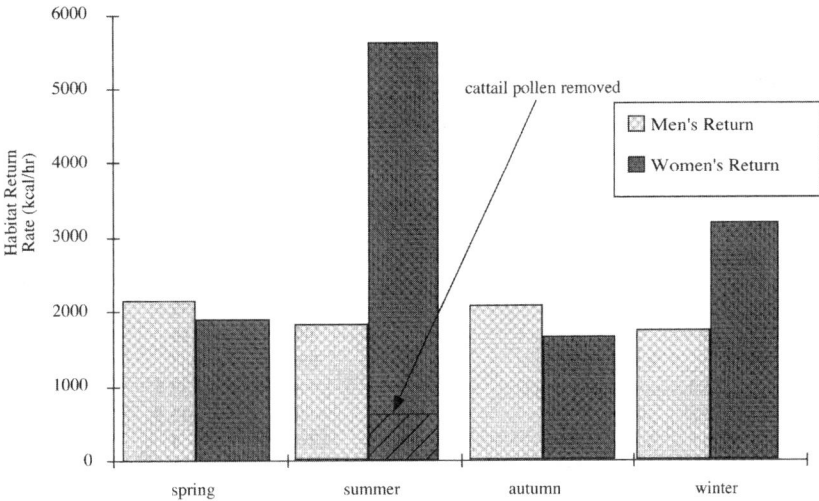

FIGURE 14.8. Highest habitat return rates by season for early Late Holocene men and women in the Carson Desert.

breadth so they could store them for winter. However, given the returns obtainable from cattail roots, which are richest in starch over winter (Madsen et al. 1997), it is difficult to envision the need for food storage in any winter when cattail roots are easily available. This suggests that it may have paid to store low-ranked resources only during bad years when marsh resources were unavailable for harvest.

Great Basin wetlands are vulnerable to periodic floods and droughts that devastate marsh productivity over a few bad years (Elston 1990; Hamilton and Auble 1993:2). Also, marsh productivity fluctuates according to internal cycles that are independent of climatic circumstances. For example, booming muskrat populations "eat out" emergent vegetation, causing mammal, fish, and bird populations to crash (Sojda and Solberg 1993:5; Weller 1981:58–65; cf. Thompson and Hallock 1988:137–138). Unpredictable, inter-annual variability of this sort may have pe-

riodically devastated marsh productivity, prompting women to widen their prey breadth for a few bad years.

Archaeological evidence supports the premise that Carson Desert foragers underwent periodic food shortages that would have widened diet breadth and intensified reliance on food stores. Cases of dental hypoplasias and cortical bone loss among skeletal samples are evidence that prehistoric children of the Carson Desert experienced occasional food shortages and nutritional stress, despite overall health and robusticity (Larsen and Kelly 1995:212). Skeletal samples also exhibit high frequencies of eburnation on joint surfaces. Osteoarthritis induced by severe mobility is the most likely cause of this pathology, but frequent consumption of stored seeds contaminated with fungal mycotoxins may have contributed to its occurrence (Brooks et al. 1990; Larsen and Kelly 1995:133). This accords with the frequent occurrence of storage pits on Carson Desert sites (Ingbar 1985;

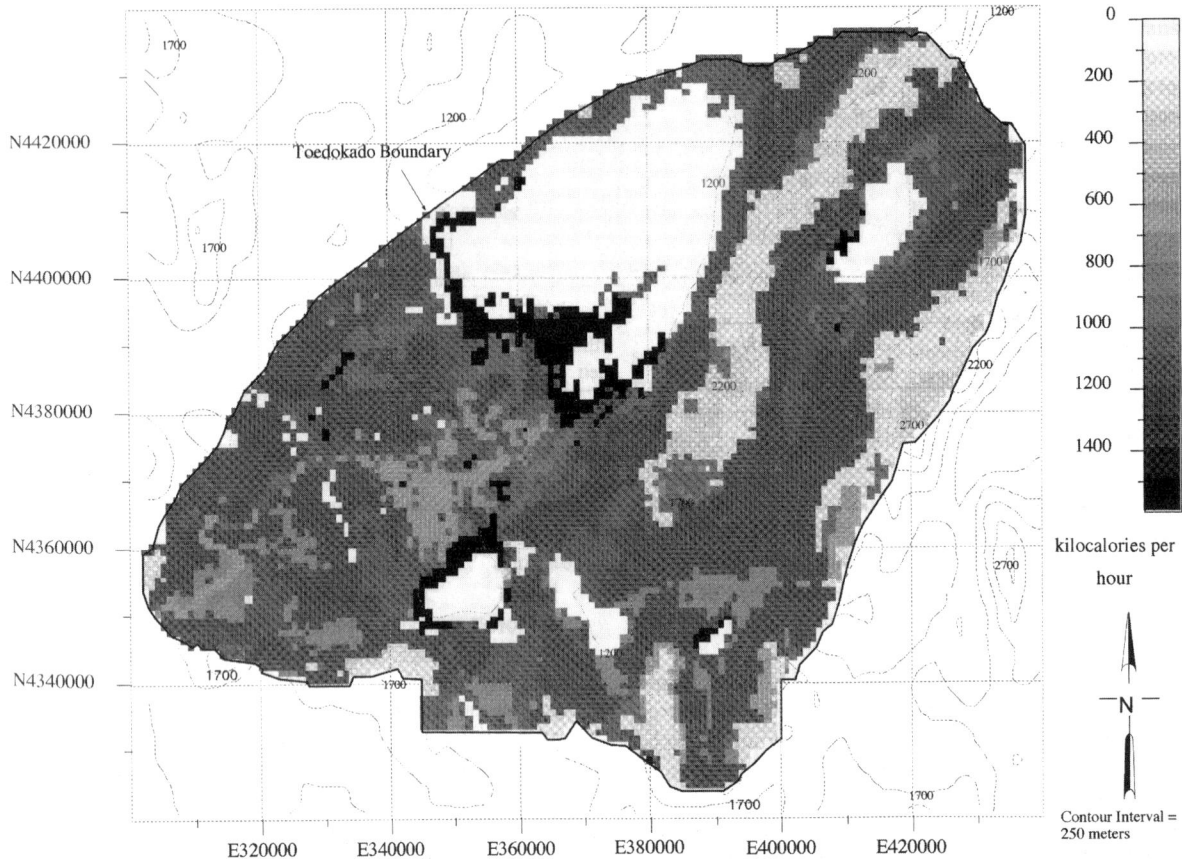

FIGURE 14.9. Early Late Holocene women's autumn foraging returns (kcal/h) in Carson Desert.

TABLE 14.7. Spearman's rank correlation coefficient test results for men's and women's habitat rankings by season.

Season	Early Late Holocene		Ethnohistoric	
	r	Z	r	Z
Spring	.66	4.25	.73	4.56
Summer	.56	3.55	.73	4.64
Autumn	−.29	−1.83	−.11	−.72
Winter	.44	2.79	.39	3.1

Kelly 2001; Raven 1988, 1990; Tuohy et al. 1987), food in which would have been prone to molding (Zeanah 1988). Coprolites often contain resources procured in different seasons, such as cattail pollen and bulrush seeds, indicating consumption of stored foods (Rhode 2003:915; Thomas 1985:386; Wigand and Mehringer 1985:117). These data suggest that Carson Desert foragers experienced periodic food shortages, despite the normal bounty yielded by Carson Desert wetlands. Whether women harvested lowest-ranked plants in response to or anticipation of food shortages, a critical foraging goal of women's foraging must have been to reliably provision their children through periods of scarcity.

Spatial Overlap Between Men's and Women's Foraging Opportunities

The third observation deriving from Tables 14.3–14.6 concerns spatial and temporal differences between men's and women's foraging opportunities. Men and women usually achieve high returns in the same habitats, but the foraging interests of men and women sometimes conflict. Table 14.7 lists Spearman's Rank Correlation Coefficients for men's and women's habitat ranking by season for ethnohistoric and early Late Holocene scenarios. In both simulations, men's and women's rankings correlate significantly during winter, spring, and summer. However, there is no significant correlation between ethnohistoric autumn rankings, whereas early Late Holocene autumn rankings approach a negative correlation at the 0.1 significance level.

Fig. 14.9 maps simulated autumn foraging return rates in the Carson Desert for women during the early Late Holocene. The figure shows that there would have been little upland competition for women's attention in autumn. The effect of Neoglacial climate on women's foraging would have been greatest in the extent and variety of profitable habitats. For example, marshes were more widespread in the Carson Desert than at the time of ethnohistoric observation, whereas there were fewer alternative habitats offering competitive foraging returns, absent pinyon in

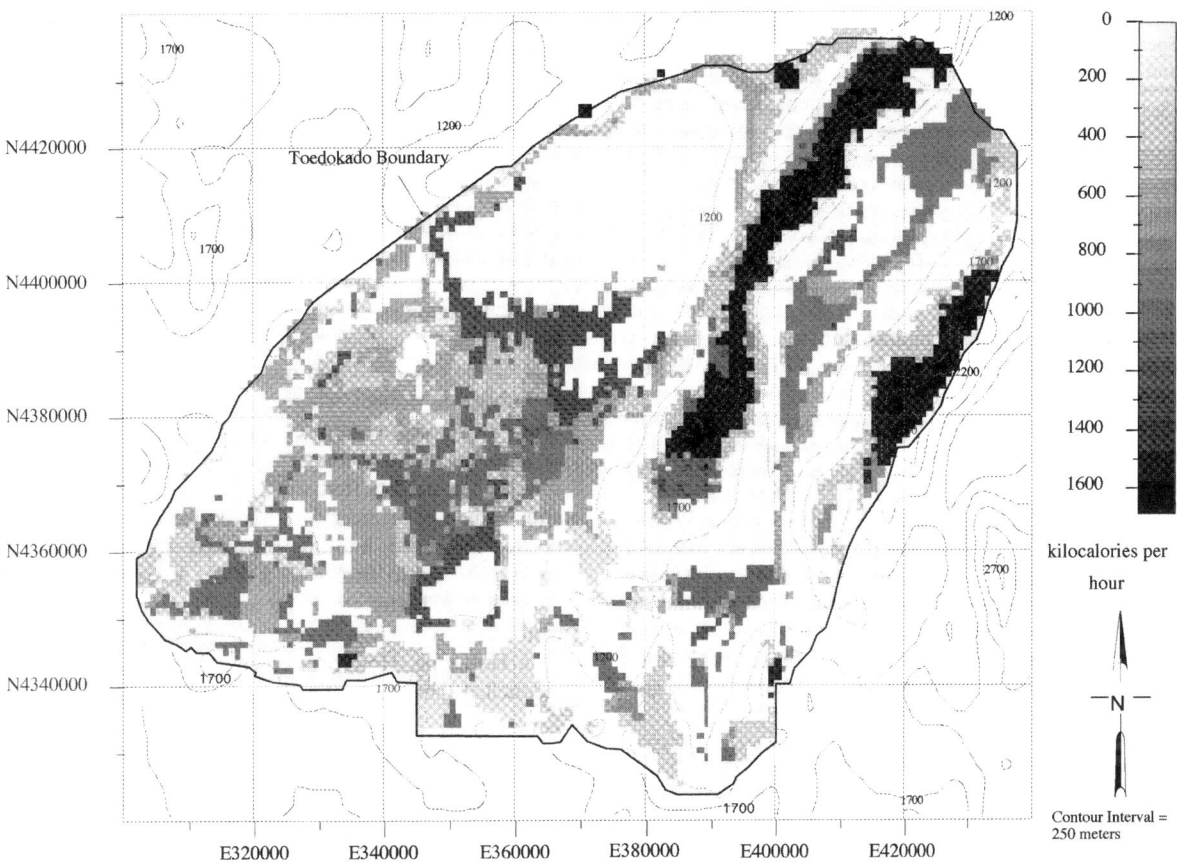

FIGURE 14.10. Early Late Holocene men's autumn foraging returns (kcal/h) in Carson Desert.

neighboring uplands. Therefore, women's optimum foraging interests should have been even more tightly focused on marshes during this period than at ethnographic observation. Fig. 14.10 shows the distribution of men's autumn returns. Note that the map is almost a reverse image of Fig. 14.9; men's best autumn hunting patches are unquestionably in the Clan Alpine and Stillwater Mountains, whereas lowland marshes are less productive.

Conflicting Foraging Goals and Patch Choice During the Early Late Holocene

The most surprising insights of the simulations are that ethnohistoric men lowered their overall foraging return rate by specializing in hunting, whereas ethnohistoric and early Late Holocene women lowered their overall foraging return rate by harvesting low-ranked plants. Obviously, the simulations may simply be wrong; daily and yearly variation in resource abundance would be likely causes of the error. Yet recall ethnographic cases of Carson Desert women hunting and men gathering that I discounted as opportunistic or cooperative in defining men's and women's prey. Occasionally, men's returns must have been so low that even Rank 2 seeds entered men's optimal prey set, whereas women experienced foraging returns so high that all but the highest-ranked plants must have fallen out of their optimal prey set. How then, can the ethnographic specialization of men as hunters and women as gatherers be explained?

Given an evolutionary perspective of sexual division of labor, it seems likely that men bypassed plant resources that were in their optimal prey set because they received greater reproductive benefits by sharing large packages of meat. In contrast, women harvested plants that fell out of their optimal prey set, because doing so ensured provisioning their offspring, particularly during episodic food shortages. Therefore, Carson Desert men and women foraged to achieve different fitness goals. Comparison of the foraging return landscapes for men and women indicates that these goals were best achieved in different habitats in autumn, when hunting opportunities were best and the need to accumulate overwinter food stores greatest. The mesic climate of the Neoglacial only aggravated this scheduling conflict. Moreover, once we understand that Carson Desert men and women pursued different kinds of prey to achieve different objectives, foraging theory predicts that the optimal responses of men and women to increased resource abundance should have been dramatically different.

Clearly, Carson Desert habitats are patches that vary in the types and proportion of resources they contain. Thus, the selection of habitats by men and women can be considered using the

TABLE 14.8. Number of seasonal habitats falling into ordinal ranges of overall caloric return in early Late Holocene and ethnohistoric simulations for men and women.

Range of Caloric Return (kcal/h)	Early Late Holocene Men	Ethno-historic Men	Early Late Holocene Women	Ethno-historic Women
>900	35	13	37	33
600–899	20	21	26	26
300–599	21	31	57	56
0–299	88	99	44	49

patch choice model (MacArthur and Pianka 1966). The model ranks different types of resource patches, predicting that foragers should prefer the most profitable patch types and that a change in resource abundance may alter the breadth of patch selection. Unlike the diet breadth model, which excludes search costs in ranking individual resources, the patch choice model includes the cost of searching within patches in determining their rank. Travel time between patches is excluded. Thus, the rank of a patch type is independent of its abundance, but not the abundance of resources within that patch type. This complicates predictions about whether patch breadth should expand or contract with greater resource abundance according to whether search or handling costs represent the major investment in harvesting those resources.

For example, consider patches that contain resources that are expensive to handle but cheap to find (such as wetland seeds). Increasing the abundance of those resources should increase the abundance of wetland patches worthwhile to harvest (i.e., lower travel time between productive wetland stands), but not diminish the need to search for stands sufficiently to raise the ranking of wetlands relative to other patch types. In this situation, foragers may harvest the same or a narrower array of patch types (emphasizing wetlands) because more examples of the higher-ranked patch types are available to forage within.

In contrast, increasing the abundance of resources that are difficult to find but cheap to handle (for example large game) will increase overall foraging returns within hunting patches, as well as increase the number of patches worthwhile for hunting. Foragers should broaden the variety of patch types they elect to hunt in, because more high-ranked patch types fall into their optimal set (cf., MacArthur and Pianka 1966).

Earlier I observed that Neoglacial climatic conditions should have increased the abundance of both men's and women's prey in the Carson Desert, improving the foraging returns of both genders. However because of differences in search and handling costs, women should have foraged more exclusively in their highest-ranked patch types (wetlands) because those habitats were more abundant. In contrast, men should have found it profitable to hunt in a wider range of habitats because the kinds of high-ranked patch types increased. Table 14.8 tallies the number of seasonal habitats in which the simulations predict men or women could achieve overall caloric foraging returns in four ordinal categories. Early Late Holocene women achieve overall foraging returns greater than 900 kcal/h in only four more habitats than ethnohistoric women, whereas early Late Holocene men achieve overall foraging returns greater than 900 kcal/h in 22 more habitats than ethnohistoric men. Thus, the simulations predict that the optimal responses of men and women to increased resource abundance were opposite; improved hunting prospects would have encouraged men to forage in a wider variety of upland habitats, but better gathering opportunities would have continued to anchor women to wetlands in the early Late Holocene. How could men and women have best reconciled these conflicting foraging interests?

Modeling Central Place Foraging Tactics in Light of Gender Conflicts

This is clearly a central place foraging issue. Ethnographic Carson Desert foragers inhabited a patchy environment and exploited dispersed resource patches from residential base camps. Both women and men logistically transported some resources to residences where they processed, stored, and consumed those resources (Thomas 1985:24–26). Thus, men and women had to decide not only where to forage, but also where to position base camps and how to exploit spatially and temporally dispersed patches from central places.

Central place foraging models often assume that foragers locate base camps to minimize travel and transport costs (Horn 1966; cf., Orians and Pearson 1977), and carrying resources imposes significant costs on pedestrian Great Basin hunter-gatherers (Jones and Madsen 1989). However, recognition that men and women faced conflicts of foraging interest poses a dilemma for modeling central place foraging tactics from the perspective of behavioral ecologists, who prefer to model the foraging cost-benefits of individuals since natural selection operates more strongly on individuals than groups. Yet, if male and female foragers in a camp pursue different sets of prey to achieve different objectives, whose individual interests are served by positioning camps to minimize travel and transport costs?

This dilemma warrants consideration of how males and females can mutually reconcile divergent foraging interests because sexual division of labor divorces foraging from consumption among hunter-gatherers. Although men and women pursue different prey to achieve their own ends, both consume resources procured by the opposite gender while offspring benefit from the resources harvested by both sexes. Thus, the moniker "optimal diet" is a misnomer in the case of human sexual division of labor; male and female foragers have overlapping diets, but the breadth of prey that men and women seek may not be determined solely by momentary optimization of caloric returns. Although both men and women should seek to maximize the returns for procuring their sets of prey by foraging in their best patches, it is not in the interest of either sex to choose a settlement strategy that precludes the opposite sex from economically foraging and

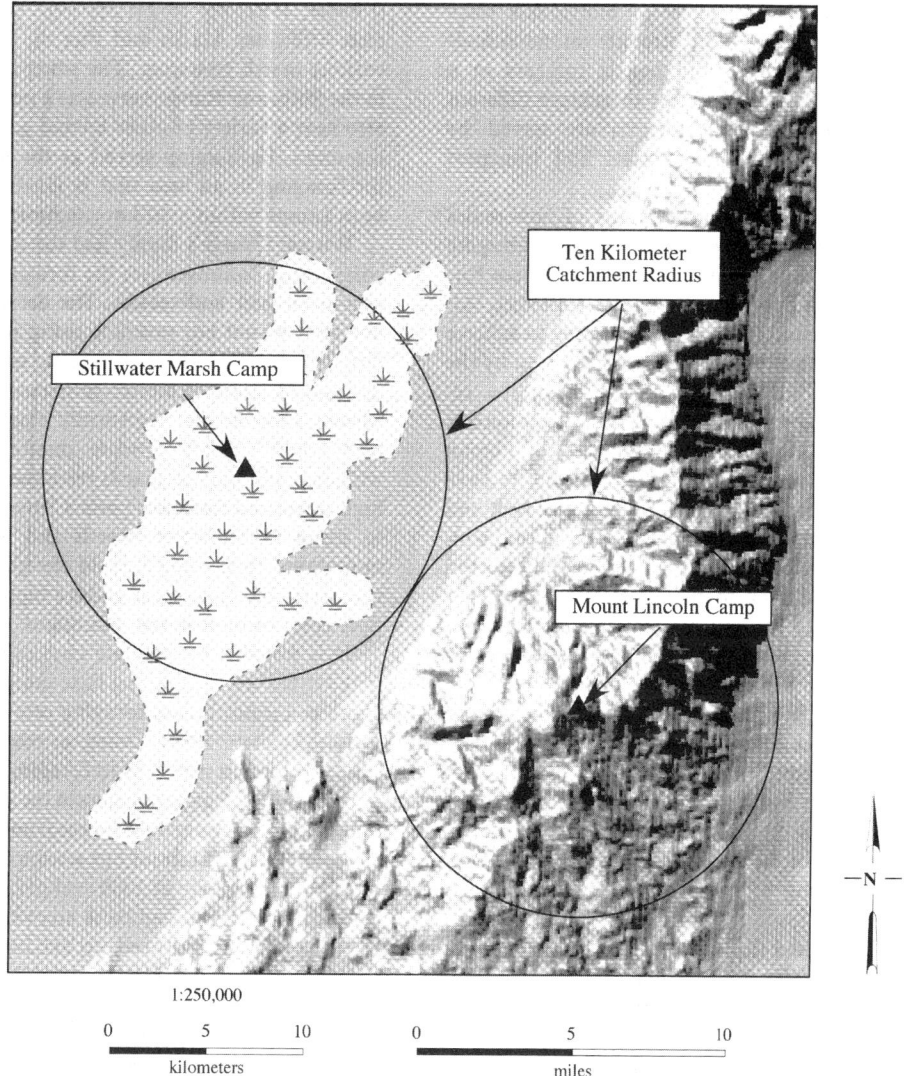

FIGURE 14.11. Two ethnographic Toedokado central place base camps showing 10 km catchment radius (after USGS Reno, Nevada; CA, 1:250,000, 1957, Revised 1971)

contributing to the caloric intake of camp residents. When foraging interests conflict, each sex faces tradeoffs between placing residences in their own best patch and imposing high travel and transport costs on the opposite gender, versus facilitating foraging of the opposite gender by living elsewhere and enduring high travel and transport costs themselves.

One way to model these tradeoffs is to assume that hunter-gatherers prefer to live at the central place location offering the highest combined rate that men and women foragers can return to consumers at a particular central place. When modeled from the perspective of caloric consumers (particularly offspring), hunter-gatherers who camp at locations offering higher combined central place foraging return rates must enjoy a selective advantage over competitors who camp only where men's or women's best foraging interests are served. Also, if women's critical constraint concerns accumulation of stores to tide over forthcoming food shortages, they should prefer to camp in the location where the intake rate of food into camp is greatest, even if this is not in women's best foraging patch. This is the simplest assumption allowing modeling of how men and women can resolve conflicts of interest through a central place foraging strategy. It treats men and women as separate foragers pursuing different sets of prey, while avoiding empirical generalizations about the relative mobility of men and women.

Fig. 14.11 illustrates the approximate locations of two ethnographic base camps used by the family of Wuzzie George, a famous ethnographic Toedokado Paiute (Fowler 1992:26–27, 35, 39).[7] One camp occurred near Stillwater Marsh and was the hub for procuring various marsh resources. The other, on Mount Lincoln in the Stillwater Range, served as a pinyon camp but was also near a variety of other upland game, seed, and root resources. Each camp served as the local central place for foraging in an area that is illustrated by encircling both camps within a 10-km catchment radius.

Wuzzie George's family had the option of occupying either camp contingent on the foraging opportunities of any given year

TABLE 14.9. Constants for calculating travel and transport costs over variable grades.

Percent Grade	Caloric Cost of Walking per km	Caloric Cost of Carrying 1 kg per km
−40	88.4	2.78
−30	72.6	1.85
−20	60.0	1.11
−10	36.6	0.42
0	50.6	0.4
10	115.2	1.32
20	169.4	3.13
30	239.8	6.11
40	325.3	10.21

and season. The decision to encamp at one spot would not preclude using resources at the alternative location through logistic mobility and sexual division of labor. If hunter-gatherers chose to camp on Mount Lincoln and logistically exploit marshes, then men could cheaply procure montane resources, but women could harvest marsh resources only by investing high travel and transport costs. Conversely, if they chose to camp in Stillwater Marsh and logistically exploit mountains, then women could cheaply procure wetland resources, but men must endure high travel and transport costs to hunt in the mountains.

It is possible to simulate circumstances that would make either site the optimal base camp by calculating the combined central place foraging return rate for each location (Zeanah 1999). Doing so requires estimation of several variables pertinent to foraging, travel, and transport costs for use as parameters in the simulation. Assume that a family of four could forage from either location for as long as 90 days (about one season). If each individual requires a net intake of 3000 kcal per day to satisfy daily caloric requirements and bank an overwinter food store, then the band of four needs to net 1,080,000 kcal during their stay at the camp. Men could procure sheep or deer in montane patches, whereas women could procure large bulrush seeds in marshes. Thus, each gender may simultaneously procure a different high-ranked resource in separate habitats. Either 25 sheep or 354 kg of large bulrush seeds could satisfy the net requirement.

For purposes of the simulation, assume that a hunter camped at Mount Lincoln must walk 1 km to reach a hunting patch, but one living at Stillwater Marsh must travel 17 km to hunt. Conversely, a forager living in Stillwater Marsh may walk 1 km to reach the gathering patch, but one living on Mount Lincoln must travel 17 km to gather. Regardless of home or destination, assume that foragers travel at speeds of 3 km/h and expend 300 kcal/h collecting and processing resources (Passmore and Durnin 1955). The maximum transportable weight of resource is 25 kg per load.

Also assume that hunters dry meat before transport to maximize the caloric values of loads of meat. Drying meat reduces the weight of meat by 60% without reducing caloric content (Lee 1979:223; UN Food and Agriculture Organization 1990:15), and thus allows a larger package of calories to be carried per 25-kg load. The additional handling cost for drying meat is assumed to be 35 h per 100 kg (Reidhead 1976:82–83), so the return rate for a 25 kg package of dried sheep or deer meat lowers to about 4640 kcal/h. Plant resources are assumed to be field processed to the extent processed by Simms (1987:104–135) in his experimental harvests of Great Basin plant resources.

Modeling transport costs in the Carson Desert also requires monitoring effects of gradient on travel and transport effort. The relief and vegetation zonation of the Carson Desert affected hunter-gatherer settlement decisions because of the relative costs of transporting a resource uphill or down in steep terrain. Therefore, this simulation assumes that slope influences the caloric costs of travel and transport. Table 14.9 lists data for estimating the effects of slope on transport costs (MacDonald 1961; see also Brannon 1992).

Finally, since hunter-gatherers should seek to maximize the combined caloric intake rate into a camp, they should prefer to take higher-ranked resources. Since large game is higher-ranked than large bulrush seeds, Carson Desert foragers should take as much large game as encounter rates will allow, and make up deficits in the net caloric requirement by procuring large bulrush seeds. This allows simulated foragers to tradeoff the cost and benefits of procuring and transporting one resource against the other, according to the abundance of the more profitable resource. Thus, the amount of the higher-ranked resource transported to a central place (large game) is the independent variable determining how much of the lower-ranked resource (bulrush seed) must also be transported.

In this scenario, central place foraging return rates are calculated for both locations at different proportional contributions of large game to the total caloric requirement. This is done by adding the travel, processing, and transport costs (in calories) required to obtain large game and bulrush seed obtained from either base camp to the net caloric requirement. Therefore, Eq. (14.1) calculates the central place foraging net return rate for any possible base camp location:

$$\frac{\left\{ R + \sum_{i=1}^{2} \left[FN_i * C_h + \left(\sum_{s=1}^{n} D_{is} * NN_i * W_s \right) + \left(\sum_{s=1}^{n} D_{if} * NN_i * L_i * T_s \right) \right] \right\}}{\left\{ \sum_{i=1}^{2} \left[FG_i + \frac{\sum_{s=1}^{n} D_{is} * NGi}{V} \right] \right\}} \quad (14.1)$$

where R is the net caloric requirement (1,080,000 kcal); FN_i is the handling time (hours) for total loads of resource i comprising net caloric requirement; FG_i is the handling time (hours) for total loads of resource i comprising net caloric requirement plus additional loads required to cover caloric costs; C_h is the caloric cost of handling resources (300 kcal/h); D_{is} is the distance of slope s to and from nearest patch of resource i (km); D_{if} is the

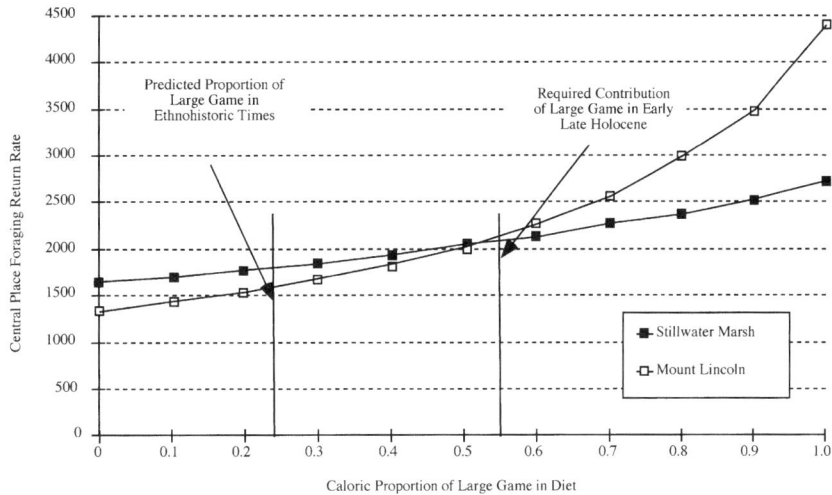

FIGURE 14.12. Central place foraging return rate for Mount Lincoln and Stillwater Marsh camps at different proportional contribution of large game to net caloric intake of camp.

distance of slope f from nearest patch of resource i (km); NN_i is the number of loads of resource i transported comprising net caloric requirement; NG_i is the number of loads of resource i transported comprising net caloric requirement plus additional loads required to cover caloric costs; W_s is the caloric costs of walking across grade s (kcal/km); L_i is the total weight of one load of resource transported (max. = 25 kg); T_s is the caloric costs of carrying load across grade s (kcal/km); V is the walking speed (3 km/h).

The formula adds the caloric costs of procuring and transporting the nearest available large game (resource i_1) and bulrush (resource i_2) seeds to the net caloric requirement, and then divides by the total hours necessary for handling and transporting the total gross caloric intake. The amount of resource i_1 garnered by hunters serves as the independent variable determining how much of the dependent variable, resource i_2, gatherers must procure. This calculates a central place foraging rate for each camp under circumstances where men choose to hunt large game in upland habitats, whereas women harvest bulrush seeds in marshes, given the particular contribution of large game to the net caloric requirement.

Applying the values presented in Tables 14.10 and 14.11 to the equation, the central place foraging return rate was calculated for different proportional contributions of large game to the caloric requirement (Fig. 14.12). Hunter-gatherers should choose the camp with the highest return rates after transport. The figure indicates that the camp on Mount Lincoln should be optimal as long as game makes up more than half of the total caloric requirement. For example, if hunters could obtain 50% of the caloric target as sheep, the Stillwater Marsh base provides a central place foraging return rate of approximately 2080 kcal/h, whereas the Mount Lincoln camp provides a slightly lower return rate of roughly 2040 kcal/h. Alternatively, if hunters could satisfy 60% of the caloric requirement with large game, then Mount Lincoln provides central place foraging returns of about 2300 kcal/h, whereas Stillwater Marsh returns only about 2150 kcal/h.

The proportion of montane large game predicted for the Carson Desert for 90, 5-h hunting days at ethnohistoric encounter rates is indicated in Fig. 14.12. At this encounter rate, sheep could supply only 24% of the net caloric requirement, thus the model predicts Stillwater Marsh camps will be optimal central places. Ethnohistoric Carson Desert bands faced with the task of simultaneously exploiting large bulrush seed and large game would obtain higher combined central place foraging return rates by camping in women's best foraging habitat. Thus, men should exhibit greater logistic mobility than women to accommodate women's settlement decisions.

This inference applies only to the ethnohistoric landscape. Certainly, the Carson Desert was capable of supporting higher densities of large game during the more mesic Neoglacial. Prehistoric men should have procured more large game when game were more abundant, because of game's high rank and reproductive benefit. Would the greater abundance of game during the early Late Holocene be sufficient to prompt a hunting-oriented subsistence-settlement strategy?

Fig. 14.12 also shows the contribution of large game to the caloric requirement if large game supplies 55% of the caloric requirement, barely enough to make the Mount Lincoln camp the optimal central place location. However, making such encounter rates sustainable would require a total population of approximately 3000 sheep, far more than the modern carrying capacity of 475 sheep estimated by the Bureau of Land Management (1991:37) and the 900 sheep hypothetically permitted by a 75% increase in forage during the early Late Holocene.

Discussion

The simulation suggests that a hunting-oriented subsistence-settlement strategy is unlikely to have existed in the Carson Desert during the early Late Holocene. However, the point is

TABLE 14.10. Distance, slope, and transport cost estimates for calculating central place foraging returns between Stillwater Marsh and Mount Lincoln.

Description	D_{is} Distance (km)	Slope (degrees)	W_s Walking Cost (kcal/km)	T_s Carrying Cost (kcal/kg/km)
Stillwater Marsh camp to sheep hunting patches on Mount Lincoln	10	1.5	50.6	—
	2	5	82.9	—
	5	15	142.3	—
	5	−15	48.3	.8
	2	−5	43.6	.4
	10	−1.5	50.6	.4
Stillwater Marsh camp to bulrush gathering patches in Stillwater Marsh	1	−1.5	50.6	—
	1	1.5	50.6	.4
Mount Lincoln camp to bulrush gathering patches in Stillwater Marsh	5	−15	48.3	—
	2	−5	43.6	—
	10	−1.5	50.6	—
	10	1.5	50.6	.4
	2	5	82.9	.86
	5	15	142.3	1.42
Mount Lincoln camp to sheep hunting patches on Mount Lincoln	1	5	82.9	—
	1	−5	43.6	.41

TABLE 14.11. Weight, time, and load size estimates required to satisfy the net caloric requirement for 25 kg loads of dried sheep and large bulrush seeds.

Proportion of Sheep	1.0	0.9	0.8	0.7	0.6	0.5	0.4	0.3	0.2	0.1	0.0
Dried sheep											
FN_1—net handling time (h)	233	209	186	163	140	116	93	70	47	23	0
FG_1—gross handling time (h)	248	223	198	173	149	124	99	74	50	25	0
NN_1—net loads of sheep	14	12	11	10	8	7	5	4	3	1	0
NG_1—gross loads of sheep for Stillwater Marsh camp	18	16	15	13	11	9	8	6	4	2	0
NG_1—gross loads of sheep for Mount Lincoln camp	15	14	13	11	10	7	6	5	3	2	0
L_1—weight per net load	24.5	23.8	22.9	24.0	22.9	24.5	22.9	20.6	22.9	17.2	0.0
Large bulrush seeds											
FN_2—net handling time (h)	0	64	127	191	254	318	381	445	509	572	636
FG_2—gross handling time (h)	0	75	150	224	299	374	449	524	598	673	748
NN_2—net loads of bulrush	0	2	3	5	6	8	9	10	12	13	15
NG_2—gross loads of bulrush for Stillwater Marsh camp	0	2	4	6	7	9	11	12	14	15	18
NG_2—gross loads of bulrush for Mount Lincoln camp	0	3	4	7	8	11	12	14	16	18	21
L_2—weight per net load	0.0	17.7	21.2	22.1	23.6	23.6	23.6	23.6	23.6	24.5	24.8

not to assert that the model settles the issue; obviously assumptions used to simulate prehistoric foraging behavior in the Carson Desert could be erroneous. It clearly is theoretically possible for large game to have been sufficiently abundant to prompt a settlement pattern emphasizing male foraging decisions, contrary to any empirical generalization that constraints to women's mobility must always tether base camp locations. However, it is a valid observation that the hunting success rates necessary to make such a strategy worthwhile seem an order of magnitude greater than those feasible in the ethnohistoric landscape or inferable for the early Late Holocene period.

Given this observation, it seems more likely that successful encounter rates with large game during the Neoglacial were insufficient to promote a hunting-oriented settlement strategy. In this scenario, women would emphasize marshes and men would concentrate on terrestrial hunting patches with more exclusivity than they did during ethnohistoric times, magnifying scheduling conflicts between men's and women's best foraging interests. The best response to this circumstance would be to increase the logistic mobility of men and tether residences to marshes, allowing early Late Holocene hunter-gatherers to exploit the best patches of both sexes simultaneously.

Exceptions to this pattern should have occurred when catastrophic events disrupted wetland productivity for a few years. At such times, hunter-gatherers would have the option of either staying in the Carson Desert by broadening their respective prey sets, or moving where higher foraging return rates were available. In the latter case, Carson Desert foragers may have cached specialized marsh-edge subsistence gear whenever flooding, drought, or internal biotic cycles made it worthwhile to abandon marsh-edge encampments in favor of camping in adjacent valleys and wetlands (Kelly 2001:292). Such events would not supplant women's need to feed children and store food, so the subsistence-settlement system would have continued to emphasize residence near women's best foraging patches.

Recent reanalysis of Hidden Cave coprolites (Rhode 2003) is consistent with the stance that women's foraging decisions likely drove the pre-1250 BP settlement strategy. Direct radiocarbon dating of the specimens shows that all predate 1250 BP, falling within two periods of time from 3800 to 3400 BP and 1900 to 1500 BP. As discussed previously, steroid analysis indicates that women probably deposited the specimens, suggesting that women sometimes foraged independently of men before 1250 BP. Cluster analysis of the dietary constituents of the coprolites identifies three constellations of wetland resources varying in the proportion of fish and bird bone, bulrush seed, and cattail pollen they contain, and a fourth set containing high proportions of dryland goosefoot and ricegrass seeds. The segregation of dryland seeds from wetland resources is consistent with the argument developed here that low-ranked seeds served as fallback foods when higher-ranked marsh resources were unavailable for women's foraging.

Evidence that it was periodic disruption of marsh productivity, rather than enhanced hunting opportunities, that drew women out of Stillwater Marsh is available in Fairview Valley, the next basin southeast of Stillwater Marsh (see Fig. 14.1). Fairview Valley is notable in the Carson Desert for the richness of ricegrass stands (Young et al. 1983; Zeanah et al. 1995; see Fig. 14.3) and the rarity of perennial water. However, annual runoff often fills the playa pan in Fairview Valley in spring and early summer when ricegrass is available (Zeanah et al. 1995). Large occupation sites with abundant ground stone tools fringe the playa (Zeanah 1996; Zeanah et al. 1995).

It seems likely that hunter-gatherers camped in Fairview Valley to harvest nearby patches of ricegrass when water stood in the pan, but under normal circumstances the marshes of the Carson Desert offered far better spring and summer foraging opportu-

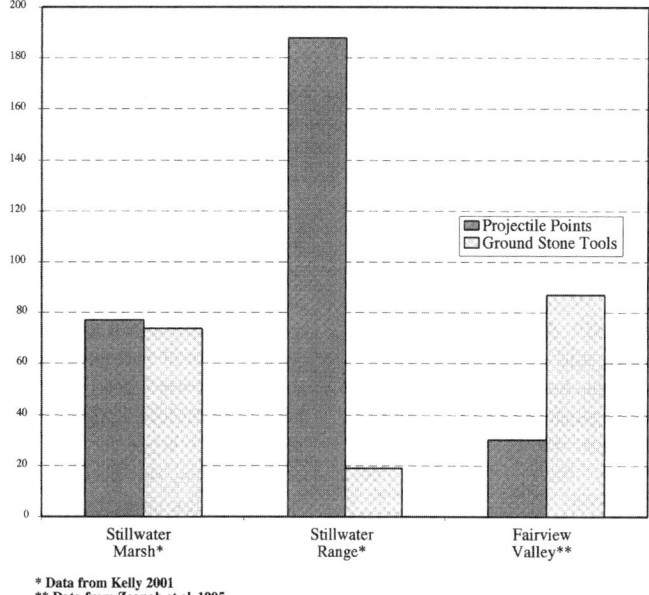

FIGURE 14.13. Comparison of projectile point and ground stone tool frequencies in three regions of the Carson Desert.

nities for women than such xeric seed stands. Less productive, but widespread, patches of ricegrass were available much closer to Stillwater Marsh (see Fig. 14.3). Moving to Fairview Valley to harvest ricegrass seeds only makes sense if marsh resources were unavailable. If so, extensive occupation in Fairview Valley indicates that occasional failures of the marsh drew Carson Desert foragers into neighboring valleys. These data complement evidence from the Stillwater Mountains where evidence of logistic hunting is common, and ground stone tools rare (Kelly 1985, 1995:22–23; 1999, 2001).

Fig. 14.13 compares the representation of projectile points and ground stone tools documented in systematic surveys of Stillwater Marsh, the Stillwater Mountains, and Fairview Valley, assuming that these artifact types serve as reliable markers of men's and women's foraging activity. Projectile points dominate in the Stillwater Mountains, whereas ground stone tools are more common in Fairview Valley.[8] The contrast is clear; women were reluctant to gather in Stillwater Range hunting patches only 20 km from their Stillwater Marsh encampments, but were occasionally willing to move more than 30 km to harvest distant, low-ranked, but prolific, seed stands in Fairview Valley.

Table 14.12 tallies the number of pre- and post-1250 BP diagnostic projectile points in the three regions. The representation of diagnostic projectile points in Fairview Valley is not significantly different from Stillwater Marsh or the Stillwater Mountains ($x^2 = 3.24$, $p = .202$, $df = 2$), and Fairview Valley yielded a higher percentage of pre-1250 BP point types (56%) than either Stillwater Marsh (32%) or the Stillwater Mountains (40%). The distribution suggests long continuity in the use of Fairview Valley (at least by men) relative to the other two regions.

TABLE 14.12. Counts of Late Holocene projectile points found in three regions of the Carson Desert.

Point Type[a]	Stillwater Mountains[b]	Stillwater Marsh[b]	Fairview Valley[c]
Pre-1250 BP	33	19	10
Post-1250 BP	50	40	8

[a] Pre-1250 BP point types include Humboldt, Gatecliff, and Elko Series. Post 1250 BP points are Eastgate, Rose Spring, Cottonwood, and Desert Series.
[b] Data from Kelly (2001).
[c] Data from Zeanah et al. (1995).

Regional Implications

These foraging simulations have profound implications for subsistence-settlement strategies in the Great Basin, but care must be taken that they are correctly understood. Although the simulations suggest hunting returns must breach a very high threshold for a hunting-oriented strategy to be worthwhile, it derives from resource distributions specific to the Late Holocene Carson Desert. The threshold would not necessarily be so high in other times and places where resources were differently arrayed. For example, elsewhere my colleagues and I (Elston and Zeanah 2002; Zeanah et al. 1995:312–326) have argued that the probable habitat mosaic of the Early Holocene Carson Desert was capable of supporting higher densities of large game in lowland distributions coinciding with women's best foraging patches. In such a situation, the foraging interests of men and women are less likely to conflict and a hunting-oriented economy is more likely. Consequently this simulation does not justify a generalization that hunting-oriented subsistence-settlement strategies never occurred in the Great Basin, but it does call on archaeologists to carefully think about women's subsistence options when considering whether such strategies occurred in the past.

It is clear that men's foraging returns are not crucial to determining when and where hunting-oriented settlement strategies may have occurred. Instead, the critical factor concerns women's requirement to reliably feed their children. This was often best achieved by residing in women's best foraging patches in the Late Holocene Great Basin, where temporal fluctuations in resource abundance often made food storage mandatory. Hunting must be sufficiently productive and reliable to overcome women's provisioning constraint for a hunting-oriented strategy to be favored. Otherwise, men should hunt logistically. Mere demonstration that higher hunting returns were feasible in the past is insufficient basis for expecting a hunting-oriented economy to have occurred. This is an important lesson in the Great Basin where archaeologists have often constructed models in which men's subsistence choices either explicitly or implicitly drive prehistoric subsistence-settlement strategies (Bettinger 1999; Bettinger and Baumhoff 1982; Butler 1986; Elston 1982, 1986; Kelly 1985; O'Connell et al. 1982; Tuohy 1974; Warren 1967).

The analysis presented here also bears implications for Hildebrandt and McGuire's (2002) recent proposal that prestige-based, large game hunting by men increased across California and the western Great Basin between 4000 and 1000 BP. As in this paper, McGuire and Hildebrandt call attention to the implications of sexual division of labor for understanding prehistoric foraging strategies and approach the issue from the perspective of human behavioral ecology. The authors point to widespread evidence that logistic hunting of large game increased about 4000 years ago, more or less simultaneously with evidence of burgeoning populations, greater sedentism, heavier reliance on food stores, and more intensive use of plants. They propose that Neoglacial foragers crossed a threshold whereby women met basic subsistence needs allowing men to specialize in hunting as a means for enhancing social prestige. Mostly consistent with their regional sweep, my model of Carson Desert prehistory bolsters Hildebrandt and McGuire's argument by contributing a detailed analysis of the foraging ecology and archaeology of a particular part of the western Great Basin.

However, it is worth pointing out two important differences between Hildebrandt and McGuire's treatment of sexual division of labor and this one. First, McGuire and Hildebrandt (2002:235) rather selectively call attention to only one faction (although a very persuasive one) in the debate among human behavioral ecologists concerning the evolution of sexual division of labor and food sharing. That camp argues that large-game hunting offers no provisioning advantage that could not be more efficiently gained by gathering, and that the evolution of hunting by males was little more than a form of sexual selection (Bliege Bird 1999; Hawkes 1990, 1991, 1996; Hawkes and Bliege Bird 2002; Hawkes et al. 2001; Smith and Bliege Bird 2000). There is another view that hunting males provide women and children with critical resources they could not otherwise easily obtain. In this view, standard optimal foraging models that intentionally reduce foraging choices to momentary caloric return rates simply do not monitor the fitness benefits of hunting. Those benefits may lie in protein rather than calories, require a lifetime of training to become proficient at acquiring, or are crucial at certain points in human life history (Gurven et al. 2000; Hill 1988; Hill and Kaplan 1993; Kaplan et al. 2000; Marlowe 2003). In this paper, I have drawn from the consensus among human behavior ecologists that men and women seek different arrays of prey in order to achieve different fitness goals without basing my analysis on which interpretation of the evolution of hunting may be correct. In contrast, Hildebrandt and McGuire depend on the correctness of the "show-off" explanation of male subsistence behavior to rationalize their interpretation of prestige hunting.

Second, both camps of human behavioral ecologists agree that, whatever the cause for its selection, specialized male strategies for procuring meat have a long evolutionary history in the genus *Homo* (Kaplan et al. 2000; O'Connell et al. 2002). This makes it critical that McGuire and Hildebrandt demonstrate why they expect "prestige-based" hunting to vary significantly within a relatively short span of time but broad expanse of space in western North America. They correctly point out that hunting and sexual division of labor are not hardwired into the human genotype and can be expected to vary in response to changes in the natural and social environment. However, the environmental changes that may have triggered such changes are not fully thought out.

This analysis of Carson Desert foraging ecology argues that intensified logistic hunting was a predictable response by men to improved foraging conditions in an environment where women's foraging opportunities also improved. But this prediction is made for a particular ecology containing a wetland that was rich in women's resources, and uplands that were impoverished of high-ranked women's resources (pinyon) but rich in game (predominately sheep) before 1250 BP. In contrast, Hildebrandt and McGuire's (2002:248–249) argument for California encompasses areas where critical men's and women's resources (sea mammals, mule deer, and acorns for example) were arrayed in dramatically different ecological distributions, and may have been *adversely* affected by the onset of the Neoglacial. Hildebrandt and McGuire might be correct that selection favored intensified, logistic hunting by males across such an ecologically diverse region, but they have yet to cast their argument in a testable manner. Instead, they have used theory developed to test hypotheses about the evolution of hunting in the Pleistocene against the behavior of modern hunter-gatherers to loosely accommodate broad regional trends in western North American prehistory.

Discussion

By debating whether the relative contributions of hunting and gathering to prehistoric diet changed significantly over time, Great Basin archaeologists (Bettinger 1977; Elston 1982, 1986; Heizer 1967; Heizer and Baumhoff 1970; Jennings 1957; Kelly 1985; O'Connell 1975; O'Connell et al. 1982; Thomas 1973; Tuohy 1974; Warren 1967) have often unwittingly considered whether men's and women's roles as food providers varied from the ethnographic record. They have come to explicitly address sexual division of labor by using Darwinian theory only over the last twenty years (Bettinger 1999; Bettinger and Baumhoff 1982; Hildebrandt and McGuire 2002; Kelly 1997; Leach 1999; Simms 1987), but have been poorly equipped with the middle range tools necessary for linking evolutionary theory with the archaeological record. As a result, some of these approaches have invoked explanations about foraging efficiency, ethnic identity, and gender prestige that are difficult to test archaeologically.

By adopting a human behavioral ecology stance and using optimal foraging theory as a research strategy, I have illustrated how central place foraging models can serve as middle range theory for investigating sexual division of labor in the archaeological record. In the Great Basin, the best archaeological evidence for variability in the contributions of hunting and gathering to prehistoric diets often comes from site distribution studies that monitor the spatial and temporal distribution of residential base camps, or excavations of such base camps themselves. Whether residential base camps occur in locations that seem most advantageous for pursuing large game or harvesting plants and small animals often serves as a primary trait distinguishing hunting from gathering-oriented settlement patterns. Since residential bases are the central places where men and women must resolve conflicts of foraging interest, central place foraging models capable of predicting the location and content of camps under different ecological scenarios are powerful tools for investigating sexual division of labor.

Using such models, I have argued that simple shifts in diet breadth are unlikely to trigger the gender-related shifts in settlement pattern and mobility strategy that Great Basin archaeologists have often envisioned in the past. Instead archaeologists should pay attention to conflicts of foraging interest between men and women, particularly as they affect women's need to provision children, when formulating and testing models of subsistence-settlement change.

Acknowledgments

Special thanks go to Kristen Hawkes whose thoughtful comments on my dissertation inspired this article. Several people have commented on various earlier drafts including Robert Elston, Kathy Heath, William Hildebrandt, Robert Kelly, Melinda Leach, Karen Lupo, and several anonymous reviewers. Their constructive criticisms have significantly improved this paper, and their efforts are greatly appreciated. Whatever mistakes remain are the fault of this author alone.

Notes

1. Archaeologists in the western United States have also envisioned subsistence-settlement strategies that were more gathering oriented than ethnographic examples (cf. McGuire and Hildebrandt 1994).
2. Most archaeologists currently reject characterization of Early Holocene foragers in the Great Basin as specialized "big game hunters" (Jones and Beck 1999:89; Simms 1988; Willig and Aikens 1988:26–27). Nevertheless, many argue that Early Holocene foragers hunted a generalized array of game while rarely harvesting seeds. This emphasis differed sufficiently from Middle and Late Holocene subsistence strategies to prompt dramatically different mobility strategies, settlement patterns and technological organization (Beck and Jones 1997:216–221; Zeanah et al. 1995; Elston and Zeanah 2002; Rusco and Davis 1986:180, 209).
3. Pinyon, one of the most productive upland resources procured by ethnographic Toedokado Paiute women (Wheat 1967), was probably absent from the Carson Desert uplands until after 1600 BP (Kelly 1995:18; 2001:36).
4. Kelly was specifically referring to the Traveler-Processor model of the Numic Spread through the Great Basin (Bettinger and Baumhoff 1982). That model posits that a gathering-oriented adaptation replaced a hunting-oriented adaptation throughout the

Great Basin during the Late Holocene, and is notable because of the emphasis it places on the role of sexual division of labor in the transition. However, the theoretical perspective of that model differs from the behavioral ecology stance taken in this paper, because it assumes sexual division of labor and sex ratios are determined by the common good of ethnic groups rather than played out among the conflicting interests of individual foragers (Zeanah and Simms 1999).

5. Foraging return rates in kcal/h were calculated using the equation developed by Winterhalder and colleagues (1989:325) for estimating resource encounter rates for a hypothetical forager searching randomly in each habitat at a set travel speed and search radius. Plant encounter rate calculations were based on the assumption that plant resources occur in 10 m^2 stands estimated from the percentage weight and cover contribution of each plant resource to the total vegetation biomass of the habitat. Small game were encountered in population units (burrows, nests, leks) that were assumed to occur in densities reported by modern game management studies in the habitats assessed by the model as being most suitable for each species. Population densities diminished proportionally to the habitat suitability score for all less attractive habitats. Large-game encounters were assumed to be with individual animals with highest population densities (as estimated from modern wildlife population estimates) occurring in the best habitats and diminishing proportionally with less suitable habitats. Large-game encounter rates were then adjusted to reflect unsuccessful hunts based on game management estimates. The foraging return rate obtainable for the optimal breadth of prey, by each sex, in each season was then calculated using the encounter rate estimates, and experimental and estimated post-encounter caloric return rates based on the formula for the diet breadth model (Schoener 1971; Simms 1987). Detailed discussion of the rationale and methodology is presented in Zeanah (1996).

6. The habitat foraging return rates and rankings in Tables 14.3–14.6 are not to be confused with the resource return rates and class ranks presented in Table 14.2, which represent the post-encounter caloric returns per unit of handling cost offered by individual resources and the order in which searching foragers should add resources to the diet. Tables 14.3–14.6 present the overall foraging return per unit of search and handling cost for a simulated hunter or gatherer seeking all resources in the optimal diet within each habitat (symbolized by an alphabetic prefix representing vegetation community and an arbitrary numeric identifier). These overall foraging return rates allow each habitat to be ranked as foraging patches (cf. MacArthur and Pianka 1966).

7. The exact locations of the two camps are not known and are only estimated from descriptions and maps provided by Fowler (1992). Despite locational ambiguity, the wetland camp certainly occurred in habitats assessed by the model as a primary women's foraging patch, whereas the mountain camp occurred in prime men's autumn territory. The arrival of pinyon in late prehistoric times also made the upland camp a suitable base for women's foraging.

8. Intermountain Research's survey of Stillwater Marsh (Raven 1990) documented frequencies of ground stone tools that are far greater than those of Kelly's (1985) survey, and overwhelm the representation of ground stone in both the Stillwater Mountains and Fairview Valley. The difference results from post-flooding exposure of numerous occupation sites and the attention paid by Intermountain Research toward identifying very small fragments of ground stone. The figures from Kelly's earlier survey were used here because circumstances of site visibility and recordation intensity were more comparable to those of the Stillwater Mountain and Fairview Valley surveys.

References

Barlow, K. R., D. Metcalfe
1996 Plant utility indices: two Great Basin examples. *Journal of Archaeological Science* 23:351–371.

Beck, C., G. T. Jones
1997 The terminal Pleistocene/early Holocene archaeology of the Great Basin. *Journal of World Prehistory* 11(2):161–236.

Bettinger, R. L.
1977 Aboriginal human ecology in Owens Valley: prehistoric change in the Great Basin. *American Antiquity* 42(1):3–17.
1999 From traveler to processor: regional trajectories of hunter-gatherer sedentism in the Inyo-Mono region, California. In *Settlement Pattern Studies in the Americas: Fifty Years Since Viru*, edited by B. R. Billman and G. M. Feinman, pp. 39–55. Smithsonian Institution Press, Washington, D.C.

Bettinger, R. L., M. A. Baumhoff
1982 The Numic spread: Great Basin cultures in competition. *American Antiquity* 47(3):485–503.

Binford, L. R.
1980 Willow smoke and dog's tails: hunter-gatherer settlement systems and archaeological site formation. *American Antiquity* 45:4–20.

Bliege Bird, R.
1999 Cooperation and conflict: the behavioral ecology of the sexual division of labor. *Evolutionary Anthropology* 8(2):65–75.

Brannon, J. A.
1992 On modeling resource transport costs: suggested refinements. *Current Anthropology* 33(1):56–60.

Brooks, S. T., M. B. Haldeman, R. H. Brooks
1988 *Osteological Analyses of the Stillwater Skeletal Series: Stillwater Marsh, Churchill County, Nevada.* US Fish and Wildlife Service Cultural Resource Series, No. 1, Portland.
1990 Unusual eburnation frequencies in a skeletal series from the Stillwater Marsh area, Nevada. In *Wetland Adaptations in the Great Basin*, edited by J. C. Janetski and D. B. Madsen, pp. 97–105. Museum of Peoples and Cultures, Brigham Young University, Provo.

Budy, E. E.
1988 Plant macrofossils. In *Preliminary Investigations in Stillwater Marsh: Human Prehistory and Geoarchaeology*, edited by C. Raven and R. G. Elston, pp. 343–356. US Fish and Wildlife Service Cultural Resource Series 1, Portland.

Bullock, M.,
1994 Appendix H. Ground stone use wear replication study. In *Behind the Argenta Rim: Prehistoric Land Use in Whirlwind Valley and the Northern Shoshone Range*, edited by C. Raven and R. G. Elston. Intermountain Research, Silver City, NV. Bureau of Land Management, Battle Mountain District on behalf of Santa Fe Pacific Gold Corporation. Ms. on file, BLM, Battle Mountain.

Butler, B. R.
1986 Prehistory of the Snake and Salmon River area. In *Great Basin*, edited by W. L. d'Azevedo, pp. 127–134. Handbook of North American Indians, vol. 11. Smithsonian Institution Press, Washington, D.C.

Chamberlin, R. V.
1911 The ethnobotany of the Gosiute Indians. *Memoirs of the American Anthropological Association* 2(5):329–405.

Couture, M. D., M. F. Ricks, L. Housley
1986 Foraging behavior of a contemporary northern Great Basin foraging population. *Journal of California and Great Basin Anthropology* 8(2):150–160.

Elston, R. G.
1982 Good times, hard times: prehistoric culture change in the western Great Basin. In *Man and Environment in the Great Basin*, edited by D. B. Madsen and J. F. O'Connell, pp. 186–206. Society for American Archaeology Papers 2, Washington, D.C.
1986 Prehistory of the western area. In *Great Basin*, edited by W. L. d'Azevedo, pp. 135–149. Handbook of North American Indians, vol. 11. Smithsonian Institution Press, Washington, D.C.
1990 Ebb and flow: Stability and persistence of marshes in the Carson Desert. Paper presented at the 55th Annual Meeting of the Society for American Archaeology, Las Vegas.

Elston, R. G., D. W. Zeanah
2002 Thinking outside the box: a new perspective on diet breadth and sexual division of labor in the Prearchaic Great Basin. *World Archaeology* 34(1):103–130.

Ember, C.
1975 Residential variation among hunter-gatherers. *Behavior Science Research* 3:199–227.

Estioko-Griffen, A., P. Griffen
1981 Woman the hunter: the Agta. In *Woman the Gatherer*, edited by F. Dahlberg, pp. 121–151. Yale University Press, New Haven.

Fowler, C. S.
1976 The processing of ricegrass by Great Basin Indians. *Mentzelia* 2:2–4.
1986 Subsistence. In *Great Basin*, edited by W. L. d'Azevedo, pp. 64–97. Handbook of North American Indians, vol. 11. Smithsonian Institution Press, Washington, D.C.
1989 *Willard Z. Park's Ethnographic Notes on the Northern Paiute of Western Nevada, 1933–1944*, vol. 1. Compiled and edited by C. S. Fowler. University of Utah Anthropological Papers No. 114, Salt Lake City.
1992 *In the Shadow of Fox Peak: An Ethnography of the Cattail-Eater Northern Paiute People of Stillwater Marsh*. US Department of the Interior, Fish and Wildlife Service, Region 1, Stillwater National Wildlife Refuge. Cultural Resource Series No. 5. US Government Printing Office, Washington, D.C.

Gurven, M., W. Allen-Arave, K. Hill, M. Hurtado
2000 It's a Wonderful Life: Signaling generosity among the Ache of Paraguay. *Evolution and Human Behavior* 21:263–282.

Hamilton, D. B., G. T. Auble
1993 *Wetland Modeling and Information Needs at Stillwater National Wildlife Refuge*. USDI Fish and Wildlife Service, Ft. Collins, CO.

Hawkes, K.
1990 Why do men hunt? Some benefits for risky strategies. In *Risk and Uncertainty in Tribal and Peasant Economies*, edited by E. Cashdan, pp. 145–166. Westview Press, Boulder.
1991 Showing off: tests of another hypothesis about men's foraging goals. *Ethology and Sociobiology* 11:29–54.
1996 Foraging differences between men and women: behavioral ecology of the sexual division of labor. In *The Archaeology of Human Ancestry: Power, Sex, and Tradition*, edited by J. Steele and S. Shennan, pp. 283–305. Routledge, London.

Hawkes, K., R. Bliege Bird
2002 Showing off, handicap signaling, and the evolution of men's work. *Evolutionary Anthropology* 11:58–67.

Hawkes, K., J. F. O'Connell, N. G. Blurton Jones
2001 Hunting and nuclear families: some lessons from the Hadza about men's work. *Current Anthropology* 42:681–709.

Heizer, R. F.
1967 *Analysis of Human Coprolites from a Dry Nevada Cave*. University of California Archaeological Survey Report 70, Berkeley, pp. 1–2.

Heizer, R. F., M. A. Baumhoff
1970 *Big Game Hunters in the Great Basin: A Critical Review of the Evidence*. Papers on the Anthropology of the Western Great Basin. Contributions of the University of California Research Facility, No. 7, Berkeley.

Hiatt, B.
1978 Woman the gatherer. In *Woman's Role in Aboriginal Society*, edited by F. Gale, pp. 4–15. Australian Institute of Aboriginal Studies, Canberra.

Hildebrandt, W. R., K. R. McGuire
2002 The ascendance of hunting during the California Middle Archaic: an evolutionary perspective. *American Antiquity* 67:231–256.

Hill, K.
1988 Macronutrient modifications of optimal foraging theory: an approach using indifference curves applied to some modern foragers. *Human Ecology* 16:157–197.

Hill, K., H. Kaplan
1993 On why foragers hunt and share food. *Current Anthropology* 34:701–710.

Hill, K., H. Kaplan, K. Hawkes, A. Hurtado
1987 Foraging decisions among Ache hunter-gatherers: new data and implications for optimal foraging models. *Ethology and Sociobiology* 8:1–36.

Hooper, L. M.
1994 Harvest and yield studies of some ethnographic food plants in the Yucca Mountain area. Poster Session presented at the 24th Biannual Great Basin Anthropological Conference, Elko, NV.

Horn, H.
1966 The adaptive significance of colonial nesting in the Brewers Blackbird (*Euphagus cyanocephalus*). *Ecology* 49:682–694.

Hurtado, A., K. Hawkes, K. Hill, H. Kaplan
1985 Female subsistence strategies among the Ache of eastern Paraguay. *Human Ecology* 13:1–28.

Ingbar, E.
1985 Analysis of Hidden Cave feature morphology. In *The Archaeology of Hidden Cave, Nevada*, edited by D. H. Thomas, pp. 307–321. Anthropological Papers 61, American Museum of Natural History, New York.

Jennings, J. D.
1957 *Danger Cave*. University of Utah Anthropological Papers No. 27, Salt Lake City.

Jones, G. T., C. Beck
1999 Paleoarchaic archaeology in the Great Basin. In *Models for the Millennium: Great Basin Anthropology Today*, edited by C. Beck, pp. 83–95. University of Utah Press, Salt Lake City.

Jones, K. T., D. B. Madsen
1989 Calculating the cost of resource transportation: a Great Basin example. *Current Anthropology* 30(4):529–533.

1991 Further experiments in native food processing. *Utah Archaeology* 14(1):68–77.

Kaplan, H., K. Hill, J. Lancaster, A. M. Hurtado
2000 A theory of human life history evolution: diet, intelligence, and longevity. *Evolutionary Anthropology* 9:156–185.

Kelly, I. T.
1932 Ethnography of Surprise Valley Paiutes. *University of California Publications in American Archaeology and Ethnology* 31: 67–210.

Kelly, R. L.
1985 Hunter-gatherer mobility and sedentism. A Great Basin study. Unpublished Ph.D. Dissertation, University of Michigan Department of Anthropology, Ann Arbor.
1990 Marshes and mobility in the western Great Basin. In *Wetland Adaptations in the Great Basin*, edited by J. C. Janetski and D. B. Madsen, pp. 259–276. Brigham Young University Museum of Peoples and Cultures Occasional Papers No. 1, Provo, UT.
1995 Hunter-gatherer lifeways in the Carson Sink: a context for bioarchaeology. In *Bioarchaeology of the Stillwater Marsh: Prehistoric Human Adaptation in the Western Great Basin*, edited by C. S. Larsen and R. L. Kelly, pp. 12–32. Anthropological Papers of the American Museum of Natural History, No. 77, New York.
1997 Late Holocene Great Basin prehistory. *Journal of World Prehistory* 11(1):1–50.
1999 Theoretical and archaeological insights into foraging strategies among the prehistoric inhabitants of the Stillwater Marsh wetlands. In *Prehistoric Lifeways in the Great Basin Wetlands*, edited by B. E. Hemphill and C. S. Larsen, pp. 117–150. University of Utah Press, Salt Lake City.
2001 *Prehistory of the Carson Desert and Stillwater Mountains: Environment, Mobility, and Subsistence in a Great Basin Wetland*. University of Utah Press Anthropological Records Number 123, Salt Lake City.

Larralde, S. L., S. M. Chandler
1980 *Archaeological Inventory in the Seep Ridge Cultural Study Tract, Uintah County, Northeastern Utah*. Nickens and Associates, Montrose, CO. Bureau of Land Management, Contract No. YA-553-CTO-1008.

Larsen, C. S., D. L. Hutchinson
1999 Osteopathology of Carson Desert foraging: reconstructing prehistoric lifeways in the western Great Basin. In *Prehistoric Lifeways in the Great Basin Wetlands*, edited by B. E. Hemphill and C. S. Larsen, pp. 184–202. University of Utah Press, Salt Lake City.

Larsen, C. S., R. L. Kelly
1995 *Bioarchaeology of the Stillwater Marsh: Prehistoric Human Adaptation in the Western Great Basin*. Anthropological Papers 77. American Museum of Natural History, New York.

Larsen, C. S., C. B. Ruff, R. L. Kelly
1995 Structural analysis of the Stillwater postcranial human remains: Behavioral implications of articular joint pathology and long bone diaphyseal morphology. In *Bioarchaeology of the Stillwater Marsh: Prehistoric Human Adaptation in the Western Great Basin*, edited by C. S. Larsen and R. L. Kelly, pp. 107–133. Anthropological Papers 77. American Museum of Natural History, New York.

Leach, M.
1999 In search of gender in Great Basin prehistory. In *Models for the Millennium: Great Basin Anthropology Today*, edited by C. Beck, pp. 182–191. University of Utah Press, Salt Lake City.

Lee, R. B.
1979 *The !Kung San: Men, Women and Work in a Foraging Society*. Cambridge University Press, Cambridge.

Madsen, D. B., L. Eschler, T. Eschler
1997 Winter cattail collecting experiments. *Utah Archaeology* 10(1): 1–19.

MacArthur, R. H., E. R. Pianka
1966 On optimal use of a patchy environment. *American Naturalist* 100:603–609.

MacDonald, I.
1961 Statistical studies of recorded energy expenditure of man. *Nutrition Abstracts and Reviews* 31(3):739–757.

Marlowe, F. W.
2003 A critical period for provisioning by Hadza men: implications for pair bonding. *Evolution and Human Behavior* 24:217–229.

Maynard Smith, J.
1978 *The Evolution of Sex*. Cambridge University Press, Cambridge.

McGuire, K. R., W. R. Hildebrandt
1994 The possibilities of women and men: gender and the California Milling Stone Horizon. *Journal of California and Great Basin Anthropology* 16:41–59.

Napton, L. K., R. F. Heizer
1970 Analysis of human coprolites from archaeological contexts, with primary reference to Lovelock Cave, Nevada. In *Archaeology and the Prehistoric Great Basin Lacustrine Subsistence Regime as Seen from Lovelock Cave, Nevada*, edited by R. F. Heizer and L. K. Napton, pp. 87–130. Contributions of the University of California Archaeological Research Facility, Berkeley.

O'Connell, J. F.
1975 The prehistory of Surprise Valley. *Ballena Press Anthropological Papers* 4:1–57.

O'Connell, J. F., K. Hawkes, K. D. Lupo, N. G. Blurton Jones
2002 Male strategies and Plio-Pleistocene archaeology. *Journal of Human Evolution* 43:831–872.

O'Connell, J. F., K. T. Jones, S. R. Simms
1982 Some thoughts on prehistoric archaeology in the Great Basin. In *Man and Environment in the Great Basin*, edited by D. B. Madsen and J. F. O'Connell, pp. 227–240. Society for American Archaeology Papers 2, Washington, D.C.

Orians, G. H., N. E. Pearson
1977 On the theory of central place foraging. In *Analysis of Ecological Systems*, edited by D. J. Horn, G. R. Stairs, and R. D. Mitchell, pp. 155–177. Ohio State University Press, Columbus.

Passmore, R., J. V. Durnin
1955 Human energy expenditure. *Physiological Review* 35:801–839.

Raven, C.
1988 Cultural features and site structure. In *Preliminary Investigations in Stillwater Marsh: Human Prehistory and Geoarchaeology*, edited by C. Raven and R. G. Elston, pp. 385–399. Cultural Resource Series, US Fish and Wildlife Service, Region 1, Portland, OR.
1990 *Prehistoric Human Geography in the Carson Desert, Part II: Archaeological Field Test of Model Predictions*. Cultural Resource Series, Number 4. US Fish and Wildlife Service, Region 1, Portland, OR.

Raven, C., R. G. Elston (Eds.)
1988 *Preliminary Investigations in Stillwater Marsh: Human Prehis-*

tory and Geoarchaeology. Cultural Resource Series 1, US Fish and Wildlife Service, Region 1, Portland, OR.

1989 *Prehistoric Human Geography in the Carson Desert, Part 1: A Predictive Model of Land-Use in the Stillwater Wildlife Management Area*. Cultural Resource Series, Number 3. US Fish and Wildlife Service, Region 1, Portland, OR.

Raymond, A. W., V. M. Parks

1989 *Surface Archaeology of Stillwater Marsh, Churchill County, Nevada*. US Fish and Wildlife Service, Stillwater Wildlife Management Area, Fallon, NV.

1990 Archaeological sites exposed by recent flooding of Stillwater Marsh, Carson Desert, Churchill County, Nevada. In *Wetland Adaptations in the Great Basin*, edited by J. C. Janetski and D. B. Madsen, pp. 33–62. Museum of Peoples and Cultures, Occasional Paper No. 1. Brigham Young University, Provo.

Raymond, A. W., E. Sobel

1990 The use of tui chub as food by Indians of the western Great Basin. *Journal of California and Great Basin Anthropology* 12(1):2–18.

Reidhead, V. A.

1976 Optimization and food procurement at the prehistoric Leonard Haag site, Southeastern Indiana: a linear programming approach. Unpublished Ph.D. Dissertation, Department of Anthropology, Indiana University.

Rhode, D.

2003 Coprolites from Hidden Cave revisited: evidence for site occupation history, diet, and sex of occupants. *Journal of Archaeological Science* 30:909–922.

Rhode, D., K. D. Adams, R. G. Elston

2000 Geoarchaeology and Holocene landscape history of the Carson Desert, western Nevada. *Geological Society of America, Field Guide* 2:45–74.

Roust, N. L.

1967 *Preliminary Examinations of Prehistoric Human Coprolites from Four Western Nevada Caves*. University of California Archaeological Survey Reports 70, Berkeley, pp. 49–88.

Rusco, M. K., J. O. Davis

1986 *Studies in Archaeology, Geology, and Paleontology at Rye Patch Reservoir, Pershing County, Nevada*. Nevada State Museum Anthropological Papers 20, Carson City.

Schoener, T. W.

1971 Theory of feeding strategies. *Annual Review of Ecology and Systematics* 2:369–404.

Schoeninger, M. J.

1995 Dietary reconstruction in the prehistoric Carson Desert: stable carbon and nitrogen isotopic analysis. In *Bioarchaeology of the Stillwater Marsh: Prehistoric Human Adaptation in the Western Great Basin*, edited by C. S. Larsen and R. L. Kelly, pp. 96–106. Anthropological Papers of the American Museum of Natural History, No. 77, New York.

1999 Prehistoric subsistence strategies in the Stillwater Marsh region of the Carson Desert. In *Prehistoric Lifeways in the Great Basin Wetlands*, edited by B. E. Hemphill and C. S. Larsen, pp. 151–166. University of Utah Press, Salt Lake City.

Shimkin, D. B., R. M. Reid

1970 Sociocultural persistence among Shoshoneans of the Carson River basin (Nevada). In *Languages and Cultures of Western North America: Essays in Honor of Sven S. Liljeblad*, edited by E. H. Swanson, pp. 172–200. Idaho State University Press, Pocatello.

Simms, S. R.

1987 *Behavioral Ecology and Hunter-Gatherer Foraging: An Example from the Great Basin*. British Archaeological Reports: International Series 381, Oxford.

1988 Conceptualizing the Paleo-Indian and Archaic in the Great Basin. In *Early Human Occupation in Far Western North America: The Clovis-Archaic Interface*, edited by J. A. Willig, C. M. Aikens, and J. L. Fagan, pp. 41–520. Nevada State Museum Anthropological Papers Number 21, Carson City.

Smith, E. A., R. L. Bliege Bird

2000 Turtle hunting and tombstone opening: public generosity as costly signaling. *Evolution and Human Behavior* 21:245–261.

Sojda, R. S., K. L. Solberg

1993 *Management and Control of Cattails*. In Fish and Wildlife Leaflet 13.4.13. USDI Fish and Wildlife Service, Washington, DC, pp. 1–7.

Steward, J. H.

1933 Ethnography of the Owens Valley Paiute. *University of California Publications in American Archaeology and Ethnology* 33:233–350.

1938 *Basin-Plateau Sociopolitical Groups*. Bureau of American Ethnology Bulletin 120, Smithsonian Institution. US Government Printing Office, Washington, D.C.

1941 *Cultural Element Distributions: XII, Nevada Shoshone*. Anthropological Records, vol. 4, No. 2. University of California Press, Berkley.

Stewart, O. C.

1941 *Culture Element Distributions XIV: Northern Paiute*. Anthropological Records, vol. 4, No. 3. University of California Press, Berkeley, pp. 361–446.

Thomas, D. H.

1973 An empirical test for Steward's model of Great Basin settlement patterns. *American Antiquity* 38:155–176.

1985 *The Archaeology of Hidden Cave, Nevada*. Anthropological Papers 61. American Museum of Natural History, New York.

Thompson, S., B. Hallock

1988 *Newlands Project Operating Criteria and Procedures, Record of Decision, Wetland Analysis*. USDI Fish and Wildlife Service, Fallon, NV.

Todt, D. L., N. Hannon

1998 Plant food resource ranking on the upper Klamath River of Oregon and California: a methodology with archaeological implications. *Journal of Ethnobiology* 18(2):273–308.

Tuohy, D. R.

1974 A comparative study of late Paleo-Indian manifestations in the western Great Basin. In *A Collection of Papers on Great Basin Archaeology*, edited by R. G. Elston and L. Sabini, pp. 99–116. Nevada Archaeological Survey Research Paper No. 8, Reno.

Tuohy, D. R., A. J. Dansie, M. B. Haldeman

1987 *Final Report on Excavations in the Stillwater Marsh Archaeological District*. Nevada State Museum Archaeological Services Report, Carson City.

United Nations, Food and Agriculture Organization

1990 *Manual on Simple Methods of Meat Preservation*. FAO Animal Production and Health Paper 9. United Nations Food and Agriculture Organization, Rome.

US Department of the Interior Bureau of Land Management

1991 *Nevada BLM Statewide Wilderness Report, Vol. IV*. Carson City and Ely Districts. USDI, Bureau of Land Management, Nevada State Office, Reno.

Warren, C. N.
1967 The San Dieguito complex: a review and hypothesis. *American Antiquity* 32(2):168–185.

Washburn, S., C. Lancaster
1967 The evolution of hunting. In *Man the Hunter*, edited by R. Lee and I. Devore, pp. 293–303. Aldine, Chicago.

Weller, M. W.
1981 *Freshwater Marshes: Ecology and Wildlife Management*. University of Minnesota Press, Minneapolis.

Wheat, M. M.
1967 *Survival Arts of the Primitive Paiutes*. University of Nevada Press, Reno.

Williams, G. C.
1975 *Sex and Evolution*. Princeton University Press, Princeton.

Willig, J. A., C. M. Aikens
1988 The Clovis-Archaic interface in far western North America. In *Early Human Occupation in Far Western North America: The Clovis-Archaic Interface*, edited by J. A. Willig, C. M. Aikens, and J. L. Fagan, pp. 325–360. Nevada State Museum Anthropological Papers Number 21, Carson City.

Wigand, P. E., P. J. Mehringer Jr.
1985 Pollen and seed analysis. In *The Archaeology of Hidden Cave, Nevada*, edited by D. H. Thomas, pp. 108–124. Anthropological Papers 61, American Museum of Natural History, New York.

Winterhalder, B., W. Bailolargeon, F. Cappelletto, I. R. Daniel Jr., C. Prescott
1989 The population ecology of hunter-gatherers and their prey. *Journal of Anthropological Archaeology* 7:280–283.

Young, J. A., R. A. Evans, B. A. Roundary
1983 Quantity and germinability of *Oryzopsis hymenoides* seeds in Lahontan sands. *Journal of Range Management* 36(1):82–86.

Zeanah, D. W.
1988 Pit features and food storage strategies in the Great Basin. Paper presented at the 21st Great Basin Anthropological Conference, Park City, UT.
1996 Predicting settlement patterns and mobility strategies: an optimal foraging analysis of hunter-gatherer use of mountain, desert, and wetland habitats in the Carson Desert. Unpublished Ph.D. Dissertation, University of Utah, Department of Anthropology, Salt Lake City.
1999 Transport costs, central place foraging, and hunter-gatherer alpine land use strategies. In *Intermountain Archaeology: Selected Papers of the Rocky Mountain Anthropological Conference*, edited by D. B. Madsen and M. D. Metcalfe, pp.1–14. University of Utah Anthropological Papers, Number 122, Salt Lake City.

Zeanah, D. W., R. G. Elston, J. A. Carter, D. P. Dugas, J. E. Hammett
1995 *An Optimal Foraging Model of Hunter-Gatherer Land Use in the Carson Desert*. Intermountain Research Report to US Department of the Navy and the US Fish and Wildlife Service.

Zeanah, D. W., S. R. Simms
1999 Modeling the gastric: Great Basin subsistence studies since 1982 and the evolution of general theory in Great Basin archaeology. In *Models for the Millennium: Great Basin Anthropology Today*, edited by C. Beck, pp. 118–140. University of Utah Press, Salt Lake City.

15

Declining Foraging Efficiency and Moa Carcass Exploitation in Southern New Zealand

Lisa Nagaoka

INTRODUCTION

Early zooarchaeological models of prey body-part transport decisions predicted that the kind and number of skeletal parts carried back to a home base depend upon the "utility" of those parts and the distance from a carcass to the home base [10, 21, 31, 32, 50, 54, 55]. For vertebrates, utility is typically specified for skeletal parts in terms of meat, marrow, and grease content [10, 11], and is more correctly termed "food utility" [39]. Elements are ranked in terms of their food utility, and decisions about body-part transport by foragers are assumed to have been made based on this ranking. In general, foragers are expected to process large animals in the field and discard low-utility body parts in order to create more transportable units [9, 12, 39, 38, 46, 47], with higher-utility elements transported back to the home base for consumption.

These early studies examined carcass butchery and transport decisions in a manner similar to those found in foraging theory models of behavioral ecology. However, only recently have researchers explicitly linked carcass exploitation to foraging theory models, providing a theoretical link to these middle-range studies [45]. In particular, patch choice models have been used to examine body-part transport decisions (e.g., [13, 16, 40, 44] and references therein). Patch choice is used to determine in which patches a forager will pursue prey and how long they should spend in each patch. However, when applied to human carcass exploitation, the scale of analysis is reduced so that the models are used to examine how human predators forage across individual carcasses, selecting and transporting elements based on the utility of each element [9, 21]. In this paper I discuss the theoretical assumptions and predictions that models based in foraging theory entail when applied to human foragers. I then apply these models to the moa assemblage from the Shag River Mouth site in southern New Zealand.

As with many other Pacific Islands, the human colonization of New Zealand led to the extinction of many native animals, the best-known examples of which are the moas (Aves: Dinornithiformes). These ten species of birds [14, 26] went extinct 100–200 years after Polynesians colonized New Zealand in the 13th century [3, 57]. Moas were much stockier in build than most other ratites, with individuals that weighed up to 250 kg. The abundance of moa remains in archaeological middens attests to the significant role that human predation played in the decline and eventual extinction of moas [1, 2, 57].

In previous research, I used foraging theory models to demonstrate how human foraging patterns responded to the decline in the availability of moas at the Shag River Mouth site [42–44]. The site is located on the east coast of the South Island of New Zealand (Figure 15.1), and was occupied from AD 1250 to AD 1450 [6]. Moas were the largest terrestrial vertebrate in New Zealand, and thus they are assumed to be high-ranked resources for human foragers. As the availability of moas declined over time at Shag Mouth, the proportion of smaller game pursued increased so that more effort was required to produce the same net gains resulting in a decrease in overall foraging efficiency. In addition, the decline in foraging efficiency was significant enough so that foragers also increased the number of resources pursued, as well as the number of patches or environments exploited. This decreased foraging efficiency was shown to result from resource depression, a reduction in the availability of high-ranked prey due to human foraging efforts [17], rather than to environmental changes or technological innovations. Thus, foragers were responding to a situation of declining resource availability that they created.

This demonstrated decline in foraging efficiency means that Shag Mouth offers a valuable dataset with which to consider carcass exploitation. Resource depression and the corresponding decline in foraging efficiency should have affected how foragers butchered, transported, and utilized moa carcasses over time. Using patch choice models of foraging theory, several expectations about carcass exploitation are generated and tested.

*Originally published in *Journal of Archaeological Science* 32(2005):1328–1338

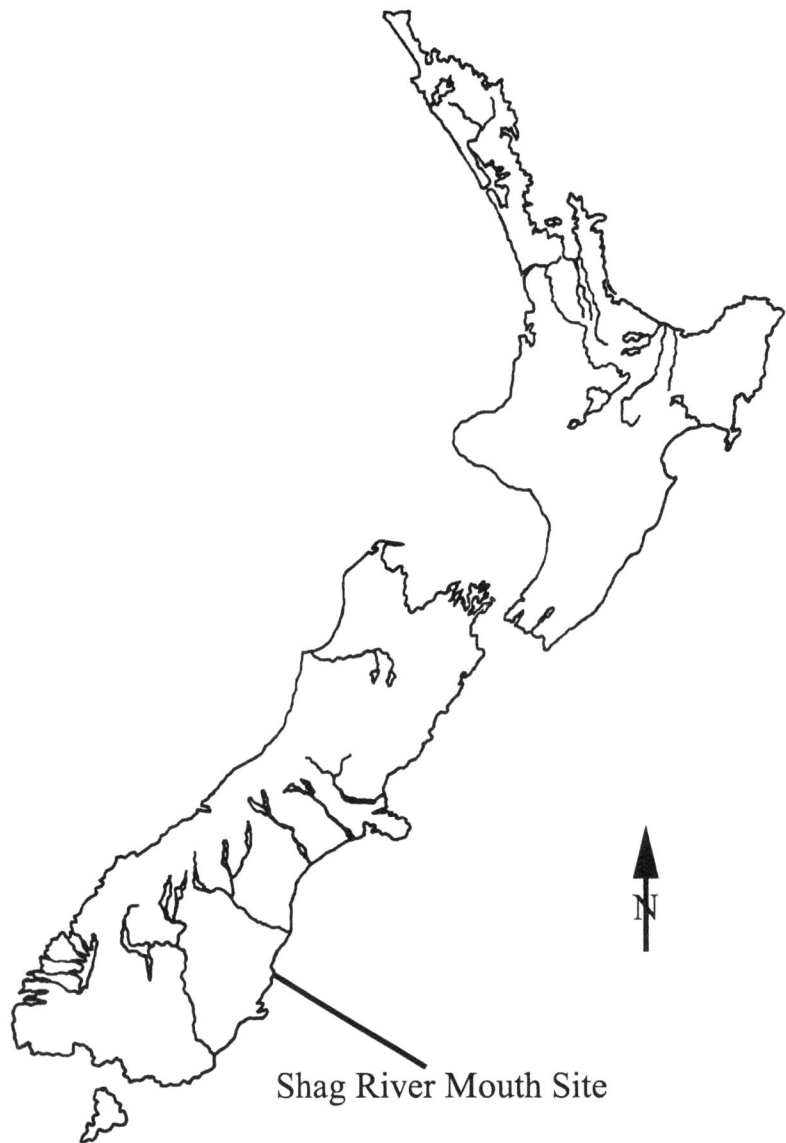

FIGURE 15.1. Location of the Shag River Mouth site in New Zealand.

FORAGING THEORY AND CARCASS TRANSPORT AND PROCESSING DECISIONS

Much of the early work on carcass exploitation used models that were similar in structure to foraging theory models. Foragers were assumed to have transported elements based on their utility, which is analogous to the net return rate used in foraging theory. Thus, the formal incorporation of carcass exploitation studies into a foraging theory framework was a relatively simple step. Initially, the specific model used to explain butchery and transport decisions was the marginal value theorem (MVT) [13, 44], which had been designed to account for foragers who consume their prey at the point of capture [17]. However, in most situations, humans are better characterized as central place foragers, who forage from a "home base" and return prey to that base for consumption [15, 16]. Central place foraging models differ from the Marginal Value Theorem in that they incorporate the additional travel costs associated with central place foraging.

Orians and Pearson [48] developed two central place models, for which foragers make choices that maximize net returns per foraging trip. The greater the distance is and the higher the transport costs are, the greater the net returns per trip should be. One problem with the application of Orians and Pearson's models to human foragers is that they were designed for non-human predators that transport their prey whole. Human predators, on the other hand, often butcher prey that are too large to carry whole, and it is these butchering decisions that are not taken into account in Orians and Pearson's models. To address this limitation, Cannon [16] created an archaeological model that incorporates Metcalfe and Barlowe's [38] research on transport and processing decisions into Orians and Pearson's central place foraging model. Cannon's central place forager prey choice model can be used for understanding both prey choice and processing/transport decisions, but I focus on the latter for the purpose of this analysis.

Cannon's model assumes that foragers will maximize the "delivery rate" when making butchering and transport decisions [16:4]. Delivery rate is similar to the net return rate, except that in addition to handling costs and transport costs, it also includes processing costs. Cannon differentiates between butchering that is associated with handling costs and butchering considered as processing costs. Handling costs incorporate the time required to transform the carcass into a load that can be carried, while the processing costs include the extra butchering done to maximize the utility of the elements transported. The assumption is that handling costs should be relatively constant across prey of the same species and size. Thus, changes in butchering practices are expected to relate to differences in the way prey are processed.

The amount of time a forager spends butchering, in particular processing a carcass, is affected by both transport distance and prey encounter rate. If prey encounter rates decrease, but transport distance remains unchanged, then field processing of prey may decline [17]. Instead, each individual carcass should be exploited more intensively so that a broader range of elements will be returned. However, foragers tend to deplete resources in a particular pattern such that transport distance increases with declining prey encounter rates [24]. In general, foragers will first exploit resources around the home base. When local areas around the central place are being exploited, transport costs are relatively low. In such cases, less field processing should occur and a relatively broad range of high and low return elements will be transported. As resource availability declines locally and foragers expand their foraging radius, distance to prey, and thus transport costs increase. In such situations, processing time is expected to increase in order to maximize the utility of the load transported. Foragers should be more selective not only about what is pursued, but also about what portions of those prey items are returned to the central place. That is, they will tend to maximize the delivery rate by processing the carcass so that the utility per load transported back to the central place is high [16].

At Shag Mouth, moa abundance declined significantly during the occupation of the site [43]. Thus, if the distribution of moas across the landscape followed the pattern described above, then foragers had to travel farther over time to exploit them. As transport distance increased, time spent processing moa carcasses should have also increased. Foragers should have become more selective about what they transported, with fewer lower utility elements being transported over time.

To determine if the delivery rate of elements transported changed over time, the mean utility or average returns per element for a given sample is examined. Mean utility is calculated by multiplying each specimen within a sample or assemblage by the corresponding utility value for that element [13]. The utility for all specimens is then summed and divided by the total number of specimens for the sample. Each sample is represented by a single value, which can then be used to statistically examine mean utility amongst samples. Samples with a high mean utility have a large proportion of high utility elements. If a broader range of elements is represented, then the mean utility for that layer will be lower. Thus, for the Shag Mouth moas, since encounter rates are known to have declined and travel costs are expected to have increased, mean utility of the elements transported is expected to have increased over time as foragers maximized their delivery rate.

Foraging Theory and Skeletal Element Breakage Decisions

In addition to changing butchery and transport patterns, if foraging efficiency declines significantly, foragers may also try to extract more calories from the skeletal elements transported back to the site through activities such as marrow and grease extraction [13]. The Marginal Value Theorem (MVT) can be used to make predictions about the nature of this resource exploitation. The MVT assumes that the time allocated to foraging within a patch will depend on the net return rate of that patch, and the average return rate for all exploited patches, taking into account the costs of moving between those patches [17]. When high-ranked prey are abundant in all patches, the average return rate is high. In this situation, foragers are expected to focus on exploiting high return prey types, moving on to the next patch as soon as the returns drop to the average for all patches. Thus, when the average return rate is high, the amount of time spent in each patch will be relatively short. The MVT also assumes that foraging behaviors affect prey abundances, often leading to declines in encounter rates over time. Thus, as patch and average return rates across all patches decline due to resource depression, more foraging time is allocated within each patch. In other words, each patch is used more intensively. In addition, patches that once had net returns below the average return rate may now be cost-effective to exploit.

Although the MVT was developed to deal with the spatial distribution of prey types [17, 48], archaeologically it has been used to examine if skeletal elements transported back to a home base were exploited more intensively over time [13]. Since exploitation of bones occurs at the site, travel costs are not an issue, thus the MVT is an appropriate model. The scale of analysis is changed so that each skeletal element is considered a patch. The prediction made from the MVT is that with declining foraging efficiency, foragers should increase the time allocated to extracting resources out of productive skeletal elements. In other words, when foraging efficiency is low, elements transported back to the home base will be used more intensively through activities such as marrow and grease extraction [13]. In addition, the number of skeletal elements exploited may expand to include lower ranked elements.

Marrow and grease extraction both require that individual elements be broken, but the nature of the fragmentation differs for each activity [49]. Building on Lyman's [33] distinction between the intensity and extent of fragmentation, Wolverton [56] developed a model that applied these concepts to identify marrow versus grease exploitation. The breakage patterns associated with marrow versus grease extraction are expected to differ because marrow and grease are distributed differently across and within elements. Marrow is typically found in certain elements, and usually only in parts of those elements (e.g., long bone shafts). To obtain marrow, the marrow cavity need only be broken open or "breached" [41:224]. Therefore, marrow extraction can be characterized by the extent of fragmentation, i.e., the proportion of specimens that were broken, which can be measured by calculating the percentage of unbroken elements (% whole). A decrease in % whole indicates that a larger proportion of skeletal elements were broken over time, suggesting that marrow extraction increased.

In contrast to marrow, grease is associated with all elements, thus a broader range of elements are likely to be exploited. In addition, grease, though it occurs in high abundance in cancellous bone, is not limited to certain portions of elements like marrow. Therefore, all portions of an element may be fragmented to extract grease. It is assumed that more intensive grease extraction is characterized by smaller bone fragments. The smaller a skeletal element is fragmented, the more grease that can be extracted from it because there is more surface area from which to draw grease. Thus, grease extraction is characterized by the intensity or degree of fragmentation [34, 56], which can be measured by the ratio of the number of fragmented specimens in the sample (NISP) to the minimum number of elements represented by those specimens (MNE) [20, 34]. NISP:MNE has previously been used to measure both marrow and grease extraction [13, 44]. However, it is a better measure of just grease extraction because it indicates fragment size [34, 56], and fragment size is not related to marrow extraction efficiency.

For the Shag Mouth moa, the MVT predicts that the time allocated to marrow and grease extraction is likely to have increased if the decline in foraging efficiency is significant enough to warrant more intensive exploitation of relatively low utility and low ranked resources. Like other flightless birds, long bones of moa had medullary cavities that contained marrow. More intensive use of marrow-bearing skeletal elements should result in a decline in the proportion of whole bones (% whole) over time. In addition, there should be variation in element breakage based on the amount of marrow available across elements. Kooyman [28] estimated the marrow cavities for tibiotarsi to be the largest with a range of 120–960 cc, while tarsometatarsi were 80–730 cc, and femura were 140–240 cc. Thus, tibiotarsi should be relatively high ranked in terms of marrow extraction. Elements such as phalanges, which have small marrow cavities are low ranked and should only be exploited after foraging efficiency has declined significantly.

Unlike marrow, grease extraction is a more time consuming process. Grease is found within the bone structure and is less accessible than marrow, thus extracting grease from elements typically entails fragmenting and boiling the bone. This process requires more energy with fewer caloric returns, therefore grease should be a lower ranked resource than marrow. As such, grease will be exploited after marrow extraction is added to foraging behaviors, and only after foraging efficiency has declined significantly to warrant the exploitation of such a low return resource.

Archaeologically, the assumption about grease extraction is that smaller fragments will allow for more grease to be extracted, and thus is an indication of more intensive use of skeletal elements. However, recent experimental research by Church and Lyman [18] suggests that the amount of grease that can be extracted from deer (*Odocoileus* sp.) long bones does not increase significantly when the bone fragments are smaller. Therefore, the assumption that increased fragmentation rates implies more intensive use of elements for grease extraction may be faulty. Church and Lyman's experiment was designed to examine grease extraction over a time scale of one hour intervals up to 14 hours. The amount of grease extracted reaches a state of diminishing returns by the third hour across all fragment sizes. However, the data for the first hour of boiling varies from 29% to 63% of the grease extracted depending on fragment size. The smaller the fragment was, the more grease was extracted in that first hour. These data suggest that if bone is boiled for a duration of less than one hour, then increased fragmentation of bone may provide more grease. The shorter boiling time may be more relevant to New Zealand where ceramic vessels were not available and boiling was done by placing hot cooking stones into wooden bowls [4]. If boiling time for grease extraction was a limiting factor such that fragment size mattered, then the MVT predicts that the amount of time that foragers spent fragmenting skeletal elements for grease extraction should have increased across time if foraging efficiency declined significantly.

Changing Processing and Transport Patterns of Moas at Shag Mouth

With resource depression and an increase in travel costs due to an expanding foraging radius, the transport of moa carcass parts should narrow through time to mainly high utility elements, i.e., the mean utility of elements per layer is expected to increase (Table 15.1). Since moas are extinct, it is impossible to derive utility indices for them directly. Instead, I use a utility index developed for kiwis (*Apteryx* spp.), ratite relatives of moas that are endemic to New Zealand [29]. Kiwis, though much smaller than moas, are similar in shape. They, like moas, are stocky and have more robust legs compared to other ratites such as ostriches, which tend to have longer, gracile legs designed for running. To derive his kiwi utility index, Kooyman [29] weighed the meat taken from each element of eight kiwi carcasses, then ranked these values and normalized them by setting the top value at 100 and adjusting the remaining values accordingly.

TABLE 15.1. Summary of expectations for carcass transport and skeletal element breakage given declining foraging efficiency.

	CONDITION	PREDICTIONS	ARCHAEOLOGICAL MEASURES
Transport	▼ in local resource availability and foraging efficiency; ▲ in transport costs	▲ in utility of elements transported per load	▲ in mean utility (▲ in high utility elements; ▼ in low utility elements)
Breakage	▼ in resource availability and foraging efficiency	▲ in extent of bone fragmentation due to marrow extraction	▼ in %whole
		▲ in intensity of bone fragmentation due to grease extraction	▲ in NISP:MNE

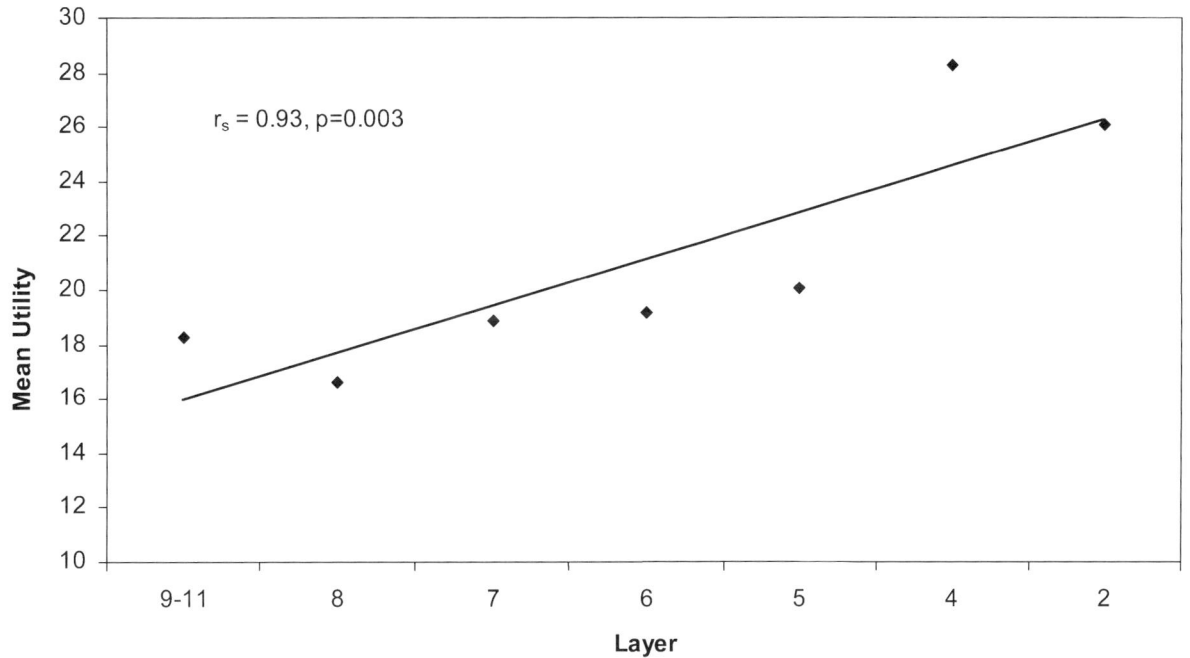

FIGURE 15.2. Mean moa utility across layers. Layer 2 is the youngest; Layers 9–11 are the oldest.

Using Kooyman's utility data and the number of identified moa specimens in the Shag Mouth assemblage, I calculated the mean utility of elements for each layer. The data are plotted in Figure 15.2. Layer 11 is the oldest of the nine cultural strata at Shag Mouth, while the youngest stratum is Layer 2. Layers 9 through 11 were combined because the size of the sample from each layer was relatively small. Over time, there was a significant increase in mean utility of moa elements transported to the site (r_s = 0.93, p = 0.003). Thus, the range of elements transported narrowed from the initial use of both higher and lower utility elements toward a focus on higher return elements later on. The narrowing of the range of elements transported suggests that distance traveled to procure moas increased over time.

An increase in processing time can be seen in changes in the relative abundance of high and low utility skeletal elements. The percentage of high utility elements (femora, tibiotarsi, fibulae, and cervical vertebrae) steadily increases as expected (Figure 15.3). In contrast, the frequency of low utility elements (phalanges, tarsometatarsi, caudal vertebrae) does not decrease constantly, but rises slightly until Layer 6 before declining (Figure 15.4). This pattern suggests that low utility elements may have been initially transported as riders with higher utility elements up until Layer 6. Later, processing time appears to have increased so that these low utility elements were removed and thus became less common in the assemblage.

Indeed, an increase in field processing time can also be seen in changes in the relative abundance of the neck elements of moas. Tracheal rings and cervical vertebrae are both neck elements. Cervical vertebrae support sizable amounts of meat as evidenced in their high utility value, however, tracheal rings, the ossified segments of the windpipe, have little to no utility. If field processing increased relative to element utility, then the proportional abundance of tracheal rings should decrease over time. Tracheal ring relative abundance at the site decreases

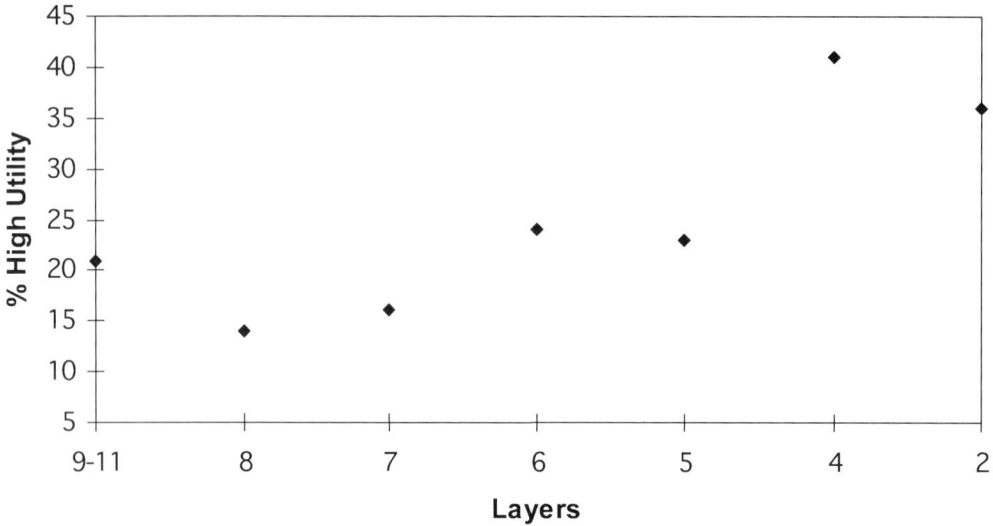

FIGURE 15.3. Relative frequency of high utility elements (femora, tibiotarsi, fibulae, cervical vertebrae) across layers.

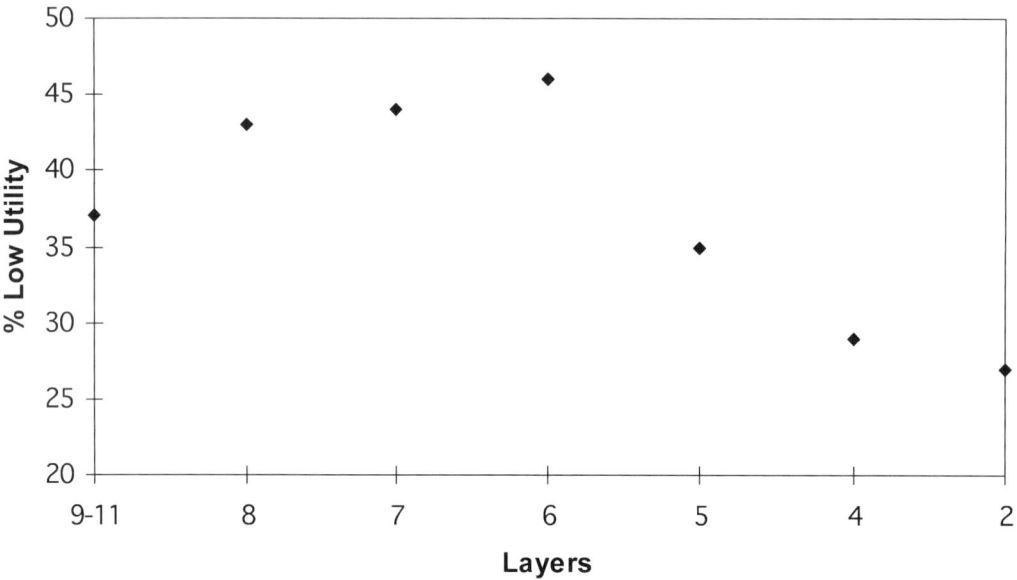

FIGURE 15.4. Relative frequency of low utility elements (phalanges, tarsometatarsi, caudal vertebrae) across layers.

significantly over time while the percentage of cervical vertebrae remains relatively constant (Figure 15.5), suggesting that pre-transport processing of moas increased over time.

The evidence suggests that foragers became more selective about what they were transporting in order to maximize net foraging returns. Increased selectivity is linked to resource depression and increased travel costs as more distant hunting grounds were used. Initially, local populations of moas were exploited and a broad range of high and low utility elements were brought back to the site. As local moa populations dwindled and people traveled farther to obtain them, the cost of transporting their remains back to the site increased. Carcasses were processed not only to create transportable packages, but also to maximize the delivery rate so that mainly high utility elements were transported back to the site.

Taphonomic factors, such as carnivore attrition or differential preservation, have been shown to affect skeletal part frequency [34, 35]. Thus, alternative explanations for the pattern of skeletal element representation must be examined. Carnivore attrition can differentially affect the survivorship of elements as well as shafts and ends of elements [36]. For moas, carnivore damage to epiphyseal ends would also significantly affect element identifiability. However, the Shag Mouth faunal assemblage shows surprisingly little carnivore damage, even though

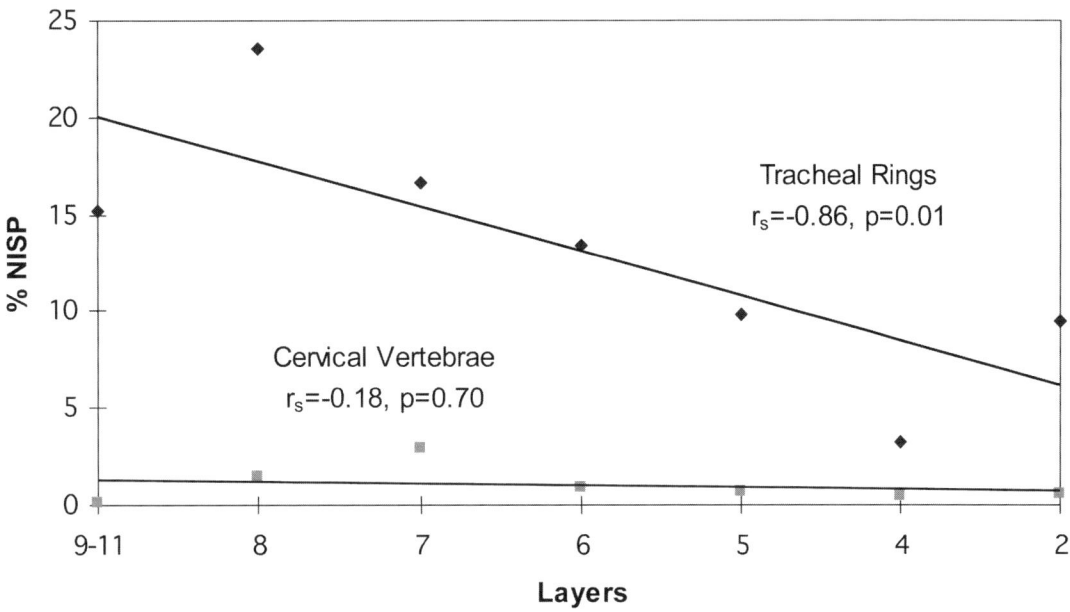

FIGURE 15.5. Relative frequency of moa cervical vertebrae and tracheal rings across layers.

domesticated dogs were known to have been present in and around the site [7]. Less than 3% of the assemblage displays evidence of carnivore gnawing. The relatively low percentage of carnivore damage at Shag Mouth and other South Island sites is not uncommon and was first noted in the 1870s by Julius von Haast [23]. Haast, the original excavator of the Shag Mouth site, went so far as to erroneously speculate that the paucity of dog gnawing indicated that dogs must not have had access to the bones, and thus were not domesticated. Whether domesticated or not, the fact remains that dog gnawing appears to have played an insignificant role in skeletal part frequency.

Bone density has also been identified as a factor that can affect skeletal element survivorship. Elements of low density are less likely to preserve than high-density elements [31, 34]. If bone density and element utility are correlated, then the pattern seen in the mean utility index may alternatively be due to differential destruction of low-density bones in the upper layers rather than to changes in transport decisions alone. To determine if bone density affects mean utility data, the skeletal element survivorship was compared to the bone density values for rheas, a ratite relative of moas [19]. Of the nine cultural layers examined, three (Layers 4, 5 and 7) had samples that showed a significant correlation between bone density and skeletal element frequencies (Table 15.2). Thus, for these layers, the skeletal element representation may be influenced by preservation bias.

To determine if bone density is affecting the pattern of mean utility, the samples are reanalyzed using high-density elements that are least likely to be underrepresented due to differential preservation [8]. The densest elements according to Cruz and Elkin's [19] rhea density data are the femur, fibula, pelvis, phalanx, tarsometatarsus and tibiotarsus. Figure 15.6 shows that

TABLE 15.2. Results of the Spearman's rank correlation analysis between bone density and % survivorship for moa elements.

LAYER	CORRELATION COEFFICIENT
2	$r_s = 0.33$, $p = 0.29$
4	$r_s = 0.70$, $p = 0.001$ [a]
5	$r_s = 0.63$, $p = 0.007$ [a]
6	$r_s = 0.41$, $p = 0.08$
7	$r_s = 0.56$, $p = 0.02$ [a]
8	$r_s = 0.30$, $p = 0.44$
9	$r_s = 0.24$, $p = 0.41$
10	$r_s = 0.40$, $p = 0.60$
11	$r_s = -0.33$, $p = 0.52$

[a] Significant correlation between density and % survivorship.

using the high-density elements, mean utility still increases significantly over time ($r_s = 0.94$; $p = 0.005$). Thus, while density is correlated with skeletal element abundance for some layers, the overall trend of increasing mean utility does not appear to be affected by differential preservation.

CHANGING BREAKAGE PATTERNS OF MOA SKELETAL ELEMENTS AT SHAG MOUTH

A conservative approach to understanding element fragmentation resulting from butchering limits the analysis of skeletal elements to those specimens showing clear evidence of human fracturing, such as green fractures, cut marks, and impact marks [e.g., 28, 29, 30]. However, this approach can exclude parts of the assemblage that might have resulted from butchering, but have no preserved marks from butchering processes on the specimens. An alternative approach is to use all specimens in the

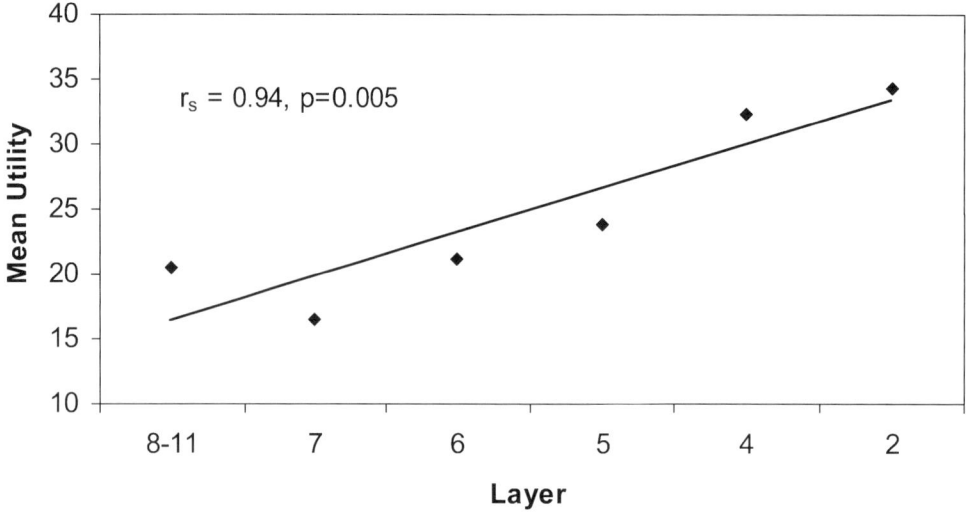

FIGURE 15.6. Mean moa utility for high density elements across layers.

assemblage and evaluate the taphonomic history of the assemblage for alternative explanations for the patterns [34, 56].

The Shag Mouth assemblage is well-preserved and its taphonomic history appears to be dominated by cultural rather than natural taphonomic processes. As mentioned above, there is very little carnivore damage on the Shag Mouth fauna. Differential preservation as indicated by bone density also does not appear to have a significant effect on the assemblage. In addition, less than 15% of the assemblage was categorized as significantly weathered on the Behrensmeyer weathering scale [7]. Instead, the presence of human impact fractures on moa bone in the Shag Mouth assemblages suggests that moa bone fragmentation is generally related to human activities. Thus for the purposes of this analysis, I am assuming that bone fragmentation was caused mainly by human butchering practices. Other alternative explanations are examined below.

Marrow Extraction

With declining moa populations, the MVT predicts that the elements transported back to the site should be used more intensively if the decline in foraging efficiency was significant (Table 15.1). One manifestation of intensification would be an increase in marrow extraction, which would result in a lower proportion of whole skeletal elements. Long bones typically contain the largest amounts of easily accessible marrow. Since moas did not have upper limbs, only moa leg elements (femur, tibiotarsus, tarsometatarsus, and phalanges) are analyzed. The percentage of whole elements (% whole) is used to examine changes in marrow extraction. Of the four skeletal elements, the samples of complete femora, tibiotarsi, and tarsometatarsi were too small to examine trends across time. Only four tarsometatarsi out of a total of 58 specimens were complete; of the 37 femora specimens, only one was whole; and none of the 290 tibiotarsi specimens were complete. Thus, the small percentage of unbroken skeletal elements across all stratigraphic layers suggests that marrow extraction from these leg elements may have been practiced regularly throughout the occupation of the site.

Phalanges are the one leg element that has a large enough sample of whole elements to examine across time. Of the limb bone elements, phalanges have small marrow cavities, thus they are expected to be low-ranked in terms of marrow utility [10], and should be exploited after other higher-ranked elements such as tibiotarsi. If foraging efficiency declined significantly over time, then low-ranked phalanges may have been used more intensively.

Figure 15.7 shows that the percentage of whole moa phalanges declines significantly over time (X^2_{trend} = 3.83, p = 0.05). The decrease in whole moa phalanges suggests that marrow extraction from low-ranked phalanges increased even though fewer phalanges were being transported back to the site over time. Thus, while marrow extraction from higher-ranked leg elements appears to have been a common practice across time, it is possible that foraging efficiency declined sufficiently to warrant more intensive marrow extraction of low-ranked elements such as phalanges. Differential preservation due to bone density does not appear to have played a role in this pattern since phalanges are relatively high-density elements.

Grease Extraction

Another form of intensification is increased grease extraction as reflected by an increase in bone fragmentation. To determine if moa skeletal elements were being used more intensively through time, the ratio of NISP to MNE is calculated using data on the sternum, pelvis, femur, tibiotarsus, and tarsometatarsus. Phalanges, vertebrae, and ribs were not considered because the element portion for these elements was not recorded, thus MNE could not be determined.

An increase in fragmentation should be represented by a corresponding increase in the NISP:MNE ratio. As Figure 15.8 shows, the change in fragmentation rate across time is not sig-

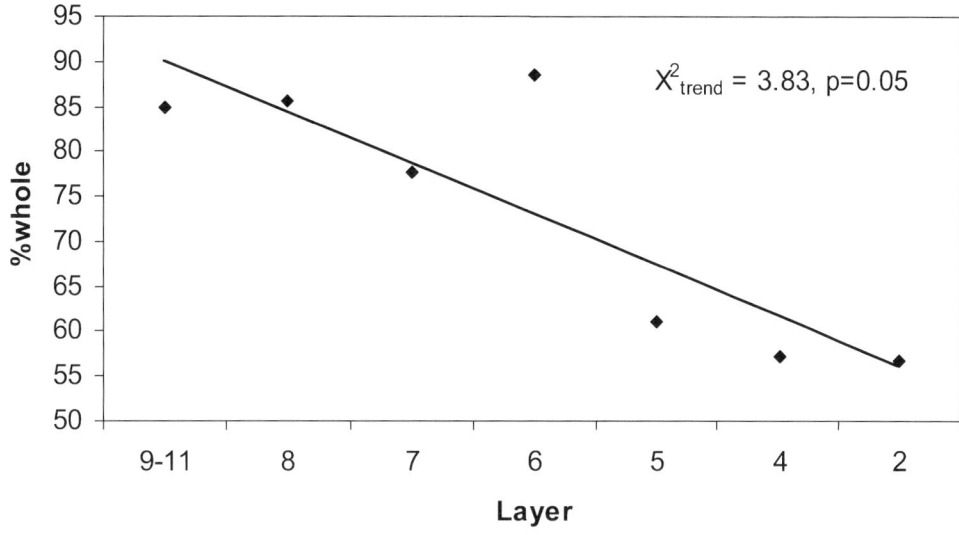

FIGURE 15.7. Percentage of unbroken moa phalanges across layers.

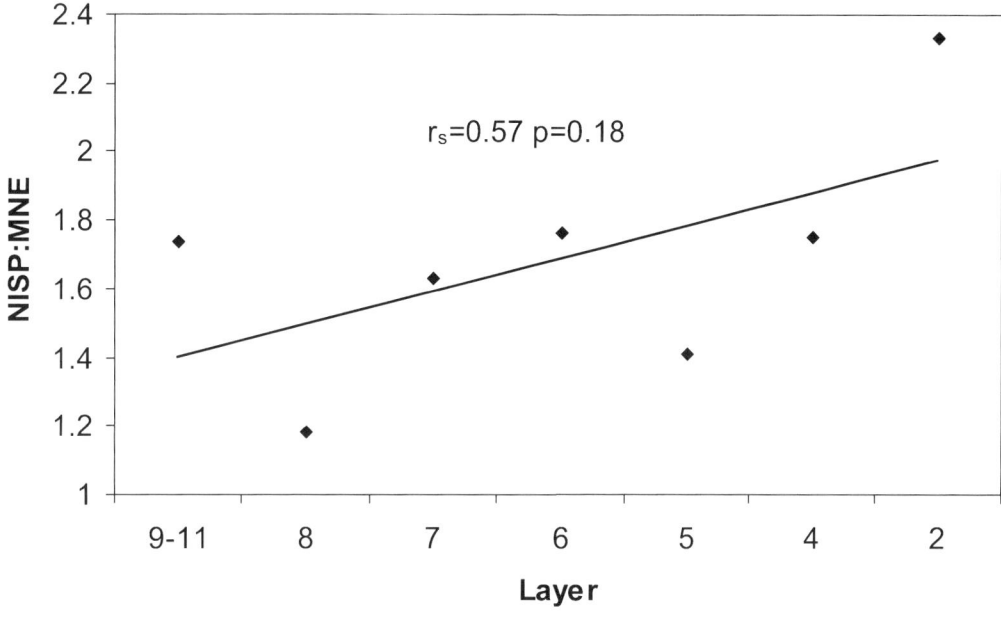

FIGURE 15.8. Moa bone fragmentation across layers.

nificant ($r_s = 0.57$, $p = 0.18$). It appears that foragers did not fragment moa skeletal elements for grease extraction more intensively over time. Thus, foraging efficiency may not have declined enough to warrant the use of this relatively low-return resource across all elements.

The fragmentation pattern, however, varies across skeletal element. In particular, the ratio of NISP to MNE for tibiotarsi increases significantly (Figure 15.9), and it appears that the slight increase in fragmentation rate seen in Figure 15.8 is driven mainly by the tibiotarsi data. Haast [23] in his original excavation noted that, unlike the femora and tarsometatarsi, most of the moa tibiotarsi that he excavated were broken. He surmised that the reason for the varying rates of fragmentation across leg elements was because of femora and tarsometatarsi "not having been thought of sufficient value to pay for the trouble of extraction" [23:94]. More recently, in his original analysis of fracture frequency data on the Shag Mouth moa assemblage, Kooyman [30] also suggested that only tibiotarsi were regularly broken. The tibiotarsus is the largest leg element, and thus it is the element that likely contained the most grease. The differential fragmentation of tibiotarsi may indicate that the decline in foraging efficiency was not large enough to direct significant effort into extracting resources like grease across all elements, but only from the one element with the greatest grease utility. The

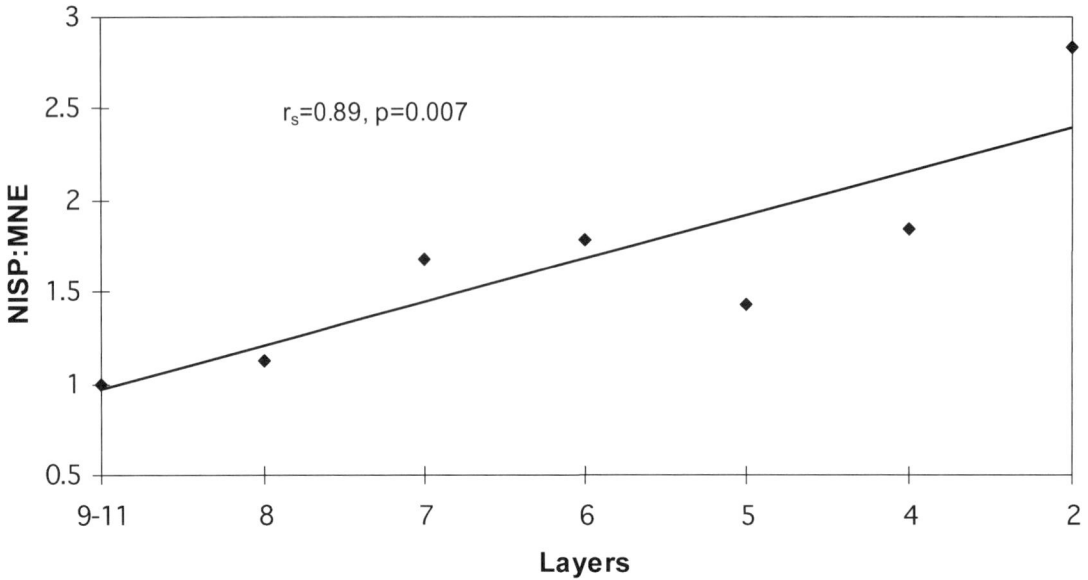

FIGURE 15.9. Fragmentation of moa tibiotarsi across layers.

apparent pattern of differential fragmentation may also be due to other factors, including differential identifiability, differential preservation, or the tool utility of particular elements.

Differential identifiability across elements may cause some skeletal elements to be underrepresented, which would affect the NISP to MNE ratio. In general, as bone fragmentation increases, it becomes more difficult to identify fragments to a particular skeletal element [37]. For moas, the bone structure of tibiotarsi is distinctive, and fragments are identifiable even when pieces are small. Other elements, however, are more difficult to differentiate unless fragments are relatively large. Thus, fragmentation of elements for grease extraction may have increased across all elements, but this increased fragmentation may have also caused a significant decline in the identifiability of fragments for all elements except tibiotarsi.

To examine the effects of identifiability on skeletal element representation, the ratio of the number of unidentifiable and identifiable specimens (NSP) to the number of specimens identified to a particular element (NISP) is calculated [22, 56]. For the Shag Mouth sample, only the leg elements are used. NISP is the number of specimens identified to the three leg elements: femur, tibiotarsus, and tarsometatarsus. NSP includes specimens identified to these elements as well as fragments that could only be identified to "leg elements". If fragmentation increased to the point where the identifiability of leg elements decreased, then the ratio of NSP:NISP should increase over time. Contrary to this expectation, the NSP:NISP ratio did not change across time significantly, indicating that rate of identifiability does not appear to have changed (Figure 15.10). Thus, differential identifiability does not appear to be the cause of differential fragmentation rates across elements.

As discussed above, taphonomic factors such as bone density can also affect skeletal element representation. If preservation conditions changed over time to favor more-robust elements, then less-dense leg elements may be underrepresented in later assemblages. However, according to the bone density data for rheas [19], tibiotarsi element portions tend to have similar density values as other leg elements. Thus, differential preservation does not appear to be a significant factor in the variability in fragmentation rates across elements.

Difference in bone breakage patterns across elements may instead be due to the value of the element as a raw material for tools rather than as food [51]. If this is the case, then the nature of the artifacts created from moa elements should dictate the amount of bone breakage. While other moa skeletal elements may have been used as raw material for tools [28], tibiotarsi are the most commonly identified element used for fishhooks and other artifacts. The medial-anterior surface of the bone is large, flat and very thick, from which large, durable artifacts could be made. The increase in tibiotarsi fragmentation relative to other leg elements could indicate that tibiotarsi were being used more intensively for tools. More intensive fragmentation may be related to changes in the size of fishhooks produced from tibiotarsi.

One-piece fishhooks required larger pieces of raw material than composite fishhooks. The shanks of the composite fishhooks were made of wood or shell, with only the point being made of bone. Thus, the manufacture of smaller composite fishhooks could have resulted in more artifacts being made from each piece of bone, increasing the fragmentation rate. If the increase in tibiotarsus fragmentation is a product of a shift in fishhook manufacture, then the frequency of composite fishhooks relative to one-piece fishhooks is expected to increase over time.

At the Shag Mouth site, the type of fishhooks made from bone changes over time (Table 15.3). Early on, the one-piece fishhooks were common, as they were across southern New Zealand

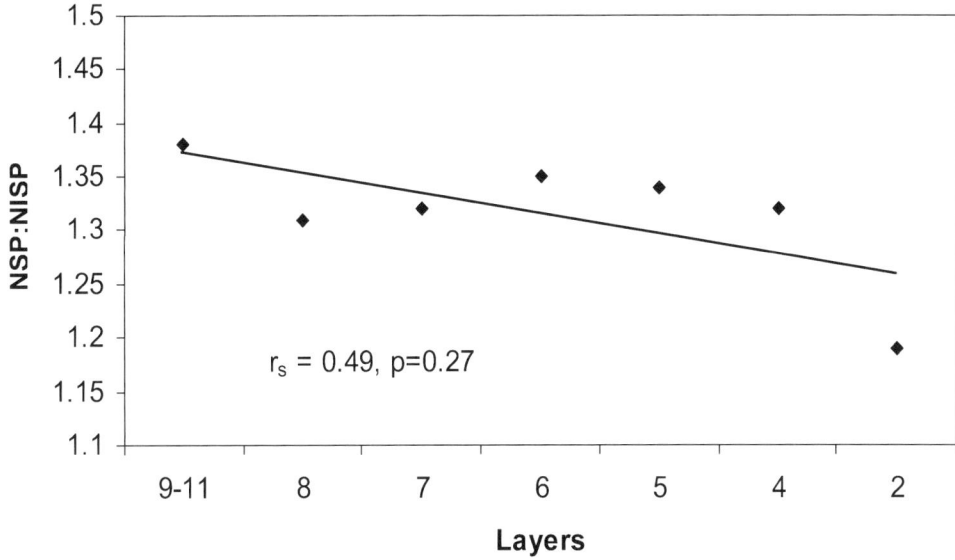

FIGURE 15.10. Identifiability of moa leg elements across layers.

TABLE 15.3. The relative abundance of one-piece and composite fishhooks and lures from the Shag Mouth site.

Layer	N	Barracouta Lure (%)	Minnow Lure (%)	Two-Piece Fishhook (%)	One-Piece Fishhook (%)
2	16	18.8	6.3	18.8	56.3
4	36	11.1	5.6	13.9	69.4
5	8			12.5	87.5
6	1				100.0
7	1				100.0
8	1				100.0
9	1				100.0

[5, 25]. Composite fishhooks and lures occur later. The advent of these later hook types is likely to be due either to changing subsistence practices and/or reduced availability of raw material resources. In either case, the increased fragmentation of tibiotarsi could be due to the manufacture of high numbers of smaller bone hooks. Additional research on changes in the frequency of fishhook manufacturing traits such as cut marks, drill holes, tab and core size is required to determine how tool utility affects moa bone breakage patterns.

There remain two possible explanations for the fragmentation pattern of moa elements at Shag River Mouth. The first is that foraging efficiency did not decline sufficiently to warrant more intensive use of all elements through grease extraction. Instead, only tibiotarsi appear to have been targeted for grease extraction because of its greater utility. Another possibility is that foraging efficiency did not decline significantly enough for foragers to intensify grease extraction in any of the long bones. Rather, any differences in the skeletal element representation are due to the elements' use as raw materials for tool manufacture. Thus, tibiotarsi may have been the focus of both grease extraction and tool manufacture.

Summary and Conclusions

The study of prehistoric carcass exploitation has been relatively limited in New Zealand. Only a few researchers have systematically examined how body parts of large animals such as moas and seals were exploited differentially [e.g., 27, 28, 30, 52, 53]. As described above, many of the patterns on moa carcass exploitation documented in this research have been noted in these earlier studies. However, unlike previous research, which studied butchering patterns at the site level, this study examines changes in butchering and bone fragmentation decisions over time. A temporal perspective has allowed for an analysis of change in carcass exploitation as moa populations were declining.

At Shag Mouth, changes in carcass exploitation are linked to declining overall foraging efficiency. Human foraging behavior at Shag Mouth led to a decline in the availability of high-ranked moas and an overall decrease in foraging efficiency [43, 44]. At the same time, the spatial distribution of moa changed. They became less available locally so that foragers had to travel farther to hunt them, and thus travel costs increased. Because of the declining encounter rates and increasing travel costs, foragers altered the manner in which they butchered and transported moas to maximize their net returns. In addition, bone breakage patterns were also affected by the decline in foraging efficiency.

More importantly, this research uses foraging theory to integrate changes in carcass exploitation with dietary changes under one analytical framework. Cannon's [16] central place prey choice model works well to set up testable expectations about carcass exploitation as transport distance increases and/or prey availability declines. Thus, butchery/transport practices and skeletal element exploitation are explained within the context of general subsistence trends, providing a better understanding of how and why carcass exploitation changes over time. The changes in moa carcass exploitation observed in this paper are directly linked to changes in prey availability, prey distribution, and overall foraging efficiency. Thus, this research builds upon previous studies that examined prey and patch choice decisions, and provides a more complete and integrated picture of resource exploitation patterns in southern New Zealand.

Acknowledgments

Thanks to Ian Smith, Helen Leach, and Atholl Anderson for access to the Shag Mouth fauna collections and database. Lee Lyman offered helpful direction on an early draft. Steve Wolverton provided thoughtful comments on the various versions of this paper. Thanks also to the four anonymous reviewers for making this a stronger paper.

References

1. A. Anderson, The extinction of moas in southern New Zealand, in: P. S. Martin, R. G. Klein (Eds.), *Quaternary Extinctions*, University of Arizona, Tucson, 1984, pp. 728–740.
2. A. Anderson, Mechanics of overkill in the extinction of New Zealand moas, *Journal of Archaeological Science* 16 (1989) 137–151.
3. A. Anderson, *Prodigious Birds*, Cambridge University, Cambridge, 1989.
4. A. Anderson, *Traditional Lifeways of the Southern Maori*, University of Otago, Dunedin, 1994.
5. A. Anderson, W. Gumbley, Fishing gear, in: A. Anderson, B. Allingham, I. Smith (Eds.), *Shag River Mouth: The Archaeology of an Early Southern Maori Village*, Australian National University, Canberra, 1996, pp. 148–160.
6. A. Anderson, I. Smith, T. Higham, Radiocarbon chronology, in: A. Anderson, B. Allingham, I. Smith (Eds.), *Shag River Mouth: The Archaeology of an Early Southern Maori Village*, Australian National University, Canberra, 1996, pp. 60–69.
7. A. Anderson, T. Worthy, R. McGovern-Wilson, Moa remains and taphonomy, in: A. Anderson, B. Allingham, I. Smith (Eds.), *Shag River Mouth: The Archaeology of an Early Southern Maori Village*, Australian National University, Canberra, 1996, pp. 200–213.
8. J. E. Beaver, Identifying necessity and sufficiency relationships in skeletal-part representation using fuzzy-set theory, *American Antiquity* 69 (2004) 131–142.
9. R. L. Bettinger, *Hunter-Gatherers: Archaeological and Evolutionary Theory*, Plenum, New York, 1991.
10. L. R. Binford, *Nunamiut Ethnoarchaeology*, Academic, New York, 1978.
11. L. R. Binford, *Bones: Ancient Men and Modern Myths*, Academic, New York, 1981.
12. D. W. Bird, R. L. B. Bird, Contemporary shellfish gathering strategies among the Meriam of the Torres Strait Islands, Australia: Testing predictions of a central place foraging model, *Journal of Archaeological Science* 24 (1997) 39–63.
13. J. M. Broughton, *Resource Depression and Intensification during the Late Holocene, San Francisco Bay: Evidence from the Emeryville Shellmound*, University of California, Berkeley, 1999.
14. M. Bunce, T. H. Worthy, T. Ford, W. Hoppitt, E. Willerslev, A. Drummond, A. Cooper, Extreme reversed sexual size dimorphism in the extinct New Zealand moa *Dinornis*, *Nature* 425 (2003) 172–175.
15. M. D. Cannon, Large mammal relative abundance in Pithouse and Pueblo period archaeofaunas from southwestern New Mexico: resource depression among the Mimbres Mogollon? *Journal of Anthropological Archaeology* 19 (2000) 317–347.
16. M. D. Cannon, A model of central place forager prey choice and an application to faunal remains from the Mimbres Valley, New Mexico, *Journal of Anthropological Archaeology* 22 (2003) 1–25.
17. E. L. Charnov, Optimal foraging, the marginal value theorem, *Theoretical Population Biology* 9 (1976) 129–136.
18. R. R. Church, R. L. Lyman, Small fragments make small differences in efficiency when rendering grease from fractured artiodactyl bones by boiling, *Journal of Archaeological Science* 30 (2003) 1077–1084.
19. I. Cruz, D. Elkin, Structural bone density of the Lesser Rhea (*Pterocnemia pennata*) (Aves: Rheidae). Taphonomic and archaeological implications, *Journal of Archaeological Science* 30 (2003) 37–44.
20. D. K. Grayson, *Quantitative Zooarchaeology*, Academic, New York, 1984.
21. D. K. Grayson, Bone transport, bone destruction, and reverse utility curves, *Journal of Archaeological Science* 16 (1989) 643–652.
22. D. K. Grayson, Alpine faunas from the White Mountains, California: Adaptive change in the prehistoric Great Basin? *Journal of Archaeological Science* 18 (1991) 483–506.
23. J. v. Haast, Notes on the Moa-hunter encampment at Shag Point, Otago, *Transactions and Proceedings of the New Zealand Institute* 7 (1874) 91–98.
24. W. J. Hamilton, K. E. F. Watt, Refuging, *Annual Review of Ecology and Systematics* 1 (1970) 263–286.
25. J. Hjarno, *Maori Fishhooks in Southern New Zealand*, Otago Museum, Dunedin (1967).
26. L. Huynen, C. D. Millar, R. P. Scofield, D. M. Lambert, Nuclear DNA sequences detect species limit in ancient moa, *Nature* 425 (2003) 175–178.
27. B. Kooyman, Moa utilisation at Owen's Ferry, Otago, New Zealand, *New Zealand Journal of Archaeology* 6 (1984) 47–57.
28. B. Kooyman, Moa and Moa Hunting: An Archaeological Analysis of Big Game Hunting in New Zealand, Ph.D. dissertation, University of Otago, Dunedin, 1985.
29. B. Kooyman, Moa procurement: communal or individual hunting? in: L. B. Davis, B. O. K. Reeves (Eds.), *Hunters of the Recent Past*, Unwin Hyman, London, 1990, pp. 327–351.
30. B. Kooyman, Moa hunting and butchery, in: A. Anderson, B. Allingham, I. Smith (Eds.), *Shag River Mouth: The Archaeology of an Early Southern Maori Village*, Australian National University, Canberra, 1996, pp. 214–222.
31. R. L. Lyman, Bone density and differential survivorship of fossil classes, *Journal of Anthropological Archaeology* 3 (1984) 259–299.
32. R. L. Lyman, Anatomical considerations of utility curves in zooarchaeology, *Journal of Archaeological Science* 19 (1992) 7–22.
33. R. L. Lyman, Relative abundances of skeletal specimens and taphonomic analysis of vertebrate remains, *Palaios* 9 (1994) 288–298.
34. R. L. Lyman, *Vertebrate Taphonomy*, Cambridge University, Cambridge, 1994.
35. C. W. Marean, N. Cleghorn, Large mammal skeletal element transport: Applying foraging theory in a complex taphonomic system, *Journal of Taphonomy* 1 (2003) 15–42.
36. C. W. Marean, L. M. Spencer, Impact of carnivore ravaging on zoo-

archaeological measures of element abundance, *American Antiquity* 56 (1991) 645–658.

37. F. Marshall, T. Pilgram, Meat versus within-bone nutrients: Another look at the meaning of body part representation in archaeological sites, *Journal of Archaeological Science* 18 (1991) 149–163.

38. D. Metcalfe, K. R. Barlow, A model for exploring the optimal tradeoff between field processing and transport, *American Anthropologist* 94 (1992) 340–356.

39. D. Metcalfe, K. T. Jones, A reconsideration of animal body-part utility indices, *American Antiquity* 53 (1988) 486–504.

40. N. D. Munro, Zooarchaeological measures of hunting pressure and occupation intensity in the Natufian, *Current Anthropology* 45 Supplement (2004) S5–S33.

41. N. D. Munro, G. Bar-Oz, Gazelle bone fat processing in the Levantine Epipaleolithic, *Journal of Archaeological Science* 32 (2005) 223–239.

42. L. Nagaoka, Using diversity indices to measure changes in prey choice at the Shag River Mouth site, Southern New Zealand, *International Journal of Osteoarchaeology* 11 (2001) 101–111.

43. L. Nagaoka, The effects of resource depression on foraging efficiency, diet breadth, and patch use in southern New Zealand, *Journal of Anthropological Archaeology* 21 (2002) 419–442.

44. L. Nagaoka, Explaining subsistence change in southern New Zealand using foraging theory models, *World Archaeology* 34 (2002) 84–102.

45. J. F. O'Connell, Ethnoarchaeology needs a general theory of behavior, *Journal of Archaeological Research* 3 (1995) 205–255.

46. J. F. O'Connell, K. Hawkes, N. Blurton Jones, Hadza hunting, butchering, and bone transport and their archaeological implications, *Journal of Anthropological Research* 44 (1988) 113–161.

47. J. F. O'Connell, K. Hawkes, N. Blurton Jones, Reanalysis of large mammal body part transport among the Hadza, *Journal of Archaeological Science* 17 (1990) 301–316.

48. G. H. Orians, N. E. Pearson, On the theory of central place foraging, in: D. J. Horn, G. R. Stairs, R. D. Mitchell (Eds.), *Analysis of Ecological Systems*, Ohio State University Press, Columbus, OH, 1979, pp. 155–177.

49. A. K. Outram, New approach to identifying bone marrow and grease exploitation: why the "indeterminate" fragments should not be ignored, *Journal of Archaeological Science* 28 (2001) 401–410.

50. D. Perkins, P. Daly, A hunter's village in Neolithic Turkey, *Scientific American* 219 (1968) 96–106.

51. N. D. Sharp, Fremont Farmers, Hunters: Faunal Resource Exploitation at Nawthis Village, Central Utah, Ph.D. Dissertation, Department of Anthropology, University of Washington, 1989.

52. I. W. G. Smith, Sea Mammal Hunting and Prehistoric Subsistence in New Zealand, Ph.D. dissertation, University of Otago, 1985.

53. I. W. G. Smith, The mammal remains, in: A. Anderson, B. Allingham, I. Smith (Eds.), *Shag River Mouth: The Archaeology of an Early Southern Maori Village*, Australian National University, Canberra, 1996, pp. 185–199.

54. M. C. Stiner, *Human Predators and Prey Mortality*, Westview, Boulder, 1991.

55. T. E. White, Observations on the butchering technique of some aboriginal peoples: no. 2, *American Antiquity* 19 (1953) 160–164.

56. S. Wolverton, NISP:MNE and % whole in analysis of prehistoric carcass exploitation, *North American Archaeologist* 23 (2002) 85–100.

57. T. H. Worthy, R. N. Holdaway, *The Lost World of the Moa*, Indiana University Press, Bloomington, IN, 2003.

PART III, SECTION 2

The Intersection of Technology, Subsistence, and Mobility

The papers in this section of *Evolutionary Ecology and Archaeology* relate aspects of technology to the subsistence and mobility patterns of prehistoric people, and they complement in many ways the papers in the preceding section, which explore issues of subsistence and mobility primarily through studies of food remains. In some cases the models used in these papers are the same ones from foraging theory that are used in papers reprinted elsewhere in this volume, though recast for application to tools and tool materials rather than food resources. In other cases, original models are developed in order to explore issues unique to the realm of technology. These papers on technology do an especially good job of illustrating the integrative nature of the EE approach, or its ability to provide theoretical links between various aspects of behavior. They also present innovative ways of thinking about the costs and benefits of technological alternatives and of measuring those costs and benefits archaeologically.

The paper by S. Kuhn provides an excellent example of the application of the EE logic that was described in the introduction to this volume. Moreover, this paper has provided an important foundation for EE-based studies of stone tools that has been built upon by many other researchers, as evidenced by some of the other works reprinted in this section. Kuhn develops a model that considers the optimal design of the lithic toolkits carried by highly mobile hunter-gatherers. His model explores the tradeoff between transport cost, which is a function of tool weight, and tool utility, which he defines in terms of the amount of usable sharp edge potentially available on the tools that a forager carries. Importantly, this model results in testable predictions based on archaeologically measurable variables, and Kuhn notes the contrasts between this approach and more traditional inductive or post hoc approaches to the study of the technological adaptations of mobile foragers.

The most notable, and counterintuitive, prediction of Kuhn's model is that the strategy that maximizes tool utility per unit mass is to carry many smaller flake tools, rather than a few larger ones. A corollary to this is that individual tools carried by highly mobile foragers may not exhibit multifunctional design characteristics: if the most economic strategy is to carry many small tools, multifunctionality can be achieved by making each of those small tools suitable for a different, relatively narrow range of tasks. This prediction contrasts with the more common assumption that the tools of mobile foragers should be designed to serve a variety of functions, and Kuhn notes that this insight might help to explain, for example, why North American Folsom lithic assemblages often consist of a diverse set of individually specialized flake tools.

This paper also shows how EE models can be useful even when their predictions are not met in specific empirical cases. Kuhn's model, which assumes a goal of maximizing the utility per unit weight of transported toolkits, predicts that cores should never be carried as part of those toolkits, but Kuhn notes that there are many ethnographic and archaeological examples of mobile hunter-gatherers doing just that. Kuhn suggests that this may be because the functional properties of cores (such as the ability to serve as hammers or choppers) provide a benefit that is not captured in the goal assumption of his model. He also notes how certain tool functions, such as use as a projectile point, may be subject to tight size constraints that override the relationship between tool utility and tool size on which his model is based. By demonstrating the limitations of the model's assumptions in this way, the process of evaluating the model leads to new knowledge about the factors that underlie stone tool design beyond what was contained in the model's starting assumptions.

The paper by C. Beck and colleagues also addresses aspects of the relationship between transport costs and stone tool manufacture, but it focuses on issues pertaining to the transport of toolstone away from the quarries where it is procured, rather than issues of tool design per se. Beck et al. consider the role that transport costs play in decisions that are made at a quarry through a detailed empirical application of Metcalfe and Barlow's (1992) model of the tradeoff between transport costs and field processing costs. In so doing, they show that the same constraints that may often influence decisions about the transport of food resources (see the applications of the Metcalfe and Barlow model that are reprinted in the previous section of this volume) can also influence decisions about the transport of raw material for tools.

The quarries that Beck et al. study are sources of the fine-grained volcanic rocks that were commonly used during the Great Basin Paleoarchaic (terminal Pleistocene/early Holocene) period to make stemmed projectile points. It appears that the manufacturing sequence for these points involved initial preparation at the quarry of bifaces, which were typically then transported for completion elsewhere. Beck et al. consider two pairs of quarry sites and residential sites, wherein the distance between quarry and associated residential site is considerably greater for one pair than the other. Their analysis shows that when transport distance was greater, bifaces were reduced to a further degree at the quarry, thereby increasing the number of potential completed points per unit weight that could be carried away, though at a cost of higher field processing time. This study presents a compelling demonstration of the role that transport costs can play in the organization of toolstone procurement. It also presents some key methodological insights that enable foraging models to be applied to lithic resources, and the authors discuss additional steps that might be taken to further advance such applications.

While transport costs may help to explain some aspects of archaeological lithic assemblages, the technologies used in many prehistoric settings were likely adapted to other, or at least additional, constraints. In a consideration of microlithic technologies in northern Asia, the paper by R. Elston and P. Brantingham that is reprinted here addresses one such constraint—subsistence risk, or the probability of experiencing a shortfall in food resources. Elsewhere, Bamforth and Bleed (1997) and Fitzhugh (2001) have also used the EE approach to elucidate relationships between technology and risk.

North Asian microblades, which date primarily from just after the Last Glacial Maximum into the early Holocene, appear to have been used mainly as components of composite projectile points that were manufactured by insetting the microblades into point shafts made of bone, antler, or ivory. Elston and Brantingham point out that the two methods of producing microblades that were used in north Asia—removing them either from "wedge-shaped" or from "boat-shaped" cores—may have varied in their costs and benefits: the use of wedge-shaped cores requires larger raw material package sizes and additional steps in the manufacturing process, but, Elston and Brantingham suggest, it may also have enabled the production of microblades that were more uniform in size and shape than those that could be produced from boat-shaped cores. Drawing on the Z-score model of foraging theory (discussed in detail by Gremillion 1996, reprinted here), Elston and Brantingham argue that prehistoric north Asian hunters made decisions about which type of core to use based on whether the context in which they found themselves favored risk-averse or risk-prone behavior. Although the data required for a detailed test of these ideas about risk and microlithic technologies are not yet available, Elston and Brantingham do specify ways in which future analyses of archaeofaunal data might provide such a test. They also discuss how a focus on the adaptive strategies that underlie lithic technologies—a focus that can be operationalized through the EE approach—might help to move lithic analysis beyond its frequently purely typological concerns.

The paper by J. Bright et al. continues the theme of linking technology to subsistence economics. These authors incorporate issues of technology into basic foraging models by considering the effects that technologies have on the handling times of resources. Bright and colleagues have developed a "tech investment" model that predicts the optimal amount of time to invest in a particular technology given such factors as the encounter rates and initial handling times for resources, and given the degree to which investment in a technology might reduce resource handling times. This model is discussed in detail in a companion theoretical paper (Ugan et al. 2003; also see Bettinger et al. 2006 for important further discussion), and in the paper that is reprinted here it is applied to late prehistoric archaeological data from the Little Boulder Basin area of the north-central Great Basin.

Among the predictions of the tech investment model is that the amount of time invested in a particular technology should vary positively with the amount of time spent handling the resources that the technology is used to procure or process: the greater the amount of time spent handling a particular resource, the higher the payoff to time invested in technologies that might reduce the handling time of that resource. The results of the authors' initial application of the tech investment model to the Little Boulder Basin archaeological record accord with this prediction, as it appears that foragers there allocated varying amounts of time to ground stone, ceramic, and flaked stone technologies in conjunction with changes in the importance of different types of resources in the subsistence economy. More specifically, it appears that as low-return resources, particularly seeds, came to comprise a larger proportion of the diet, greater amounts of time were invested in technologies that could reduce the handling times for those resources, (i.e., milling stones and ceramics), and less time was allocated to technologies used in the procurement of higher-return large mammals (i.e., bifacial flaked stone tools).

Like the other papers in this section, this paper pays appropriate attention to the need for methods of measuring costs and benefits with artifactual data, and Bright et al. present some robust ways of doing so. More important, this paper and the tech investment model on which it is based also open up a potentially very productive avenue for future research into the linkages between technological change and changes in the use of subsistence resources. By showing why "technological intensification" might follow from instances of resource intensification (such as those considered in papers in the preceding section of this volume), this work has important implications for our understanding of major developments in human prehistory.

In sum, the papers in this section have helped to advance our understanding of the following important issues in the study of technology:

1. Stone tool design features that are adaptive for a highly mobile way of life.
2. The role that transport costs can play in decisions about the steps in the tool manufacturing process that are carried out at quarry vs. residential sites.
3. The general importance of transport costs as a constraint for prehistoric populations.
4. Tool design features that are adaptive in situations that favor either risk-averse or risk-prone behaviors.
5. Ways in which technologies might change in response to changes in the diet (such as technological intensification in conjunction with resource intensification).
6. General ways in which the distributions of artifact types over space and time might be explained in terms of adaptive strategies.

References

Bamforth, D. B., P. Bleed
1997   Technology, Flaked Stone Technology, and Risk. In *Rediscovering Darwin: Evolutionary Theory in Archaeological Explanation*, edited by C. M. Barton and G. A. Clark, pp. 267–290. Archaeological Papers of the American Anthropological Association No. 7. American Anthropological Association, Arlington, Va.

Bettinger, R. L., B. Winterhalder, R. McElreath
2006   A Simple Model of Technological Intensification. *Journal of Archaeological Science* 33:538–545.

Fitzhugh, B.
2001   Risk and Invention in Human Technological Evolution. *Journal of Anthropological Archaeology* 20:125–167.

Metcalfe, D., K. R. Barlow
1992   A Model for Exploring the Optimal Tradeoff between Field Processing and Transport. *American Anthropologist* 94:340–356.

Ugan, A., J. Bright, A. Rogers
2003   When is Technology Worth the Trouble? *Journal of Archaeological Science* 30:1315–1329.

16

A Formal Approach to the Design and Assembly of Mobile Toolkits

Steven L. Kuhn

Archaeologists interested in prehistoric hunter-gatherers have become concerned with identifying the effects of mobility on technology. One consequence of high levels of residential mobility is extensive reliance on the limited battery of artifacts that people can carry along with them as they move about the landscape. Because dependence on these "mobile toolkits" is at least partly contingent on patterns of land use, learning to recognize their distinctive characteristics can help archaeologists gain insights into prehistoric foraging adaptations using lithic technological data.

It is widely accepted that mobile toolkits should be composed so as to minimize transport cost, at the same time ensuring that the component artifacts are as durable and multifunctional as possible. This paper explores the implications of these basic expectations in terms of attributes that are potentially measurable in the archaeological record. The discussion centers on a simple model that relates the transport cost, durability, and potential versatility of artifacts to a single variable—artifact size. Relations between these three characteristics are treated as a kind of optimization problem. The analysis poses two simple questions. Presupposing that mobile toolkits are designed to maximize durability and functional versatility, while simultaneously minimizing weight (1) should people carry cores or tools/tool blanks, and (2) should they transport a few large artifacts or a larger number of small ones?

What Are Mobile Toolkits, and What Has Been Said about Them?

As used here, the term "mobile toolkit" refers to artifacts that mobile individuals keep with them most or all of the time, implements that are subject to virtually continuous transport. The term corresponds to what Binford (1979) refers to as "personal gear." Familiar examples of mobile toolkits exist in modern life. Each of us owns a series of artifacts—identification and credit cards, a pen, perhaps a notebook, agenda, penknife, cosmetic kit, or vial of medication—that remains on or near our person at all times. Owners of automobiles generally employ another such toolkit, consisting of a jack, lug wrench, and other tools, and sometimes basic medical or survival equipment. These artifacts are quite distinct from, and treated differently than, the vastly larger array of material that fills our homes and offices. Transported toolkits are available wherever the user finds herself/himself, whereas it is necessary to be at a particular location in order to make use of most other artifacts and facilities.

Mobile toolkits are an important component in the technologies of all foraging populations. Raw materials are seldom ubiquitous, and even when they are, it is often neither practical nor convenient to stop and make tools as needs arise. As a consequence, it is advantageous to keep a limited inventory of implements on hand at all times, as a hedge against unpredictable but unavoidable exigencies. The sizes and contents of mobile toolkits vary extensively, according to environment, mode of transport, and other factors. For example, Aboriginal hunters in the Western Desert of Australia generally carry very few versatile implements with them (Gould 1969:76), while Nunamiut Eskimos habitually pack a substantial quantity of gear when making extended trips by dogsled (Binford 1977). Nonetheless, all mobile groups or individuals carry at least a limited array of artifacts with them at all times.

The extent to which foragers must rely on what they can carry along with them also varies. All mobile hunter-gatherers have at least a minimal transported toolkit. At the same time, all rely to some extent on both expediently manufactured tools and "curated" artifacts that are stored or cached at fixed locations. The frequency and scale of residential mobility (Binford 1977, 1979; Kelly 1988; Kuhn 1989, 1992; Shott 1986), the scheduling of resource availability (Bleed 1986; Torrence 1983), and the natural distribution of usable raw materials (Bamforth 1986, 1991) all play a role in determining how much of the total technological inventory must be transported about, as opposed to being either opportunistically produced or strategically disposed among different locations.

*Originally published in *American Antiquity* 59(1994):426–442

From an archaeological perspective, it is important to recognize that the observation that an artifact was transported does not necessarily ensure that it was part of a mobile toolkit. Tool users may move artifacts and raw materials *to* heavily visited places, keeping them there for future use. Such items may well show evidence of having been moved, often over appreciable distances, but they were not perpetually carried from place to place.[1] As a result, recognizing the archaeological consequences of a dependence on mobile toolkits requires going beyond simple evidence of transport.

Working from both archaeological and ethnographic observations, a number of researchers have commented on factors likely to constrain the manufacture and assembly of continuously transported gear. It is widely held that the greatest limitation on mobile toolkits is portability (e.g., Gould 1969:76; Marshall 1976:367) or transport cost. Nelson (1991:76) argues concisely that "Transportable designs accommodate the constraints of mobility and anticipate future needs.... The specific function of a tool, in the task sense, may influence the form less than do the exigencies of transporting it." At the same time, the gear that tool-users carry with them must be sufficient to cope with a broad and changeable set of circumstances, and must last until there is an opportunity to replace it. In keeping with these basic premises, most researchers agree that strategies for putting together mobile toolkits should tend to optimize their potential usefulness relative to weight, the primary determinate of transport cost.

Although optimizing *utility* with respect to transport cost may be a general concern in designing and assembling transported gear, there is more than one possible means to this end. In order to keep weight down, the artifacts that people carry around with them should be few in number (Shott 1986; Tindale 1965:159; Torrence 1983:13), and the tools themselves should be small and light (Ebert 1979; Nelson 1991). At the same time, in order to maximize utility, such tools should be "flexible" or multifunctional (e.g., Bleed 1986:741; Nelson 1991:73–4; Shott 1986:20), a supposition supported by ethnological observations (e.g., Gould 1980:70; Smyth 1878:132). Because opportunities to replace the components of transported toolkits may be few and far between, constituent artifacts should be durable and inherently "maintainable" (Bleed 1986; Kelly 1988:721), and should show evidence of frequent reworking.

While all of these propositions about the design of transported artifacts are reasonable, it is often difficult to translate them into predictions about the kinds of artifacts archaeologists study. After all, how does one measure the flexibility or maintainability of a stone scraper or projectile point? Most frequently, researchers simply offer post hoc arguments to the effect that exotic artifacts, or artifacts associated with evidence of frequent movement, are in fact portable, highly flexible, etc. (see review in Nelson [1991]). Although such inductive strategies can aid in identifying the range of potential responses to the exigencies of producing transported toolkits, they do not furnish useful general predictions. For example, in western North America, the use of bifacial-core technology seems to have been strongly correlated with the frequency and scale of residential mobility (Kelly 1988; Kelly and Todd 1988). While this is good evidence for suggesting that some property of bifacial reduction was advantageous to Paleoindian and later populations in coping with the demands of a mobile lifestyle, it only begs the question of why highly mobile Australian foragers (see White and O'Connell 1982:116) and (presumably) mobile Middle and Upper Paleolithic hunter-gatherers in western Europe so infrequently exploited bifacial-core-reduction technologies.

Most attempts to actually measure the hypothetical properties of interest are also post hoc in nature. For example, it has been suggested that the "flexibility" or multifunctionality of tools can be gauged by counting the number of retouched edges, utilized employable units ("EUs"), or distinct patterns of use wear present on artifacts (e.g., Shott 1986). While undeniably interesting, such observations speak more to artifact use histories than to tool design. Artifacts moved around by mobile foragers would almost by definition have stayed in circulation longer than tools made and abandoned in the same place, and it would not be surprising to find that they were employed in a greater variety of contexts. The simple fact that a tool was used in several different ways does not necessarily imply that multifunctionality was an important consideration in its design.

A Model of Transport Cost and Utility

Rather than being concerned with explaining a specific set of archaeological observations, this study asks what transported stone tools should be like if the prevailing ideas about design constraints on them are valid. It focuses particularly on the essential opposition between portability on one hand, and durability and/or functional versatility on the other. By itself, an overwhelming concern with minimizing transport cost would probably result in toolkits composed of the smallest possible quantities of very lightweight artifacts. In contrast, the need for durable, flexible implements to meet a variety of unpredictable requirements would seem to favor the selection of larger implements, capable of supporting multiple edges, either simultaneously or serially. Interrelations between artifact size, transport cost, and potential utility can thus be approached as a problem in optimization, wherein beneficial increases in portability (i.e., smaller tools) have potentially detrimental implications for artifact use lives and functional versatility, and vice versa. The optimal solutions to the dilemma of assembling groups of artifacts light enough to be portable but large enough to cope with a variety of needs can provide general predictions about how toolmakers ought to behave, if the common suppositions about mobile toolkits are correct. Such predictions can aid in interpreting archaeological patterning. Even more importantly, however, they can help to evaluate the validity of our basic assumptions.

This study is based on a highly simplified model of toolkit

design and assembly in a forager technology. The hypothetical inventory of technological options includes only two types of stone artifacts—unifacially retouched tools (or tool blanks) and cores. It is postulated that the "goal" of designing mobile toolkits is to minimize the risk of being caught without serviceable tool edges in an environment where raw-material availability is uncertain. It is also assumed that there are some limits on how much technological material people can practically carry around. After all, every kilo of tools or raw material people carry with them means one kilo less of food, clothing, or shelter. The most beneficial (the optimal) artifact design is thus the one that produces the greatest potential utility relative to the cost of transporting it. Given this optimal goal, the analysis is designed to explore two questions: What kinds of artifacts should people carry with them (tools/blanks or cores), and how large should they be?

Like Jones and Madsen (1989:530), I assume that transport cost for all artifacts is directly proportional to mass. Since only a single type of raw material is considered, mass can in turn be treated as a simple function of volume. Although the actual energy expended in carrying a unit mass over a given distance varies according to the gradient and nature of the terrain traversed (Brannan 1991), such considerations are overly fine grained for a model of this type.

Unlike transport cost, the utility of artifacts can be measured in terms of more than one currency or dimension. In the following analyses, this important variable is quantified in two somewhat distinct ways. At the outset, potential utility is equated with the possibility of producing fresh, usable edges from an artifact. A retouched tool that is too small and too worn to be used has a utility of zero, while a large, fresh flake blank with extensive potential for resharpening or reworking has a high potential utility. Similarly, a large core would have a much higher potential utility than a small flake tool. Subsequently, a somewhat more complex situation is modeled, in which utility is a product of both the potential for producing fresh edges, and the lengths of those edges. Other factors, such as functional properties, are purposefully excluded for the time being; their influence will become evident where the model's predictions do not conform with archaeological and ethnographic observations about how human beings actually behaved.

To simplify the analysis, cores are treated as though they were of cubic proportions (length = width = thickness). Tools/blanks are treated as though they were square in plan view, and always thinner than they are long or wide (i.e., length = width > > thickness). While the ratios of the primary dimensions of tool blanks vary widely among archaeological assemblages, there is little doubt that the length, width, and thickness of flakes are strongly correlated in pieces produced by a given method of core reduction. These arbitrary and somewhat unrealistic conventions about the shapes of artifacts do not significantly affect the model. Assuming cores to be lenticular instead of cubic, or making flakes ovoid instead of square, would not al-

TABLE 16.1. Equations used to calculate transport cost and potential utility.

Artifact Class	Equation	
Transport cost (= mass)		
Tools	$\sim f(\text{length}^3)$	
Cores	$\sim \text{length}^3$	
Potential utility		
Tools	$\sim \text{length} - \text{minimum size}$	(1)
Tools	$\sim \text{length}(\text{length} - \text{minimum size})$	(2)
Cores	$\sim C(\text{length}^3 - \text{minimum size}^3)$, (where $0 < C < 1$)	

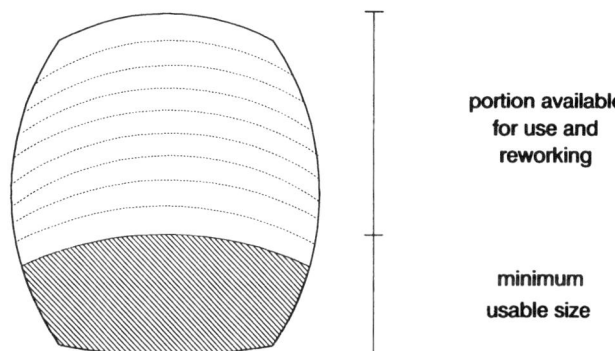

FIGURE 16.1. Schematic illustration of tool blank, showing minimum size and portion available for use.

ter the basic association between transport cost and potential utility. It would simply make the calculations more tedious. For the purposes of this analysis, the fixed nature of the artifact proportions is more significant than the actual shapes of things.

Table 16.1 presents the equations used to estimate transport cost and utility for the hypothetical tools and cores. Because the energy expended when carrying artifacts is directly proportional to mass, transport cost for cores is calculated as proportional to the cube of their lengths. The lengths and widths of flakes are assumed always to be equal, but relative thickness is allowed to vary. The variable f represents an arbitrary proportion of artifact thickness to length, and is always less than 1.0. In the following analyses, f is allowed to vary from .25 to .10, which seems to be a reasonable range based on comparisons with actual archaeological assemblages. Thus, transport cost for flake tools is calculated as a fraction (between .25 and .10) of length to the third power.

The calculation of potential utility is more complex, and is different for retouched tools and cores. The potential utility of an ideal, marginally retouched tool is initially taken to be equivalent to the number of fresh edges that can be produced through successive reworking of one (or both) margin(s). Hence, utility is determined by a single dimension, either length or width. In practice, however, the entire length of a tool is not available to be transformed into fresh edges. As illustrated schematically in Figure 16.1, flake tools are assumed to have a minimum usable

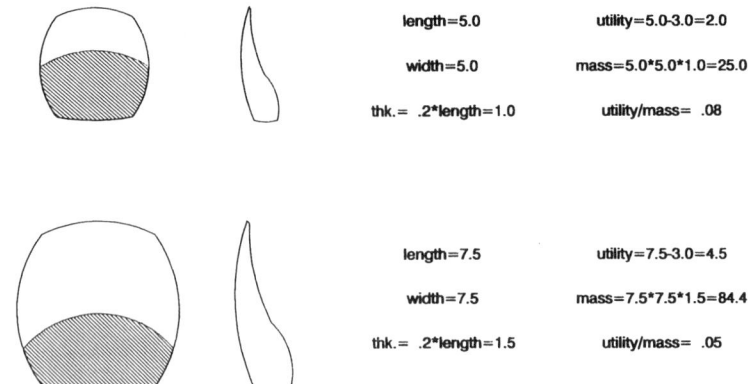

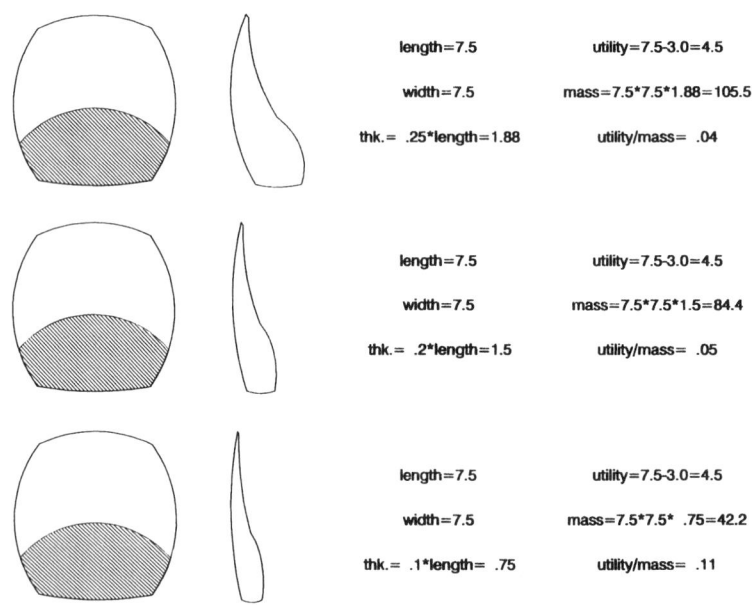

FIGURE 16.2. Examples of mass and utility calculations for tool blanks of different sizes.

FIGURE 16.3. Examples of mass and utility calculations for tool blanks of different thicknesses.

size, below which it is impossible to hold them in the hand or in a socket. A more realistic approximation of utility sets it proportional to one linear dimension minus this minimum size (Table 16.1, Equation 1). Only "plan view" dimensions are important; an endscraper 10 cm long and 2 cm thick is considered to have the same potential utility as one the same length but only 1 cm thick. Since the discussion deals only with unifacial tools, this assumption would seem to be valid, although it might not hold for bifacial tools.

The potential usefulness of a core is assumed to depend on its volume rather than one (or two) of its linear dimensions. As with tools, the entire volume of a core is not available to be turned into usable artifacts. Cores are assumed to contain a minimum, "irreducible" unit, a block of stone too small to produce functional blanks. The volume of this unusable portion is fixed at the third power of the minimum serviceable length for tools. Furthermore, no core is a perfectly efficient source of functional edges. Raw material is lost in preparing platforms, regularizing the face of detachment, and coping with errors and faults, and not every flake produced is large enough or of the correct form to be used as a tool. As just discussed, each flake produced also "contains" a small, unusable portion. The variable C is used to adjust for the wastage of cores as a source of edges. C, which can be viewed as a coefficient of production efficiency, is allowed to vary between .10 and .025.[2]

ARTIFACT UTILITY AND TRANSPORT COST

In the following analyses, utility and transport cost are calculated for artifacts of varying sizes and shapes using the equations shown in Table 16.1 above. Figures 16.2 and 16.3 illustrate sample

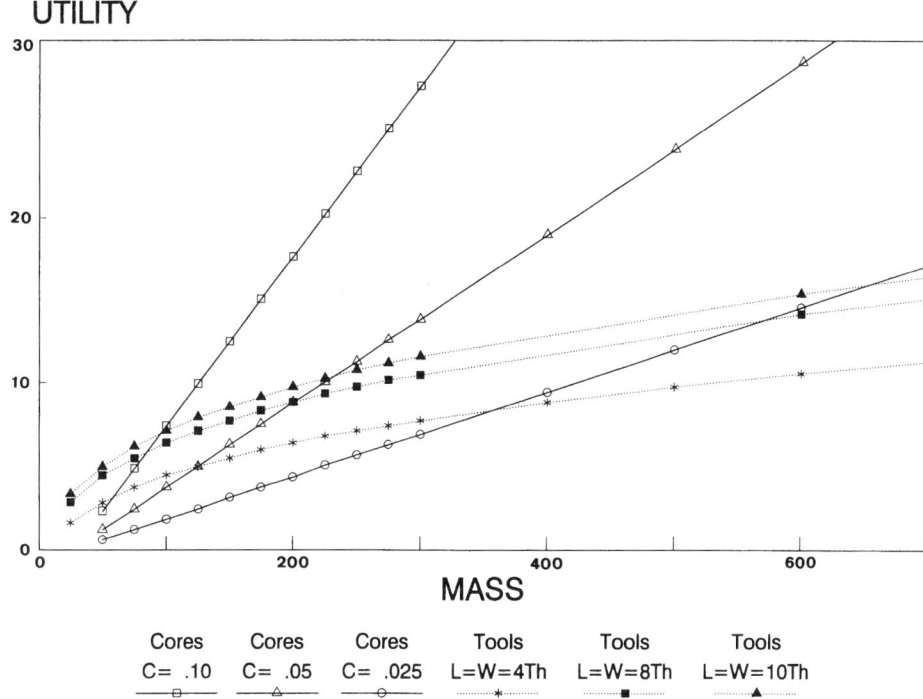

FIGURE 16.4. Relation between potential utility and mass for tools and cores. Coefficient of production efficiency (C) and thicknesses of tool blanks vary.

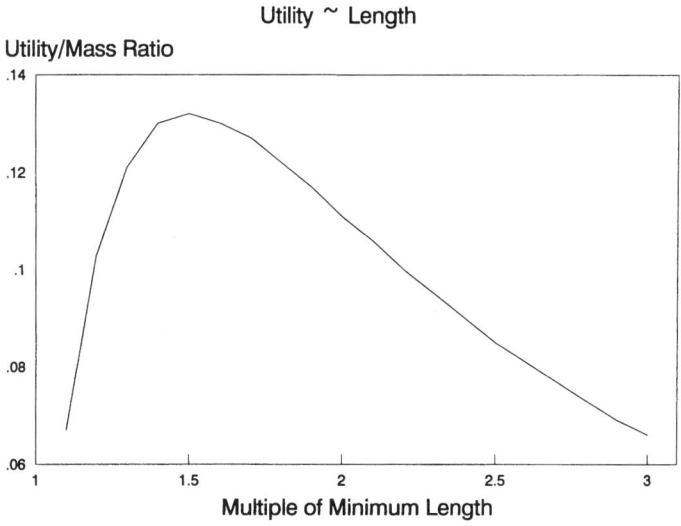

FIGURE 16.5. Utility/mass ratios for retouched tools plotted against length, expressed as a multiple of minimum usable size. Utility ~ length − minimum size.

calculations for flake tools of varying lengths and thicknesses, respectively. The minimal usable size for artifacts is fixed arbitrarily at three units of length; consequently, the minimum volume for cores capable of producing usable tools is three to the third power, or 27 cubic "units." In Figures 16.4–16.7 below, the x-axes of the plots measure either mass—equivalent to transport cost—or artifact size, expressed as a multiple of the minimum usable length. The y-axes plot either utility or the ratio of utility per unit mass. The goal in presenting each graph is to show the most efficient technological option, providing the greatest utility relative to the cost of transport (mass). In some graphs, the relative thicknesses of tool blanks or the coefficients of production efficiency for cores are allowed to vary. It would be impossible to illustrate utility and transport cost for all possible values of these two variables, so a limited number of cases are shown, representing extremes within the total range of possibilities.

Figure 16.4 is a plot of artifact utility against mass for idealized tools and cores of varying sizes. Each point on each curve represents a single artifact with a mass corresponding to the value on the x-axis. The three tool curves (dotted lines) represent different length/width/thickness ratios, while the three core curves (solid lines) represent different values of the coefficient of production efficiency. The most efficient options, those affording the highest utility per unit of mass, are the curves that come closest to the upper left corner of the graph.

The most striking aspect of Figure 16.4 is that the potential utility of cores increases as a linear function of mass, while the utility of tools quickly reaches a point beyond which large increments in weight yield minimal gains in utility. The curvilinear shapes of the tool curves reflect the fact that tool utility is calculated as a simple function of length (or width), whereas mass is proportional to volume, or length to the third power (length * width * thickness, where the three dimensions are highly correlated). In contrast, *both* the utility and the transport cost of cores are proportional to volume. Decreasing the relative thickness of tools markedly increases the ratio of utility to mass, and very low values of C cause cores to gain utility more slowly. Nonetheless, it is always true that relative gains in utility for retouched tools decline as size increases, whereas for cores the ratios between the two variables remain constant.

One lesson that could be drawn from Figure 16.4 is that, if one had to meet all needs with a single transported item, it should be a core, unless core technology was extremely inefficient (a very low value of C), or the total mass of the transported toolkit was very small (the area at the extreme left edge of the graph). More importantly, however, the plot demonstrates that, with respect to the postulated requirements of transported toolkits, increasing the sizes of tools is not a uniformly economical alternative for achieving higher levels of potential utility.

Fortunately, of course, the use of ever-larger tools or cores is not the only option for adding to the potential utility of the transported toolkit. Instead of packing one very large implement, one could carry many smaller flakes or retouched tools, those with more favorable utility-to-mass ratios. This alternative bypasses the problem caused by the autocorrelation between linear dimensions of flakes, as several small tools placed end to end can be visualized as one very long, narrow (or very short, broad) artifact, with a lengthy working edge and/or great potential for successive renewal. The question nonetheless remains as to how large or small these blanks should be.

The nonlinear nature of the relation between utility and mass for retouched tools indicates that there is some optimum size for such artifacts, a point at which the greatest utility per unit weight is achieved. Figure 16.5 plots the ratio of utility to mass for hypothetical tool blanks of different sizes (on the y-axis), expressed (on the x-axis) as multiples of the minimum usable length. The most efficient size is indicated by the highest point on the curve, or the highest utility/mass ratio. If utility is assumed to be proportional to artifact length—to the potential for resharpening or renewal—the most favorable ratio is obtained with surprisingly small artifacts. The most efficient artifacts would be about 1.5 times the minimum usable length, although severe declines are not encountered until tools become significantly larger. Note, however, that the utility/mass ratio of artifacts three times the minimum length is equivalent to that of artifacts only 1.1 times the minimum useful size.

In the context of assembling mobile toolkits, then, it would be most efficient to employ a large number of artifacts about one and one-half times their minimum usable size, rather than a single large tool with the same potential utility. Figure 16.6 shows that, as might be expected, the strategy of transporting a number of relatively small tools and blanks produces a linear relation between utility and mass. The plot is similar to Figure 16.4, except that each point on the three "tool curves" represents an assemblage of several tools/blanks of the specified dimensions (1.2, 1.5, or 2.0 times minimum size), with *a composite* mass corresponding to the position on the y-axis, rather than a single large artifact. The curves for cores are identical to those in Figure 16.4, although they appear flatter because the x-axis has been extended. As would be inferred from Figure 16.5, tools and blanks 1.5 times minimum usable size (the square markers) yield the greatest utility per unit mass. Moreover, points on the dotted lines are always higher on the y-axis at a given position on the x-axis than are points on the solid lines. This implies that a toolkit composed of several retouched tools or blanks of optimal, or even suboptimal, size will always provide greater utility than a core of equivalent mass.[3] The simple reason is that the heights of the core curves along the y-axis are always modified by the coefficient of production efficiency C, which is always less than 1.0. In other words, because cores are never perfectly efficient at turning mass into utilizable flakes, it is always preferable to carry retouched tools or blanks, so long as the latter are kept within a reasonable size range.

To this point, we have assumed that the utility of artifacts is determined only by their potential to yield fresh edges. This supposition may not always be justified. Intuitively, it seems that bigger tools ought to be more effective for at least some tasks. It may well be possible to butcher an entire deer using a single stone flake no more than a few centimeters long, but anyone who has tried to free a rusty bolt with a short-handled screwdriver knows that the size of an implement is important in determining how much pressure or leverage can be brought to bear on the task at hand. Longer edges may also be more effective for certain jobs. It is no accident that professional butchers prefer 8–10-inch blades over paring knives, even for relatively fine work.

It is a relatively simple matter to modify the mass/utility equations to reflect a situation in which overall artifact size is important for more than its role in determining renewability or resharpening potential. If total tool size confers functional advantages, then it might be more realistic to say that the utility of tools is proportional to two linear dimensions, the length of the edge as well as the dimension perpendicular to the edge (equiva-

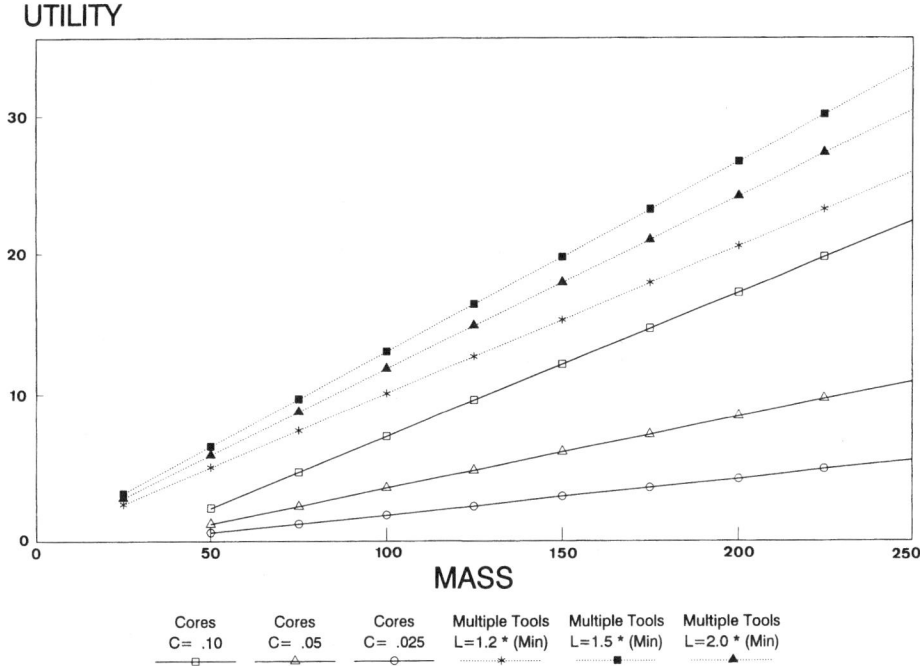

FIGURE 16.6. Relationship between potential utility and mass for toolkits consisting of multiple small tools, and for cores with different values of the coefficient of production efficiency (C).

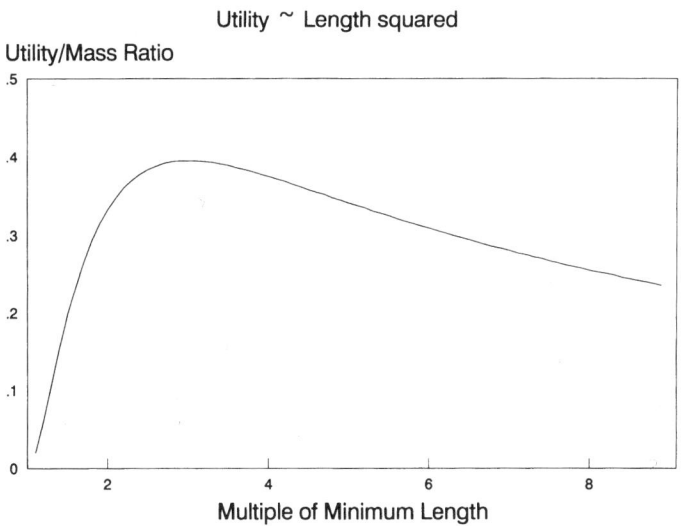

FIGURE 16.7. Utility/mass ratios for retouched tools, plotted against length, expressed as a multiple of minimum usable size. Utility ~ length − minimum size.

lent to the potential for resharpening or renewal). The appropriate equation is shown in Table 16.1 (Equation 2).

Changing how utility is calculated would have little effect on the contrasts between cores and tools in the associations among mass, size, and utility, as described in Figure 16.4. While the utility of tools would now increase as a function of the square of their length, mass still increases with volume (or as a function of the third power of length). Therefore, utility continues to exhibit declining returns relative to the increasing mass as tool sizes increase, although the falloff is not as pronounced as before.

At the same time, utility per unit mass is now a function of the square of tool length, so the coefficients of core efficiency, calculated based on the utility/mass ratio of the optimal tool form (see Note 2), would be higher than in Figure 16.4, thereby raising the core curves on the y-axis. However, making tool utility proportional to the square of length *does* have pronounced effects on calculations of the optimal size for transported tools. Figure 16.7 is similar to Figure 16.5 above, except that utility is now calculated as proportional to length ∗ (length − minimum size). Here again, the potential utility per unit mass increases dramatically for artifacts larger than the minimum usable size, although the falloff is more gradual as artifacts get very large. With the new equation, the highest point on the curve, indicating the optimal utility/mass ratio, occurs for artifacts that are three times the minimum size. Impressionistically, at least, this would seem to be a more reasonable size range than the figure of 1.5 times minimum usable length arrived at previously.

Some Implications of the Model

The results obtained to this point can be summarized as follows. If the transported toolkit consists of a single object, and a tool-user desires to maximize its utility relative to the cost of transport, a core should always be carried. Changing the shapes of tool blanks, making them thinner relative to linear dimensions, increases the utility/mass ratio of transported tools somewhat, but does not alter the fact that gains in utility provided by very large flake tools do not compensate for their increased weight. The best *overall* strategy for maximizing utility per unit mass is to carry many smaller flake tools, as long as they are within a certain size range. Depending on how one assesses utility, flakes and tools between 1.5 and 3 times the minimum usable size provide the most economical option.

Although the model of artifact form and utility presented here is comparatively simple and abstract, it has a number of interesting implications. Perhaps the most surprising concerns the assumed value of multifunctional artifacts in transported toolkits. Because it is more economical to carry several small tools than one big one, it may be unrealistic to expect multifunctionality to be an important concern in designing assemblages of transported stone tools. If multifunctional tools must be relatively large, either to support multiple working edges or so that they can be worked successively into different forms (see Nelson 1991:71–74), there is simply more payoff to packing several smaller, more functionally "specialized" implements. Thus, we should not necessarily expect that assemblages created by relatively mobile groups heavily dependent on transported toolkits will be monotonous, containing few functionally distinct stone-tool forms. This principle could explain the existence of many apparently specialized flaked tools in Folsom Paleoindian chipped-stone assemblages (Frison and Bradley 1980:64–72; Frison and Stanford 1982:45–55), thought to have been created by foragers practicing very high levels of residential mobility (Kelly and Todd 1988). The same principle might also account for the apparent lack of association between numbers of "functionally specialized" tools and changing mobility strategies during the Archaic period at the Koster site (Lurie 1990:55).

It is difficult to dispute ethnographic observations that high residential mobility is often associated with comparatively restricted diversity in artifact assemblages (Hitchcock 1982:370–380; Shott 1986). There are two problems with applying these observations directly to archaeological lithic assemblages, however. First, while there is no doubt that residential stability allows people to keep a wider variety of functionally specialized artifacts at hand, it is not necessarily the case that the transported tools used by more mobile people must be designed to fill *all* the functions of the larger stocks of artifacts employed by more sedentary groups. Some "potential" needs may simply go unmet where mobility restricts the size of the toolkit, while others may be met using available implements on an improvisational basis (e.g., Binford 1979; Nelson 1991). I would argue that functional versatility becomes an option only when artifacts must otherwise be large for functional reasons, a situation discussed below.

Another implication of this model concerns the effects of raw-material size on the dimensions of transported artifacts. Although it is disadvantageous from a purely economic standpoint to carry very large tools, it is beneficial to have implements somewhat larger than the minimum usable size. If there are no raw-material-imposed restrictions on artifact size, transported tools should be comparatively small (e.g., Ebert 1979). On the other hand, in cases where extremely small raw materials were exploited, we might expect evidence that toolmakers took special efforts to maximize the sizes of transported tools so as to approach the optimum. Of course, artifacts made for use and abandonment without transport could be as small or as large as was convenient.

Pontinian Mousterian assemblages from coastal Italy (Kuhn 1994) represent a case in point. Most of the stone artifacts in these assemblages are made on flakes from very small flint pebbles. Because of the raw materials, most potential tool blanks were smaller than the minimal optimal size for transported tools. However, under conditions of relatively high mobility, when Mousterian populations would have been most reliant on transported artifacts, it appears that special measures were taken to maximize the resharpening potential of flake blanks. In strata where faunal patterns and other evidence attest to the exploitation of highly dispersed resources, land-extensive foraging patterns, and relatively ephemeral use of caves, toolmakers depended on primarily centripetal (radial) and bipolar methods of core reduction (Kuhn 1994; Stiner and Kuhn 1992). Both of these methods yielded comparatively large flakes, with relatively great potential for renewal, from the available raw materials (Kuhn 1992, 1994). In contrast, evidence for somewhat diminished mobility, more prolonged occupations, and less extensive reliance on transported artifacts in other Mousterian strata is associated with core technologies yielding smaller, narrow flakes with relatively little potential for resharpening (Kuhn 1994; Stiner and Kuhn 1992).

The graphs presented in Figures 16.4–16.8 imply that, while very large tools will always be inefficient as components in mobile toolkits, shifting techniques of manufacture to adjust the geometric proportions of flakes or the way a given volume of raw material is exploited can result in more favorable utility/weight ratios, particularly for tool blanks. The use of alternative techniques of core reduction or blank production may also be quite visible archaeologically, more so than contrasts in the sizes of transported artifacts that are recovered only after being extensively reworked and reduced. In this regard, it is interesting that the products of Levallois cores, both blanks and the tools made on them, were transported more frequently and over longer distances than other artifacts in many European Mousterian contexts (e.g., Geneste 1989; Jaubert et al. 1990; Rensink et al. 1991). Levallois artifacts are especially thin (compared to length and width), so it is not surprising to find that they were preferentially selected to serve as components of transported toolkits. Similarly, among the many splendid qualities attributed to bifacial tools and cores (e.g., Kelly 1988; Nelson 1991) should be added the fact that bifacial technology permits an unequaled degree of control over all three linear artifact dimensions. Other considerations aside, bifacial reduction may be ideal for making transported artifacts simply because it allows one to most closely approach the optimal weight/utility ratios.

An Anomaly: the Role of Cores in Mobile Toolkits

Human technologies, like any other type of behavior, are complex and multicausal phenomena. The forms stone tools take, the techniques used to manufacture and maintain them, and their dispositions across space are contingent on a vast number of factors. Yet this analysis has proceeded from the assumption that two considerations—transport cost and potential utility—guide the design and assembly of transported toolkits. The value of this kind of simple model is that it yields "sharper," more easily rejected predictions than complicated models, which can be "fudged" to fit almost any outcome (Jefferys and Berger 1992). Because its implications are relatively clear-cut, a very simple model can show how and under what circumstances the basic premises underlying it do not hold and how these assumptions must be modified in order to better predict real phenomena. The greatest utility of such models is often realized by showing how they *fail* to account for contemporary and prehistoric behavior.

Optimizing the utility of artifacts relative to their weight is arguably the single most important consideration in making up assemblages of artifacts that people carry around with them. Based on this premise, the model predicts that the most economical toolkits would consist of numerous relatively small implements, between 1.5 and 3.0 times their minimum usable sizes. The fact that finished tools and/or tool blanks appear to have been the items most frequently transported and maintained by mobile populations in a range of contexts (e.g., Bamforth 1985; Ellis 1989; Hoffman 1991:349; Kozlowski 1991; Reher 1991:279; Rensink et al. 1991) would seem to furnish general support for this notion. By this same token, if the maximization of utility per unit mass were the only important consideration, we would *never* expect to find *cores* as parts of transported toolkits. It is true that using the most homogeneous raw material available (e.g., Goodyear 1989) or the most efficient flake production technology can increase the utility of cores per unit weight.

Similarly, preparation of cores at the point of procurement can reduce the amount of wastage people must carry (e.g., Metcalfe and Barlow 1992:352). Nonetheless, cores will never be perfectly efficient at converting mass to usable flakes, so it would seem most rational to reduce cores into tool blanks prior to carrying them off. Yet in spite of their demonstrable inefficiency, it is clear from ethnographic and archaeological observations that at least some cores were regularly transported as part of "personal gear" by mobile populations in a variety of contexts (e.g., Binford 1979:276; Holen 1991; Kelly 1988; Smyth 1878:132), albeit not as frequently as finished tools or tool blanks.

A number of arguments might be offered to explain the occasional inclusion of cores in mobile toolkits. Cores could serve as a kind of "sheath" for undetached flakes, and striking off flakes only as needed would ensure that the edges were fresh and undamaged. However, while it is true that sharp flake margins are easily dulled by incidental contact with other artifacts or containers, there are a variety of low-cost tactics for protecting them. Stone-tool users in Australia wrapped spear heads in soft bark (Thompson 1949:65), or even carried flake knives wrapped in their hair (Tindale 1965:147). Transported cores would also provide the option of producing a variety of flake or blank forms (e.g., flakes with acute or obtuse edges) as needed. But the energetically optimal solution is to carry many small tools anyway, so it would be just as practical to take along a variety of ready-made specialized implements or blanks.

A more satisfactory explanation for the "inefficient" inclusion of cores in transported toolkits derives from the unique *functional* properties of cores as tools. Large blocks of stone with thick edges can fill a variety of needs for which smaller flake tools are entirely inadequate. Cores can serve as hammers, anvils, pestles, or pounders, or as chopping tools or meat cleavers, purposes that require a certain mass. Frison (1974, 1978:313–314) notes that heavy-duty choppers or cleavers are extremely useful, if not indispensable, for butchering large animals like bison. If the raw materials are readily available, large, heavy-duty tools are often made of tough, granular, and fairly common materials such as metamorphic rocks (e.g., Gould et al. 1971:163). Even low-quality stone is uncommon in some regions, however, and it would be beneficial to carry around at least one relatively massive artifact.

Although it is risky to make generalizations about large regions, it is possible that variation in the distribution of stone resources may explain the unusually heavy emphasis on bifacial-core technology among some mobile hunter-gatherer groups in western North America, mentioned above. What is most distinctive about the New World cases is not the existence of bifacially modified tools, as these are known from many other regions and time periods, including both Pleistocene Europe and Australia. More remarkable is the frequent use of large bifacial cores. In many parts of western North America, particularly the plains and grasslands, stones of any sort are uncommon. Where one could not rely on finding even poor-quality stone,

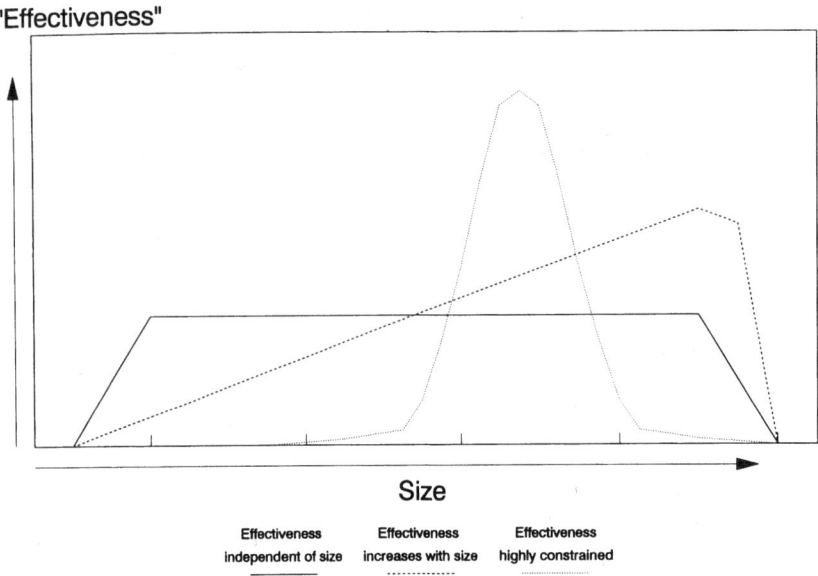

FIGURE 16.8. Hypothetical relations between artifact sizes and functional effectiveness.

there would be a distinct advantage to having at least one substantial implement on hand. And, as many authors have noted, bifacial cores are a particularly cost-efficient source of flakes and preforms for tools, as well as being good heavy-duty tools (Kelly 1988; Nelson 1991). In contrast, stone of one kind or another is essentially ubiquitous across much of Europe, and apparently in at least some parts of desert Australia as well (Gould et al. 1971). In more stone-rich areas, where people could be assured of finding stone adequate for at least heavy-duty implements, core technology might be more strictly oriented toward producing flakes with functionally or economically favorable properties, rather than toward increasing the transportability of the cores themselves.

Conclusion

This paper explores the implications of a simple model of how transported toolkits and their components should be designed. The model is founded on the assumption that the predominant concern in making artifacts subject to more or less continuous transport is maximizing their potential utility, relative to the cost of transport. Initially, the utility of blanks and retouched tools is measured in terms of the potential for producing fresh edges; subsequently, the lengths of those edges are taken into account. Setting up a few simple conventions about artifact geometry and morphology permits one to assess the most economical way to assemble transported toolkits, within the limitations of the fundamental assumptions that underlie the model. Arguably the biggest "surprise" arising from this analysis is the conclusion that the highest utility/mass ratios are obtained by carrying many relatively small tools. This conclusion stands in apparent contradiction to the widespread notion that the elements of transported toolkits should be designed to be multi-functional (e.g., Bleed 1986; Gould 1980; Nelson 1991:73–74; Shott 1986). It is also somewhat inconsistent with the observations that people in the past *did* move both cores and fairly large tools around the landscape with some frequency.

If this model contradicts certain notions about how artifacts *ought to be* designed, it may be that these ideas are somewhat unrealistic. After all, the model is a simple mathematical actualization of the basic assumptions about tool multifunctionality. The lesson of this analysis is that, if optimization of utility relative to transport cost is the only strategic "goal," then *generalized* multi-functionality is not a relevant design consideration, so long as it requires making tools larger. Artifacts that comprise mobile toolkits may well show evidence of having been used in a wider variety of tasks, if only because they were more often employed in situations where there was little else on hand to choose from. But this reflects the context of tool use and not design.

On the other hand, evidence that prehistoric and recent foraging populations often carried cores as part of transported toolkits implies that it is not always appropriate to assess artifact utility exclusively in terms of the potential to produce new edges (i.e., in terms of tool length or core volume). Larger size may confer unique functional advantages to some artifacts. In the case of transported cores, it is argued that their large mass permits their use in heavy pounding or chopping tasks, roles that small flake tools simply cannot fill.

It has been assumed throughout this analysis that the sizes of artifacts are free to vary widely in response to the demands of portability. While this might be true for most stone tools, it does not necessarily hold for all artifacts. Figure 16.8 illustrates three hypothetical associations between the size of an artifact and its effectiveness in a particular application. Here, we are speaking purely of functional characteristics, not strategic or economic attributes. The solid line represents a situation in which large artifacts are not significantly more useful than small ones, within reasonable limits. This situation corresponds to the initial premises of this discussion, when artifact utility was assessed strictly in terms of the potential to produce new edges. Under such conditions, the most efficient transport option is to carry many small tools that are approximately 1.5 times the minimum usable length. The dashed line represents a situation in which larger artifacts *are* more effective tools, at least up to the point where they become unwieldy. The situation represented by the dashed line is equivalent to setting artifact utility as a function of both edge length and a tool's potential for renewal. As demonstrated above, reformulating the basic model to account for two artifact dimensions when calculating utility produces a larger and seemingly more realistic estimate—about three times minimum usable size—for optimal elements of transported toolkits.

The first two scenarios probably account for the majority of stone tools, particularly those implements employed in relatively "high tolerance" manufacture and maintenance activities. However, the functional demands of some applications may impose much more stringent upper and lower limits on artifact sizes. The third (dotted) line in Figure 16.8, showing a very sharp and limited peak in effectiveness within a narrow size range, represents this situation. Tightly constrained size limits may be particularly relevant to the design of artifacts like containers and weapons. A simple wooden spear two or three times its minimum usable length would be quite unwieldy, if not completely useless. Similarly, an effective arrow or dart point must be designed for penetrating power, and cannot be too heavy for the shaft. In this kind of "low tolerance" application, an artifact's renewal potential and transport cost are probably eclipsed in the realm of artifact design by the requirements of making a practical, dependable weapon.

When the functional constraints of a particular task make adjusting artifact sizes impractical, we would expect to see a shift to alternative approaches to "lightening the load." If, and only if, tools must be kept within narrow size parameters, then modifying their designs to accommodate multiple functions could well be a fruitful strategy. Another option that exists within lithic technology is to change the geometry of artifacts, altering the relation between the functional dimensions (length and width) and total volume (or mass). Tactics of stone-tool manufacture, like bifacial reduction, which allow for a high degree of control of overall artifact shape, should be used most frequently to make implements otherwise subject to tight functional constraints on size. It is probably no accident that bifacial-reduction technology is used to make large Late Paleoindian projectile points and/or knives, while scrapers, drills, and other "manufacture and maintenance" tools are more often made on simple flake blanks (e.g., Frison and Stanford 1982; Gramly 1990).

One could ask whether the small energetic costs or benefits of carrying a particular kind of tool or toolkit would be sufficient to influence the apparent choices of real people under most conditions. After all, stone tools probably constituted a relatively small component of what mobile foragers actually toted from place to place. The response to this rhetorical question is that we do not know for sure. However, it is reasonable to think that transport costs could have been important in determining what people carried with them and what they left behind. Moreover, understanding the basic economics of human technologies can greatly expand what one can learn from studying stone artifacts. The simple model discussed here furnishes a set of basic predictions about how people should behave, provided that a number of underlying assumptions are legitimate. The observations cited above suggest that the premises of the model are often valid, but that we also need to account for factors like the purely functional advantages of size in certain contexts. Comparing the model's predictions more extensively to archaeological or ethnographic evidence is one potential avenue for recognizing even more fundamental limitations in how we think humans behave.

Acknowledgments

Although it was my idea to write this paper, the extent to which it is cogent and intelligible is largely attributable to the advice and critical comments of Mary Stiner, the editorial staff of *American Antiquity,* and four anonymous reviewers.

Notes

1. Metcalfe and Barlow (1992) provide an interesting discussion of the economics of short-range, one-way transport of both foodstuffs and lithics.
2. The values of the coefficient of production efficiency for cores (C) are calculated relative to the utility/mass ratio for retouched tools with a length 1.5 times the minimum usable artifact size (here fixed at 3 units of length). As shown in this paper, this is the most economical size for transported artifacts. Where length = width = 8 ∗ thickness, idealized tools of this size yield approximately .13 units of utility (length) per unit mass. For a value of C equal to .10, about 77 percent of the usable mass of the core would be transformed into tool blanks of the optimal size, while for a value of C equal to .025, around 19 percent of a core's mass would be transformed into such blanks. Although I could find no published quantitative experimental data on the conversion of core mass into flakes of a particular size, these values seem reasonable as the upper and lower limits on core efficiency.
3. In contrast to the case of tools/tool blanks, there is no payoff in transporting several small cores rather than one large one, since both utility and transport cost for cores are proportional to volume. In fact, carrying several small cores presents a slight disadvantage, since each core would "contain" a small, irreducible portion of equal size.

References

Bamforth, D.
1985 The Technological Organization of Paleoindian Small Group Bison Hunting on the Llano Estacado. *Plains Anthropologist* 32:1–16.

1986 Technological Efficiency and Stone Tool Curation. *American Antiquity* 51:38–50.

1991 Technological Organization and Hunter-Gatherer Land Use: A California Example. *American Antiquity* 56:216–34.

Binford, L. R.

1977 Forty-Seven Trips: A Case Study in the Character of Archaeological Formation Processes. In *Stone Tools as Cultural Markers,* edited by R. V. S. Wright, pp. 24–36. Australian Institute of Aboriginal Studies, Canberra.

1979 Organization and Formation Processes: Looking at Curated Technologies. *Journal of Anthropological Research* 35:255–273.

Bleed, P.

1986 The Optimal Design of Hunting Weapons. *American Antiquity* 51:737–747.

Brannan, J. A.

1991 On Modeling Resource Transport Costs: Suggested Refinements. *Current Anthropology* 33:56–60.

Ebert, J. I.

1979 An Ethnographical Approach to Reassessing the Meaning of Variability in Stone Tool Assemblages. In *Ethnoarchaeology,* edited by C. Kramer, pp. 59–74. Columbia University Press, New York.

Ellis, C. J.

1989 The Explanation of Northeastern Paleoindian Lithic Procurement Patterns. In *Eastern Paleoindian Lithic Resource Use,* edited by C. J. Ellis and J. C. Lothrop, pp. 139–164. Westview Press, Boulder, Colorado.

Frison, G. C.

1978 *Prehistoric Hunters of the High Plains.* Academic Press, New York.

Frison, G. C. (editor)

1974 *The Casper Site: A Hell Gap Bison Kill.* Academic Press, New York.

Frison, G. C., B. Bradley

1980 *Folsom Tools and Technology at the Hanson Site, Wyoming.* University of New Mexico Press, Albuquerque.

Frison, G. C., D. J. Stanford

1982 *The Agate Basin Site: A Record of the Paleoindian Occupation of the Northwestern High Plains.* Academic Press, New York.

Geneste, J-M.

1989 Les industries de la Grotte Vaufrey: Technologie du débitage, économie et circulation de la matiére premiére. In *La Grotte Vaufrey a Cenac et Saint-Julien (Dordogne). Paleoenvironnement, Chronologie et Activités Humaines,* edited by J-P. Rigaud, pp. 441–519. Mémoires de la Société Préhistorique Française Tome XIX. Paris.

Goodyear, A.

1989 A Hypothesis for the Use of Cryptocrystalline Raw Materials Among Paleoindian Groups of North America. In *Eastern Paleoindian Lithic Resource Use,* edited by C. Ellis and J. Lothrop, pp. 1–10. Westview Press, Boulder, Colorado.

Gould, R.

1969 *Yiwara: Foragers of the Australian Desert.* Charles Scribner's Sons, New York.

1980 *Living Archaeology.* Cambridge University Press, New York.

Gould, R. A., D. A. Koster, A. H. L. Sontz

1971 The Lithic Assemblage of the Western Desert Aborigines of Australia. *American Antiquity* 36:149–169.

Gramly, R. M.

1990 *Guide to the Paleoindian Artifacts of North America.* Persimmon Press, New York.

Hitchcock, R.

1982 *The Ethnoarchaeology of Sedentism: Mobility Strategies and Site Structure Among Foraging and Food Producing Groups in the Eastern Kalahari Desert, Botswana.* Unpublished Ph.D. dissertation, Department of Anthropology, University of New Mexico, Albuquerque.

Hoffman, J.

1991 Folsom Land Use: Projectile Point Raw Material as a Key to Mobility. In *Raw Material Economies Among Prehistoric Hunter-Gatherers,* edited by A. Montet-White and S. Holen, pp. 335–356. Publications in Anthropology No. 19. University of Kansas, Lawrence.

Holen, S.

1991 Bison Hunting Techniques and Lithic Acquisition Among the Pawnee: An Ethnohistoric and Archaeological Study. In *Raw Material Economies Among Prehistoric Hunter-Gatherers,* edited by A. Montet-White and S. Holen, pp. 399–411. Publications in Anthropology No. 19. University of Kansas, Lawrence.

Jaubert, J., M. Lorblanchet, H. Laville, R. Slott-Moller, A. Turq, J-P. Brugal

1990 *Les Chasseurs d'Aurochs de La Borde: Un Site du Paléolithique Moyen (Livernon, Lot).* Documents d'Archéologie Française No. 27. Editions de la Maison des Sciences de l'Homme, Paris.

Jefferys, W. H., J. O. Berger

1992 Ockham's Razor and Bayesian Analysis. *American Scientist* 80:64–72.

Jones, K. T., D. B. Madsen

1989 Calculating the Cost of Resource Transportation: A Great Basin Example. *Current Anthropology* 30:529–534.

Kelly, R.

1988 The Three Sides of a Biface. *American Antiquity* 53:717–734.

Kelly, R., L. C. Todd

1988 Coming into the Country: Early Paleoindian Hunting and Mobility. *American Antiquity* 53:231–244.

Kozlowski, J.

1991 Raw Material Procurement in the Upper Paleolithic of Central Europe. In *Raw Material Economies Among Prehistoric Hunter-Gatherers,* edited by A. Montet-White and S. Holen, pp. 187–196. Publications in Anthropology No. 19. University of Kansas, Lawrence.

Kuhn, S. L.

1989 Hunter-Gatherer Foraging Organization and Strategies of Artifact Replacement and Discard. In *Experiments In Lithic Technology,* edited by D. Amick and R. Mauldin, pp. 33–47. BAR International Series 528. British Archaeological Reports, Oxford.

1992 On Planning and Curated Technologies in the Middle Paleolithic. *Journal of Anthropological Research* 48:185–214.

1994 *Mousterian Lithic Technology and Raw Material Economy: A Case Study.* Princeton University Press, Princeton, in press.

Lurie, R.

1990 Lithic Technology and Mobility Strategies: The Koster Site Middle Archaic. In *Time, Energy, and Stone Tools,* edited by R. Torrence, pp. 46–56. Cambridge University Press, Cambridge.

Marshall, L.

1976 Sharing, Talking and Giving: Relief of Social Tensions among the !Kung. In *Kalahari Hunter-Gatherers: Studies of the !Kung and Their Neighbors,* edited by R. Lee and I. DeVore, pp. 349–371. Harvard University Press, Cambridge, Massachusetts.

Metcalfe, D., K. R. Barlow
1992 A Model for Exploring the Optimal Trade-off Between Field Processing and Transport. *American Anthropologist* 94:340–356.

Nelson, M.
1991 The Study of Technological Organization. In *Archaeological Method and Theory,* vol. 3, edited by M. Schiffer, pp. 57–100. University of Arizona Press, Tucson.

Reher, C.
1991 Large Scale Lithic Quarries and Regional Transport Systems on the High Plains of Eastern Wyoming: Spanish Diggings Revisited. In *Raw Material Economies Among Prehistoric Hunter-Gatherers,* edited by A. Montet-White and S. Holen, pp. 161–168. Publications in Anthropology No. 19. University of Kansas, Lawrence.

Rensink, E., J. Kolen, A. Spieksma
1991 Patterns of Raw Material Distribution in the Upper Pleistocene of Northwest and Central Europe. In *Raw Material Economies Among Prehistoric Hunter-Gatherers,* edited by A. Montet-White and S. Holen, pp. 141–160. Publications in Anthropology No. 19. University of Kansas, Lawrence.

Shott, M.
1986 Settlement Mobility and Technological Organization: An Ethnographic Examination. *Journal of Anthropological Research* 42:15–51.

Smyth, R. B.
1878 *Aborigines of Victoria: With Notes Relating to the Habits of the Natives of other Parts of Australia and Tasmania.* 2 vols. Government Printing Office, Melbourne.

Stiner, M. C., S. L. Kuhn
1992 Subsistence, Technology, and Adaptive Variation in Middle Paleolithic Italy. *American Anthropologist* 94:306–339.

Tindale, N. B.
1965 Stone Implement Making Among the Nakako, Ngadadjara and Pitjandjara of the Great Western Desert. *Records of the South Australian Museum* 15:131–164.

Thompson, D. F.
1949 *Economic Structure and the Ceremonial Exchange Cycle in Arnhem Land.* MacMillan and Co., Melbourne.

Torrence, R.
1983 Time Budgeting and Hunter-Gatherer Technology. In *Hunter-Gatherer Economy in Prehistory,* edited by G. Bailey, pp. 11–22. Cambridge University Press, Cambridge.

White, J. P., J. F. O'Connell
1982 *The Prehistory of Australia, New Guinea and Sahul.* Academic Press, New York.

☒ 17 ☒

Rocks are Heavy

Transport Costs and Paleoarchaic Quarry Behavior in the Great Basin

*Charlotte Beck, Amanda K. Taylor, George T. Jones,
Cynthia M. Fadem, Caitlyn R. Cook, and Sara A. Millward*

Archaeological study of quarry workshops has been sporadic at best, even though they have been of interest to archaeologists for over 100 years (e.g., Holmes 1891, 1897, 1919). This lack of treatment is likely due to the many problems that seem to be inherent in dealing with these sites (Singer 1984). Their areal extent alone, often many hectares, can be daunting, but added to this factor are the vast numbers of artifacts (thousands or even millions) that are generally present (Singer and Ericson 1977). Ericson (1984:2) reflects that while the investigation of quarries should be a primary focus in the study of peoples who relied upon stone tools, prehistoric archaeologists have seemingly avoided analyses of such sites due to "technical and methodological limitations imposed by a shattered, overlapping, sometimes shallow, nondiagnostic, undatable, unattractive, redundant, and at times voluminous material record." Essentially, because lithic procurement sites are often used over long periods of time by different groups of people, quarry records are frequently characterized by mixed technologies and artifact types.

Although quarries often are complex records of human behavior, both temporally and spatially, they represent an important aspect of prehistoric land use and adaptation. Quarries are sources of lithic raw material, referred to here as toolstone, a significant component of most prehistoric technologies. The acquisition of toolstone thus assumes a theoretical importance on a level with that of food and other resources and must be considered as such if the nature of prehistoric human behavior is to be fully understood.

Quarry assemblages can exhibit considerable variation (e.g., Gramly 1980; Holmes 1897; Johnson 1984; Raab et al. 1979), although the explanations for this variation differ. Holmes (1897), for example, related various kinds of quarry assemblages to differences in environmental setting. Raab et al. (1979) argue that the variation in assemblages from quarries reflects the number and kinds of different activities performed. These explanations, of course, are not necessarily mutually exclusive. Johnson (1984:225) states that "the activities represented in a quarry site assemblage depend on how the site articulates with the rest of the prehistoric settlement system," which includes the environmental context of the quarry as well as the activities performed there.

In this paper, we address the issue of quarry assemblage variability as it relates to the overall settlement and subsistence strategy. Specifically, we examine the costs and benefits of particular quarry behaviors in relation to the cost of transporting the resulting products upon departure, and how these behaviors are reflected in particular quarry assemblages. We explore the conditions under which it is beneficial to invest more effort in reduction and manufacture at the quarry and those under which it is better to transport raw material to a residential site for this purpose. We use as an example two Great Basin dacite quarries[1] and their associated residential sites[2] from eastern and central Nevada (Figure 17.1). All four sites were utilized almost exclusively by populations during the terminal Pleistocene and early Holocene, referred to here as the Paleoarchaic (Beck and Jones 1997:162).

The Great Basin Paleoarchaic

The earliest well-documented occupation of the Great Basin begins about 11,500–11,000 rcy BP, about the same time substantial Paleoindian records appear in other parts of North America. In the Great Basin, however, this early occupation is represented by an archaeological record comprised mainly of lithic artifacts resting on the surface, and it is thus difficult to date precisely, but we do know that it persists until ca. 8000–7500 rcy BP. Although fluted points are present in the Great Basin, they have yet to be dated and they commonly occur with large-stemmed point forms that are known to have also been present in the terminal Pleistocene. These latter forms, however, persist until the mid-Holocene, or ca. 7500 rcy BP. Thus, there is little to distin-

*Originally published in *Journal of Anthropological Archaeology* 21(2002):481–507

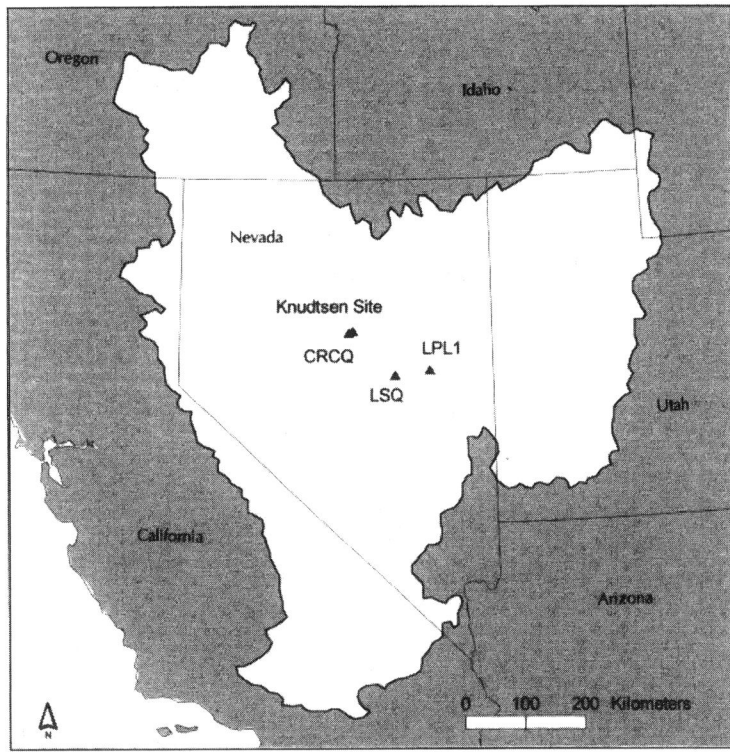

FIGURE 17.1. Map showing locations mentioned in the text.

guish the earliest Great Basin records in the terminal Pleistocene from those that followed in the early Holocene (and are generally termed "Archaic") (Simms 1988). For this reason, following Willig (1989), we use the term "Paleoarchaic" to refer to the Great Basin terminal Pleistocene/early Holocene archaeological record (see Beck and Jones 1997).

Despite more than six decades of research, archaeologists still do not know a great deal about Paleoarchaic subsistence. A realistic, if general, picture can be drawn from site location data, the paleoenvironmental record, and the meager but significant subsistence records from dry rockshelters and open sites (Beck and Jones 1997). Paleoarchaic sites typically occupy high relict landforms, including Pleistocene lake features (e.g., spits, beaches) as well as alluvial terraces, in valley bottoms; sites in upland areas are rare, a pattern markedly contrasting with the mid- and late Holocene record. It seems clear, based on these landform associations, that Paleoarchaic groups established residential/activity sites adjacent to wetlands, i.e., shallow lakes, marshes, springs and seeps, and active streams. This settlement pattern persists throughout the span of Paleoarchaic occupation despite evidence of substantial climatic changes. Following a relatively moist Younger Dryas (ca.11,200–10,100 rcy BP [Madsen 1999:77]), when paleoenvironmental records indicate that substantial surface water was available, valleys became increasingly dryer during the early Holocene (Grayson 1993), although moisture apparently remained high relative to modern conditions. With drying and presumably a parallel loss of preferred resources, Paleoarchaic diet breadth appears to have expanded, focusing increasingly on smaller mammals, birds, and fish (Beck and Jones 1997). Finally, near the close of the early Holocene, increased and substantial plant use is signaled by widespread appearance of groundstone artifacts.

Paleoarchaic mobility appears to have been substantial and wide-ranging. For example, in the central Great Basin populations traveled along a north-south trajectory from the northeastern to southern sections of Nevada, a distance of more than 450 km (Jones et al. 2002). Similar patterns are emerging from the western Great Basin, also indicating wide-ranging movement in a north-south direction (Graf 2001:126–133). Given these large territorial ranges, raw material procurement was most likely embedded in the Paleoarchaic subsistence schedule. As little archaeological evidence other than stone tools remains of this system, however, lithic procurement and manufacturing waste are the principle clues to Paleoarchaic subsistence strategies and other aspects of their lifeways.

The Paleoarchaic toolkit shares many similarities with that of Paleoindians in the Plains and Southwest; it includes various kinds of scrapers, gravers, notched tools, wedges, hammerstones, and abraders (Beck and Jones 1997). An additional component of this toolkit is the crescent (Figure 17.2e), which rarely occurs in Paleoindian toolkits outside the Great Basin (but see Daugherty 1956:247–249; Frison and Bradley 1999:34). The primary component of the Paleoarchaic toolkit, however, is the large stemmed point (Figure 17.2a–d). These points, which may have functioned both as weapon tips on thrusting spears and as knives (Beck and Jones 1993, 1997), and the flake debris from

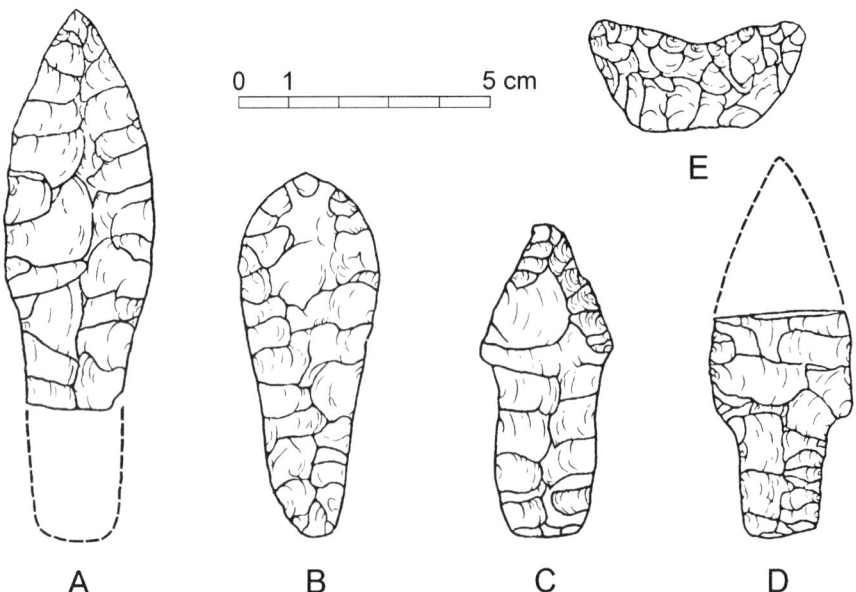

FIGURE 17.2. Paleoarchaic long stemmed points (A–D) and crescent (E). Types are as follows: (A) Cougar Mountain; (B) heavily reharpened Cougar Mountain; (C) Lake Mohave; (D) Parman; (E) "Butterfly Crescent" (see Beck and Jones, 1997 for original type definition and citations).

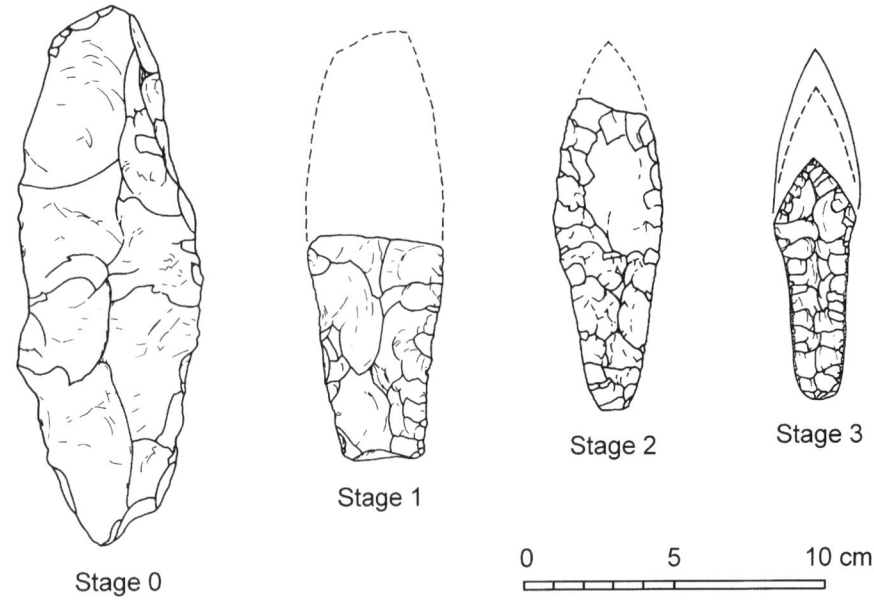

FIGURE 17.3. Biface reduction sequence represented in Great Basin Paleoarchaic sites. Definitions for biface stages (Table 17.2) are based on Callahan (1979).

their manufacture, dominate most Paleoarchaic assemblages. In most of the Great Basin, these points are made primarily of fine-grained volcanics, including andesites, dacites, and basalts (Basgall 1993a,b, 2000; Beck and Jones 1990b); in the northern Great Basin they are more often made of obsidian (Amick 1995), which is readily available in that area. They are rarely, however, made from chert. These points represent the final product of a biface reduction system that is standardized to such an extent that it is readily identifiable, even in assemblages containing later material (Figure 17.3).

It has been suggested by a number of scholars that bifacial technologies, such as those used by Paleoindians of North America, serve well in situations of high mobility and where a range of tool needs would select for flexibility (Goodyear 1989; Kelly 1988:719–720, 2001:67–68; Kelly and Todd 1988:237; Nelson 1991:66–76; Torrence 1989:63). Bifacial cores, while designed to enable production of a bifacial tool like a projectile point, constitute serviceable tools in themselves and provide for flakes of predictable morphologies to serve as expedient tools or blanks for further reduction. Bifacial technologies are not represented universally among mobile populations and are poorly known in ethnographic accounts (Kelly 1988, 2001). Still, as Hiscock (1994:277) suggests, where they evolved, bifacial technologies seem to have represented a response to the uneven distribution of lithic sources across the landscape (see also, Andrefsky 1994; Bamforth 1986:23–24; Kelly 2001:68–72; Kuhn 1994:40), serving as a tactic to reduce the risk of being caught unprepared, either without the proper tools or without a sufficient supply of toolstone.

It is certainly the case that bifaces served as the central component of the transportable toolkit of Paleoarchaic peoples of the Great Basin, as they did for Paleoindians elsewhere in North America. This is evidenced not only by the prevalence of bifaces in Paleoarchaic assemblages but also by the fact that the majority of flakes with platforms in these assemblages are biface reduction flakes.[3] In our work in eastern Nevada, for example, we have collected nearly 18,000 unmodified flakes from 15 surface sites. Only 5687 (32.2%) of these flakes have platforms but nearly 80% of these (n = 4415) are biface reduction flakes.[4]

In our studies, we have also noted that Paleoarchaic biface assemblages vary considerably with respect to the extent of reduction undertaken at different locales (Beck and Jones 1988 1990b, 1993, 1994b). In some cases, primarily middle-to-late reduction debris is predominant while in others, only early-to-middle reduction is evident. Here we investigate the variation in biface assemblages with respect to the extent of reduction undertaken at different locales using the predictions of central place foraging models, which explain variation in the field processing of resources as a function of travel time. With respect to lithic assemblages, variation in the reduction stages represented in different assemblages, especially those from quarries, is explained by the anticipated travel distance and the cost of transporting toolstone.

Central Place Foraging Models

In 1980, Binford introduced the notions of residential and logistical mobility in relation to how foragers move across the landscape and how they exploit resources. Residentially mobile foragers generally forage on a daily basis from their residential base, returning to that base at the end of the day; the residential base is moved as resources within the foraging radius are depleted. Logistical mobility, on the other hand, involves travel to procurement sites that are too far from the residential base to return the same day; in these cases small subgroups travel from the residential base to these locations and may remain there for several days or weeks. Whether organized residentially or logistically, foragers must transport resources from procurement sites back to their residential base. Although the need to transport resources from distant procurement sites has long been recognized as a conditioning variable in butchering and field processing techniques (e.g., White 1953, 1954), more recently archaeologists have attempted to quantify the costs of resource transport and how these costs may affect the types and quantities of food remains found in archaeological sites (e.g., Barlow and Metcalfe 1996; Bettinger et al. 1997; Binford 1978; Bird and Bliege Bird 1997; Drennan 1984a,b; Jones and Madsen 1989; Metcalfe and Barlow 1992; Metcalfe and Jones 1988; O'Connell et al. 1988, 1990; Rhode 1990). These studies have demonstrated that field processing reduces the amount of unusable material, thereby reducing transport costs of the load being carried back to the residential base. As a result, low-utility portions of a resource, such as nut hulls or certain animal parts, may be rare or absent at the residential site. The important implication is, of course, that the type and quantity of resources represented in the archaeological record of the residential base may not bear a direct relationship to the importance of those resources in the diet (Bird and Bliege Bird 1997:52–53).

Transport costs have been addressed using central place foraging models (e.g., Orians and Pearson 1979; Stephens and Krebs 1986), primarily with respect to differential transport of animal body parts (e.g., Binford 1978; Metcalfe and Jones 1988; O'Connell et al. 1988, 1990), but more recently for other resources as well (Barlow and Metcalfe 1996; Bettinger et al. 1997; Metcalfe and Barlow 1992). "Central place models posit that, holding technology and field conditions constant, optimal foraging behavior changes as a function of increasing round-trip travel time from central place to foraging location and back" (Bettinger et al. 1997:888). Generally, as the distance between central place and foraging location increases, field processing of resources becomes more cost-effective. The most familiar of these models, and probably the best suited to archaeological use (Bettinger et al. 1997:888), is that developed by Metcalfe and Barlow (1992) and Barlow and Metcalfe (1996). This model explores the trade-offs that foragers face when transporting resources comprised of components of different utility over long distances (Bird and Bliege Bird 1997:44). In the simplest case, a resource "package" is comprised of only two components, one of which has low or no utility. Metcalfe and Barlow (1992:342) offer nuts as an example, which consist of nutmeat contained in a valueless hull. A forager must decide whether removing the hulls at the procurement site is economically more beneficial than transporting unshelled nuts back to the residential base. The trade-off here is between the time spent processing in the field and the cost of transporting the unprocessed resource from the field site. If, for example, the hull equals half of the nut volume, then twice as much useful nutmeat can be transported in each load if the hulls are removed before transport, decreasing the amount of travel time expended per unit of useful material; on the other hand, field processing increases the amount of time expended per unit of useful material at the procurement site (Bettinger et al. 1997:888). If the distance to be traveled is great, the benefit of field processing is obvious in that fewer trips will

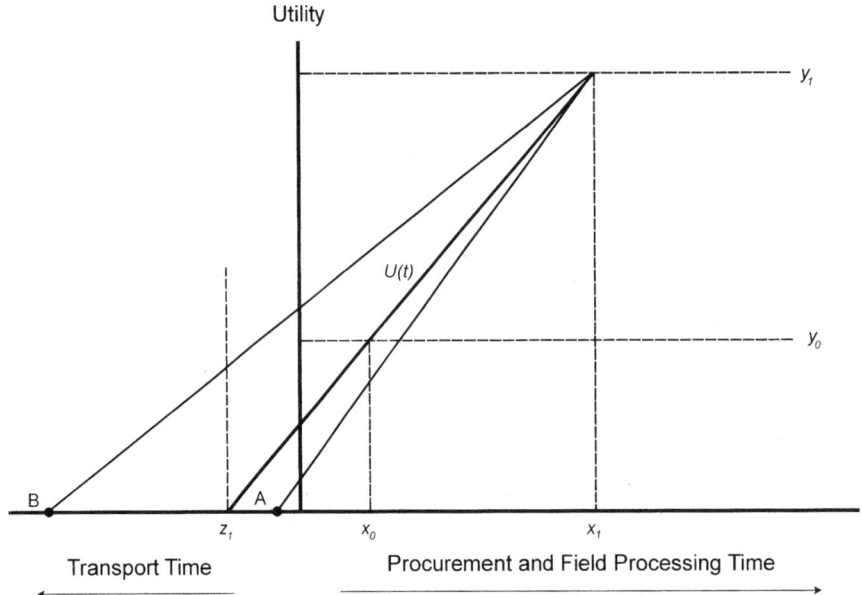

FIGURE 17.4. Central Place Foraging Model describing the relationship between field processing and resource utility. In this model a resource has two components. Removal of the low utility component incurs time costs while increasing the utility of the resource for transport. An increase in field processing time from x_0 to x_1 will increase resource utility from y_0 to y_1. The utility function $U(t)$ represents the relationship between the cost of field processing a resource and the increased utility of a load of that resource due to processing. The slope of this function predicts the travel time, z_1, at which field processing becomes cost-effective. (After Metcalfe and Barlow (1992, 346).)

have to be made, but if the distance is short, it may be more cost-effective to make the extra trips and process the nuts at the residential base.

The primary assumption of central place foraging models is that foragers will make economically efficient decisions concerning field processing and transport costs; but implied in this assumption is that "either processing time in camp has no cost, or no processing in camp is required for consumption or use" (Metcalfe and Barlow 1992:345; see also, Bettinger et al. 1997: 888). Metcalfe and Barlow (1992:345) go on to say, however, that "for estimating the order in which parts should be culled during field processing (for resource packages consisting of more than two components), and the *relative* (emphasis in original) travel times at which they should be removed and discarded, it is only necessary to assume that processing time in camp is less costly per unit than field processing time, or less time is required to process in camp than in the field, or both."

The decision of whether or not to spend time field processing a resource depends on (1) the field processing time for that resource; (2) the gain in resource utility due to field processing; and, (3) the distance to the central place (Bettinger et al. 1997: 888). Figure 17.4 presents a model of the two-component resource discussed above in which field processing is a discrete process; that is, the component of low utility is removed in the field, increasing the overall utility of the transported load. In exploring their model, Metcalfe and Barlow use a scale of proportional utility where utility is scaled from 1.0 to 0.0; 1.0 indicates the highest possible utility, 0.0 no utility at all (Metcalfe and Barlow 1992:354). For example, if nuts are composed of 30% nutmeat and 70% unusable shell, the nutmeat has a utility of 1.0 and the shell has a utility of 0.0. Therefore, an unprocessed load has a proportional utility of 0.3, while that of a processed load is 1.0.

The utility function $U(t)$ shown in Figure 17.4 represents the relationship between the cost of field processing a resource and the increased utility of a load of that resource due to processing; an increase in field processing time from x_0 to x_1 will increase resource utility from y_0 to y_1. The slope of this function predicts the travel time, z_1, (and thus travel distance) at which field processing becomes cost-effective (Bird and Bliege Bird 1997:44). There is an inverse relationship between the benefit derived from field processing a resource and the travel time at which field processing that resource becomes worthwhile (Metcalfe and Barlow 1992:347).

In testing this model among the Meriam of the Torres Strait Islands, Australia, Bird and Bliege Bird (1997) find good agreement between the predictions of the model and their observations. In their study, Bird and Bliege Bird observed procurement and processing of two-component resources: several species of shellfish comprised of high-utility flesh and low/no-utility shell. They find that, in general, the Meriam field process the low utility components of shellfish for transport "in a manner

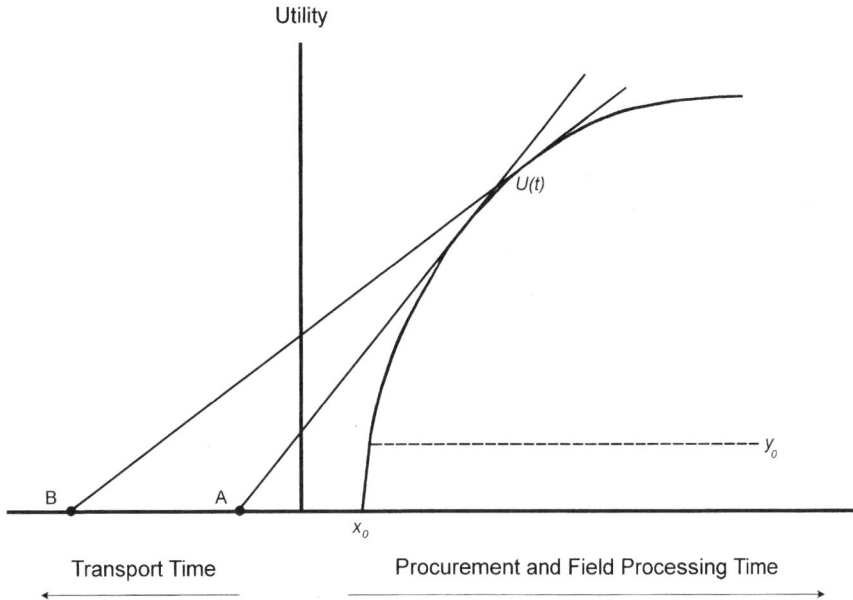

FIGURE 17.5. Central Place Foraging Model in which the utility function U(t) is not discrete but continuous and differentiable. This function describes a case of "diminishing returns" in which the gain in utility per increment of processing decreases with increased field processing time. The slope of a line tangent to the curve predicts the travel time at which a certain level of field processing is cost-effective. Points A and B represent two hypothetical residential camps. (After Metcalfe and Barlow, 1992, 349.)

that maximizes the rate at which they deliver edible flesh to a central locale" (Bird and Bliege Bird 1997:52). They also observe that shellfish species that are relatively difficult to field process, for which field processing adds little to the proportion of usable meat, or are collected near the residential base or temporary camp, were not field processed but transported whole (Bird and Bliege Bird 1997:52).

As Metcalfe and Barlow (1992:348) note, however, "utility functions for some resources may be continuous and differentiable," as depicted in Figure 17.5. "The question here is not whether a low-utility component of a resource is expected to be transported, but rather how much of it will be transported" (Metcalfe and Barlow 1992:348). As is the case in Figure 17.4, the slope of a line tangent to the curve in Figure 17.5 predicts the travel time at which a certain level of field processing becomes cost-effective. This model predicts that if foragers must travel to point B, then they will process the resource to a greater degree in the field than if they were traveling to point A.

The procurement of food resources, however, rarely occurs in the absence of technology (Grayson and Cannon 1999:146) and thus the costs and benefits of raw material acquisition, processing, and transport must figure into the overall evaluation of a particular foraging strategy. As Kelly (2001:68) notes, to spend time acquiring and processing toolstone at the quarry, a forager must give up the opportunity to do something else, which can exact a cost. In presenting their model, Metcalfe and Barlow

(1992:341) suggest that the same trade-off between field processing and transport that had been evident for some time regarding butchering and field processing prey "should be evident for other resources procured in packages that include useful and useless (or less useful) parts." Although Metcalfe and Barlow concentrate their efforts on food resources, they note the model's applicability to toolstone, a "packaged" resource (1992:352). Several researchers have begun to explore the use of the Metcalfe and Barlow model to investigate procurement, processing, and transport costs of lithic toolstone (e.g., Elston 1990, 1992a:153–174, 775–801; Kelly 2001:65–76). We draw on these studies below in our use of this model in an attempt to explain variability in Paleoarchaic biface assemblages at quarry and associated residential sites.

MODELING BIFACE REDUCTION AND TRANSPORT COSTS

Figure 17.5 shows a utility function that increases with increasing processing effort but at a decreasing rate. Biface reduction has been described as a continuous process (e.g., Ingbar et al. 1989; Magne 1985; Muto 1971; Raab et al. 1979; Shott 1996b; Sullivan and Rozen 1985) and presumably is well modeled by such a utility function. In practical terms, however, most archaeologists segment this continuum into a series of stages (e.g., Bloomer et al. 1992; Bolduroian 1990; Bradbury and Carr 1999; Johnson 1981, 1984, 1989; Magne and Pokotylo 1981; Nami 1999); it is certainly the case that, prehistorically, mobile populations often

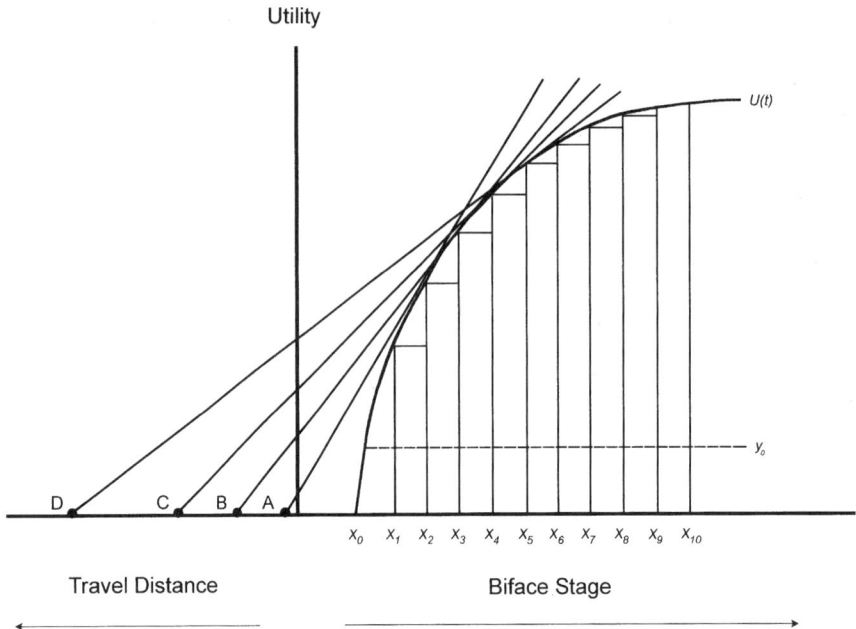

FIGURE 17.6. Central Place Foraging Model describing biface reduction revised from the continuous and differentiable model depicted in Figure 17.5. In this case, field processing is a discrete, multi-step process. The x-axis to the right represents discrete biface stages; the x-axis to the left depicts travel distance, which is implied by travel time (see Bettinger et al., 1997; Bird and Bliege Bird, 1997; Jones and Madsen, 1989). A line tangent to the curve predicts the travel distance at which reduction of a biface to a particular stage is cost-effective. Points A–D represent hypothetical residential bases.

reduced bifaces to various degrees short of the finished product for transport (Pecora 2001:175). The stages of reduction can thus be represented as points along the utility curve $(x_0, x_1; \ldots; x_n)$, as depicted in Figure 17.6. The slope of a line tangent to the curve at any of these points predicts the travel time, and therefore travel distance, at which field processing to that particular stage is worthwhile.

Using an optimal load size (that is, the maximum volume or weight that can be economically carried by an individual [Bettinger et al. 1997:892; Jones and Madsen 1989:529–530]), the travel time at which field processing a resource will increase its utility for transport can be calculated (see Bettinger et al. 1997; Metcalfe and Barlow 1992 for mathematical details). The data necessary for this calculation include procurement time for a load of unprocessed resources, the time required to procure and process a load of resources, the utility of the unprocessed load, and the utility of the processed load. The utility of plant and animal food resources is measured in calories per unit weight or volume,[5] while travel time, procurement time, processing time, and optimal load size are estimated on the basis of ethnographic descriptions or actualistic studies (e.g., Barlow and Metcalfe 1996; Bettinger et al. 1997; Binford 1978; Bird and Bliege Bird 1997; Bunn et al. 1988; Jones and Madsen 1989; O'Connell et al. 1988, 1990; Rhode 1990; see Grayson and Cannon (1999) for a discussion of the difficulties involved in arriving at these and other measures used in optimal foraging models). Comparable estimates for the acquisition and processing of toolstone,

however, are not available. Ethnographic observations, for instance, rarely include information concerning the procurement and processing of toolstone at a quarry, especially procurement and processing times. Some work has been done on the amount of toolstone used by an individual over the period of a year (e.g., Gould 1977; Hayden 1977), but no data of which we are aware exist on optimal load size for an individual procurement event. As Kelly (2001:67) notes, understanding the costs related to toolstone procurement, processing, and transport is hindered by the fact that "there are no peoples left who make extensive use of stone tools." Whereas the hunting and butchering of animals as well as differential transport of animal parts can still be observed among some hunter-gatherer groups (although in a modern context), the data concerning these activities with respect to toolstone will have to be generated experimentally.

Elston and his colleagues (Carambelas and Raven 1991; Elston 1992a) conducted several actualistic experiments using toolstone from the Tosawihi quarries in northeastern Nevada. To the extent possible, we use the results of these experiments in our discussions below. These results, however, do not give us the necessary data to calculate specific travel times for toolstone types at which field processing becomes cost-effective. We must, instead, examine archaeological cases to see if they conform to the *qualitative* predictions (Kaplan and Hill 1992) of the model in Figure 17.6, using the experimental results as supportive data. We begin by examining if our four assemblages are appropriately modeled by the same utility function.

Toolstone Utility

In simplest terms, the measure of toolstone utility would seem to be the amount of usable stone in a core, but defining what is usable stone is problematic. Bettinger et al. (1997:888), in a hypothetical example, suggest that in the manufacture of a biface, half of the core ends up as waste (and thus unusable) material; therefore a raw cobble has half the utility per unit weight as the final product, the biface. The Tosawihi experiments suggest that as much as 98% of the extracted raw material ends up as waste (Elston 1992a:787). In reality, however, toolstone utility is likely somewhat more complex than Bettinger et al. (1997) suggest. "Sources may be variable in their utility depending upon the form of different technological organizations" (Ingbar 1994: 55), and thus toolstone utility is related to a number of different factors.

First, some of what is termed waste, in fact, may be usable material and serve as expedient tools or blanks for additional reduction (Elston 1992a:787). Thus the utility of an unflaked cobble of toolstone relates to the intention for the final product(s), that is, whether the cobble is being reduced only to produce blanks, to produce a finished biface for a particular function, or is to be reduced to a certain stage for transport. In the latter case, the biface may serve as a core from which flakes can be derived for use as informal tools or as blanks, but also can itself be used as is or further reduced into a formal tool.

Second, there is not necessarily a one-to-one relationship between core and biface. Depending upon the toolstone package size, i.e., boulder versus cobble, that package may produce many biface blanks or may actually *be* the biface blank. Therefore, toolstone utility will vary according to the size of the toolstone package.

A third factor that may affect toolstone utility concerns the functional requirements of the tool being manufactured. Some tools require particular kinds of raw material. Archaeological patterns, for instance, suggest that a preference was shown for high-quality cherts when manufacturing fluted points, although they were sometimes made from obsidian as well (Beck and Jones 1990b; Goodyear 1989; Kelly 1988; Kelly and Todd 1988). These tools, however, were never made from fine-grained volcanics, such as andesite, dacite, or basalt (Beck and Jones 1990b, 1997). Given this apparent functional requirement, an artisan will choose to travel a long distance to obtain suitable raw material and pass up unsuitable materials closer to hand. Thus, utility in this case is related to the tool being manufactured; that is, dacite will never have high utility when the goal is the manufacture of fluted points.

Finally, the quality of the raw material will affect toolstone utility. Fracture mechanics, even within material types (e.g., chert), can vary considerably and thus affect search, procurement, and processing costs. The Tosawihi experiments suggest that assaying and processing costs increase substantially for low-quality toolstone (Elston 1992a:787; see also, Kelly 2001:69)

It is important, however, to distinguish between the utility of the *toolstone* and the utility of the *tool*. A number of authors have addressed the latter issue (e.g., Elston 1992b:40–42; Gould and Saggers 1985:129–134; Kelly 2001:68–72; Kuhn 1994; Shott 1996a:270–273), suggesting various ways in which this variable might be quantified, from the number of useful strokes of a tool per unit weight or volume of used stone (Gould and Saggers 1985:131) to the potential for producing fresh edges combined with the length of those edges (Kuhn 1994:429). Shott (1996a: 270) suggests employing two measures, maximum utility, which is equal to the average maximum use-life of specimens in a tool class, and realized utility, the actual period of use of an individual specimen. There is an important difference, however, between the *resource* (e.g., pinon nuts, toolstone) and the *use* of the resource (e.g., food, tools). When considering the acquisition of toolstone at a quarry, factors such as the number of strokes that are eventually accomplished with a tool are not relevant; what *is* relevant is simply optimizing the amount of useful stone that can be economically transported from the quarry. Given the complexities mentioned above, then, how might this quantity be measured?

The factors outlined above that likely have an effect on toolstone utility are:
1. the intended intermediate and final tool morphologies;
2. the size of the toolstone package;
3. the functional requirements of the intended product; and,
4. the quality of the toolstone.

It is obvious from these factors that different kinds of toolstone (e.g., chert versus dacite) can have different utilities and thus a single "currency" is not appropriate for all. By holding the type of toolstone constant (i.e., only considering dacite), then two sources of variation are largely eliminated, the functional requirements of the intended product and toolstone quality. As stated earlier, Paleoarchaic stemmed points are almost always manufactured from fine-grained volcanic toolstones, except in those areas where obsidian is plentiful; they are rarely made from chert. In the case considered here, a comparison is made between two dacite quarries where material is of comparable quality and where the manufacturing trajectory leads to a single final product—large stemmed points; thus the qualities that make chert attractive for certain tools are irrelevant.

The problem caused by difference in toolstone package size can be approached through experimental studies that focus on finding the average number of blanks of a certain size that can be produced from a single package. Considering quarries where the package size is similar, as is the case for the two dacite quarries considered below, eliminates this problem altogether. We are left, then, with the most difficult source of variation, of course, the intended product(s) of toolstone reduction.

Intent in prehistoric lithic reduction is often unknowable. As stated above, however, the reduction sequence for Paleoarchaic stemmed points is highly standardized (Figure 17.3) and identifiable even in temporally mixed assemblages. This biface trajectory dominates the assemblages at both of the dacite quarries considered here; expended and broken fragments of finished stemmed points are also present, representing specimens broken when very nearly completed or exhausted pieces made of another source material that were discarded. Therefore we are confident that the toolstone at both quarries can be modeled using the same utility function.

The question remains, however, is the biface reduction sequence considered here validly modeled by the diminishing returns utility curve as suggested above? "Flintknapping is fundamentally a reductive process" (and as such,) "places predictable and repetitive size constraints on the byproducts produced" (Ahler 1989b:89; see also, Ahler 1989a:205–210). As reduction continues, "the maximum possible size as well as the average size of the flake byproducts should decrease progressively as the tool itself becomes progressively smaller" (Ahler 1989b:89; see also, Ahler 1989a:205–210; Stahle and Dunn 1982:86). Newcomer (1971) demonstrated this relationship experimentally over 30 years ago, and subsequent studies have supported his conclusions (e.g., Ingbar et al. 1989:124; Pecora 2001:179–180; Stahle and Dunn 1982, 1984). It is true that small flakes will likely dominate numerically throughout the reduction process; however, individual large flakes will weigh considerably more than individual small flakes and thus will dominate the average flake weight in stages where they are most numerous (Ahler 1989b: 90; Stahle and Dunn 1982:86–87).

The reduction in flake weight, however, is not linear but curvilinear (e.g., Ahler 1989b; Newcomer 1971:92; Stahle and Dunn 1982:89), with weight diminishing quickly during the earliest stages of reduction, but slowing considerably as reduction proceeds. Taking the simplest case, that a single cobble is used to produce a single biface, then toolstone utility can theoretically be measured in terms of the weight of the finished biface (e.g., Elston 1992a:791). Using the Bettinger et al. (1997) example from above, that 50% of the core ends up as waste, then toolstone utility should increase with the removal of each waste flake. Because the flakes removed in initial reduction will be, on the average, of the greatest weight, then the increase in utility per unit of weight will be greatest with initial reduction, and decline fairly quickly in conjunction with flake weight, just as suggested by the diminishing returns utility curve in Figure 17.5. This relationship is supported by the Tosawihi experiments (e.g., Elston 1992a:792).

In sum, central place foraging models provide explanations for certain kinds of variation in resource procurement and processing. Specifically, whether or not a forager chooses to process a resource in the field to increase the utility of that resource for transport depends on (1) field processing time, (2) the increase in resource utility due to processing, and (3) the distance to the central place. The models predict that there is a travel time for each resource beyond which it is profitable to spend time field processing to eliminate low-utility components. As our discussion suggests, biface reduction at the two Paleoarchaic quarries considered here can validly be modeled by the diminishing returns utility curve in Figure 17.5; analytically, however, it is more useful to segment this continuum into a set of stages, as modeled in Figure 17.6.

Although we cannot provide specific travel times for the cases presented here, we can use the model in Figure 17.6 to predict directional tendencies (Kaplan and Hill 1992) in the assemblages from the two quarry and residential site pairs in which the distance between the members of each pair is substantially different. In one case the quarry is only 9 km from the residential site, while in the other case this distance is 60 km. The Tosawihi experiments suggest that when the distance between quarry and residential site is less than 10 km, high proportions of early-stage bifaces and debitage are expected at the quarry, but if this distance is greater than 10 km, a greater proportion of middle-to-late stages will dominate (Elston 1992a:798). Therefore, we would expect considerably more reduction of bifaces at the quarry in the latter case than in the former.

The Central Great Basin Database

The four assemblages examined in this analysis were collected as part of fieldwork done by Beck and Jones over the last 15 years in eastern and central Nevada (e.g., Beck and Jones 1988, 1990a,b, 1993, 1994a,b; Jones and Beck 1990; Jones et al. 1997; Jones et al. 1996; Jones et al. 2003; Huckleberry et al. 2001). The work was initiated with the purpose of studying Paleoarchaic land use in these areas.

The Cowboy Rest Creek Quarry is located in Grass Valley, central Nevada (Figure 17.1). This quarry occurs on an extensive alluvial fan containing abundant large cobbles. Data were collected from two areas of high artifact density, the first (Locality 1) in 1999 and the second (Locality 2) in 2001. At Locality 1, all bifaces within a 50 × 50 m area were collected with point provenience using an electronic total station; debitage and other artifacts were collected from four randomly selected grids within this 2500 m² area. Several thousand artifacts were collected from this quarry, but currently, total artifact counts are available only for bifaces (n = 58).

At Locality 2, which lies approximately 1 km northwest of Locality 1, analysis was done in the field. All bifaces—as well as all artifacts of chert ($n = 47$) and obsidian ($n = 4$) and two possible hammerstones—within an approximately 150 × 250 m area were flagged. Each artifact was given a point provenience and in-field analysis was completed for each of the 613 dacite bifaces.

Lying approximately 9 km to the northeast of Cowboy Rest Creek Quarry is the Knudtsen Site, which occurs on an east-west oriented spit reaching out onto the valley floor. This site is represented by pockets of artifacts extending almost the entire length of the spit, perhaps for 3 km. Artifact density is highest at the eastern end of the spit and this is where we focused our work. Two separate localities, termed Knudtsen 1 and Knudtsen 2, were collected in 1999. All bifaces at Knudtsen 1 ($n = 215$), an area of 200 × 140 m, were given point provenience while a 10% sample of the area was randomly selected for grid collection.

Knudtsen 2 was a dense artifact concentration that contained an extraordinary number of artifacts within a 60 × 100 m area. Again, bifaces (*n* = 656) were collected with point provenience while all other artifacts within the 6000 m² area (*n* = 59,806) were collected within 2 × 2 m units.

Given the proximity of Knudtsen to Cowboy Rest Creek Quarry, we have made the assumption that these bifaces are made from Cowboy Rest Creek dacite. Upon reviewing the ethnographic literature, Kelly (1995:133) suggests that a 20–30 km round-trip is the maximum distance that foragers will travel in daily food collection efforts. The round-trip distance between Knudtsen and Cowboy Rest Creek Quarry (18 km) lies below this maximum, suggesting the quarry lies within the foraging range of Knudtsen and that most of the dacite from this site was carried there from the Cowboy Rest Creek Quarry. In addition, the Cowboy Rest Creek dacite is very uniform and exhibits a distinctive red weathering surface, both of which are evident in the dacite bifaces from Knudtsen, adding to our confidence that the material at this site is from the Cowboy Rest Creek Quarry. Finally, the exhausted state of obsidian bifacial tools at Knudtsen indicates that people arrived at that site with largely depleted toolkits in need of refurbishment and Cowboy Rest Creek Quarry was the nearest at hand.

The Little Smoky Quarry is located in the southeastern portion of Little Smoky Valley in eastern Nevada (Figure 17.1) and was first described by Price (1989). This quarry also coincides with an extensive alluvial fan containing dacite cobbles. Collection was made in a 480 m² area (Locality 1), selected because of its high artifact density (Beck and Jones 1994b). A total of 309 artifacts were collected, 130 of which are bifaces. A second area (Locality 2) was examined in 2001; as in the case of Cowboy Rest Creek Quarry, Locality 2 analysis was done in the field and the artifacts were not collected. This area was also chosen because of high artifact density and lies adjacent to the previous collection area. All bifaces within an 1800 m² area were analyzed. Of the 210 bifaces analyzed, 200 are made of dacite; the remaining 10 are made from obsidian (*n* = 7) and chert (*n* = 3).

Although extensive systematic archaeological surveys have not been conducted in northern Little Smoky Valley, the work that has been completed in the valley has not identified any Paleoarchaic site comparable to the Knudtsen Site. A site approximately 15 km north of Little Smoky Quarry, Black Point, exhibits a small Paleoarchaic assemblage but is dominated by late Archaic artifacts (Price and Johnston 1988). While there are a number of Paleoarchaic sites to the southeast in Railroad Valley (Zancanella 1988), which lies about 65 km from Little Smoky Quarry, we have not analyzed the raw materials from these sites. A good quality dacite source, however, occurs on the Duckwater Indian Reservation, which lies between Railroad Valley and Little Smoky Quarry; we suspect the majority of the Railroad Valley assemblages are comprised of this material.

The closest substantial Paleoarchaic site to Little Smoky Quarry is Limestone Peak Locality 1 (henceforth referred to as Limestone Peak-L1), located ca. 60 km to the east of Little Smoky Quarry in southwestern Jakes Valley (Figure 17.1). In 1991 a total of 6932 artifacts were collected from the site surface. Although this assemblage contains 268 fine-grained volcanic bifaces, only those that could be identified as having come from Little Smoky Quarry (*n* = 54) were used in the present analysis. Little Smoky Quarry dacite exhibits a distinctive pattern of large phenocrysts (ca. 1 mm in diameter), spaced about 5–10 mm apart. None of the other 20 dacite and andesite sources we have examined from this region and characterized geochemically (Jones et al. 1997) possess this morphology, and therefore we are confident of our sample selection.

A total of 1925 dacite bifaces from four sites were examined in this analysis. Table 17.1 shows the number of bifaces represented in each of the site assemblages.

TABLE 17.1. Total number of dacite bifaces analyzed at each of the four sites studied.

Site Name	Site Type	Location	No. of Bifaces
Cowboy Rest Creek-Loc. 1	Quarry	Grass Valley	58
Cowboy Rest Creek-Loc. 2	Quarry	Grass Valley	613
Knudtsen-Locality 1	Residential	Grass Valley	215
Knudtsen-Locality 2	Residential	Grass Valley	656
Little Smoky Quarry-Locality 1	Quarry	Little Smoky Valley	130
Little Smoky Quarry-Locality 2	Quarry	Little Smoky Valley	200
Limestone Peak Locality 1	Residential	Jakes Valley	54

Analytical Protocol

Our arguments center on the extent of biface reduction undertaken at each of the quarries and residential sites. For this evaluation, we employ two measures of reduction stage: a traditional stage classification based on that devised by Callahan (1979) and a biface thinning index proposed by Johnson (1981). Traditional stage classifications are often criticized as problematic because they segment what are believed to be theoretical continua (Teltser 1991:366). There is some indication, however, that there are sometimes empirical disjunctions in the reduction continuum, especially as the knapper changes from hard to soft hammer percussion (see Pecora 2001), but these disjunctions do not necessarily occur at the same point in every case. Therefore stage assignment can still be somewhat arbitrary. The advantage of stage classifications is that they are generally based on multiple criteria and thus assignment of an artifact to a stage is not based on a single variable.

The Johnson Thinning Index has the advantage of measuring reduction along a continuum but is based on only two criteria:

TABLE 17.2. Biface stage determinants (after Callahan, 1979).

STAGE	FEATURES
0	Large biface with irregular shape and low symmetry; few very widely and/or variably spaced flake scars; very wide edge offset (very sinuous); very thick and irregular cross-section
1	Large biface with irregular shape; widely and/or variably shaped flakes; wide offset; thick and irregular cross-section
2	Large biface with semi-regular and symmetrical shape; closely and/or semi-regularly spaced flake scars; edge offset moderate; cross-section semi-regular
3	Regular, symmetrical biface; closely and/or quite regularly spaced flake scars (pressure flaking often present); offset close (little edge sinuosity); cross-section thin and regular; later edge grinding evident on haft

weight and plan area. These two independent approaches, however, complement one another and act to limit bias that may be introduced by simply using one or the other.

Stage Classification

We began our laboratory analyses by attempting to determine the dominant patterns of reduction for the quarry and associated site assemblages, identifying each biface according to its stage of manufacture. In theory, a biface is processed through several stages of reduction before it reaches the form of the final product—in our case, a stemmed point. The classification used here creates four stages of reduction (Figure 17.3, Table 17.2), modified from those of Callahan (1979:10–11). Stage definitions are based on form, number and shape of flake scars, edge sinuosity, and thickness.

Each of the 1925 bifaces in the sample was examined by two analysts, and these results were compared to test the reliability of the classification. Analysts consistently agreed on stage assignments for the Cowboy Rest Creek Quarry bifaces, but there was some disagreement regarding assignments for the other assemblages. In reviewing the results it was evident that one analyst relied more heavily on edge sinuosity to make stage assignments while the second analyst emphasized flaking patterns. Indeed, these features do not precisely change in tandem (that is, sinuosity may decrease more quickly than symmetry in one case while the reverse may be true in another). As a consequence, even though each analyst agreed on the assignment of those bifaces exhibiting the modal character of each stage, they judged some bifaces differently. To correct for the differences, the assignments of each biface from each assemblage were averaged, creating a set of intermediate stages and increasing the number of classes to seven.[6]

The Johnson Thinning Index

As an alternative to stage classification, Johnson (1981) developed a thinning index (JTI) for identification of the trend in biface reduction from early-stage to late-stage manufacture. The index, which is computed as the ratio between weight and plan surface area, is based on the hypothesis that a manufacturer would maximize biface surface area while minimizing thickness; thus, early-stage bifaces are considerably thicker than later-stage bifaces. As Johnson (1981:13) points out, however, "small, early stage bifaces are likely to have the same thickness as large, nearly complete bifaces." To control for the effect of difference in initial blank size, Johnson suggests that the ratio of weight to "plan view area" is a more useful way to compare overall thinning and, therefore, stage of manufacture (Johnson 1981:13). In theory, weight decreases at a faster rate than plan area and thus the decrease in the JTI is not linear but curvilinear, with values leveling off as reduction nears completion. Conceivably, a number of variables can be measured to track size changes in biface manufacture but many, such as length, cannot be applied to broken specimens. An advantage of the JTI is that this ratio can be applied equally well to whole and fragmentary bifaces.

The JTI was computed for each biface in the quarry and residential site assemblages by weighing it and calculating its plan view area. Weight was determined in the laboratory using a standard digital balance; in the field, weight was measured using a mechanical balance. Area was determined for the bifaces at Knudtsen, Limestone Peak-L1, Cowboy Rest Creek Quarry, Locality 1, and Little Smoky Quarry, Locality 1 according to the method described by Johnson (1981). Each of these bifaces was placed on a standard office photocopier and copied. Because the light projects the shape of the biface onto a two-dimensional surface, this method reduces error. Each biface was then compared to its respective image to ensure accuracy. A transparent 0.5 cm² grid was placed over the copied image of each biface. Where edges were irregular, measurement units were approximated to fit the curve. The resultant count of squares was converted to cm².

For field-based analyses at Cowboy Rest Creek Quarry, Locality 2 and Little Smoky Quarry, Locality 2 bifaces were carefully traced onto graph paper (0.5 cm²) and the same method as above was followed for calculating their plan view areas and the JTI.

As stated above, two different collections were made at the two quarries and at the Knudtsen site. Before combining each with its respective assemblage, we must examine if they are statistically identical. In this, as well as subsequent analyses, two statistical approaches are used to make comparisons: stage assignments are compared using the two-sample Kolmogorov-Smirnov (K-S) test and JTI values are compared using t tests.

The distribution of biface stages in the two pairs of quarry assemblages is shown in Figure 17.7. In general the assemblage pairs are quite similar. This observation is confirmed by a two-sample K-S test, which fails to detect significant differences between the two localities at either Cowboy Rest Creek Quarry or Little Smoky Quarry (Table 17.3). Likewise, when JTI values from the two localities at Little Smoky Quarry (Figure 17.8) are compared by t test (separate variance test), the two assemblages

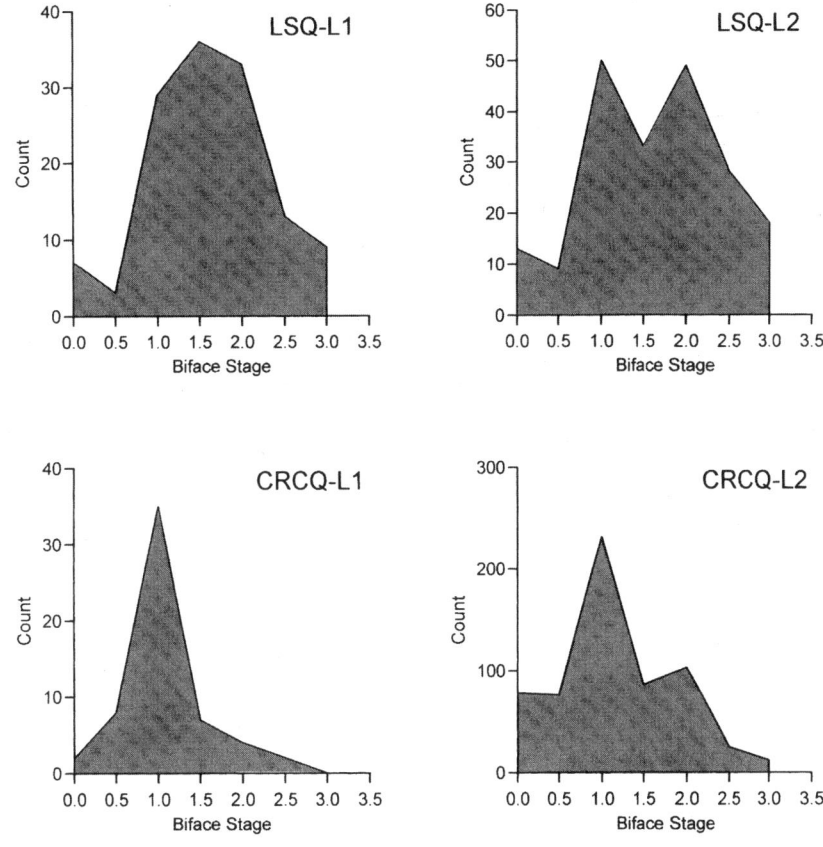

FIGURE 17.7. Profile summary graphs of biface stages represented in the Little Smoky Quarry, Localities 1 and 2 (LSQ-L1, LSQ-L2) and Cowboy Rest Creek Quarry, Localities 1 and 2 (CRCQ-L1, CRCQ-L2) assemblages.

are statistically identical. The test between the two localities at Cowboy Rest Creek Quarry, however, indicates that the two assemblages are statistically dissimilar. Based on these results, we have chosen to combine the Little Smoky Quarry assemblages in later analyses but will treat the Cowboy Rest Creek Quarry assemblages separately.

The Knudtsen Site assemblages were also compared with respect to biface stage and JTI (Table 17.3). In regards to the distribution of stages, the two assemblages cannot be differentiated statistically. A comparison of the JTI means, however, indicate a significant difference between Knudtsen 1 and Knudtsen 2. Thus, we treat the two Knudtsen localities separately in subsequent analyses.

Transport Distance and Quarry Behavior

The central place foraging model in Figure 17.6 describes and explains the relationship between the effort expended at a lithic quarry and the distance from that quarry to a foraging or residential site. This model predicts that foragers will invest greater or lesser time processing toolstone at a quarry depending on how far they intend to travel upon their departure. In the case of the assemblages explored here, those distances are, in one case, 9 km, and in the other, 60 km. The model in Figure 17.6 predicts a greater degree of field processing would be expected at

TABLE 17.3. Statistical tests between different collection samples at the two quarries and at the Knudtsen site.

Comparison	Separate Variances t Test of JTI			K–S Test Biface Stage
	t	df	p	p
Little Smoky Quarry, Localities 1 and 2	−1.693	227.8	0.092	0.936
Cowboy Rest Creek Quarry, Localities 1 and 2	−2.090	103.4	0.039	0.223
Knudtsen 1 and Knudtsen 2	2.027	268.4	0.044	0.930

Little Smoky Quarry before traveling the 60 km to Limestone Peak-L1 than at Cowboy Rest Creek Quarry, which is only 9 km from Knudtsen. As stated earlier, the Tosawihi experiments suggest that if the quarry is less than 10 km from the residential site, foragers will likely conduct minimal field processing and thus transport a large portion of low-utility material to the residential site (Elston 1992a:798). We expect, therefore, the assemblages at the two quarries to differ substantially with respect to the stages of reduction represented; that is, Little Smoky Quarry should

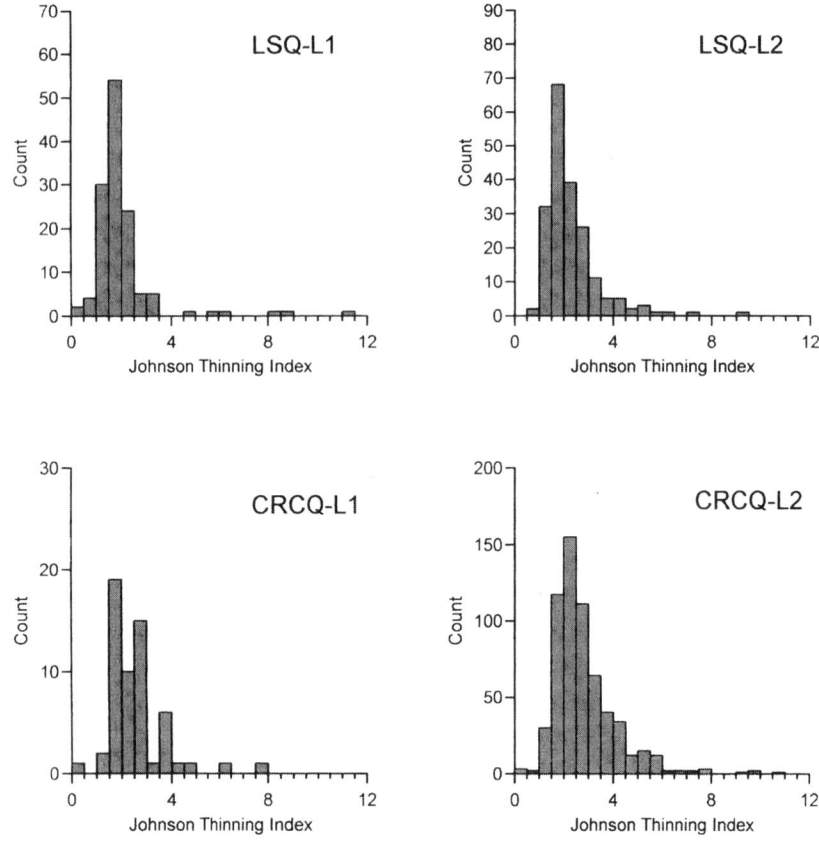

FIGURE 17.8. Histograms of Johnson Thinning Index (JTI) values for bifaces in the Little Smoky Quarry, Localities 1 and 2 (LSQ-L1, LSQ-L2) and Cowboy Rest Creek Quarry, Localities 1 and 2 (CRCQ-L1, CRCQ-L2) assemblages.

TABLE 17.4. Biface stages represented in the Little Smoky Quarry and Cowboy Rest Creek Quarry assemblages.

Biface stage	Little Smoky Quarry		Cowboy Rest Creek Quarry, Locality 1		Cowboy Rest Creek Quarry, Locality 2		Cowboy Rest Creek Quarry, Localities 1 and 2	
	No.	%	No.	%	No.	%	No.	%
0	20	6.1	2	3.4	78	12.7	80	11.9
0.5	12	3.6	8	13.8	76	12.4	84	12.5
1.0	79	23.9	35	60.3	233	37.9	268	39.9
1.5	69	20.9	7	12.1	86	14.5	93	13.9
2.0	82	24.9	4	6.9	103	16.8	107	16.0
2.5	41	12.4	2	3.4	25	4.8	27	4.0
3.0	27	8.2	0	0.0	12	2.0	12	1.8
Total	330	100.0	58	100.0	613	100.0	671	100.0

exhibit a greater number of bifaces at later stages of reduction than Cowboy Rest Creek Quarry. This is exactly what we find.

Figure 17.7 shows the distribution of biface stage classes at both quarries and illustrates how each quarry differs from the other. The Little Smoky Quarry assemblage is more heavily weighted towards middle and late-stage bifaces than either of the Cowboy Rest Creek Quarry samples. As Table 17.4 shows, nearly two-thirds (64.3%) of the bifaces at Cowboy Rest Creek Quarry fall in the first three reduction stages while only a third (33.6%) of Little Smoky Quarry bifaces represent stages 0–1.

This pattern is also evident in the distribution of the JTI (Figure 17.8). JTI values decrease with increasing reduction and, as Table 17.5 shows, the Little Smoky Quarry mean for this variable is much smaller than that at Cowboy Rest Creek Quarry. These differences are confirmed by statistical comparison (Table 17.6).

It seems clear from these results that Paleoarchaic knappers staged lithic tool manufacture differently at these two workshop localities, carrying reduction further at Little Smoky Quarry than at Cowboy Rest Creek Quarry. It follows that their companion residential sites should exhibit complementary re-

TABLE 17.5. Johnson Thinning Index statistics for Little Smoky Quarry and Cowboy Rest Creek Quarry.

Statistic	Little Smoky Quarry	Cowboy Rest Creek Quarry, Locality 1	Cowboy Rest Creek Quarry, Locality 2	Cowboy Rest Creek Quarry, Localities 1 and 2
N	327	58	613	671
Mininum	0.14	0.15	0.37	0.15
Maximum	11.02	7.66	27.86	27.86
Mean	2.18	2.60	2.98	2.94
SD	1.25	1.17	2.27	2.20

TABLE 17.6. Statistical tests between Little Smoky and Cowboy Rest Creek Quarry assemblages.

Comparison	Separate Variances t Test of JTI			K–S Test Biface Stage
	t	df	p	p
Little Smoky Quarry and Cowboy Rest Creek Quarry, Locality 1	−2.494	81.5	0.015	0.000
Little Smoky Quarry and Cowboy Rest Creek Quarry, Locality 2	−6.930	937.3	0.000	0.000
Little Smoky Quarry and Cowboy Rest Creek Quarry	−6.980	974.8	0.000	0.000

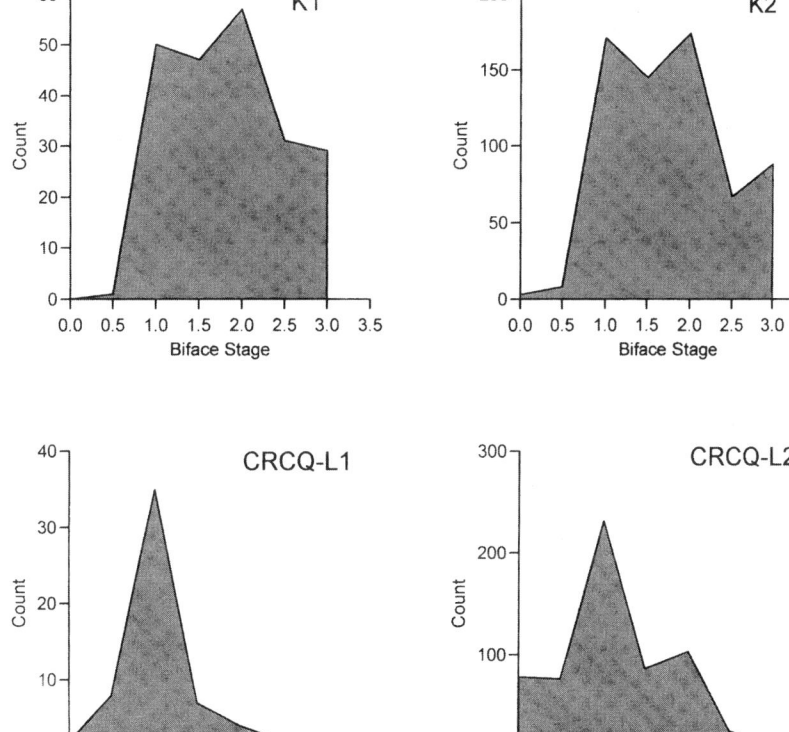

FIGURE 17.9. Profile summary graphs of biface stages represented in the Knudtsen, Localities 1 and 2 (K1, K2) and Cowboy Rest Creek Quarry, Localities 1 and 2 (CRCQ-L1, CRCQ-L2) assemblages.

duction signatures. That is, the biface assemblage at each residential site should be dominated by later-stage products than its associated quarry (Elston 1992b; Johnson 1989:20–22; Metcalfe and Barlow 1992:352). It does not necessarily follow, however, that Knudtsen and Limestone Peak-L1 will differ from one another in the same way as their respective quarries differ. The assemblage composition at these two sites will be governed as much by the on-site activities and the anticipated travel distance upon leaving as by the material brought from the quarry. This is a complex issue not easily dealt with here but a problem for future research.

A comparison of the distribution of biface stages confirms that the Knudtsen assemblages contain more later-stage bifaces than those from Cowboy Rest Creek Quarry (Figure 17.9; Table 17.7). This difference is mirrored by the JTI means from Knudtsen and Cowboy Rest Creek Quarry (1.20 and 2.94, re-

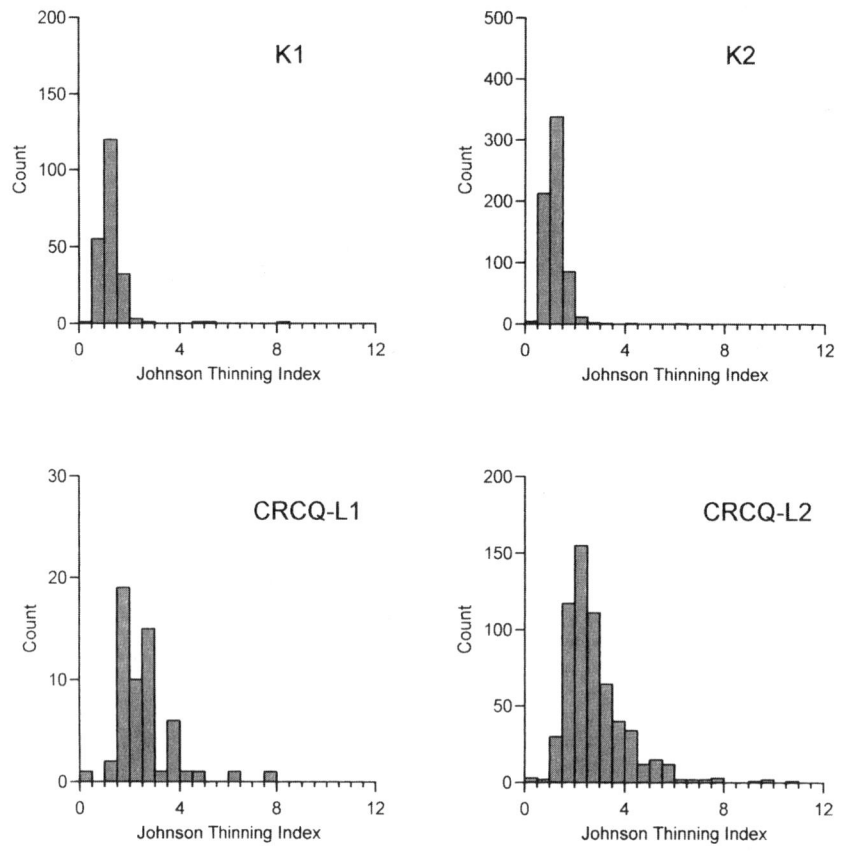

FIGURE 17.10. Histograms of Johnson Thinning Index (JTI) values for bifaces in the Knudtsen, Localities 1 and 2 (K1, K2) and Cowboy Rest Creek Quarry, Localities 1 and 2 (CRCQ-L1, CRCQ-L2) assemblages.

TABLE 17.7. Biface stages represented in all four assemblages.

BIFACE STAGE	COWBOY REST CREEK QUARRY		KNUDTSEN		LITTLE SMOKY QUARRY		LIMESTONE PEAK-L1	
	No.	%	No.	%	No.	%	No.	%
0	80	11.9	3	0.5	20	6.1	0	0.0
0.5	84	12.5	9	1.0	12	3.6	3	5.5
1.0	268	39.9	221	25.4	79	23.9	14	25.9
1.5	93	13.9	192	22.1	69	20.9	17	31.5
2.0	107	16.0	231	26.5	82	24.9	13	24.1
2.5	27	4.0	98	11.3	41	12.4	5	9.3
3.0	12	1.8	117	13.4	27	8.3	2	3.7
Total	671	100.0	871	100.0	330	100.0	54	100.0

spectively; Figure 17.10; Table 17.8). Both the differences in biface stages and those in JTI values are statistically significant (Table 17.9). Finally, a pilot study of debitage from Knudtsen ($n = 247$) and Cowboy Rest Creek Quarry ($n = 183$), focusing on the analysis of size, platform configuration, percentage of dorsal cortex, and the number of dorsal scars, indicates the debitage from the quarry was derived from earlier reduction stages than that from the residential site (Kessler 2002), supporting our findings here.

Turning to Little Smoky Quarry and Limestone Peak-L1, as shown in Table 17.7 and Figure 17.11, there is little difference in

TABLE 17.8. Johnson Thinning Index statistics for all four assemblages.

STATISTIC	COWBOY REST CREEK QUARRY	KNUDTSEN	LITTLE SMOKY QUARRY	LIMESTONE PEAK-L1
N	671	871	327	54
Minimum	0.15	0.08	0.14	0.51
Maximum	27.86	8.39	11.02	3.61
Mean	2.94	1.20	2.18	1.38
SD	2.20	0.50	1.25	0.48

TABLE 17.9. Statistical tests between Little Smoky and Cowboy Rest Creek Quarries and their associated residential/activity sites.

COMPARISON	SEPARATE VARIANCES t TEST OF JTI			K–S TEST BIFACE STAGE
	t	df	p	p
Cowboy Rest Creek Quarry, Locality 1 and Knudtsen 1	−8.21	67.6	<0.001	<0.001
Cowboy Rest Creek Quarry, Locality 1 and Knudtsen 2	−9.17	58.3	<0.001	<0.001
Cowboy Rest Creek Quarry, Locality 2 and Knudtsen 1	−16.50	812.7	<0.001	<0.001
Cowboy Rest Creek Quarry, Locality 2 and Knudtsen 2	−19.27	650.2	<0.001	<0.001
Cowboy Rest Creek Quarry and Knudtsen 1	−20.07	722.5	<0.001	<0.001
Cowboy Rest Creek Quarry and Knudtsen 2	−20.38	718.8	<0.001	<0.001
Cowboy Rest Creek Quarry, Locality 1 and Knudtsen	−9.01	58.4	<0.001	<0.001
Cowboy Rest Creek Quarry, Locality 2 and Knudtsen	−18.98	653.1	<0.001	<0.001
Cowboy Rest Creek Quarry and Knudtsen	−20.07	722.5	<0.001	<0.001
Little Smoky Quarry and Limestone Peak-L1	8.42	196.5	<0.001	0.900

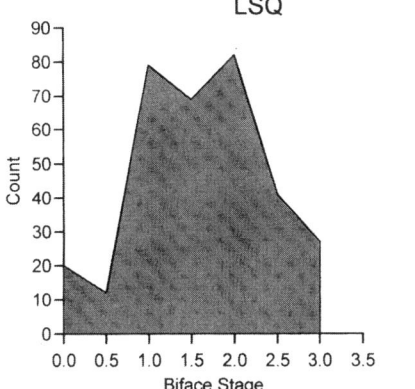 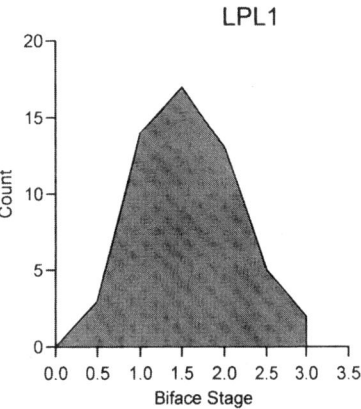

FIGURE 17.11. Profile summary graphs of biface stages represented in the Little Smoky Quarry (LSQ) and Limestone Peak Locality 1 (LPL1) assemblages.

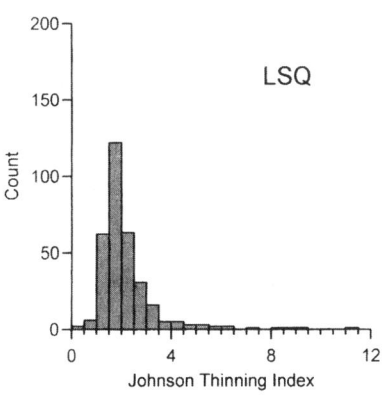 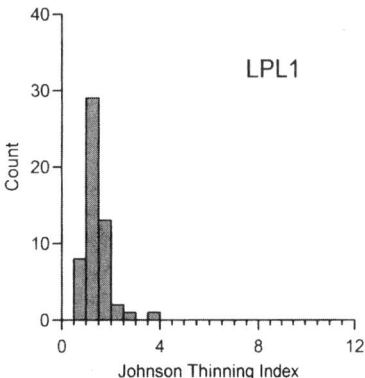

FIGURE 17.12. Histograms of Johnson Thinning Index (JTI) values for bifaces in the Little Smoky Quarry (LSQ) and Limestone Peak Locality 1 (LPL1) assemblages.

the biface stages represented at these two sites (p = 0.900, Table 17.9). In contrast, however, the JTI values of these assemblages differ in the manner expected (Figure 17.12, Table 17.9), with the large majority of values at Limestone Peak-L1 (92.6%) being ≤2.0 while 40.9% of those at Little Smoky Quarry are >2.0. This significant difference seems all the more important because the distribution of JTI is curvilinear and tends to level off as biface manufacture nears completion and does not discriminate well between assemblages of predominantly later stages.

In summary, we predicted that field processing of dacite at Little Smoky Quarry would have proceeded further than at Cowboy Rest Creek Quarry because the distance to be traveled after leaving Little Smoky Quarry (to Limestone Peak-L1) is much greater (60 km) than that between Cowboy Rest Creek Quarry and Knudtsen (9 km). Consequently, Little Smoky Quarry should contain a more complete range of biface stages than Cowboy Rest Creek Quarry. Further, Knudtsen and Limestone Peak-L1 should exhibit biface assemblages that are later in reduction stage than those at their associated quarries, Cowboy Rest Creek and Little Smoky, respectively, because reduction continued at the residential sites.

Our predictions are met with respect to the differences between the two quarries and between Cowboy Rest Creek Quarry and the Knudtsen site. In the comparison between Little Smoky Quarry and its associated residential site, Limestone Peak-L1, there is no difference in the biface stages represented. The significant difference in JTI means, however, suggests that reduction at Limestone Peak-L1 did proceed further there than at Little Smoky Quarry.

Discussion and Conclusion

We have presented an analysis of biface assemblages from two dacite quarries and associated residential sites using the central place foraging model proposed by Metcalfe and Barlow (1992). Our aim was to evaluate if variability in biface reduction at these different sites was a function of transport costs. The model application has been successful, but we realize there is a long way to go in this endeavor. To this point, central place foraging models have been used almost exclusively to investigate the relationship between the processing and transport of food resources and have been most successfully applied in ethnographic situations, where behavior can be observed and time budget analysis can be directly applied to assess the costs of alternative behaviors. Their application to decisions regarding toolstone acquisition are in their infancy.

Central place foraging models are applicable to instances of economizing behavior. This would suggest that these models are appropriate to studies of lithic tool manufacture and transport, and the success of our analysis appears to confirm this claim. In fact, we were successful in spite of the fact that archaeological assemblages represent "time-averaged" behaviors, which will serve to mask these relationships unless redundant behaviors yield strong signatures. Still, there may be questions concerning the suitability of these models for examining behavioral variability regarding lithic toolstone. This concern, termed dynamic sufficiency by Lewontin (1974:11), regards whether the model contains "enough and the right variables to account for the field in question" (Dunnell 1980:85). Although our success here suggests these models, in fact, *do* have dynamic sufficiency, the results could be fortuitous. For future studies the starting point, then, is to continue to evaluate whether the variables in these models are sufficient, and this can be done by shifting those variables held constant and those allowed to vary.

In the case presented in this paper, for instance, we have held toolstone quality, package size, and intended product constant in order to examine transport distance and its effects on assemblage composition. Under the right empirical circumstances, transport distance could be held constant instead, while allowing other factors to vary. Consider, for example, two sources for which quality differs, quality being measured in terms of isotrophy (i.e., homogeneity), sharpness of edges, durability, and workability (Elston 1992b:35). The utility curves for these two materials will differ. Because of increased assaying and processing times for the poorer quality material, it would take more time, and thus effort, to reduce a blank of the same size to, say, a Stage 3 biface than it would for the better quality material. We would predict, then, that biface manufacture would be carried out to different stages at these material sources despite both lying equal distances from residential sites. These costs, at least in an average way, can be estimated through repeated experiments.

Regarding the quantitative measurability of the variables in these models in the archaeological record, or what Lewontin (1974:11) refers to as empirical sufficiency, there are some serious obstacles. To apply the Metcalfe and Barlow model in its entirety, we must know: (1) the optimal load size for a particular type of toolstone; (2) the procurement time for an unprocessed load; (3) the time required to procure and process to different points along the reduction continuum for a processed load; (4) the utility of an unprocessed load; and, (5) the utility of loads processed to different points along the reduction continuum. The Tosawihi experiments (Elston 1992b) have provided some initial data on procurement and processing times, but these data pertain to quarrying activities that involved considerable effort to excavate raw material from quarrying pits. At surface quarries such as those considered here the initial effort is expended in search and assaying time, and thus experiments must be done that focus specifically on these circumstances. Clearly, a great deal of experimental work is needed to derive estimates of acquisition and processing costs.

The most significant obstacle to a more rigorous, quantitative application of this model, however, lies in establishing quantitative estimates of toolstone utility. Economists "use utility as a *descriptive* (emphasis in original) tool: they measure utility by observing what people choose, not by independent criteria" (Stephens and Krebs 1986:105). In models concerning food resources, utility is more directly measured as caloric return. Neither option can be applied to measure toolstone utility, although the proportion of caloric return attributable to technology is an appealing heuristic. But in all practical senses, if measurable at all, toolstone utility as measured by caloric return would be subject to widely varying estimates and therefore of dubious value. Toolstone utility, then, probably should be conceived in a different manner.

Using the concept of proportional utility as discussed above (e.g., Metcalfe and Barlow 1992:354), we might assign a finished biface a utility of 1.0 and the original blank a value less than 1.0, depending upon the amount of waste removed in the reduction process (see also, Bettinger et al. 1997:888). But, of course, this requires that we know the weight of the original blanks reduced to form the biface. A more important question, it seems, is what constitutes a finished biface? If the intent is to take reduction to a form such as a projectile point, and there is no need for the flakes produced in the process, then it is the point that will have a utility of 1.0. But if the intent is to only partially reduce a biface so that it can serve as a core, then it reaches a maximum utility earlier in the reduction sequence. However, as numerous authors who have written about bifacial core technology have pointed out, the advantage of such a technology is in both the finished product *and* in the flakes removed. It is precisely this factor that makes utility so difficult to measure. Which component contributes the most to utility? We may be able to go some distance towards solving this problem with studies that begin with cores of different sizes and in which bifaces are manufactured with different intents: finished tool, creation of blanks, creation of bifacial cores. Using a measure of size, such as weight, edge length, or area, a utility value can be assigned to the original blank, depending on the final product. In one case, the intended product may be a bifacial core, and thus reduction is taken to a particular point, at which the utility is 1.0. In another case, the intended product may also be a bifacial core, but the production of blanks as well, and therefore the number of flakes of a certain size can be counted, their relative contribution to the final weight determining their proportion of the utility.

In this type of procedure, however, the archaeological record must play an important role. In a certain type of reduction trajectory, for example, the average flake size we must consider for the manufacture of tools must be determined. That is, if the purpose is to use flakes that result from the manufacture of a bifacial core or projectile point, there should be an optimal size below which a flake is not likely to be used. This can only be determined empirically, and although these estimates may not be specifically correct in every situation, the uncertainty of such a determination will likely be no greater than the uncertainties involved in estimates of food resource return rates, whether derived ethnographically or experimentally. As Grayson and Cannon (1999:146) point out, "both ethnographically and experimentally derived return rates have an unknown and unknowable relationship to return rates that characterized the past." Grayson and Cannon (1999:145) also point out that one way around this problem has been to convert resource return rates to return rate classes; as a result, the models rely only on ordinal return rate estimates. The same could be done for toolstone utility estimates or "reduction classes," as discussed above.

In the end the study presented here may seem overly simplistic. It does, however, demonstrate that central place foraging models can be useful in trying to understand variability in lithic assemblages that has by default been interpreted as functional differences in sites. In spite of the problems they see with the implementation of optimal foraging models to archaeological circumstances, Grayson and Cannon (1999:150) state that "foraging theory provides the best, if not the only, means currently available to archaeologists for examining interactions between people and their environments within an evolutionary framework." We are sure that many archaeologists would dispute this bold claim. Nevertheless, foraging theory in general, and central place foraging models in particular, offer fruitful insights into the reasons for assemblage variability. What we have done here is the easy part; the hard part is yet to come.

Acknowledgments

An earlier version of the paper was presented as a poster at the 66th Annual Meeting of the Society for American Archaeology in New Orleans, LA. We would like to thank Michael Cannon, Donald Grayson, David Madsen, and two anonymous reviewers for their helpful comments on previous versions of the paper.

Notes

1. Butler and May (1984:185) define true quarrying as "actual excavation to locate and remove the raw material from its geological matrix." Such quarries are relatively rare in the Desert West (but see Elston and Raven [1992]); more often, toolstone occurs in cobble/boulder form on the surface. Here we use the term "quarry" in the more general sense, to refer to areas of prehistoric toolstone procurement, whether that procurement involves actual extraction from rock or sediment matrix or simply cobble testing and collection.
2. The term "residential site" is used here in the sense meant by Binford (1980) at which the entire group camped, presumably for some extended period of time, and where a variety of activities took place. These sites generally exhibit a wide range of artifactual forms used in these different activities.
3. A biface reduction flake is defined here as any flake removed from a bifacial core. These flakes generally have: (1) a faceted platform, (2) an acute external platform angle, (3) a lip at the intersection of the platform and ventral surface, and (4) a concave dorsal surface/arched longitudinal axis. This class encompasses the more specific category of bifacial thinning flakes (see Whittaker 1994:185–187).
4. It is true, as Ahler (1989b:87) notes, that "flakes might appear to be technology-specific, such as bifacial thinning flakes and bipolar flakes, can be produced by multiple techniques." Ahler's experimental results presented in Table 1 (Ahler 1989b:88), however, show that of the 91 biface thinning flakes (a more specific category than our biface reduction flakes), 88 (96.7%) were produced through biface edging and thinning, resulting in an error of 3.3%. If we reduce the total number of biface reduction flakes by 3.3% (from 4415 to 4269), our assemblage is still overwhelmingly dominated by these flakes (75.1%).
5. That the only (or even primary) payoff of all food resources is in energy has been challenged by a number of researchers, especially with respect to meat and hunting (e.g., Bliege Bird et al. 2001; Hawkes et al. 2001) but the issue is debated (e.g., Gurven et al. 2000; Hill and Kaplan 1993; Kaplan et al. 2000; see also, "Comments" following Hawkes et al. 2001). We do not enter this debate here, but accept the assumption of central place foraging models that energy is a significant benefit of food resources.

6. It would be inappropriate to use an average of two ordinal categories for statistical comparison. For example, if a teacher is evaluated on a scale of 1–6, it is inappropriate for his/her superior to suggest his/her teaching is better or worse because his/her "average" is higher or lower than another's. In the present case, however, we are simply using the average to designate an additional category in between the originally defined stages. In no case were the assignments made by the two analysts more than one stage apart (e.g., Stage 1 versus Stage 2) and thus the averaged category (Stage 1.5) indicates only that the biface has been reduced to a greater degree than Stage 1 but to a lesser degree than Stage 2. Statistical comparisons are not made between biface stages.

References

Ahler, S. A.
1989a Experimental knapping with KRF and Midcontinent Cherts: overview and applications. In *Experiments in Lithic Technology*, edited by D. S. Amick and R. P. Mauldin, pp. 199–234. BAR International Series 528. British Archaeological Reports, Oxford.
1989b Mass analysis of flaking debris: studying the forest rather than the tree. In *Alternative Approaches to Lithic Analysis*, edited by D. O. Henry and G. H. Odell, pp. 85–118. Archaeological Papers of the Anthropological Association no. 1.

Amick, D. S.
1995 Raw material selection patterns among paleoindian tools from the Black Rock desert, Nevada. *Current Research in the Pleistocene* 12:55–57.

Andrefsky, W., Jr.
1994 Raw-material availability and the organization of technology. *American Antiquity* 59:21–34.

Bamforth, D. B.
1986 Technological efficiency and tool curation. *American Antiquity* 51:38–50.

Barlow, K. R., D. Metcalfe
1996 Plant utility indices: two Great Basin examples. *Journal of Archaeological Science* 23:351–371.

Basgall, M. E.
1993a *Archaeology of the Nelson Basin and Adjacent Areas, Fort Irwin, San Bernardino County, CA.* Prepared for the US Army Corps of Engineers, Los Angeles District, by Far Western Anthropological Research Group, Davis, CA.
1993b Early Holocene Prehistory of the North-Central Mojave Desert. Unpublished Ph.D. Dissertation, Department of Anthropology, University of California, Davis.
2000 Patterns of toolstone use in late-Pleistocene/early-Holocene assemblages of the Mojave desert. *Current Research in the Pleistocene* 17:4–6.

Beck, C., G. T. Jones
1988 Western Pluvial Lakes Tradition occupation in Butte Valley, Nevada. In *Early Human Occupation in Far Western North America: The Clovis-Archaic Interface*, edited by J. A. Willig, C. M. Aikens, and J. L. Fagan, pp. 273–301. Nevada State Museum Anthropological Papers no. 21.
1990a Late Pleistocene/early Holocene occupation in Butte Valley, Eastern Nevada: three season's work. *Journal of California and Great Basin Anthropology* 12:231–261.
1990b Toolstone selection and lithic technology in early Great Basin prehistory. *Journal of Field Archaeology* 17:283–299.
1993 The multi-purpose function of Great Basin stemmed series points. *Current Research in the Pleistocene* 10:52–54.
1994a Dating surface assemblages using obsidian hydration. In *Dating in Surface and Exposed Contexts*, edited by C. Beck, pp. 47–76. University of New Mexico Press, Albuquerque.
1994b On-site analysis as an alternative to collection. *American Antiquity* 59:304–315.
1997 The terminal Pleistocene/early Holocene archaeology of the Great Basin. *Journal of World Prehistory* 11:161–236.

Bettinger, R. L., R. Malhi, H. McCarthy
1997 Central place models of acorn and mussel processing. *Journal of Archaeological Science* 24:887–899.

Binford, L. R.
1978 *Nunamiut Ethnoarchaeology*. Academic Press, New York.
1980 Willow smoke and dogs' tails: hunter-gatherer settlement systems and archaeological site formation. *American Antiquity* 45:4–20.

Bird, D. W., R. L. Bliege Bird
1997 Contemporary Shellfish gathering strategies among the Meriam of the Torres Strait islands, Australia: testing predictions of a central place foraging model. *Journal of Archaeological Science* 24:39–63.

Bliege Bird, R., E. A. Smith, D. Bird
2001 The hunting handicap: costly signaling in male foraging strategies. *Behavioral Ecology and Sociobiology* 50:9–19.

Bloomer, W. W., K. Ataman, E. Ingbar, M. W. More
1992 Bifaces. In *Archaeological Investigations at Tosawihi, A Great Basin Quarry*, edited by R. G. Elston and C. Raven, pp. 81–131. Prepared for the Bureau of Land Management, Elko Resource Area, by Intermountain Research Reports, Silver City, NV.

Boldurian, A. T.
1990 *Lithic Technology at the Mitchell Locality of Blackwater Draw: A Stratified Folsom Site in Eastern New Mexico*. Plains Anthropologist, Memoir 24.

Bradbury, A. P., P. J. Carr
1999 Examining stage and continuum models of flake debris analysis: an experimental approach. *Journal of Archaeological Science* 26:105–116.

Bunn, H. T., L. E. Bartram, E. M. Kroll
1988 Variability in bone assemblage formation from Hadza hunting, scavenging, and carcass processing. *Journal of Anthropological Archaeology* 7:412–457.

Butler, B. M., E. E. May
1984 Introduction to Part III: Quarry and Workshop Sites. In *Prehistoric Chert Exploitation: Studies from the Midcontinent*, edited by B. M. Butler and E. E. May, p. 185. Center for Archaeological Investigations, Occasional Paper No. 2.

Callahan, E.
1979 *The Basics of Biface Knapping in the Eastern Fluted Point Tradition: A Manual for Flint Knappers and Lithic Analysts*. Archaeology of North America v. 7.

Carambelas, K., S. Raven
1991 Actualistic Quarrying Experiments at Tosawihi. Paper presented at the 56th meeting of the Society for American Archaeology, New Orleans.

Daugherty, R. D.
1956 Archaeology of the Lind Coulee Site, Washington. *Proceedings of the American Philosophical Society* 100:223–278.

Drennan, R. D.
1984a Long-distance movement of goods in the Mesoamerican Formative and Classic. *American Antiquity* 49:27–43.
1984b Long-distance transport costs in pre-Hispanic Mesoamerica. *American Anthropologist* 86:105–112.

Dunnell, R. C.
1980 Evolutionary theory and archaeology. In *Advances in Archaeological Method and Theory*, vol. 3, edited by M. B. Schiffer, pp. 35–99. Academic Press, New York.

Elston, R. G.
1990 A cost-benefit model of lithic assemblage variability. In *The Archaeology of James Creek Shelter*, edited by R. G. Elston and E. Budy, pp. 153–163. University of Utah Anthropological Papers no. 115.
1992a Economics and strategies of lithic production at Tosawihi. In *Archaeological Investigations at Tosawihi, A Great Basin Quarry*, edited by R. G. Elston and C. Raven, pp. 775–801. Prepared for the Bureau of Land Management, Elko Resource Area, by Intermountain Research Reports, Silver City, NV.
1992b Modeling the economics and organization of lithic procurement. In *Archaeological Investigations at Tosawihi, A Great Basin Quarry*, edited by R. G. Elston and C. Raven, pp. 31–47. Prepared for the Bureau of Land Management, Elko Resource Area, by Intermountain Research Reports, Silver City, NV.

Elston, R. G., C. Raven (Eds.)
1992 Archaeological *Investigations at Tosawihi, A Great Basin Quarry*. Prepared for the Bureau of Land Management, Elko Resource Area, Intermountain Research Reports, Silver City, NV.

Ericson, J. E.
1984 Toward the analysis of lithic production systems. In *Prehistoric Quarries and Lithic Production*, edited by J. E. Ericson and B. A. Purdy, pp. 2–9. Cambridge University Press, Cambridge.

Frison, G., B. Bradley
1999 *The Fenn Cache: Clovis Weapons and Tools*. One Horse Land and Cattle Company, Santa Fe, NM.

Goodyear, A. C.
1989 A hypothesis for the use of cryptocrystalline raw materials among Paleoindian groups of North America. In *Eastern Paleoindian Lithic Resource Use*, edited by C. Ellis and J. Lothrop, pp. 1–10. Westview Press, Boulder.

Gould, R. A.
1977 Ethnoarchaeology; or, where do models come from? In *Stone Tools as Cultural Markers*, edited by R. V. S. Wright, pp. 162–168. Australian Institute of Aboriginal Studies, Canberra.

Gould, R. A., S. Saggers
1985 Lithic procurement in central Australia: a closer look at Binford's idea of embeddedness in archaeology. *American Antiquity* 50:117–136.

Graf, K. E.
2001 Paleoindian Technological Provisioning in the Western Great Basin. Master's Thesis, Department of Anthropology and Ethnic Studies, University of Nevada, Las Vegas.

Gramly, R. M.
1980 Prehistoric industry at the Mt. Jasper mine, Northern New Hampshire. *Man in the Northeast* 20:1–24.

Grayson, D. K.
1993 *The Desert's Past: A Natural Prehistory of the Great Basin*. Smithsonian Institution, Washington, D.C.

Grayson, D. K., M. D. Cannon
1999 Human paleoecology and foraging theory in the Great Basin. In *Models for the Millennium: Great Basin Anthropology Today*, edited by C. Beck, pp. 141–151. University of Utah Press, Salt Lake City.

Gurven, M., K. Hill, A. Hurtado, R. Lyles
2000 Food transfers among Hiwi foragers of Venezuela: tests of reciprocity. *Human Ecology* 28:171–214.

Hawkes, K., J. F. O'Connell, N. G. Blurton Jones
2001 Hunting and nuclear families. *Current Anthropology* 42:681–695.

Hayden, B.
1977 Stone tool functions in the western desert. In *Stone Tools as Cultural Markers*, edited by R. V. S. Wright, pp. 178–188. Australian Institute of Aboriginal Studies, Canberra.

Hill, K., H. Kaplan
1993 On why male foragers hunt and share food. *Current Anthropology* 34:701–706.

Hiscock, P.
1994 Technological responses to risk in Holocene Australia. *Journal of World Prehistory* 8:267–292.

Holmes, W. H.
1891 Aboriginal novaculae quarries in Garland County, Arkansas. *American Anthropologist* 4:313–316.
1897 *Stone Implements of the Potomac-Chesapeake Tidewater Province*. Bureau of American Ethnology, Bulletin 15. Smithsonian Institution, Washington, D.C.
1919 *Handbook of Aboriginal American Antiquities*. Part I. The Lithic Industries. Bureau of American Ethnology Bulletin 60.

Huckleberry, G., C. Beck, G. T. Jones, A. Holmes, M. D. Cannon, S. D. Livingston, J. M. Broughton
2001 Terminal Pleistocene/early Holocene environmental change at the Sunshine Locality, North-Central Nevada, USA. *Quaternary Research* 55:303–312.

Ingbar, E. E.
1994 Lithic material selection and technological organization. In *The Organization of North American Prehistoric Chipped Stone Tool Technologies*, edited by P. J. Carr, pp. 45–56. International Monographs in Prehistory. Archaeological Series 7.

Ingbar, E. E., M. L. Larson, B. A. Bradley
1989 A non-typological approach to debitage analysis. In *Experiments in Lithic Technology*, edited by D. S. Amick and R. P. Mauldin, pp. 67–99. BAR International Series 528.

Johnson, J. K.
1981 *Yellow Creek Archaeological Project* v. 2. Tennessee Valley Authority Publications in Anthropology no. 2.
1984 Measuring prehistoric quarry site activity in northeastern Mississippi. In *Prehistoric Chert Exploitation: Studies from the Midcontinent*, edited by B. M. Butler and E. E. May, pp. 225–235. Center for Archaeological Investigations Occasional Paper No. 2. Southern Illinois University at Carbondale.
1989 The utility of production trajectory modeling as a framework for regional analysis. In *Alternative Approaches to Lithic Analysis*, edited by D. O. Henry and G. H. Odell, pp. 119–133. Archaeological Papers of the American Anthropological Association no. 1.

Jones, G. T., C. Beck
1990 An obsidian hydration chronology of late Pleistocene/early Holocene surface assemblages from Butte Valley, Nevada. *Journal of California and Great Basin Anthropology* 12:84–100.

Jones, G. T., D. G. Bailey, C. Beck
1997 Source provenance of andesite artifacts using non-destructive XRF analysis. *Journal of Archaeological Science* 24:929–943.

Jones, G. T., C. Beck, E. Jones
2002 Lithic source use and paleoarchaic foraging territories in the Great Basin. *American Antiquity*, in press.

Jones, G. T., C. Beck, F. L. Nials, J. J. Neudorfer, B. Brownholtz, H. Gilbert
1996 Recent archaeological and geoarchaeological investigations at

the Sunshine Locality, Long Valley, Nevada. *Journal of California and Great Basin Anthropology* 18:48–63.

Jones, K. T., D. B. Madsen
1989 Calculating the cost of resource transportation: a Great Basin example. *Current Anthropology* 30:529–534.

Kaplan, H., K. Hill
1992 The evolutionary ecology of food acquisition. In *Evolutionary Ecology and Human Behavior*, edited by E. A. Smith and B. Winterhalder, pp. 167–201. Aldine de Gruyter, New York.

Kaplan, H., K. Hill, J. Lancaster, A. M. Hurtado
2000 A theory of human life history evolution: diet, intelligence, and longevity. *Evolutionary Anthropology* 9:156–185.

Kelly, R. L.
1988 The three sides of a biface. *American Antiquity* 53(4):717–734.
1995 *The Foraging Spectrum: Diversity in Hunter-Gatherer Lifeways.* Smithsonian Institution Press, Washington, D.C.
2001 *Prehistory of the Carson Desert and Stillwater Mountains: Environment, Mobility, and Subsistence in a Great Basin Wetland.* University of Utah Anthropological Papers No. 123.

Kelly, R. L., L. C. Todd
1988 Coming into the country: early Paleoindian hunting and mobility. *American Antiquity* 53:231–244.

Kessler, R. A.
2002 Trash: A Study of Great Basin Paleoarchaic Debitage. Poster presented at the 67th meeting of the Society for American Archaeology, Denver.

Kuhn, S. L.
1994 A formal approach to the design and assembly of mobile toolkits. *American Antiquity* 59(3):426–442.

Lewontin, R. C.
1974 *The Genetic Basis of Evolutionary Change.* Columbia University Press, New York.

Madsen, D. B.
1999 Environmental change during the Pleistocene-Holocene transition and its possible impact on human populations. In *Models for the Millennium: Great Basin Anthropology Today*, edited by C. Beck, pp. 75–82. University of Utah Press, Salt Lake City.

Magne, M. P.
1985 *Lithics and Livelihood: Stone Tool Technologies of Central and Southern Interior B.C.* Archaeology Survey of Canada, Mercury Series No.133. Ottawa.

Magne, M. P., D. Pokotylo
1981 A pilot study in bifacial lithic reduction sequences. *Lithic Technology* 10:34–47.

Metcalfe, D., K. R. Barlow
1992 A model for exploring the optimal trade-off between field processing and transport. *American Anthropologist* 94:340–356.

Metcalfe, D., K. T. Jones
1988 A reconsideration of animal body-part utility indices. *American Antiquity* 53:486–504.

Muto, G. R.
1971 A stage analysis of the manufacture of stone tools. In *Great Basin Anthropological Conference 1970: Selected Papers*, edited by C. M. Aikens, pp. 109–118. University of Oregon Anthropological Papers no. 1.

Nami, H.
1999 The Folsom biface reduction sequence: evidence from the Lindenmeier Collection. In *Folsom Lithic Technology: Explorations in Structure and Variation*, edited by D. S. Amick, pp. 82–97. International Monographs in Prehistory. Archaeological Series 12.

Nelson, M. C.
1991 The study of technological organization. In *Archaeological Method and Theory*, vol. 3, edited by M. B. Schiffer, pp. 57–100. The University of Arizona Press, Tucson.

Newcomer, M. H.
1971 Some quantitative experiments in handaxe manufacture. *World Archaeology* 3:85–94.

O'Connell, J. F., K. Hawkes, N. Blurton Jones
1988 Hadza hunting, butchering, and bone transport and their archaeological implications. *Journal of Anthropological Research* 44:113–161.
1990 Reanalysis of large mammal body part transport among the Hadza. *Journal of Archaeological Science* 17:301–316.

Orians, G. H., N. E. Pearson
1979 On the theory of central place foraging. In *Analysis of Ecological Systems*, edited by D. J. Horn, R. D. Mitchell, and G. R. Stairs, pp. 154–177. Ohio State University Press, Columbus.

Pecora, A. M.
2001 Chipped stone tool production strategies and lithic debris patterns. In *Lithic Debitage: Context, Form, Meaning*, edited by W. Andrefsky, Jr., pp. 173–190. University of Utah Press, Salt Lake City.

Price, B. A.
1989 *Archaeological Survey of Seismic Test Lines in Chevron Exploration and Production Services' Pancake Prospect, White Pine, Eureka, and Nye Counties, Nevada.* Bureau of Land Management Cultural Resources Report 6-956. Reno, Nevada.

Price, B. A., S. E. Johnston
1988 A model of late Pleistocene and early Holocene adaptation in Eastern Nevada. In *Early Human Occupation in Far Western North America: The Clovis-Archaic Interface*, edited by J. A. Willig, C. M. Aikens, and J. L. Fagan, pp. 231–250. Nevada State Museum Anthropological Papers 21, Carson City.

Raab, L. M., R. F. Cande, D. W. Stahle
1979 Debitage graphs and archaic settlement patterns in the Arkansas Ozarks. *Midcontinental Journal of Archaeology* 4:167–182.

Rhode, D.
1990 On transportation costs of Great Basin resources: an assessment of the Jones-Madsen model. *Current Anthropology* 31:413–419.

Shott, M.
1996a An exegesis of the curation concept. *Journal of Anthropological Research* 52:259–280.
1996b Stage versus continuum in the debris assemblage from production of a fluted biface. *Lithic Technology* 21:6–22.

Simms, S. R.
1988 Conceptualizing the Paleoindian and Archaic in the Great Basin. In *Early Human Occupation in Far Western North America: The Clovis-Archaic Interface*, edited by J. A. Willig, C. M. Aikens, and J. L. Fagan, pp. 41–52. Nevada State Museum Anthropological Papers 21, Carson City.

Singer, C. A.
1984 The 63-kilometer fit. In *Prehistoric Quarries and Lithic Production*, edited by J. E. Ericson and B. A. Purdy, pp. 35–48. Cambridge University Press, Cambridge.

Singer, C. A., J. E. Ericson
1977 Quarry analysis at Bodie Hills, Mono County, California: a

case study. In *Exchange Systems in Prehistory*, edited by T. K. Earle and J. E. Ericson, pp. 171–188. Academic Press, New York.

Stahle, D. W., J. E. Dunn
1982 An analysis and application of size distribution of waste flakes from the manufacture of bifacial tools. *World Archaeology* 14:84–97.
1984 *An Experimental Analysis of the Size Distribution of Waste Flakes from Bifacial Reduction*. Technical Report No. 2. Arkansas Archaeological Survey, Fayetteville.

Stephens, D. W., J. R. Krebs
1986 *Foraging Theory*. Princeton University Press, Princeton.

Sullivan III, A. P., K. C. Rozen
1985 Debitage analysis and archaeological interpretation. *American Antiquity* 50:755–779.

Teltser, P. A.
1991 Generalized core technology and tool use: a Mississippian example. *Journal of Field Archaeology* 18:363–375.

Torrence, R.
1989 Retooling: Towards a behavioral theory of stone tools. In *Time, Energy, and Stone Tools*, edited by R. Torrence, pp. 57–66. Cambridge University Press, Cambridge.

Whittaker, J. C.
1994 *Flintknapping: Making & Understanding Stone Tools*. University of Texas Press, Austin.

White, T. E.
1953 Observations on the butchering techniques of some aboriginal peoples, no. 2. *American Antiquity* 19:160–164.
1954 Observations on the butchering technique of some aboriginal peoples, nos. 3, 4, 5, and 6. *American Antiquity* 19:254–264.

Willig, J. A.
1989 *Paleo-archaic Broad Spectrum Adaptations at the Pleistocene-Holocene Boundary in Far Western North America*. Ph.D. Dissertation, University of Oregon. University Microfilms, Ann Arbor.

Zancanella, J. K.
1988 Early lowland prehistory in southcentral Nevada. In *Early Human Occupation in Far Western North America: The Clovis-Archaic Interface*, edited by J. A. Willig, C. M. Aikens, and J. L. Fagan, pp. 251–271. Nevada State Museum Anthropological Papers 21, Carson City.

18

The Effect of Handling Time on Subsistence Technology

Jason Bright, Andrew Ugan, and Lori Hunsaker

Many anthropologists and archaeologists have dismissed evolutionary ecology as inappropriate for the study of humans, often on the basis of the qualitative statement that we are somehow inherently different from our non-human counterparts. To us, this view seems better expressed as a question rather than a dismissive statement. Can observations from the animal world inform how we think about humans in useful ways?

Most applications of evolutionary ecology to archaeology use the prey choice model as a point of departure to address changes in prehistoric subsistence strategies (e.g., Stiner et al. 2000). Part of the attraction of the model comes from the potential universality of its application to prehistoric subsistence decisions in any region and at any time (O'Connell 1995), and the fact that the evidence of past meals is a common component of the archaeological record worldwide. Here, we take a step away from prehistoric subsistence economy to summarize and outline what we call the "Tech Investment Model", which addresses the technological elaborations that often accompany prehistoric subsistence transitions.

To begin, we demonstrate how processing costs affect a resource's profitability, in order to highlight the potential of this model and others like it. We then summarize the model, and suggest how archaeologists might use it, by applying five coarse predictions to a locally important transition in the Great Basin of western North America. Much more important than the particular model or the local case, however, is the attention to general relationships between the costs of technological innovation and its effects on an individual's foraging returns. We suggest that this approach complements many existing explanations of technological change. We conclude by showing how the model can address other technological transitions, such as the transition between Middle and Upper Paleolithic technologies.

*Originally published in *World Archaeology* 34(2002):164–181

Two Foraging-Related Models
A Classic Foraging Model

Throughout the 1970s and 1980s, ecologists often turned to the prey choice model (Stephens and Krebs 1986:17–24) to study the foraging behavior of animals. The prey choice model is used to predict whether a forager should exploit a resource upon encounter or pass it up in favor of searching for other prey. Applications of the model often assume the goal of the forager is to maximize its net rate of energy gain for a foraging bout.

The decision to take or not take a resource is based in part upon the post-encounter return rate of the prey item. A return rate is calculated by dividing the caloric value of the item by the time spent handling it (i.e., calories/hr). Handling time has two components. Time spent acquiring prey once it has been encountered, such as chasing down a deer after it has been hit with an arrow, is referred to as pursuit time. Time spent turning the acquired prey into a meal, like butchering a deer or grinding corn into flour, is called processing time.

Resources vary widely in their handling times (Table 18.1). For example, an Alyawara woman needs about 6.5 hours to handle one kilogram of *Acacia anuera* seed, but only 15 minutes to handle a kilogram of *Ipomea costata* tuber (O'Connell and Hawkes 1981:123). Because handling time is a critical component of the post-encounter return rate of an item, the return rates for prey vary widely as well. Return rates for acacia seeds are lower than for the tuber resource because of the lengthy handling seeds require, even though the seeds contain more energy per unit volume.

The relationship between handling time and return rate holds true for Great Basin resources as well. The game and pollen resources presented in Table 18.1 all yield fewer calories per kilogram than the nut and seed resources shown in the table, but

TABLE 18.1. Handling times and return rates of several Australian and Great Basin prey.

Location	Name	Kind of Prey	Energy Yield (Cals/kg)	Handling (Hr/kg)	Kcals/Hr	Source
Australia	*Ipomea costata*	tuber	1563	0.25	6252	1
Australia	*Solanum centrale*	fruit	2992	0.5	5984	1
Australia	*Amphibilarus* sp.	lizard	1050	0.25	4200	1
Australia	*Vigna lanceolata*	tuber	862	0.75	1724	1
Australia	*Xyleutes* sp.	witchity grub	2600	1.75	1486	1
Australia	*Acacia coriacea*	seed	3551	>5.25	<676	1
Australia	*Acacia anuera*	seed	3778	6.5	580	1
Australia	*Acacia cowleana*	seed	3589	6.5	552	1
Great Basin	*Ovis canadensis*	large game	1258	0.04	31450	2
Great Basin	*Antilocapra americana*	large game	1258	0.04	31450	2
Great Basin	*Lepus* sp.	medium game	1078	0.07	15400	2
Great Basin	*Thomomys* sp.	small game	1078	0.10	10870	3
Great Basin	*Typha latifolia*	pollen	1040	0.37	2750	3
Great Basin	*Pinus monophylla*	nuts	6560	7.3	900	4
Great Basin	*Echinochloa crusgalli*	seed	3510	5.0	702	3
Great Basin	*Allenrolfea occidentalis*	seed	2300	18.4	150	3

Sources: O'Connell and Hawkes (1981:123); Simms (1987); Simms (1985); Barlow and Metcalfe (1996).
Note: For prey reported with a range of handling time and return rates, we have reported the shortest reported handling time and therefore the highest return rate. Simms (1985) and Barlow and Metcalfe (1996) are in general agreement concerning *Pinus monophylla* and *Allenrolfea occidentalis*.

they also take less time to handle, and so rank higher than available seed resources.

Any woman relying on seeds for a substantial portion of her calories will spend a great deal of time processing them. If she reduces the time she spends handling the seeds, she improves the post-encounter returns for that resource. She might reduce the time she spends handling seeds by using larger or better stones that allow for more seeds to be processed more efficiently, or by making a wooden tray designed to catch flour and protect against loss. But there is an important trade-off here. She must *spend* time making or improving a tool in order to *reduce* the time she spends handling prey. This is a problem that archaeologists have been wrestling with for more than ten years (e.g., Russell 1989:653), and a problem that every human forager faces.

Summary of the Tech Investment Model

The tech investment model is a fine-grained mathematical model that articulates the relationship between time invested in manufacture and reductions in handling time. The model makes several assumptions. It assumes that a unit of time invested into tool manufacture cannot be invested into another activity, and that each unit of time invested (e.g., time spent shaping grinding-stone surfaces) decreases handling time of some resource by an equal amount. It also assumes that a forager's goal when manufacturing subsistence tools is to maximize his or her net rate of energy gain.

Because we only summarize the model in order to highlight potential applications, we forgo formal presentation here. The full derivation is available upon request (Ugan and Rogers 2000), and is in preparation for publication (Ugan et al. n.d.).

The model demonstrates that the optimal amount of time to invest in tool production is a product of three variables, which are outlined below:

1. *Total time spent searching for food* The longer someone searches for food, the greater the amount of food they can acquire. For instance, a woman occupying a camp for ten days has less time to search for prey, and therefore has less prey to handle, than someone occupying a camp for a month. We expect the first forager to invest less in a particular technology, all else being equal.

2. *Encounters with a resource* Imagine two foragers, one who encounters many acacia seeds in a month and one who encounters few. Because the frequencies at which the two prey are encountered and taken are different, we expect the foragers to invest in seed-handling technology differently, even though the base handling time of the resource is the same.

3. *The base handling time of a resource* This last point refers directly to how different resources vary in the handling time they require, as outlined in Table 18.1. As long as seeds take more time to process than tubers, a woman stands to gain more by investing in a seed technology than a tuber technology, all else being equal.

We draw upon points 1–3, and several independent lines of evidence, to explain technological changes in the Little Boulder Basin Area (LBBA) of north-central Nevada. We do this by generating some rather robust predictions about the nature and direction of technological change in the LBBA, in order to demonstrate the basic utility of the model. The time period in question covers a transition to a broader diet, decreased residential

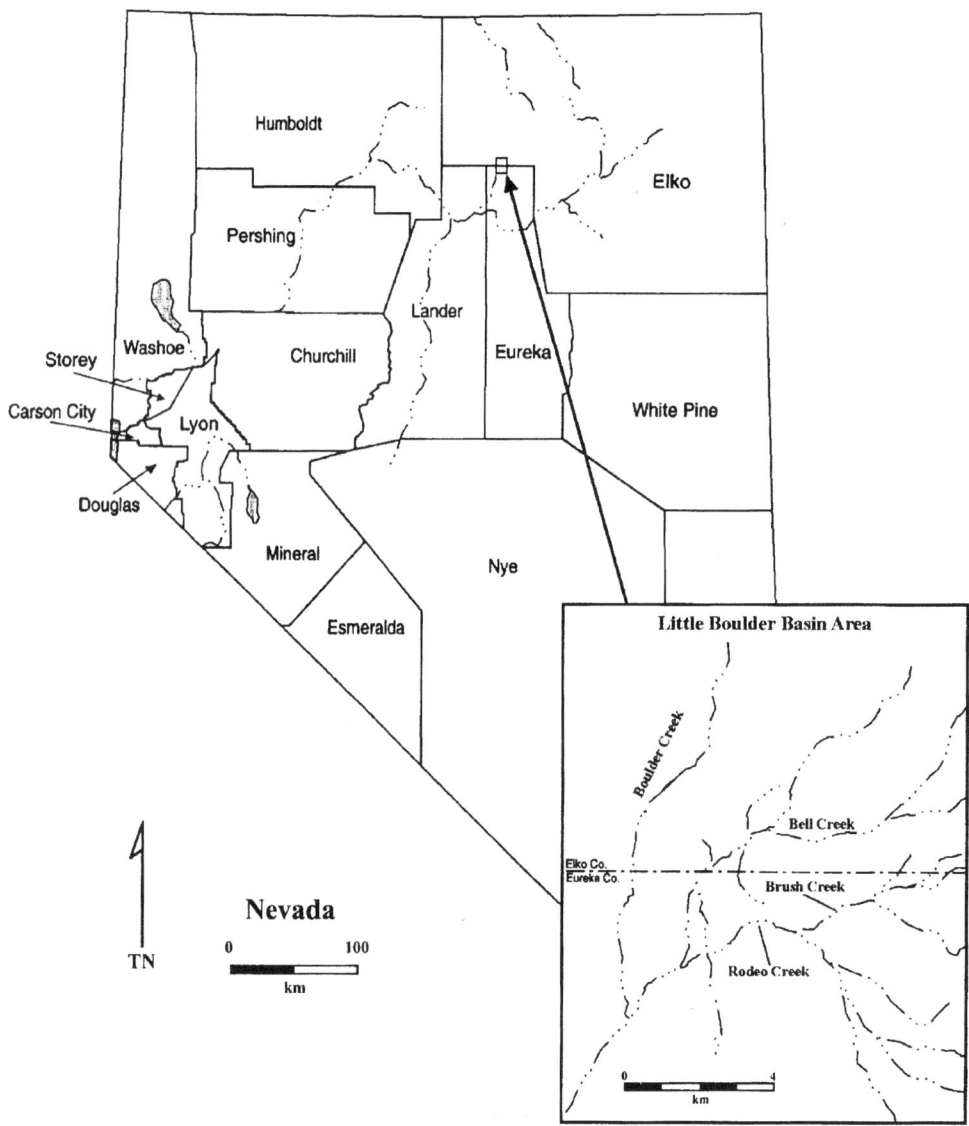

FIGURE 18.1. Map of Nevada, showing the location of the Little Boulder Basin Area (modified from Schroedl 1995:1).

mobility and technological innovation in many places across the Great Basin (Bettinger and Baumhoff 1982; Simms 1983). In the LBBA, longer residence times are indicated during later occupations, based upon changes in the size of occupations and the array of artifact types (Point 1). The contents of firehearths suggest decreasing encounter rates with larger game (Point 2) and a shift to diets that include a significant amount of seed resources (Point 3). Each of these phenomena has implications for changes in subsistence-related technology.

Test Case: The Little Boulder Basin Area

Background

The archaeology of the LBBA is presented in several volumes (La Fond and Jones 1995; La Fond et al. 1995; Schroedl 1995, 1996, 1997, 1999; Stratford and La Fond 1995; Tipps 1996; Tipps and Stratford 1996; see also Bright and Ugan 1999; Ugan and Bright 2001). In total, close to forty sites have been excavated in a roughly 100 square km area of north central Nevada (Figure 18.1).

Many sites in the LBBA are multi-component palimpsests; they present the remains of many occupations during many different temporal periods. Because many artifacts are not temporally diagnostic, we limit our discussion of technological change to artifacts that were recovered from single-component occupations. Sometimes these occupations represent individual sites; in other cases they are discrete components of larger sites. Occupations prior to cal. AD 700 are scarce and ephemeral, and are dealt with only minimally here.

Prey Choice

Analysis of over 250 firehearths (eighty-one radiocarbon-dated) provides important information for understanding LBBA prey choice through time (Bright 1999a). LBBA hearths can generally be divided into two groups based on size (Table 18.2). Size differences are significant in every dimension ($p < .001$). All prey items found in hearths were divided into four categories based on experimental return rates and body size. These categories are high, medium, and low return rate mammals, and seeds as a separate, lower-ranked category. Table 18.3 lists the taxa included in each group.

Hearth contents are summarized in Table 18.4. Large hearths contain equal or greater numbers of identified bone specimens (NISP) from high and medium return mammals, but significantly fewer specimens of low return animals. Twice as many large hearths contain the bones of high and medium return rate mammals, but twice as many small hearths contain the remains of low return mammals. Small hearths also contain significantly more seeds than large hearths do ($p = .032$, two-tailed t-test). The ratio of small hearths to large ones that contain seeds is about two or three to one. We take this as evidence that hearth size is related to prey choice. The larger the prey, the larger the hearth.

Other lines of evidence also suggest an association between hearth size and resource type. Milling stones commonly occur within five meters of a firehearth (Birnie 1996) and in most cases it is a small, seed hearth. In addition, over 90 per cent of all pottery sherds are found next to small seed hearths (Bright 1999b). Chipped stone tools thought to be used for butchering large game, on the other hand, are more common on sites with higher frequencies of large firehearths (Bright 1999a).

If we accept that hearth type and prey items are even moderately correlated as outlined above, then an interesting temporal pattern appears. Figure 18.2 indicates that large hearths appear early (c. cal. 2250 BC) and increase in frequency throughout the temporal sequence. Small hearths, however, appear only after cal. AD 700, and not in number until cal. AD 1200. Then they increase in frequency through time until both hearth types are represented equally at the end of the sequence. We note that our sample of firehearths could be affected by preservation issues, as both kinds of hearths are relatively few in number before cal. AD 700. Nonetheless, the increase in the relative frequency of small hearths is still evident.

This means that, after cal. AD 1200–1300, LBBA foragers take relatively less large game and spend more time acquiring seed resources. Therefore, technologies that serve to cut the handling time associated with seeds and small game should show greater investment after cal. AD 1300, while those that serve to cut handling times of larger game should show less.

TABLE 18.2. Dimensions of large and small firehearths.

	THICKNESS (cm)		WIDTH (cm)		LENGTH (cm)	
SIZE	N	MEAN	N	MEAN	N	MEAN
Large	67	14.6 ± 5.4	120	62.3 ± 21.6	112	53.0 ± 16.0
Small	44	9.3 ± 4.2	104	51.7 ± 24.9	102	44.3 ± 22.2

TABLE 18.3. Return rate categories for LBBA prey.

HIGH	MEDIUM	LOW	SEEDS (LOWEST)
Bison	Jackrabbit	Squirrel	Mormon needlegrass
Mountain sheep	Cottontail	Mouse	Ditchgrass
Antelope	Badger	Vole	Bluegrass
Deer	Marmot		Goosefoot
	Gopher[a]		

[a] Although gophers are small, experimental return rates are reported as similar to other medium-sized animals (Simms 1987). Gophers are included as small in another publication (Ugan and Bright 2001), and the general trends in LBBA prey choice do not change. Prey for which no published rates are available are grouped with other animals of similar size.

Variation in Subsistence Technology

In order to evaluate the proposed relationship between changing foraging opportunities and changes in technology, we examine evidence for differential investment in three kinds of tools: milling stones, ceramics, and chipped stone tools. Every archaeologist is familiar with the range of variation one can find in any of these tool categories. Milling stones vary from locally procured flat, unmodified slabs, to heavily troughed or shaped metates and mortars (e.g., Adams 1993, 1999; Basgall 1987; Moratto 1984; Russell 1989). In western North America, pottery includes crude, sun-baked pots (e.g., Kelly 1964:77, 186) and intricately painted, carefully smoothed vessels (e.g., Colton and Hargrave 1937). Pots can vary dramatically in terms of wall thickness, temper size and surface improvements (e.g.,

TABLE 18.4. Hearth contents in the LBBA (from Bright 1999a).

Seed Resources

HEARTH SIZE	n	MEAN SEED COUNT	% WITH SEEDS	% WITH > 14 SEEDS
Large	122	4.3 ± 11.9	32.0	5.7
Small	82	24.3 ± 101.5	66.0	18.0

Game Resources

HEARTH SIZE	n	MEAN NISP LARGE	MEAN NISP MEDIUM	MEAN NISP SMALL	% WITH LARGE	% WITH MEDIUM	% WITH SMALL
Large	28	0.9 ± 1.5	9.4 ± 23.8	2.1 ± 4.2	42.9	42.9	35.7
Small	14	1.9 ± 7.0	5.0 ± 13.6	43.6 ± 123.6	25.0	25.0	67.9

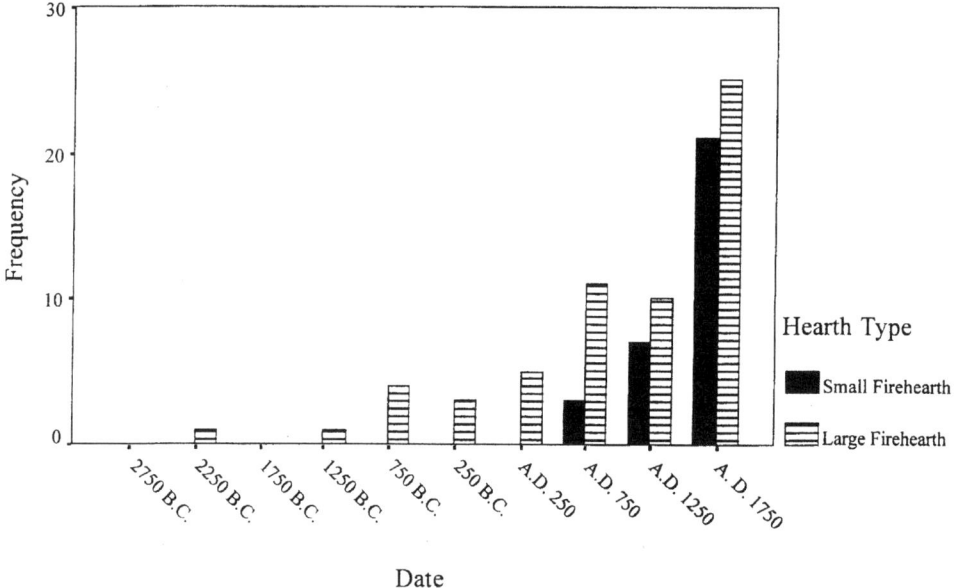

FIGURE 18.2. The temporal distribution of calibrated radiocarbon dated firehearths in the Little Boulder Basin Area.

Braun 1983; Bright and Ugan 1999; Simms et al. 1997; Whalen 1994:70–91). Chipped stone tools range from simple, unmodified flakes knocked off locally available rocks to finely flaked, symmetrical bifaces made from distantly acquired materials (e.g., Kelly 1985; Parry and Kelly 1987). In the sections that follow, we propose a number of hypotheses concerning how investment in these three subsistence-related technologies should vary across the cal. AD 1300 threshold in the LBBA.

Predictions and Results

Combining the floral and faunal data from the two sizes of firehearths in Table 18.4 and the temporal distribution of those hearths in Figure 18.2, it appears that seed resources appear early, but initially represent only a minor contribution to local diets. They become common after cal. AD 1200, when the small hearth type appears in number. Larger game is exploited throughout the sequence, but makes a relatively small contribution to local diets after AD 1300. A series of hypotheses follow from these general patterns.

Milling Stones

We include manos, metates, mortars and pestles as milling stones. These tools reduce handling times in several ways. By reducing food particle size and allowing indigestible fibrous elements to be removed, milling reduces cooking times and allows more food to be processed by the gastrointestinal tract (Stahl 1989). Variation in the size and form of milling stones affects how quickly these benefits can be realized by dictating the amount of a resource that can be ground at once and by protecting against loss. Improvements in milling efficiency can be achieved by using raw materials of appropriate coarseness, even if this includes non-local materials acquired at some distance, and by shaping stones to increase milling area and to better retain ground materials (Hard 1990; Horsfall 1987; Russell 1989; Stahl 1989). Intentional shaping and acquisition of non-local raw materials may therefore indicate increasing investment on the part of a forager. Although grinding stones can be used for a variety of tasks (e.g., hide tanning) we think their most common use in the LBBA is as a seed-processing tool. Almost all grinding stones are recovered from within a few meters of a firehearth (Birnie 1996), most commonly the small, seed hearths.

Hypothesis 1: Since milling stones can be used with minimal initial costs (a naturally occurring flat, unmodified rock found nearby) and because some milling is necessary to make seeds digestible, we expect to see milling stones appear as soon as seed resources enter diets. Investment, however, should be minimal. The earliest LBBA milling stones should be made from locally available material, and show no investment in terms of intentional shaping.

Seeds initially appear in limited quantities ca. 2250 BC (Coulam 1996). One milling stone made of locally available stone (Birnie 1996) is present in an occupation which dates between 2200 and 850 BC and is made from locally available material. Contrary to our prediction, it is intentionally shaped.

Hypothesis 2: Milling stones should increase in frequency at AD 1200–1300, when the small hearths appear. Foragers should invest time producing more efficient milling stones in order to offset increases in the time they spend handling seed resources. Therefore, post-AD 1300 milling stones should consist of better suited, non-local material and show evidence of intentional shaping.

Table 18.5 summarizes milling stone data across the AD 1300

threshold. Two-thirds of LBBA milling stones date to post-1300, and non-local material is used only post-1300. Finally, 75 per cent of the post-1300 assemblage is intentionally shaped.

Ceramics

Clay vessels offer a number of technological advantages over other technologies (Arnold 1985:128–44). A forager is able to handle a variety of resources with ceramic vessels, including seeds. She is able to reduce handling time of a resource by processing more of it at once, and she can control this effect as she chooses by altering the size of the ceramic vessel. She may also extract a nutritional benefit otherwise unavailable (e.g., Reid 1990). Cooking improves digestibility and detoxifies some resources (e.g., starch). Moist heat (e.g., boiling) enhances digestibility more than dry heat (Stahl 1989:181). Thus, ceramic cooking vessels have an important effect on reducing handling time. By controlling vessel quality, a forager can control the use-life of the vessel, altering the duration over which these benefits are realized.

In the Great Basin, clay and temper need to be procured and ground, clay must be coiled and the coils must be added together and smoothed out, all before the pot is fired (Baldwin 1950). Spending time in any of these activities improves resistance to thermal shock, spalling, cracking and shrinkage (Braun 1983:118–19; Bright and Ugan 1999; Bright et al. 1998; Kingery et al. 1986:768–813; Kirchner 1979:1–12; Schiffer at al. 1994:197; Simms et al. 1997; Skibo and Schiffer 1987:93), as well as efficiency of heat transfer and resistance to thermal and mechanical shock (Reid 1990). All of these factors affect the utility and performance of the vessel.

Ceramic vessels are more costly than the most basic milling stones. Generally speaking, the tech investment model suggests that such high start-up costs are economically more acceptable when they represent a small fraction of the total time spent handling a resource. This is more likely to occur later in time in the LBBA, when the archaeology suggests longer residence times, and when seeds make a larger contribution to local diets. Once adopted, investment in ceramic manufacture should continue to increase as the total amount of resource handled increases.

We think that ceramic vessels are most commonly used in tandem with grinding stones for handling seed resources. In the LBBA, 90 per cent of ceramics are recovered directly beside a small, seed hearth (Bright 1999a, 1999b).

Hypothesis 3: If it is the case that ceramic vessels represent more time invested than basic milling stones, they should appear later in time, when seeds make such a large contribution to diets that it pays to go through the extra effort.

Figure 18.3 presents calibrated radiocarbon dates associated with seven different ceramic vessels. These dates are taken from radiocarbon-dated firehearths with good associations between hearth and sherds. Though the beginnings of some two-sigma ranges reach back before cal. AD 1200, earliest mid-points clus-

TABLE 18.5. Changes in groundstone technology in the LBBA.

TIME	COUNT (%)	NON-LOCAL MATERIAL	INTENTIONALLY SHAPED
Before AD 1300	5 (38%)	0%	20%
After AD 1300	8 (62%)	25%	75%

TABLE 18.6. Investment in LBBA pottery.

VARIABLE	BROWN	GRAY
Wall thickness (mm)	5.73 ± 1.3 (n = 234)	5.00 ± 0.9 (n = 56)
Temper size (mm)	1.53 ± 1.1 (n = 257)	0.47 ± 0.4 (n = 58)
Surface preparation	13% (n = 257)	27.6% (n = 58)

ter at cal. AD 1200 and most of the two-sigma ranges fall post-1200. Thus, ceramic technology does not appear in the LBBA until handling times associated with taking seed resources represent such a significant cost that it pays to improve handling efficiency by using costly ceramic vessels.

Hypothesis 4: Since the relative contribution of seeds to local diets continues to increase after AD 1300, investment in ceramic manufacture should increase as well.

Investment data on brown- and greywares in the LBBA (Bright 1999b; Bright and Ugan 1999) reveal that greywares exhibit greater investment than brownwares do, in that they have thinner walls, finer temper and are more often smoothed and polished on their surfaces (Table 18.6). We have no evidence of functional differences between the two types of wares, so we predict that greywares, then, ought to appear later than brownwares. Three out of four vessels that date between cal. AD 1200 and 1600 are brownware vessels, while two out of three vessels that date post cal. AD 1600 are greyware vessels (Figure 18.3).

Chipped Stone Tools

Chipped stone tools have a wide range of uses, many of which are directly or indirectly related to the procurement and processing of larger game. Stone tools can also be used to procure plants and small animals, but the Great Basin ethnographic record suggests they were not commonly used this way. Small game were often acquired with snares, traps or sticks, while seed-beaters, winnowing trays, basketry, pottery and milling stones were more common plant-processing technologies (Steward 1938).

We rely on a simple distinction between formal/bifacial tools versus expedient/core-flake. We do this because this distinction is commonly made in North American lithic analyses, and because most archaeologists will accept that formal/bifacial tools represent more time invested in manufacture than simple flake tools. Examples of formal tools include projectile points of all types, knives and simple bifaces. Expedient tools include

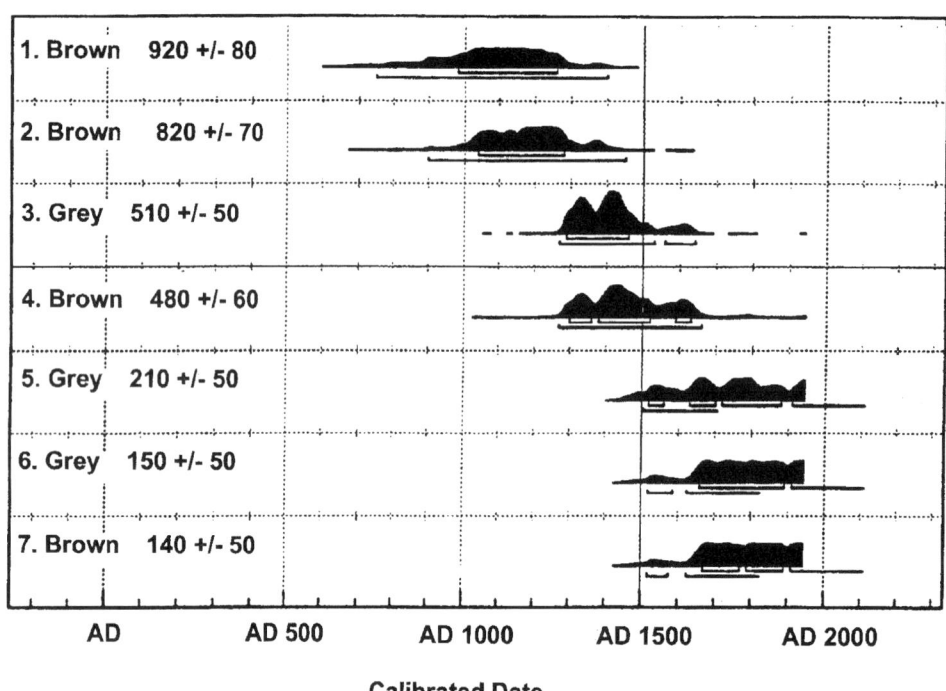

1. Site 26Eu1483, Rock-filled Firepit 5. Beta - 69150.
2. Site 26Eu1483, Firepit with Rocks 3. Beta - 69152.
3. Bootstrap Bench, Firepit 2. Beta - 109932.
4. Round Mountain Camp, Firepit 1. Beta - 96777.
5. Site 26Ek4688, Firepit 2. Beta - 132867.
6. Site 26Ek6487, Firepit 2. Beta - 129155.
7. Site 26Ek4690, Firepit 1. Beta - 69176.

FIGURE 18.3. Distribution of calibrated radiocarbon dates associated with brown- and grayware vessels in the Little Boulder Basin Area.

scrapers, choppers, denticulates and modified flake tools. Because over 90 per cent of all toolstone found in LBBA archaeological assemblages comes from the nearby Tosawihi quarry (La Fond 1996), distance to source is not an important factor.

Hypothesis 5: The reduced emphasis on large and medium game after AD 1300 should be accompanied by a shift away from higher-cost bifacial tools in favor of more expedient core/flake technology.

Between AD 700 and 1300, formal, bifacial tools constitute 96.8 per cent of all chipped stone tools while expedient core/flake tools make up only 3.2 per cent. After AD 1300, there is a shift to 85.1 per cent bifacial and 14.9 per cent core/flake. These data are generally consistent with Hypothesis 5.

DISCUSSION

These results support the claim that LBBA foragers respond to changes in diet by trading-off investment in one technology for investment in another. Specifically, foragers seem to be trying to reduce the handling times that various kinds of prey require. When game constitutes a larger portion of the local diet, LBBA foragers invest more time into making the kinds of tools that cut the time they spend handling that game. But, when seeds make a larger contribution, they relax the time investment they put into game-related technologies and invest more time into making the kinds of technology that help them handle seeds.

Although simplistic in a way, these results suggest that the relationship between handling time and technological investment is worth attention, for at least two reasons. First, the changes in LBBA technology presented here are often taken as evidence of population replacement in the Great Basin in general (see papers in Madsen and Rhode 1994), or changes in raw material availability (Andrefsky 1994), gender roles (Gero 1991) or mobility (Kelly 1985). While aspects of all of these are clearly important to technological change, the issue of handling time is likewise important to each of them.

For example, Gero (1991) describes the role of "expedient" stone tools in activities commonly performed by women. If the kinds of prey women commonly handle with stone tools present few or no handling problems, it may be little surprise that women do not spend time making high-investment "formal tools." However, if we can generalize that women's resources commonly include plants, we might expect them to spend more time improving grinding stones or seed-beaters.

Because efficiently exploiting prey is a problem all foragers face, handling time problems are also common to all foragers, everywhere. Therefore, paying attention to the relationship between handling-time and technological change complements existing models of technological change.

Second, this model can be applied to technological changes in many settings across the world, as archaeologists from any region can identify locally important transitions in technological innovation. The time period we address (c. AD 1200–1300) is one such transition in many areas of the Great Basin (Bettinger and Baumhoff 1982; Madsen and Rhode 1994). Others might include the Australian Small Tool Tradition (Mulvaney and Kamminga 1999:44–5, 230–1), the appearance of the first pottery vessels in the Japanese Palaeolithic (Reynolds and Barnes 1984), or the spread of Iron Age technologies across southern Africa (Schmitt 1996).

Another important episode of technological innovation is captured in the sequence from Middle Palaeolithic flake technologies to Upper Paleolithic blades, burins and scrapers (Klein 1999:522–4). Stiner et al. (2000) present a detailed description of subsistence transitions in Italy and Israel across the Middle to Upper Palaeolithic transition. They demonstrate that Middle Palaeolithic hunters preferred slow-moving or sessile prey types, like tortoises and shellfish. These kinds of prey present a certain set of handling problems. Although they may be easy to catch, hard shells prevent hunters from processing them easily. This means that the pursuit component of handling time is low, but the processing component is high. Some amount of technology might be dedicated to the processing problems these prey present.

In the Upper Palaeolithic, hunters supplement tortoise and shellfish with more mobile prey, like birds, rabbits and hares. These prey use escape as a defense mechanism instead of a hard shell. This poses a different set of handling problems that emphasize pursuit rather than processing. A simple prediction from the tech investment model is that Middle Palaeolithic technologies should be more focused on reducing the processing component of handling time, and that the elaboration of Upper Palaeolithic technology is somehow tied to reducing pursuit time.

In order to apply the model to cases like this, we need to ask, "What are the gains that come from spending time making different kinds of tools?" It may well be the case that an edge need be only so sharp for a given task, and that spending more time sharpening it further is a waste. As yet, we lack a framework for understanding how any unit of time invested in tools translates into some decreased handling time.

To begin building such a framework we need to know at least two things. First, we need to know the costs and benefits of investing more time into different subsistence technologies. Fortunately, this topic has been studied at length for many common artifact classes (e.g., Adams 1999; DeBoer and Lathrap 1979; Ellis 1997; Knecht 1996; Rye and Evans 1976; Stahl 1989).

Second, we also need to know more about how prey are handled, and the effects of varying technological investment on different tasks. For example, pulling an apple from a tree does not require any technology at all. Taking five hundred apples from a tree does not necessarily require any technology either, but it does take a lot of time. How much time should one spend building a tool that helps acquire apples twice as fast, or five times as fast? By studying the direct effects of technology on the handling time of various prey, we can develop a strong basis for understanding variation in subsistence technology.

Conclusions

A formal model, based in a body of theory usually applied to non-human animals, is used to derive hypotheses about the nature of variation in subsistence technology. We present technological decisions as being motivated by the single goal of improving return rates by cutting handling time, but also point out that decisions in this regard constitute trade-offs of one thing over another. The hypotheses we generated from the model have helped us understand changes in the nature of Great Basin milling stones, ceramics, and chipped stone tools.

Over time the kinds of prey that LBBA foragers may handle with chipped stone tools represent a smaller proportion of their diet, and are therefore handled less often, so foragers spend less time making stone tools. At the same time, seeds become more important to local diets. Accordingly, foragers invest more time into making tools that can be used to handle seeds. In both cases, changes can be understood as a response to the handling time these prey require.

Acknowledgments

This paper is a summary of a symposium presented at the 27th Great Basin Anthropological Conference in Ogden, Utah. In addition to the authors, presenters included Peter Ainsworth, Kristen Hawkes, Greg Miller, James O'Connell and Alan Rogers. Jack Broughton and Steve Kuhn acted as discussants. We gratefully thank all involved, including those members of the audience who approached us afterwards with helpful insights. Many people graciously read and commented on several drafts of this article, and to them we extend our thanks. Alan Schroedl at P-III Associates, Inc. (Salt Lake City) has kindly allowed us access to the LBBA data, some of which were unpublished at the time of this writing. We thank Stephen Shennan for the invitation to participate in this volume, and two reviewers for their helpful comments on an earlier draft. We also thank everyone who managed to find us two-and-a-half seasons of Senators' games during the preparation of the symposium and this article.

References

Adams, J. L.
1993   Toward understanding the technological development of manos and metates. *Kiva* 58:331–44.
1999   Refocusing the role of food-grinding tools as correlates of

subsistence strategies in the U. S. southwest. *American Antiquity* 64:475–98.

Andrefsky, W., Jr.
1994 Raw-material availability and the organization of technology. *American Antiquity* 59:21–34.

Arnold, D. E.
1985 *Ceramic Theory and Culture Process*. Cambridge University Press, Cambridge.

Baldwin, G. C.
1950 The pottery of the Southern Paiute. *American Antiquity* 16:50–6.

Barlow, K. R., D. Metcalfe
1996 Plant utility indices: two Great Basin examples. *Journal of Archaeological Science* 23:351–71.

Basgall, M. E.
1987 Resource intensification among hunter-gatherers: acorn economies in prehistoric California. *Research in Economic Anthropology* 9:21–52.

Bettinger, R. L., M. A. Baumhoff
1982 The Numic Spread: Great Basin cultures in competition. *American Antiquity* 47:485–503.

Birnie, R. I.
1996 Groundstone technology in Little Boulder Basin. In *Open Site Archaeology in Little Boulder Basin: 1993–1994, Data Recovery Excavations in the North Block Tailings Impoundment Area, North-Central Nevada*, edited by A. R. Schroedl, pp. 707–28. Cultural Resources Report 5039-01-9601, Vol. II. P-III Associates, Salt Lake City.

Braun, D. P.
1983 Pots as tools. In *Archaeological Hammers and Theories*, edited by J. A. Moore and A. S. Keene, pp. 107–34. Academic Press, New York.

Bright, J. R.
1999a Little Boulder Basin Area thermal feature analysis. In *Open-Site Archaeology: 1996 Bootstrap Data Recovery Excavations, North-Central Nevada*, edited by A. R. Schroedl, pp.340–62. Cultural Resources Report 5082-01-9804, P-III Associates, Salt Lake City.
1999b Little Boulder Basin Area pottery: modeling pottery manufacture and use among hunter-gatherers. In *Open-Site Archaeology: 1996 Bootstrap Data Recovery Excavations, North-Central Nevada*, edited by A. R. Schroedl, pp. 363–76. Cultural Resources Report 5082-01-9804, P-III Associates, Salt Lake City.

Bright, J. R., A. Ugan
1999 Ceramics and mobility: assessing the role of foraging behavior and its implications for culture-history. *Utah Archaeology* 12:17–29.

Bright, J. R., S. R. Simms, A. Ugan
1998 Residential mobility and ceramic variation in Utah's West Deserts. Poster presented at 26th Great Basin Anthropological Conference. Manuscript in possession of authors.

Colton, H. S., L. L. Hargrave
1937 *Handbook of Northern Arizona Pottery Wares*. Museum of Northern Arizona Bulletin 11. Museum of Northern Arizona, Flagstaff.

Coulam, N.
1996 Plant utilization in Little Boulder Basin. In *Open Site Archaeology in Little Boulder Basin: 1993–1994, Data Recovery Excavations in the North Block Tailings Impoundment Area, North-Central Nevada*, edited by A. R. Schroedl, pp. 729–53. Cultural Resources Report 5039-01-9601, Vol. II, P-III Associates, Salt Lake City.

DeBoer, W. R., D. W. Lathrap
1979 The making and breaking of Shipibo-Conibo ceramics. In *Ethnoarchaeology: Implications of Ethnography for Archaeology*, edited by C. Kramer, pp. 102–38. Columbia University Press, New York.

Ellis, C. J.
1997 Factors influencing the use of stone projectile tips. In *Projectile Technology*, edited by H. Knecht, pp. 37–74. Plenum Press, New York.

Gero, J. M.
1991 Genderlithics: women's roles in stone tool production. In *Engendering Archaeology: Women and Prehistory*, edited by J. M. Gero and M. W. Conkey. Blackwell, Oxford.

Hard, R. J.
1990 Agricultural dependence in the mountain Mogollon. In *Perspectives on Southwestern Prehistory*, edited by P. E. Minnis and C. L. Redman, pp. 135–49. Westview Press, Boulder, CO.

Horsfall, G.
1987 A design theory perspective on variability in grinding stones. In *Lithic Studies among the Highland Maya*, edited by B. Hayden, pp. 332–77. University of Arizona Press, Tucson.

Kelly, I. T.
1964 *Southern Paiute Ethnography*. Anthropological Papers No. 69 University of Utah Press, Salt Lake City.

Kelly, R. L.
1985 Hunter-gatherer mobility and sedentism: a Great Basin study. Unpublished PhD dissertation, University of Michigan.

Kingery, W. D., H. K. Bowen, D. R. Uhlman
1976 *Introduction to Ceramics*. Wiley, New York.

Kirchner, H. P.
1979 *Strengthening of Ceramics*. Marcel Dekker, New York.

Klein, R. G.
1999 *The Human Career: Human Biological and Cultural Origins*, 2nd edn. University of Chicago Press, Chicago.

Knecht, H.
1997 Projectile points of bone, antler and stone: experimental explorations of manufacture and use. In *Projectile Technology*, edited by H. Knecht, pp. 191–212. Plenum Press, New York.

La Fond, A. D.
1996 Chipped stone technology in Little Boulder Basin. In *Open Site Archaeology in Little Boulder Basin: 1993–1994, Data Recovery Excavations in the North Block Tailings Impoundment Area, North-Central Nevada*, edited by A. R. Schroedl, pp. 673–706. Cultural Resources Report 5039-01-9601, Vol. II, P-III Associates, Salt Lake City.

La Fond, A., J. B. Jones
1995 *Data Recovery Excavations at the Yaha Site: An Open Prehistoric Camp Site Along Rodeo Creek, Northern Eureka County, Nevada*. Cultural Resources Report 5010-01-9301. P-III Associates, Salt Lake City.

La Fond, A., B. L. Tipps, M. K. Stratford
1995 *Data Recovery Excavations at Site 26EU1494*. Cultural Resources Report 5028-01-9510. P-III Associates, Salt Lake City.

Madsen, D. B., D. Rhode
1994 *Across the West: Human Population Movement and the Expansion of the Numa*. University of Utah Press, Salt Lake City.

Moratto, M. J.
1984 *California Archaeology*. Academic Press, Orlando, FL.

Mulvaney, J., J. Kamminga
1999 *Prehistory of Australia*. Smithsonian Institution Press, Washington, D.C.

O'Connell, J. F.
1995 Ethnoarchaeology needs a general theory of behavior. *Journal of Archaeological Research* 3:205–55.

O'Connell, J. F., K. Hawkes
1981 Alyawara plant use and optimal foraging theory. In *Hunter-Gatherer Foraging Strategies: Ethnographic and Archaeological Analyses*, edited by B. Winterhalder and E. A. Smith, pp. 99–125. University of Chicago Press, Chicago.

Parry, W. J., R. L. Kelly
1987 Expedient core technology and sedentism. In *The Organization of Core Technology*, edited by J. Johnson and C. Morrow, pp. 285–304. Westview Press, Boulder, CO.

Reid, K. C.
1990 Simmering down: a second look at Ralph Linton's "North American Cooking Pots." In *Hunter-Gatherer Pottery of the Far West*, edited by J. M. Mack. Anthropological Papers No. 73. Nevada State Museum, Carson City.

Reynolds, T. E. G., G. L. Barnes
1990 The Japanese Paleolithic: a review. *Proceedings of the Prehistoric Society* 50:49–63.

Russell, K. W.
1989 Groundstone. In *Kayenta Anasazi and Navajo Ethnohistory on the Northwestern Shonto Plateau: The N-16 Project*, edited by A. R. Schroedl, pp. 649–89. P-III Associates, Salt Lake City.

Rye, O. S., C. Evans
1976 *Traditional Pottery Techniques of Pakistan: Field and Laboratory Studies*. Smithsonian Contributions to Anthropology No. 21. Smithsonian Institution Press, Washington, D.C.

Schiffer, M. B., J. M. Skibo, T. C. Boelke, T. C. Neupert, M. Aronson
1994 New perspectives on experimental archaeology: surface treatments and thermal response of the clay cooking pot. *American Antiquity* 59:197–217.

Schmitt, P. R. (ed.)
1996 *The Culture and Technology of African Iron Production*. Gainesville: University Press of Florida.

Schroedl, A. R.
1995 *Open Site Archaeology in Little Boulder Basin: 1992 Data Recovery Excavations in the North Block Heap Leach Facility Area, North-Central Nevada*. Cultural Resources Report 492-02-9317. P-III Associates, Salt Lake City.
1996 *Open Site Archaeology in Little Boulder Basin: 1993–1994, Data Recovery Excavations in the North Block Tailings Impoundment Area, North-Central Nevada*. Cultural Resources Report 5039-01-9601, Vols I–II. P-III Associates, Salt Lake City.
1997 *Data Recovery Excavations at Site 26EK6232, Eureka County, Nevada*. Cultural Resources Report 5063-01-9703. P-III Associates, Salt Lake City.
1999 *Open-Site Archaeology: 1996 Bootstrap Data Recovery Excavations, North-Central Nevada*. Cultural Resources Report 5082-01-9804. P-III Associates, Salt Lake City.

Simms, S. R.
1983 Comments on Bettinger and Baumhoff's explanation of the "Numic Spread" in the Great Basin. *American Antiquity* 48:825–34.
1985 Acquisition cost and nutritional data on Great Basin resources. *Journal of California and Great Basin Anthropology* 7:117–26.
1987 *Behavioral Ecology and Hunter-Gatherer Foraging: An Example from the Great Basin*. Oxford: BAR International Series 381.

Simms, S. R., A. Ugan, J. R. Bright
1997 Plain-ware ceramics and residential mobility: a case study from the Great Basin. *Journal of Archaeological Science* 24:779–92.

Skibo, J. M., M. B. Schiffer
1987 The effects of water on processes of ceramic abrasion. *Journal of Archaeological Science* 14:83–96.

Stahl, B.
1989 Plant-food processing: implications for dietary quality. In *Foraging and Farming: The Evolution of Plant Exploitation*, edited by D. R. Harris and G. C. Hillman, pp. 172–94. Unwin Hyman, London.

Stephens, D. W., J. R. Krebs
1986 *Foraging Theory*. Princeton University Press, Princeton, NJ.

Steward, J. H.
1938 *Basin-Plateau Aboriginal Sociopolitical Groups*. Bureau of American Ethnology Bulletin 120. Smithsonian Institution, Washington, D.C.

Stiner, M. C., N. D. Munro, T. A. Surovell
2000 The tortoise and the hare: small-game use, the broad spectrum revolution and paleolithic demography. *Current Anthropology* 42:39–73.

Stratford, M. K., A. D. La Fond
1995 *Data Recovery Excavation at Site 26EU2124*. Cultural Resources Report 5046-01-9518. P-III Associates, Salt Lake City.

Tipps, B. L.
1996 *Open Site Archaeology Near Upper Boulder Creek: Data Recovery Excavations at Sites 26EK5270, 26EK5271, and 26EK5274 in the East Development Area, Elko County, Nevada*. Cultural Resources Report 5032-01-9609. P-III Associates, Salt Lake City.

Tipps, B. L., M. K. Stratford
1996 *Data Recovery Excavations at Site 26EU1505, Eureka County, Nevada*. Cultural Resources Report 5072-01-9606. P-III Associates, Salt Lake City.

Ugan, A., J. Bright
2001 Measuring foraging efficiency with archaeological faunas: the relationship between relative abundance indices and foraging returns. *Journal of Archaeological Science* 28:1309–21.

Ugan, A., A. Rogers
2000 Modeling the relationship between foraging returns, the optimal diet and technological investment. Paper presented at the 27th Great Basin Anthropological Conference, Bend, Oregon.

Ugan, A., A. Rogers, J. Bright
n.d. The Tech Investment Model. Manuscript in possession of authors.

Whalen, M. E.
1994 *Turquoise Ridge and Late Prehistoric Residential Mobility in the Desert Mogollon Region*. Anthropological Papers No. 118. University of Utah Press, Salt Lake City.

19

Microlithic Technology in Northern Asia

A Risk-Minimizing Strategy of the Late Paleolithic and Early Holocene

Robert G. Elston and P. Jeffrey Brantingham

Microblade technology was important in hunter-gatherer adaptations throughout northern Asia from the latter part of the late Pleistocene through the Pleistocene/Holocene transition and beyond (Chard 1974; Chen 1984; Chen and Wang 1989; Derev'anko 1998; Gai 1985; Goebel 1995; Goebel et al. 2000; Goebel and Slobodin 2001; Hoffecker et al. 1993; Kobayashi 1970; Kuzmin and Orlova 1998; Lu 1998; Seong 1998). To date, researchers in the region have focused primarily on issues of culture history, in which questions of culture origin and typology are foremost (Seong 1998). Consequently, we are faced with a welter of type lists, divided and subsumed into archaeological cultures that never quite account for everything—always some assemblage or date falls outside the general pattern. The failure of the archaeological record to conform to such categories is not a reflection of our excavation or typological skills. Rather, it is the inevitable result of variability in past human behavior that is not categorical, but situational. While the distribution and variability of microlithic technology remain important variables to track, we direct attention to issues involving the role of microlithic technology in adaptive strategies. In other words, what were the major problems microlithic technologies solved for north Asian hunter-gatherers? Answering this question will help explain the emergence, spread, and persistence of microlithic technology in northern Asia.

We approach the question from the vantage point of artifact design (Bamforth and Bleed 1997; Bleed 1987; Ellis 1997; Knecht 1993, 1997) and risk analysis employing the Z-score model (Bettinger 1991; Stephens 1981; Stephens and Charnov 1982; Winterhalder et al. 1999). First we discuss the function of Asian microblades and outline the general costs and benefits of three aspects of microlithic technology: (1) the advantages of organic points with microblade insets over simple organic points and flaked stone points, (2) the relative efficiency of biface and microblade production, and (3) the relative advantages of wedge-shaped and split-pebble microcores in terms of the Z-score model. We conclude with a review of the role of microlithic technology as a strategy to cope with increasing variability in late Pleistocene environments of northern Asia and, finally, suggest further lines of inquiry.

THE FUNCTION OF ASIAN MICROBLADES

Late Paleolithic Asian microlithic technology was employed to produce stone insets for the margins of composite tools made of organic materials such as bone, antler, and ivory. No Late Paleolithic organic components of composite weapons have been recovered in China, but Late Paleolithic examples of inset projectile points are known from Siberia (Chard 1974:fig. 1.16; Derev'anko 1998:figs. 87, 127). Both composite points and knives have been recovered from Neolithic sites in Siberia and northeast China (Chard 1974:figs. 2.14, 2.16; Guo 1995:fig. 1.11; Lu 1998:92). Modified, end-hafted microblades were also sometimes used as scrapers, drills, knives, and projectile points (Elston et al. 1997; Gai 1985; Lu 1998:92; Madsen et al. 1998). There is no evidence that inset tools were ever used as sickles in the Chinese Neolithic, where this function was performed by serrated implements of ground stone, bone, or shell (Chang 1986:figs. 49, 52, 126; Shih 1992:figs. 8, 9; Underhill 1997:fig. 4).

Thus, we follow Lu (1998) and others in assuming that most Asian pre-Neolithic microblades were intended to be used as marginal insets for projectile points made of bone, antler, or ivory. In this chapter, we refer to such points without insets as *organic points* (Knecht 1997) and to organic points with insets as *inset points*. The costs and benefits of inset points are best understood in comparison with the alternatives: simple organic points and flaked stone points.

PERFORMANCE OF ORGANIC AND FLAKED STONE POINTS

The performance characteristics of organic and flaked stone points are known from experimental replication and use, and as reflected in the ethnographic record. Both aspects have been

---

*Originally published in *Thinking Small: Global Perspectives on Microlithization*, edited by R. G. Elston and S. L. Kuhn, pp. 103–116. Archaeological Papers of the American Anthropological Association No. 12, American Anthropological Association, Arlington, VA (2002).

covered recently in papers by Heidi Knecht (1993, 1997) and Christopher Ellis (1997). Their findings are summarized briefly as follows (Table 19.1).

Flaked stone points are relatively fragile and often fail catastrophically in use. Breakage is particularly common when they hit the ground or impact trees or rocks. Importantly, stone points apparently become brittle and easily damaged in severe cold. Spears and arrows with stone points are composite weapons in which the stone component is expendable (Bamforth and Bleed 1997:129) and probably expected to fail. Some ethnographic hunters believe that stone points are more lethal when they break up on penetration, because the fragments increase tissue damage and bleeding. Compared to the costs of fabricating shafts, with their nocks or sockets, fletching, and hafts, flaked stone points are cheap and quick to make. The stone-tipped projectile is designed to allow replacement of a damaged point without having to create an entirely new weapon. The use of a foreshaft may offer additional protection to the main shaft, as well as increased ease in point replacement. However, making foreshafts for stone points and fitting the points to them are additional costs in manufacture, and the use of a foreshaft produces a weaker weapon than one with a solid shaft.

Organic points are made by splitting the bone or antler and shaping it by scraping and grinding. Organic points are more durable than stone points, less likely to break in use, and more easily repaired if dulled or broken. The time required to make an organic point is greater than that for most flaked stone points. However, organic points have longer use-lives, and one may spend more time making a series of expendable stone points than making one organic point (Knecht 1997). Organic points cause less-lethal wounds than stone points with their sharp cutting edges that promote tissue damage and bleeding.

As summarized in Table 19.2, Ellis (1997) has evaluated the factors influencing choice of stone or organic points in a survey of over 100 ethnographic cases. He found that stone arrow and spear points are used almost exclusively for large and dangerous game (bears and large herd animals) and in warfare where killing the enemy is the object. However, the use of thrusting spears with stone points tended to be limited to situations in which there was little danger to the user or in which a number of replacement weapons were available. Organic points were used when wielding thrusting spears against smaller herd animals, against dangerous animals when no replacement weapons were available, when multiple thrusts were to be given in warfare, on arrows used in cold weather (when stone points are brittle and easily broken), and when throwing spears in heavy underbrush.

Ellis's review of ethnographic data suggests that various factors including ambient temperature and prey type condition the selection and use of organic or flaked stone points. The primary factor underlying these situational decisions appears to be the cost or risk associated with weapon failure (Bamforth and Bleed 1997; Torrence 1989). When the potential cost is relatively low, stone points tend to be used. When the cost of failure is high,

TABLE 19.1. Performance characteristics of organic and flaked stone points.

Organic Points	Stone Points
Material limits morphology	More varied morphology possible
Considerable time to manufacture	Can be made relatively quickly
Cause less tissue damage	Cause great tissue damage, bleeding
More durable: • unaffected by cold weather • rarely damaged in transport • rarely break in use	Less durable: • more brittle in cold weather • easily damaged in transport • invariably break on missed shot • frequently shatter when penetrating prey
Easily repaired/resharpened	Difficult or impossible to repair
Less likely to be used with foreshaft	More often used with foreshaft
Reusable	Expendable

TABLE 19.2. Ethnographic situational use of organic and flaked stone points.

Organic Points Used	Stone Points Used
On thrusting spears without foreshafts against herd animals	When hunting large game
Against large/dangerous animals without replacement weapons	When attacking dangerous solitary animals (bears)
On thrusting spears used in hand-to-hand fighting	On thrusting spears when danger to user is slight
On arrows used in cold weather	For killing enemies at a distance
On spears cast in heavy brush	

reliable weapons are called for (*sensu* Bleed 1987), and organic points are more likely to be employed.

Costs and Benefits of Inset Points

Creating inset points involves manufacturing, using, and maintaining a complex toolkit in which we might find hard and soft hammers, pressure tools, holding devices, anvils, splitting wedges, grinders, scrapers, slotting tools (burins, gravers, saws), and mastic (Flenniken 1987; Tabarev 1997). Toolkits for *either* flaked stone or organic points contain fewer items.

The use of inset points in a weapons system is costly. It requires a large investment to learn how to produce microblades of a consistent width and thickness and how to do it quickly enough to meet demands in field conditions (Bamforth and Bleed 1997:132). Making microblades takes more time than producing bifaces from comparable amounts of raw material (Flenniken 1987). Making appropriate insets from microblade blanks and installing them in the margins of slotted organic points requires still more time. It may also be costly to learn to create slots of consistent width and depth and time consuming to produce them.

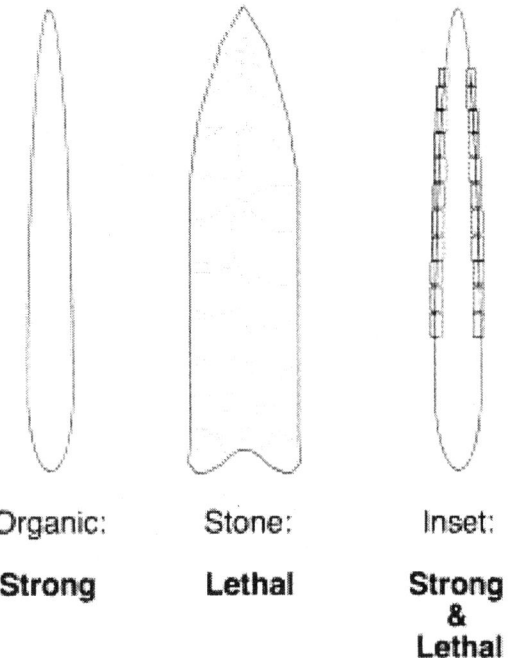

FIGURE 19.1. Organic, stone, and inset organic points compared.

TABLE 19.3. Experimental microblade and biface production based on data in Flenniken (1987:122–123).

REDUCTION STAGE	MEAN WEIGHT (g)	PREVIOUS STAGE	% OF PREVIOUS STAGE
Microblades			
Initial flake	not given		
Stage 2 biface	219.1		
Prepared bifacial core (w/ski spall removed)	158.8	Stage 2 biface	72.48
Exhausted microblade core	33.4	prepared core	21.03
Prepared microblade core wt. minus exhausted core wt.	125.4		
Average of 101 usable microblades per core reduction	not given		
Microblades per reduction of prepared core (equal to 41.6% of prepared core wt.)	66.06	Stage 2 biface	30.15
Insets produced from microblades	33.0	Stage 2 biface	15.06
Bifaces			
Initial flake	not given		
Stage 2 biface (ca. the same as for microblade trajectory)	219.1		
Completed biface	64.6	Stage 2 biface	29.5

Ellis (1997) does not cite ethnographic cases of inset point use. However, his review of ethnographic uses of organic points and flaked stone points suggests that the best design features of both are combined in inset points. Inset organic points can cause tissue damage and bleeding similar to that caused by stone points with less likelihood of breaking; they are both strong and lethal (Figure 19.1). Inset points can be used more than once, eliminating the need for frequent tip replacement and use of foreshafts. At the same time, inset points are highly maintainable, requiring only replacement of lost or broken insets and occasional tip resharpening. Batches of insets can be produced in anticipation of need (Bamforth and Bleed 1997:132; Kuhn 1994).

As we have suggested, however, the cost of producing inset points is much higher than that for weapons with organic or flaked stone tips. The payoff for the increased cost of inset weapons is their performance in situations in which the cost of failure is catastrophic and/or deadly: when they must be used multiple times without replacement, when the prey is a key resource that must be taken during a limited time, when the prey is a dangerous animal or a human enemy to be engaged at close quarters, or when the weapon is used in very cold environments.

Relative Efficiency of Bifaces and Microblades in Production of Cutting Edge

After raw material is procured, the first step in preparation of virtually all varieties of wedge-shaped cores involves some degree of bifacial reduction (Bleed 2002). Thus, we must assume that all prehistoric knappers who made microblades from wedge-shaped cores could have produced bifaces to use as tools if they had chosen to do so. In some cases, bifaces used as tools and points do accompany microblades in lithic assemblages (Dyuktai Cave, Siberia; Uenodaira and Araya, Japan; Hutouliang and Pigeon Mountain Basin, China); in others, unifacially modified flakes (Afontova Gora-2, Siberia) and macroblades (Kokorevo-2, Siberia) serve the same functions as bifacial knives and points.

While this variability is no doubt multidimensional, we explore a few aspects of it by comparing the costs and benefits of producing bifaces and microblades. Our analysis makes use of experimental data (Table 19.3) provided by Flenniken (1987). To compare time and raw material consumption of the two techniques, Flenniken experimentally produced microblades from replicate Dyuktai bifacial cores, and, as well, made finished bifaces. As blanks to start the reduction process, he used large flakes of comparable size for both the wedge-shaped cores and bifaces. Flenniken (1987:122–132) concluded from his experiment that production of bifaces is the least economical use of toolstone because, while usable microblades (average 101 per core) comprised 41.6 percent of prepared core weight, finished bifaces comprised only 29.5 percent of blank weight. However, our analysis suggests this conclusion may be incorrect because Flenniken apparently did not calculate wastage of each technique as a proportion of the original weight of the raw material blank, nor did he include the final production step of microblade production.

Flenniken (1987) does not report the mean weights of the flake blanks used as the initial starting points of reduction. We are able to estimate, however, the weights of key products and by-products by using the mean weights of Stage 2 bifaces as a starting point and working back through the process using the proportions of products and debitage reported. Flenniken reports a mean weight of 219.1 gm for the Stage 2 biface "blanks" used to make microblade cores. We assume that when Flenniken refers to "blanks" for the biface trajectory, he also means Stage 2 bifaces. This assumption is important and reasonable given (1) the statement that the reduction process started in each case with flakes of comparable size and (2) that the reported starting weights for microblade cores (Stage 2 biface) and formal bifaces ("blank") are equivalent. Comparisons based on these assumed equivalencies indicated that as much as 69.85 percent of the initial raw material is wasted in microblade core preparation, reduction, and maintenance; 30.15 percent of the initial raw material is converted into microblades. The production of completed bifaces uses 70.5 percent of the initial raw material; 29.5 percent of the initial blank weight is retained in the final biface.

This comparison suggests that there is *no* clear advantage in raw material economy between the two techniques. However, there are further potential costs and benefits associated with these technologies. Importantly, a handful of microblades is merely a bundle of stone splinters, not a finished tool. Insets, commonly made by snapping off both ends of the microblade, comprise about one-third to one-half of the intact microblade weight. Thus, 33 gm of snapped microblade insets from 66 gm of microblades per core in Flenniken's experiment is only 15.06 percent of the initial blank weight, less than half the yield of finished bifaces. Flenniken also points out that microblades cost more in time to produce; he took 51 minutes on average to make about 100 microblades, or about twice the time per core (exclusive of heat treatment) used to make each biface. Thus, compared with microblade production, biface production appears to take less time and to make a more economical use of raw material.

At the same time, we are aware that bifaces and microblades may represent different technological modalities that are not directly comparable in terms of a general economic currency such as raw material weight. Recall that the focus of our discussion has been on the relative performance characteristics of inset and flaked stone points. Flaked stone points are so effectively lethal because of their *cutting edges*. If we assume that length of usable cutting edge is the currency to be maximized (Kuhn 1994), and production time is the cost to be minimized, we now have an equivalent basis for comparison. Flenniken's experimental bifaces averaged about 15 cm in length, thus yielding a maximum of 30 cm of cutting edge (less if hafted) per biface. The mean microblade yield per core was 101. Assuming each microblade could provide an inset 2 cm long, the average core would provide 202 cm of cutting edge, nearly seven times that of the experimental biface for just twice the cost in time and the same cost in raw material.

Nevertheless, one might argue that the ability to resharpen a biface adds utility. While it is true that a biface used as a scraper or knife can be resharpened many times over a long use-life, a bifacial projectile point is likely to catastrophically fail after at most a few uses (Rondeau 1996). Most bifacial stone points probably could not be resharpened enough to match the 202 cm of inset cutting edge utility produced by microblade core reduction in the previous example. Additional utility may accrue from employing a biface as a tool prior to its service as a microblade core, as was sometimes the case in Japan (Bamforth and Bleed 1997) and China (Lu 1998), and by using biface thinning flakes and ski spalls as tools or tool blanks. While such recycling can reduce raw material demands, and can provide additional cutting edge (cf. Kelly 1988; Kuhn 1994), it is not edge of the same uniform quality provided by insets, and it does not serve the same purpose as projectile point edges. Such cost and benefit analysis seems ripe for further exploration through experimental replication and modeling.

Relative Efficiency of Boat-Shaped and Wedge-Shaped Microcores

Asian microblade cores are frequently divided into a large number of types (Chen and Wang 1989; Morlan 1976), though Seong (1998) has made a significant contribution by reducing these bewildering type lists to four basic variants. For the purposes of this chapter, we will distinguish between two basic approaches: boat-shaped cores made on split pebbles or thick flakes, and wedge-shaped cores made on bifacial blanks (Figure 19.2, A, B) (Kobayashi 1970).

Microblades can be produced from boat-shaped cores in a few steps (Morlan 1976). One primary advantage of boat-shaped cores is that they are usually produced from small, pebble-sized toolstone packages locally available in many geomorphic contexts. The use of a minimal number of technical steps in preparing boat-shaped cores appears to allow a large proportion of the pebble blanks to be turned into microblades (Seong 1998:272–274). Fewer steps in preparation may also reduce the risk of core failure since fewer things can go wrong. In contrast, bifacially prepared wedge-shaped cores require large toolstone packages, which may not be available locally. Preparation of a wedge-shaped core involves minimally the initial preparation of a biface, removing one margin (ski spall) to form the platform, adjusting the platform margins to fit the holding device, and then removing a crested blade to start the working face (cf. Figure 19.2, A; Bleed 2002; Flenniken 1987; Tabarev 1997).

Although we presently lack supporting experimental data, we believe that the advantage of the wedge-shaped core over the boat-shaped core lies in greater uniformity of the microblades produced. Microblades from wedge-shaped cores should yield microblades with less variation in width and thickness, less ventral and lateral curvature, and more units with a trapezoidal cross section—all qualities that enhance the efficiency of microblades as components of slotted organic armatures.

For example, consider a collection of microblades recovered from QG3, a Mesolithic archaeological site in Pigeon Mountain

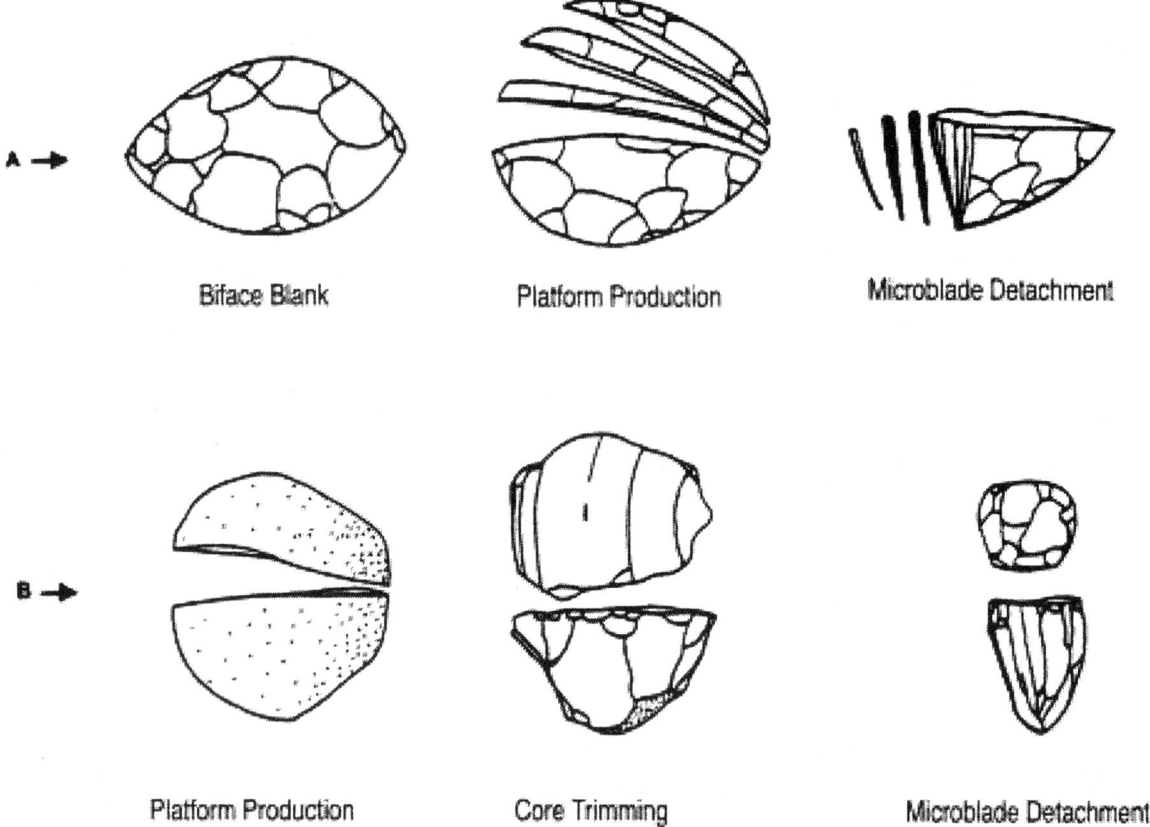

FIGURE 19.2. Asian microblade cores. A, Wedge-shaped made from biface blank; B, boat-shaped made on split-pebble blank (from Kobayashi 1970, used with permission, University of Wisconsin Press).

TABLE 19.4. Width and thickness of production and maintenance microblades from site QG3, Ningxia, China.

	n	Mean Width	σ	Mean Thickness	σ
Production	81	6.00	1.79	1.56	0.69
Maintenance	130	5.23	1.58	1.76	1.68

Basin, Ningxia, China (Elston et al. 1997; Madsen et al. 1998). These are divided into two classes: production microblades to be used as insets and maintenance microblades detached to maintain optimum core geometry and adjust core radius. The ratio of production to maintenance microblades in the QG3 sample suggests that for every production microblade, more than one maintenance microblade was removed. The former are wider and thinner, with more parallel margins and less ventral curvature, and some are retouched. Note in Table 19.4 that the thickness mean and standard deviation(s) are less for production microblades than for maintenance microblades from QG3. This suggests a concern of the QG3 knappers was to constrain production microblade thickness to a narrow, slot-fitting standard.

The manner in which wedge-shaped cores help constrain microblade variability is illustrated in a simplified model of core geometry (Figure 19.3) that demonstrates the relationships between three key variables: the radius of the core working face (from which microblades are detached) and the width and thickness of detached microblades.

Imagine that the working face of the core is a cylinder (Figure 19.3, a), which in plan view (looking down on the platform) approximates the arc of a circle with radius r (Figure 19.3, b). The width of a microblade detached from the core approximates a chord c, with microblade thickness x. The intersection of x and c forms a right angle.

By the Pythagorean theorem:

$$r^2 = (r - x)^2 + \left[\frac{c}{2}\right]^2 \qquad (19.1)$$

Rearranging equation 1 for c:

$$c = 2\sqrt{2(rx) - x^2} \qquad (19.2)$$

Core face maintenance is necessary on all microblade cores because microblade removal tends to increase r locally (Figure 19.3, c). Maintenance should be less efficient on cores with wide working faces where r can freely increase, and more efficient on wedge-shaped cores whose narrow working faces help prevent

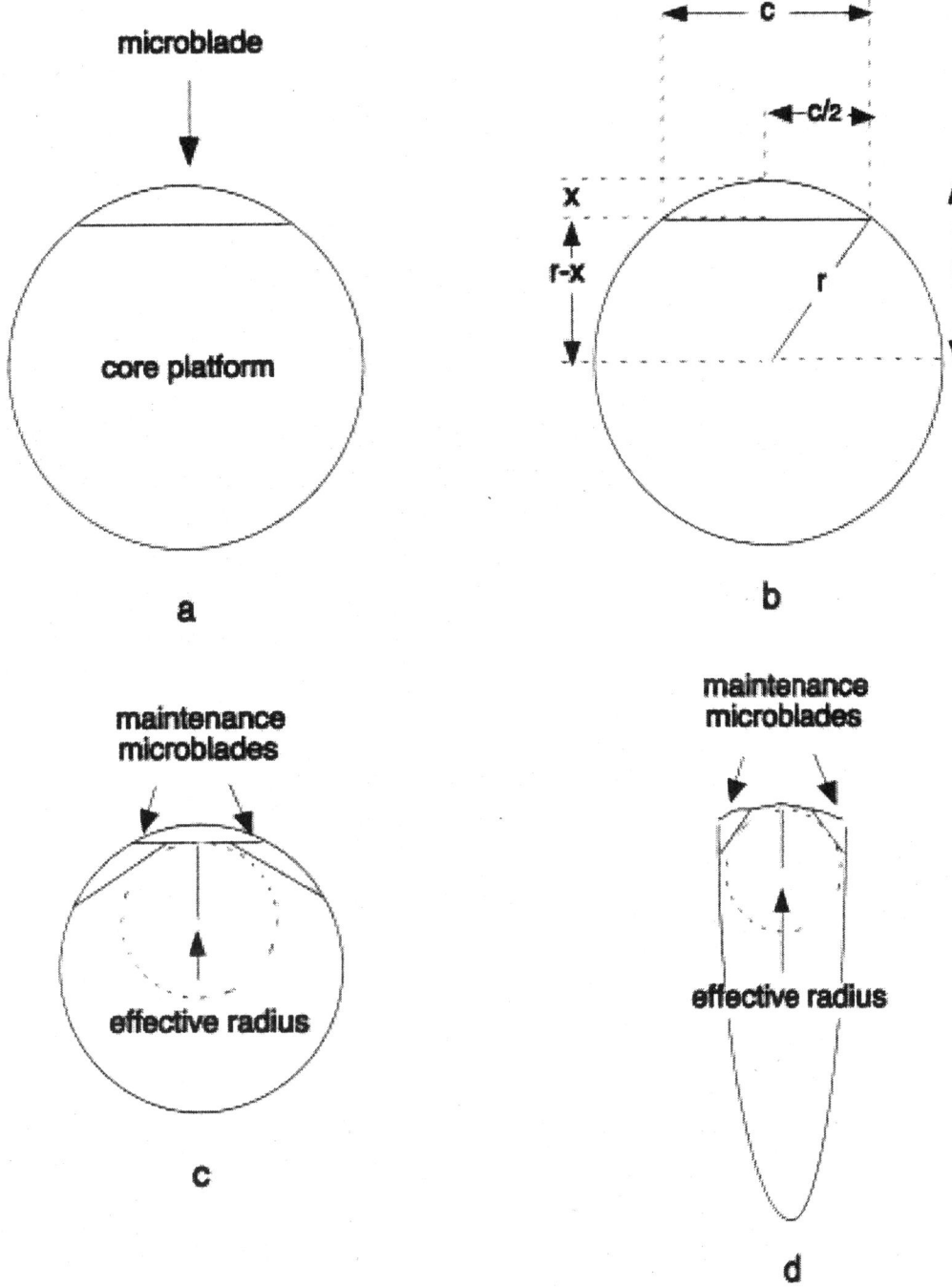

FIGURE 19.3. Simplified model of core geometry. a, Core platform and microblade; b, geometric relationships; c, reducing effective radius with maintenance microblades; d, small effective radius on wedge-shaped core.

r from increasing beyond a certain value (Figure 19.3, d). The wedge-shaped core will require removal of less material (smaller maintenance microblades) for maintenance.

To illustrate the relationship between c and x, Figure 19.4 is a plot of values for microblade width c by equation 19.2, where thickness x is constant (using the QG3 production microblade mean of 1.56 mm) and r varies between 2 mm and 40 mm. As r increases, so does width c, but at a decelerating rate. Thus, when r is small (below about 10 mm) small adjustments to the core have a large effect on microblade width (c).

Rearranging equation 19.1 for r:

$$r = \frac{1}{2}\left[\frac{x^2 + \frac{1}{4}c^2}{x}\right] \qquad (19.3)$$

Production microblades from site QG3 have a mean width of 6.00 mm and mean thickness of 1.56 mm (Table 19.4). By equation 19.3, it appears that the QG3 knappers employed microcores with a working face radius r of about 3.7 mm, somewhat larger than a BB shot. Since this value is much smaller than the width of any cores recovered from the site, we assume that

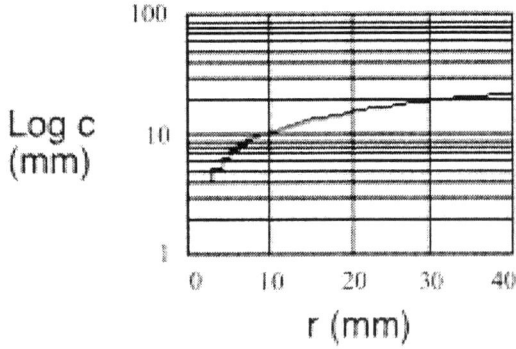

FIGURE 19.4. Step plot of core radius (r) and log microblade width.

the ancient knappers must have created local regions where the *effective* radius was small, by removal of maintenance blades and by otherwise relieving the platforms of production microblades before their detachment. On wide cylindrical cores (Figure 19.3, c), this process would have produced more variable results and wasted more material than on narrow wedge-shaped cores (Figure 19.3, d).

RISK AND MICROBLADE TECHNOLOGY

The archaeological record suggests the relationship between boat-shaped and wedge-shaped cores is not developmental or evolutionary. In Japan, the first microcores to appear are boat-shaped, but these are later supplanted by wedge-shaped cores (Aikens and Higuchi 1982; Andrefsky 1987; Kobayashi 1970); however, in Korea, wedge-shaped cores appear first, followed by boat-shaped cores (Seong 1998). In China, boat-shaped cores are earliest, but remain in use when wedge-shaped cores appear (Chen 1984; Chen and Wang 1989; Lu 1998). In Siberia, the first microblades were made on miniaturized prismatic or conical blade cores; later, true microcores included both boat-shaped and wedge-shaped forms (Derev'anko 1998; Goebel 1995; Goebel et al. 2000; Goebel and Slobodin 2001; Hoffecker et al. 1993). In this light, variation in the proportions of boat-shaped to wedge-shaped cores in lithic assemblages appears to reflect a number of situational factors. In certain times and places, the need for uniformity is paramount; in other circumstances, it is not. Some of this variation is due to raw material availability (Seong 1998), but much of it may be a response to risk.

Winterhalder et al. (1999:302) define risk as "unpredictable variation in the outcome of a behavior," with fitness or utility consequences. When such consequences are nonlinear, the behavior can be analyzed with one of a family of risk-sensitive models, including the Z-score model (Bettinger 1991; Stephens and Charnov 1982; Winterhalder et al. 1999). Risk-sensitive analysis requires the assignment or estimation of the probability outcomes for each behavior and specification of outcomes and their values as utility or fitness. Most studies of risk sensitivity have shown both animals and humans avoid risk; risk-prone behavior with regard to food selection is invariably a response to dire shortfall.

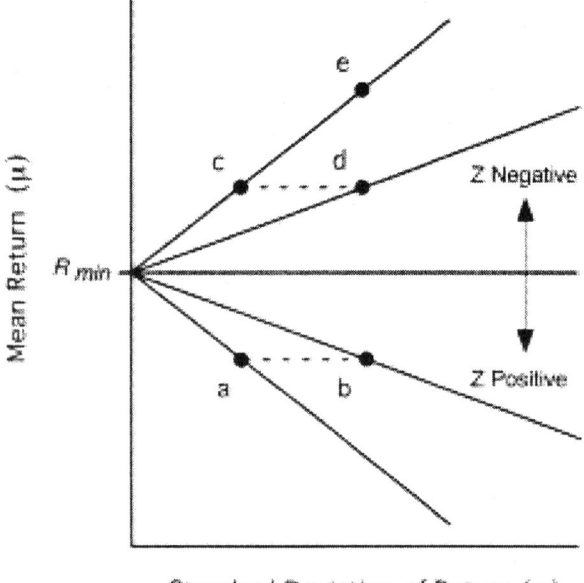

FIGURE 19.5. Isovalue map for risk-sensitive Z-score model (after Winterhalder et al. 1999:fig. 2).

The Z-score model calculates the probability of a resource shortfall as the distance from the shortfall threshold (r_{min}) to the mean (μ) and standard deviation (σ) of the expected return:

$$Z = \frac{(r_{min} - \mu)}{\sigma} \quad (19.4)$$

When the mean expected return is below r_{min}, a more variable behavior (higher Z) will increase the probability of surpassing the shortfall threshold and consequently reduce overall risk. In Figure 19.5, for example, behaviors a and b share means below r_{min}, but the model predicts that behavior b (with its greater σ) will be preferred. Conversely, when the mean expected return is above r_{min}, Z will be negative, and a less variable behavior (lower σ) will be less risky. In Figure 19.5, where the means of behaviors c and d are both above r_{min}, behavior c, with its smaller Z, will be preferred over behavior d. Because points falling on a particular line have the same Z, a forager should be indifferent to choices among them (i.e., behaviors c and e in Figure 19.5).

In the case of microblade technology, it seems likely that wedge-shaped and boat-shaped cores each have a different probability distribution in yield of production microblades. While we cannot yet specify the probability outcomes, we strongly suspect that wedge-shaped cores reduce variability in microblade output. The Z-score model offers a generalized framework for predicting when precision and uniformity may be necessary in microblade production.

These relationships are modeled in Figure 19.6, where the vertical axis represents continuous values of possible returns; shortfall thresholds of hunting returns below which starvation becomes increasingly likely (r_{min} a–c) are indicated by dashed horizontal lines; and means and standard deviations of microblade output from wedge-shaped cores (W) and boat-shaped cores (B) are represented by dots and lines. Here, behaviors W and B have the same mean (μ), but the standard deviation (σ) of B is greater. For present purposes, let us assume that different technological behaviors influence the outcome (μ and σ) of hunting success. In situations in which expected returns are above the threshold, and the cost of failure is high (hunting for winter meat far from base camp), hunters can be expected to make choices that reduce behavioral variation around the expected mean. This situation is graphically represented in Figure 19.6 when the shortfall threshold is equal to $r_{min}a$, and the means of both W and B lie above it. Production from wedge-shaped cores (W) is the better choice, because there is a chance that output from boat-shaped cores (B) will fall below $r_{min}a$. In this situation, boat-shaped cores should rarely, if ever, be used. Variation and risk can be further reduced by making wedge-shaped cores part of the transported hunt toolkit, thus ensuring the availability of the highest-quality replacement parts. While hunters could manufacture batches of microblades before the hunting trip, their transport introduces the possibility of dulling and breakage (Ellis 1997:58). Furthermore, the transported supply may turn out to be insufficient to meet all the contingencies of the hunt. In many (but not all) situations, therefore, transporting cores will do more to reduce variance in return.

What if these same hunters find they need more microblades than can be produced from their supply of wedge-shaped cores? In Figure 19.6, this situation pertains when the shortfall threshold is equal to $r_{min}b$. Both W and B share equal means that lie below $r_{min}b$, but the model predicts that the hunters should try making microblades from whatever material is available, including boat-shaped cores on split pebbles (B). This tactic at least stands a chance of exceeding the threshold ($r_{min}b$). The proportion of usable microblades to waste matters relatively little in this context in which high variability in output can be tolerated with little cost.

Figure 19.6 also helps us to imagine situations in which the shortfall threshold ($r_{min}c$) is below the mean of either microblade production technique. In this case, foragers may be indifferent to whether boat-shaped or wedge-shaped cores are used from the standpoint of risk reduction, but other contingencies may apply. For example, if prey structure varies seasonally, production in anticipation of critical hunts could be less important in some seasons. Or perhaps certain classes of individuals tend to use the technology with the more variable output, such that we might see boat-shaped cores used to train children in microblade production techniques or perhaps boat-shaped cores used by women to produce tools for their own needs around camp.

The point is not whether any of these scenarios truly reflect prehistoric behavior, but that risk-sensitivity models, including the Z-score model, help frame lithic variability differently, and more productively, than does a focus on typology. In principle, the implementation of the Z-score model in the study of microblade assemblage variability is straightforward. The model

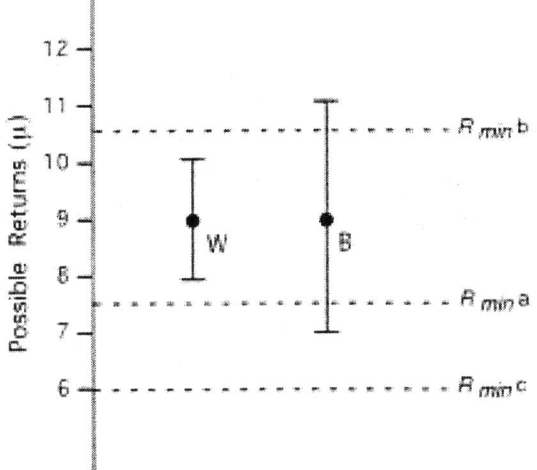

FIGURE 19.6. Choice model for microblade production using wedge-shaped cores (W) and boat-shaped cores (B) in relation to hypothetical production shortfall thresholds ($r_{min}a$–c). Dots and vertical lines for W and B are means and standard deviations. For $r_{min}a$, choice of W will lower risk of falling below threshold; for $r_{min}b$, choice of B will increase the chance of meeting the threshold. For $r_{min}c$, neither W nor B will increase risk of falling below the threshold; the choice between them is more likely to be made in consideration of factors other than risk.

suggests that risk should condition variability in microblade dimensions, raw materials, core dimensions, and production technology. It should be possible to estimate the relative risk inherent in different prehistoric situations (cf. Bleed, 2002) and arrange them along an ordinal scale, perhaps represented at one end by time-limited hunts of large animals such as caribou, and on the other by summer occupation of residential bases. Testing such hypotheses requires metric measurements of microblades and microblade production debris from multiple assemblages of sufficient size and lack of bias. Unfortunately, assemblages from northern Asia meeting the above criteria are not yet abundant. Future studies will depend on consistent application of fine screening and attention to metric measurements in assemblage analyses.

THE EMERGENCE OF MICROLITHIC TECHNOLOGY IN NORTHERN ASIA

Although we argue that the use of inset points in Asia is contingent and situational, we are presently able to match their use only with the most general environmental parameters. Microlithic technology is strongly associated with latitude (Bamforth and Bleed 1997; Bettinger et al. 1994; Derev'anko 1998; Elston 2002; Goebel 2002; Kuzmin and Orlova 1998; Kuzmin and Tankersley 1996; Lu 1998, 1999; Morlan 1970), occurring from about 33°N to above 70°N, and from far western China and Siberia to the Pacific coast, Sakhalin Island, and the islands of Japan. This northerly distribution suggests that Asian microlithic technology and inset weapons are in part a solution to problems of provisioning through long, harsh winters when resources are less abundant and more difficult to access, and when failure to procure sufficient resources has fatal consequences (Bamforth and Bleed 1997; Torrence 1983, 1989).

This hypothesis may be further supported by the timing of the appearance of microlithic technology summarized by Goebel (2002). North Asian hunter-gatherers may have started experimenting with microlithic technology during the last glacial maximum (LGM), although it did not become an important part of toolkits until post-LGM times, most likely between 17,000 to 18,000 BP. Many of the skills and much of the knowledge required to initiate the production of microblades were inherent in the macroblade technologies of the Siberian early Upper Paleolithic, but not in China except for the brief incursion at Shuidonggou (Brantingham 1999; Madsen et al. 2001). The development of indirect percussion, or precision pressure flaking techniques using specialized indenters, may have allowed knappers to surpass a major threshold in workable toolstone package sizes. These innovations and the development of microlithic technologies in general were likely keyed to strategies for reducing foraging risk.

It is impossible at present to do more than present the above scenario as a hypothesis requiring further testing. The faunal record from China for this period is very poor. There is more faunal information from Siberia (Goebel 2002) that implicates microblade use with reindeer hunting. However, the detailed faunal analyses needed to assess diet breadth, body part utility, prey population dynamics, or seasonality do not presently exist.

What is common over this region in the late Pleistocene is climate, which was cold, dry, and increasingly variable after the LGM. While mean annual temperatures gradually increased following the LGM, this did not make long winters any easier to bear for humans. Furthermore, warming caused changes in the distribution of prey animals as the "mammoth fauna" (Powers 1996) retreated northward and remaining species rearranged themselves. For example, in China large lakes reappeared in the deserts of the Ordos and Tengger (Madsen et al. 1998), but we suspect that if large game was attracted to these "oases," hunting pressure would have soon reduced numbers of species and prey abundance. This may also have been the case in Siberia at this time for which Powers (1996) describes the environment as "hyperzonal"; that is, lacking the vegetation zones we see today, but divided into myriad small patches of steppe, forest, and tundra. While this helps account for the wide range of species present in some late Pleistocene archaeological sites, its likely effect on hunters would have been to increase mobility, since small "good patches" could be quickly depleted. If human populations were growing at the same time, this would have further stressed the dwindling faunal resource base and increased the variability in hunting returns. Microlithic technology can be viewed as part of a strategy of intensifying large-game hunting to help cope with the increasing seasonal variance and unpredictability of key resources.

Large-game hunting was important through the Pleistocene/Holocene transition, and the use of inset points continued as well, suggesting that animals remained key resources. However, now shortages and high variation in prey availability were not just seasonal, but continually present. Beginning sometime after 14,000 BP, technology was adopted allowing the efficient use of lower-ranked resources such as seeds, signaled by the appearance of grinding tools and pottery in southern Siberia, Mongolia, and China (Elston et al. 1997; Fairservis 1993; Kuzmin et al. 1997; Lu 1999; Maringer 1950; Zhuschchikhovskaya 1997). Where it could be practiced, Neolithic agriculture finally solved this set of subsistence problems, and microblades faded out of general use soon thereafter (Lu 1999). Microblades tended to persist only in the parts of northern Asia and in the New World where agriculture was late or unfeasible; where hunting to provision over winter in a patchy environment remained a risky, but key, subsistence strategy. Of course, as this volume makes abundantly clear, tactics designed to cope with risks of cold-weather large-game hunting do not explain microlithic technologies and composite armatures in other parts of the world where it was not cold and large animals were not hunted (i.e., southwest Asia, Australia). Nevertheless, foragers face other risks (time-limited critical resources, human enemies, and so on) that may be reduced using microliths and composite armatures.

Conclusion

Our goal in this chapter was to illustrate the value of thinking about microlithic technology in terms of adaptive strategy and risk management, rather than in terms of origins and typology. We believe that this approach will help us comprehend the variable role played by microlithic technology in north Asian hunter-gatherer economies. Formal models of risk and utility (such as the Z-score model, transport cost models, toolstone procurement models, and tool utility models) in conjunction with experimental replication will illuminate lithic assemblage variation at local and regional scales (Elston and Zeanah 2002; Kuhn 1994; Metcalfe and Barlow 1992; Zeanah et al. 1995). These models are, however, unlikely to generate fully testable hypotheses without much better faunal data than presently exist. Although most of the Siberian faunal assemblages remain to be analyzed using contemporary zooarchaeological methods, at least the collections exist. This is not the case for China where many fewer sites of the late Pleistocene are known and fewer still have even species lists. We hope that by demonstrating the utility of this kind of approach, more archaeologists will become interested in applying it.

Acknowledgments

Thanks to Ted Goebel, Andre Tabarev, and Slava Kuzmin for much-needed data and guidance with regard to microlithic technology in the Siberian Paleolithic, to Robert Bettinger for advice about the Z-score model, to Robert Clerico and Ian Agol for their help with the mathematics of the core/microblade model, and to Jim O'Connell, Drew Ugan, and anonymous reviewers for their helpful comments. Research in Pigeon Mountain Basin, China, was supported by the National Geographic Society and National Science Foundation.

References

Aikens, C. M., T. Higuchi
1982 *Prehistory of Japan*. Academic, New York.

Andrefsky, W., Jr.
1987 Diffusion and Innovation from the Perspective of Wedge Shaped Cores in Alaska and Japan. In *The Organization of Core Technology*, edited by J. K. Johnson and C. A. Morrow, pp. 13–43. Westview, Boulder.

Bamforth, D. B., P. Bleed
1997 Technology, Flaked Stone Technology, and Risk. In *Rediscovering Darwin: Evolutionary Theory and Archeological Explanation*, edited by C. M. Barton and G. A. Clark, pp. 109–139. Archeological Papers of the American Anthropological Association No. 7. American Anthropological Association, Arlington, Va.

Bettinger, R. L.
1991 *Hunter-Gatherers: Archaeological and Evolutionary Theory*. Plenum, New York.

Bettinger, R. L., D. B. Madsen, R. G. Elston
1994 Prehistoric Settlement Categories and Settlement Systems in the Alashan Desert of Inner Mongolia, PRC. *Journal of Anthropological Archaeology* 13:74–101.

Bleed, P.
1987 The Optimal Design of Hunting Weapons: Maintainability or Reliability. *American Antiquity* 51(4):737–747.

2002 Cheap, Regular, and Reliable: Implications of Design Variation in Late Pleistocene Japanese Microblade Technology. In *Thinking Small: Global Perspectives on Microlithization*, edited by R. G. Elston and S.L Kuhn, pp. 95–102. Archaeological Papers of the American Anthropological Association No. 12. American Anthropological Association, Arlington, Va.

Brantingham, P. J.
1999 Astride the Movius Line: Late Pleistocene Lithic Technological Variability in Northeast Asia. Ph.D. dissertation, Department of Anthropology, University of Arizona, Tucson.

Chang, K. C.
1986 *The Archaeology of Ancient China*. 4th ed. Yale University Press, New Haven.

Chard, C. S.
1974 *Northeast Asia in Prehistory*. University of Wisconsin Press, Madison.

Chen, C.
1984 The Microlithic of China. *Journal of Anthropological Archaeology* 1(3):79–115.

Chen, C., X. Wang
1989 Upper Paleolithic Microblade Industries in North China and Their Relationships with Northeast Asia and North America. *Arctic Anthropology* 26(2):144–145.

Derev'anko, A. P., ed.
1998 *The Paleolithic of Siberia*. Urbana: University of Illinois Press.

Ellis, C. J.
1997 Factors Influencing the Use of Stone Projectile Tips. In *Projectile Technology*, edited by H. Knecht, pp. 37–74. Plenum, New York.

Elston, R. G.
2002 Toward Understanding the Spatial and Temporal Distribution of Microlithic Technology in Northeast Asia and Eastern Beringia. In *Entering America: Northeast Asia and Beringia before the Last Glacial Maximum*, edited by D. B. Madsen. University of Utah Press, Salt Lake City.

Elston, R. G., C. Xu, D. B. Madsen, K. Zhong, R. L. Bettinger, J. Li, P. J. Brantingham, H. Wang, J. Yu
1997 New Dates for the North China Mesolithic. *Antiquity* 71:985–993.

Elston, R. G., D. Zeanah
2002 Modeling Pre-Archaic Diet Breadth and Patch Choice in the Great Basin of North America. *World Archaeology* 34(1):103–130.

Fairservis, W. A.
1993 *The Archaeology of the Southern Gobi–Mongolia*. Carolina Academic, Durham, N.C.

Flenniken, J. J.
1987 The Paleolithic Dyuktai Pressure Blade Technique of Siberia. *Arctic Anthropology* 24(2):117–132.

Gai, P.
1985 Microlithic Industries in China. In *Paleoanthropology and Paleolithic Archaeology in the People's Republic of China*, edited by R. Wu and J. Olsen, pp. 225–242. Academic, New York.

Goebel, T.
1995 The Record of Human Occupation of the Russian Subarctic and Arctic. *Byrd Polar Research Center Miscellaneous Series* M-335:41–46.

2002 The "Microblade Adaptation" and Recolonization of Siberia during the Late Upper Pleistocene. In *Thinking Small: Global Perspectives on Microlithization*, edited by R. G. Elston and S. L. Kuhn, pp. 117–131. Archaeological Papers of the American Anthropological Association No. 12. American Anthropological Association, Arlington, Va.

Goebel, T., S. B. Slobodin
1999 The Colonization of Western Beringia: Technology, Ecology, and Adaptations. In *Ice Age Peoples of North America: Environments, Origins, and Adaptations of the First Americans*, edited by R. Bonnichsen and K. L. Turnmire, pp. 104–155. Oregon State University Press, Corvallis.

Goebel, T., M. R. Waters, I. Buvit, M. V. Konstantinov, A. V. Konstantinov
2000 Studenoe-2 and the Origins of Microblade Technologies in the Transbaikal, Siberia. *Antiquity* 74:567–575.

Guo, D. S.
1995 Hongshan and Related Cultures. In *The Archaeology of Northeast China: Beyond the Great Wall*, edited by S. M. Nelson, pp. 21–64. Routledge, New York.

Hoffecker, J. F., W. P. Powers, T. Goebel
1993 The Colonization of Beringia and the Peopling of the New World. *Science* 259:46–53.

Kelly, R. L.
1988 The Three Sides of a Biface. *American Antiquity* 53(4):717–734.

Knecht, H.
1993 Early Upper Paleolithic Approaches to Bone and Antler Projectile Technology. In *Hunting and Animal Exploitation in*

the Later Paleolithic and Mesolithic of Eurasia, edited by G. L. Peterkin, H. M. Bricker, and P. Mellars, pp. 33–47. Archeological Papers of the American Anthropological Association No. 4. American Anthropological Association, Arlington, Va.

1997 Projectile Points of Bone, Antler, and Stone: Experimental Explorations of Manufacture and Use. In *Projectile Technology*, edited by H. Knecht, pp. 191–212. Plenum, New York.

Kobayashi, T.
1970 Microblade Industries in the Japanese Archipelago. *Arctic Anthropology* 7(2):38–58.

Kuhn, S. L.
1994 A Formal Approach to the Design and Assembly of Mobile Toolkits. *American Antiquity* 59(3):426–442.

Kuzmin, Y. V., A. J. T. Jull, Z. S. Lapshina, V. E. Medvedev
1997 Radiocarbon AMS Dating of the Ancient Sites with Earliest Pottery from the Russian Far East. *Nuclear Instruments and Methods in Physics Research B* 123:496–497.

Kuzmin, Y. V., L. A. Orlova
1998 Radiocarbon Chronology of the Siberian Paleolithic. *Journal of World Prehistory* 12(1):1–53.

Kuzmin, Y. V., K. B. Tankersley
1996 The Colonization of Eastern Siberia: An Evaluation of the Paleolithic Age Radiocarbon Dates. *Journal of Archaeological Science* 23(4):577–585.

Lu, T. L. D.
1998 The Microblade Tradition in China: Regional Chronologies and Significance in the Transition to Neolithic. *Asian Perspectives* 37(1):85–112.
1999 *The Transition from Foraging to Farming and the Origin of Agriculture in China*. BAR International Series, 774. British Archaeological Reports, Oxford.

Madsen, D. B., R. G. Elston, X. Chen, R. L. Bettinger, Z. Kan, J. Brantingham, A. L. Jingjen
1998 The Loess/Paleosol Record and the Nature of the Younger Dryas Climate in Central China. *Geoarchaeology* 13(8):847–869.

Madsen, D. B., J. Li, P. J. Brantingham, X. Gao, R. G. Elston, R. L. Bettinger
2001 Dating Shuidonggou and the Upper Paleolithic Blade Industry in North China. *Antiquity* 75:706–716.

Maringer, J.
1950 *Contribution to the Prehistory of Mongolia*. Stockholm: Statens Etnografiska Museum.

Metcalfe, D., K. R. Barlow
1992 A Model for Exploring the Optimal Trade-Off between Field Processing and Transport. *American Anthropologist* 94(2):340–356.

Morlan, R. E.
1970 Wedge-Shaped Core Technology in Northern North America. *Arctic Anthropology* 7(2):17–37.
1976 Technological Characteristics of Some Wedge-Shaped Cores in Northwestern North America and Northeast Asia. *Asian Perspectives* 19(1):96–106.

Powers, W. R.
1996 Siberia in the Late Glacial and Early Postglacial. In *Humans at the End of the Ice Age: The Archaeology of the Pleistocene-Holocene Transition*, edited by L. G. Straus, B. V. Eriksen, J. M. Erlandson, and D. R. Yesner, pp. 229–242. Plenum, New York.

Rondeau, Michael F.
1996 When Is an Elko? In *Stone Tools: Theoretical Insights into Human Prehistory*, edited by George H. Odell, pp. 229–243. Plenum, New York.

Seong, C.
1998 Microblade Technology in Korea and Adjacent Northeast Asia. *Asian Perspectives* 37(2):245–278.

Shih, X.
1992 The Discovery of the Pre-Yangshao Culture and Its Significance. In *Pacific Northeast Asia in Prehistory*, edited by C. M. Aikens and S. N. Rhee, pp. 125–132. Washington State University Press, Pullman.

Stephens, D. W.
1981 The Logic of Risk-Sensitive Foraging Preferences. *Animal Behavior* 26:628–629.

Stephens, D. W., E. L. Charnov
1982 Optimal Foraging: Some Simple Stochastic Models. *Behavioral Ecology and Sociobiology* 10:252–263.

Tabarev, A. V.
1997 Paleolithic Wedge-Shaped Microcores and Experiments with Pocket Devices. *Lithic Technology* 22(2):139–149.

Torrence, R.
1983 Time Budgeting and Hunter-Gatherer Technology. In *Hunter-Gatherer Economy in Prehistory*, edited by G. Bailey, pp. 11–22. Cambridge University Press, Cambridge.
1989 Retooling: Toward a Behavioral Theory of Stone Tools. In *Time, Energy and Stone Tools*, edited by R. Torrence, pp. 57–66. Cambridge University Press, Cambridge.

Underhill, A. P.
1997 Current Issues in Chinese Neolithic Archaeology. *Journal of World Prehistory* 11(2):103–160.

Winterhalder, B. P., F. Lu, B. Tucker
1999 Risk-Sensitive Adaptive Tactics: Models and Evidence from Subsistence Studies in Biology and Anthropology. *Journal of Archaeological Research* 7(4):301–348.

Zeanah, D., J. A. Carter, R. G. Elston, J. E. Hammett
1995 *An Optimal Foraging Model of Hunter-Gatherer Land Use in the Carson Desert*. Report to U.S. Fish and Wildlife Service and U.S. Department of the Navy. Intermountain Research, Silver City, Nev.

Zhuschchikhovskaya, I.
1997 On Early Pottery-Making in the Russian Far East. *Asian Perspectives* 36(2):159–174.

PART IV

Food Production Strategies: Origins, Spread, and Variation

Certainly one of the most significant developments in human prehistory was the development of agricultural economies in the millennia following the Last Glacial Maximum. Archaeologists have long been interested in issues related to the origins and spread of agriculture, and such issues have been addressed by scholars using the EE approach for about as long as it has been active in archaeology (e.g., Keegan 1986; Layton et al. 1991; Hawkes and O'Connell 1992). In recent years, EE-based explorations of the development of agriculture have blossomed fully, resulting in the publication of a volume devoted entirely to this area of inquiry (Kennett and Winterhalder 2006; see also, for example, Hard and Roney 2005).

The papers reprinted in this fourth part of *Evolutionary Ecology and Archaeology* are examples of seminal applications of EE models to archaeological data pertaining to agricultural origins, and as such they provide important footing for the more recent research noted above. They also illustrate the insights that EE theory provides into the origins and spread of agriculture and into the variability in the degree of investment in agriculture that is evident in the global archaeological record. Additional important insights, as well as a useful discussion of the relationship between EE and "selectionist" approaches to the evolution of agriculture, are provided in the influential theoretical work of Winterhalder and Goland (1997).

This section begins with a paper by K. J. Gremillion, which itself begins by making a compelling case for the use of Darwinian theory in the study of agricultural origins. Gremillion then discusses how the prey (or diet choice) and Z-score models of foraging theory, as well as an opportunity cost model of time allocation, can help to explain developments such as the initial adoption of domesticated crops and subsequent increases in the amount of time invested in producing those crops. This discussion includes an important consideration of the potential relevance of both efficiency and risk to decisions about the use of domesticates, and Gremillion also points out that the factors that underlie the initial adoption of crops may be quite different from those that underlie later increases in their importance.

Gremillion's clear presentation of the models that she discusses, and of the ways in which these models can be applied to decisions that individuals must make when adopting and practicing agriculture, have provided an important foundation for subsequent theoretical development (e.g., Barlow 1997, 2006; Cannon 2001).

Gremillion presents two case studies from eastern North America, one of which involves the use of maize, which was farmed to some extent in parts of the east at least as early as AD 1, but which did not become a major component of the diet until the period between AD 800 and 1000. During this period, maize became the dominant food resource across more and more of the east, virtually replacing many of the indigenous domesticates that had been grown in some places to that point. Gremillion applies the time allocation model to this development, noting that maize may have been comparatively unimportant at first because the process of growing it may have provided low returns relative to its opportunity costs. Such opportunity costs arise from the fact that time devoted to growing maize is time that cannot be spent on other activities like tending to other crops, foraging for wild resources, or pursuing nonsubsistence activities. Gremillion points out that the optimal amount of time to allocate to maize farming would have increased either if its opportunity costs declined—for example, owing to a decline in the productivity of alternative resources—or if the returns that it provided increased—for example, through the introduction of more profitable varieties of maize, the development of more efficient technologies for producing a crop, or the development of more effective means of storage or exchange.

Gremillion evaluates which of these specific factors might best explain the increase in the importance of maize that occurs across eastern North America between AD 800 and 1000. She notes, for example, that there is virtually no archaeological evidence for the introduction of a more productive variety of maize across the east at the time in question, and instead proposes that the best-supported hypothesis is that the labor costs of maize declined due to the development of more effective technologies.

Such points about the factors that might lead to increasing investment in maize agriculture are developed further in the paper by Barlow that is reprinted in this section of the volume, though some of Barlow's specific conclusions contrast with those of Gremillion, as we discuss below.

S. R. Simms and K. W. Russell provide an example of the ethnoarchaeological research that has long played a large role in the EE approach to archaeology (see, e.g., O'Connell 1995), and they also derive clear implications from such research for our understanding of the archaeological record pertaining to the origins of agriculture in southwestern Asia. These authors first discuss an ethnographic study that was directed at estimating return rates for the harvesting of wheat and barley using traditional hand-harvesting techniques. In doing so, they point out the important distinction that must be kept in mind between gross agricultural yields and net return rates, or yield obtained per unit time (see also Winterhalder and Goland 1997). Simms and Russell also compare their ethnographic results for hand-harvesting to experimentally derived results for tool-assisted harvesting using a variety of sickle types. They then calculate overall return rates—which incorporate time spent tilling, sowing, threshing, and milling in addition to harvesting—for wild wheat and for cultivated wheat produced with a variety of technologies.

From their ethnographic and experimental results, Simms and Russell derive implications for the study of the origins of agriculture in southwest Asia. These include the observation that, because hand-harvesting can provide return rates comparable to those obtained using early sickles, the appearance of sickles in the archaeological record may not necessarily mark the beginning of cereal cultivation. Simms and Russell suggest instead that sickles may simply have enabled the expansion of cultivation from areas with sandy soils into areas with silty or clayey soils, where hand harvesting is less effective, and/or they may have enabled the extension of the harvesting period, thereby allowing higher yields to be obtained. In either case, the authors note, the appearance of sickles in the archaeological record would indicate the extensification or intensification of agriculture, rather than its initial appearance. Simms and Russell also point out how the rough equivalence of the return rates for wild wheat and hand-harvested cultivated wheat could have opened up a wide range of subsistence alternatives to the late Pleistocene occupants of southwest Asia: once overall foraging efficiency declined to a point where it was economical to include grass seeds—whether wild or domesticated—in the diet, foragers may have been able to choose from a variety of possible subsistence strategies the one that was best suited to the particular constraints of environment and mobility that they faced.

Although detailed tests of the hypotheses that Simms and Russell propose remain to be conducted, their work illustrates very well the power of EE theory for generating insights into the selective factors involved in the transition from foraging to farming. Only by focusing on the return rates of various behavioral or technological alternatives do the implications that Simms and Russell derive become apparent, and only through the application of EE theory does it become clear that return rates are the variable on which archaeologists should focus when developing explanations for the evolution of agriculture (see Winterhalder and Goland 1997). Simms and Russell also develop some important methodological innovations for the study of the beginnings of agriculture, which Barlow builds upon in her paper that we discuss below. Though brief, the paper by Simms and Russell is truly an important milestone in EE-based research into food production strategies.

Also very innovative is the paper by M. S. Alvard and L. Kuznar. This paper is unique among those in this volume in that it draws on a foraging theory model in which the goal is maximization of productive efficiency over the very long term (years or decades). Both the production of domesticated crops and animal husbandry—the specific issue that Alvard and Kuznar address here—are activities that provide delayed returns, and long-term maximization is the appropriate goal to assume when this is the case. Indeed, as Alvard and Kuznar illustrate, EE can even help us understand the conditions under which individuals should devote effort to activities with delayed, rather than immediate, returns. Key to this issue is the rate at which delayed returns are discounted, and Alvard and Kuznar thus use a time-discounting model in their consideration of animal husbandry (also see Tucker 2006 for an application of a time-discounting model to horticulture). Alvard has used this approach elsewhere in ethnographic studies of prey conservation (or lack thereof) by hunter-gatherers, and in this paper Alvard and Kuznar extend this line of research by defining animal husbandry as conservation, or the payment of a short-term cost (foregoing immediate harvest) in exchange for a long-term benefit (higher returns realized at some point in the future).

Alvard and Kuznar introduce the model that they develop to apply to the issue of animal husbandry by discussing general conditions under which prey conservation should be expected to occur. In doing so, they draw on related research topics—particularly territoriality and collective action—that have been pursued very fruitfully by ethnographers using the EE approach (e.g., Dyson-Hudson and Smith 1987; Hawkes 1992). The model that Alvard and Kuznar present is designed to determine when an important condition favoring husbandry will be met: specifically, when the long-term benefit to be obtained from husbanding an animal population—or the benefit obtained from harvesting it at its maximum sustainable yield (MSY)—will be higher than the short-term cost that doing so imposes. Building on the work of Clark (1990), Alvard and Kuznar argue that small-bodied taxa are more likely to be husbanded than large-bodied taxa, all else being equal. They go beyond this general prediction, however, by developing through allometric analysis an equation to predict the maximum prey body size that it pays to husband, given a certain discount rate. Using a range of discount rates for humans that are well-supported both theoretically and empirically, they propose an "economically husbandable" maximum body size of about 40 kg.

Alvard and Kuznar note that 40 kg is on the order of the body size of sheep and goats, the taxa first domesticated in southwest Asia. They point out that an earlier foraging theory-based approach to the issue of animal domestication in this part of the world (Russell 1988) predicted that the largest-bodied ungulates such as cattle and camels should have been domesticated first, not the smallest-bodied, and they propose that the incongruity between that prediction and the archaeological record was the result of a failure to consider the importance of long-term benefits and time-discounting to the husbandry issue. Alvard and Kuznar suggest that the later domestication of larger-bodied taxa may have occurred due to a reduction in human population growth rates and/or an increase in sedentism, either of which may have led to a reduction in the discount rate that people applied (Rogers 1994).

The final paper in this section, by K. R. Barlow, builds on issues discussed in the previous papers and derives some very important implications for our understanding of variability over space and time in the intensity of agricultural production in the American Southwest. Barlow begins by discussing the Fremont archaeological culture of the eastern Great Basin and northern Colorado Plateau, which is characterized by extreme geographic diversity in the importance of agriculture relative to hunting and gathering. Barlow makes the important point that prehistoric farming groups are best viewed as exhibiting a continuum of strategies, with low-intensity horticulture and high mobility at one end and labor-intensive agriculture and high sedentism at the other. She proposes that variability along this continuum can be explained in terms of variability in the costs and benefits of farming and foraging in different settings, and she operationalizes such a cost-benefit approach through the application of foraging theory.

Using ethnographic data from Latin America, Barlow derives estimates of return rates for maize agriculture in a variety of settings and involving a variety of agricultural technologies and strategies. Several important points emerge from this analysis, perhaps most notable of which is the observation that the return rates for maize agriculture are highly constrained by labor costs, particularly those imposed by post-harvest processing. This implies that increases in agricultural yields, such as may have resulted from improvements in climate or the development of more productive varieties of maize, may not provide a sufficient explanation for increases over time in the importance of agriculture relative to hunting and gathering. Rather, Barlow proposes that the abundance of high-return wild resources is more important in determining the optimal degree of investment in agriculture. She points out that low-intensity farming produces return rates comparable to those of many wild plant resources that are common in the Great Basin and the Southwest, while labor-intensive agriculture produces return rates comparable to those of small seeds, which are among the lowest-return resources in these regions. Using the logic of the prey model, Barlow concludes that it is economical to practice intensive agriculture only when rates of encounter with high-return wild resources are low.

This conclusion contrasts with Gremillion's conclusion, discussed above, that the increase in the importance of maize that occurred in eastern North America between AD 800 and 1000 was due to the development of more effective technologies that reduced the labor costs associated with maize agriculture. This raises some issues that are certainly worthy of consideration in future research. Gremillion recognizes that declines in the productivity of wild resources might have led individuals to allocate increasing amounts of time to agriculture, but she proposes that the "more effective technologies" hypothesis is better supported for the east. However, this may be the case simply because tests of the "declining wild resource productivity" hypothesis have yet to be conducted in eastern North America, and such tests might lead to an important new understanding of the causes of agricultural intensification in this part of North America. On the other hand, further research into the effects of technological changes on the return rates of agriculture is also in order, both for eastern North America and the Southwest, and we note that such research could benefit from application of the tech investment model discussed by Bright et al. in their paper that is reprinted in the previous section of this volume.

It thus seems clear that, in the future, the EE approach can continue to provide important insights into the transitions from foraging to farming that occurred virtually worldwide beginning in the terminal Pleistocene. The insights that have already been provided by the works reprinted in this section of *Evolutionary Ecology and Archaeology* include the following:

1. The importance of farming relative to foraging, as well as the precise farming strategies that were employed in a particular setting, likely varied continuously over space and time in response to local environmental and technological constraints.
2. The return rates of various behavioral or technological alternatives, which incorporate the costs of agriculture, are more relevant to decisions about the adoption or intensification of agriculture than are the gross yields provided by those alternatives.
3. Transitions from foraging to farming involved a shift from short-term goals to long-term goals and the payment of an immediate cost in exchange for a delayed benefit.
4. Different factors may have been involved in the initial domestication or adoption of crops and subsequent increases in their importance.
5. Intensive agriculture may have developed due to a variety of factors—including declines in the productivity of wild resources, the introduction of more profitable crop varieties, and the development of more efficient technologies or more effective means of storage or exchange—and the factors involved may have varied from case to case.

6. Domesticates may have been incorporated into prehistoric diets either because they improved productive efficiency or because they reduced risk.
7. Low-return crops may have been incorporated into the diet without replacing other resources, while the incorporation of high-return crops may often have resulted in lower-return resources being dropped from the diet.
8. The appearance of certain technologies in the archaeological record—such as sickles in southwest Asia—may indicate the intensification of agriculture, rather than its initial development.
9. Husbandry pays off when the reproductive rate of an animal population is greater than the rate at which future returns are discounted; this is more likely to be the case for smaller-bodied animals, and the order in which animals were domesticated in southwest Asia is understandable in light of this.

References

Barlow, K. R.
1997 Foragers that Farm: A Behavioral Ecology Approach to the Economics of Corn Farming for the Fremont Case. Ph.D. dissertation, Department of Anthropology, University of Utah, Salt Lake City.
2006 A Formal Model for Predicting Agriculture among the Fremont. In *Behavioral Ecology and the Transition to Agriculture*, edited by D. J. Kennett and B. Winterhalder, pp. 87–102. University of California Press, Berkeley.

Cannon, M. D.
2001 Large Mammal Resource Depression and Agricultural Intensification: An Empirical Test in the Mimbres Valley, New Mexico. Ph.D. dissertation, Department of Anthropology, University of Washington, Seattle.

Clark, C. W.
1990 *The Optimal Management of Renewable Resources* 2nd ed. Wiley-Interscience, New York.

Dyson-Hudson, R., E. A. Smith
1978 Human Territoriality: An Ecological Reassessment. *American Anthropologist* 80:21–41.

Hard, R. J., J. R. Roney
2005 The Transition to Farming on the Río Casa Grandes and in the Southern Jornada Mogollon Region. In *The Late Archaic Across the Borderlands: From Foraging to Farming*, edited by B. J. Vierra, pp. 141–186. University of Texas Press, Austin.

Hawkes, K.
1992 Sharing and Collective Action. In *Evolutionary Ecology and Human Behavior*, edited by E. A. Smith and B. Winterhalder, pp. 269–300. Aldine de Gruyter, New York.

Hawkes, K., J. F. O'Connell
1992 On Optimal Foraging Models and Subsistence Transitions. *Current Anthropology* 33:63–65.

Keegan, W.
1986 The Optimal Foraging Analysis of Horticultural Production. *American Anthropologist* 88:92–107.

Kennett, D. J., B. Winterhalder
2006 *Behavioral Ecology and the Transition to Agriculture*. University of California Press, Berkeley.

Layton, R. H., R. A. Foley, E. Williams
1991 The Transition between Hunting and Gathering and the Specialized Husbandry of Resources. *Current Anthropology* 32:255–274.

O'Connell, J. F.
1995 Ethnoarchaeology Needs a General Theory of Behavior. *Journal of Archaeological Research* 3:205–255.

Rogers, A. R.
1994 Evolution of Time Preference by Natural Selection. *The American Economic Review* 84:460–481.

Russell, K.
1988 *After Eden: The Behavioral Ecology of Early Food Production in the Near East and North Africa*. BAR International Series 391. British Archaeological Reports, Oxford.

Tucker, B.
2006 A Future Discounting Explanation for the Persistence of a Mixed Foraging-Horticulture Strategy among the Mikea of Madagascar. In *Behavioral Ecology and the Transition to Agriculture*, edited by D. J. Kennett and B. Winterhalder, pp. 22–40. University of California Press, Berkeley.

Winterhalder, B., C. Goland
1997 An Evolutionary Ecology Perspective on Diet Choice, Risk, and Plant Domestication. In *People, Plants, and Landscapes: Studies in Paleoethnobotany*, edited by K. J. Gremillion, pp. 123–160. University of Alabama Press, Tuscaloosa.

Diffusion and Adoption of Crops in Evolutionary Perspective

Kristen J. Gremillion

Optimization models developed within evolutionary ecology provide a useful framework for analyzing patterns of acceptance and utilization of new crops by food-producing populations. The deterministic version of the diet choice model predicts that new crops of relatively low profitability are likely to enter the diet if highly ranked resources are in short supply, but will be ignored if such items are abundant. A stochastic version of the same model (the "z score" model) predicts adoption when the chance of energetic shortfall is very low and addition of new foods (even if they are of poor quality) may minimize risk by reducing variance in yields. Similar environments may consequently elicit strikingly different responses to new resources depending on which goal (e.g., risk minimization or efficiency maximization) guides decision-making. A case study, the introduction of the peach to eastern North America, illustrates how these models might be enlisted to develop and evaluate adaptation-based explanations of crop adoption. The opportunity cost model is used to predict optimal time allocation to different tasks as a function of the relationship between rising opportunity costs and diminishing returns. Manipulation of the model suggests that high opportunity costs may have limited investment in maize production following the initial introduction of this crop to eastern North America. Both increased productivity of maize and improvement in labor efficiency are likely causes of its eventual establishment as an economic staple. These findings illustrate the heuristic value of evolutionary ecology models as tools for investigating the origins and development of agricultural systems, including the diffusion of specific crops.

Anthropological interest in the origins of agriculture in recent years has been expressed in theoretical exploration of the subject as well as archaeological documentation of the prehistoric development of farming economies and the spread of particular crop plants (Cowan and Watson 1992; Gebauer and Price 1992; Harris and Hillman 1989; Rindos 1984; Smith 1989, 1992a, 1992b). The domestication of locally available wild plants by foraging populations has perhaps generated the liveliest debate. Less effort is expended on explaining the adoption of new crops by populations whose commitment to food production is already established, although efforts to document the timing of such crop introductions are central to culture historical reconstruction and are a part of archaeological research programs worldwide. Perhaps because they are thought to be subsumed under the explanation of food production in general, the causal processes that determine whether a new crop will be accepted and what subsistence role it will play thereafter are seldom explicitly addressed.

I contend that the issue of crop adoption can yield important insights into the cultural and ecological processes that drive agricultural change. I also argue that the origins of agricultural systems and their transformation are fundamentally evolutionary problems and demand consideration from a Darwinian perspective. Both because of its emphasis on the evolution of adaptive behaviors in specific environmental contexts and its systematic methodology, evolutionary ecology offers a valuable strategy for formulating explanations for the adoption and utilization of introduced crops. In the ensuing discussion, I present two general models that can be used to predict what patterns of utilization of such crops are likely to emerge under a selective regime that favors optimizing strategies of food acquisition. I then discuss the applicability of these models to specific cases from eastern North America. Finally, I summarize methodological pitfalls and advantages of using optimal foraging models to explain agricultural innovation and change.

Natural Selection, Evolutionary Ecology, and Agricultural Change

Darwin recognized that the process of domestication represents a kind of selection in which human beings, rather than forces of nature, determine the reproductive success of individuals (Darwin 1859). Since his time, evolutionary analysis of plant

*Originally published in *Journal of Anthropological Archaeology* 15(1996):183–204

domestication and agricultural change in prehistory has diversified along disciplinary lines. A strongly Darwinian approach characterizes research into crop evolution and genetics, which has close ties to plant population biology, physiology, and plant geography and ecology (e.g., Doggett 1965; Donald and Hamblin 1983; Harlan et al. 1973; Wilson 1990a, 1990b). The process of natural selection also plays a pivotal role in recent anthropological investigations of plant domestication and the development of food production as a form of coevolution (Crites 1987; Gremillion 1989a; O'Brien 1987; Rindos 1984; Smith 1992a) and in archaeological analyses of change in crop morphology (Cowan 1996; Gremillion 1993a, 1993b; King 1985, 1987; Smith 1985a, 1985b). Approaches borrowed from evolutionary ecology and microeconomic theory have also been employed to further understanding of agricultural origins and change in prehistory (Gardner 1992; Keene 1981; Reidhead 1981; Winterhalder and Goland 1996; Yesner 1981). On the other hand, as several researchers have pointed out (e.g., Dunnell 1980; O'Brien and Holland 1990), many archaeologists who attack this problem using evolutionary concepts have shown little interest in natural selection as a mechanism promoting change in human behavior (e.g., Flannery 1968, 1973; Struever 1968). Such studies emphasize systemic function over cause and offer no explicit framework within which the evolutionary consequences of subsistence decisions can be predicted (Smith and Winterhalder 1992).

The failure to incorporate natural selection into explanations of agricultural change explicitly seems a tactical error, given the power and generality of this theory. Although it is only one of many evolutionary processes, natural selection can act upon all types of phenotypic variability, whether genetically or culturally transmitted (Boyd and Richerson 1985; Durham 1982). The robustness, elegance, and simplicity of the theory of natural selection make it perhaps the most axiomatic general explanatory principal in modern biology and one that should be as useful to archaeologists as it is to paleobiologists.

The process of natural selection operates in the same fashion whether the variants in question are genetically or culturally based (or of complex determination). In order for selection to occur, there must be variation between individuals that is heritable, and possession of different variants must affect the survival and reproductive chances of individuals (Boyd and Richerson 1985:173). Differential reproductive success ensures that genetic variants associated with successful phenotypes will be passed on to future generations. Information that is encoded and transmitted culturally rather than genetically is subject to a parallel process (Boyd and Richerson 1985:11; O'Brien and Holland 1990: 35). Possession of a trait may also enhance both reproductive success *and* the ability of an individual to effectively transmit her or his own inventory of cultural information. For instance, certain subsistence techniques result in improved nutrition and consequently larger families, thereby conferring greater opportunity for cultural and biological parentage and the transmission of information about those techniques. Furthermore, the phenotypic flexibility of humans allows for rapid behavioral adjustment to environmental conditions. This flexibility and at least some of the "decision rules" that shape its expression in particular cases are themselves likely to represent evolved traits that have been established as a result of natural selection (Boone 1994; Boyd and Richerson 1985:11, 132–133).

Darwinian analysis is faced with the problem of how to link the theory of natural selection to empirical phenomena. One approach that has been quite successful in connecting theory and data has been developed primarily within the field of evolutionary ecology and falls into the general category of optimization analysis (Smith and Winterhalder 1992:50–51). Often optimization analysis takes the form of the development of relatively simple models that predict how behavior will vary in different environmental contexts, given certain constraints. These models employ the assumptions that optimization (generally defined as a function of cost-benefit relationships) tends to be favored by selection and that measurable variables (such as energetic efficiency) are robust correlates of fitness. Model predictions can then be compared with ethnographic or archaeological data by way of hypothesis testing in order to refine understanding of relationships between key variables. Models of this kind are not intended to be perfectly realistic representations of real-world situations, nor is their construction an end in itself; rather, they are heuristic tools used to organize the investigative journey from abstract theory to hypothesis testing and explanation (Winterhalder and Smith 1992:13–14).

Optimization models of foraging behavior that combine microeconomic and ecological theory can be adapted to formulate predictions about the adoption and use of novel resources. Two models widely used in human evolutionary ecology, the diet choice model and the opportunity cost model of time allocation, can be employed to predict which food items should be sought and how much time should be allocated to the pursuit or production of each. These models provide a framework for addressing these questions: Under what circumstances should selection favor the widespread adoption of a new plant resource? And, how will selection regulate the investment of time and/or energy that an individual devotes to exploiting such a resource?

Two Optimization Models and their Implications for Crop Diffusion

The Diet Choice Model

A major assumption of many optimal foraging models is that natural selection favors individuals who maximize net energy intake while foraging (Winterhalder 1981; Kaplan and Hill 1992). Energetic efficiency enhances fitness by permitting maximization of energy capture or by freeing time for non-foraging activities necessary to survival and reproduction. Benefits accruing to the forager are usually measured by selecting some material commodity, such as calories and nutrients, to act as a currency. Total costs include both direct energetic and time costs of exploitation (resource costs) and the costs incurred by foregoing some other activity that could have been carried out alternatively (opportunity costs) (Kaplan and Hill 1992).

One of the simplest of foraging models is the diet choice model, which has been used to analyze human foraging behavior in a variety of environmental settings (Stephens and Krebs 1986; Winterhalder 1981; Kaplan and Hill 1992). The diet choice model envisions a world in which foragers move through the environment, searching for all prey types simultaneously. Prey types are ranked according to profitability (net energy acquired after accounting for pursuit and handling costs). The model predicts that a forager will add resources to the diet until the profitability of the next-highest ranked item is lower than the net acquisition rate, or NAR (a function of the total time and energy costs and benefits, including searching for prey) of continuing to search for and pursue only the higher–ranked items. Any resource not included in this optimal resource set should be consistently ignored (not pursued when encountered). An important implication of the model is that as highly ranked resources become less abundant, search costs increase (reducing the NAR of items in the diet) such that lower-ranked items are added.

In its deterministic form, the diet choice model assumes the equivalency of options that produce the same *average* return rate (Stephens 1990; Winterhalder 1990). However, individuals may discriminate between choices that offer a fixed outcome and those that are variable. Suppose there are two feeding options with the same mean but different variance (assuming that the food yields of both are normally distributed). Which option should be preferred? The response most likely to benefit the individual can be predicted using the mathematically derived "z-score model" of food choice, which is based on the properties of the normal curve (Stephens 1990). The essential elements of the z-score model can be illustrated graphically (Figure 20.1). The probability of failing to acquire some minimum amount of food needed for survival (m) is represented by the area under the curve and to the left of m; conversely, the probability of acquiring more than the required amount is proportional to the area to its right. Thus, if needs (m) are well below the mean expected yield, risk of failure to meet those needs is minimized by choosing the low-variance alternative (Option 1 in Figure 20.1). However, if the required amount greatly exceeds average expected returns, the high-variance option (Option 2 in Figure 20.1) is the only choice of the two that offers any chance (albeit slim) of survival.

The fitness benefits of a particular mix of resources may thus be a function not of average energy acquisition rates exclusively, but also of their variance. Optimization in this context often means counterbalancing risk and inefficiency. Consideration of variability in yield and its effects upon the benefits of different feeding options can be used to add realism to the diet choice model (Winterhalder 1990). In applying the z-score model to diet breadth, risk is defined as the probability of yield drop-

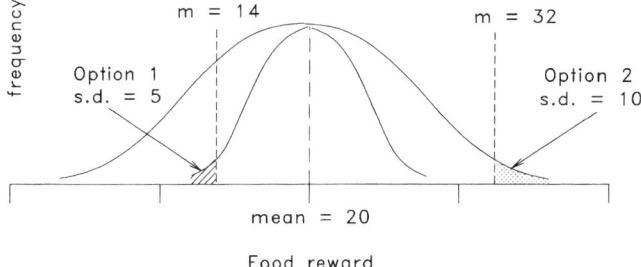

FIGURE 20.1. Graphic representation of the z score model (Stephens 1990; Winterhalder 1990). The two curves represent the normally distributed food rewards associated with alternative feeding options that have the same average returns but different standard deviations. The variable m represents some minimum yield required for survival. If $m = 14$, the low-variance option (Option 1) minimizes the area of the curve to the left of m (diagonal hatching) and thus minimizes risk (in mathematical terms, this option minimizes the z score of m). However, if $m = 32$, the high-variance option (Option 2) maximizes the probability of obtaining the required amount of food (dot hatching).

ping below some fixed minimum requirement m (Winterhalder 1990:72). In order to explore the relationships among diet breadth, foraging efficiency, and risk, Winterhalder (1990) ran a simulation of diet selection that incorporated stochastic variation in prey encounter rate and pursuit costs. Repeated runs of the simulation were used to calculate a mean and standard deviation for the normally distributed NAR of diets including varying numbers of prey items. In this simulation, total variance in yield (as measured by the standard deviation of the net acquisition rate, or NAR) declines as resources are added, whereas the NAR itself begins to decline after reaching a peak at the energetically optimal diet breadth of three items. This energetically optimal diet also minimizes risk, provided that m is near the average expected yield. However, where expected average returns are much higher than m, risk is minimized by broadening the diet and thereby reducing variance (even though some decline in average efficiency results). Conversely, where m is high compared to average returns, the strategy most likely to result in survival is to choose a diet with high variance that has a better chance of meeting minimum requirements (likewise in exchange for some loss of energetic efficiency). Thus, in many cases there is a trade-off between maximizing average returns and reducing risk by choosing a resource mix with either high or low variance (depending on the value of m).

The deterministic version of the diet choice model can be used to hypothesize how variation in the profitability rank of a new food item and resource abundance might result in contrasting patterns of adoption (Figure 20.2). A newly introduced resource of high profitability will enter the diet if its inclusion raises average return rates (cells a and c). However, the pattern of adoption (by replacement of or addition to items already in use) varies depending on environmental conditions. If highly ranked resources are abundant, search costs are low and a relatively narrow diet will result. Under these conditions, replacement is

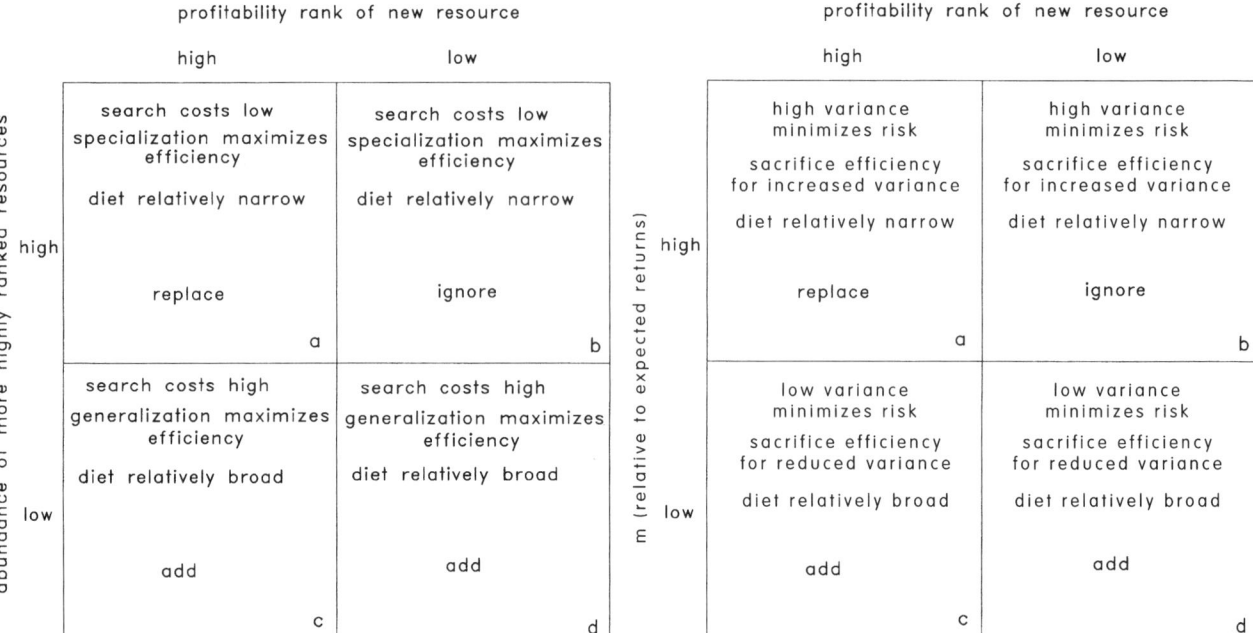

FIGURE 20.2. Predicted behavior with respect to novel resources of varying profitability assuming the goal of efficiency maximization. Outcomes are determined by the effect of different options upon average rates of return. Each cell posits one of the following possible outcomes: (a) replace an existing resource with the new one, (c) and (d) add the new resource to the existing diet; or (b) ignore the new resource. Note that cell entries are limited to the *most likely* outcome predicted by the model.

FIGURE 20.3. Predicted behavior with respect to novel resources of varying profitability assuming the goal of risk minimization. Outcomes are determined by the effect of different options upon the probability of failing to meet minimum requirements (m). Each cell posits one of the following possible outcomes: (a) replace an existing resource with the new one, (c) and (d) add the new resource to the existing diet, or (b) ignore the new resource. Note that cell entries are limited to the *most likely* outcome predicted by the model.

more likely than expansion of the diet to include the new item, because expansion will probably result in some loss of efficiency (cell a). If highly ranked resources are instead in short supply, search time is greater and the diet correspondingly broad (cell c). In this environmental context, new resources of high rank are likely to be added to the existing resource set. The fate of items of low rank at the margin of the optimal diet is less certain. If highly ranked resources are abundant, a resource of low profitability would probably remain beyond the margins of the optimal diet and consequently would be ignored (cell b). However, the same resource is likely to be added under conditions of scarcity (cell d), although it will be ignored if its return rates fall below the average for resources already in use.

The deterministic version of the diet choice model specifies the fate of the new resource and its relationship to other foods already in use assuming that the average rate of return is the most appropriate currency and that anticipated returns neither greatly exceed nor fall seriously short of minimum requirements. The z-score model of diet choice introduces the effects of variance upon the utility of different dietary options given contrasting expectations of success in food acquisition (Figure 20.3). Although increasing the number of items in the diet does not necessarily reduce variance, this may often be the case (Stephens 1990:30–31), as it is in Winterhalder's (1990) simulation. Thus, a risk-sensitive forager who expects resources to be abundant relative to need may opt for a broad (low variance) diet, as will an energy maximizer when high-quality resources are scarce. If the required minimum (m) is well below expected returns, reduced variance in yield is advantageous and new resources should be added to the diet, since under these conditions some loss of efficiency can be tolerated (cells c and d). If needs instead greatly exceed expected returns, diet breadth should contract and become limited to the most highly ranked resources. Under these circumstances, a new resource is likely to replace one or more existing foods if its profitability is high enough to offset the disadvantages of high variance in yield (cell a). If the new resource is low in quality and m is relatively high, it should be ignored (cell b).

Comparison of the two models suggests that adoption of novel resources can be expected to occur under conditions of either scarcity or abundance, although different pathways and goals may be involved in each case. In a resource-rich environment (one in which highly ranked resources are plentiful), a relatively narrow diet will maximize efficiency. Under these conditions, a new item is likely to replace one or more traditional foods, but only if it is profitable (Figure 20.2, cells a and b). A resource-rich environment may also be one in which the required minimum return rate is much lower than the expected average

(Figure 20.3, cells c and d). Given a goal of risk minimization, acceptance of the new item in this context is less contingent upon its profitability and should occur by addition rather than replacement. In parallel fashion, a resource-poor environment may be one in which a narrow diet composed of high-quality foods minimizes risk (in which case the new resource should be ignored unless it is highly profitable) and a relatively broad diet maximizes efficiency (in which case it is likely to be added even if it is of low quality).

The Opportunity Cost Model

Implicit in the diet choice model is the principle that the benefits of subsistence options can only be assessed relative to alternatives; by foregoing one activity in favor of another, the individual incurs *opportunity costs*. Because it can be used to predict how optimal time allocation for a resource or activity changes given different cost/benefit relationships, the opportunity cost model is a useful tool for analyzing the factors underlying the role of a new resource within a subsistence system. Most often, opportunity cost models are used to analyze tradeoffs between foraging and other fitness-enhancing activities (Winterhalder 1983; E. Smith 1987); however, it applies equally to the optimization of time allocation to mutually exclusive food acquisition tasks. This model proposes that the value of any activity is a function of the amount of time devoted to it. Opportunity costs rise with increasing time allocation because alternative fitness-enhancing activities (those that are neglected) become more valuable. Rising opportunity costs are accompanied by a diminishing marginal rate of return as needs are met and fitness benefits reach a plateau. Opportunity costs always rise along with time allocation, assuming that activities are mutually exclusive and that the time available for all activities is finite. Eventually a point will be reached at which opportunity costs exceed the total payoff. The optimal amount of time allocated to an activity is that which maximizes the difference between these two variables (Hames 1992) (Figure 20.4).

The implications of this general model are of interest for generating explanations for the subsistence role played by newly acquired resources. One potentially important observation is that the rate of return accruing to an activity is not identical at all levels of time allocation. This is so because the marginal rate of return (increase in output that results from an additional unit of input) diminishes with increasing effort as needs are filled and as the productive capability of the environment declines. For example, although the farmer who places additional land into production of a high-yielding crop variety can anticipate high net returns of calories, the actual utility of exercising such an option is limited by the farmer's nutritional needs as well as opportunities to convert the harvest into other commodities once those needs are met. The eventual leveling of the benefits curve thus reflects the fact that no matter how energetically efficient it is, an activity whose rewards cannot be utilized is a poor choice for further investment of effort. Furthermore, intensification of

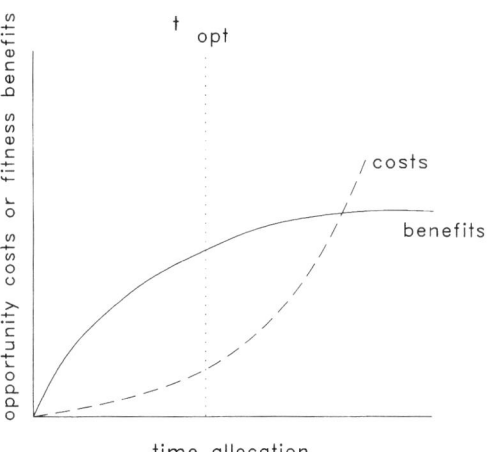

FIGURE 20.4. The opportunity cost model. The solid line represents diminishing returns, the dashed line represents rising opportunity costs. Optimal time allocation for an activity maximizes the distance between the two curves (adapted from Hames 1992: Fig. 7.1).

production by the same hypothetical farmer may result in rapid depletion of soil fertility, simultaneously decreasing yields. The returns curve therefore contains information about both profitability (energy acquired per unit time, assuming time is an adequate surrogate for the costs of acquisition) and actual benefits to the individual. This interpretation of the returns curve further implies that, in the absence of fitness-enhancing alternatives (in other words, given low opportunity costs), time invested in an activity whose rate of return is relatively low can still have substantial benefits.

Case Studies

The models presented above offer a logical and systematic way to develop hypothetical explanations for patterns of crop diffusion by predicting optimal responses to crop introductions in different environmental contexts. In the discussion to follow, I consider how these general models enhance understanding of particular cases of crop adoption.

The Diet Choice Model and Historic Crop Introductions in the Southeastern United States

European colonists introduced a number of new plant species, including crops, to the southeastern United States beginning in the sixteenth century (Crosby 1986; Ruhl 1993; Sauer 1971; Reitz and Scarry 1990; Viola and Margolis 1991). Initially, a surprisingly small number of these plants made their way into native diets. Before European settlement, native populations of the interior Southeast continued to follow a largely traditional mode of subsistence that centered around the cultivation of maize and other crops, seasonal acquisition of mast and fruits, hunting of deer and other mammals, and fishing (Gremillion 1989b, 1993c; Holm 1987). Assuming that the acceptance of plant species by these populations was selective, its patterning illuminates the factors that determined which novel resources were readily

TABLE 20.1. Nutritional data for selected fleshy fruits (quantity per 100 grams of edible portion).[a]

TAXON	CALORIES	PROTEIN (grams)	VITAMIN C (milligrams)
Persimmon (*Diospyros virginiana*)	127	.8	66
Elderberry (*Sambucus* sp.)	72	2.6	36
Grape (*Vitis* sp.)	69	1.3	4
Plum (*Prunus* sp.)	66	.5	0
Blueberry (*Vaccinium* sp.)	62	.7	14
Blackberry (*Rubus* sp.)	58	1.2	21
Cherry, sour (*Prunus* sp.)	58	1.2	10
Groundcherry (*Physalis* sp.)	53	1.9	11
Maypops[b] (*Passiflora incarnata*)	40	2.0	21
Peach (*Prunus persica*)	38	.6	7
Strawberry (*Fragaria* sp.)	37	.7	59

[a] Data from Watt and Merrill (1975), except where otherwise noted.
[b] Data from Gardner (1992:105).

adopted and which were ignored. The evolutionary ecology models discussed above (especially the two versions of the diet choice model) offer additional insights into the problem.

Three Old World crops whose use during the Historic period have been well documented archaeologically are the peach (*Prunus persica*), the cowpea (*Vigna unguiculata*), and the watermelon (*Citrullus lanatus*). Elsewhere (Gremillion 1993c) I have observed that the shared characteristics of Old World crops that were incorporated into native Southeastern economies during Historic times point the way toward an explanation of subsistence change and stability following European contact. All three are weedy, meaning that they often escape from cultivation and thrive under anthropogenic disturbance without requiring large investments of human labor. Thus, they are all relatively cheap to produce, so that their initial cultivation need not have been costly in terms of time and energy outlay. Also contributing to their attractiveness is the fact that all three species had native analogues whose products and cultivation requirements were similar. Thus, incorporating these novel crops did not entail any reorganization of subsistence activities.

The two versions of the diet choice model can be used to refine this argument and add additional depth to the proposed explanation. To illustrate, I take the introduction of the peach into the interior Southeast as an example. Peach, native to Asia, gained wide acceptance amongst southeastern groups. Peach trees require minimal tending, bear fruit within 3 to 5 years after planting, and often germinate spontaneously (Sheldon 1978). Although modern peach orchard management techniques include grafting, pruning, and fertilizing, fruit production is only enhanced by, and not dependent on, these activities (Childers 1954; Gould 1918). Brought to the New World as early as Columbus' second voyage, peaches were introduced to the coastal Southeast early in the sixteenth century and became a common component of Spanish mission gardens along the Atlantic and Gulf coasts of North America (Reitz and Scarry 1985; Sheldon 1978). After 1620, peaches become increasingly visible archaeologically in the Southeastern interior (Gremillion 1993c). Unlike many crops introduced by the Spanish, peaches flourished in the climate of the Southeast and spread rapidly (Ruhl 1993; Reitz and Scarry 1985:55).

I evaluate the usefulness of the diet choice model for explaining patterns of crop adoption by comparing its predictions with the actual effects of introduction of the peach upon native southeastern subsistence as documented archaeologically and historically. In order to do so, it is necessary to assess characteristics of the environment that are relevant to the model, such as resource profitability, resource abundance, and the relationship between required and anticipated return rates. A comparison of the caloric and nutritional characteristics of several fleshy fruit taxa known to have been used in the Southeast during Historic times indicates that peaches rank relatively low in kilocalories per 100 g of edible portion (Watt and Merrill 1975) (Table 20.1). Although this ratio does not take costs into account, it is likely that processing costs (including any preparation for storage) were similar for different fleshy fruit taxa. Planting of peach trees near settlements would have enhanced their profitability by reducing transportation and travel costs and perhaps also by increasing their productivity. However, this gain in benefits would have been accompanied by increased production costs associated with activities such as pruning and clearing (if practiced). Given these considerations, peach is unlikely to have been among the most profitable fleshy fruits available to southeastern populations.

The relative abundance of fleshy fruit taxa is more difficult to estimate in the absence of precise environmental data. Anthropogenic disturbance of woody vegetation due to settlement, fires, and agriculture was common in the Historic Southeast (Gremillion 1989b:147; Hammett 1992). The open habitats thus created were quickly colonized by heliophilic, disturbance-loving plants, among them many fleshy fruit-bearing taxa (Horn 1974; Yarnell 1982). Given a disturbance regime of this sort, alternative resources were potentially quite dense near human settlements. Planting and tending of native fruit trees such as persimmons in Historic times (see for example Van Doren 1928: 57, 272, 280) probably increased their density beyond that of non-managed populations. However, assessments of abundance should also take into account the effects of predation and more severe forms of environmental disturbance.

Because of the lack of a sound basis for categorizing the southeastern environment as being either rich or poor in fleshy fruit resources, I consider the implications of both scenarios for the model's predictions. If more highly ranked fleshy fruits were in fact abundant, then the diet choice model predicts that peach (as a resource of relatively low profitability) should have been ignored (Figure 20.2, cell b). If these profitable items were instead relatively scarce, the model predicts that a new resource such as peach should have been added to the diet (depending on its return rate relative to the average for the existing resource set) (Figure 20.2, cell d). The first outcome (ignore) disagrees with the archaeological and historic evidence for the rapid adoption of peaches in the Southeast. The second option (add) is consistent with archaeological evidence, which indicates that peaches joined, rather than replaced, traditional foods (Gremillion 1989b; Chapman and Shea 1981). Historical accounts of "orchards" in which native persimmons and plums were grown alongside peaches further attest to the additive use of crops that were adopted from European sources (Van Doren 1928:57, 272, 280). Thus, if native fleshy fruits of higher rank were scarce, the deterministic version of the diet choice model succeeds in predicting the adoption of peach and its integration into a relatively broad diet. If, however, such resources were abundant, then the model performs poorly and some refinement of the model or of estimates of relevant costs and benefits is warranted.

The z-score model offers an alternative that considers the effects of crop adoption upon risk. Assuming that peach was introduced as a resource of low profitability, this model predicts that it should have been ignored (if need was significantly higher than expected average returns) or added (if need was instead considerably lower). There is reason to suppose that the latter situation is more likely. Suppose that m is defined as the food energy needed to meet the minimum daily requirement for ascorbic acid, which is available primarily in fleshy fruits and greens (although organ meats are also an important source). Further assume that individuals must meet their ascorbic acid needs exclusively through consumption of fleshy fruits. It is estimated that 10 mg/day of ascorbic acid will cure and prevent scurvy (Bureau of Nutritional Sciences 1983). Given a fleshy fruit diet of the 11 taxa in Table 20.1, this need would be met by an average of 44 g of fruit. This quantity of fresh fruit could have been easily acquired during the summer and fall, when natural productivity is at its maximum. Its availability would have been further extended into the winter, judging by historic accounts of the drying of fruit for storage (Gremillion 1989b:87–88). Greens would have become available in early spring. Thus, the return rate from fleshy fruits required to meet ascorbic acid needs (m) should have been easily met by fresh or stored fruit throughout most of the year, and perhaps by fresh greens during the spring. Under such conditions (m low relative to anticipated returns; new resource of low profitability), the z-score model predicts addition of peaches to the diet, as seems to have in fact occurred in the Southeast. This outcome can be expected regardless of the role of ascorbic acid in the diet, provided that the relationship between required and expected energetic returns from fleshy fruits has been correctly estimated.

The use of evolutionary ecology models in this case refines understanding of the relationships between variables beyond that provided by the more inductive, common-sense interpretation offered in Gremillion (1993c). It proposes an explanation for why peaches were added to the existing suite of crops rather than replacing one or more traditional resources. The replacement option would have been advantageous only if peach had proven to be highly profitable compared to fleshy fruits already in the diet. If peach was instead as low in rank as its caloric and nutritional composition suggest, its use may have acted to increase energetic efficiency (if more profitable resources were scarce) or to minimize risk (if returns were expected to greatly exceed requirements).

If proposed as an explanation for the adoption of peach by southeastern groups during the Historic period, this analysis is subject to a number of valid criticisms. For example, the relevant environmental variables (such as costs and benefits of fleshy fruit harvesting and production) are not realistically estimated. Seasons of availability and the consequences of storage have been ignored, and only a small subset of the total potential resource set has been considered. Consequently, hypotheses to explain the adoption of this crop cannot be rigorously evaluated. However, the discussion of model predictions does serve to illustrate how evolutionary ecology models might be used as frameworks for constructing explanations of patterns of crop adoption in the past.

Indigenous and Introduced Crops in Eastern North America

Recent research in eastern North America has allowed the refinement of the history of crop introductions and reveals a complex pattern of agricultural development (Fritz 1990; Smith 1989, 1992a, 1992b; Watson 1989). Several indigenous annual species were first domesticated during the second and late third millennia BC (Asch and Asch 1985; Crites 1993; Smith and Cowan 1987). After ca. 500 BC, agricultural economies based on these native plants became widespread in the Midwest and Midsouth. Maize, a native of Mesoamerica, was introduced to eastern North America before AD 1 (Riley et al. 1994). In the Midwest and Midsouth, maize seems to have been initially incorporated as a garden crop into a long-established system of relatively small-scale farming based on indigenous crops such as sumpweed (*Iva annua*), sunflower (*Helianthus annuus*), and chenopod (*Chenopodium berlandieri*) (Fritz 1990:397; Gremillion and Yarnell 1986; Watson 1989:560). However, stable carbon isotope ratios in human bone from the eastern United States indicate that maize was not a significant dietary component until after ca. AD 800–1000 (Ambrose 1987; Bender et al. 1981; Buikstra et al. 1987; Lynott et al. 1986). In the midcontinental region, both quantity and ubiquity of maize remains on archaeological sites are correspondingly low until after about AD 800 (Asch and Asch 1985; Johannessen 1984). Thus, in the Midwest maize seems to have played only a minor subsistence role for at least 600 years following its initial introduction (Smith 1992b). The interior Southeast, where indigenous crops seem not to have been as important, displays a different archaeological pattern; there, maize rose to dominance as a food crop between AD 900 and 1000, shortly after its initial appearance in the archaeological record (Scarry 1986). Although some of the indigenous crops (such as sunflower) continued to be grown and used up until the contact period in parts of eastern North America, they lost

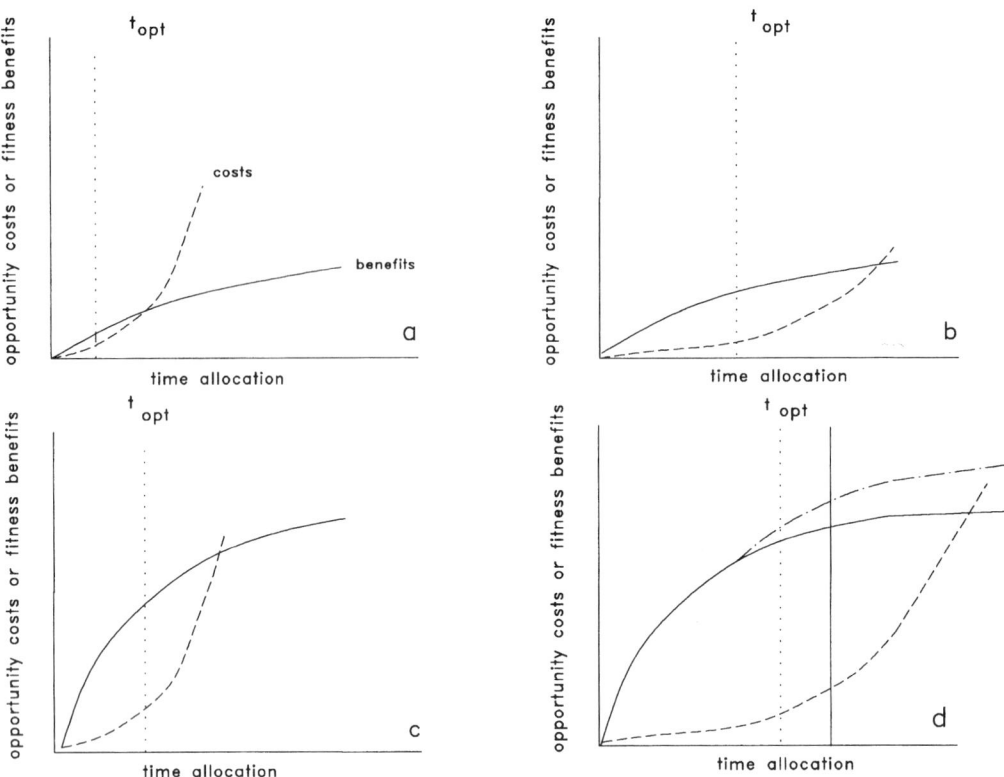

FIGURE 20.5. Hypothetical changes in opportunity cost and diminishing returns for maize: (a) initial scenario representing rapidly rising opportunity costs coupled with a quickly leveling return curve; (b) effects of lowering the opportunity cost curve on optimal time allocation, with returns held constant; (c) effects of raising the profitability (net return rate) of maize production on optimal time allocation, with opportunity costs held constant; (d) both lowering opportunity costs and raising return rates on optimal time allocation; the dot-and-dash curve represents an increase in the marginal rate of return, and the solid vertical line marks the new optimal time allocation. Solid curves represent diminishing returns; dashed curves represent opportunity costs. Optimal time allocation maximizes the distance between the two curves (vertical lines).

much of their former importance and in some localities (such as the middle Ohio Valley) seem to have virtually disappeared from native gardens and fields (Wagner 1987).

At the center of current debates about the rise of food producing economies in eastern North America is the question of why maize became important when it did. Gardner's (1992) linear programming results suggest that (based on yields of historic varieties) maize is so profitable to produce that it should rapidly become important among populations maximizing net energy acquisition subject to nutritional constraints. In light of such findings, as well as the centrality of maize agriculture to subsistence economies encountered at contact, the long period of limited use following its initial introduction is intriguing. Factors suggested to account for this temporal lag include the inferiority of the varieties of maize initially available and the labor requirements associated with maize production.

The opportunity cost model can assist in determining which variables are likely to have had the greatest effects upon the allocation of time and effort to maize production. In its simplest form, this model proposes that the optimal amount of time allocated to use of a resource is a function of the relationship between increasing opportunity costs and diminishing returns. Suppose that the earliest maize grown in the Eastern Woodlands was in fact relatively unproductive and consequently offered poor caloric returns per unit of effort expended on planting, cultivating, harvesting, and processing. These proposed relationships between opportunity costs and diminishing returns are proposed in simple graphical form in Figure 20.5a. Total benefits remain low, and opportunity costs rise sharply as time is taken away from more beneficial activities. By manipulating these hypothetical curves, it is possible to propose what kinds of real-world changes would be required to push the vertical dotted line towards the right, in the direction of greater time allocation. The introduction or development of a more productive maize variety or an improvement in technological efficiency peculiar to (or differentially benefiting) maize production—even with no change in opportunity costs—would both have had the effect of raising optimal time allocation somewhat (Figure 20.5c). (For purposes of illustration, opportunity costs are kept constant, although in reality they would be likely to decline as maize production became more efficient). Alternatively, some increase in optimal time allocation can be expected to occur as a result of the elimination of profitable alternative activities (Figure 20.5b). For example, depletion of game would have reduced the costs of diverting effort from hunting. In this case, increased investment in maize farming would have been beneficial even without any change in its profitability. Finally, a major increase in optimal time allocation to maize production can be expected when both the returns curve and the opportunity cost curve are modified simultaneously (Figure 20.5d). However, increased profitability does not remove the constraints that cause benefits to diminish as more time is allocated. The point at which the marginal rate of return approaches zero may remain the same even if the returns curve acquires a steeper slope. For greater effort to be favored, the ceiling on potential benefits of maize production would have to be lifted. Such a change would be most likely to occur if mechanisms such as storage and exchange were available to extend benefits beyond those associated with immediate subsistence needs.

The scenario represented by Figure 20.5c (an increase in the profitability of maize agriculture) is plausible in light of botanical and archaeological evidence, but different pathways to this outcome are not equally well supported. Net returns may change as a function of either an increase in benefits or a decrease in costs. The assertion that early forms of maize were relatively unprofitable, and that improved varieties were needed to make greater investment in production worthwhile, is offered some support by the observation that maize requires more care than its weedy predecessors and tends to rapidly deplete soil fertility (Lopinot 1992). However, the yields of historic varieties of maize per unit of cultivated land are quite high compared to those of most of the native crops (Rindos and Johannessen 1991:43; B. Smith 1987). Furthermore, empirical support for the introduction of a morphologically distinctive (and presumably productively superior) maize variety at the critical time is virtually nonexistent. The contemporaneous success of maize in the Southwest also argues against the assertion that maize was poorly adapted to temperate climates at the time of its introduction to the Eastern Woodlands (Fritz 1992). Unless humid conditions posed further adaptive problems for early eastern maize that contributed to low yields, it seems likely that a reduction in labor costs contributed significantly toward the proposed increase in the profitability of maize production. The association of high quality hoes made of Mill Creek chert with the increase in presence of maize on archaeological sites in the Southeast after about AD 750 (Smith 1986:54) may be evidence of such technological improvements.

Effective means of storing maize would have removed constraints on the potential benefits of its production, enhancing the value of increased effort (Figure 20.5d). A correlation between increased production of maize and its developing role as a stored commodity is supported by the construction of large storage pits in many midwestern localities after ca. AD 800 (Smith 1986:54). The frequent occurrence of both kernel and cob remains in early Mississippian components in the American Bottom region, along with high ratios of maize remains to native crops seeds, offers further support for the establishment of maize as a storable staple (Lopinot 1992:75). However, the fact that maize is actually more convenient to store than the native starchy seed crops (Lopinot 1992:55) casts doubt on the interpretation that absence of appropriate technology limited the potential of maize as a storable resource.

Exchange is another mechanism whereby limits on the marginal utility of maize production might have been lifted. The Mississippi period is well known for its geographically widespread trade in exotic goods such as pearls, marine shell, and other raw materials and finished products. However, similar networks had been in place at the time maize was first introduced to the Eastern Woodlands. For the exchange value of maize to have made a difference in optimal time allocation, demand for maize as a commodity would have needed to arise or intensify. Unsynchronized food shortages across a broad region represent one potential source of demand. However, stresses of this kind faced by Mississippian populations appear to have been met primarily by redistributive networks and centralization of authority (Muller and Stephens 1991).

Reduced opportunity costs associated with foregoing other subsistence activities (such as mast harvesting, cultivation of native crops, and hunting) are even less compelling as a cause for increased effort devoted to maize agriculture. Late summer and early fall, harvest time for maize and other crops, is a time of great resource abundance in the Eastern Woodlands. Wild fruits and nuts as well as crops reach their peak availability, and considerable effort had to be invested in processing storable plant foods whose payoff would not be realized for months. Barring a serious shortage of some customarily exploited resource, it seems unlikely that a poor-quality variety of maize would have been utilized in quantity because of an absence of competing demands on time.

This discussion of model predictions in light of archaeological evidence has several implications for understanding why maize played such a limited role for hundreds of years following its introduction, and why the dramatic shift in its utilization occurred when it did. One of these is that the initial rejection or adoption of a crop and its economic role have different consequences and thus are likely to require different kinds of explanations. Assuming it was introduced as a resource of relatively low profitability, maize would have been adopted by agricultural populations that maintained a broad resource base, either due to resource depletion or as a means of reducing risk. However, opportunity costs associated with maize farming would have limited investment in its production. A sharp upswing in that investment is most likely to have resulted from an improvement in labor efficiency, perhaps accompanied by an increase in productivity as the plant became better adapted to the humid temperate climate. Exchange networks and storage technology predate the intensification of maize farming; thus, their absence cannot be held responsible for limiting the realizable benefits of maize production. Unless rival subsistence activities were severely curtailed, declining opportunity costs alone are unlikely to have driven up time allocation to maize production.

Once maize was established as a major crop after ca. 1000 AD, the role of other plant resources changed. Winterhalder and Goland (1996) point out that if maize was initially low-ranked but high in density, population growth should result, causing alternative resources of high rank to be depleted and consequently diminish in importance. The optimal diet, however, should remain relatively broad. If high in rank, the elevation of maize to the role of dietary staple should be accompanied by a contraction of diet breadth. These two scenarios parallel variation in Late Prehistoric subsistence systems of the eastern United States (Rossen and Edging 1987). Mississippian maize farmers who occupied the fertile, broad river valleys of the Midwest and Midsouth retained a broad resource base including migratory waterfowl, fishes and other aquatic species, large and small mammals, wild plant foods, and indigenous crops (Smith 1985c). In contrast, Fort Ancient populations of the middle Ohio Valley appear to have had narrower diets that were more heavily focused on maize (Wagner 1987). Perhaps these differences reflect separate evolutionary pathways for the emergence of maize-based economies in response to regional differences in the productivity of maize and the effectiveness of agricultural technology.

The development of specialized economies based on maize can also be viewed as a response to risk. The z-score model predicts that a change in the relationship between expected average food acquisition and minimum requirements should result in contraction or expansion of diet breadth. Narrowing of the diet to emphasize the most highly ranked resources (even at the cost of reduced efficiency) would be the predicted response to the appearance of a large deficit between expected and required returns. Anticipated food shortage may help to explain the virtual abandonment of native crops in favor of maize after ca. AD 1000 in some areas (Smith 1992b). However, such a response is unlikely to occur except in cases of a considerable gap between needs and returns. Furthermore, strategies such as food sharing or field dispersion are more effective means of reducing risk in that they do not entail such large sacrifices of energy efficiency (Winterhalder 1990; Winterhalder and Goland 1996). This factor would have placed serious constraints on the value of dropping some resources as a way of reducing risk, except in the most extreme cases of food shortage.

Discussion

I argue that research programs aimed at understanding changes in agricultural systems in prehistory can benefit from the development and testing of hypotheses based on evolutionary ecology models. Following this path involves overcoming a number of admittedly formidable obstacles that demand refinement of methods for assessing the goodness of fit between models and data as well as adjustment of general models to better approximate the complexity of human decision-making and its consequences. Despite these concerns, evolutionary ecology models offer the promise of generating new insights into old problems from the standpoint of a coherent body of theory.

Any discussion that emerges from a theoretical foundation inevitably raises issues of testability and confirmation. Are the insights into crop diffusion offered by the use of evolutionary ecology likely to be empirically demonstrated, or are they merely plausible stories with a new twist? Testing requires accurate reconstruction of subsistence behavior using archaeological data, a task that continues to occupy the best minds in many specialties and requires further refinement. Archaeologists lack the ethnographer's advantage of direct observation of environmental and behavioral variability. The nature of the archaeological data base heightens the problem of equifinality that is always encountered as part of hypothesis testing. That is, even if data and hypothesis are congruent, it is often difficult to rule out the operation of alternative causes that might produce the same archaeological pattern. However, given adequate consideration of such alternatives and continued methodological refinement, there is no reason why testing should not come to play an increasingly important role in the application of evolutionary ecology to archaeological problems.

The chief virtue of the simplest of foraging models, their applicability to a wide variety of situations, becomes increasingly problematic when used to account for variation in behaviorally complex organisms. One remedy is refinement of foraging models to reflect more realistically the complexity of interactions between human populations and their resources. For example, the diet choice model assumes that the search for prey is random (that is, prey items are encountered in proportion to their density). This assumption underlies the prediction that as highly ranked prey become less abundant, the efficiency of foraging is lowered, and the diet expands to include lower-ranked items. It is clearly violated by human actors, particularly in the case of managed resources. The locations of harvestable stands are well known to foragers and farmers alike, so that search costs are often nil (Gardner 1992; Keegan 1986). Consequently, a reduction in abundance of highly ranked crops does not necessarily lower overall foraging efficiency by increasing the costs of searching for prey. The simplicity of the z-score model is similarly problematic in that it assumes that individuals act alone. In fact, as Winterhalder (1990) has demonstrated, manipulating diet breadth to minimize the possibility of resource shortfalls is limited in its effectiveness by the tradeoff between energetic efficiency and risk reduction. Because they sidestep this constraint, food sharing and (for farming populations) field dispersion may be superior means of reducing risk, although they entail energetic costs of their own (Goland 1993).

However, consideration of variables not explicitly accounted for in the models is not precluded by the evolutionary ecology approach. For example, familiarity with crops similar to those newly introduced can be modeled as cultural information whose existence modifies cost-benefit functions. In other words, prior knowledge about related plants reduces the costs associated with any novel resource and thereby enhances its profitability, making its entry into the optimal diet more likely. Conversely, reluctance to adopt agricultural innovations can also be linked to the energetic inefficiency that typically characterizes initial experiments with new subsistence techniques (Bettinger and Baumhoff 1982:489). Refusal to replace traditional crops with newer, more productive varieties may reflect an understanding of the relationship between variance in yield and risk in particular environmental contexts. Rather than representing irrational conservatism, these decisions can be seen instead as responses that are, at least in some cases, both logical and adaptive. Evolutionary ecology approaches also can, and should, take into account historical aspects of environmental variability, such as the economic exchanges that provide or prevent access to exotic food resources.

One of the major strengths of optimization analysis in evolutionary ecology, its emphasis on the tendency of natural selection to result in optimal decision-making by human individuals, has also been cited as a weakness. Central to most models used in evolutionary ecology is the assumption that selection is the primary force promoting adaptive behavior, and that optimization can best be measured in terms of material benefit (which is correlated with fitness). Well-reasoned defenses of the optimization assumption and its use in evolutionary ecology have been given elsewhere (Smith 1983; Smith and Winterhalder 1992). However, suppose this assumption has been violated in the case of the peach introduction example. Is it then justifiable to conclude that people eagerly acquired and grew peaches simply because they tasted good, regardless of their energetic returns or risk-reduction value? Certainly this interpretation (which is equivalent to stating that people did what they wished to do) may be accurate, and yet if true it adds little to our understanding of the patterning known to exist in human behavior both past and present. It also leaves unanswered the question of why the preference for sweet-tasting fruits exists, a question that quickly returns us to the evolutionary arena. Despite the vast flexibility of human behavior (including perhaps the ability to behave in nonadaptive ways), it is reasonable for investigators to begin under the assumption that selection has played a major role in shaping many or even most of the food acquisition behaviors that are reflected archaeologically.

Even given these caveats, the foregoing analysis of crop diffusion using optimization models has proved to be a rich source of propositions linking environmental variables to different patterns of crop acquisition. These propositions offer some fresh vantage points from which to consider the problem of diffusion and adoption of crops and suggest directions for hypothesis development. The most important of these can be summarized as follows:

1. The adoption of new crops may not depend upon their energetic contribution or their effect on overall food acquisition efficiency exclusively, but may also reflect the impact of their utilization upon risk. Similar environments may consequently elicit strikingly different responses to new resources depending on which criterion (e.g., risk minimization or efficiency maximization) guides decision-making.
2. A population with an already effective adaptation (that is, one in which returns are high relative to anticipated needs) may adopt relatively unprofitable resources in order to minimize risk of failures. Thus, acceptance of new crops, even if they are of low quality, need not indicate subsistence stress.
3. The addition of novel items to the resource set and their establishment as staples have different implications for fitness and demand distinct explanations. For example, the effects of a change in diet breadth upon risk or efficiency of food acquisition may explain the initial adoption of a new resource, but will be unable to account for increasing allocation of effort to its production.
4. Time spent on a crop that is inefficient to produce may have substantial fitness benefits in the absence of profitable alternatives (that is, when opportunity costs are low).
5. Once adopted, a new resource that couples low profitability with high opportunity costs will play only a minor dietary role. Optimal allocation of time to producing such a crop will remain low until its profitability is enhanced and/or constraints on the benefits of additional time allocation are removed. Improved management techniques, the development of more productive varieties, storage technology, and a rise in exchange value are likely candidates for producing these changes.

These findings indicate that evolutionary ecology models have several advantages as research tools for investigating the origins and development of agricultural systems, including the diffusion of specific crops. The generality of many of the models frequently used by evolutionary ecologists allows their application to diverse research questions. Use of a simple graphical format allows for nonquantitative exploration of the behavioral consequences of modifying variables even in the absence of knowledge of their precise values. Elements of the models can be manipulated in ways that reveal previously unsuspected relationships between variables, adding new insights to problems in a way that post-hoc interpretation is unlikely to accomplish. Such insights often emerge during the testing phase of analysis, in which model predictions are compared with real-world observations. In this context, the lack of congruence between data and model prediction does not represent a failure to explain and does not represent wasted effort.

Rather, the points of disagreement provide the basis for refining the models and hypotheses that have been developed. In this way, the basic approach of model-building within an evolutionary ecology framework can be adapted and refined to create analytic tools appropriate for the task of developing explanations for the diffusion and adoption of crops.

Acknowledgments

I am grateful to Paul Gardner, Carol Goland, Bruce Winterhalder, and one anonymous reviewer for their helpful comments and encouragement. My interest in the diffusion of peaches and other Old World crops was inspired in part by my involvement as a graduate student with the archaeological research conducted by the Research Laboratories of Anthropology, University of North Carolina, Chapel Hill between 1982 and 1989. Despite these debts, I alone am responsible for the content of this article.

References

Ambrose, S. H.
1987   Chemical and isotopic techniques of diet reconstruction in eastern North America. In *Emergent Horticultural Economies of the Eastern Woodlands,* edited by W. F. Keegan, pp. 87–107. Occasional Papers No. 7, Center for Archaeological Investigations, Southern Illinois University, Carbondale.

Asch, D. L., N. B. Asch
1985   Prehistoric plant cultivation in west-central Illinois. In *Prehistoric Food Production in North America,* edited by Richard I. Ford. Anthropological Papers No. 75, University of Michigan Museum of Anthropology, Ann Arbor.

Bender, M. M., D. A. Baerreis, R. L. Steventon
1981   Further light on carbon isotopes and Hopewell agriculture. *American Antiquity* 46:346–353.

Bettinger, R. L., M. A. Baumhoff
1982   The Numic spread: Great Basin cultures in competition. *American Antiquity* 47:485–503.

Boone, J. L.
1994   *Is It Evolution Yet?: A Critique of "Darwinian Archaeology."* Paper presented at the Annual Meeting of the Society for American Archaeology, Anaheim.

Boyd, R., P. J. Richerson
1985   *Culture and the Evolutionary Process.* University of Chicago Press, Chicago.

Buikstra, J. E., J. Bullington, D. K. Charles, D. C. Cook, S. Frankenberg, L. W. Konigsberg, J. P. Lambert, L. Xue
1987   Diet, demography, and the development of horticulture. In *Emergent Horticultural Economies of the Eastern Woodlands,* edited by William F. Keegan, pp. 67–85. Occasional Papers No. 7, Southern Illinois University, Carbondale.

Bureau of Nutritional Sciences
1983   *Recommended Nutrient Intakes for Canadians.* Department of National Health and Welfare, Ottawa.

Chapman, J., A. B. Shea
1981   The archaeobotanical record: Early Archaic period to contact in the lower Little Tennessee River valley. *Tennessee Anthropologist* 6:61–84.

Childers, N. F., editor
1954   *Mineral Nutrition of Fruit Crops.* Somerset Press, Somerville, New Jersey.

Cowan, C. W.
1996   Evolutionary changes associated with the domestication of *Cucurbita pepo*: Evidence from eastern Kentucky. In *People, Plants and Landscapes: Studies in Paleoethnobotany,* edited by Kristen J. Gremillion. University of Alabama Press, in press.

Cowan, C. W., P. J. Watson, editors
1992   *The Origins of Agriculture: An International Perspective.* Smithsonian Institution Press, Washington, D.C.

Crites, G. D.
1987   Human-plant mutualism and niche expression in the paleoethnobotanical record: A Middle Woodland example. *American Antiquity* 52:725–740.
1993   Domesticated sunflower in fifth millennium BP temporal context: New evidence from middle Tennessee. *American Antiquity* 58:146–148.

Crosby, A. W.
1986   *Ecological Imperialism: The Biological Expansion of Europe, 900–1900.* Cambridge University Press, Cambridge.

Darwin, C.
1859   *The Origin of Species.* Random House, New York.

Doggett, H.
1965   Disruptive selection in crop development *Nature* 206:279–280.

Donald, C. M., J. Hamblin
1983   The convergent evolution of annual seed crops in agriculture. *Advances in Agronomy* 36:97–143.

Dunnell, R. C.
1980   Evolutionary theory and archaeology. In *Advances in Archaeological Method and Theory,* edited by Michael B. Schiffer, Volume 3, pp. 35–99. Academic Press, New York.

Durham, W. H.
1982   Interactions of genetic and cultural evolution: Models and examples. *Human Ecology* 10:289–323.

Flannery, K. V.
1968   Archaeological systems theory and early Mesoamerica. In *Anthropological Archaeology in the Americas,* edited by Betty J. Meggers, pp. 67–87. Anthropological Society of Washington, Washington, D.C.
1973   The origins of agriculture. *Annual Review of Anthropology* 2:271–310.

Fritz, G. J.
1990   Multiple pathways to farming in precontact eastern North America. *Journal of World Prehistory* 4:387–435.
1992   "Newer", "better" maize and the Mississippian emergence: A critique of prime mover explanations. In *Late Prehistoric Agriculture: Observations from the Midwest,* edited by William I. Woods, pp. 19–43. Studies in Illinois Archeology No. 8, Illinois Historic Preservation Agency, Springfield.

Gardner, P. S.
1992   Diet optimization models and prehistoric subsistence change in the Eastern Woodlands. Unpublished Ph.D. dissertation, Department of Anthropology, University of North Carolina, Chapel Hill.

Gebauer, A. B., T. D. Price, editors
1992   *Transitions to Agriculture in Prehistory.* Prehistory Press, Madison, Wisconsin.

Goland, C.
1993   *Field Scattering as Agricultural Risk Management: A Case Study from Cuyo Cuyo, Department of Puno, Peru.* Mountain Research and Development, International Mountain Society and United Nations University, University of California Press, Berkeley.

Gould, H. P.
1918   *Peach-growing.* Macmillan, New York.

Gremillion, K. J.
1989a   The development of a mutualistic relationship between humans and maypops (*Passiflora incarnata* L.) in the southeastern United States. *Journal of Ethnobiology* 9:135–155.

1989b   Late prehistoric and historic period paleoethnobotany of the North Carolina Piedmont. Unpublished Ph.D. dissertation, University of North Carolina, Chapel Hill.
1993a   The evolution of seed morphology in domesticated *Chenopodium*: An archaeological case study. *Journal of Ethnobiology* 13:149–169.
1993b   Crop and weed in prehistoric eastern North America: The *Chenopodium* example. *American Antiquity* 58:496–508.
1993c   Adoption of Old World crops and processes of cultural change in the Historic Southeast. *Southeastern Archaeology* 12:15–20.

Gremillion, K. J., R. A. Yarnell
1986   Plant remains from the Westmoreland–Barber and Pittman–Alder sites, Marion County, Tennessee. *Tennessee Anthropologist* 11:1–20.

Hames, R.
1992   Time allocation. In *Evolutionary Ecology and Human Behavior*, edited by Eric Alden Smith and Bruce Winterhalder, pp. 203–236. de Gruyter, New York.

Hammett, J. E.
1992   Ethnohistory of aboriginal landscapes in the southeastern United States. *Southern Indian Studies*, Volume 41.

Harlan, J. R., J. M. J. de Wet, E. G. Price
1973   Comparative evolution of cereals. *Evolution* 27:311–325.

Harris, D. R., G. C. Hillman (Editors)
1989   *Foraging and Farming: The Evolution of Plant Exploitation.* Unwin Hyman, London.

Holm, M. A.
1987   Faunal remains from the Wall and Fredricks sites. In *The Siouan Project: Seasons I and II,* edited by R. S. Dickens, Jr., H. T. Ward, and R. P. S. Davis, Jr., pp. 237–258. Monograph Series No. 1, Research Laboratories of Anthropology, University of North Carolina, Chapel Hill.

Horn, H. S.
1974   The ecology of secondary succession. *Annual Review of Ecology and Systematics* 5:25–37.

Johannessen, S.
1984   Paleoethnobotany. In *American Bottom Archaeology,* edited by Charles J. Bareis and James W. Porter, pp. 197–214. University of Illinois Press, Urbana.

Kaplan, H., K. Hill
1992   Evolutionary ecology of food acquisition. In *Evolutionary Ecology and Human Behavior,* edited by E. A. Smith and B. Winterhalder, pp. 167–202. Aldine de Gruyter, New York.

Keegan, W. F.
1986   The optimal foraging analysis of horticultural production. *American Anthropologist* 88:92–107.

Keene, A. S.
1981   Optimal foraging in a nonmarginal environment: A model of prehistoric subsistence strategies in Michigan. In *Hunter-Gatherer Foraging Strategies,* edited by Bruce Winterhalder and Eric Alden Smith, pp. 171–193. University of Chicago Press, Chicago.

King, F. B.
1985   Early cultivated cucurbits in eastern North America. In *Prehistoric Food Production in North America,* edited by Richard I. Ford, pp. 73–98. Anthropological Papers No. 75, Museum of Anthropology, University of Michigan, Ann Arbor.
1987   The evolutionary effects of plant cultivation. In *Emergent Horticultural Economies of the Eastern Woodlands,* edited by William F. Keegan, pp. 51–66. Occasional Paper No. 7, Center for Archaeological Investigations, Southern Illinois University, Carbondale.

Lopinot, N. H.
1992   Spatial and temporal variability in Mississippian subsistence: The archaeobotanical record. In *Prehistoric Agriculture: Observations from the Midwest,* edited by William I. Woods, pp. 44–94. Studies in Illinois Archaeology No. 8, Illinois Historic Preservation Agency, Springfield.

Lynott, M. J., T. W. Boutton, J. E. Price, D. E. Nelson
1986   Stable carbon isotopic evidence for maize agriculture in southeast Missouri and northeast Arkansas. *American Antiquity* 51:51–65.

Muller, J., J. E. Stephens
1991   Mississippian sociocultural adaptation. In *Cahokia and the Hinterlands,* edited by Thomas E. Emerson and R. Barry Lewis, pp. 297–310. University of Illinois Press, Urbana.

O'Brien, M. J.
1987   Sedentism, population growth, and resource selection in the Woodland Midwest: A review of coevolutionary developments. *Current Anthropology* 28:177–197.

O'Brien, M. J., T. D. Holland
1990   Variation, selection, and the archaeological record. In *Archaeological Method and Theory,* edited by Michael B. Schiffer, Volume 2, pp. 31–79. University of Arizona Press, Tucson.

Reidhead, V. A.
1981   *A Linear Programming Model of Prehistoric Subsistence Optimization: A Southeastern Indiana Example.* Prehistory Research Series 6, Indiana Historical Society, Indianapolis.

Reitz, E. J., C. M. Scarry
1985   *Reconstructing Historic Subsistence with an Example from Sixteenth-Century Spanish Florida.* Special Publications Series No. 3, Society for Historical Archaeology, Ann Arbor.
1990   Herbs, fish, scum, and vermin: Subsistence strategies in sixteenth-century Spanish Florida. In *Columbian Consequences, Volume 2: Archaeological and Historical Perspectives on the Spanish Borderlands East,* edited by David Hurst Thomas, pp. 343–354. Smithsonian Institution Press, Washington, D.C.

Riley, T. R., G. R. Waltz, C. J. Bareis, A. C. Fortier, K. E. Parker
1994   Accelerator mass spectrometry (AMS) dates confirm early *Zea mays* in the Mississippi River Valley. *American Antiquity* 59:490–497.

Rindos, D.
1984   *The Origins of Agriculture: An Evolutionary Perspective.* Academic Press, New York.

Rindos, D., S. Johannessen
1991   Human-plant interactions and cultural change in the American Bottom. In *Cahokia and the Hinterlands,* edited by Thomas E. Emerson and R. Barry Lewis, pp. 35–45. University of Illinois Press, Urbana.

Rossen, J., R. Edging
1987   East meets west: Patterns in Kentucky Late Prehistoric subsistence. In *Current Archaeological Research in Kentucky: Volume One,* edited by David Pollack, pp. 225–234. Kentucky Heritage Council, Frankfort.

Ruhl, D.
1993   Old customs and traditions in new terrain: Sixteenth- and seventeenth-century archaeobotanical data from La Florida. In *Foraging and Farming in the Eastern Woodlands,* edited by

C. Margaret Scarry, pp. 255–283. University Presses of Florida, Gainesville.

Sauer, C. O.
1971 *Sixteenth Century North America.* University of California Press, Berkeley.

Scarry, C. M.
1986 Changes in plant procurement and production during the emergence of the Moundville chiefdom. Unpublished Ph.D. dissertation, University of Michigan, Ann Arbor.

Sheldon, E. S.
1978 Childersburg: Evidence of European contact demonstrated by archaeological plant remains. *Southeastern Archaeological Conference Special Publication* 5:28–29.

Smith, B. D.
1985a The role of *Chenopodium* as a domesticate in pre-maize garden systems of the eastern United States. *Southeastern Archaeology* 4:51–72
1985b *Chenopodium berlandieri* ssp. *jonesianum*: evidence for a Hopewellian domesticate from Ash Cave, Ohio. *Southeastern Archaeology* 4:107–133.
1985c Mississippian patterns of subsistence and settlement. In *Alabama and the Borderlands: From Prehistory to Statehood,* edited by R. Reid Badger and Lawrence A. Clayton, pp. 64–79. University of Alabama Press, Tuscaloosa.
1986 The archaeology of the southeastern United States: from Dalton to DeSoto, 10,500 BP–500 BP. In *Advances in World Archaeology,* edited by F. Wendorf and A. Close, Volume 5, pp. 1–92. Academic Press, Orlando.
1987 The economic potential of *Chenopodium berlandieri* in prehistoric eastern North America. *Journal of Ethnobiology* 7:29–54.
1989 Origins of agriculture in eastern North America. *Science* 246:1566–1571.
1992a *Rivers of Change: Essays on Early Agriculture in Eastern North America.* Smithsonian Institution Press, Washington, D.C.
1992b Prehistoric plant husbandry in eastern North America. In *The Origins of Agriculture: An International Perspective,* edited by C. W. Cowan and P. J. Watson, pp. 101–120. Smithsonian Institution Press, Washington, D.C.

Smith, B. D., C. W. Cowan
1987 Domesticated *Chenopodium* in prehistoric eastern North America: New accelerator dates from eastern Kentucky. *American Antiquity* 52:355–357.

Smith, E. A.
1983 Anthropological applications of optimal foraging theory: a critical review. *Current Anthropology* 24:625–651.
1987 On fitness maximization, limited needs, and hunter-gatherer time allocation. *Ethology and Sociobiology* 8:73–85.

Smith, E. A., B. Winterhalder
1992 Natural selection and decision making: Some fundamental principles. In *Evolutionary Ecology and Human Behavior,* edited by Eric Alden Smith and Bruce Winterhalder, pp. 25–60. de Gruyter, New York.

Stephens, D. W.
1990 Risk and incomplete information in behavioral ecology. In *Risk and Uncertainty in Tribal and Peasant Economies,* edited by Elizabeth Cashdan, pp. 19–46. Westview, Boulder, Colorado.

Stephens, D. W., J. R. Krebs
1986 *Foraging Theory.* Princeton University Press, Princeton.

Struever, S.
1968 Woodland subsistence-settlement systems in the Lower Illinois valley. In *New Perspectives in Archaeology,* edited by Sally R. Binford and Lewis R. Binford, pp. 285–312. Aldine, Chicago.

Van Doren, M., Editor
1928 *Travels of William Bartram.* Dover, New York.

Viola, H. J., C. Margolis, Editors
1991 *Seeds of Change.* Smithsonian Institution Press, Washington, D.C.

Wagner, G.
1987 Uses of plants by the Fort Ancient Indians. Unpublished Ph.D. dissertation, Department of Anthropology, Washington University, St. Louis.

Watson, P. J.
1989 Early plant cultivation in the Eastern Woodlands of North America. In *Foraging and Farming: The Evolution of Plant Cultivation,* edited by D. R. Harris and G. C. Hillman, pp. 555–571. Unwin Hyman, London.

Watt, B. K., A. Merrill
1975 *Handbook of Nutritional Contents of Foods.* Dover, New York.

Wilson, H. D.
1990a *Quinua* and relatives (*Chenopodium* sect. *Chenopodium* subsect. *Cellulata*). *Economic Botany* 44 (Supplement):92–110.
1990b Gene flow in squash species. *Bioscience* 40:449–455.

Winterhalder, B.
1981 Optimal foraging strategies and hunter–gatherer research in anthropology: Theory and models. In *Hunter-Gatherer Foraging Strategies,* edited by B. Winterhalder and E. A. Smith, pp. 13–35. University of Chicago Press, Chicago.
1983 Opportunity cost foraging models for stationary and mobile predators. *American Naturalist* 122:73–84.
1990 Open field, common pot: Harvest variability and risk avoidance in agricultural and foraging societies. In *Risk and Uncertainty in Tribal and Peasant Economies,* edited by Elizabeth Cashdan, pp. 67–88. Westview, Boulder, Colorado.

Winterhalder, B., C. Goland
1996 An evolutionary ecology perspective on diet choice, risk, and plant domestication. In *People, Plants, and Landscapes: Studies in Paleoethnobotany,* edited by Kristen J. Gremillion. University of Alabama Press, in press.

Winterhalder, B., E. A. Smith
1992 Evolutionary ecology and the social sciences. In *Evolutionary Ecology and Human Behavior,* edited by Eric Alden Smith and Bruce Winterhalder, pp. 3–24. Aldine de Gruyter, New York.

Yarnell, R. A.
1982 Problems of interpretation of archaeological plant remains of the Eastern Woodlands. *Southeastern Archaeology* 1:1–7.

Yesner, D. R.
1981 Archaeological applications of optimal foraging theory: Harvest strategies of Aleut hunter-gatherers. In *Hunter-Gatherer Foraging Strategies,* edited by Bruce Winterhalder and Eric Alden Smith, pp. 148–170. University of Chicago Press, Chicago.

Bedouin Hand Harvesting of Wheat and Barley

Implications for Early Cultivation in Southwestern Asia

Steven R. Simms and Kenneth W. Russell

The food-producing transition is increasingly seen not from the vantage of the Neolithic or as a progression from one evolutionary type to another but as a set of alternative foraging strategies with selective advantages and disadvantages that vary with the socioecological circumstances (Layton, Foley, and Williams 1991:255). This view focuses attention on the constraints placed upon foragers by different circumstances. We present data on one such constraint resulting from ethnographic study among the Bedul Bedouin of Jordan. At the time of the study (1986), many Bedul cultivated wheat and barley using local seed stocks, and tillage, and rainfall farming conditions. More interesting, they harvested their fields by hand, only rarely using sickles. We were therefore able to compare the cost of hand harvesting with that of harvesting with various types of sickles known from the past. Harvesting by hand proved less costly than the use of early sickles, and hand harvesting of cultivated cereals was similar in cost to the harvesting of wild cereals despite the investment in field preparation. These findings have implications for the recognition of food production in the archaeological record.

The Bedul Bedouin and Cereal Cultivation

The Bedul Bedouin are the traditional inhabitants of Petra, Jordan, a locale best-known as a major urban center during Nabataean-Roman times (Figure 21.1). During intermittent fieldwork from 1986 to 1994, the Petra Ethnoarchaeological Project focused on the ethnography, ethnohistory, and ethnoarchaeology of the Bedul Bedouin (Simms and Russell 1997). In the first years of our study, many Bedul retained a traditional life despite the tourism associated with Petra. They perceived change as harmful to their limited ties to the tourist economy, and to maintain this aspect of their livelihood they held to traditional lands and practices longer than most Bedouin in Jordan. Bedul ethnohistory and their contemporary situation are described elsewhere (Russell 1993, Simms and Kooring 1996).

In addition to their pastoral subsistence activities, the Bedul traditionally cultivated fields of wheat, barley, and indigenous tobacco throughout the Petra region. They also kept small truck gardens in restricted locales. Their agricultural fields were encountered by early travelers, and it appears that cultivation was part of the Bedul economy from at least the 19th century (McKenzie 1991, Russell 1993). Archaeological evidence from fields at Petra suggests that some of them may date to the medieval period (Simms and Russell 1997:chap. 2).

In many respects, Bedul cereal cultivation involved ancient techniques. A government survey in southern Jordan in 1988 found that 88% of the farmers were using a local variety (or varieties) of wheat known broadly as *qatma safra* (El-Hurani 1988: 41, 42, table 52). This unimproved, mass-selected hard-wheat landrace, relatively low in gluten quality and content but high in protein (Salim 1961:12, 14, table 7), was cultivated by the Bedul. Tillage was traditionally performed using animal-traction ards. Harvested grain was threshed using animal traction on stone and dirt threshing floors located near agricultural fields. Winnowing was done in the wind using wooden forks. Of special interest here is the method of harvest: Dry or slightly green tillers of grain were gathered together with a sweep of the hand and broken off with a short jerk backwards accompanied by a downward tilt of the wrist. This action proceeded until both hands were full of small bundles of grain, at which point the bundles were laid on the ground in piles. Modern mechanized equipment has replaced traditional implements, and both tillage and threshing are generally done now by tractor. In 1986 and 1988, traditional practices continued to be employed in remote areas or in locations of difficult access for tractors and other mechanized equipment.

The areas in which Bedul agricultural fields were historically located (Figure 21.1) include the ruins of the ancient city of Petra, the southern half of the Petra Valley, the lower Wadi Beida drainage and its surrounding plateau, and several outlying areas more than 10 km from Petra. The Bedul also cultivated pockets

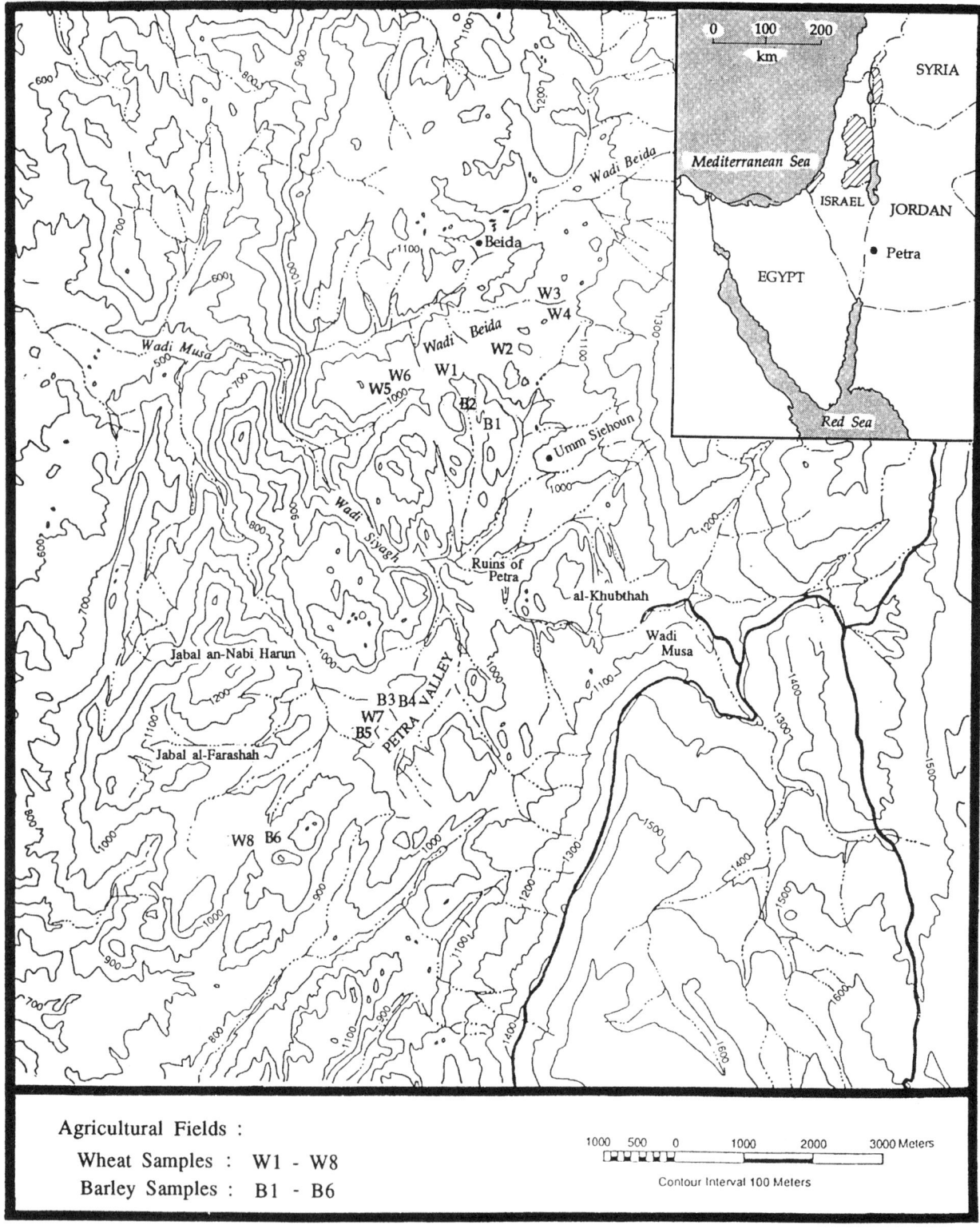

FIGURE 21.1 The study area, showing Bedul fields and sources of grain samples.

of arable land on the sandstone massifs surrounding the Petra Valley. The well-known Natufian and Pre-Pottery Neolithic B (12,000–8,000 BP) site of Beida (e.g., Kirkbride 1966, 1968) is only 4 km north of the heart of Petra. Bedul fields are thus, in terms of topography, geology, and, in some respects, cultivation practices, analogous to prehistoric ones.

THE COSTS OF CULTIVATION

The wooden ards employed in southwestern Asia and throughout the arid and semiarid regions of the world have changed little since prehistoric times (Avitsur 1965; Hopfen 1969:47–55; Russell 1988:38–40, 118; Varisco 1982; Watson 1979:74–75), and the ards used by the Bedul are no exception. (Palmer and Russell

[1993] review the traditional ards used in Jordan.) Of course, the earliest tillage was by hand with the help of digging sticks, hoes, and other such implements. Russell (1988:109–34) compiled a detailed survey of ethnographic and historical accounts to compute the labor costs of cereal cultivation at various levels of technology. He found that hand tillage suitable for cultivating wheat and barley requires approximately 35 hours per dunam, an area of 1,000 m² or 2.25 acres (Russell 1988:115). This value is employed here for modeling early cultivation employing hand tillage. Bedul informants commonly reported that one could till three dunams in an 8–10-hour day with a team of donkeys. This is equivalent to 2.7–3.3 hrs./du. These observations are comparable to data on ard tillage from other ethnographic and historic sources (Russell 1988:122–23, tables 25 and 27). For complete field preparation using an ard, a value of 20 hrs./du is used in the calculations here (Russell 1988:124).

In May 1986 we obtained quantitative data on wheat and barley yields from Bedul fields. A representative square meter of the crop in a field was totally harvested, and both clean seed (hand-cleaned by the researchers) and straw/chaff yields were measured using hand-held scientific scales. In three instances, two samples were taken in order to quantify obvious yield differences in the same field as a result of microtopographic and ecological variations. The results are presented in Table 21.1. The Bedul considered 1986 to be an average to good year, noting that 1983 had been exceptional. Wheat and barley yields and informant evaluations about the quality of the harvest among Bedouin farmers in the Negev desert in the late 1950s as reported by Mayerson (1960:18) are consistent with our findings for the Bedul. Additional consistent comparisons of the Bedul yields with others from the region can be found in Russell (1988:71, table 3; 112, table 16).

Comparisons of the nutritional composition and energetic value of cultivated wheat (Feldman 1976:121; Aykroyd and Doughty 1970:18) and wild einkorn collected in Turkey by Harlan (1967:198) suggest that the general values have not significantly changed as a result of domestication. The energy value of the wild einkorn harvested by Harlan was 3,567 kcal/kg. Modern red winter wheat analyzed at the same time was found to have an energy value of 3,474 kcal/kg. Modern wheat contains more carbohydrates but slightly less protein and fat than wild einkorn. Analyses of two wheat samples from Bedul fields produced an average of 3,511 kcal/kg. The average of two barley samples was 3,313 kcal/kg. The nutritional composition of the Bedul samples indicated a higher protein content than is typical of modern domesticated wheats. For the calculations here, we rounded the energy values to 3,500 kcal/kg for wheat and 3,300 kcal/kg for barley.

In May and June 1986, quantitative data on the labor costs of harvesting wheat and barley by hand were recorded in fields on the plateau south of the lower Wadi Beida drainage and in the southern Petra Valley. During this period, fields were being harvested by the Bedul in groups ranging in size from the few members of a single household to groups of over 20 individuals from related families. The actual harvesting was done by adult males and females, with children, adolescents, and the elderly assisting by carrying bundles of grain to larger piles.

To obtain quantitative data, a representative focal person was chosen and timed by stopwatch, recording the time elapsed for each square meter harvested. The purpose and nature of the recording were not revealed to the focal person, and the recorder pretended to be observing another harvester. The measurement of the number of square meters covered by a harvester was done with a tape measure after the completion of the timed observation. In four cases, harvesting labor was determined in fields for which grain yields had been previously established. Returns are based on the yield of clean grain and exclude tilling, sowing, and processing costs. The results of these four cases are presented in Table 21.2. (We made additional observations but exclude them here because they represent instances in which the timing was interrupted or the yields from the field under observation could not be obtained. These incomplete results did, however, fall within the ranges recorded during the complete experiments.)

Subjects 2 and 3 harvested from sandy soils, while subject 4 harvested from silty, clayey loam. The standard method did not work for subject 4, who had to jerk the stalks from the hard soil a few tillers at a time. The resulting differences in the labor required are readily apparent. Subjects 2 and 3 worked at comparable paces, resulting in expenditures of 10.7 and 10.8 hrs./du, even though the density of crops being harvested was 40 and 64 kg/du respectively. In contrast, subject 4 invested 44.7 hrs./du. In addition to occasional expletives, subject 4 repeatedly expressed his desire for a sickle.

The important and often misunderstood relationship between efficiency—a rate of return (e.g., energy/time)—and abundance (productivity or yield/area) is apparent here. Subject 4 was working in a field with an exceptionally high density

TABLE 21.1. Grain and Straw/Chaff Yields (kg) for Wheat (W) and Barley (B) Fields at Petra in May 1986.

Sample	Location	Yield/m² Grain	Yield/m² Straw/Chaff	Yield kg/du Grain	Yield kg/du Straw/Chaff
W1	L. Wadi Beida	.09	.201	90	201
W2	L. Wadi Beida	.158	.343	158	343
W3	L. Wadi Beida	.066	.117	66	117
W4	L. Wadi Beida	.04	.094	40	94
W5	L. Wadi Beida	.033	.04	33	40
W6	L. Wadi Beida	.248	.422	248	422
W7	S. Petra Valley	.095	.174	95	174
W8	S. Petra Valley	.064	.079	64	79
B1	Ras Mu'aesrah	.072	.108	72	108
B2	Ras Mu'aesrah	.06	.095	60	95
B3	S. Petra Valley	.044	.049	44	49
B4	S. Petra Valley	.135	.110	135	110
B5	S. Petra Valley	.67	.098	67	98
B6	S. Petra Valley	.05	.058	50	58

TABLE 21.2. Return Rates for Bedul Males Hand Harvesting Wheat and Barley near Petra in May and June 1986.

DATA	SUBJECT			
	1	2	3	4
Subject's age	45–48	45	20	30
Cereal	Barley	Wheat	Wheat	Wheat
Field location	S. Petra Valley	L. Wadi Beida	S. Petra Valley	L. Wadi Beida
Sample	B5	W4	W8	W6
Yield (kg/du)	67	40	64	248
No. of m² timed	20	4	12	4
Avg. minutes/m²	1.1	0.64	0.65	2.68
Yield (kg/hr)	3.66	3.74	5.94	5.58
Return in kcal/hr	12,078	13,090	20,790	19,530
Return rate (hrs./du)[a]	18.3	10.7	10.8	44.7

Note: Calculated for barley at 3,300 kcal and wheat at 3,500 kcal.
[a] Return rate excludes tilling, sowing, and processing costs.

of wheat because it was located in the trough of a shallow drainage where water accumulated. In this portion of the field, he was also working in very hard ground. His increased time per dunam was likely due to the higher density of wheat. He thus produced a greater yield while investing more effort per unit of area. However, his return rate, a measure of effort per unit of time (time that could have been spent doing other things), was only comparable to that of the other subjects. His complaints about the increased effort stemmed from the difficulty of hand harvesting in hard ground. His circumstances and response suggest that the use of sickles may, in part, be related to soil density and raise the question of the relationship between the return rates of hand harvesting and harvesting with sickles.

The possibility of a correlation between hand versus sickle harvesting and cultivation of cereals on sandy versus dense soils was suggested by Bohrer (1972). Comparison of Bedul hand harvesting with harvesting using prehistoric sickles supports this hypothesis. Comparative reaping costs of various prehistoric sickles were organized and reported by Russell (1988:115–18, table 20), drawing mainly on replication studies by Steensberg (1943) on barley and oats and Korobkova (1981) on wheat. These experiments employed replicate sickles of simple and composite forms, the latter using both ancient and modern lithic blade inserts. From this work it is possible to compare the return rates from hand harvesting with those from the use of simple early lithic sickles, sickles with flint microblade inserts, and various bronze and iron sickles. The trend in these data is toward increasing harvesting efficiency across sickle types through time. With the most ancient forms of sickles reaping costs 34 hrs./du (Russell 1988:117). Sickles with lithic blade inserts reduced costs to 20 hrs./du (p. 118), and early metal sickles were of similar efficiency.

Comparison of these figures with Bedul reaping costs for barley (18.3 hrs./du) shows that hand harvesting is as efficient as harvesting with advanced lithic and early metal sickles. This relationship is stronger in the case of wheat grown on sandy soils, with Bedul return rates of 10.7–10.8 hrs./du being significantly higher than for harvesting with all early sickles and competitive with harvesting with early iron sickles (Russell 1988:116–17). The return rate (energy/time) for the subject harvesting wheat growing in hard soil was similar to that for hand harvesting, but his productivity (energy/area) would have been improved with the use of a sickle (thus intensifying production). However, with the typically lower density of wheat in sandy soils a sickle would have been only a third to half as efficient as hand harvesting. Intensification increases productivity at the expense of efficiency by adding labor and, eventually, non-human energy to the system. This raises the question what selection pressures would lead to this seeming irony.

COMPARISONS

A variety of available data can be combined with the findings reported here to compare the overall return rates for harvesting wild forms of wheat (Harlan 1967, O'Connell and Hawkes 1981, Russell 1988) with hand tillage and hand harvesting of domesticated wheat (this study), hand tillage and hand harvesting using early lithic sickles, and finally animal-traction tillage and advanced lithic/early metal sickles. Data on processing costs can also be included to calculate return rates for the entire cultivation and preparation process (O'Connell and Hawkes 1981, Wright 1994) (table 21.3). For wild wheat procurement, tilling and sowing are, of course, omitted. For cultivated wheat, the aforementioned analyses of Russell (1988) identify the costs of hand and ard tillage, sowing, and threshing. Wright (1994) carefully details processing costs by examining the various forms of milling necessary to make cereals edible. She reports considerable variability in the time required to process groats, and therefore we use a suite of her samples (3–15) bridging the processing steps of dehusking and groat preparation, yielding an average of 0.9 hr/kg. We suggest that this value approximates the midpoint of what was likely a wide range of processing costs. Further, it compares well with the processing value of 1.0 hr/kg used by O'Connell and Hawkes (1981), which was based on direct observations of foragers grinding hard seeds. Our goal is not so much to find the "real" return for early cultivators as to identify potentially robust relationships.

The return rates for wild wheat (1,986 kcal/hr) and hand-tilled and hand-harvested cultivated wheat (1,815 kcal/hr) are similar. This counterintuitive similarity is probably due to the offsetting of wild wheat's higher harvesting cost (because of its weak rachis and smaller seeds) by the absence of the need to expend energy in field preparation. At the same time, the costs of intensification associated with cultivation (tilling and sowing) are offset by characteristics of domestication including larger seeds, less loss of seed because of a tough rachis, and, under favorable circumstances, the increased density of plants resulting from field preparation. The employment of early lithic sickles

TABLE 21.3. Return Rates for Alternative Methods of Wheat Procurement.

METHOD	COST (hrs./du)					RETURN RATE (kcal/hr)
	TILLING	SOWING	HARVESTING	THRESHING	MILLING	
Wild wheat	—	—	45.5[a]	5	37.6[b]	1,986
Cultivated wheat, hand-tilled and hand-harvested	35	8	10.8[c]	5	37.6	1,815
Hand-tilled and harvested with early lithic sickles	35	8	34	5	37.6	1,463
Hand-tilled and harvested with advanced lithic sickles	35	8	20	5	37.6	1,870
Ard-tilled and harvested with early metal sickles	20	8	20	5	37.6	1,932

Note: Energy content of wheat = 3,500 kcal/kg. All calculations for methods involving cultivation assume a yield of 50 kg/du. Estimates for tilling, sowing, harvesting, and threshing from Russell (1988:113–24) unless otherwise noted.
[a] Wild wheat return rate of 1.1 kg/hr (Harlan 1967) scaled to 50 kg/du for comparison with cases of cultivation.
[b] Wright (1994:246, fig. 4). We use an average of her samples 3–15, representing dehusking and groat preparation. These samples reflect the greatest degree of overlap in costs for this level of preparation, ranging from 0.5 to 1.3 hrs./kg.
[c] Hand harvesting cultivated wheat is from direct observation of Bedul.

reduces the return rate to 1,463 kcal/hr, and this loss is not overtaken until the advent of advanced sickles employing lithic inserts in the Pottery Neolithic period.

CONCLUSIONS AND IMPLICATIONS

Hand harvesting versus harvesting with sickles. Harvesting cereals by hand is the oldest and most enduring method throughout the Old World (Bohrer 1972), and experimental studies have shown that it is effective (Anderson 1991, Harlan 1967). The efficiency of hand harvesting compared with harvesting with early sickles suggests that there is no necessary link between the appearance of sickles in the archaeological record and the earliest cereal cultivation. Early sickles may have been used for purposes other than harvesting grasses/cereals (see Anderson 1991; Unger-Hamilton 1989, 1991). The advantages of sickles lay in opportunities for intensification rather than increased efficiency. Early cereal cultivation in southwestern Asia, North Africa, Central Asia, and Europe took place in the light, sandy soils (Sherratt 1980:315; 1983:98) that occurred in pockets in the Levantine forests and in larger expanses in surrounding areas. As cultivation expanded into areas of heavier soils during the Neolithic, the advantages of sickles increased. Nevertheless, Natufian-age sickles are known from areas characterized by sandy soils. If sickles were used here to harvest grass seed, their primary advantage at this early date would have been in extending the harvest. Microwear study (Unger-Hamilton 1989, 1991) indicates that Natufian sickles were employed on green wheat, which would have enabled the harvest to begin early. As the grain stands dried, they would have continued to be harvested at the higher return rate afforded by hand harvesting. Thus, greater productivity would have been obtained at the expense of efficiency. In sum, sickles indicate intensification, a very different behavioral pattern from initial cultivation.

Diet breadth, resource choice, and intensification. The proposition that diets expanded to include seed resources from the Upper Paleolithic through the Epipaleolithic in southwestern Asia is supported by the 19,000 BP dates for wild wheat and barley at Ohalo II (Kislev, Nadel, and Carmi 1992). Seed resources are relatively high in cost (O'Connell and Hawkes 1981, Simms 1987), and once diets began to expand to include high-cost resources a host of these expensive but storable resources became available. The similarity in return rates between wild wheat and hand-harvested cultivated wheat indicates that when the diet includes expensive resources, subtle variation in circumstances may favor one or another strategy. Perhaps this is why some researchers infer general foraging while others see specialization in nuts, acorns, marsh resources, or wild cereals (e.g., Henry 1989, Shipek 1989, Bar-Yosef and Belfer-Cohen 1992, Olszewski 1993, McCorriston 1994).

Conceptualizing the move to food production in terms of alternative adaptive strategies means that with adequate data on the return rates of the alternative resources available we should be able to predict which strategies would be favored under different circumstances. For example, consider the impact on Epipaleolithic foragers of different mobility patterns: Those with higher mobility would be expected to harvest wild wheat (or nuts, or acorns), perhaps store some, and move on. Less mobile foragers exploiting areas that include cereal patches on a recurrent basis would, through repeated use, begin the shift to cultivation and increased storage (or intensification of acorn collection). Selection for stronger territorial claims should increase with tethered seasonal rounds and repeated use of patches. The management of cereal patches by cultivation would be one expression of territoriality. Thus, the differences between Levantine forests, desert plains, and wetland locales are probably important to understanding why some people continued as foragers, some processed acorns, and others embarked upon a trajectory only retrospectively referred to as cereal "farming." That such variability in seasonal round and tempo and mode of mobility existed even as early as the Epipaleolithic Geometric Kebaran complex seems well demonstrated in recent syntheses for the region (Henry 1989, 1995). Variability in adaptive strategies would be expected over time as well, and the conditions of selection may be especially dynamic during periods of rapid climatic change such as the Younger Dryas.

The whys of the food-producing transition revolve more around such questions as what would select for intensification focused on two cereal grasses rather than other resources such as acorns and what conditions would account for the cases in which agriculture was rejected or delayed. The move toward food production began with a broadening of the diet in the Upper Paleolithic and Epipaleolithic, opening a suite of forager strategies across space and through time. Given the similarity in return rates between wild wheat and hand-harvested cultivated wheat, early cultivation could have been one of these. The rapid appearance of sickles and "village" trappings in the Early Natufian period marks intensification, and it seems that this too was spatiotemporally variable. One implication of a broadening diet is a decline in foraging efficiency. Hawkes and O'Connell (1992: 64) observe that, "where diet is broad and handling represents the bulk of foraging effort, improvements in handling efficiency would have large effects." Thus, when foragers are exploiting seeds, whose high cost is largely due to harvesting and processing (O'Connell and Hawkes 1981, Simms 1987, Wright 1994),

the development of techniques to reduce these costs would give them an advantage. By the Neolithic period, the improvements in sickle technology discussed here may have accomplished this, but this was long after first cultivation. As for understanding why some people continued as foragers, it is possible to expand upon Hawkes and O'Connell's observation. Intensification applied to resources already in the diet (e.g., seeds) need not represent an increase in efficiency, at least initially. Rather, the effort directed toward intensification need be competitive only with the alternative of adding more expensive (lower-ranked) resources to the diet. Both strategies—early cultivation and foraging for expensive wild resources—would be expected to occur for a while.

At this point, these inferences can only be considered hypotheses, albeit with the prospect of testing. A fuller exploration will require data on the costs and benefits of various resources. Foraging theory (Stephens and Krebs 1986) holds great promise for these issues but has not been fully exploited in southwestern Asia (but see Russell 1988, Layton, Foley, and Williams 1991, Hawkes and O'Connell 1992, Winterhalder and Goland 1993). In order to overcome the limitations of culture-historical and post-hoc explanation typical of the literature on the food-producing transition, theoretical models will be essential. There is a need for more data on the costs associated with different behaviors, and we hope to contribute to that in a small way here. No less important will be a bridging of literatures and dialog, because in testing the predictions of theoretical models archaeological data will also be essential.

Acknowledgments

Kenneth Russell passed away in the field in Jordan in 1992, when most of this work was in progress or rough-draft form. Continuation of the project is in his memory. Funding was provided by the L. S. B. Leakey Foundation, a U.S. Information Agency Fellowship, the American Center for Oriental Research, Amman, and Utah State University, Logan. We gratefully acknowledge the support of the Department of Antiquities of Jordan. We thank K. Renee Barlow and James F. O'Connell, University of Utah, for good comments, an anonymous referee for helpful perspective, and Marina Hall for editorial assistance. Article scope, emphasis, and any errors remain the responsibility of the senior author.

References

Anderson, P. C.
1991   Harvesting of wild cereals during the Natufian as seen from experimental cultivation and harvest of wild einkorn wheat and microwear analysis of stone tools. In *The Natufian Culture in the Levant*, edited by O. Bar-Yosef and F. Valla, pp. 521–55. International Monographs in Prehistory, Archaeological Series 1.

Avitsur S.
1965   *The Native Ard of Eretz-Israel: Its History and Development.* Aveshalom Institute for Homeland Studies, Tel-Aviv.

Aykroyd, W. R., J. Doughty
1970   *Wheat in Human Nutrition.* Food and Agriculture Organization, Rome.

Bar-Yosef, O., A. Belfer-Cohen
1992   From foraging to farming in the Mediterranean Levant. In *Transitions to Agriculture in Prehistory*, edited by A. B. Gebauer and T. D. Price, pp. 21–48. (Monographs in World Archaeology 4.) Prehistory Press, Madison.

Bohrer, V. L.
1972   On the relation of harvest methods to early agriculture in the Near East. *Economic Botany* 26:145–55.

Feldman, M.
1976   Wheats. In *Evolution of Crop Plants*, edited by N. W. Simmonds, pp. 120–28. Longman, London.

Harlan, J. R.
1967   A wild wheat harvest in turkey. *Archaeology* 20:197–201.

Hawkes, K., J. F. O'Connell
1992   On optimal foraging models and subsistence transitions. *Current Anthropology* 33:63–66.

Henry, D. O.
1989   *From Foraging to Agriculture: The Levant at the End of the Ice Age.* University of Pennsylvania Press, Philadelphia.
1995   *Prehistoric Cultural Ecology and Evolution: Insights from Southern Jordan.* Plenum Press, New York.

Hopfen, H. J.
1969   *Farm Implements for Arid and Tropical Regions.* Food and Agriculture Organization Agricultural Development Paper 91.

El-Hurani, M. H.
1988   *Report on the Wheat Baseline Data Survey Conducted in 1988.*

Jordanian National Agricultural Development Project Publication 10.

Kirkbride, D. V. W.
1966 Five seasons at the Pre-Pottery Neolithic village of Beidha in Jordan. *Palestine Exploration Quarterly* 98:8–72.
1968 Beidha: Early Neolithic village life south of the Dead Sea. *Antiquity* 42:263–74.

Kislev, M. E., D. Nadel, I. Carmi
1992 Epipaleolithic (19,000 BP) cereal and fruit diet at Ohalo II, Sea of Galilee, Israel. *Review of Palaeobotany and Palynology* 73:161–66.

Korobkova, G. F.
1981 Ancient reaping tools and their productivity in light of experimental tracewear analysis. In *The Bronze Age Civilization of Central Asia: Recent Soviet Discoveries*, edited by P. L. Kohl, pp. 325–49. M. E. Sharpe, Armonk, N.Y.

Layton, R., R. Foley, E. Williams
1991 The transition between hunting and gathering and the specialized husbandry of resources: A socio-ecological approach. *Current Anthropology* 32:255–74.

McCorriston, J.
1994 Acorn eating and agricultural origins: California ethnographies as analogies for the ancient Near East. *Antiquity* 68:97–107.

McKenzie, J.
1991 The Beduin at Petra: The historical sources. *Levant* 23:139–45.

Mayerson, P.
1960 *The Ancient Agricultural Regime of Nessana and the Central Negeb*. Colt Archaeological Institute, London.

O'Connell, J. F., K. Hawkes
1981 Alyawara plant use and optimal foraging theory. In *Hunter-Gatherer Foraging Strategies: Ethnographic and Archaeological Analyses*, edited by B. Winterhalder and E. A. Smith, pp. 99–125. University of Chicago Press, Chicago.

Olszewski, D. I.
1993 Subsistence ecology in the Mediterranean forest: Implications for the origins of cultivation in the Epipaleolithic southern Levant. *American Anthropologist* 95:420–35.

Palmer, C., K. W. Russell
1993 Traditional ards of Jordan. *Annual of the Department of Antiquities of Jordan* 37:37–53.

Russell, K. W.
1988 *After Eden: The Behavioral Ecology of Early Food Production in the Near East and North Africa*. British Archaeological Reports International Series 391.
1993 Ethnohistory of the Bedul Bedouin of Petra. *Annual of the Department of Antiquities of Jordan* 37:15–35.

Salim, M. H.
1961 *Investigations on Improvement of Wheat and Barley Production in Jordan 1952–1960*. Ministry of Agriculture Research Bulletin 1.

Sherratt, A.
1980 Water, soil, and seasonality in early cereal cultivation. *World Archaeology* 11:313–30.
1983 The secondary exploitation of animals in the Old World. *World Archaeology* 15:90–104.

Shipek, F. C.
1989 An example of intensive plant husbandry: The Kumeyaay of Southern California. In *Foraging and Farming: The Evolution of Plant Exploitation*, edited by D. R. Harris and G. C. Hillman, pp. 159–70. Unwin Hyman, London.

Simms, S. R.
1987 *Behavioral Ecology and Hunter-Gatherer Foraging: An Example from the Great Basin*. British Archaeological Reports International Series 381.

Simms, S. R., D. Kooring
1996 The Bedul Bedouin of Petra, Jordan: Traditions, tourism, and an uncertain future. *Cultural Survival Quarterly* 19:22–25.

Simms, S. R., K. W. Russell
1997 *Ethnoarchaeology of the Bedul Bedouin of Petra, Jordan: Implications for the Food Producing Transition, Site Structure, and Pastoralist Archaeology*. Utah State University Contributions to Anthropology 22.

Steensburg, A.
1943 *Ancient Harvesting Implements: A Study in Archaeology and Human Geography*. Nationalmuseets Skrifter Arkaeologisk-Historisk Raekke I.

Stephens, D. W., J. R. Krebs
1986 *Foraging Theory*. Princeton University Press, Princeton.

Unger-Hamilton, R.
1989 The Epi-Palaeolithic southern Levant and the origins of cultivation. *Current Anthropology* 30:88–103.
1991 "Natufian plant husbandry in the southern Levant and comparison with that of the Neolithic periods: The lithic perspective." In *The Natufian Culture in the Levant*, edited by O. Bar-Yosef and F. Valla, pp. 483–519. International Monographs in Prehistory, Archaeological Series 1.

Varisco, D. M.
1982 The ard in highland Yemeni agriculture. *Tools and Tillage* 4:158–84.

Watson, P. J.
1979 *Archaeological Ethnography in Western Iran*. Viking Fund Publications in Anthropology 57.

Winterhalder, B., C. Goland
1993 On population, foraging efficiency, and plant domestication. *Current Anthropology* 34:710–15.

Wright, K. I.
1994 Ground-stone tools and hunter-gatherer subsistence in Southwest Asia: Implications for the transition to farming. *American Antiquity* 59:238–63.

22

Deferred Harvests

The Transition from Hunting to Animal Husbandry

Michael S. Alvard and Lawrence Kuznar

The subsistence transition that occurred at the end of the Pleistocene has been of continuing interest to anthropologists (Binford 1968; Byrd 1994a; Childe 1952; Flannery 1965; Harris 1977, 1998; Henry 1995; Price and Brown 1985; Winterhalder and Goland 1997; Zeder 1994; and others). During this transition, humans moved from a full-time hunting and gathering way of life to one based on food production. This Neolithic transition was a watershed shift in the subsistence strategy for our species. Associated with the transition were a number of well-documented components including sedentism (Bar-Yosef and Belfer-Cohen 1989; Brown 1985; Harris 1977; Liebermann 1993), increased population density (Cohen 1977; Harris 1977), as well as the development of political, social, and economic complexity (Brown 1985; Byrd 1994b; Harris 1977; Price and Brown 1985; Wright 1978).

A major feature of the transition was the change in our relationship to other organisms. Humans shifted from being hunters and gatherers to resource husbanders. We went from exploiting the somatic potential of other organisms to co-opting and increasing their reproductive potential. An important aspect of this change involved the timing of resource use. Most hunter-gatherers focus on short-term returns, while food producers have a more farsighted perspective (Woodburn 1982). This difference has major implications for how individuals make subsistence decisions. In this article, we focus on the shift from hunting to animal husbandry by applying an evolutionary ecology approach. From the perspective of evolutionary theory and rational choice theory in economics (see Smith and Winterhalder 1992), animal husbandry presents a number of interesting theoretical problems. Models to explain why the transition occurred, as well as when and where it did, have tended to focus around climate change, resource stress, and population pressures as the key causal variables (e.g., Binford 1968; Childe 1952; Cohen 1977; Flannery 1973; Harris 1977, 1998; Hole 1996; McCorriston and Hole 1991; Price and Brown 1985; Rosenberg 1990). This is interesting because, as we will argue below, animal husbandry involves deferring benefits at a time when, according to many arguments, resources were becoming scarce (Cohen 1977:40; Hecker 1984; Kent 1988). While in hindsight animal husbandry seems like a logical long-term solution, husbandry in the short term was a costly proposition to its early practitioners.

Foraging Theory, Game Conservation, and Traditional Hunters

To understand the economics of husbandry from a hunter's point of view, it is useful to review work on resource conservation. Recent empirical work has produced a growing consensus that resource conservation in traditional hunting economies is not as common as previously thought (Alvard 1993, 1994, 1995, 1998; Hames 1987, 1991; Low 1996; Low and Heinen 1993; Ruttan and Borgerhoff Mulder 1999; Smith 1983). Original confusion resulted because many subsistence behaviors are *apparently* conservative when they are associated with sustainable harvests or the harvests are biased in ways that are consistent with genuine conservation. Such outcomes are termed *epiphenomenal conservation* (Hunn 1982).

To avoid this sort of confusion, Alvard (1995, 1998) developed an operational definition of *conservation* to allow empirical testing using foraging theory. Foraging theory predicts subsistence decisions by assuming that hunters maximize foraging return rates independent of long-term depletion effects (for a review see Kaplan and Hill 1992). *Conservation* is defined as subsistence behavior where the conserver checks his or her level of resource use to some point below what would be fitness maximizing in the short term. The costs of this restraint are exchanged for future long-term, sustainable benefit. Using the short-term cost criterion alone, behavior that has unintended conservation-like consequences but no short-term costs can be rejected as conservation.

While the basic foraging models describe foragers as short-

*Originally published in *American Anthropologist* 103(2001):295–311

term maximizers, the resource conservation models describe foragers as long-term maximizers (Alvard 1993, 1998). Slobodkin (1961, 1968; see also Errington 1946) argued that a "prudent predator" would harvest those members of the prey population with the lowest reproductive value. *Reproductive value*, defined by Fisher (1958), is a function of age and is the relative number of offspring that remain to be born to an individual of age *x*, taking into account the probability that the individual will live to age *x*. Reproductive value is low among young animals because mortality for most mammals is very high in the young and many do not attain reproductive age. Critical to the argument developed below, resource conservation predicts intraspecific selectivity similar to a husbanding or management strategy. To maximize long-term return rates, harvests are biased toward males and low reproductive value individuals. In fact, one way that archaeologists show that prehistoric people were husbanding animals is by examining the sex and age distribution of faunal remains. Animal husbandry is revealed with a characteristic mortality signature consisting of a disproportionate number of young males (Redding 1981).

Alvard tested the foraging and conservation hypotheses among the Piro of the Peruvian Amazon. This work indicates that hunters killed prey types predicted by foraging theory regardless of the prey species' vulnerability, reproductive value, or state of local depletion (Alvard 1993, 1994, 1995). Indeed, hunters harvested a number of species at rates greater than the maximum sustainable yield (MSY) (Alvard et al. 1997). Work with Wana blowgun hunters and trappers of Indonesia indicates similar results (Alvard 2000). In short, the data indicate that Piro and Wana hunters maximize their short-term return rates in spite of potential negative long-term consequences. Numerous other studies agree (Hames 1987:96; Hames and Vickers 1982:374; Kaplan and Hill 1985:236; Smith 1991:256; Winterhalder 1981:97). In contrast, pastoralists practice a conserving strategy by slaughtering young male animals who neither have high reproductive value nor are necessary for the reproduction of a herd, leaving individuals with higher reproductive value behind to reproduce more prey (see Barth 1961:8; Cribb 1991:29; Dahl and Hjort 1976; Kuznar 1991, 1995; Redding 1984). The lack of prey conservation among hunters and its arguable presence among husbanders (see below) indicates that the differences between these strategies can shed light on significant shifts in human decision making that accompanied the Neolithic transition.

Why Resource Conservation by Subsistence Hunters Is Rare

While truly altruistic conservation is unexpected, evolutionary theory does not rule out conservation per se. Rogers (1991) points out that conservation may or may not evolve depending on both its short-term and long-term costs and benefits. Three factors are hypothesized to pattern the costs and benefits of conservation (Alvard 1998). The first relates to issues of ownership, private property, and territoriality. A number of researchers have suggested that conservation is unlikely in traditional horticultural and hunter-gatherer societies because the resources they use are often open access (Hames 1987, 1991; Smith 1983). No one owns an open-access resource, and everyone has the right to consume it (see Schlager and Ostrom 1992). Individuals are unlikely to provide a public good by altruistically limiting their own harvest of such resources if others who have not necessarily sacrificed anything are free to share the benefits (Olson 1965; see also Boone 1992; Hawkes 1992). These sorts of problems have been termed *collective action problems*, or *Tragedy of the Commons problems*, (Acheson 1989; Hardin 1968, 1993; Olson 1965) and have long been a topic of study by resource economists (Tietenberg 1996).

Ownership helps prevent competitors from free-riding and can help to solve the collective action problem. Resource owners are more motivated to conserve because ownership increases the probability that they can realize the benefits of their short-term sacrifice. Ownership, however, implies resource defense and territoriality. If territorial defense costs are high and private ownership difficult, resources remain open access and conservation is less likely because of collective action problems (Alvard 1998). Ingold (1980:5; see also Harris 1996:448; Reed 1984:2) has addressed the issue of common versus restricted access in his analysis of the development and character of reindeer exploitation. He noted that the relationship between hunter and prey is essentially that of predation whereas the relationship between a pastoralist and livestock is essentially that of protection and defense (Ingold 1980:27). Dyson-Hudson and Smith (1978), in their analysis of territoriality, note that resources that are both predictable and *locally* abundant are likely to be defended; a closely managed animal herd fits their definition of a defendable resource perfectly.

Second, for a resource to be economically defendable and for conservation to provide benefits, the resource must have sufficient value. Relative scarcity adds to the marginal value of a resource, assuming there is demand (Tietenberg 1996). If a resource is abundant and not limiting, an additional unit of resource has less value to a consumer than if the resource is scarce. If resources have less value, the returns from the defense required to make them a private good decline, and conservation is less likely (Beckerman and Valentine 1996; Tietenberg 1996). The corollary is that the *less* abundant a resource, the greater its value, the greater the return for defense and ownership, and the *more* likely conservation will pay off.

Lastly, if the resource's opportunity costs are high, then conservation is also unlikely. In the case of conservation, the opportunity costs accrue from not exploiting the resource immediately. To understand opportunity costs, it is useful to introduce the economic concept of discounting. Much evidence shows that people tend to prefer present consumption to future consumption (e.g., Loewenstein and Elster 1992). The discount rate is the economic measure of the rate at which current income is valued over future income (Fisher 1930). Time preference refers to the tendency to favor certain schedules of resource consumption over others. The rate that future benefits are discounted and measured in terms of present value is the discount rate.

The concept of discounting is important for understanding both resource conservation and animal husbandry. Time preferences are thought to exist for two reasons (Fisher 1930). The first is because of some probability that future benefits will not be realized. The second is because compounding gain is lost with delayed consumption. For example, a hunter may discount a future goat and prefer to kill it today because its future benefits may not be realized; the goat may be killed and eaten by someone else, or perhaps the hunter may die. Even if the future goat is 100% assured, however, the hunter may still discount it because present use of the goat could result in higher long-term benefits. This can occur if the current benefits compound over time. In modern market economies, cash resources can be put into a bank with compounding gains equal to the interest rate. In nonmarket, biological systems, the resource can be invested in a growing population of descendants. A hunter could use a goat to feed his offspring or use it to obtain additional mating opportunities.

Animal Husbandry Is Prey Conservation

We argue that animal husbandry is an example of resource conservation as we defined above. Animal husbandry can be considered conservation because the criterion of restraint is met. Husbanded animals are prey that are not pursued upon encounter. According to the strict prey choice model, foragers are predicted to always pursue prey that are in the optimal diet and ignore those that are not (Stephens and Krebs 1986). The optimal diet is the suite of prey that when pursued upon encounter will maximize the rate of return. A short-term cost is paid for husbandry (the animals are not immediately pursued, killed, and eaten), and the benefits (the animal and its descendants) are deferred to some future point. Others have also noted the conceptual similarities between conservation and animal husbandry (Russell 1988:15).

Defined theoretically as conservation, animal husbandry is predicted to arise and persist in competition with a hunting strategy under conditions that favor conservation. First, the prey resources would have to be private rather than open-access goods. Next, for the costs of defense to be justified, the prey resources would have to have increased value. Finally, the discounted deferred returns from husbandry must be higher than the short-term returns from hunting.

The first two conditions were discussed in the classic economic treatment of property rights by Demsetz (1967). From that discussion one can see how increased competition for valuable game resources can lead to the development of private property and the territoriality required to avoid the free-rider problem of conservation. Demsetz cites the work of Leacock (1954) and Speck (1915) on native North American hunting territories and the fur trade. Leacock argued that territoriality was minimal before the development of the fur trade of the 1700s. The value of game, especially beaver, increased dramatically with the trade, as did the scale of hunting. As Demsetz (1967:352) notes, the geographic evidence collected by Leacock indicated an unmistakable correlation between early centers of fur trade and the oldest and most complete development of private hunting territories.

While increased resource value and territoriality are required, they are not sufficient conditions for successful conservation or husbandry to develop, as the near extinction of the beaver during the fur trade demonstrates (see Krech 1999:chap. 7 and references). Even if resources are valued to the point that territorial defense pays, it may not be in the best interest of the hunter to husband if the opportunity costs of conserving are too high (see also Acheson 1989:364). The discounted deferred returns from husbandry must be higher than the short-term returns from hunting. In the last section of this paper, we will discuss this last prediction in more detail.

Discounting and the Opportunity Costs of Animal Husbandry

Rogers (1991) developed a model to examine how conservation might evolve if resource inheritance (in this case territories) was uncertain. He imagined two strategies. The Conserve strategy pays a short-term cost in reduced fertility but increases the fertility of those who inherit the territory. The Prodigal strategy does not conserve and hence does not pay the cost of conservation but bequeaths poor territories to descendants. In Rogers' model, individuals need territories in order to breed, and fertility is greater in better territories. Rogers obtains the counterintuitive result that conservation is *not* favored as inheritance approaches certainty. The problem is that the conserver's population is growing, but the resources (in this case a territory) do not. This is a classic Malthusian conundrum (see Wood 1998 for a recent review). Rogers concludes that this is a difficult hurdle to surmount.

One way to mitigate (but not completely avoid) this problem is to imagine that the territories in Rogers' model are biological resources—perhaps a herd of ungulates. If the reproductive rate of the herd is greater than the discount rate, it may pay to conserve the herd. If tended properly, a parent can manage a growing herd of ungulates so that each child receives as much as was originally held by the parent. As long as the reproductive rate of the "territory" keeps pace with that of the conservers, there are always animals for subsequent generations to inherit.

One might ask, however, where will all the animals graze? Eventually the pasture will deplete, the growth rate of the herds will drop below the discount rate, and conservation will no longer be the best strategy. How long a system of conservation like this can maintain before collapsing into exploitation remains a theoretical and empirical question. The dilemma of

TABLE 22.1. Symbols used in text.

Symbol	Parameter
N	Density (#/km²)
K	Carrying capacity (#/km²)
r_{max}	Maximum intrinsic rate of increase
r	Intrinsic rate of increase
H	Harvest rate
W	Body mass (kg)
W*	Maximum body size to be husbanded
s	Standing biomass (N * W)
i	Discount rate
b_1	Coefficient for the r_{max}-body mass allometric equation
b_2	Coefficient for the density-body mass allometric equation
b_3	Coefficient for the biomass-body mass allometric equation
A_1	Constant for the r_{max}-body mass allometric equation
A_2	Constant for the density-body mass allometric equation

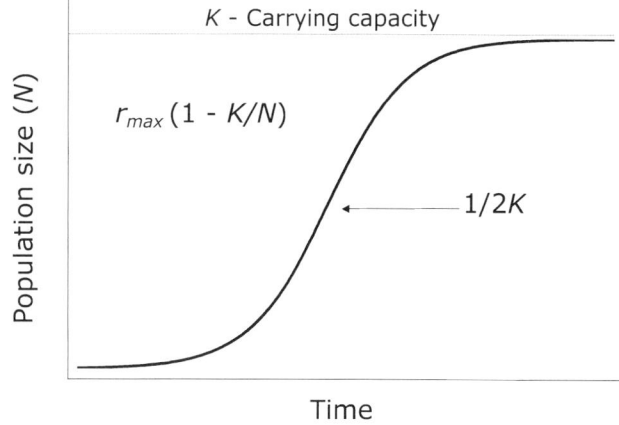

FIGURE 22.1. Model of logistic population growth. At low population densities, growth is rapid, exponential, and reaches a maximum at 1/2K. Since population growth is the greatest, 1/2K is the population size that provides the MSY. After this point, density-related factors take hold, and growth begins to slow until density equals the carrying capacity (K), births equal deaths, and growth is zero.

feeding growing herds in the interest of one's progeny potentially explains the predatory expansionism typical of pastoral societies (Bailey 1980; Barfield 1989; Khazanov 1994).

How fast would a herd have to grow for its conservation to pay off? This question can be answered by examining Clark's (1990) work with MSY models (see Table 22.1 for a description of the parameters used in this paper). These models calculate the maximum number of animals that can be removed from a population on a regular, sustainable basis without driving the population extinct (Caughley 1977). Standard mammalian MSY models assume density-dependent logistic population growth with the highest rate of growth occurring at intermediate density levels (Figure 22.1). In such a context, to obtain the largest sustainable harvest from an animal population, harvesters should maintain the population at a density (N) near

$$N = 1/2 * K \quad (1)$$

and harvest at the rate of

$$H = 1/2 * r_{max} \quad (2)$$

This yields a maximum annual harvest in animals per year of

$$MSY = r_{max} K/4 \quad (3)$$

where K = carrying capacity and r_{max} = the maximum intrinsic rate of increase (Caughley 1977).

Clark (1990) criticized these models because they assume a zero discount rate. That is, the resources harvested at some distant point in the future are implicitly assumed to be worth as much as resources harvested today. In other words, MSY models assume that resources harvested 10, 20, or 30 years from now have the same value as resources harvested today. Clark argues that MSY models do not take into account the opportunity costs of *not* harvesting above the maximum sustainable yield. These costs accrue from not consuming the resource in the short term. At a more basic level, the issue is one concerning the proper criterion of optimality (Bulmer 1994). MSY models assume *a priori*

that sustainability is the goal on which harvesters base their decisions. From an evolutionary as well as rational economic view, sustainability is only a viable tactic if it leads to greater fitness in the former case or greater profits in the latter. Clark concluded that for a maximum sustainable yield strategy to be favored, the reproductive rate of the resource population must be greater than the discount rate—the rate of return from the best current alternative investment. If the discount rate is higher than the reproductive rate of the prey species, the optimal choice is to harvest the resource as rapidly as possible and invest the capital in the current investment with the higher payoff. Clark's often-cited example explains why economically rational whalers often overharvest whales rather than hunt them sustainably. Because whale reproductive rates are often lower than the interest rate, economically rational whalers can receive a higher return by harvesting whales to the point of extinction than they can get from harvesting whales at the MSY indefinitely.

Should the Piro Conserve Collared Peccaries or Spider Monkeys?

Using Clark's reasoning we can pose the same theoretical question of Piro hunters (see also Alvard 1998). The Piro, whom we mentioned earlier, hunt a variety of game including collared peccaries (*Tayassu tajacu*) and spider monkeys (*Ateles paniscus*). Collared peccaries are piglike ungulates that weigh around 20 to 25 kg. This species provides the greatest amount of hunted meat in the Piro diet. Black spider monkeys are medium-sized primates of the family Cebidae with adult body weights of between 6 and 10 kg (Alvard 1993). These species vary in reproductive rate and body size—both factors that determine the payoff to husbanding. If meat is valued sufficiently and territoriality

TABLE 22.2. The data to calculate the maximum sustainable yield. The lump sum column is the return if the entire population was killed. The last column presents rate of return that must be found for the lump sum to be invested and yield a return as high as the MSY.

Species	K (per 314 km²)	r_{max}	MSY (rK/4)	Body Wt. (kg)	MSY (kg)	Lump Sum (kg)	Rate
Collared peccary	3,714	0.84	780	22	17,160	81,708	21%
Spider monkey	4,191	0.08	84	9	756	6,804	2%

provides sufficient assurance that future benefits can be realized, which species would be more profitable to husband, collared peccaries or spider monkeys?

Alvard (1998) calculated the maximum sustainable yield per year for both species (Table 22.2). The carrying capacity is calculated for an arbitrarily chosen area of 314 km², an area with a radius of 10 km—the maximum distance a Piro hunter can travel away from the village and return before dark. The population at MSY is $1/2K$, or 1,857 for peccaries and 2,095 for monkeys, but the MSY are significantly different—780 peccaries and 84 spider monkeys. In a common currency, the MSY strategy returns 17,160 kg a year for peccaries and only 756 kg a year for spider monkeys.

For husbanding to be favored, future productivity from husbanding must be greater than what could be had from investing the animal resources into reproduction today at interest rate (i). An extreme exploit strategy: kill all in one year, returns a lump sum of 81,708 kg of peccaries and 37,710 kg for spider monkeys. For the exploit strategy to be favored over a husband strategy for peccaries, the lump sum must be invested in a current investment that has an annual return greater than the MSY:

$$Ki > rK/4 \quad (4)$$

or to simplify,

$$i > r/4 \quad (5)$$

For peccaries this obtains if i = 21%. For spider monkeys, exploit pays if the current alternative returns only 2% or more. Other things being equal, this comparison shows that peccaries are much more likely to be conserved than spider monkeys. This is because the opportunity cost of conserving peccaries is lower than it is for spider monkeys. It would take 50 years at the MSY rate to recoup the lump sum for spider monkeys but only 4.7 years for peccaries.

In spite of the peccaries' high reproductive rate, the Piro do not conserve them. Why? Unlike modern whalers, Piro hunters do not have credit markets in which to invest their peccary profits. They do invest, however, in their own reproduction in the evolutionary sense. One can imagine two options for Piro hunters. The first is for hunters to harvest as much peccary in the present as would maximize their short-term fertility and survivorship. They can do this by investing the meat directly into offspring or into mating opportunities (Hawkes et al. 1995; Trivers 1972). The second option is for hunters to harvest peccaries at a lower but perhaps more sustainable level and pay a short-term fertility and survivorship cost in exchange for the long-term benefits associated with sustainable resources.

Apparently, the returns from peccary conservation would be lower than what hunters obtain from unrestrained peccary harvesting. Why? First, peccaries are a costly-to-defend, open-access resource. Restraint on the part of any one hunter provides a public good enjoyed by all hunters. Second, data show that the Piro peccary harvest is currently less than the MSY. That is, the current harvest *is* already sustainable (Alvard et al. 1997). Peccaries are abundant relative to demand, and the Piro enjoy a diet flush with meat (Alvard 1993). Finally, the Piro may have a short-term time preference (a high discount rate) with respect to peccaries and other prey resources (Alvard 1998).

Animal Husbandry

With this background, we can now address the transition that occurred in human history at the Pleistocene/Holocene boundary. In the Old World, the transition defines the boundary between the Paleolithic and the Neolithic (~12,000 yr. ago). This is the period when the first plants and animals were being husbanded for food production. In the New World, the transition occurred independently a few thousand years later (Wing 1983). Indeed, independent transitions from foraging to food production occurred at different locations around the globe. Sheep and goats were the first animals domesticated, possibly as early as 11,000 years ago, and certainly by 9,000 years ago at sites such as Zawi Chami, Shanidar, Jarmo, Tepe Sarab, Ganj Dareh, and Tepe Ali Kosh in the foothills of the Zagros mountains near the present-day border of Iran and Iraq (Flannery 1969:86; Hole 1989:97, 1996:273; Mason 1984; Meadow 1989; Redding 1984: 239; Ryder 1984). Subsequently, they were introduced to the Levant and Anatolia around 9000 BP during the Prepottery Neolithic B phase (Harris 1998:8; Levy 1992; Uerpmann 1989). Larger animals, such as cattle, pigs, horses, and camels, were domesticated later. Horses and camels were domesticated relatively late in prehistory (post-5500 BP and 5000 BP respectively), after profound economic, social, and ecological changes had altered people's lives in southern Asia (Ben-Shun 1989; Bokonyi 1984; Levine 1999; Mason 1984; Zarins 1992). Our model, in its present form, cannot include all of the essential complexities of such late domestication events. Our model explains the initial domestication of small stock such as sheep and goats. Later, we will consider the conditions that would have altered the parameters of our model to make larger animals attractive as conservable resources.

A number of researchers have offered explanations of why and how the transition to animal husbandry occurred (Flannery

1969, 1973; Hole 1989, 1996; Reed 1984; Uerpmann 1989). A recent attempt to understand animal husbandry from a behavioral ecology perspective is the work of Russell (1988). Russell used an approach incorporating optimal foraging theory's prey choice model to predict the order in which prey species were first husbanded. Like other models, Russell envisions population pressures driving the system. In his scenario, greater human population density led to prey depletion and depressed hunting returns. When hunting return rates dropped to lower than what could have been obtained from husbandry, the transition occurred. In this sense, husbanded prey and hunted prey represent two distinct prey types.

Russell collected quantitative data on productivity and labor costs for the husbanding of camels, cattle, sheep, and goats and calculated the respective return rates for each species, averaged over one year (Russell ignores pigs). With herds of 100 animals, camels have the highest return (665 kilocalories per hr), followed by cattle (615 kcal/hr), sheep (203 kcal/hr), and goats (109 kcal/hr). Based on these results, Russell predicted that as human population density increased and hunting returns declined, prey species would be husbanded in the order of their profitability: first cattle and camels, then sheep and goats. As mentioned earlier, however, the archeological record indicates domestication occurred in just the opposite order. The smaller species and less profitable (at least in the short term) types were husbanded before the larger ones. Why was this the case?

Russell claims that the archeological record is misleading (1986:341). A more satisfying answer is that Russell's approach exposes some ambiguity in his prey choice model. The critical issue is the length of time over which the models assume foraging return rates are expected to be maximized. The time scale over which Russell's model assumes individuals will optimize is too myopic. He assumes a 0% discount rate (Kagel et al. 1986). Husbanding, however, is not a strategy where payoffs are measured in the short term. The conservative nature of husbanding requires an approach that examines the long-term payoffs of alternative strategies. It is true, as Russell demonstrated, that the short-term return rates for husbanding are higher for larger-bodied animals compared to smaller-bodied animals if herds are of equal size (say 100 head). It takes time for herds to grow, however, and the payoffs occur at some point in the future. Herds are biological populations that can grow, and different species have different rates of increase. These differences affect the value of different species as potentially husbanded resources. Table 22.3 presents estimates of the maximum intrinsic rate of increase (r_{max}) for cattle, sheep, goats, and camels, based on data from modern species presented in Dahl and Hjort (1976). Herds of sheep and goats with r_{max} values between 0.6 and 0.7 can grow significantly faster than herds of cattle or camels that have r_{max} values less than half as large (0.22–0.25).

Figure 22.2 shows the effect of herd growth rates on relative return rates. Because herds of sheep and goats grow much faster than herds of cattle or camels, after 70 years return rates

TABLE 22.3. Data from Dahl and Hjort (1976) used to iteratively calculate r_{max} from Cole's (1954) equation.

SPECIES	ANNUAL BIRTHRATE	AGE AT FIRST REPRODUCTION	AGE AT LAST REPRODUCTION	r_{max}
Camel	0.5	3	23	0.225
Goat	2.0	1	9	0.700
Sheep	1.5	1	9	0.600
Cattle	0.5	2	13	0.255

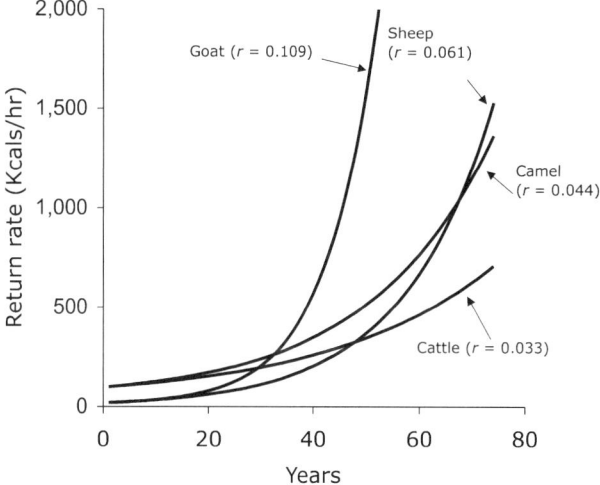

FIGURE 22.2. Effect of herd growth on return rates to husbanding. Herd size increases with time according to the reproductive rate of each species. Because of economies of scale, return rates increase accordingly. Since sheep and goats reproduce much faster than cattle or camels, over time return rates for these animals rapidly match, then surpass, return rates for the larger-bodied species. Herd growth is estimated here using the intrinsic rate of increase. Herds begin with a size of two individuals. Original return rate data for husbanding from Russell (1986).

are higher for sheep and goats. Also important is that absolute herd yield (kg per year) also increases much faster for sheep and goats. Starting with a population of two individuals, goats reach a herd size of 100 in 24 years—for sheep in 40 years. A herder would have to wait 55 years to achieve a herd of 100 camels and 72 years for a herd of 100 cattle. Return rates, as calculated by Russell, and the absolute herd yield are higher for goats after only one human generation and for sheep after only 2.5 human generations, compared to cattle and camels. This result suggests that their higher rates of population growth were one reason why sheep and goats were husbanded before camels and cattle.

These results beg a number of questions, however. How does a strategy of husbanding these animals compare to a hunting strategy? Herds of smaller animals grow more quickly, and most of them can be kept at higher densities than larger-bodied animals. Is this important? To answer these questions, below we develop a model of animal husbandry that predicts the range of body sizes that should be husbanded.

The Model

The model contains a number of assumptions. Foremost, we assume that the transition to husbandry occurred under conditions of increasing population density. We also assume that populations became more sedentized and less mobile. And finally, we assume that the benefits to husbanding depend on the value of animal resources to individual fitness. Evidence in support of these assumptions will be provided in the discussion section. At low human density or high prey abundance, hunters can more easily obtain optimal quantities, and resources have diminishing returns on fitness. In this context, there is no incentive for hunters to pay the short-term costs of conserving if time and effort can be better spent on other fitness-enhancing activities and long-term benefits are small (Alvard 1998; Kaplan and Hill 1992; Smith 1987). Recall that this is one of the reasons offered above to explain why Piro hunters do not husband peccaries. During the transition period, we expect competition for resources to have increased with human density (and/or depleting resources) and prey to have become more valuable. Fitness returns no longer diminish because prey become increasingly rare. Full-time hunters would find that return rates for hunting begin to drop as competition increases and animals become scarce. The point at which a hunter might consider husbanding and not hunting a scarce prey item would depend on the resource's value and its discount rate.

Allometry

There are two major reasons for discounting—uncertainty over future benefits and the opportunity costs associated with lost compounding gain. We will reserve the question of uncertainty and its impact on husbanding for later and will focus now on the issue of compounding gain. The future discounted value of husbanded animal resources varies according to the type of animal. As noted by Clark (1990), the prey's reproductive rate is especially critical for determining whether it is more economical for hunters to maintain the population at the MSY or exploit it as rapidly as possible.

Reproductive rate varies with a number of parameters, but body size explains much interspecific variance. In fact, body size correlates with many of the characteristics that determine the cost and benefits of husbanding (Peters 1983; Schmidt-Nielson 1984). As we will show, larger animals are bigger packages and, all other things being equal, will be preferred in the short term simply because they represent more resource. They reproduce more slowly, however, than smaller-bodied animals. In addition, larger animals live at lower densities but at higher biomass (kg per unit area) than small-bodied types (Figure 22.3). These allometric relationships between body size, density, and reproductive rate provide a powerful analytic tool for examining the costs and benefits of animal husbandry. The first step to determine the benefits from husbanding, and the best animals to husband, will be to calculate the maximum sustainable yield for a range of prey body sizes. The MSY represents the benefits to husband-

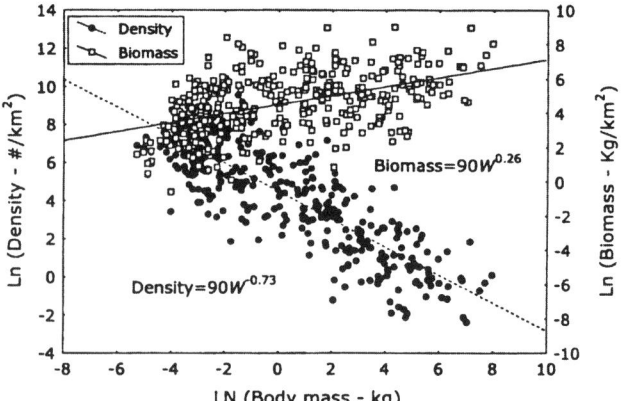

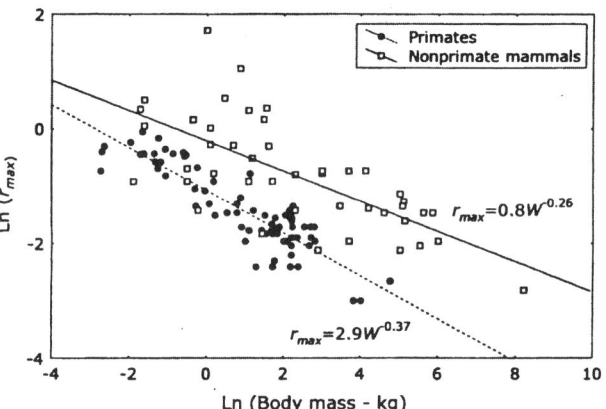

FIGURE 22.3. Allometric relationships of body mass for mammals. The top figure plots both density (number of individuals per km²) and biomass (density times body weight per km²) as a function of body mass for a sample of 368 mammalian herbivore species (data from Damuth 1987). The bottom figure plots r_{max} as a function of body weight for a sample of 72 primate species (Ross 1992) and 40 nonprimate mammal species (Henneman 1983).

ing. This will be compared to what could be obtained from hunting the same animals.

Many morphological, physiological, and life history traits vary between species with body size according to the following relationship (Peters 1983):

$$X = AW^b \qquad (6)$$

W is body mass, A is the intercept, b is the slope of the line for the linear relationship of the log-transformed variables:

$$\log(X) = \log(A) + b[\log(W)] \qquad (7)$$

The most well-known allometric relationship is described by Kleiber's rule, which states that metabolic rate is a function of body mass where $b \cong 0.75$ power (Kleiber 1961).

To estimate the MSY (kilograms of biomass harvested per year per square kilometer) for potentially husbanded species of different sizes, we use three variables and two well-documented allometric life history relationships. The variables are the maximum intrinsic rate of increase (r_{max}), density (N), and body mass (W). r_{max} is a measure of a species' maximum rate of population growth in an environment where resources are not limiting. Density is the number of individuals per unit area. Body mass is weight in kilograms for adults.

Across a variety of taxa, the data indicate that larger-bodied species reproduce slower. There is a general negative allometric relationship between body size and reproductive rate with the general form

$$r_{max} = A_1 W^{b_1} \qquad (8)$$

As discussed below, r_{max} is one of a number of life history variables that are predicted to scale allometrically to the approximate power of –0.25 of body weight (Charnov 1993). A variety of data sets show slopes between $b_1 = -0.25$ and $b_1 = -0.37$ (Charnov 1993; Henneman 1983; Robinson and Redford 1986; Ross 1992). For example, for a sample of 40 nonprimate mammal species, Henneman (1983) found that $\ln(r_{max}) = -0.27 * \ln(W) - 0.11$ or, $r_{max} = 0.9 W^{-0.27}$. A is a constant that varies with trophic level and taxon. For example, primates as a group show the same slope but a lower A value than nonprimate mammals (Charnov 1993). After controlling for body size, primates as a group reproduce slower than other types of mammals, but small primates still reproduce faster than large primates in a manner predicted by equation 7. Using covariance analysis, Charnov (1993) found a common slope of –0.31 for Henneman's (1983) data set of nonprimate mammals and Ross's (1992) primate data set (Figure 22.3).

Larger-bodied organisms are also usually found at lower population densities than smaller-bodied animals, following a general negative allometric relationship of the form:

$$N = A_2 W^{b_2} \qquad (9)$$

Using a sample size of 368, Damuth (1987) found the relationship for mammalian herbivores to be $\ln(N) = -0.73 * \ln(W) + 4.5$ or $N = 90 W^{-0.73}$. Inversely, standing biomass (density times body mass divided by unit area) *increases* with body size. In this case, $b_3 = 0.26$. Figure 22.3 demonstrates these two allometric relationships.

These are useful relationships because they can be used to calculate MSY as an allometric function of body size. Recall that MSY theory shows that yield is maximized at approximately MSY $= (K r_{max})/4$. If it is assumed that the densities (N) reported by Damuth (1987) are at carrying capacity, maximum sustainable yields in kilograms can be calculated as

$$\text{MSY} = (N W r_{max})/4 \qquad (10)$$

Because density (N) and reproductive rate (r_{max}) are related allometrically to body mass (W), equations 8 and 9 can replace them in equation 11, to get

$$\text{MSY} = (A_2 W^{b_2}) * (A_1 W^{b_1}) * W / 4 \qquad (11)$$

or,

$$\text{MSY} = (90 W^{-0.73}) * (0.9 W^{-0.31}) * W / 4 \qquad (12)$$

Assuming $A_2 = 90$, $b_2 = -0.73$ (Damuth 1987), and $A_1 = 0.9$ (Henneman 1983), $b_1 = -0.31$ (from Charnov [1993] for primates and nonprimate mammals), the MSY is between 22.7 kg per km² per year for an animal that weighs 50 kg and 14.6 kg per km² per year for an animal that weighs 3,000 kg. Note the large difference in body size but that the MSY does not differ proportionally. Charnov (1993), in fact, argues that MSY should theoretically be *independent* of body mass. This result obtains because of the nature of the allometric relationships between r_{max}, density, and body size. Smaller species reproduce faster and live at higher density but are small bodied; these effects nearly cancel one another out using the exponents from Henneman and Damuth. The exponents suggested by Charnov (1993) for the density and r_{max} allometric equations ($b_1 = -0.25$ and $b_2 = -0.75$) sum to equal W^{-1} or $1/W$. This cancels exactly when multiplied by the W term in equation 11. Consequently, MSY is independent of body mass and dependent primarily on the constants A_1 and A_2:

$$\text{MSY} = A_1 A_2 / 4 \qquad (13)$$

Using these exponents and the constants $A_2 = 90$ and $A_1 = 0.9$ from above, the MSY for any body-sized species is approximately 20.2 kg per km². This estimate, however, is too high. The estimate used to calculate r_{max} from Henneman (1983) and Ross (1992) is derived from Cole's (1954) formula. Cole's equation uses age at first reproduction, annual fecundity, and age at last reproduction, but ignores mortality. Charnov (1993) provides a correction for this overestimation. He notes the invariant relationship between the product of age at first reproduction and mortality, the product of age at first reproduction and yearly fecundity, and body size for mammals (for details see Charnov 1993:124). For nonprimate mammals, he calculates $r_{max} = 0.4 W^{-0.25}$. Using $A_1 = 0.4$ in equation 13 reduces the MSY for any body-sized species to approximately 9 kg per km². This is a more realistic estimate of the yield.

While MSY is the same across body sizes, standing biomass, s (density multiplied by body mass), scales positively with body size. The larger the species, the greater the standing biomass to the approximate power of positive 0.25. This means that husbanding larger-bodied animals requires much more "capital" in terms of standing biomass to achieve the same MSY than does husbanding smaller-bodied animals. The relationship indicated in Figure 22.3 shows that for a mammal the size of a camel (~500 kg), 450 kg of biomass are standing per km², while for a mammal the size of a sheep (~30 kg) only 217 kg are standing per km².

This has important implications for a husbanding strategy. Recall that the delayed benefits of husbanding are discounted by the opportunity costs—the costs of not using the resource immediately. The standing biomass represents the resources that husbanders do not consume. The lower the discount rate compared to the returns expected from husbanding, the more likely husbanding will pay off. Since the amount of biomass *not* being used is greater for larger animals, the opportunity costs of husbanding are greater for larger animals. The opportunity costs are measured by multiplying the standing biomass (s—this represents what could be obtained from a hunting strategy) times the discount rate (i).

$$s i > K r_{max} / 4 \qquad (14)$$

Density cancels, to leave
$$i > r_{max}/4 \quad (15)$$
This is exactly the same result obtained with equation 5. In this case, the carrying capacity is equal to the standing biomass ($K = s$). If the opportunity costs (standing biomass multiplied times the discount rate) are greater than the productivity that could be obtained from husbanding (the MSY), hunters are predicted to exploit that body-sized species rather than husband it.

It might be argued that we have overestimated the opportunity cost of husbanding because there are diminishing returns to hunted resources; that is, the cost of forgoing the 100th ungulate is less than forgoing the first. It should be noted that we assume increasing population density, resource shortages, and competition at the time of the transition. Competition pushes the system down to the linear end of the diminishing returns curve. In other words, there is no 100th ungulate. Game is rare, each animal is valuable, and the cost of forgoing a large-bodied animal is high.

Setting the Discount Rate

Knowing the evolutionary discount rate is important for understanding the time depth that humans should take into account when the costs and benefits of subsistence and reproductive decisions occur at some point in the future. An evolutionary-based discount rate was formulated deductively by Rogers (1994), who hypothesized an evolved time preference by natural selection. He concluded that the evolved human time preference, or discount rate (i), should depend critically on three factors: (p), the rate of population growth, (R), the average relatedness between the individual and his offspring, and the generation length, (T):
$$i = (-\ln R)/T + p \quad (16)$$
Delayed benefits that accrue to descendants are discounted because the longer benefits are deferred, the less related descendants become. At some point in the future, direct descendants become as related to ego as ego is related to the average person in the present population. For example, after 240 years, or 9 generations (generation time in this case is 30 years), the coefficient of relatedness between the descendant receiving the husbanded animals and their ancestor is $R = 0.002$. For comparison, the average degree of relatedness between individuals in the Piro village of Diamante is $R = 0.027$ (Alvard 1998). The average degree of relatedness for 12 Yanomamo villages is $R = 0.086$ (Chagnon 1988). Assuming $R = 0.5$ between parent and offspring, after 156 years the direct descendants of an individual Piro in Diamante today would be as related to ego, on average, as any two individuals chosen randomly from the village in the present.

If the average coefficient of relatedness between parent and offspring is 0.5, and one generation is about 30 years, the discount rate is 2.4% plus the rate of population growth according to Rogers. Future reproduction is additionally discounted by the rate of population growth because delayed reproduction contributes a smaller proportion of the larger future population than does current reproduction if the population is growing. Rogers assumes that the rate of population growth must have been zero on average prehistorically and thus did not contribute to the evolved time preference. He concludes with an average discount rate of 2.4%.

Recent demographic work with the Ache, a hunting and gathering group in Paraguay, suggests that perhaps a higher rate is justified. Using the best demographic data to date from a precontact foraging period, Hill and Hurtado (1996) found that the Ache have had an annual growth rate of over 2.5% for at least the last 100 years. They argue that hunters and gatherers in general may be characterized by periods of rapid growth interspersed with population crashes. In this scenario, long-term, average population growth remains near zero, but individuals are exposed to a selective environment where most years are characterized by nonzero growth. If that were the case prehistorically, selection would favor a time preference greater than 2.4%—in the case of the Ache, as high as 5% (Alvard 1998). Indeed, it is now apparent that humans experienced much environmental variability in their evolutionary past that could have affected population growth rates (Potts 1996). This predicts the possibility of adaptive plasticity with respect to how schedules of reward are preferred (Hansson and Stuart 1990).

Calculation of the Maximum Body Size to Husband

As mentioned above, for husbanding to be favored, its future benefits must be discounted and still be greater than the returns expected from hunting. Recall equation 15, where $i > r_{max}/4$ for hunting to pay off. For husbanding to pay, the inverse must be true, $i < r_{max}/4$. To calculate the body size below which husbanding pays (W^*), we replace r_{max} with its body mass function (equation 8) and simplify to get
$$W^* = e^{[\ln(4i)-A]/b} \quad (17)$$
Note that prey density is not a factor here. This is because a change in density affects both the MSY and the opportunity costs as measured by standing biomass. Standing biomass increases as density increases, but so does MSY and the effects cancel. W^* is, however, sensitive to the r_{max} function and the discount rate. If we follow Charnov (1993) and assume that $b_1 = -0.25$, $A_1 = 0.4$, and the discount rate is 2.4%, which corresponds to zero population growth on the part of the humans, the maximum body size to husband is 309 kg. If a discount rate of 4% is used, the maximum body size to husband falls to around 40 kg. A 4% discount rate corresponds to a 1.6% population growth rate. The results for a variety of ranges of reasonable discount rates are presented graphically in Figure 22.4.

The predicted body size falls within the range of probable body sizes for the species first domesticated. Data on modern domesticated sheep show much variability but bracket the predicted size. As examples of the extremes, adult male Nilotic sheep weigh between 11 and 25 kg, while male Lohi sheep from India weigh up to 68 kg (Dahl and Hjort 1976). Goat sizes vary as well, but again within the predicted range. Indian Jumnapai male goats can weigh up to 79 kg, while East African Mubende male goats weigh 20 kg (Dahl and Hjort 1976). We discuss the subsequent domestication of larger animals below.

Discussion

We have applied principles derived from evolutionary ecology, conservation biology, economics, and archaeology in a cross-disciplinary approach to address a classic question within the field of anthropology—the transition from hunting to animal husbandry. By operationally defining animal husbandry as conservation, we could predict that husbandry likely arose during conditions that favored resource conservation. Conservation is more likely when goods are privately owned/territorially defended rather than open access, when resources are sufficiently valued/scarce to justify the costs of defense, and when the opportunity costs of restraint are low. Even if resources are highly valued and well defended, however, husbanding may not be the best strategy if the opportunity costs are too high. As we have shown, prey body size predicts much of the variance in the parameters of interest (reproductive rate, density, and biomass) for measuring the benefits and opportunity costs of husbanding alternative prey types.

The first two conditions that favor conservation were not discussed to any great extent in this article, but current archaeological research on the period of the Neolithic transition in the Middle East indicates that the circumstances were conducive to resource conservation. Human populations were growing and becoming more sedentary, perhaps as a result of resource stress provoked by global cooling during the Younger Dryas period (ca. 11,000 to 10,000 BP—Bar-Yosef 1998; Byrd 1994a; McCorriston and Hole 1991; Sherratt 1997). In a number of complex but generally confirmed models of domestication, archaeologists and ethnographers have established that population growth of forager bands leads to the packing of people into a finite environment (Aldenderfer 1998; Binford 1987; Haaland 1995; Harris 1977; Hitchcock 1982; Kuznar 1989; Price and Brown 1985). Packing necessarily leads to reduced mobility and the territorial defense of resources as suggested by Dyson-Hudson and Smith (1978). Reduced mobility reduces constraints on women's fertility, which in turn increases population growth (Binford and Chasko 1976; Haaland 1995; Harpending and Wandsnider 1982; Kuznar 1989; Reed 1984). The net result creates the conditions of increased population density, increased pressure upon resources, and the consequent increase in resource value essential to our model. The packing/sedentism model has been useful for explaining the origins of sedentism, cultural complexity, and food production in the Middle East (Bar-Yosef and Belfer-Cohen 1992; Flannery 1973; Harris 1969, 1977, 1998), East Africa (Close and Wendorf 1992; Haaland 1995), Andean South America (Aldenderfer 1998; Kuznar 1989, 1995; MacNeish 1983), and in Eastern North America (Brown 1985; Watson 1989).

As one expects from work that resides on the margins of

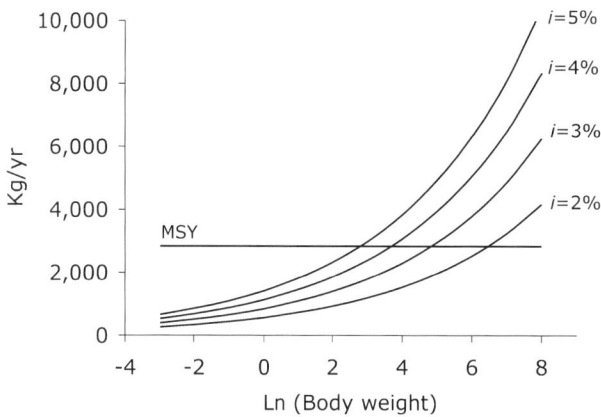

FIGURE 22.4. Annual harvest in kilograms per 314 km² as a function of body size for both the husbanding strategy (MSY) and the hunting strategy. MSY is 2,844 kg/yr and is independent of species' body size. The returns from the hunting strategy assume an extreme exploit strategy where total biomass is harvested and invested at the discount rate. Hunting returns increase with prey body weight because larger animals are found at greater biomass. The MSY line and hunting line cross at W^*. This is where the benefits from husbanding and its opportunity costs are equal. Above W^*, the opportunity costs of husbanding are too high, and the prey type should be pursued and consumed upon encounter. W^* depends on the discount rate. Returns from hunting and the corresponding W^* values are plotted for four values of i. For 2%, $W^* = 640$ kg; for 3%, $W^* = 126$ kg; for 4%, $W^* = 40$ kg; and for 5%, $W^* = 16$ kg. As mentioned above, 314 km² is an area with a radius of 10 km—the approximate distance a hunter can travel from a central place and return before dark.

many fields, more questions were raised than answered, and there are many opportunities for fruitful future investigation. For example, why did hunting continue to coexist alongside husbandry? In the manner of most models, this article takes a relatively complex phenomenon and simplifies it. The transition from hunting and gathering to food production was not a simple one, however. For example, strong evidence shows that both wild and domesticated animal resources were used concurrently long after the initial incorporation of husbanded resources in the economies of the ancient Middle East (Stein 1989; Zeder 1994) and the Andes (Olivera 1998). Ethnographers commonly observe that pastoralists, reluctant to slaughter and eat their valuable livestock, actively hunt wild herbivores as sources of meat. Examples are found in pastoral settings and among people as diverse as those of the South American Andes (Custred 1977; Flannery et al. 1989; García 2001; Kuznar 1995), Tibet (Ekvall 1968; Goldstein and Beall 1990), East Africa (Evans-Pritchard 1940; Klima 1970; Robertshaw and Collett 1983), Siberia (Ingold 1980), Europe (Creighton and Segui 1998), the Middle East (Barth 1961; Cribb 1991), and the Navajo (Downs 1972; Luckert 1975). The persistence of hunting might be expected for two reasons. First, our model predicts that hunting remains the optimal choice for larger prey because the opportunity costs of husbanding them were too high. Cattle, pigs, and camels were not husbanded until at least a millennium after sheep and goats. Below we discuss a number of possibilities that may have dropped the opportunity costs of husbanding these animals to the point that hunting them was no longer as profitable (Russell 1988). Second, the analysis above uses predicted values of r_{max}, density, and biomass. There is much scatter around the regression lines in Figure 22.3. One would expect variation within species as well. For foragers exploiting local prey populations whose reproductive rates were lower than the predicted values, the benefits to husbanding would also have been lower, and hunting would have been the optimal strategy. Zeder (1994:119) reports what she terms a remarkable dichotomy between sites where domesticates and wild animal resources dominated. Sites with a predominance of wild fauna are located in marginal arid environments where husbanding may not have provided sufficient returns.

Why were gazelles or other animals of the appropriate size not husbanded? While goats and sheep fall in the range of predicted body size, so do a number of other prey present in the Middle East at the time of the transition. One hypothesis is that other medium-sized ungulates were not husbanded because of behavioral characters that made these species more difficult to manage (Baskin 1974). These extra costs could decrease the return rates from husbanding these species significantly, independent of their body size. Gazelles (*Gazella gazella*) were the favored prey of the Natufians and other groups who lived in the Levant region of the Middle East between 10,000 and 8,000 BC (Byrd 1989). Gazelle behavioral characters make them difficult animals to defend in large numbers and, hence, more costly to conserve. They cannot be herded in large groups or driven long distances as can sheep and goats (Clutton-Brock 1981). This may have been the reason that a number of other prey were also never husbanded, including roe deer (*Capreolus capreolus*—22 kg), ibex (*Capra ibex*), and various equids (*Equus* spp.).

Larger types, such as cattle and camels, were eventually incorporated into the suite of domesticates (in terms of our model, W^* increased). What may have caused larger-bodied prey to be husbanded? There are a number of possibilities. According to our model, one way that W^* would increase is if people's time preference became less myopic, i.e., the discount rate lowered. Recall that there are two important parameters that Rogers (1994) uses to estimate the discount rate: population growth and the average coefficient of relatedness [R] between parents and offspring. Holding the coefficient of relatedness constant at $R = 0.50$, if human population growth slowed, the opportunity costs of husbanding large game would decline. Such population crashes have been hypothesized for foragers as mentioned above. At zero population growth (a discount rate of 2.4%), W^* increases to 309 kg. Inbreeding would also increase W^*. Rogers assumes $R = 0.5$, which gives the 2.4% in the absence of population growth. The average coefficient of relatedness between parents and offspring will increase if endogamy becomes more common (cousin marriages for example), perhaps because of increased sedentism. In fact, sedentism is cross-culturally correlated with increased group cohesiveness and territorial defense of resources (Charles and Buikstra 1983; Kuznar in press). If R increases to 0.6 in the absence of population growth, the discount rate drops to 1.8%, and W^* increases to 976 kg.

Another idea to explain the move to larger game is the risk reduction hypothesis suggested by Mace (1993). Larger animals like camels and cattle reproduce more slowly but are better able to cope with drought. Mace shows that keeping herds of smaller animals like sheep and goats promises better short-term returns in terms of herd growth, but such a strategy is riskier because these animals are susceptible to drought. During a protracted transition from hunting to husbanding, such risk could have been managed by mobility or by switching to hunting. Later, more sedentary husbanders unable to move or hunt profitably might have been willing to pay the higher opportunity costs of husbanding larger animals if such a strategy reduced risk of shortfall. In fact, the need for reducing risk in food supply is hypothesized for increasingly sedentary people (Harris 1998; McCorriston and Hole 1991; Price and Brown 1985; Sherratt 1997; Stein 1989), and general risk aversion is well established among pastoralists ethnographically (De Boer and Prins 1989; Fratkin 1991; Kuznar 1991, 2000; Mace 1993).

Finally, a strategy of milk and blood extraction, part of a secondary products revolution (Sherratt 1981), may have eventually dropped the opportunity costs of husbanding large animals to the point that hunting them was no longer as profitable (Russell 1988). It is unclear when this may have occurred during the process of animal husbandry, although such intensification is common among many extant husbanders (Dyson-Hudson and Dyson-Hudson 1980).

In conclusion, we have provided an analytical model that provides baseline predictions of the timing of initial animal husbandry and the optimal size of animal to be domesticated. The model, in which we suggest that animal husbandry is a special case of resource conservation, is based on optimal foraging assumptions rooted in evolutionary theory. Because of the model's groundedness in well-established theory and its focus on a few key variables, its parameters can be altered to provide explanations and testable predictions of the domestication of various animals. Using this model, we were able to explain the initial domestication of sheep and goats approximately 10,000 years ago, and by considering factors extraneous to the model, such as human demography, risk reduction considerations, and human protection of ungulates from predation, we were able to suggest subsequent domestication events. We do not claim that this model explains all domestication, nor do we consider all potentially important factors in domestication. However, the model does provide an analysis of several key factors and allows for both explanation of archaeological data as well as suggests future lines of inquiry. These analytical tools not only allow for a better understanding of past domestication events but also bring into focus the nature of conservation and broaden our perspective on common property phenomena.

REFERENCES

Acheson, J.
1989 Management of Common Property Resources. In *Economic Anthropology*, edited by S. Plattner, pp. 351–378. Stanford University Press, Stanford, CA.

Aldenderfer, M.
1998 *Montane Foragers: Asana and the South-Central Andean Archaic*. University of Iowa Press, Ames.

Alvard, M.
1993 Testing the Ecologically Noble Savage Hypothesis: Interspecific Prey Choice by Piro Hunters of Amazonian Peru. *Human Ecology* 21:355–387.
1994 Conservation by Native Peoples: Prey Choice in a Depleted Habitat. *Human Nature* 5:127–154.
1995 Intraspecific Prey Choice by Amazonian Hunters. *Current Anthropology* 36:789–818.
1998 The Evolutionary Ecology of Resource Conservation. *Evolutionary Anthropology* 7:62–74.
2000 The Impact of Traditional Subsistence Hunting and Trapping on Prey Populations: Data from the Wana of Upland Central Sulawesi, Indonesia. In *Hunting for Sustainability in Tropical Forests*, edited by J. Robinson and E. Bennett, pp. 214–230. Columbia University Press, New York.

Alvard M., J. Robinson, K. Redford, H. Kaplan
1997 The Sustainability of Subsistence Hunting in the Neotropics. *Conservation Biology* 11:977–982.

Bailey, L. R.
1980 *If You Take My Sheep: The Evolution and Conflicts of Navajo Pastoralism, 1630–1868*. Westernlore Publications, Pasadena, CA.

Barfield, T.
1989 *The Perilous Frontier: Nomadic Empires and China*. Blackwell, Cambridge.

Barth, F.
1961 *Nomads of South Persia: The Basseri Tribe of the Khamesh Confederacy*. Little, Brown and Company, Boston.

Bar-Yosef, O.
1998 The Natufian Culture in the Levant, Threshold to the Origins of Agriculture. *Evolutionary Anthropology* 6:159–177.

Bar-Yosef, O., A. Belfer-Cohen
1989 Origins of Sedentism and Farming Communities in the Levant. *Journal of World Prehistory* 3:447–498.
1992 From Foraging to Farming in the Mediterranean Levant. In *Transitions to Agriculture in Prehistory*, edited by A. B. Gebauer and T. D. Price, pp. 21–48. Prehistory Press, Madison, WI.

Baskin, L.
1974 Management of Ungulate Herds in Relation to Domestication. In *The Behavior of Ungulates and Its Relationship to Management*, edited by V. Geist and F. Walther, pp. 530–541. International Union for the Conservation of Nature, Morges, Switzerland.

Beckerman, S., P. Valentine
1996 On Native American Conservation and the Tragedy of the Commons. *Current Anthropology* 37:659–666.

Ben-Shun, C.
1989 The Domestic Horse of Pre-Ch'in Period in China. In *The Walking Larder: Patterns of Domestication, Pastoralism, and Predation*, edited by J. Clutton-Brock, pp. 105–107. Unwin Hyman, London.

Binford, L.
1968 Post-Pleistocene Adaptations. In *New Perspectives in Archeology*, edited by S. Binford and L. Binford, pp. 313–344. Aldine, Chicago.
1987 The Reluctant Shift from Hunting to Horticulture in North America. *Social Science* 72:44–47.

Binford, L. R., W. J. Chasko
1976 Nunamiut Demographic History: A Provocative Case. In *Demographic Anthropology: Quantitative Approaches*, edited by E. B. W. Zubrow, pp. 63–143. University of New Mexico Press, Albuquerque.

Bokonyi, S.
1984 Horse. In *Evolution of Domesticated Animals*, edited by I. L. Mason, pp. 162–173. Longman, London.

Boone, J.
1992 Competition, Conflict and the Development of Social Hierarchies. In *Evolutionary Ecology and Human Behavior*, edited by E. Smith and B. Winterhalder, pp. 301–337. Aldine de Gruyter, New York.

Brown, J.
1985 Long-Term Trends to Sedentism and the Emergence of Complexity in the American Midwest. In *Prehistoric Hunter-Gatherers: The Emergence of Cultural Complexity*, edited by T. D. Price and J. A. Brown, pp. 201–231. Academic Press, New York.

Bulmer, M.
1994 *Theoretical Evolutionary Ecology*. Sinauer, New York.

Byrd, B.
1989 The Natufian: Settlement Variability and Economic Adaptations in the Levant at the End of the Pleistocene. *Journal of World Prehistory* 3:159–197.
1994a From Early Humans to Farmers and Herders—Recent Progress on Key Transitions in Southwest Asia. *Journal of Archaeological Research* 2:221–253.
1994b Public and Private, Domestic and Corporate. The Emergence of the Southwest Asian Village. *American Antiquity* 59:34.

Caughley, G.
1977 *Analysis of Vertebrate Populations*. John Wiley and Sons, London.

Chagnon, N.
1988 Life Histories, Blood Revenge, and Warfare in a Tribal Population. *Science* 239:985–992.

Charles, D., J. Buikstra
1983 Archaic Mortuary Sites in the Central Mississippi Drainage: Distribution, Structure, and Behavioral Implications. In *Archaic Hunters and Gatherers in the American Midwest*, edited by J. Phillips and J. Brown, pp. 117–145. Academic Press, New York.

Charnov, E.
1993 *Life History Invariants: Some Explorations of Symmetry in Evolutionary Ecology*. Oxford University Press, Oxford.

Childe, V. G.
1952 *Man Makes Himself*. Watts, London.

Clark, C. W.
1990 *The Optimal Management of Renewable Resources*. 2nd edition. Wiley-Interscience, New York.

Close, A., F. Wendorf
1992 The Beginnings of Food Production in the Eastern Sahara. In *Transitions to Agriculture in Prehistory*, edited by A. B. Gebauer and T. D. Price, pp. 63–72. Prehistory Press, Madison, WI.

Clutton-Brock, J.
1981 *Domesticated Animals from Early Times*. University of Texas Press, Austin.

Cohen, M.
1977 *The Food Crisis in Prehistory: Overpopulation and the Origins of Agriculture*. Yale University Press, New Haven, CT.

Cole, L.
1954 The Population Consequences of Life History Phenomena. *Quarterly Review of Biology* 29:103–137.

Creighton, O. H., J. R. Segui
1998 The Ethnoarchaeology and Post-Abandonment Behavior in Pastoral Sites: Evidence from Famorca, Alacant Province, Spain. *Journal of Mediterranean Archaeology* 11:31–52.

Cribb, R.
1991 *Nomads in Archaeology*. Cambridge University Press, Cambridge.

Custred, G.
1977 Las Punas de los Andes Centrales. In *Pastores de Puna: Uywamichiq Punarunakuna*, edited by Jorge Flores-Ochoa, pp. 55–86. Instituto de Estudios Peruanos, Lima.

Dahl, G., A. Hjort
1976 *Having Herds: Pastoral Herd Growth and Household Economy*. Liber Tryck, Stockholm.

Damuth, J.
1987 Interspecific Allometry of Population Density of Mammals and Other Animals: The Independence of Body Mass and Population Energy Use. *Biological Journal of the Linnean Society* 31:193–246.

De Boer, W. F., H. H. T. Prins
1989 Decisions of Cattle Herdsmen in Burkina Faso and Optimal Foraging Models. *Human Ecology* 17:445–464.

Demsetz, H.
1967 Toward a Theory of Property Rights. *American Economic Review* 57(2):347–359.

Downs, J.
1972 *The Navajo*. Holt, Rinehart and Winston, New York.

Dyson-Hudson, R., N. Dyson-Hudson
1980 Nomadic Pastoralism. *Annual Review of Anthropology* 9:15–61.

Dyson-Hudson, R., E. A. Smith
1978 Human Territoriality: An Ecological Reassessment. *American Anthropologist* 80:21–41.

Ekvall, R.
1968 *Fields on the Hoof: Nexus of Tibetan Nomadic Pastoralism*. Holt, Rinehart and Winston, New York.

Errington, P.
1946 Predation and Vertebrate Populations. *Quarterly Review of Biology* 21:145–177.

Evans-Pritchard, E. E.
1940 *The Nuer*. Oxford University Press, Oxford.

Fisher, I.
1930 *Theory of Interest*. Kelley and Millman, New York.

Fisher, R.
1958 *The Genetical Theory of Natural Selection*. 2nd edition. Oxford University Press, Oxford.

Flannery, K.
1965 The Ecology of Early Food Production in Mesopotamia. *Science* 147:1247–1256.
1969 Origins and Ecological Effects of Early Domestication in Iran and the Near East. In *The Domestication and Exploitation of Plants and Animals*, edited by P. J. Ucko and G. W. Dimbleby, pp. 72–100. Duckworth, London.
1973 The Origins of Agriculture. *Annual Review of Anthropology* 2:271–310.

Flannery, K., J. Marcus, R. Reynolds
1989 *The Flocks of the Wamani: A Study of Llama Herders on the Punas of Ayacucho, Peru*. Academic Press, New York.

Fratkin, E.
1991 *Surviving Drought and Development: Ariaal Pastoralists of Northern Kenya*. Westview Press, Boulder, CO.

García, L. Clara
2001 Women at Work: A Present Archaeological View of Azul Pampa Herding Culture (North West Argentina). In *Ethnoarchaeology of Andean South America: Contributions to Archaeological Method and Theory*, edited by L. Kuznar, pp. 202–220. International Monographs in Prehistory, Ann Arbor, MI.

Goldstein, M., C. Beall
1990 *Nomads of Western Tibet: The Survival of a Way of Life*. University of California Press, Berkeley.

Haaland, R.
1995 Sedentism, Cultivation, and Plant Domestication in the Holocene Middle Nile Region. *World Archaeology* 22:157–174.

Hames, R.
1987 Game Conservation or Efficient Hunting? In *The Question of the Commons*, edited by B. McCay and J. Acheson, pp. 92–107. University of Arizona Press, Tucson.
1991 Wildlife Conservation in Tribal Societies. In *Biodiversity: Culture, Conservation and Ecodevelopment*, edited by B. Oldfield and J. Alcorn, pp. 172–199. Westview Press, Boulder, CO.

Hames, R., W. Vickers
1982 Optimal Diet Breadth Theory as a Model to Explain Variability in Amazonian Hunting. *American Ethnologist* 9:357–378.

Hansson, I., C. Stuart
1990 Malthusian Selection of Preferences. *The American Economic Review* 80:529–544.

Hardin, G.
1968 The Tragedy of the Commons. *Science* 162:1243–1248.
1993 *Living within Limits: Ecology, Economics, and Population Taboos*. Oxford University Press, Oxford.

Harpending, H., L. Wandsnider
1982 Population Structures of Ghanzi and Ngamiland !Kung. In *Current Developments in Anthropological Genetics*, edited by M. Crawford and J. Mielke, pp. 29–50. Plenum Press, New York.

Harris, D.
1969 Agricultural Systems, Ecosystems and the Origins of Agriculture. In *The Domestication and Exploitation of Plants and Animals*, edited by P. J. Ucko and G. W. Dimbleby, pp. 1–15. Aldine, New York.
1977 Settling Down: An Evolutionary Model for the Transformation of Mobile Bands into Sedentary Communities. In *The Evolution of Social Systems*, edited by J. Friedman and M. Rowlands, pp. 401–417. Duckworth, London.
1996 Domesticatory Relationships of People, Plants and Animals. In *Redefining Nature: Ecology, Culture and Domestication*, edited by R. Ellen and K. Fukui, pp. 437–463. Berg, Oxford.
1998 The Origins of Agriculture in Southwest Asia. *Review of Archaeology* 19:5–11.

Hawkes, K.
1992 Sharing and Collective Action. In *Evolutionary Ecology and Human Behavior*, edited by E. Smith and B. Winterhalder, pp. 269–300. Aldine de Gruyter, New York.

Hawkes, K., A. R. Rogers, E. L. Charnov
1995 The Male's Dilemma: Increased Offspring Production Is More Paternity to Steal. *Evolutionary Ecology* 9:662–677.

Hecker, H. M.
1984 A New Look at Childe's Oasis-Propinquity Theory. In *Animals and Archaeology, Vol. 3. Early Herders and Their Flocks*, edited by J. Clutton-Brock and C. Grigson, pp. 133–144. BAR International Series, 202. British Archaeological Reports, Oxford.

Henneman, W.
1983 Relationship among Body Mass, Metabolic Rate and the Intrinsic Rate of Natural Increase in Mammals. *Oecologia* 56:104–108.

Henry, D.
1995 *Prehistoric Cultural Ecology and Evolution*. Plenum Press, New York.

Hill, K., M. Hurtado
1996 *Ache Life History: The Ecology and Demography of a Foraging People*. Aldine, New York.

Hitchcock, R. K.
1982 Patterns of Sedentism among the Basarwa of Botswana. In *Politics and History in Band Societies*, edited by E. Leacock and R. Lee, pp. 223–267. Cambridge University Press, Cambridge.

Hole, F.
1989 A Two-Part, Two-Stage Model of Domestication. In *The Walking Larder: Patterns of Domestication, Pastoralism, and Predation*, edited by J. Clutton-Brock, pp. 97–104. Unwin Hyman, London.
1996 The Context of Caprine Domestication in the Zagros Region. In *The Origins and Spread of Agriculture and Pastoralism in Eurasia*, edited by D. Harris, pp. 263–281. Smithsonian Institution Press, Washington, D.C.

Hunn, E.
1982 Mobility as a Factor Limiting Resource Use in the Columbia Plateau of North America. In *Resource Managers: North American and Australian Hunter-Gatherers*, edited by N. Williams and E. Hunn, pp. 17–43. Westview Press, Boulder, CO.

Ingold, T.
1980 *Hunters, Pastoralists, and Ranchers: Reindeer Economies and Their Transformations*. Cambridge University Press, Cambridge.

Kagel, J., L. Green, T. Caraco
1986 When Foragers Discount the Future: Constraint or Adaptation? *Animal Behavior* 34:271–283.

Kaplan, H., K. Hill
1985 Food Sharing among Ache Foragers: Tests of Explanatory Hypotheses. *Current Anthropology* 26:223–246.
1992 The Evolutionary Ecology of Food Acquisition. In *Evolutionary Ecology and Human Behavior*, edited by E. Smith and B. Winterhalder, pp. 167–202. Aldine de Gruyter, New York.

Kent, J.
1988 El Más Antiguo Sur: Una Revision de la Domestication de los Camélidos Andinos. In *Coloquio V. Gordon Childe: Estudios Sobre la Revolución Neolítica y la Revolución Urbana*, edited by L. Manzanilla, pp. 181–197. Universidad Nacional Autónoma de México, Mexico City.

Khazanov, A.
1994 *Nomads and the Outside World*. 2nd edition. University of Wisconsin Press, Madison.

Kleiber, M.
1961 *The Fire of Life*. John Wiley and Sons, New York.

Klima, G.
1970 *The Barabaig: East African Cattleherders*. Holt, New York.

Krech, S.
1999 *The Ecological Indian: Myth and History*. Norton, New York.

Kuznar, L.
1989 The Domestication of South American Camelids: Models and Evidence. In *Ecology, Settlement and History in the Osmore Basin, Peru*, edited by D. Rice, Charles Stanish, and Phillip Scarr, pp. 167–182. BAR International Series, 545. British Archaeological Reports, Oxford.
1991 Transhumant Pastoralism in the High Sierra of the South Central Andes: Human Responses to Environmental and Social Uncertainty. *Nomadic Peoples* 28:93–104.
1995 *Awatimarka: The Ethnoarchaeology of an Andean Herding Community*. Harcourt Brace, Fort Worth, TX.
2000 Application of General Utility Theory for Estimating Value in Non-Western Societies. *Field Methods* 12:334–345.
2003 Sacred Sites and Profane Conflicts: The Use of Burial Facilities and Other Sacred Locations as Territorial Markers—Ethnographic Evidence. In *Theory, Method and Practice in Modern Archaeology*, edited by Robert R. Jeske and Douglas D. Charles, pp. 269–286. Praeger, Westport, Ct.

Leacock, E.
1954 *The Montagnais "Hunting Territory" and the Fur Trade*. American Anthropologist Memoir, 78. American Anthropological Association, Menasha, WI.

Levine, M.
1999 Botai and the Origins of Horse Domestication. *Journal of Anthropological Archaeology* 18:29–78.

Levy, T.
1992 Transhumance, Subsistence, and Social Evolution in the Northern Negev Desert. In *Pastoralism in the Levant: Archaeological Materials in Anthropological Perspectives*, edited by O. Bar-Yosef and A. Khazanov, pp. 65–82. Prehistory Press, Madison, WI.

Liebermann, D.
1993 The Rise and Fall of Seasonal Mobility among Hunter-Gatherers: The Case of the Southern Levant. *Current Anthropology* 34:599–632.

Loewenstein, G., J. Elster, eds.
1992 *Choice Over Time*. Russell Sage Publications, New York.

Low, B.
1996 Behavioral Ecology of Conservation in Traditional Societies. *Human Nature* 7:353–379.

Low, B., J. Heinen
1993 Population, Resources and Environment: Implications of Human Behavioral Ecology for Conservation. *Population and Environment* 15:7–41.

Luckert, K. L.
1975 *The Navajo Hunter Tradition*. University of Arizona Press, Tucson.

Mace, R.
1993 Nomadic Pastoralists Adopt Subsistence Strategies That Maximise Long Term Household Survival. *Behavioural Ecology and Sociobiology* 33:329–334.

MacNeish, R. S.
1983 The Ayacucho Preceramic as a Sequence of Cultural Energy-Flow Systems. In *Prehistory of the Ayacucho Basin, Peru, Vol. 4:*

The Preceramic Way of Life, edited by R. MacNeish, R. Vierra, A. Nelken-Turner, R. Lurie, and A. Cook, pp. 236–280. University of Michigan Press, Ann Arbor.

Mason, I. L.
1984   Camels. In *Evolution of Domesticated Animals*, edited by I. L. Mason, pp. 106–115. Longman, London.

McCorriston, J., F. Hole
1991   The Ecology of Seasonal Stress and the Origins of Agriculture in the Near East. *American Anthropologist* 93:46–69.

Meadow, R.
1989   Prehistoric Wild Sheep and Sheep Domestication on the Eastern Margin of the Middle East. In *Early Animal Domestication and Its Cultural Context*, edited by P. Crabtree, D. Campana, and K. Ryan, pp. 24–36. University Museum of Archaeology and Anthropology, University of Pennsylvania, Philadelphia.

Olivera, D.
1998   Cazadores y Pastores Tempranos de la Puna Argentina. In *Past and Present in Andean Prehistory and Early History*, edited by S. Ahlgren, A. Munoz, S. Sjodin, and P. Stenborg, pp. 153–180. Etnografiska Museet, Göteborg.

Olson, M.
1965   *The Logic of Collective Action: Public Goods and the Theory of Groups*. Harvard University Press, Cambridge, MA.

Peters, R.
1983   *The Ecological Implications of Body Size*. Cambridge University Press, Cambridge.

Potts, R.
1996   *Humanity's Descent: The Consequences of Ecological Instability*. William Morrow, New York.

Price, T. D., J. A. Brown
1985   Aspects of Hunter-Gatherer Complexity. In *Prehistoric Hunter-Gatherers: The Emergence of Complexity*, edited by T. D. Price and J. A. Brown, pp. 3–20. Academic Press, New York.

Redding, R.
1981   Decision Making in Subsistence Herding of Sheep and Goats in the Middle East. Ph.D. dissertation, Anthropology Department, University of Michigan.
1984   Theoretical Determinants of a Herder's Decisions: Modeling Variation in the Sheep/Goat Ratio. In *Animals and Archaeology, Vol. 3. Early Herders and Their Flocks*, edited by J. Clutton-Brock and C. Grigson, pp. 223–241. BAR International Series, 202. British Archaeological Reports, Oxford.

Reed, C.
1984   The Beginnings of Animal Domestication. In *Evolution of Domesticated Animals*, edited by I. Mason, pp. 1–6. Longman, London.

Robertshaw, P. T., D. P. Collett
1983   The Identification of Pastoral Peoples in the Archaeological Record: An Example from East Africa. *World Archaeology* 15:67–78.

Robinson, J., K. Redford
1986   Intrinsic Rate of Natural Increase in Neotropical Forest Mammals: Relationship to Phylogeny and Diet. *Oecologia* 68:516–520.

Rogers, A.
1991   Conserving Resources for Children. *Human Nature* 2:73–82.
1994   Evolution of Time Preference by Natural Selection. *American Economic Review* 84:460–481.

Rosenberg, M.
1990   The Mother of Invention: Evolutionary Theory, Territoriality and the Origins of Agriculture. *American Anthropologist* 92:399–415.

Ross, C.
1992   Environmental Correlates of the Intrinsic Rate of Natural Increase in Primates. *Oecologia* 90:383–390.

Russell, K.
1986   Ecology and Energetics of Early Food Production in the Near East and North Africa. Ph.D. dissertation, Anthropology Department, University of Utah.
1988   *After Eden: The Behavioral Ecology of Early Food Production in the Near East and North Africa*. BAR International Series, 391. British Archaeological Reports, Oxford.

Ruttan, L., M. Borgerhoff Mulder
1999   Are East African Pastoralists Conservationists? *Current Anthropology* 40:621–652.

Ryder, M. L.
1984   Sheep. In *Evolution of Domesticated Animals*, edited by I. L. Mason, pp. 63–84. Longman, London.

Schlager, E., E. Ostrom
1992   Property Rights Regimes and Natural Resources: A Conceptual Analysis. *Land Economics* 68:249–262.

Schmidt-Nielson, K.
1984   *Scaling: Why Is Body Size So Important?* Cambridge University Press, Cambridge.

Sherratt, A.
1981   Plough and Pastoralism: Aspects of the Secondary Products Revolution. In *Pattern of the Past: Studies in Honor of David Clarke*, edited by I. Hodder, G. Isaac, and N. Hammond, pp. 261–305. Cambridge University Press, Cambridge.
1997   Climatic Cycles and Behavioural Revolutions: The Emergence of Modern Humans and the Beginning of Farming. *Antiquity* 71:271–287.

Slobodkin, L.
1961   *Growth and Regulation of Animal Populations*. Holt, Rinehart and Winston, New York.
1968   How to Be a Predator. *American Zoologist* 8:43–51.

Smith, E. A.
1983   Anthropological Applications of Optimal Foraging Theory: A Critical Review. *Current Anthropology* 24:625–651.
1987   Optimization Theory in Anthropology: Applications and Critiques. In *The Latest on the Best: Essays on Evolution and Optimality*, edited by J. Dupre, pp. 201–249. MIT Press, Cambridge, MA.
1991   *Inujjuamiut Foraging Strategies: Evolutionary Ecology of an Arctic Hunting Economy*. Aldine, New York.

Smith, E. A., B. Winterhalder
1992   Natural Selection and Decision Making: Some Fundamental Principles. In *Evolutionary Ecology and Human Behavior*, edited by E. A. Smith and B. Winterhalder, pp. 25–60. Aldine de Gruyter, New York.

Speck, F.
1915   The Basis of American Indian Ownership of Land. *Old Penn Weekly Review*, January 16:491–495.

Stein, G.
1989   Strategies of Risk Reduction in Herding and Hunting Systems of Neolithic South Anatolia. In *Animal Domestication and Its Cultural Context*, edited by P. Crabtree, D. Campagna, and

K. Ryan, pp. 87–97. University of Pennsylvania University Museum, MASCA, Philadelphia.

Stephens, D., J. Krebs
1986 *Foraging Theory*. Princeton University Press, Princeton, NJ.

Tietenberg, T.
1996 *Environmental and Resource Economics*. 4th edition. Scott-Foresman, New York.

Trivers, R.
1972 Parental Investment and Sexual Selection. *In Sexual Selection and the Descent of Man, 1871–1971*, edited by B. Campbell, pp. 136–179. Aldine, Chicago.

Uerpmann, H.
1989 Animal Exploitation and the Phasing of the Transition from the Palaeolithic to the Neolithic. In *The Walking Larder: Patterns of Domestication, Pastoralism, and Predation*, edited by J. Clutton-Brock, pp. 91–96. Unwin Hyman, London.

Watson, P. J.
1989 Early Plant Cultivation in the Eastern Woodlands of North America. In *Foraging and Farming: The Evolution of Plant Exploitation*, edited by D. Harris and G. Hillman, pp. 555–571. Unwin Hyman, London.

Wing, E.
1983 Domestication and Use of Animals in the Americas. In *Domestication, Conservation, and Use of Animal Resources*, edited by L. Peel and D. Tribe, pp. 21–38. Elsevier, New York.

Winterhalder, B.
1981 Foraging Strategies in the Boreal Forest: An Analysis of Cree Hunting and Gathering. In *Hunter-Gatherer Foraging Strategies*, edited by B. Winterhalder and E. Smith, pp. 66–98. University of Chicago Press, Chicago.

Winterhalder, B., C. Goland
1997 An Evolutionary Ecology Perspective on Diet Choice, Risk and Plant Domestication. In *People, Plants and Landscapes: Studies in Paleoethnobotany*, edited by K. Gremillion, pp. 123–160. University of Alabama Press, Tuscaloosa.

Wood, J.
1998 A Theory of Preindustrial Population Dynamics: Demography, Economy and Well-Being in Malthusian Systems. *Current Anthropology* 39:99–135.

Woodburn, J.
1982 Egalitarian Societies. *Man* 17:431–451.

Wright, G. A.
1978 Social Differentiation in the Early Natufian. In *Social Archaeology: Beyond Subsistence and Dating*, edited by C. Redman, pp. 201–223. Academic Press, New York.

Zarins, J.
1992 Pastoral Nomadism in Arabia: Ethnoarchaeology and the Archaeological Record—A Case Study. In *Pastoralism in the Levant: Archaeological Materials in Anthropological Perspectives*, edited by O. Bar-Yosef and A. Khazanov, pp. 219–240. Prehistory Press, Madison, WI.

Zeder, M.
1994 After the Revolution: Post-Neolithic Subsistence in Northern Mesopotamia. *American Anthropologist* 96:97–126.

23

Predicting Maize Agriculture among the Fremont

An Economic Comparison of Farming and Foraging in the American Southwest

K. Renee Barlow

Fremont assemblages are found throughout much of Utah (Figure 23.1). The Fremont culture region extends from Idaho south to the Colorado River, and from the Green River in northwestern Colorado to the Great Basin in eastern Nevada. Fremont sites are contemporary with archaeological assemblages in the Anasazi, Hohokam, Mogollon, Sinagua, and Patayan regions, and most sites date to between AD 700 and 1200 (Marwitt 1986; Massimino and Metcalfe 1999; Talbot and Wilde 1989). The characteristics that distinguish Fremont material culture from other Southwestern traditions include a local variety of 8- to 14-rowed dent maize, often hafted on sticks (Figure 23.2); ceramics that are usually plain grayware but sometimes decorated with appliqué, indentations or painted designs (upper left); small, regionally distinctive projectile-point types; a single-rod-and-bundle basket construction technique; large, "Utah-type" trough metates with a distinctive shelf and secondary grinding depression (lower right); ground-stone balls; leather moccasins; broad-shouldered anthropomorphic clay figurines (lower left) and rock-art figures (upper right) with elaborate headdresses, necklaces, and earrings (Adams 1994; Aikens 1966; Cutler and Blake 1970; Madsen 1989; Marwitt 1970, 1986; Morss 1931; Winter and Hogan 1986; Winter and Wylie 1974). Corn cobs dating from approximately 100 BC to AD 500 have been recovered from a few locations (e.g., Geib and Bungart 1989; Jennings 1980; Madsen 1989; Wilde et al. 1986; Winter and Wylie 1974), and there is increasing evidence that maize may have been a primary source of food in the diets of some individuals by this time (e.g., Coltrain 1996; Geib 1993; Talbot and Richens 1996; Wilde et al. 1986). However, assemblages with Formative characteristics are comparatively late in the Fremont area, well after major changes in diet and land use are indicated in Basketmaker Anasazi assemblages (cf. Matson 1991; Matson and Chisolm 1991; Smiley 1994). Apparently prehistoric inhabitants of the Fremont region did not produce assemblages with distinctive Fremont characteristics until some 600 years after maize was introduced locally (for similar arguments about the transition to maize agriculture elsewhere in the Southwest, see Minnis 1992 and Wills 1995).

Fremont architectural features include pithouses, granaries, and multiple-room surface structures constructed of mud and pole, adobe, or stone masonry. Many residential sites are located on alluvial sediments conducive to crop production. Substantial habitation and storage facilities suggest sedentism during at least some portions of the year. Evidence of irrigation features at some locations further suggests the possibility of intensive agriculture (Metcalfe and Larrabee 1985; Morss 1931; Talbot and Richens 1996). Maize remains and grayware ceramics are ubiquitous, and large trough-shaped metates (Figure 23.2, lower right) and loaf-shaped manos are often highly worn, indicating intensive processing of cereal grains (Hard et al. 1996). In addition, isotopic ratios in human skeletal remains from Backhoe Village, Evans Mound, and Caldwell Village are comparable to those from burials in the Anasazi region (Coltrain 1993; Decker and Tieszen 1989; Matson and Chisholm 1991). Clearly, maize farming was an important component of Fremont subsistence in some areas.

Excavated sites important to interpretations of Fremont lifeways (Figure 23.1) include the Bear River sites, the Great Salt Lake Wetland sites, Willard Mounds, Hogup Cave, and Grantsville near the Great Salt Lake (Aikens 1966, 1967; Judd 1926; Sharrock and Marwitt 1967; Simms 1999; Steward 1933a); Nephi Mounds, Pharo Village, Topaz Slough, the Bennett Site, Nawthis, Baker Village, Backhoe Village, and Five-finger Ridge in the Sevier drainage (Barlow 1997:18–19; Madsen 1979; Marwitt 1968; Metcalfe and Heath 1990; Shearin 2000; Simms 1986; Talbot and Wilde 1989; Wilde and Soper 1994); Paragonah, Evans Mound, and Median Village in southwestern Utah (Berry 1974; Dodd 1982; Judd 1919; Marwitt 1970); and Whiterocks

*Originally published in *American Antiquity* 67(2002):65–88

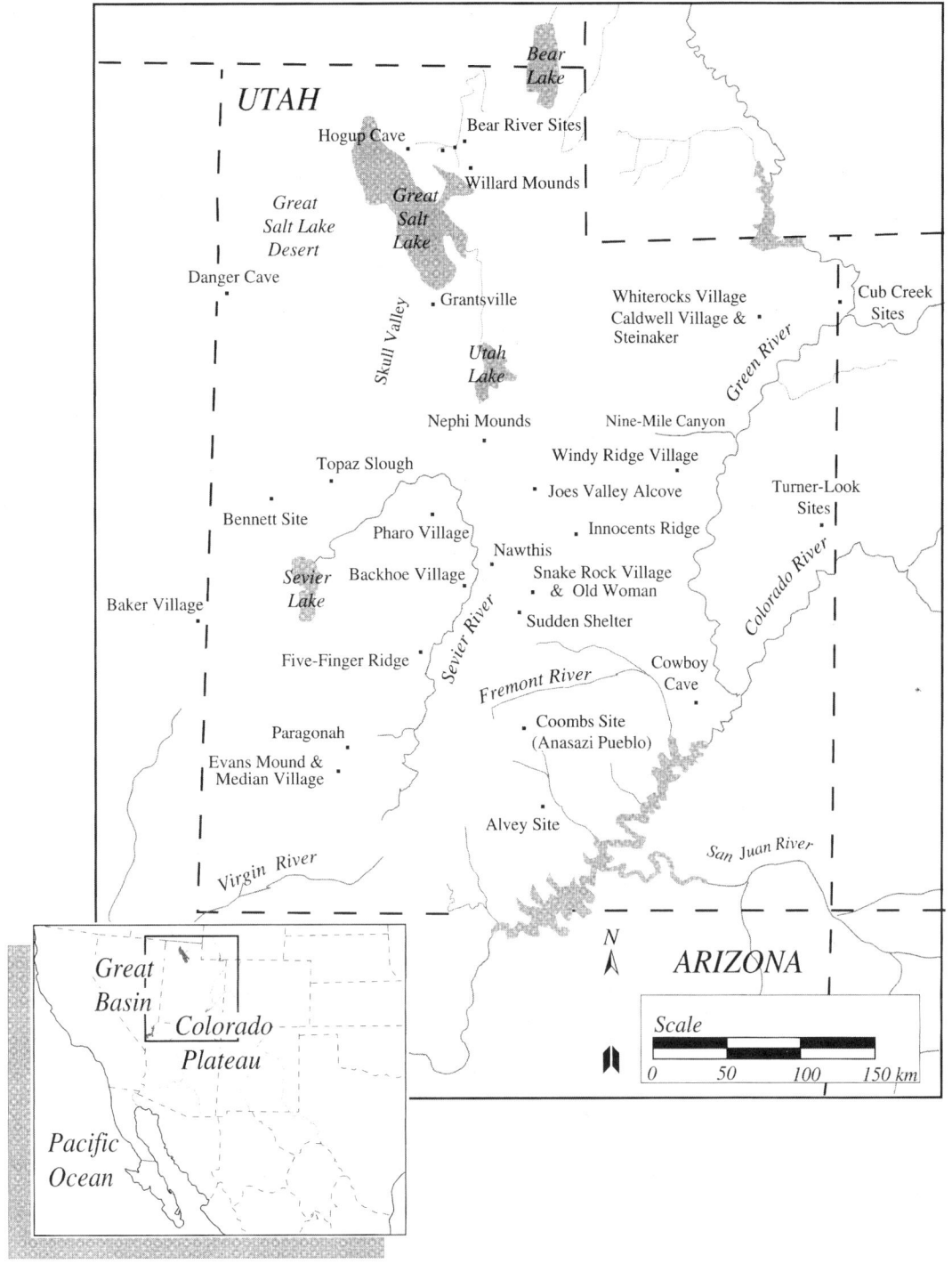

FIGURE 23.1. Map of archaeological sites in the Fremont culture area (after Marwitt 1986).

Village, Caldwell Village, Steinaker Gap, the Cub Creek sites, Nine Mile Canyon, Windy Ridge Village, Joes Valley Alcove, Innocents Ridge, the Turner-Look sites, Snake Rock Village, the Old Woman site, Triangle Cave, the Alvey site, and Cowboy Cave along the Green River drainage of the northern Colorado Plateau (Aikens 1967; Ambler 1966; DeBloois et al. 1979; Geib 1996; Jennings 1980; Madsen 1975; Miller and Matheny 1990; Schroedl and Hogan 1975; Talbot and Richens 1996; Taylor 1957; Wormington 1955). Nearly all assemblages include maize and plain gray pottery, but the frequencies of decorated Fremont wares, Utah-type metates, stone balls, figurines, and other unique Fremont artifact types vary between sites and geographical subregions, sometimes dramatically. While some sites contain dozens of pit structures or surface rooms, many others yield only rock art, the remains of an isolated granary or hearth, or a few surface artifacts. Still others are rockshelters with hearths

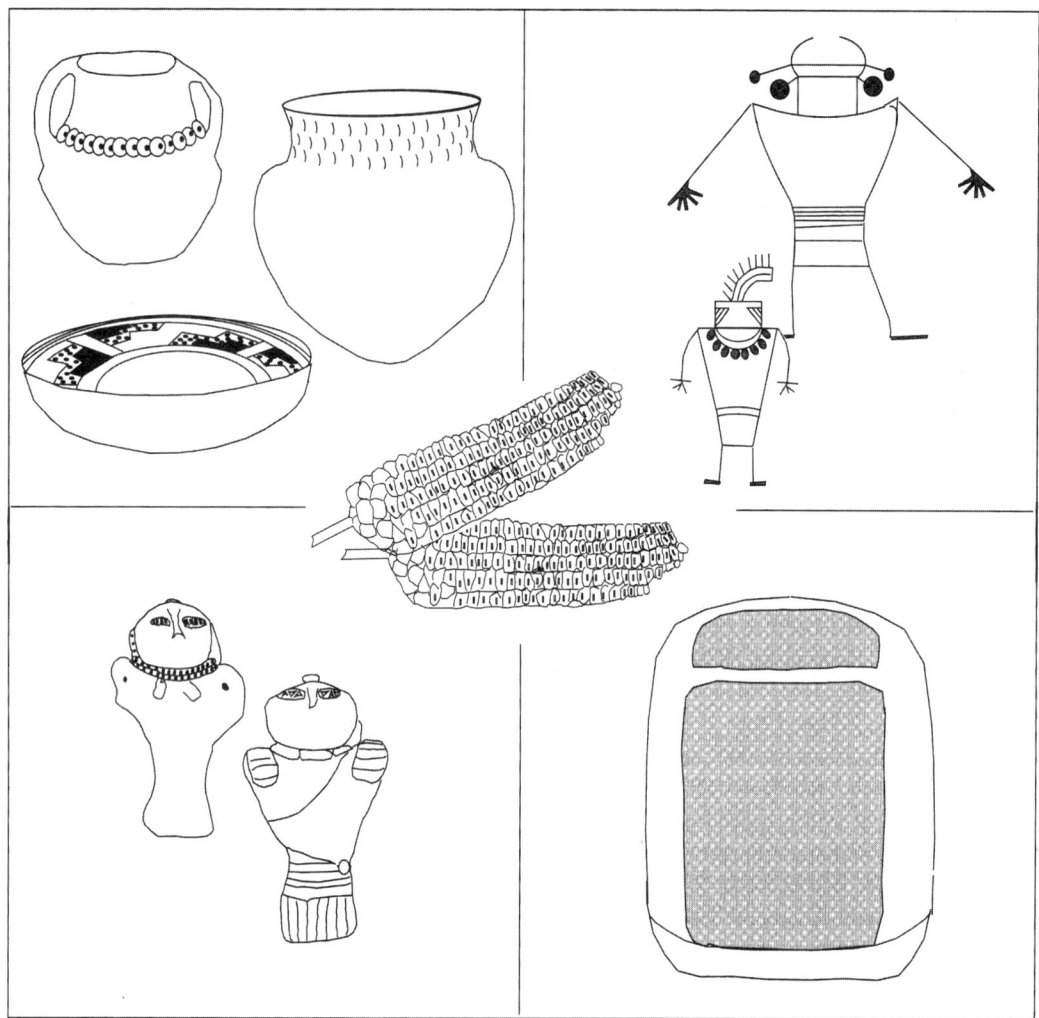

FIGURE 23.2. Fremont material culture (Judd 1926; Madsen 1989; Marwitt 1986; Morss 1931).

and storage cists or open sites that consist of the remains of highly ephemeral features similar to the wickiups used by Paiute and Shoshone-speaking Native Americans through the late 1800s (e.g., Morss 1931; Simms 1986).

The Fremont Problem

In spite of its similarities to other prehistoric Southwestern traditions, the most remarkable aspect of Fremont material culture is the diversity it presents in the importance of maize farming relative to hunting and gathering (Barlow 1997; Madsen 1989; Madsen and Simms 1998; Simms 1999). Originally designated "Puebloan" in recognition of assemblage characteristics like those in the Anasazi, Hohokam, Mogollon, and Patayan regions (Judd 1926), and long identified as the "northern periphery" of Southwest material culture (Cordell 1984; Jennings and Norbeck 1955; Steward 1933a, 1933b; Wormington 1955), Fremont was also the first regional variant named in the newly defined Southwest "Pueblo" tradition in the early 1900s (Barlow 1997). Of the disparate sites, assemblages, and culture regions now subsumed within the Fremont archaeological tradition, Noel

Morss originally proposed the name "Fremont" for Formative sites along the Fremont and Green River drainages, noting that this distinctive culture relied on both agriculture and hunting, and was much more "primitive" and mobile than other Pueblo II cultures (Morss 1931). Similarly, Jesse D. Jennings initially suggested the label "Desert-culture Horticulture" to describe what he felt was largely an Archaic lifeway in the eastern Great Basin but with the addition of maize horticulture (Jennings and Norbeck 1955). Each recognized that ecology strongly influenced the archaeological record of these foragers and farmers, and each attempted to reconstruct the roles of migration, local adaptation, and the diffusion of technology in shaping the archaeological record of the Fremont. Perhaps better known for his work at Pueblo Bonito in Chaco Canyon, Neil Judd conducted six field seasons of reconnaissance and excavation in Utah between 1915 and 1920 (Cordell 1984; Plog 1979). After investigating dozens of sites, he concluded that the material remains represented:

homes of ancient Indians who were agriculturists after a fashion, and in consequence chiefly sedentary in their habits. But these prehistoric dwellings and the cultural objects found in them are not always the same. We note changes in the type of habitation as we pass from one valley to another, and changes in the character of the utensils employed daily by the occupants of those habitations.… That culture may have originated within the drainage of the Rio San Juan, or it may have come into being on the northern and western borders of the Great Interior Basin. But it is Puebloan in fact; it is definitely and directly related to those pre-Pueblo and Pueblo cultures represented by the prehistoric ruins of northern Arizona, New Mexico and Colorado [Judd 1926:152].

In contrast with the Anasazi, Mogollon, and Hohokam regions, the primary source of inter-assemblage variability in the Fremont area is spatial rather than temporal (Barlow 1997; but see Talbot and Wilde 1989). Proposed developmental sequences are rare and are only described in a few sites after about AD 800 or 1000. These consist of the addition of corrugated or decorated ceramic wares, a shift from circular to rectangular pithouses, and the addition of coursed adobe or masonry surface structures. Such developments are broadly similar to those seen between AD 700 and 1100 throughout much of the Southwest (Marwitt 1970; compare with Cordell 1984; Kidder 1927). Most Fremont sites located in the vicinity of the Great Salt Lake (Figure 23.1) are very different from those in southern Utah, the Uintah Basin, or the Green River drainage. Areal classifications of Fremont material culture, or proposed "variants" based on geography and the spatial distribution of culture elements, are numerous, and Aikens (1966), Marwitt (1970), and Madsen (1979, 1989) have produced excellent summaries. Inter-assemblage variation is so great that local archaeologists have had difficulty consistently determining the range of sites, assemblage types, and even geographical areas to include within the definition of Fremont. In some cases, apparently contemporaneous sites within close proximity yield as much variation in assemblage composition as is common between Basketmaker and Pueblo sites in the Anasazi area, or even broadly between Fremont sites versus the Anasazi, Mogollon, or other culture regions (e.g., Ambler 1966; Barlow 1997; Madsen 1989; Marwitt 1970).

This diversity is now thought to indicate fundamental variation in prehistoric subsistence practices, particularly the importance of maize agriculture relative to hunting and collecting wild foods (Madsen 1979, 1989). Archaeologists are concerned with identifying whether prehistoric inhabitants modified their economic focus from year to year, or whether several contemporaneous strategies of diet and land use were maintained by groups occupying the region simultaneously (Berry 1974; Madsen and Simms 1998; Simms 1986, 1999; Simms et al. 1997; Zeanah and Simms 1999). In fact, "understanding adaptive diversity" has been the principal axiom guiding local studies during the last decade (Madsen and Simms 1998; Simms 1986; Zeanah and Simms 1999). In a comprehensive examination of the behavioral implications of Fremont material remains, Madsen and Simms (1998) argue that maize agriculture was never the defining core of Fremont behavioral systems. They propose that while the importance of maize differed from site to site, and was certainly a subsistence focus at some locations, "it is evident that resource choice and settlement patterns were not universally centered around farming at any time during the Fremont period" (Madsen and Simms 1998:296).

It is suggested here that the archaeological record of the Fremont simply represents a continuum of variation in the economic importance of maize and in the degree to which people inhabited central places or employed residential mobility to exploit maize fields and wild food procurement areas. The farmers and foragers who produced Fremont artifacts probably spent only a limited amount of time in maize fields during the growing season (perhaps similar to that spent in slash-and-burn agriculture in Latin America). However, local farming strategies during the 800-year Fremont period likely varied from horticultural to full time, from foragers harvesting maize from small stands scattered among wild resource procurement areas to farmers spending much of the growing season cultivating a number of different fields. Rather than viewing farming as a cultural complex or way of life that individuals either embraced or rejected, Fremont agriculture is seen here simply as a composite of economic behaviors associated with the production of food that varied in intensity and complexity from case to case. Using the concepts of human behavioral ecology, I propose that maize horticulture was highly ranked relative to most plant foods within local Archaic adaptations, but that decreasing energetic yields for hunting and gathering was the stimulus for increased investment in maize fields and increased sedentism at some sites between AD 700 and 1100. In the following analysis I examine whether spatial variation in the costs and benefits of various farming activities, coupled with differences in costs and benefits of collecting readily available indigenous plant foods, are key factors structuring intersite variability in material remains. Contrary to conventional wisdom, this analysis leads to the prediction that the abandonment of farming villages ca. AD 1300–1350 in the Fremont area was the result of increased availability of alternative, higher-ranked wild foods, and not the result of decreasing maize harvests per se.

The Benefits and Costs of Farming

An economic assessment of farming and foraging in the prehistoric Southwest requires a methodology that allows archaeologists to compare the two modes of subsistence. Anthropologists who employ optimal-foraging theory to assess diet choice among modern foragers and horticulturalists have shown that differences in the food energy gained during time spent collecting and processing various resources strongly influence subsistence patterns (e.g., Bayham 1979; Hawkes et al. 1982; Keegan 1986; Kelly 1995; O'Connell and Hawkes 1981; Simms 1987; Smith

1983; Winterhalder 1981; Winterhalder and Smith 1992). Mathematical models predict behavioral strategies that will maximize the rate of food energy gained per unit energy expended to collect and consume it.

Based on the theory of natural selection, the underlying tenet of optimal-foraging theory is that individuals who employ efficient strategies in hunting and food collection produce more offspring or more successful offspring because they gain extra resources to invest in their family, offspring, or mates. Optimal-foraging models are powerful tools for anticipating the behaviors of individuals and, surprisingly, are sometimes even more accurate in predicting the collective diets of people operating as hunting or foraging groups. The same principle is applied here to the Fremont case. Maize farming is seen simply as a series of activities that yield an expected or average return, like hunting or collecting wild seeds. Foragers are expected to invest time in a particular farming activity (e.g., planting maize, weeding or watering maize fields, or harvesting the ripened ears) rather than foraging (e.g., hunting antelope, deer, or jackrabbits, or collecting pine nuts or wild seeds) only when doing so will result in an overall increase in the net energy gained during all time spent hunting, gathering wild foods, and farming. People should spend their time in agricultural activities, pursuing and collecting available wild foods, or both, depending upon which strategy will yield the most food gained during all time invested in subsistence pursuits.

Within this behavioral framework, the time costs and energetic yields of maize farming are compared to those for collecting wild foods available in the eastern Great Basin and northern Colorado Plateau. Horticultural or agricultural strategies that yield a net increase in harvested food energy are expected to become more frequent relative to subsistence practices that are less productive.

Potential agricultural activities include various types of land preparation, the intentional sowing of indigenous or domesticated crops, weeding, fertilizing, moving or watering young plants, harvesting, transporting and storing crops, and processing them for consumption. Among modern subsistence agriculturists, these activities are often scheduled within a much larger complex of behaviors that may include hunting, collecting wild plants, manufacturing crafts, working for wages, and a variety of other important economic and social pursuits. Table 23.1 reports data detailing the average harvests gained in maize fields (bu/acre) and the labor costs associated with maize production (hr/acre) for maize fields in 29 Latin American communities. These data are from published observations of maize farming in Zinacantan, Chiapas (Cancian 1965), Panajachel, Guatemala (Tax 1963), and northwest Guatemala (Stadelman 1940), where farmers continued to cultivate maize using simple technology (Barlow 1994, 1997, 1999). Data for 106 cases are divided into three ecological settings: high elevation, or *tierra fria*; temperate zones, or *tierra templada*; and lowland fields that are often near rivers in tropical or semitropical ecological situations and referred to as "hot-country" or *tierra caliente*.

Maize Harvests

Without the aid of draft animals and animal or chemical fertilizers, maize harvests in Guatemala and southern Mexico range from approximately 10–50 bu/acre (Table 23.1). Most full-time farmers produce 12–33 bu/acre annually. Household gardens and newly cleared fields in wooded areas generally yield more abundant harvests than other field types, but the overall ecological settings of communities appear to have the greatest influence on harvests. Yields are generally highest in lowland tropical situations and river deltas, and lowest in high-elevation settings. Annual harvests of 40–50 bu/acre were reported in Santa Ana Huista and San Ramon in the *tierra caliente* and, surprisingly, also in household gardens in the *tierra fria* community of Concepcion. The poorest maize harvests were obtained in the *tierra fria* communities of San Rafael La Independencia and San Juan Ixcoy in northwest Guatemala.

Agricultural Costs

Farming costs in Table 23.1 are reported as person-hours per acre (hr/acre) in agricultural activities, and include time spent clearing land of trees, brush, or old cornstalks, sometimes burning brush or spreading composted plant material over fields and tilling it into the soil, planting maize, reseeding, cultivating, and weeding young plants, and harvesting maize. Although some fields are not tilled prior to planting, others are tilled one, two, or three times. To calculate return rates that approach those achieved by prehistoric agriculturists, only cases reporting exclusive use of hand tools (e.g., machetes, axes, hoes, and digging sticks) were used (Barlow 1997). Data for the few fields that were plowed or fertilized using draft animals were excluded from this analysis. It is important to caution that most of these estimates are based on the average number of workdays reported by farmers, not actual work times observed by anthropologists. Nevertheless, the variation is remarkable, and although all cases represent full-time, subsistence maize farmers, the amount of time invested in fields varies dramatically between communities and field types. In these cases, time spent working fields with hand tools ranges from 133–772 hr/acre annually. There is no single, average field-investment strategy that typifies maize farming.

Cost/benefit data were also calculated for five different types of maize fields located in a high-elevation valley in the Peruvian Andes (Table 23.2). Hastorf reports that these people, although not full-time maize farmers, work an average of 105 person-days per hectare in maize fields (Hastorf 1993:124–131). Harvest yields were reported in the gross kilocalories produced per hectare in "bench, irrigated valley, fertile lowland, intensive hillside and extensive hillside" fields. Conversion factors of 9 hours per person-day, 85,680 kilocalories per bushel of maize, and .4 hect-

TABLE 23.1. Annual Harvest Yields (bu/acre) and Average Field Investments (bu/acre) Associated with Maize Farming in Latin America.

COMMUNITY	WOODED		HEAVY BRUSH		SOD		GRASSLAND		CORNSTALK		GARDEN		FLOODPLAIN	
	BU/ACRE	HR/ACRE	BU/ACRE	HR/ACRE	BU/ACRE	HR/ACRE	BU/ACRE	HR/ACRE	BU/ACRE	HR/ACRE	BU/ACRE	HR/ACRE	BU/ACRE	HR/ACRE
High-Elevation Fields (*tierra fría*)														
Todos Santos C.	—		16.5	314	16.5	plow	16.5	480	16.5	314	29	335		
Santa Eulalia	29	688	16.5	386	16.5	552	12	616	16.5	386	—			
San Mateo Ixtatan	29	639	16.5	348	16.5	514	16.5	639	16.5	390	25	419		
San Rafael L. I.	—		12	315	10	475	10	558	—		25	401		
Concepcion	33	640	16.5–33	286–308	20	457	—		20	291	45	324		
San Juan Atitan			21	310	12	461	16.5	552	—		29	407		
San Sebastian C.			16.5	407	16.5	490	12	558	16.5	366	25	395	16.5	318
Santiago Ch.			16.5	247–284	16.5	205–242			12.5	230–281	—			
San Pedro S.	25	692	20	384			16.5	455	16.5	372	33	415		
San Juan Ixcoy	16.5	560	12	344			12	557	16.5	391	—			
San Martin Cu.			16.5	317					16.5	359	—			
San Miguel Acatan	33	358	16.5	358	12	377–442	16.5	441	16.5	358	33	445		
La Libertad			16.5	325			16.5	325	16.5	plow	—			
San Ildefonso Ix.			25	359–422							—			
Colotenango			16.5	372	12	730–772			16.5	plow		—		
Temperate-Zone Fields (*tierra templada*)														
San Pedro Necta	—		16.5	325	16.5	574	12	393	16.5	325	25	359		
El Quetzal	25	574	25	408	25	491	25	450	25	408				
Barillas	16.5	553	16.5	304	16.5	plow				plow				
El Injerto	25	408	25	325						plow				
Jacaltenango	22	425	16.5	251					14	248				
San Antonio Huista	33	538	33	289	16.5	428	16.5	428	25	276				
La Democracia			16.5	232	33	plow	12	159	15	229				
Cuilco	33	628	33	192	33	plow	33	545		plow				
Panajachel	20	455							20	384			30	334
Lowland-Fields (*tierra caliente*)														
Santa Ana Huista	33	416	25	313	40	454	40	plow	33	229				
Nenton			21	329	21	329	21	448	21	282				
Amelco	33	414	33	331					33	276				
San Ramon	50	348	50	265					50	265				
Zinacantan			25	143–157					16	133				
Average	28	521	22	311	19	470	18	475	20	309	30	389	23	326
St. Dev.	8	118	8	67	8	144	8	119	8	71	7	41	—	—
n	16	16	29	32	18	17	17	16	23	22	9	9	2	2

Note: Data from Cancian 1965; Stadelman 1940; Tax 1963

TABLE 23.2. Harvest Yield Data and Caloric Return Rates for Farming Maize in Peru.

	"Bench" Fields	"Irrigated Valley" Fields	"Fertile Lowland" Fields	"Intensive Hillside" Fields	"Extensive Hillside" Fields
Bushels of Corn Per Hectare	55	18	38	4	7
Bushels of Corn Per Acre	22	7	15	2	3
Caloric Return Rates for Time Spent in Fields and Processing Activities (Kcal/Hr)	1412	902	1247	307	493

Note: data from Hastorf (1993).

ares per acre were employed to convert these data to bu/acre and hr/acre for comparison with data from Guatemala and Mexico (Barlow 1997:114–116). Maize harvests in this region are generally poor, with "intensive" and "extensive" hillside fields yielding only 2–3 bu/acre, and 7 bu/acre in irrigated valley fields.

Caloric Return Rates

Table 23.3 reports energetic return rates for farming maize in terms of kilocalories per hour (from Barlow 1997). Each estimate is the food energy gained during time spent clearing, hoeing, planting, hilling, weeding and harvesting maize fields, and shelling dried maize and grinding it into meal. These rates were calculated simply by determining the food energy gained from maize per acre of cultivated land, and dividing that by the sum of hours per acre spent in each field activity during the growing season and the total time required to process the harvested maize:

Maize Farming Return Rate (kcal / hr)

$$\frac{[x \, bu/acre] \, [25.2 \, kg/bu] \, [3550 \, kcal/kg]}{\sum y \, hr/acre + [x \, bu/acre \cdot 43.55 \, hr/bu]} \quad (23.1)$$

where $x \, bu/acre$ = reported bushels of shelled, dried maize kernels harvested per acre (Table 23.1); $25.2 \, kg/bu$ = the average weight (kg) of shelled, dried maize kernels per bushel (from data in Burns 1983:59; Castetter and Bell 1942; Stadelman 1940:133); $3550 \, kcal/kg$ = the caloric value of shelled, dried maize kernels per kilogram (FAO 1953; see Barlow 1997:198 for discussion); $\sum y \, hr/acre$ = the sum of time spent in all field activities (e.g., clearing, planting, weeding, harvesting) per acre during the agricultural cycle (Table 23.1, from data reported in Barlow 1997 and Stadelman 1940); and $43.55 \, hr/bu$ = the time required to grain, pound, and grind one bushel of the harvested, dried maize into meal using stone manos and metates (from Adams 1989; Barlow 1997; Cancian 1965).

For these cases, average maize yields and field times (Tables 23.1 and 23.2) were employed for the variables x and y. However, where available, observed harvest yields and field times should be substituted. In calculating the 43.55 hr/bu post-harvest processing costs, graining costs were estimated from rates obtained by Tzotzil-speaking Mayan farmers in Chiapas, Mexico (Cancian 1965). Grinding-cost estimates were borrowed from rates achieved during maize-grinding experiments in an investigation of ground stone assemblages in the American Southwest (Adams 1989). This results in total processing costs of 1.73 hours per kilogram of harvested maize. This estimate is comparable to others calculated for processing cereals and seed foods (Barlow et al. 2000; Simms 1987; Wright 1994). If correct, it suggests that when yields ranged from 10 to 50 bu/acre, Latin American farmers using stone manos and metates spent an estimated 435–2,000 hr/acre, or 1–5 hours per day, processing maize for consumption (Barlow 1997:133; compare with Hard et al. 1996). In these cases, processing accounts for 45–90 percent of all time spent farming. This point underscores the importance of post-harvest processing costs in calculating the economic benefits of agricultural activities.

Discussion

In Latin America, maize agriculture using only simple hand tools produces a gross energetic gain of approximately 300–1,800 kcal/hr with average maize harvests of approximately 3–50 bushels per acre (Figure 23.3). Economically, this is clearly within the range of wild plant foods available to the Fremont, many of which are common in archaeological sites and were important to the Paiute and Shoshone-speaking inhabitants of the region at the time of European contact. In Figure 23.3, caloric return-rate estimates (kcal/hr) are plotted against average annual maize harvests (bu/acre). The vertical columns of points at 12, 16.5, 25, and 33 bushels per acre (bu/acre) represent cases where people reported differing amounts of time in maize fields, but the same "usual" annual maize harvests. The gray, curving lines represent sets of maize fields in Panajachel, Guatemala, Zinacantan, Chiapas, and the Upper Mantaro Valley in Peru, where one community-wide, average-work-effort estimate was reported, but variation in maize harvests was measured.

These data highlight several important aspects of the economics of maize farming and are particularly interesting in light of hypotheses concerning the spread and demise of Formative archaeological traits in the Fremont region. First, expected economic success for farming varies not only with harvest yields, but also with time invested in field activities. A strong, positive relationship between energetic returns and harvest yields would be indicated by a linear or increasing function. Instead, the observed relationship in Figure 23.3 is a *diminishing* returns curve. At best, maize agriculture in Latin America yields about 1700–1800 kcal/hr whether harvests are 10 or 50 bu/acre. The highest returns are associated with field investments of approxi-

TABLE 23.3. Caloric Return Rates for Farming Maize in Latin America Using Hand Tools.

	WOODLAND/HEAVY BRUSH	SOD	GRASS	CORNSTALK	GARDENS	FLOODPLAIN
High Elevation (*Tierra Fria*)						
Todos Santos C.	1430	plow	1232	1430	1624	
Santa Eulalia	1330–1336	1162	943	1336	—	
San Mateo Ixtatan	1364–1385	1197	1087	1331	1484	
San Rafael L. I.	1281	983	901	—	1501	
Concepcion	1421–1692	1348		1540	1763	
San Juan Atitan	1534	1091	1162	—	1554	
San Sebastian C.	1311	1221	993	1361	1508	1424
Santiago Ch.	1473–1530	1538–1599		1356–1445	—	
San Pedro S.	1256–1426		1257	1353	1594	
San Juan Ixcoy	1155–1239		994	1330	—	
San Martin Cu.	1426			1371	—	
San Miguel Acatan	1371–1645	1112–1193	1273	1371	1568	
La Libertad	1414		1414		—	
San Ildefonso Ix.	1502–1569				—	
Colotenango	1353	829–857			—	
Temperate Zone (*Tierra Templada*)						
San Pedro Necta	1414	1142	1172	1414	1545	
El Quetzal	1345–1494	1416	1453	1494		
Barillas	1161–1444					
El Injerto	1494–1582					
Jacaltenango	1424–1522			1461		
San Antonio Huista	1495–1710	1288	1288	1639		
La Democracia	1553		1575	1521		
Cuilco	1430–1812		1489			
Panajachel	1339			1416		1640
Lowlands (*Tierra Caliente*)						
Santa Ana Huista	1593–1596	1630		1772		
Nenton	1511	1511	1379	1571		
Amelco	1595–1669			1723		
San Ramon	1771–1832			1832		
Zinacantan	1801–1822			1747		
Average	1486	1242	1226	1491	1571	1532
s	167	241	205	158	85	—
n	48	18	16	22	9	2

Note: Data from Cancian 1965; Stadelman 1940; Tax 1963.

mately 100–250 hr/acre and are found at the top of the graph. Those points in the lowest portion indicate investments of approximately 400–800 hr/acre. The effects of work effort on return rates are most dramatic in regions with low-to-moderate expected harvests. For example, when production averages 12 bu/acre, farmers may experience energetic returns of 700–1650 kcal/hr, depending on how much time is invested in fields. Doubling the expected harvest to 24 bu/acre (a very high yield for subsistence farmers without the aid of chemical or animal fertilizers or plows), anticipated energetic returns increase to 1200–1800 kcal/hr. At 50 bu/acre, expected returns vary from 1770–1830 kcal/hr. Minimum expected returns increase significantly with better harvests because they are limited by high field costs, which are amortized over greater yields. However, maximum energetic gains change very little because maximum rates are constrained by high post-harvest processing costs.

Second, the effects of climate change on caloric return rates will be most dramatic in regions with very low annual maize harvests. The gray curving lines in Figure 23.3 represent changes in return rates with increasing harvests when the same field-investment strategy is maintained. Again, these are based on actual data from cases in Zinacantan, Panajachel, and the Upper Mantaro Valley. The lowest curve represents maize farming in Peru with an annual field investment of 383 hr/acre and maize harvests of 2–22 bu/acre (Table 23.2). This is a moderate field-investment strategy, with most Latin American maize farmers

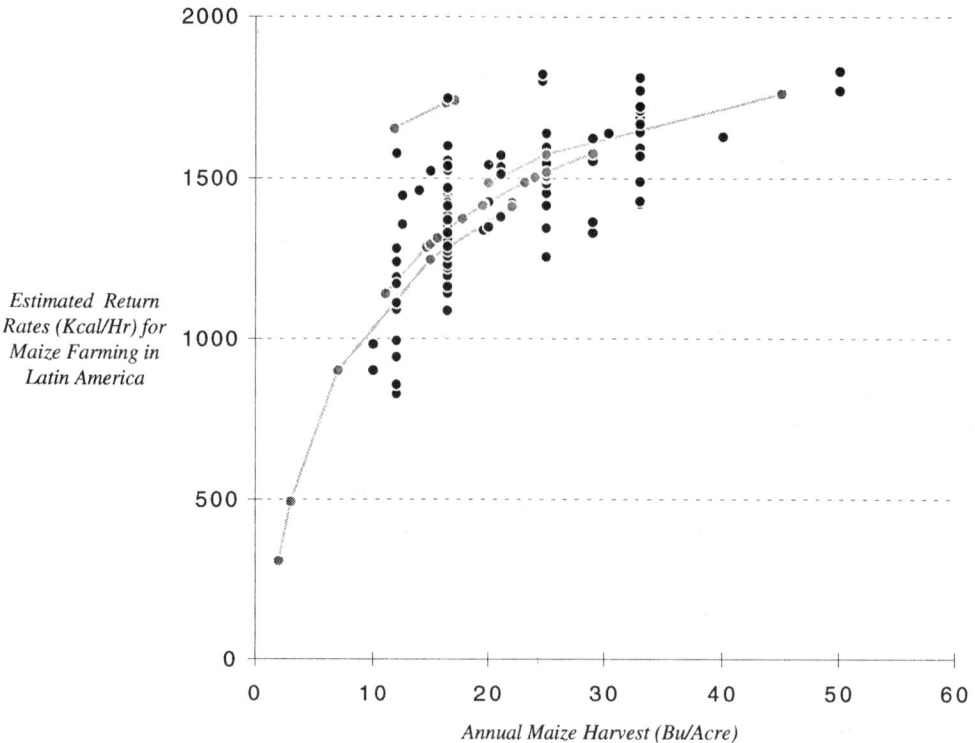

FIGURE 23.3. Caloric return rates for maize agriculture with hand tools in Latin America.

spending 200 to 600 hr/acre. The two overlapping curves near the center represent sets of fields in Panajachel with field investments of 334 and 384 hr/acre (Barlow 1997:109–110). The short gray curve near the top of the graph indicates three cases in Chiapas with annual investments of 104 hr/acre, and harvests of 12–17 bu/acre (Barlow 1997:113). This represents a low investment, "slash-and-burn" agricultural strategy. These functions suggest that farmers experience smaller and smaller marginal gains with increased harvest yields even if no additional field costs (not even time spent harvesting and husking the additional grain) were associated with higher yields (see Barlow 1997:108–117). They are particularly interesting here because they indicate maximum expected increases or decreases in return rates when extrinsic factors (e.g., differences in soils or changes in growing season precipitation or temperatures) influence expected annual harvests. For example, energetic returns could increase from about 1000–1300 kcal/hr if greater growing season precipitation or increased temperatures resulted in maize harvest increases of approximately 10–15 bu/acre. It is in these settings that spatial and temporal variation in climate or ecological setting would likely have the greatest impact on the overall success of various farming strategies.

Not surprisingly, the highest observed energetic returns in Latin America are associated with "slash-and-burn" agriculture, in which brush-covered or wooded land is cleared and/or burned (Barlow 1997; compare Boserup 1965). Maize is planted and harvested, but very little effort is invested in fields during the growing season. In low-investment agricultural strategies like these, return rates of 1650 kcal per hour may be obtained for yields as low as 12 bu/acre (Figure 23.3). This rate is comparable to the rates achieved in many fields yielding 16, 20, 30, and even 40 bu/acre.

Based on these data, I propose that prehistoric foragers and subsistence farmers in the Fremont region experienced only modest increases in economic success by investing additional time in field activities. Rather than intensifying efforts to produce an extra 2–5 bu/acre, Fremont farmers would have maximized energetic returns by investing only small-to-moderate amounts of labor in individual fields. Increasingly intensive field strategies should be associated with increasingly lower marginal returns and lower overall economic success. Potential crop yields alone are insufficient predictors of the intensity of investments in farming activities among maize agriculturists. In fact, in areas where average yields are consistently high, farmers will always do better to adopt an *extensive* strategy of planting more fields, rather than intensifying labor investments in a few fields. Consequently, hypothesized changing precipitation regimes, new varieties of maize, or the development of local farming strategies are not necessarily complete explanations of transitions to Formative lifeways in the American Southwest or the apparent abandonment of large portions of the region by full-time farmers after AD 1250. Figure 23.3 further suggests that if greater sedentism is associated with more intensive field investment, the appearance of fully "Formative" assemblage charac-

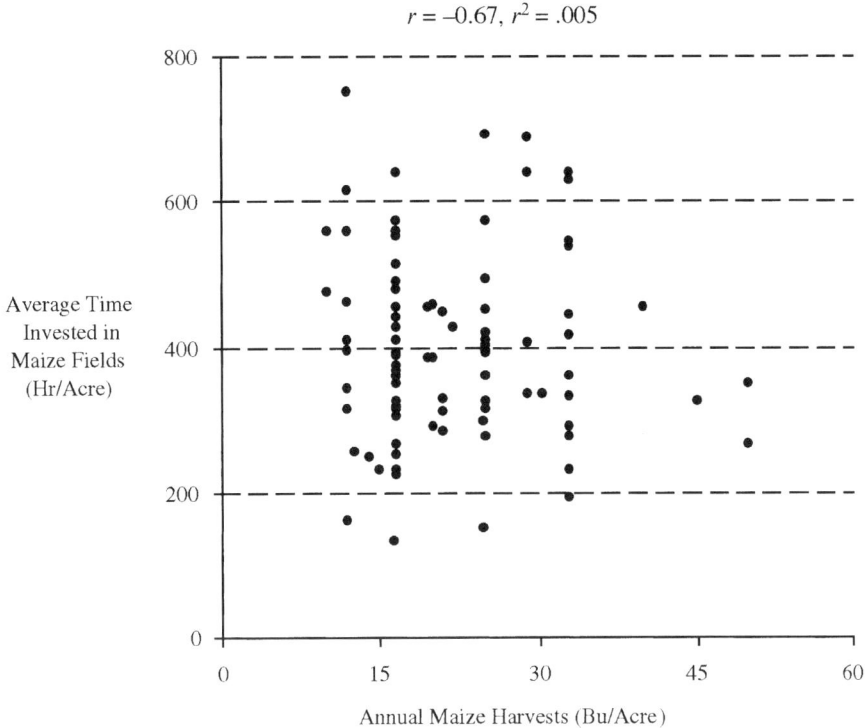

FIGURE 23.4. The relationship between time spent in maize fields and average annual maize harvests.

teristics may be indicative of greater economic stress during time spent in food procurement and production, not necessarily increasing success in farming.

Figure 23.4 illustrates the results of a preliminary test of the common archaeological hypothesis that farmers intensified maize production when greater harvests were possible. The relationship between time spent in fields (hr/acre) and harvest yields (bu/acre) among subsistence farmers in Latin America is plotted using the data presented in Table 23.3. Instead of the expected positive correlation, the points are scattered throughout the range of the graph. The data actually yield a slightly negative regression statistic of −.067, with an r^2 value of .005. Values in this range generally do not indicate a significant relationship. The most interesting information conveyed in the graph is simply that most modern subsistence maize farmers in Latin America gain between 10–40 bu/acre, and farmers generally spend between 200 and 600 hr/acre (or 22–68 workdays) in their maize fields during the growing season.

Note that the greatest time investments in Figure 23.4 are actually associated with fields that yield low-to-moderate annual maize harvests. Rather than spending more time in fields that produce fantastic amounts of maize, farmers tended to gain greater energetic returns when they minimized investments in moderate-to-rich agricultural settings. The most time appears to have been invested in locations where community averages for all fields were relatively poor, rather than in those locations that were producing "enough" maize. This suggests that increasing investment in farming is most common in communities with the lowest overall economic success.

Farming Versus Foraging

Time invested in maize agriculture should be strongly influenced both by the increases in maize yields expected with various field activities and the potential economic gains associated with local hunting-gathering opportunities. If so, associations between maize harvests and prehistoric agricultural strategies may be a great deal more complex than anticipated by simple "opportunity" or "necessity" models (see discussions in Hard 1986; Hard and Merrill 1992; Hitchcock and Ebert 1984; Minnis 1985; Wills 1995). To explore this relationship for the Fremont case, estimates of the costs and benefits of maize farming are compared to those for hunting and collecting indigenous resources available to prehistoric foragers and farmers in the Great Basin and northern Colorado Plateau. In addition to maize, the remains of many wild foods are found in archaeological assemblages throughout the Fremont region. Figure 23.5 illustrates energetic return rates for both. Most of the values for indigenous taxa are taken from post-encounter resource rankings in Simms' (1987) landmark study of the behavioral ecology of foraging in the Great Basin. However, estimates for cattail roots, pine nuts, local acorns, and Indian rice grass have been modified with supplementary data from other published studies (Barlow and Metcalfe 1996; Barlow et al. 2000; Jones and Madsen 1991; Madsen et al. 1997; see discussion in Barlow 1997:180, 200).

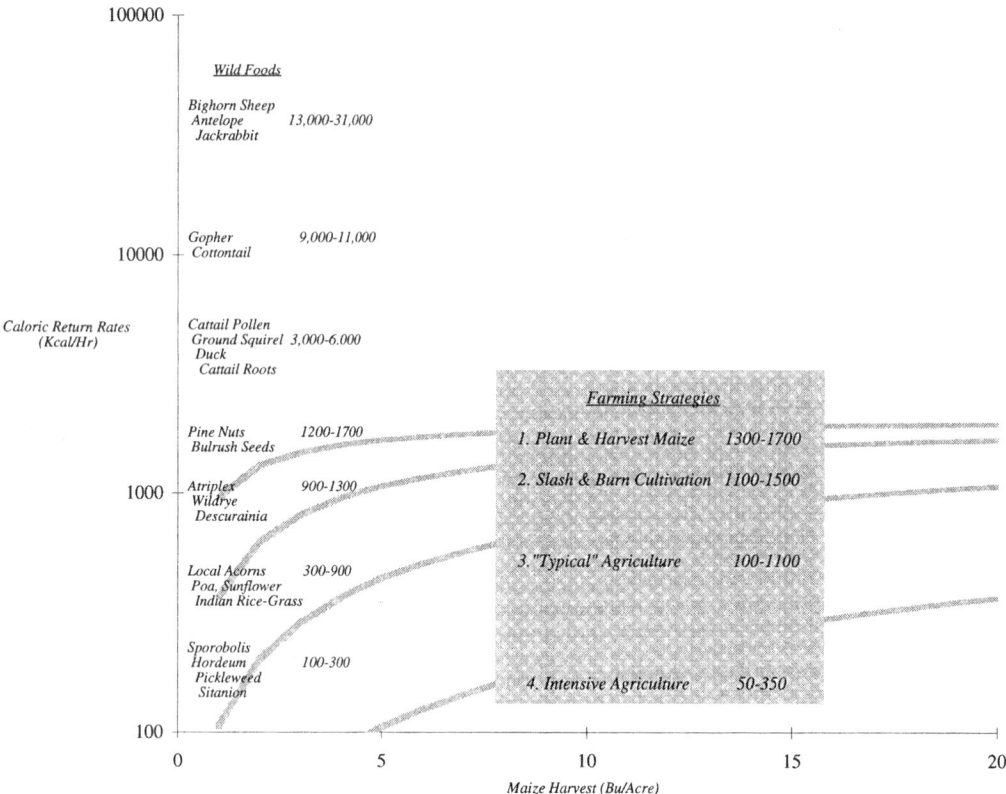

FIGURE 23.5. Comparison of caloric return rates for foraging and farming in the Great Basin and American Southwest (from data in Barlow 1997; Barlow and Metcalfe 1996; Barlow et al. 2000; Jones and Madsen 1991; Madsen et al. 1997; Simms 1987).

Wild resources are listed in rank order with those yielding the highest gross caloric return rates at the top, corresponding to the scale (Kcal/hr) on the y-axis. These return rates are based on reported caloric gains (kcal) per hour required to pursue, collect, and process each resource after it has been encountered by the forager (after Simms 1987); search times are not included. Resource rankings are entirely independent of the abundance or frequency of potential foods in the foraging environment. For ease of examination, the y-axis is scaled logarithmically, such that higher values are compressed while lower values are spread out.

The diet-breadth model predicts that the highest-ranked foods (e.g., bighorn sheep, antelope, jackrabbit) will always be pursued when encountered. With estimated post-encounter gains of 13,000–31,000 kcal/hr, lower-ranked potential foods should be ignored when local wild sheep, antelope, and jackrabbits are so dense that a forager cannot take a step without encountering one of them. Any additional resource would lower overall energetic returns. On the other hand, when these resources are scarce the forager may spend so much time searching for them that overall return rates with search time included are lower than the post-encounter returns for hunting gophers and cottontails. When overall return rates fall to 9,000–11,000 kcal/hr, the forager will increase overall energetic return rates by hunting all five potential prey items. Resources like cattail pollen, ground squirrels, and ducks should be included when local foraging rates drop to approximately 3,000–6,000 kcal/hr, and so on (Simms 1987).

Estimated return rates for maize agriculture (kcal/hr) are plotted alongside those for wild foods using four gray curves. As with the curves in Figure 23.3, these functions plot return rates with increasing maize harvests but no additional field time. Note that maize harvests on the x-axis apply only to agriculture, while wild foods are ranked independent of their own abundance. Although a continuum of variation in field investment is actually expected, the four strategies are designed to simulate varying levels of investment seen among modern foragers, horticulturists, and subsistence farmers. For this discussion, the four strategies represent the potential range of prehistoric farming tactics using digging sticks, hoes, manos, and metates:

1. Plant and harvest maize investing only 50 hr/acre, or about 5 to 6 person-days/acre;
2. Slash-and-burn agriculture investing 200 hr/acre, or 20 days/acre;
3. "Typical" subsistence agriculture with field investments of 400 hr/acre; and

4. An intensive maize agricultural strategy of 800 hr/acre that includes soil preparation before planting, tending and weeding young plants, and various field activities throughout the growing season.

Rates are calculated for harvests of 1–20 bushels per acre, and include estimated maize processing costs in addition to projected work effort in fields. The range of potential harvests was chosen based on Burns' (1983) reconstructed maize harvests in southwestern Colorado during the 700-year period from AD 650–1350. Using tree-ring data, one of Burns' data series predicted annual prehistoric maize yields of approximately 12–13 bu/acre, with a range of 6–21 bu/acre (Barlow 1997:183–190). This range actually approximates harvests in the Mesa Verde region of the Anasazi culture area. It is suggested that this range may represent maximum yields in the Fremont region, as few locations harbor the rich soils and relatively high summer precipitation that characterize the Colorado study area.

Based on these data, maize farming with simple technology is economically comparable to collecting and processing a variety of eminently storable wild seeds and nuts. In southern Mexico and Guatemala, where today maize farming continues to be a common subsistence strategy, fields with harvests of 6–20 bu/acre yield energetic returns of approximately 800–1700 kcal/hr. Modern fields in Peru with harvests of less than 6 bu/acre yield returns of about 300–500 kcal/hr. These estimates indicate that maize agriculture in Latin America generally yields much lower energetic returns than hunting game animals, but overlaps a wide range of wild plants available in the Great Basin and Southwest (see Jones and Madsen 1991; Kelly 1995; Simms 1987). In the Fremont area, high-ranked plants like pinyon pine nuts (*Pinus monophylla*) and bullrush seeds (*Scirpus*) should have become important when overall foraging returns averaged about 1,200–1,700 kcal/hr (other varieties of pinyon available in the Southwest, like *Pinus edulis* and *Pinus cemboides*, yield smaller, thicker-hulled seeds and consequently may yield lower energetic returns). A variety of local berries like wolfberry (*Lycium*), chokecherry (*Prunus*), hackberry (*Celtis*), and skunkbush (*Rhus*), and perhaps some high-ranked native root foods like yampa (*Perideridia*), bitterroot (*Lewisia*), and camas (*Camassia*) may also be included in this group, but comparable data on the costs and benefits of their exploitation are not yet available. Important local plant foods in other regions of the desert Southwest, such as mesquite, agave, and some large acorn-yielding species (Barlow et al. 2000; Diehl 1997), may also yield return rates in this range. These foods are particularly interesting in light of the caloric return-rate estimates for maize agriculture, as they overlap projected returns for time spent just planting maize with digging sticks and harvesting the ripe ears.

Many small seeds typical of historic Native American plant-food collections in the Great Basin and Southwest, and also common in archaeological sites (e.g., saltbush, wildrye, tanseymustard, blue grass, wild sunflower, and Indian rice grass and small local acorn-bearing species like *Quercus gambelii* and *Q. turbinella*), yield energetic returns comparable to "slash-and-burn" or "typical" maize investment strategies common among modern Latin American subsistence farmers. The lowest-ranked seeds (e.g., drop seed, pickleweed, squirreltail grass, and foxtail barley) are economically comparable to "intensive" maize agriculture in Latin America, but many of these were dropped from Native American subsistence practices by the early-to-mid-1900s (e.g., Barlow and Metcalfe 1996).

Initially, a maize cultivation strategy that includes minimal field investment is predicted among hunter-gatherers collecting high-ranked local plant foods like pine nuts and bullrush seeds. Greater field investments are only expected with decreases in the availability of higher-ranked wild foods, and consequently decreasing return rates for foraging efforts. Intensive maize agriculture should be associated with very broad diets, in the sense that even the lowest-ranked indigenous seeds are collected. The diet-breadth model does not speak directly to the issue of dietary importance, but predicts that all resources that are "in" the optimal diet should be pursued or collected at every encounter. In effect, foragers and farmers operating near the bottom of the chart in Figure 23.5 should work much harder than those at the top, since they spend more time pursuing opportunities that yield increasingly lower energetic returns. Conversely, low-ranked seeds and intensive farming practices should be abandoned whenever the encounter rates of higher-ranked wild foods increase, or whenever opportunities to engage in less intensive farming systems arise.

Expecting Farming among the Fremont

Southwestern archaeologists often assume either that farming yields so much food that opportunities to farm structured the timing and intensity of agriculture (e.g., Berry 1974, 1982; Minnis 1992; Van West 1994), or that farming is so "expensive" that the abundance of wild foods determined when and where maize was adopted to supplement hunting and gathering (e.g., Hard 1986; Madsen 1982; Wills 1988). The analyses here suggest that both views include elements important in understanding prehistoric transitions to agriculture. Maize fields can yield large quantities of harvestable food energy. However, a variety of alternative local wild foods yield energetic benefits that are comparable to maize farming, even in optimal conditions in Latin America. The inclusion or exclusion of maize farming should therefore not vary simply with anticipated harvests, but also with the kinds of wild foods available (Barlow 1997:159–97). In fact, there is no economic basis for the position that maize farming intrinsically provided more food or a more stable subsistence for prehistoric people than could be obtained by hunting and gathering.

1. **"Plant-and-Harvest" Strategy.** The proposed minimal "plant-and-harvest" strategy in Figure 23.5 captures the very lowest level of field investment possible with maize horticulture.

Simply planting maize using a digging stick without prior field preparation requires approximately 25 hr/acre and low estimates for harvesting maize fields by hand range from 20–28 hr/acre (Barlow 1997). Although not seen in ethnographic cases in Latin America, the hypothesized 50 hr/acre "plant-and-harvest" strategy simulates one type of maize horticulture reported in southern Utah and northern Arizona in the early 1900s (Kelly 1976) and may approximate maximum potential energetic returns for the initial adoption of maize into Archaic hunter-gatherer systems (e.g., Flannery and Reynolds 1986; Minnis 1992; Wills 1988). According to a San Juan Paiute informant, maize or "kumi" was planted in a suitable location with a stick. A small ditch was dug by hand to divert water from a nearby spring and the planted area was left until harvest time (Kelly 1976:192). A variety of wild foods were collected and stored during the summer.

Low-investment agriculture is generally associated with low absolute harvests, but estimated energetic gains of approximately 1,300–1,700 kcal/hr are anticipated with annual harvests of approximately 2–5 bu/acre. This strategy is not unlike that reported for some bushmen groups in the Kalahari desert who cultivate maize, melons, and other foods within a system of hunting-and-gathering and high residential mobility (Hitchcock and Ebert 1984), and perhaps also tropical foragers like the Hiwi of Venezuela who obtain most of their food energy from hunting and collecting wild taxa (Hurtado and Hill 1989:305). By scattering small plots of maize throughout the foraging environment, hunter-gatherers may increase overall energetic returns without substantial changes in diet, foraging practices, or settlement patterns (compare Cordell 1984; Minnis 1985).

The diet-breadth model predicts that prehistoric foragers in the Fremont region should have adopted low-investment maize horticulture whenever high-ranked plant foods like pine nuts were included in diets, or when foraging yielded average returns of about 1,300–1,700 kcal/hr. High residential mobility and the use of small maize patches in numerous locations with conditions "suitable" for maize horticulture are expected. In similar land-use systems among Southern Paiute and Kalahari horticulturists, wild foods and cultigens alike were cached in storage facilities near numerous collection areas and cultivation patches (Hitchcock and Ebert 1984; Kelly 1976). The diet-breadth model also predicts the continued use of high-ranked plants such as pine nuts, bullrush seeds, and perhaps locally available berries and root foods. Other, similarly high-ranked wild and cultivated plant foods should also be included, but many low-ranked seeds should be ignored unless overall energetic returns are low enough to warrant substantial agricultural investments. Storage structures and processing/camping facilities near maize fields and wild-food collection areas are expected, but not seasonal or long-term habitations. This expectation is consistent with recent arguments that the nascent spread of agriculture may have been rapid and difficult to track archaeologically (e.g., Smiley 1994), initially involving only a slight dietary shift with the addition of cultigens to indigenous land-use systems. It is suggested that Cowboy Cave (Jennings 1980), many Archaic cave sites in the Southwest (Smiley 1994), Formative-age Fremont sites like Topaz Slough and the Bennett Site in the west desert (Barlow 1997; Shearin 2000; Simms 1986), the remains of campsites in Skull Valley (Smith 1994), rockshelter sites with storage cists along the Fremont River (Morss 1931), and numerous sites in the Uintah Basin (Marwitt 1986) may be associated with similar levels of field investment.

2. Slash-and-Burn Strategy. "Slash-and-burn," or shifting cultivation systems, appear to yield the greatest energetic returns among full-time subsistence farmers in Latin America today. Maize farmers in Santiago, Jacaltenango, La Democracia, Cuilco, Santa Ana Huista, San Ramon, and Zinacantan employ slash-and-burn strategies in some fields, investing approximately 100–250 hr/acre for harvest yields of 12–50 bu/acre and gaining energetic returns of 1400–1800 kcal/hr (Barlow 1997). Slash-and-burn strategies are most common in tropical lowlands or ecological situations that favor maize and facilitate the growth of dense indigenous vegetation. Field investment is generally limited to land-clearing and soil preparation before planting, without substantial investments in fields during the growing season. In the Fremont region, similar field investments in rare, relatively high-yield locations might have produced harvests of approximately 5–15 bu/acre, with foragers that farm gaining energetic returns of approximately 1100–1500 kcal/hr. Temporal and spatial shifts from foraging/horticulture land-use systems to slash-and-burn agriculture are predicted with decreases in the availability of important, high-ranked plant foods like pine nuts. The collection of additional local foods like saltbush or shadscale (*Atriplex*), wildrye (*Elymus*), and tansey mustard (*Descurainia*) are expected. Sedentism during the growing season would not be necessary, but foragers are expected to invest more heavily in habitations and storage facilities near field locations. Several weeks or more of occupation are expected in the spring, with more time in autumn to harvest and store the maize, and perhaps during the winter months to consume harvested produce or transport it to a winter habitation. Local shifts from "plant-and-harvest" maize horticulture to "slash-and-burn" field-investment strategies may actually represent the most dramatic transition in diet and lifeways that occurred during the prehistoric period (compare Wills and Huckell 1994).

A number of "Formative" pithouse villages throughout the Southwest may have been associated with agricultural investment strategies similar to those used in shifting cultivation systems in tropical Latin America. Sites in the northern Tucson Basin (Roth 1992) may represent a local, early development of this kind. Farmers are not expected to have invested large amounts of time in any single field location. As foraging return rates diminished, prehistoric farmers initially would have gained higher economic returns by investing minimal efforts in additional maize fields in "suitable" locations, rather than intensifying labor in one or a few fields. Suggested sites associated with "slash-and-burn" agricultural locations include the SU site (Martin 1941; Wills 1989, 1996), the Cienega Creek sites (Diehl 1997; Huckell 1995), Basketmaker II sites on Cedar Mesa (Dohm 1994), Shabik'eschee Village (Wills and Windes 1989), and Evans Mound and Five-finger Ridge in the Fremont region (Berry 1974; Dodd 1982; Talbot and Wilde 1989). Although these clearly are substantial occupations, the model suggests their repeated use by farmers cultivating maize in nearby fields and numerous other locations, in addition to hunting and collecting wild foods. Smaller Fremont habitations like the Bear River sites, Innocents Ridge, Backhoe Village, some temporal components of Nawthis Village, and perhaps many small sites dating to later Pueblo or Classic periods in the Southwest, may also represent the locations of short-term habitations associated with locations suitable for low-investment, "slash-and-burn" agriculture and/or the collection of wild foods (e.g., Kohler 1992; Sullivan 1987).

3. **"Typical" Latin American Investment Strategy.** Greater farming investments, or intensification, are only expected with decreasing overall return rates, suggesting decreases in the availability of higher-ranked foods and better farming patches (sensu Hard 1986; Janetski 1997; Madsen 1979). The amount of time spent cultivating maize should vary with the range of alternative subsistence pursuits available during the growing season and aspects of the local ecology (e.g., soils, precipitation) that influence the marginal gains for various farming activities. Features of the archaeological record that indicate intensive farming efforts should be associated with evidence of decreases in the availability of less-costly foods, and may be accompanied by other indicators of decreasing foraging efficiency or increasing nutritional stress.

"Typical" subsistence agriculture in Latin America includes field investments of approximately 400 hr/acre. Farmers in Chiapas, Guatemala, and Peru spent 300 to 500 hr/acre in about 60 percent of the cases examined in this study (Tables 23.2 and 23.3). In reality, farmers without the aid of draft animals, irrigation systems, or chemical fertilizers generally invest 30–50 person-days per acre laboring in maize fields during the growing season. Annual maize harvests range from 2–50 bu/acre, and farmers gain energetic returns of 300–1700 kcal/hr, depending on harvest yields and level of field investment. If "typical" cultivation systems in southwestern Colorado yielded average harvests of 12–13 bu/acre and energetic returns of 700–900 kcal/hr from AD 650 to AD 1350, farming in the Fremont region likely yielded even lower returns (see discussion in Barlow 1997:183–190). Nevertheless, with typical Latin American investment levels and predicted maize harvests of 2–20 bu/acre, the model anticipates energetic returns of 300–1100 kcal/hr. This suggests that in the Fremont area there may have been little advantage for increasing field investments two-fold over slash-and-burn agriculture. If, for example, a Fremont farmer could increase average harvest yields from 12–15 bu/acre by spending 400 rather than 200 hr/acre in fields during the growing season, a gain of 3 additional bu/acre per the additional 200 hr/acre required to achieve it would yield a marginal gain of about 800 kcal/hr. Overall energetic returns for time spent farming would drop from approximately 1400 to 950 kcal/hr. Of course, the actual marginal gain would depend on the particular field activity in question (e.g., tilling the soil, irrigation, weeding) and its effect on harvest yields given local conditions for growing maize. In some areas, time spent irrigating or weeding may result in substantial increases; in others, improvements for the same activities would be much less dramatic. In general, however, increasing field investments will be indicative of decreasing foraging and farming returns, and should be associated with increasing sedentism during the growing season.

"Typical" field investments are predicted in the Fremont region only when opportunities for collecting higher-ranked foods or engaging in more profitable activities are limited. Again, nutritional stress is expected with evidence of increasing sedentism and investments in maize fields. Farmers should continue to collect pine nuts, bullrush seeds, berries, and other high-ranked plants during the growing season, but the range of species collected should increase to include locally available bluegrass (*Poa*), wild sunflower (*Helianthus*), Indian rice grass (*Oryzopsis*), and similar resources. Given the frequency of this type of strategy among modern maize farmers, it is likely that many historic and prehistoric pueblos throughout the American Southwest are associated with farming investments of approximately 400 hr/acre. Fremont sites that may be indicative of "typical" farming strategies include Nephi Mounds, Baker Village, Snake Rock Village, Caldwell Village, and the Turner-Look Sites and perhaps some occupations at Evans Mound, Nawthis Village, and Five-finger Ridge. Although adequate cost/benefit data are not yet available, it may be that indigenous species of amaranth (*Amaranthus*) and chenopod (*Chenopodium*) also yield energetic returns similar to wild sunflower, Indian rice grass, and maize farming with "typical" levels of field investment. If so, comparable estimates of energetic yields for these foods may allow archaeologists to develop inferences about larger patterns of diet, farming strategies, and overall nutritional welfare based on their presence and apparent dietary importance in Southwestern archaeological sites.

If prehistoric farmers operated under constraints similar to modern subsistence maize farmers, they may have employed several field-investment strategies simultaneously in different ecological settings (e.g., Table 23.1). Scattered granaries positioned in obscure locations or on escarpments high above canyon floors, like those in Nine-mile Canyon and along the Green River and its tributaries, may be associated with "plant-and-harvest" or "slash-and-burn" levels of investment along river channels, even if they were part of a larger system of land use that included greater labor investments and seasonal or long-term habitation structures at some locations.

4. Intensive Farming Investments. The "intensive" agricultural strategy predicted in Figure 23.5 represents the maximum level of field investment observed among modern full-time subsistence farmers. This strategy is only expected when overall returns for foraging and farming are very low. Although field investments of 500–700 hr/acre are not uncommon in high elevation, wooded fields in Latin America, investments of 700–800 hr/acre are extremely rare. Only maize farmers in the high-elevation, *tierra fria* community of Colotenango invest at this level, and then only in one type of field. Perhaps not surprisingly, average energetic returns of approximately 850 kcal/hr are among the poorest for maize farming in Guatemala. Intensive maize agriculture is only rarely expected in the Fremont region and Greater Southwest, and should be associated with use of the lowest-ranked wild food plants. To the extent that these plants were naturally abundant in local foraging environments, they may constitute a major dietary component. However, extreme nutritional stress is indicated, and intensive farming practices should be dropped from local subsistence strategies whenever energetic return rates for foraging or farming increase (sensu Boserup 1965). It may be that the complex irrigation systems at Chaco Canyon (Vivian 1990) and the "pebble-mulch gardens" of the Rio Grande basin (Lightfoot and Eddy 1995) are indicative of such levels of field labor. It is suggested here that few, if any, Fremont sites represent field-investment strategies of greater than 600 hours per acre. Sites associated with field locations requiring major land-clearing activities, like those in heavy brush or woodland settings, would be the most likely candidates.

A Gender Issue

These return-rate estimates also raise questions about the role of women in transitions to agriculture in the American Southwest. Potential male/female differences in the relative advantages of farming versus foraging have been ignored in this discussion. Instead, attention has focused exclusively on economic success as measured by average return rates and the circumstances when time spent producing food would result in more food acquired overall. However, it is a truism that men and women operating in the same foraging environment focus on different sets of resources. Although time/energy studies among the Ache of eastern Paraguay and the Alyawara of arid Australia indicate that the range of potential wild foods included in the diets of hunter-gatherers generally meet predictions of the diet-breadth model (Hawkes et al. 1982; Kaplan and Hill 1992; O'Connell and Hawkes 1981), variation in the types of resources taken by female and male foragers indicate marked gender differences in foraging goals. For the most part, the foraging behaviors of women appear to be most strongly conditioned by the economics of provisioning offspring, while men's strategies may be influenced more by potential social benefits associated with capturing large resources that are widely shared. Whereas the variance associated with hunting game animals results in the capture of one large "package" of food energy that is returned to camp and quickly distributed among all individuals, resources like berries, root foods, palm starch, and seeds can usually be collected by any individual on a daily basis (Hawkes 1990). As a consequence, women generally distribute the latter foods primarily to children and other family members.

Maize is most similar to the kinds of foods taken by women foragers worldwide. Although farming may produce a very large patch of food energy in a predictable location, the expected energetic returns for spending the time required to do so are much lower than those calculated for pursuing, capturing, skinning, gutting, and butchering animals. Even though heavy field work such as fertilizing, hoeing, tilling, and plowing with draft animals are "traditionally" male activities, women in Latin America often plant and harvest maize, sometimes engage in field labor during the growing season, and perform nearly all the necessary maize-processing tasks. To the extent that the time required to collect and process agricultural products for consumption is considered labor, prehistoric women were likely the driving force that enabled "agricultural revolutions." In most contexts in the American Southwest, the initial adoption of maize into Archaic foraging strategies may simply have been an extension of womens' everyday foraging decisions. Whether it was most often used to provision offspring or attract potential mates is beyond the scope of this discussion. The planting and harvesting of small patches of maize, like the scattering of wild seeds in the Great Basin (e.g., Steward 1938) or the cultivation of melons and maize in the Kalahari desert (Hitchcock and Ebert 1984), would likely result in resource collection, storage, and distribution tactics similar to those used for other plant foods. Consequently, it is hypothesized that initially women foragers, not men, would have gained the greatest economic advantage for adding maize and other cultigens to Archaic food collection strategies.

A Proposed Archaeological Test

In the American Southwest, the expansion of prehistoric farmers, as indicated by the frequency and spatial distribution of habitation sites with Formative assemblage characteristics, is generally thought to have been positively and strongly correlated with ecological situations that favored high harvest yields. This assumption is also prevalent in hypotheses developed to explain the increase in farmers or farming technology in the Fremont culture region, as well as spatial variation in the importance of maize and the apparent demise of Fremont material culture after AD 1350. Whether this latter transition is best accounted for by local or regional abandonments, the replacement or absorption of people that produced Fremont assemblages by immigrating, Numic-speaking foragers, or an adaptive shift back to full-time hunting and gathering, most archaeologists assume that the ultimate cause is decreasing maize yields, rather than any ecological or technological factor that might have influenced efficiency in hunting or collecting wild plant foods. However, "typical" maize agriculture in Latin America is low ranked relative to nearly all Great Basin hunting opportunities, most insect foods, and many fruits and nuts. In general, it yields energetic returns comparable to moderate- and low-ranked plant foods collected by women worldwide (e.g., Hawkes et al. 1982; O'Connell and Hawkes 1981; Simms 1987; Winterhalder 1981; cf. Kelly 1995). Furthermore, in the ethnographic cases reviewed here, there is no obvious increase in field time with greater average maize yields, and in fact, no apparent relationship exists between the two variables. Time invested in maize fields does not increase with average harvests (Figure 23.4). This result at least suggests that the amount of time farmers spend in agricultural activities is influenced by factors other than the abundance of maize at harvest time. Even in locations where harvests are relatively high, the inclusion or exclusion of maize in diets is expected to vary not simply with maize harvests, but also with the kinds of wild foods available locally and average return rates for alternative subsistence practices. Greater investment in maize farming, or "intensification," is only expected with overall *decreases* in energetic efficiency.

Fremont archaeological assemblages may provide an important test of this hypothesis. Agricultural intensification, manifested by greater investment in agricultural features and habitation and storage structures, should be associated with a broader diet (compare Diehl 1997). In locations where acorns, bluegrass, sunflower, rice grass and chenopod were abundant, such resources should be represented in summer habitations by large quantities of seeds, chaff, or other processing wastes. In addition, the logistical and intensive use of nearby patches of pine nuts, acorns, and seeds should be indicated by collection sites with processing facilities like fire hearths, ground stone, and perhaps storage architecture (e.g., Barlow and Metcalfe 1996; Barlow et al. 2000). Increasing complexity in site structure, including greater accumulations of secondary refuse, general intra-assemblage variability, and micro-refuse patterns that suggest increasingly specialized use of areas within habitations, are expected with greater lengths of time spent in fewer residences (e.g., Metcalfe and Heath 1990; O'Connell 1987; Schiffer 1972). Conversely, "slash-and-burn" maize agriculture, or low-to-moderate field-investment strategies, are expected in richer settings (sensu Madsen 1979). Foragers employing this type of farming strategy should have had greater opportunities to encounter higher-ranked wild foods, including both game animals and "preferred" or higher-ranked plants. Even though maize would have been a major part of subsistence in locations favoring high yields, diets should have been narrow in the sense that a smaller number of low-ranked taxa were included. Greater residential mobility, or a greater number of residences associated with groups of fields and wild food collection areas, are predicted. Less investment in architecture is expected at each location, and farming foragers should have been more robust and healthy than counterparts that engaged in intensive cultivation practices.

Hard et al. (1996), Simms et al. (1999), and Odell (1998) have recently suggested additional lines of archaeological evidence that may correlate with differences in agricultural investment. Hard et al. (1996) argued that inter-assemblage variation in mano size, differences in the ubiquity of maize in human coprolites and bulk soil samples, and differences in human bone chemistry (coupled with variation in the proportion of wild C_3 to C_4/CAM plant remains in macrofossil assemblages) all indicate increasing reliance on maize agriculture. A comparison of the surface area of ground stone would provide an interesting test for the Fremont case. Once again, however, it is argued here that the increased use of certain wild plant foods, particularly abundant small seeds and other foods that require extensive processing, should accompany intensified agricultural effort. Using bone chemistry, it may also be difficult to sort out the confounding issues of diet versus labor investment among subsistence agriculturalists, since some "slash-and-burn" farmers with high-yield fields in Chiapas spend very little time in maize fields during the growing season (104 hr/acre). Yet these farmers consume as much or more maize annually than those investing 600–700 hr/acre in fields in northwest Guatemala or 380 hr/acre in Peru. Other evidence from plant macrofossil assemblages, such as the ubiquity of low-ranked seeds in bulk soil samples or the number of low-ranked taxa represented in cultural contexts relative to those available in local prehistoric environments, might be an indicator of intensity in field labor.

Similarly, degree of investment in ceramic wares measured through analyses of vessel wall thickness and temper size may provide corroborative evidence about the degree of sedentism among Fremont farmers and foragers. Simms et al. (1987) have taken a first step in developing a methodology to quantify differences in sedentism by measuring the amount of effort invested in the production of pottery. To the degree that these differences correlate with investment in agricultural fields, they may aid in defining regional and temporal patterns in Fremont subsistence. Although largely untapped as a source of information about prehistoric behavior, Fremont lithic assemblages may also provide important insights about degree of investment in agriculture. Odell (1998) found decreases in the use of bifacial technology with increasing sedentism during the Holocene in the Illinois River Valley. Although not a statistically significant pattern, some documentation for increases in the kinds of tools and usewear expected with increasing agricultural efforts (e.g., hoes, hoe-sharpening flakes, chopping, and wedging usewear) was also recovered from at least one Middle Woodland site.

An auxiliary and more rigorous test of the behavioral implications of the foraging/farming model would employ ethnographic observations of modern subsistence maize farmers that utilize relatively simple farming technology and continue to collect wild plant foods (Barlow 1999). Scan surveys and focal follows of individuals engaged in field labor, as well as documentation of harvest yields and processing costs, would permit robust estimates of the costs and benefits of maize farming relative to alternative foods. By observing relationships between actual differences in maize harvests and time spent in various farming activities among subsistence farmers, archaeologists would also be in a much better position to understand the costs and benefits of various types of field-investment strategies. At a minimum, this will require quantifying variance in work effort and harvest yields of maize fields in close proximity, as well as groups of fields in differing environments. When studied within the context of modern opportunities, we can then support or falsify hypotheses about the causes of increasing or decreasing field investments with behavioral observations. Recent time/energy studies have proven the usefulness of this approach to understanding synchronic variation in diet, time spent foraging, and time spent in agricultural activities among modern foragers and horticulturists (e.g., Johnson and Behrens 1982; Keegan 1986; Lee 1968; O'Connell and Hawkes 1981; Hawkes et al. 1982; Hurtado and Hill 1989). Additional studies of this kind among small-scale subsistence agriculturists (cf. Wilken 1987:7–14) will likely provide important insights about field-investment strategies among modern foragers and farmers as well as prehistoric foragers that farmed.

Acknowledgments

I dedicate this manuscript to my good friend and mentor Duncan Metcalfe. I am also grateful to the Utah Museum of Natural History for support in printing and copying the manuscript, especially Kathy Kankainen. I wish to thank Bob Hard, Chip Wills, editor Tim Kohler, and anonymous reviewers for constructive criticism on earlier versions of this paper. As always, I thank Caprielle and Aaron for their support during many hours spent writing and at the library, for accompanying me to archaeological sites in Utah, Arizona, California, New Mexico, and Colorado, and for backpacking with me in the canyons, forests, and maize fields of the Sierra Tarahumara.

References

Adams, J. L.
1989 Experimental Replication of the Use of Ground Stone Tools. *The Kiva* 54:261–271.

Adams, K. R.
1994 A Regional Synthesis of *Zea mays* in the Prehistoric American Southwest. In *Corn and Culture in the Prehistoric New World*, edited by S. Johannessen and C. A. Hastorf, pp. 273–302. Westview Press, Boulder.

Aikens, C. M.
1966 *Fremont-Promontory-Plains Relationships in Northern Utah*. University of Utah Anthropological Papers 82. University of Utah Press, Salt Lake City.
1967 *Excavations at Snake Rock Village and the Bear River No. 2 Site*. University of Utah Anthropological Papers 87. University of Utah Press, Salt Lake City.

Ambler, J. R.
1966 Caldwell Village and Fremont Prehistory. Unpublished Ph.D. dissertation, Department of Anthropology, University of Colorado, Boulder.

Barlow, K. R.
1990 Differential Transport of Plant Parts: Experimental Data on Several Great Basin Plant Resources. Paper presented at the 22nd Great Basin Anthropological Conference, Reno.
1994 A Cost/Benefit Comparison of Farming and Foraging. Paper presented at the 24th Great Basin Anthropological Conference, Elko.
1997 *Foragers That Farm: A Behavioral Ecology Approach to the Economics of Corn Farming for the Fremont Case*. Ph.D. dissertation, University of Utah, Salt Lake City. University Microfilms, Ann Arbor.
1999 Analysis of the Costs and Benefits of Maize Farming among the Tarahumara. Proposal submitted to the National Science Foundation.

Barlow, K. R., M. Heck, J. F. O'Connell
2000 More on Acorn Eating During the Natufian: Expected Patterning in Diet and the Archaeological Record of Subsistence. In *Hunter-Gatherer Archaeobotany*, edited by S. L. R. Mason and J. G. Hather. Institute of Archaeology, London, in press.

Barlow, K. R., D. Metcalfe
1996 Plant Utility Indices: Two Great Basin Examples. *Journal of Archaeological Science* 23:351–371.

Bayham, F.
1979 Factors Influencing the Archaic Pattern of Animal Utilization. *The Kiva* 44:219–235.

Berry, M. S.
1974 The Evans Mound: Cultural Adaptation in Southwestern Utah. Unpublished M. A. thesis, Department of Anthropology, University of Utah, Salt Lake City.
1982 *Time, Space and Transitions in Anasazi Prehistory*. University of Utah Press, Salt Lake City.

Boserup, E.
1965 *The Conditions of Agricultural Growth: the Economics of Agrarian Change Under Population Pressure*. Aldine Publishing, Chicago.

Burns, B. T.
1983 *Simulated Anasazi Storage Behavior Using Crop Yields Reconstructed from Tree Rings: AD 652–1968*. Ph.D. dissertation, Department of Anthropology, University of Arizona. University Microfilms, Ann Arbor.

Cancian, F.
1965 *Economics and Prestige in a Maya Community: the Religious Cargo System in Zinacantan*. Stanford University Press, Stanford, California.

Castetter, E. F., W. H. Bell
1942 *Pima and Papago Indian Agriculture*. University of New Mexico Press, Albuquerque.

Coltrain, J. B.
1993 Fremont Corn Agriculture: A Pilot Stable Carbon Isotope Study. *Utah Archaeology* 1993:49–55.
1996 Stable Carbon Isotope Analysis. In *Steinaker Gap: An Early Fremont Farmstead*, edited by R. K. Talbot and L. D. Richens, pp. 11–122. Brigham Young University Museum of Peoples and Cultures Occasional Papers No. 2. Provo, Utah.

Cordell, L.
1984 *Prehistory of the Southwest*. Academic Press, San Diego.

Cutler, H. C., L. W. Blake
1970 Appendix I: Corn from the Median Village Site. In *Median Village and Fremont Culture Regional Variation*, edited by J. D. Jennings and N. B. Mikkelson. University of Utah Anthropological Papers 95. University of Utah Press, Salt Lake City.

DeBloois, E. I., D. F. Green, J. Wylie
1979 Joes Valley Alcove: an Archaic-Fremont Site in Central Utah. Manuscript on file, Manti-LaSal National Forest, Price.

Decker, K. W., L. L. Teizen
1989   Isotopic Reconstruction of Mesa Verde Diet from Basketmaker III to Pueblo III. *The Kiva* 55:33–47.

Diehl, M.
1997   Rational Behavior, the Adoption of Agriculture, and the Organization of Subsistence during the Late Archaic Period in the Greater Tucson Basin. In *Rediscovering Darwin: Evolutionary Theory and Archaeological Explanation*, edited by C. M. Barton and G. A. Clark, pp. 251–265. Archaeological Papers of the American Anthropological Association No. 7, Arlington, Virginia.

Dodd, W. A., Jr.
1982   *Final Year Excavations at the Evans Mound Site*. University of Utah Anthropological Papers 106, University of Utah Press, Salt Lake City.

Dohm, K. M.
1994   The Search for Anasazi Village Origins: Basketmaker II Dwelling Aggregation on Cedar Mesa. *The Kiva* 60:257–276.

Flannery, K. V., R. G. Reynolds
1986   Simulating Foraging and Early Agriculture in Oaxaca. Part VII. In *Guila Naquitz*, edited by K. V. Flannery, pp. 433–507. Academic Press, Orlando.

FAO (Food and Agriculture Organization of the United Nations)
1953   *Maize and Maize Diets: A Nutritional Survey*. FAO Nutritional Studies 9. FAO, Rome.

Geib, P. R.
1993   New Evidence for the Antiquity of Fremont Occupation in Glen Canyon, South-Central Utah. In *Proceedings of the First Biennial Conference on Research in Colorado Plateau National Parks*, edited by P. G. Rowlands, C. van Riper III, and M. K. Sogge. Transactions and Proceedings Series NPS/NRNAU/NRTP-93/10, National Park Service Natural Resources Publication Office, Denver.
1996   Early Formative Occupancy of Glen Canyon. In *Glen Canyon Revisited*, edited by P. R. Geib, pp. 78–97. University of Utah Anthropological Papers 119. University of Utah Press, Salt Lake City.

Geib, P. R., P. W. Bungart
1989   Implications of Early Bow Use in Glen Canyon. *Utah Archaeology* 1989:32–47.

Hard, R. J.
1986   Ecological Relationships Affecting the Rise of Farming Economies: A Test for the American Southwest. Unpublished Ph.D. dissertation, Department of Anthropology, University of New Mexico, Albuquerque.

Hard, R. J., W. Merrill
1992   Mobile Agriculturalists and the Emergence of Sedentism: Perspectives from Northern Mexico. *American Anthropologist* 94: 601–620.

Hard, R. J., R. P. Mauldin, G. R. Raymond
1996   Mano Size, Stable Carbon Isotope Ratios, and Macrobotanical Remains as Multiple Lines of Evidence of Maize Dependence in the American Southwest. *Journal of Archaeological Method and Theory* 3:253–318.

Hastorf, C. A.
1993   *Agriculture and the Onset of Political Inequality Before the Inka*. Cambridge University Press, Cambridge.

Hawkes, K.
1990   Why Do Men Hunt? Benefits for Risky Choices. In *Risk and Uncertainty in Tribal and Peasant Economies*, edited by E. Cashdan, pp. 145–166. Westview Press, Boulder.

Hawkes, K., K. Hill, J. F. O'Connell
1982   Why Hunters Gather: Optimal Foraging and the Ache of Eastern Paraguay. *American Ethnologist* 9:379–398.

Hitchcock, R. K., J. I. Ebert
1984   Foraging and Food Production among Kalahari Hunter/Gatherers. In *From Hunters to Farmers: The Causes and Consequences of Food Production in Africa*, edited by J. Clarke and S. Brandt, pp. 328–348. University of California Press, Berkeley.

Huckell, B. B.
1995   *Of Marshes and Maize: Preceramic Agricultural Settlements in the Cienega Valley, Southeastern Arizona*. Anthropological Papers of the University of Arizona No. 59, University of Arizona Press, Tucson.

Hurtado, A. M., K. R. Hill
1989   Seasonality in a Foraging Society: Variation in Diet, Work Effort, Fertility, and Sexual Division of Labor among the Hiwi of Venezuela. *Journal of Anthropological Research* 46:293–346.

Janetski, J. C.
1997   Fremont Hunting and Resource Intensification in the Eastern Great Basin. *Journal of Archaeological Science* 24:1075–1088.

Jennings, J. D.
1980   *Cowboy Cave*. University of Utah Anthropological Papers 104. University of Utah Press, Salt Lake City.

Jennings, J. D., E. Norbeck
1955   Great Basin Prehistory: A Review. *American Antiquity* 21:1–10.

Johnson, A., C. A. Behrens
1982   Nutritional Criteria in Machiguenga Food Production Decisions: a Linear-Programming Analysis. *Human Ecology* 10:167–189.

Jones, K. T., D. B. Madsen
1991   Further Experiments in Native Food Procurement. *Utah Archaeology* 1991:68–77.

Judd, N. M.
1919   *Archeological Investigations at Paragonah, Utah*. Smithsonian Miscellaneous Collections Volume 70, Number 3. Smithsonian Institution, Washington.
1926   *Archaeological Observations North of the Rio Colorado*. Bureau of American Ethnology Bulletin 82. Smithsonian Institution, Washington, D.C.

Kaplan, H., K. Hill
1992   The Evolutionary Ecology of Food Acquisition. In *Evolutionary Ecology and Human Behavior*, edited by E. A. Smith and B. Winterhalder, pp. 167–201. Aldine de Gruyter, New York.

Keegan, W. F.
1986   The Optimal Foraging Analysis of Horticultural Production. *American Anthropologist* 88:92–107.

Kelly, I. T.
1976   Southern Paiute Ethnography. In *Paiute Indians II*, pp. 11–224. Garland Publishing, New York.

Kelly, R. L.
1995   *The Foraging Spectrum: Diversity in Hunter-Gatherer Lifeways*. Smithsonian Institution Press, Washington, D.C.

Kidder, A. V.
1927   Southwestern Archeological Conference. *Science* 66:489–491.

Kohler, T. A.
1992   Field Houses, Villages, and the Tragedy of the Commons in the Early Northern Anasazi Southwest. *American Antiquity* 57:617–635.

Lee, R. B.
1968   What Hunters Do for a Living, or, How to Make Out on

Scarce Resources. In *Man the Hunter*, edited by R. B. Lee and I. DeVore, pp. 3–12. Aldine Publishing Company, Chicago.

Lightfoot, D. R., F. W. Eddy
1995 The Construction and Configuration of Anasazi Pebble-Mulch Gardens in the Northern Rio Grande. *American Antiquity* 60:459–470.

Madsen, D. B.
1975 *Three Fremont Sites in Emery County, Utah*. Antiquities Section Selected Papers 1:1–28, Division of State History, Salt Lake City.
1979 The Fremont and the Sevier: Defining Prehistoric Agriculturalists North of the Anasazi. *American Antiquity* 44:711–722.
1982 Get It Where the Gettin's Good: A Variable Model of Great Basin Subsistence and Settlement Based on Data from the Eastern Great Basin. In *Man and Environment in the Great Basin*, SAA Papers No. 2, edited by D. B. Madsen and J. F. O'Connell, pp. 207–226. Society for American Archaeology, Washington, D.C.
1989 *Exploring the Fremont*. Utah Museum of Natural History, Salt Lake City.

Madsen, D. B., S. R. Simms
1998 The Fremont Complex: a Behavioral Perspective. *Journal of World Prehistory* 12:255–336.

Madsen, D. B., L. Eschler, T. Eschler
1997 Winter Cattail Collecting Experiments. *Utah Archaeology* 1997:1–19.

Martin, P. E.
1941 *The SU Site, Excavation at a Mogollon Village, Western New Mexico. Second Season*. Anthropological Series, Field Museum of Natural History 32(2), Chicago.

Marwitt, J. P.
1968 *Pharo Village*. University of Utah Anthropological Papers 91. University of Utah Press, Salt Lake City.
1970 *Median Village and Fremont Culture Regional Variation*. University of Utah Anthropological Papers 95. University of Utah Press, Salt Lake City.
1986 Fremont Cultures. In *Great Basin*, edited by W. L. D'Azevedo, pp. 161–172. Handbook of North American Indians Vol. 11, W. C. Startevant, general editor, Smithsonian Institution, Washington, D.C.

Massimino, J., D. Metcalfe
1999 New Form for the Formative. *Utah Archaeology* 1999:1–16.

Matson, R. G.
1991 *The Origin of Southwestern Agriculture*. University of Arizona Press, Tucson.

Matson, R. G., B. Chisholm
1991 Basketmaker II Subsistence: Carbon Isotopes and Other Dietary Indicators from Cedar Mesa, Utah. *American Antiquity* 56:444–459.

Metcalfe, D., K. M. Heath
1990 Microrefuse and Site Structure: The Hearths and Floors of the Heartbreak Hotel. *American Antiquity* 55:781–796.

Metcalfe, D., L. V. Larrabee
1985 Archaeological Evidence for Fremont Irrigation. *Journal of California and Great Basin Anthropology* 7:244–254.

Miller, P. W., D. G. Matheny
1990 The Nine Mile Canyon Survey: Amateurs Doing Archaeology. *Utah Archaeology* 1990:122–133.

Minnis, P. E.
1985 *Social Adaptation to Food Stress: A Prehistoric Southwestern Example*. University of Chicago Press, Chicago.
1992 Earliest Plant Cultivation in the Desert Borderlands of North America. In *The Origins of Agriculture: An International Perspective*, edited by C. W. Cowan and P. J. Watson, pp. 121–141. Smithsonian Institution Press, Washington, D.C.

Morss, N.
1931 *The Ancient Culture of the Fremont River in Utah: Report on the Explorations Under the Claflin-Emerson Fund, 1928–29*. Papers of the Peabody Museum, Vol. XII, No. 3. Peabody Museum of American Archaeology and Ethnology, Harvard University, Massachusetts.

O'Connell, J. F.
1987 Alyawara Site Structure and its Archaeological Implications. *American Antiquity* 52:74–108.

O'Connell, J. F., K. Hawkes
1981 Alyawara Plant Use and Optimal Foraging Theory. In *Hunter-Gatherer Foraging Strategies: Ethnographic and Archaeological Analyses*, edited by B. Winterhalder and E. A. Smith, pp. 99–125. University of Chicago Press, Illinois.

O'Dell, G. H.
1998 Investigating Correlates of Sedentism and Domestication in Prehistoric North America. *American Antiquity* 63:553–571.

Plog, F.
1979 Prehistory: Western Anasazi. In *Southwest*, edited by A. Ortiz, pp. 108–130. Handbook of North American Indians vol. 9, W. C. Sturtevant, general editor. Smithsonian Institution, Washington D.C.

Roth, B. J.
1992 Sedentary Agriculturists or Mobile Hunter-Gatherers? Recent Evidence of the Late Archaic Occupation of the Northern Tucson Basin. *The Kiva* 57:291–314.

Schiffer, M. B.
1972 Archaeological Context and Systemic Context. *American Antiquity* 37:156–165.

Schroedl, A. R., P. F. Hogan
1975 Innocents Ridge and the San Rafael Fremont. Antiquities Section Selected Papers 1:31–66. Division of State History, Salt Lake City.

Sharrock, F. W., J. P. Marwitt
1967 *Excavations at Nephi, Utah, 1965–1966*. University of Utah Anthropological Papers 88. University of Utah, Salt Lake City.

Shearin, N.
2000 42Md1052: Final Report of Test Excavations at the Bennett site, 1993. Report submitted to the Utah Division of State History, Antiquities Section.

Simms, S. R.
1986 New Evidence for Fremont Adaptive Diversity. *Journal of California and Great Basin Anthropology* 8:204–216.
1987 *Behavioral Ecology and Hunter-Gatherer Foraging*. BAR International Series 381, Oxford.
1999 Farmers, Foragers, and Adaptive Diversity: the Great Salt Lake Wetlands Project. In *Prehistoric Lifeways in the Great Basin Wetlands: Bioarchaeological Reconstruction and Interpretation*, edited by B. Hemphill, B. Larsen and C. Larsen, pp. 22–47. University of Utah Press, Salt Lake City.

Simms, S. R., A. Ugan, J. R. Bright
1997 Plain-Ware Ceramics and Residential Mobility: A Case Study from the Great Basin. *Journal of Archaeological Science* 24:779–792.

Smiley, F. E.
1994 The Agricultural Transition in the Northern Southwest: Patterns in the Current Chronometric Data. *The Kiva* 60:165–202.

Smith, E. A.
1983 Anthropological Applications of Optimal Foraging Theory: A Critical Review. *Current Anthropology* 24:625–651.

Smith, S. J.
1994 Fremont Settlement and Subsistence Practices in Skull Valley, Northeastern Utah. *Utah Archaeology* 1994:51–68.

Stadelman, R.
1940 *Maize Cultivation in Northwestern Guatemala*. Contributions to American Anthropology and History, vol. 6, Carnegie Institution of Washington, Washington, D.C.

Steward, J. H.
1933a *Early Inhabitants of Western Utah: Part I—Mounds and House Types*. Bulletin of the University of Utah, Volume 23, Number 7. University of Utah Press, Salt Lake City.
1933b *Archaeological Problems of the Northern Periphery of the Southwest*. Museum of Northern Arizona Bulletin 5. Northern Arizona Society of Science and Art, Flagstaff.
1938 *Basin-Plateau Aboriginal Sociopolitical Groups*. Smithsonian Institution Bureau of American Ethnology Bulletin 120. Washington, D.C.

Sullivan, A. P.
1987 Seeds of Discontent: Implications of a "Pompeii" Archaeobotanical Assemblage for Grand Canyon Anasazi Subsistence Models. *Journal of Ethnobiology* 7:137–153.

Talbot, R. K., L. D. Richens (editors)
1996 *Steinaker Gap: An Early Fremont Farmstead*. Brigham Young University Museum of Peoples and Cultures Occasional Papers No. 2. Provo, Utah.

Talbot, R. K., J. D. Wilde
1989 Giving Form to the Formative: Shifting Settlement Patterns in the Eastern Great Basin and Northern Colorado Plateau. *Utah Archaeology* 1989:3–18.

Taylor, D. C.
1957 *Two Fremont Sites and Their Position in Southwestern Prehistory*. University of Utah Anthropological Papers 29. University of Utah Press, Salt Lake City.

Tax, S.
1963 *Penny Capitalism: A Guatemalan Indian Economy*. University of Chicago Press, Chicago.

Van West, C. R.
1994 *Modeling Prehistoric Agricultural Productivity in Southwestern Colorado: A GIS Approach*. Washington State University Department of Anthropology Reports of Investigations 67. Washington State University Department of Anthropology, Pullman, and Crow Canyon Archaeological Center, Cortez.

Vivian, R. G.
1990 *The Chacoan Prehistory of the San Juan Basin*. Academic Press, San Diego.

Wilde, J. D., R. A. Soper
1994 *Baker Village: A Preliminary Report on the 1991, 1992, and 1993 Archaeological Field Seasons at 26WP63, White Pine County, Nevada*. Brigham Young University Museum of Peoples and Cultures Technical Series No. 94-11, Provo, Utah.

Wilde, J. D., D. E. Newman, A. E. Godfrey
1986 *The Late Archaic/Early Formative Transition in Central Utah: Pre-Fremont Corn from the Elsinore Burial Site 42Sv2111, Sevier County, Utah*. Brigham Young University Museum of Peoples and Cultures Technical Report No. 86-20. Provo, Utah.

Wilken, G. C.
1987 *Good Farmers: Traditional Agricultural Resource Management in Mexico and Central America*. University of California Press, Berkeley.

Wills, W. H.
1988 *Early Prehistoric Agriculture in the American Southwest*. School of American Research Press, Santa Fe.
1989 Patterns of Prehistoric Food Production in West-Central New Mexico. *Journal of Anthropological Research* 45:139–157.
1995 Archaic Foraging and the Beginning of Food Production in the American Southwest. In *Last Hunters-First Farmers: New Perspectives on the Prehistoric Transition to Agriculture*, edited by T. D. Price and A. B. Gebauer, pp. 215–242. School of American Research Press, Santa Fe.
1996 The Transition from the Preceramic to Ceramic Period in the Mogollon Highlands of Western New Mexico. *Journal of Field Archaeology* 23:335–358.

Wills, W. H., B. B. Huckell
1994 Economic Implications of Changing Land-Use Patterns in the Late Archaic. In *Themes in Southwest Prehistory*, edited by G. J. Gumerman, pp. 3–52. School of American Research Press, Santa Fe.

Wills, W. H., T. C. Windes
1989 Evidence for Population Aggregation and Dispersal during the Basketmaker III Period in Chaco Canyon, New Mexico. *American Antiquity* 54:347–369.

Winter, J. C., P. F. Hogan
1986 Plant Husbandry in the Great Basin and Adjacent Northern Colorado Plateau. In *Anthropology of the Desert: Essays in Honor of Jesse D. Jennings*, edited by C. J. Condie and D. D. Fowler, pp. 117–144. University of Utah Anthropological Papers 110. University of Utah Press, Salt Lake City.

Winter, J. C., H. G. Wylie
1974 Paleoecology and Diet at Clydes Cavern. *American Antiquity* 39:303–315.

Winterhalder, B.
1981 Foraging Strategies in the Boreal Environment: An Analysis of Cree Hunting and Gathering. In *Hunter-Gatherer Foraging Strategies,* edited by B. Winterhalder and E. A. Smith, pp. 66–98. University of Chicago, Chicago.

Winterhalder, B., E. A. Smith
1992 Evolutionary Ecology and the Social Sciences. In *Evolutionary Ecology and Human Behavior,* edited by E. A. Smith and B. Winterhalder, pp. 3–23. Aldine De Gruyter, New York.

Wormington, M.
1955 *A Reappraisal of the Fremont Culture*. Proceedings of the Denver Museum of Natural History 1. Denver Museum of Natural History, Denver.

Wright, K. I.
1994 Ground-Stone Tools and Hunter-Gatherer Subsistence in Southwest Asia: Implications for the Transition to Farming. *American Antiquity* 59:238–263.

Zeanah, D. W., S. R. Simms
1999 Modeling the Gastric: Great Basin Subsistence Studies since 1982 and the Evolution of General Theory. In *Models for the Millennium: Great Basin Anthropology Today*, edited by C. Beck, pp. 118–140. University of Utah Press, Salt Lake City.

PART V

Cooperation and Competition in Complex Societies

In the latter half of the Holocene, complex societies characterized by high degrees of economic specialization and interdependence, social stratification, and centralized authority began to develop in many parts of the world. Like the origins of agriculture, the development of complex societies has long been a major research focus of archaeologists. Unlike the origins of agriculture, however, issues related to the development and maintenance of social complexity have only begun to be addressed by archaeologists employing the EE approach, and the number of empirical applications in this area is therefore comparatively small.

Two such applications are reprinted in *Evolutionary Ecology and Archaeology*, and these illustrate the kinds of issues related to social complexity that are amenable to analysis using EE theory, particularly issues involving cooperation and competition. We note that other empirical applications do exist (e.g., Fitzhugh 2003; Kantner 1996; Kennett 2005), as does important theoretical work (e.g., Boone 1992, 1998, 2000; Boone and Kessler 1999). Given the limited amount of EE-based research into complex societies that has occurred to date, this is perhaps the area in which the potential for the approach to make important new contributions is the greatest.

The paper by T. A. Kohler and C. R. Van West that is reprinted here develops a model to predict the appearance of food-sharing networks, and the authors apply this model to archaeological data from the Anasazi region of the American Southwest. While the Anasazi societies that they consider are not "complex societies" in the usual archaeological usage of the term, they do exhibit some of the important features of social complexity, and the issue on which Kohler and Van West focus—the development of exchange networks—is very relevant to understanding the evolution of complex societies. To explore the development of food-sharing networks, Kohler and Van West use a model of risk-sensitive foraging (also see Elston and Brantingham 2003 and Gremillion 1996, both reprinted here), which considers how the marginal value of food varies among producers and consumers under different sets of environmental conditions. They use this model to predict that the sharing of food among households should occur when and where mean productivity is high but variability in productivity from year to year and across space is also high.

Kohler and Van West test this prediction against the archaeological record of the northern Anasazi region by using tree-ring data to identify periods when their model predicts that food-sharing networks should be either present or absent, and by then considering how common various archaeological indicators of food-sharing are during those periods. The results of this analysis generally agree with their prediction, as evidence of community- or regional-scale integration is most apparent during periods of high mean productivity and high interannual and spatial variability. In addition, however, they find that variability in human population size apparently also drove food-sharing in a way not accounted for by their model. This leads them to consider ways in which such variability in population size may have altered the utility function, the foundation upon which their risk-sensitivity model—and the decisions made by individuals in the region—is based.

Kohler and Van West's use of EE provides an explanation for many of the most prominent characteristics of prehistoric Puebloan societies in the Southwest, such as the presence of aggregated communities with public architecture. They argue that these characteristics emerged as a result of individual households making decisions in accordance with their own self-interest about whether or not to participate in food-sharing networks.

Whereas Kohler and Van West focus on issues of cooperation that are central to the development of social complexity, the paper by F. Neiman considers counteracting processes of competition that are also very relevant. The specific empirical focus of Neiman's paper is the so-called Maya "collapse," which has long been a major topic of research in Mesoamerican archaeology. The main source of evidence for a Maya "collapse" has always been the cessation of monument construction in the Maya lowlands, and Neiman begins his paper by developing a compelling explanation for the construction of monumental architecture—traditionally used by archaeologists as a key indicator of social complexity—based in principles of costly signaling theory. This body of theory was originally developed by Zahavi (e.g., Zahavi and Zahavi 1997), and it has been recast in the mathematical form that characterizes the EE approach by Grafen (1990a, 1990b). Costly signaling theory has seen extensive ethnographic application (see overview in Hawkes and Bliege Bird 2002), as well as some additional archaeological use (e.g., McGuire and Hildebrandt 2005).

Neiman proposes that Maya monuments, built by competing elites, served as trustworthy signals of the competitive ability of their builders. Such signals benefit both the signaler and the receiver—and thus constitute an evolutionarily stable strategy—by enabling them to avoid more costly physical confrontations. However, for this strategy to be stable, the signals must provide an accurate representation of the competitive ability of the signaler, and they must therefore require payment of a cost that inferior competitors cannot afford. Neiman argues that Maya monuments exhibit design features that are consistent with the idea that they served as such costly signals: for example, they were truly costly and impossible to counterfeit.

The statistical analysis of terminal monument dates that Neiman presents shows a trend in the cessation of monument construction that began across a swath of the central Maya lowlands and progressed outwards from there to the northeast and southwest. Neiman proposes that this trend is most consistent with an "ecological disaster" hypothesis, and he suggests that such a disaster, driven by soil erosion, led to a decline in the number of competing elites and in the range of competitive ability among them. In turn, he argues, this led to a decline in the intensity of costly signaling and the cessation of monument construction. Neiman also derives conclusions about the movement of individuals away from centers affected by declining ecological conditions and about the average size of Maya polities; these conclusions gain from insights provided by costly signaling theory just as his analysis of the trend in terminal monument dates does. Given this, it is appropriate that Neiman concludes his paper, and this volume, by discussing the benefits to the scientific process that derive from the use of explicit theory of the sort that EE models provide.

As noted above, applications of EE to issues of social complexity are in their infancy, and we anticipate much progress in this area in the future. The papers by Kohler and Van West and by Neiman that are reprinted here provide examples of the kinds of issues relevant to the study of complex societies that can be addressed with the EE approach. The substantive points about world prehistory that can be derived from the papers in this section include:

1. The presence or absence of exchange networks can be understood in terms of self-interested action by individuals in general, and in terms of sensitivity to risk in particular.
2. Exchange systems in the prehistoric Puebloan Southwest, evidenced by such features of the archaeological record as road systems and public architecture, may have developed and disintegrated, respectively, in response to environmental conditions that alternately favored risk-averse and risk-prone behavior.
3. The construction of monumental architecture—a traditional archaeological hallmark of complex societies—can be understood as costly signaling by competing elites.
4. The cessation of monument construction that marks the Maya collapse was likely the result of environmental degradation, which removed the conditions that favored intense costly signaling.

References

Boone, J. L.
1992   Competition, Conflict, and Development of Social Hierarchies. In *Evolutionary Ecology and Human Behavior*, edited by E. A. Smith and B. Winterhalder, pp. 301–338. Aldine de Gruyter, New York.
1998   The Evolution of Magnanimity: When is it Better to Give than to Receive? *Human Nature* 9: 1–21.
2000   Status Signaling, Social Power, and Lineage Survival. In *Hierarchies in Action: Cui Bono?*, edited by M. W. Diehl, pp. 267–290. Center for Archaeological Investigations, Occasional Papers 27. Southern Illinois University, Carbondale.

Boone, J. L., K. L. Kessler
1999   More Status or More Children?: Social Status, Fertility Reduction, and Long-Term Fitness. *Evolution and Human Behavior* 20:257–277.

Fitzhugh, B.
2003   *The Evolution of Complex Hunter-Gatherers: Archaeological Evidence from the North Pacific*. Kluwer-Plenum, New York.

Grafen, A.
1990a   Sexual Selection Unhandicapped by the Fisher Process. *Journal of Theoretical Biology* 144:473–516.
1990b   Biological Signals as Handicaps. *Journal of Theoretical Biology* 144:517–546.

Hawkes, K., R. Bliege Bird
2002   Showing Off, Handicap Signaling, and the Evolution of Men's Work. *Evolutionary Anthropology* 11:58–67.

Kantner, J.
1996   Political Competition among the Chaco Anasazi of the American Southwest. *Journal of Anthropological Archaeology* 15:41–105.

Kennett, D. J.
2005   *The Island Chumash: Behavioral Ecology of a Maritime Society*. University of California Press, Berkeley.

McGuire, K. R., W. R. Hildebrandt
2005   Re-Thinking Great Basin Foragers: Prestige Hunting and Costly Signaling during the Middle Archaic Period. *American Antiquity* 70:695–712.

Zahavi A., A. Zahavi
1997   *The Handicap Principle: A Missing Piece of Darwin's Puzzle*. Oxford University Press, Oxford.

24

The Calculus of Self-Interest in the Development of Cooperation

Sociopolitical Development and Risk among the Northern Anasazi

Timothy A. Kohler and Carla R. Van West

INTRODUCTION

One clear trend that can be discerned in the last 15 years of otherwise-rather-protean Southwestern archaeology is a growing recognition that at any given time, demographic, productive, and organizational strategies were quite variable, even within comparatively small regions. Within a general trend of increasing economic intensification, strategies and tactics of production (Leonard and Reed 1993) could change quickly and in ways that are not, apparently, easily predictable. Perhaps it is part of the process of normal science to begin by emphasizing the most obvious patterns in time and space—as exemplified for the Anasazi by the Pecos classification—and only then, to focus on understanding the reasons for the variability that remains, unexplained and initially even undescribed, around those modes. In any case, we have moved from an almost complete focus on mean, or normative, behavior to a greater concern with variability around those norms.

Part of this process of elaborating normative thought has been to reexamine the archaeological record for information on how climatic or environmental *variability* (as opposed to average conditions) has affected prehistoric behavior. Of the great deal of work done in this area, that of Braun and Plog (1982) is perhaps the most widely cited. Their central thesis is that "*sustained increases in the intensity of integration will occur as concomitants of sustained increases in region-wide uncertainty or risks arising from environmental change*"[1] (emphasis in original). Their focus is on *regional* connectedness (presumably including interaction and cooperation), measured especially through stylistic traditions; environmental risks are measured primarily through the proxy of increasing local populations, which are held to increase risk through restricting mobility after a certain level of population is reached. In a similar vein, Upham (1984) has argued that *food exchange* among large late prehistoric sites in *local* clusters is related to the management of production risk in high-risk/high-diversity environments. Implicit in such a model is that food exchange should increase in times of increasing risk (we will be using risk here to mean variability in agricultural production, as measured by the standard deviation, rather than as the probability of falling below some acceptable level of income). In general these approaches assume some sort of positive linear relationship between increasing variability and increasing "integration," usually making a further connection between increasing connectedness, or integration, and increasing opportunities for emergence of hierarchical sociopolitical systems.

The present chapter builds on this earlier work to formulate a new framework for analyzing and predicting the behavior of producers in various contexts of productive shortage and abundance. We begin by assuming that nearly all agricultural production among the northern Anasazi was at the household level. By drawing on microeconomic theory, evolutionary ecology, and recent studies in the behavioral ecology of foragers, we then develop a context to predict when networks of food exchange among households ought to appear, based solely on the self-interest of the households involved. Data on production variability in the Mesa Verde Region between AD 901 and 1300—recently developed by Van West (1994)—are examined in this context, and these sections draw heavily on Van West and Kohler (1995). We then try to argue that networks of food sharing provide a foundation for more general elaboration of sociopolitical complexity in cases where production provides a surplus. In the final sections of this chapter we examine the implications of this work and point to some of the fruitful avenues for research that it suggests.

---

*Originally published in *Evolving Complexity and Environmental Risk in the Prehistoric Southwest*, edited by J. A. Tainter and B. B. Tainter, pp. 169–196, Addison-Wesley, Reading MA (1996).

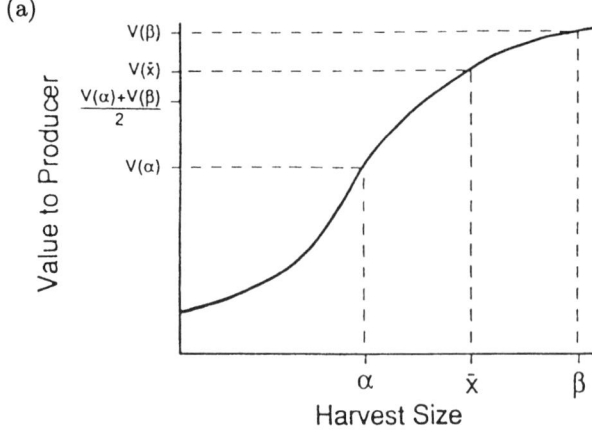

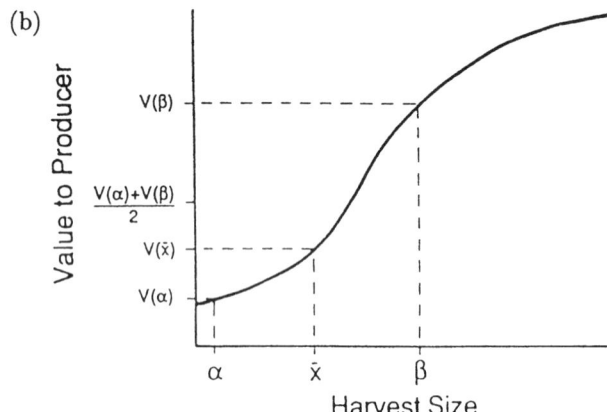

FIGURE 24.1. (a) Good-year economics. Units of production on the x-axis; units of value to the producer of a certain level of harvest on the y-axis. The variability notated by α and β could represent different plots in the same year, or different years in a series of generally good years. V(x̃) represents the average value realized after pooling; [V(α) + V(β)]/2 represents the average value realized by not pooling. After Smith (1988). (b) Bad year economics. In contrast to good years, in bad years V(x̃) (the expected value of pooling) is less than [V(α) + V(β)]/2 (the expected value of not pooling).

Utility Functions and Risk Sensitivity

As archaeologists have been searching for such linear relationships between risk and behavior over the last 12 years, studies of animal foraging systems during the same period have moved in different directions. An exemplary early study (Caraco et al. 1980) drew on utility theory to explain why juncos sometimes prefer a risky foraging strategy (involving a variable number of seeds), and sometimes a risk-free foraging strategy (involving a fixed number of seeds), when the long-term payoffs from both were the same. Utility theory was developed in economics in the mid- to late nineteenth century, although Daniel Bernoulli, in the 1730s, was already postulating that the value of an additional dollar to a person was inversely proportional to his current wealth (Rima 1972:182).

The concept of marginal utility expresses the subjective value or want-satisfying power of an additional unit of a given good to a particular user. The importance an individual attaches to an additional unit of a particular good depends in part on its relative scarcity. The larger the supply of a given commodity, the smaller will be its relative significance at the margin (Rima 1972:182–183).

A *utility function* relates some amount of goods (on its abscissa, or x-axis) to some psychological utility or value for those goods to a consumer (on the ordinate, or y-axis). Caraco et al. (1980) noticed that when birds were deprived of food, they were more likely to choose a variable reward, whereas they favored a fixed reward when their energy budget was positive. Deprivation (or satiety), they concluded, changed the shape of the utility function that guided junco foraging "decisions" so that it was not linear in form (see also Krebs and Kacelnik 1991; Stephens 1981). Decisions that systematically discriminate between different strategies that have the same expected (mean) value but differing degrees of variance are said to be risk-sensitive (Kaplan et al. 1990; Stephens 1990).

Consider the utility function graphed in Figure 24.1(a). It represents one possible relationship between total output of a foraging or production system as measured in units such as bushels or kilocalories (on the x-axis) and the *value* or *utility* of that production to the producer (on the y-axis). (An equivalent analysis, in the language of behavioral ecology, could be made by considering fitness outcomes rather than utility on the y-axis.) If the relationship were a straight line, then the difference in value between, say, 10 and 20 bushels of corn would be the same as that between 30 and 40. There is, however, reason to expect that this relationship is often of the general sigmoid form graphed here (see Smith 1988:236; Smith and Boyd 1990: 170–172). By this we do not suggest that we can define a specific curve for this (or any) archaeological context, but our conclusions will strengthen the conjecture that the appropriate curve belongs to the general, sigmoid-shaped family of functions. In such a relationship, the value of marginal production increases more quickly than production itself when few units are being produced, but more slowly when many units are being produced. So the difference in value between 10 and 20 bushels of corn may be either more—or less—than the difference between 30 and 40 bushels, depending on how the inflections on the utility function interact with these production levels on the x-axis. Slightly more formally, the sigmoid curve graphing total utility starts off with zero slope, has an increasing first derivative (which is the marginal utility) until the second derivative reaches zero, and then a diminishing marginal utility as the first derivative asymptotically approaches zero. One important emphasis of this chapter is that the household decision to share, or to hoard, should be affected by whether household production is above, or below, the major inflection point where the second derivative equals zero, beyond which marginal utility begins to decrease.

In the lower left-hand portion, or upward-concave portion, of the total utility curve, then, the marginal utility of another bushel is always greater than the last. The flat, initial segment of the curve may be thought of as indicating the presence of an "outside option"; under extremely adverse conditions—below some threshold required to meet reciprocal obligations or avoid starvation, or below the minimum threshold required to make a living as a farmer—defection into the countryside to pursue a foraging strategy, or to join a different group, must have been a possibility at most times. The diminishing marginal value segment of the curve (concave downwards, in the upper right-hand portion of the graph) is based on a great deal of economic research in the tradition discussed by Rima above. In the present case, for subsistence farmers, we presume that an additional bushel of maize in a very large harvest is not as valuable as an additional bushel of maize in a moderate harvest, since very large harvests present storage problems; since maize degrades to some extent during storage; and since transportation for exchange is not free. Yet very large harvests probably occur, given lucky combinations of planting and harvesting conditions, even if farmers plant only enough to get adequate harvests under average conditions. Sebastian (1991) makes the interesting argument that the ninth-century pithouse-to-pueblo transition marks the development of a strategy of *overproduction* (in average and good years) and an emphasis on storage, rather than foraging, to get through lean years.

An important characteristic of these curves is the way in which values of means taken on the *x*-axis differ from the mean values of the unaveraged yields taken on the *y*-axis. If mean production falls within the diminishing marginal value segment of the curve (as in Figure 24.1(a)), then the mean value of having chosen a risky gain (simplified here as $[V(\alpha) + V(\beta)]/2$ is *lower* than the mean value of having chosen a risk-free gain $V(\bar{x})$.[2] At high production levels, each household will optimize the value of its production by consuming at the mean production level (say, for its village) rather than averaging the value of consuming randomly-varying harvests through hoarding and storage (because, in good years, $V(\bar{x}) > [V(\alpha) + V(\beta)]/2$). In other words, households should avoid risk during a series of good years. Conversely, if we slide the $\bar{x}$ and variance indications α and β well to the left on the *x*-axis, the mean value of production is increased by consuming the average of the stochastic variability (because, as in Figure 24.1(b), $V(\bar{x}) < [V(\alpha) + V(\beta)]/2$). In bad years, households should be risk-seeking in their production/consumption strategies.

There are two major dimensions to variability in agricultural production for the Anasazi. First, holding space constant, there is variability from year to year in the production of any given plot due to climatic fluctuation (and other factors); second, within any year, there is spatial variability in the productivity of various agricultural plots. Many behaviors have been linked to attempts to avoid these risks. Two important strategies are storage, to buffer the year-to-year variability, and pooling food among producers, to buffer spatial variability within any year.[3] Since stored food may also be subject to pooling, and pooled resources may not be immediately consumed, it might be expected that these conceptually distinct responses to temporal and spatial variability might in practice be correlated. The terms "pooling" and "sharing" are used as synonyms here to encompass both the restricted sharing and the more general pooling of resources distinguished by Hegmon (1989). Although the precise sharing rules in place are of great interest, all we need to assume about those rules at this point is that (1) the system provides for those producing above the local mean to give more to those who produced below the mean than vice versa; and (2) that reciprocal obligations are recognized and can generally be met. The number of sharing households necessary to achieve dramatic reductions in income variance is related to the degree of correlation in production among households, and may in practice be fairly small (Hegmon 1989; Winterhalder 1986).

To Share or Not to Share

The model to be examined here is quite simple, although some of its predictions are not intuitively obvious. The decision variable to be analyzed is whether or not a household should pool its agricultural production prior to consumption. A number of assumptions are made.[4] In periods characterized by relatively high mean production, behaviors involving pooling of production ought to be attractive, both for the prodigious households (since the value of production they give away through sharing in a good year will be less to them than the value of the production they expect to receive through sharing in a bad year) and for the less fortunate (for the same reasons).[5] If these same periods are also subject to relatively high year-to-year fluctuation (we will be measuring temporal and spatial variability below in terms of their standard deviations), the difference between the value of the mean post-pooling consumption rate ($V[\bar{x}]$) and the value of the mean non-pooling consumption rate ($[V(\alpha) + V(\beta)]/2$) is accentuated (compare Figures 24.2(a) and (b)). So periods with high mean production, coupled with high annual fluctuation in production, ought to be especially favorable for the development of pooling behaviors. Finally, and for the same reasons, periods having high spatial variability in production should also tend to favor risk-averse (pooling) behaviors. As long as these conditions persist, and as long as the *distribution* of yields (*not* the yields themselves) is more or less similar for all households, the system of sharing will seem to be to the advantage of all, and ought to persist.[6]

On the other hand, periods when mean production is quite low ought to discourage sharing (and favor defection from any ongoing system of sharing). (The term "defection" is borrowed from game theory [e.g., Axelrod 1984:8] as shorthand for the decision not to go along with some system of cooperation.) This is because the value (on the *y*-axis) of the mean risky (non-pooling)

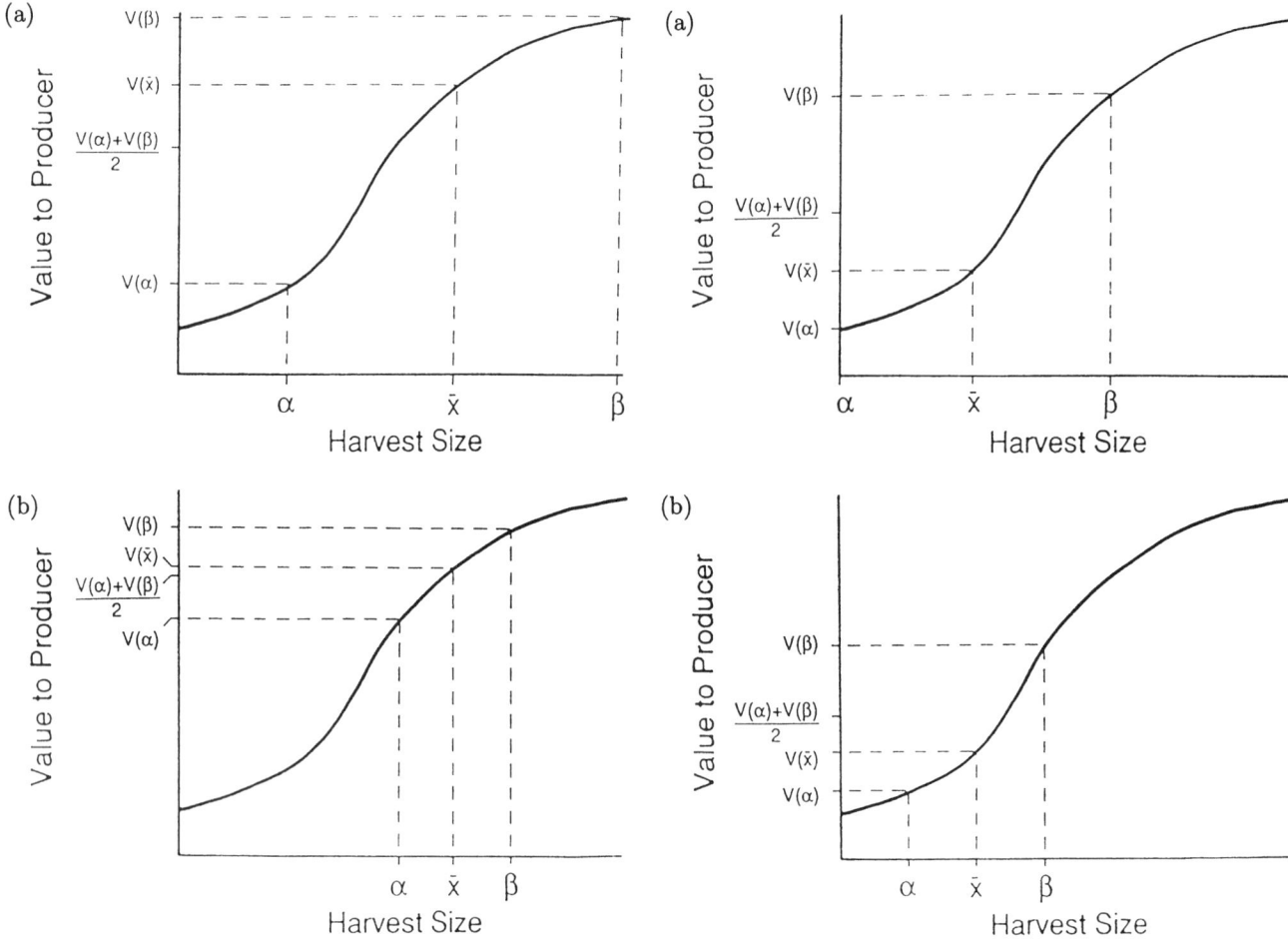

FIGURE 24.2. In good years, increasing production variance (through time or space) increases the relative value of pooling when average production is held constant. (a) A series of years with a relatively high mean production and relatively high variance results in relatively large differences between the expected value of sharing versus not sharing. (b) A series of years with the same mean but relatively low variance results in less difference between the expected value of sharing versus not sharing. Notation as in Figure 24.1.

FIGURE 24.3. In bad years, increasing production variance (through time or space) increases the relative value of not pooling when average production is held constant. (a) A series of years with a relatively low mean production and relatively high variance results in relatively large differences between the expected value of sharing versus not sharing. (b) A series of years with the same mean but relatively low variance results in less difference between the expected value of sharing versus not sharing. Notation as in Figure 24.1.

consumption is higher than the value of the mean risk-free (i.e., post-pooling) consumption. Moreover—and this point seems to go against traditional archaeological intuition—high temporal or spatial variability in periods of low mean production will exaggerate the difference between the values of the mean risky and the mean risk-free consumption rates. In periods of low mean production, the relative value of defection compared with sharing is greatest when temporal or spatial variability is *greatest* (compare Figures 24.3(a) and (b); see Hegmon 1989:93 for a related point, expressed in the currency of risk reduction rather than utility maximization). Another way to say all this is that when yields are generally poor, and have been for some time, the prodigious (but still hungry) household is discouraged from sharing by the fact that the production it gives away (in a slightly less lean year) is greater in value than the production it can hope to receive (in a year that is even worse).

Summary of the Model for Predicting Development of Food Sharing among Households

Pooling of food (a form of risk-averse behavior, and perhaps a critical element of cooperative behavior in general) is most likely to develop in periods of high mean productivity, high variability in productivity from year to year, and great differences in productivity across space. In periods characterized by low mean productivity but high temporal and/or spatial variability, pooling is not in the best interests of the participants, and is expected to break down, if present, or not to develop. Because we cannot directly observe utility functions for any society, especially a prehistoric one, it is comforting to learn that Sebastian (1991:111) has marshalled ethnographic evidence to show that small-scale agricultural societies do tend to share in good times, and hoard in lean times, a behavior which makes sense only if their implicit utility functions resemble those assumed here. In the next two

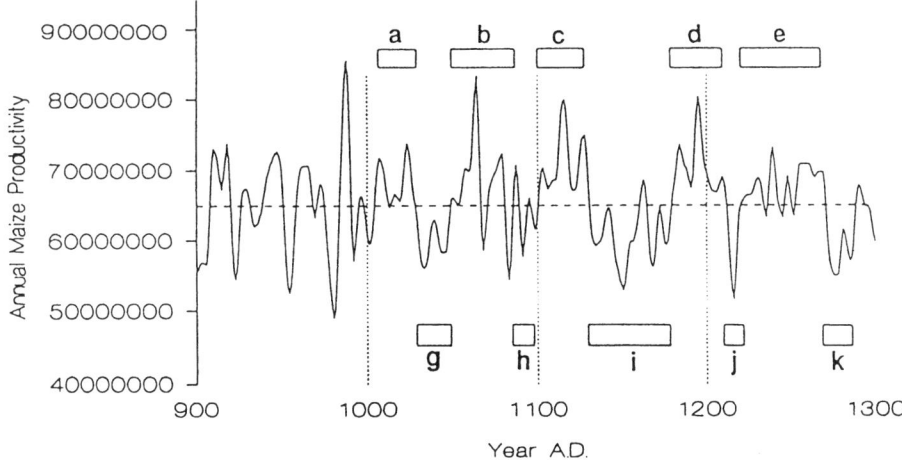

FIGURE 24.4. Smoothed annual estimates of total maize productivity in kilograms. Periods identifiable as favorable for, or unfavorable for, pooling are identified by bars above and below the series, respectively.

METHODS FOR EVALUATING MODEL PERFORMANCE
IDENTIFYING PERIODS OF INTEREST

sections we discuss how periods that meet these conditions are defined, and how "pooling behavior" (and its demise) might be recognized in the archaeological record.

Because the shape of the utility function entails that the expected value of pooling will exceed the expected value of not pooling only when the mean production is relatively high, it was first necessary to identify periods of relatively high mean production in the record. Such periods should be relatively long in order to have some chance of being able to recognize them in the archaeological record, and to allow decision-makers to have some idea as to the relative payoff for pooling versus defection in order to be able to make decisions on that basis. This would probably be impossible on the basis of very few years with given conditions.

Periods with relatively high mean production were found by smoothing Van West's (1994:133) estimates for total maize production for a study region encompassing about 1500 km² in Southwest Colorado. These estimates were developed for each year from AD 901–1300, at a spatial scale of 4 ha., and are sensitive to soil depth classes, elevation, and soil moisture (as retrodicted through estimates of Palmer Drought Severity Indices, calibrated from the relationship between modern temperature and precipitation data and recent tree-ring indices). Van West (1994) describes in detail how these paleoproductivity estimates are derived, and the environments and prehistory of the study area. The smoothed series of production values is displayed in Figure 24.4. This series appears to encompass five periods (from 24 to 50 years in length) of relatively high average production, identified by solid bars above the graph, and five (somewhat shorter) periods of relatively stable low means, identified by bars below the graph. The periods of below-normal production, expected to be unfavorable for pooling, range in length from 10 to 50 years; many of the unfavorable periods are rather short. This left 117 years (AD 901–1005 and 1289–1300) which were not part of either a stable high or low period and which are considered neutral in terms of the model.

The next step was to compute measures of temporal and spatial variation in production within each of these defined periods. This is necessary because the model predicts that pooling will be most attractive in periods with relatively high means that also exhibit high temporal and spatial variability in production. Temporal variability was measured here as the standard deviation around the mean production (per hectare per year) for each period. Spatial variability within each year was likewise measured by the per hectare standard deviations, and these annual measures were averaged for all the years within each defined period to form a measure of average spatial variability in each period.

For each period, the mean production and these measures of temporal and spatial variation are displayed in Table 24.1. (The accuracy of these figures can be roughly estimated for the historic period, but not for the prehistoric period, and it is likely that we present more significant digits in this table than is warranted.) In the first column of this table, the periods are ranked according to the best estimate of the overall attractiveness of pooling. For the set of periods in which pooling is expected to develop, this ranking is achieved by first ranking the scores in columns 2, 3, and 4 from high to low (with a rank of 1 assigned to high positive scores). Then each period is assigned an overall rank by taking the median of these three ranks. Thus, the ranks for period [c] are 1, 3, and 1, yielding a median of 1,[7] the highest rank for any period. We used ranks, rather than standardizing these three values as z-scores and taking their mean, because of some disjunction between what we would like to measure (the achieved production per household, given some particular

TABLE 24.1. Periods with differential advantages for pooling, ordered by median rank for mean total maize productivity and measures of temporal and spatial variability.

PERIOD[1] IN YEARS AD (MEDIAN RANKS)	MEAN ANNUAL MAIZE PRODUCTIVITY DURING PERIOD × 100 (RANK WITHIN GROUP)	STANDARD DEV. FOR ANNUAL MAIZE PRODUCTIVITY DURING PERIOD × 100 (RANK WITHIN GROUP)	MEAN STANDARD DEVIATION IN PRODUCTION ACROSS REGION DURING PERIOD (RANK WITHIN GROUP)	VALUE OF POOLING AS PREDICTED BY MODEL
[c] 1100–1129 (1)[2]	71,356 (1)	13,702 (3)	305 (1)	strongly positive
[d] 1180–1211 (2)	70,797 (2)	14,893 (2)	303 (2)	|
[a] 1006–1029 (3)	68,118 (3)	13,683 (4)	299 (3)	|
[b] 1049–1088 (4)	68,109 (4)	16,000 (1)	294 (5)	|
[e] 1222–1271 (5)	66,418 (5)	11,772 (5)	296 (4)	weakly positive
all 117 years not included in a favorable or unfavorable period	64,666	14,284	289	approx. neutral
[h] 1089–1099 (1)	59,607 (1)	10,961 (3)	275 (1)	weakly negative
[g] 1030–1048 (3)	59,046 (3)	10,687 (1)	277 (3.5)	|
[i] 1130–1179 (3.5)	59,433 (2)	12,130 (5)	277 (3.5)	|
[j]1212–1221 (4)	58,333 (4)	11,866 (4)	276 (2)	|
[k] 1272–1288 (5)	58,033 (5)	10,797 (2)	280 (5)	strongly negative
901–1300[3]	64,925	13,937	290	—

[1] The bracketed letters in front of each period identify the periods in the plot of smoothed regional production values (Figure 24.4).
[2] The median ranks of periods in this analysis differ slightly from those in Van West and Kohler (1995), where measures of temporal and spatial variability were based on residuals from regressions against productivity measures.
[3] Means for the entire series, for comparison.

distribution of population and fields in any given year) and what we are actually measuring (an estimate of potential production across the entire landscape).[8]

The periods in which the expected value of defection is greater than that of sharing were then ranked according to the same logic. The greatest rewards for not sharing are connected with low mean production (the lowest is assigned a rank of 5 in Table 24.1, column 2), high temporal variation (the highest is assigned a rank of 5 in column 3 of the same table), and high spatial variation (the highest is assigned a rank of 5 in column 4). Then the median of these three ranks was used to assign to each period an overall attractiveness for sharing behavior. For example, period [k] with ranks of 5, 2, and 5 received a median rank of 5, identifying it as the least favorable period for pooling in the 400-year record. This is the period from AD 1272 through 1288.

Identifying Pooling Behavior in the Archaeological Record

To examine the predictions of this model we must be able to monitor the growth and demise of cooperative behavior in the northern Anasazi Southwest, particularly those cooperative behaviors that might involve food sharing. The vast literature on integration, reciprocity, redistribution, political complexity, and aggregation is all germane in attempting to identify what these behaviors might be. Our tactic in choosing measures that ought to be involved with food sharing was to choose as many indices as possible in the full realization that none may be a pure measure of the dimension. We consider all to be at least weakly involved with the dimension of interest.

The measures selected are listed in the first column of Table 24.2. More discussion of how these facets of the archaeological record might be involved with food sharing can be found for aggregation at the site level in, for example, Glassow (1977:206); for growth and aggregation at the community level in Orcutt et al. (1990) and Sebastian (1991); for great kivas and triwall structures in Plog (1974:127); for reservoirs in Haase (1985); and so forth. Ford (1972) provides a general perspective on the importance of the movement of food in contemporary Tewa ritual and society. Our underlying premise is that all of the cooperative behaviors implied by these collective features—but especially aggregation itself—are built on the foundation of food sharing and provide a theater where continual face-to-face contacts reinforce mutual obligations and make free-riding easier to detect. Other measures of increased interaction, such as higher intraregional rates of exchange of regionally produced ceramic and lithic materials, should also be expected in periods in which regional production potentials favored development of pooling. Unfortunately, we know of no studies that describe the volumes of any intraregional flows of materials with enough temporal precision to be useful in testing the present model.

Table 24.2 lists only one "positive" test implication for the periods in which sharing is expected to break down—the dissolution of aggregated sites. Of course, we also expect to see no evidence in these periods of the behaviors we connect with sharing.

The temporal precision of our expectations exceeds the temporal precision of a good deal of the archaeological record. For this reason the tree-ring dated sites in the study area (Van West and Kohler 1995: Table 4) are especially valuable for testing the expectations. We add to the group of tree-ring dated sites

TABLE 24.2. Initial test of the Pooling Model.

	High Positive				← Predicted Strength of Pooling Advantage →					High Negative	
	1100–1129	1180–1211	1006–1029	1049–1088	1222–1271	Neutral years	1089–1099	1030–1048	1130–1179	1212–1221	1272–1288
Expectations for Periods in which pooling is Expected:											
Growth/aggregation at site level (e.g., more rooms indicate more residents)	++	+	+	+	+++	?	?	?	+	?	—
Growth/aggregation at community level (e.g., more sites indicate more members)	++	+	+	+	++	?	?	?	?	?	—
Great Kivas	7	2	?	1	5	?	?	?	?	?	0?
Reservoirs	+	+	?	?	++	?	?	?	+	?	–?
Great Houses	10	?	?	1	3	?	?	?	+	?	0
Roads	5	1	0	?	1	?	?	?	+	?	0
Enclosing Walls/Interior Plazas	1	?	0	0	6	?	?	?	?	+	0?
Triwalled and Biwalled Structures	1?	0	0	0	5	?	?	?	?	?	0?
Foundation of hamlets associated with the establishment/growth of aggregates	++	+	++	+	+++	?	?	?	—	?	—
Expectations for Periods in which Defection is Expected:											
Breakup of aggregates	?	?	?	?	?	+[a]	?	?	+	?	+
Overall strength of evidence for pooling (rank relative to other periods)	1	3	5	4	2	not ranked	9	9	6	7	9

[a] This refers to the breakup of the mid-ninth-century Pueblo I villages (as in the Dolores Archaeological Project area) which may take place slightly before AD 900.

TABLE 24.3. Performance of competing models for development of cooperative behaviors in the archaeological record.

Period (AD)	Deg. of Integration Seen in Record (Rank Based on Data on Table 24.2[1])	Deg. of Integration Predicted by Risk-Sensitivity Model (Table 24.1)	Deg. of Integration Predicted by Risk Abundance	Deg. of Integration Predicted by Buffering Model[2]	Deg. of Integration Predicted by Risk-Buffering Model with Abundance Removed[3]
1100–1129	1	1	1	1.5[4]	8
1222–1271	2	5	5	5	6.5
1180–1211	3	2	2	1.5[4]	4
1049–1088	4	4	4	3	5
1006–1029	5	3	3	4	2.5
1212–1221	6	9	9	7.5[4]	2.5
1130–1179	8.5[4]	8	7	6	6.5
1030–1048	8.5[4]	7	8	9	9
1089–1099	8.5[4]	6	6	10	10
1272–1288	8.5[4]	10	10	7.5[4]	1
T_b correlation coefficients with column 2 data		0.60	0.60	0.71	−0.07
T_b correlation coefficients with "risk-sensitivity" model			0.96	0.61	−0.16

[1] In making this ranking, we assumed, as suggested above, that the "+" marks in the 1130–1179 column of Table 24.2 are due primarily to problems in dating sites that are in fact slightly earlier; that is why the ranks here differ slightly from those in Table 24.2. Only more research at the questionable sites could settle this problem.
[2] This ranking was calculated after removing the linear relationship of the measures of temporal and spatial variability (columns 3 and 4 of Table 24.1) with a measure of annual maize productivity (column 2, Table 24.1) through regression. The resultant residuals are termed measures of "relative" temporal and spatial variation by Van West and Kohler (1995) and are displayed in columns 4 and 5 of their Table 3.
[3] Calculated by ranking the periods based on the average of the ranks for temporal and spatial variability (using data from columns 3 and 4 of Table 24.1), taking ties into account as appropriate.
[4] Ties.

another group of sites for which probable peaks of occupation can reasonably be derived from ceramic materials, and tabulate those items of public architecture from this larger set of sites that we wish to use as indices of increased sharing of resources (see Van West and Kohler 1995:Table 5). Finally, in Table 24.2, these data are tabulated against the periods identified under our model (in Table 24.1) as either rewarding cooperative food sharing, or rewarding defection.

Results

The general pattern of the record is strongly in the directions anticipated by the model. The period between 1100 and 1129—in which the expected value of cooperative behaviors is highest—coincides with the local buildup of the "Chacoan system." The mid-1200s, when the "terminal" aggregation takes place at canyonhead sites such as Sand Canyon, is also, correctly, predicted to be a time when cooperative behaviors are expanding or stable. The breakup of the local Chacoan-related system (between 1130 and 1179) and the final abandonment of the region (in the 1270s or 1280s) are likewise found in those periods in which defection ought to be advantageous, according to the model. Finally, although we did not include it as a formal expectation for periods of defection in Table 24.2, cannibalism perhaps represents the ultimate in negative reciprocity. The occupations of the four sites within or near our study area for which White (1992) finds strong evidence for cannibalism generally coincide with periods for which defection is predicted. The Grinnell site (AD 1135–1150, White 1992:376) is occupied entirely within period [i]; 5MT3 at Yellow Jacket (AD 1025–1050, White 1992:378) is occupied almost entirely within period [g]; 5MTUMR-2346 is poorly dated but the probable occupation in the first half of the twelfth century generously overlaps period [i]. Perhaps least likely to be in conformity with our prediction is the cannibalism at 5MT1 in the Yellow Jacket complex, probably dating to the late AD 900s and early 1000s, apparently overlapping neutral, favorable, and unfavorable periods, although possibly terminating in unfavorable period [g].

Some of the apparent failures in prediction may reflect weaknesses in our ability to date the archaeological record precisely. We suggest that many of the "+" signs in the 1130–1179 period in fact pertain to sites actually belonging to the immediately preceding period of more favorable conditions. Three of the periods in which we predict defection (1030–1048, 1089–1099, and 1212–1221) are simply too short to identify with confidence in the record.

How does this model perform relative to some obvious rivals for predicting the timing of increase in cooperative behaviors? In Table 24.3 we examine two such models—one in which simple abundance (high productivity) is used as a predictor of cooperative behaviors (column 4), and a second, "risk-buffering" model, similar to that espoused by Braun and Plog, where periods of greatest temporal and spatial variability are predicted to have most evidence for cooperative behaviors (column 5).

The performance of all three models is quite good, and probably undifferentiable, given the vagaries of the data. The correlation between the predictions of our model (labeled the "risk-sensitivity model" in column 3 of Table 24.3) and the archaeological record as we read it, is 0.60 as measured by Kendall's Tau-b (T_b); between the simple abundance model, and the record, 0.60; and between the "integration as risk-buffering" model, and the record, 0.71.

In fact the predictions of all three models are very similar, as can be appreciated from the last row of Table 24.3, which displays the correlation coefficients among the predictions. In this landscape, the spatial variance in production across the study area in any year is highly correlated with the mean annual (per hectare) productivity. Areas of thin rocky soils in the study area do poorly almost all the time, whereas production in areas of deep loess soils is very responsive to recent and current precipitation. It should also be noted that *periods* in which mean productivity is relatively high tend to display relatively high year-to-year variability, probably because of the mean/variance effect, in which high means entrain high variances, but also because there is a floor below which productivity rarely falls, so that periods of relatively low productivity (with few years departing radically from that floor) automatically display low year-to-year variance. Taken together, these two effects dictate that periods of high temporal and spatial variability (which according to the risk-buffering model ought to favor increased integration) are almost automatically periods of high abundance, *and* periods when the calculus of self-interest discussed above would predict food sharing. That the risk-buffering model performs well because of this effect, and not because of variability per se, can be seen in the last column of Table 24.3, where we subtract the linear dependence that our measures of temporal and spatial variability have on raw production. So recast, this model performs very poorly against the archaeological data. In future comparisons of these models, it would be desirable to select more carefully the spatial frame so that areas of relatively low productivity are not included; this would minimize the mechanical correlation between high productivity and temporal and spatial variability in productivity. However, it is inescapably a characteristic of these landscapes that these factors are correlated to some extent.

In assessing the performance of the risk-sensitivity model, it is important to remember that the household-level decisions that it predicts are being made on the basis of perceived *per household* values. Thus, the regional productivities (by which we predict the household-level decisions) ought to be corrected for regional population sizes. We have not been able to do so, although such corrections would be possible for some subsets of the study area. With such a (negative) correction to income

TABLE 24.4. Differences between predictions of risk-sensitivity model and archaeological data, examined against rough population index. Shaded cells indicate much less cooperative behavior than predicted; outlined cells indicate much more than predicted.

Period (AD)	Difference Between Predicted and Observed Rank (Positive Values Indicate More Cooperative Behavior Than Predicted)	Ordinal Index of Probable Regional Population Density (1 = Lowest)[1]
1006–1029	−2	1
1030–1048	−1.5	1
1049–1088	0	2
1089–1099	−2.5	1
1100–1129	0	2
1130–1179	−.5	1
1180–1211	−1	2
1212–1221	3	2
1222–1271	3	3
1272–1288	1.5	1

[1] These are impressionistic. Relatively accurate population estimates can be made only for specific portions of the project area (especially the Dolores Archaeological Project area, Mockingbird Mesa, Upper Sand Canyon, and Lower Sand Canyon). These estimates, tabulated in Van West and Kohler (1995:Table 1) show that, generally, population increases and decreases are not synchronized within the study area, further complicating the task of making regional estimates.

(on the x-axis of our utility functions) the relative value of pooling (on the y-axis) would likewise tend to decrease, and this effect should be more marked in times of high population than in times of low population. Therefore, when populations are high, a given level of production and degree of spatial and temporal variability should result in *less* likelihood for food sharing and related cooperative behaviors than in times of low population, according to the model.

In Table 24.4 we examine our predictions for this effect. What we find is the opposite of what we expect. The marked cells indicate periods for which the model and the data disagree the most. Large positive residuals (indicating more cooperative behavior in the record than predicted; outlined in Table 24.4) are found in the context of relatively high populations. On the other hand, the negative residuals (indicating less cooperation than predicted; stippled in Table 24.4) are found mostly in the context of relatively low population levels. Unless we assume that the decision-making model, or the data, are seriously flawed, high population levels are promoting the desirability of cooperation in some way that more than counteracts the negative bias against cooperation that they should entrain by their effects on per-household income.

Discussion

There are several ways in which the apparently strong effect of population size on promoting cooperation could be explained. One contributing factor might be that during this time population never reached a level in which per-household income was severely affected. This possibility is difficult to reject, although it has been possible to demonstrate that in particular situations, households have certainly had to work harder to maintain an acceptable level of income in times of high population and aggregation. For example, Kohler et al. (1986) were able to show that in the Dolores area, average (per household) distance to fields must have increased from only about 0.2 km, under conditions of low population and dispersion, to over 1.6 km, when population size was high and people lived in large villages (see also Orcutt et al. 1990).

However, even if per-household production was not diminished when populations levels were high, this by itself cannot explain what appears to be the *positive* effect that high populations have on cooperative behaviors (at best it would explain the absence of a negative effect). Where else can we look for some cause?

One place is in the nature of the landscape itself, with its rich array of natural and social features. The effects of large, sedentary populations of farmers on the biotic landscape are now beginning to be understood (Floyd and Kohler 1990; Kohler 1992b; Kohler and Matthews [1988]; Speth and Scott 1989). Wild-food depletion in conjunction with high population levels increased Anasazi commitment to and dependence upon agriculture. Under such conditions the landscape would have begun to fill up with marks of ownership (Adler 1990; Kohler 1992a), particularly where agriculture was most reliable and productive. In periods of both low population and low production, defection into an open landscape retaining abundant wild resources was more attractive than in periods of low production coupled with high population. We suggest that population levels—because of their effect on the landscape and on the required level of agricultural intensification—in effect changed the shape of the utility function to make sharing more attractive when population was high, and less attractive when population was low. From a game-theoretic perspective, the value of defection relative to cooperation was lower in populous landscapes. The dilemma of the 1270s/1280s in the study area was that the biotic and social characteristics of the landscape necessitated a sharing adaptation at the same time as climatic characteristics were making such cooperative behaviors increasingly unattractive. The abandonment of the Four Corners area may be due more to the undesirability of dispersing into a landscape that had lost its attractiveness for the dispersion/disintensification option, than to the absolute production levels that could have been achieved.

Our simple model predicts that cooperative behaviors such as aggregation can develop in the absence of marked increase in population, and at any population level. This prediction may in fact agree with the circumstances of aggregation in the Zuni (Stone 1992) and Taos (Crown et al. 1990) areas. If our interpretation of the weaknesses of the simple model is correct, however, the productive characteristics that favor pooling would have to be correspondingly more exaggerated to have the same effect in a sparsely populated region as in a densely populated region (see

a parallel argument in Plog et al. 1988). The population history and level of the region are key contexts within which the utility model that influences decision-making must operate.

A second important area in which this simple model is almost certainly too simple is in its ignorance of extraregional productive characteristics. Presumably at least part of the population history of our study area can be explained by its attractiveness relative to other regions, in the same way as some population movements within the study area have been explained (Schlanger 1988). If so, this in turn will affect the shape of the utility function (or, in game-theoretic terms, the nature of the payoff matrix for cooperating versus defecting) within the study area.

Finally, to lop off some portion of the Southwest from the whole for analysis is to assume that developments in each subregion were independent. Clearly our study area is influenced by the development of the Chacoan system outside of its borders, and probably by other external events and systems that are not so obvious. This study area is one of convenience, justifiable mostly on the basis of the high-quality data that makes it possible to formulate and test, in very preliminary fashion, the present model. Nevertheless, our general success suggests that the formation and dispersion of villages in this time and place is largely conditioned by local factors, although the form and particularities of villages and their attendant social lives are undoubtedly affected by complex, local and nonlocal, considerations.

Summary and Implications

In this chapter we work toward a method for predicting the development of cooperative behaviors/food-sharing among small-scale horticulturalists. The model behind the method focuses on sharing of food. Food sharing is not only one of the most fundamental forms of human cooperation but is symbolically implicated in more elaborate cooperative endeavors and may form the foundation for their elaboration. Our goal in this chapter has not been to define the links between food sharing and the various cooperative behaviors that leave more visible residues in the archaeological record. Nevertheless, examination of the performance of the model suggests that its utility goes well beyond the prediction of food sharing to the prediction of a large set of cooperative behaviors.

The model put forward here predicts that the effect of variance (risk) in production on cooperative systems involving food sharing will depend on whether these risks are situated in a context of relatively high, or relatively low, production. Cooperative behaviors are most valuable in the context of high production coupled with high temporal and spatial variability. Such behaviors are least likely in circumstances of low production coupled with high spatial and temporal variability. Essentially, the present method requires knowing something about the shape of the utility function before any predictions can be made about the relative value of cooperation versus defection. We have assumed that years of production substantially below the mean are low enough so that they will be located in the increasing-marginal-value segment of the curve, and that years well above the mean are within the decreasing-marginal-value segment of the curve. These assumptions require continued scrutiny. Provisionally, the success of the predictions they allow tends to confirm their general accuracy.

It has been realized by others that villages and expansive regional systems in the later northern Southwest tended to form during periods that were favorable to agricultural production (see, for example, Orcutt et al. 1990; Sebastian 1991). In this chapter we build theory to explain *why* this should happen; this theory also successfully predicts when such cooperative systems should fall apart. One implication of this approach is that to turn climatic variability into a variable that is effective in human affairs, we first have to pass this variability through two filters—the first connecting climatic variability with production of economically significant materials (which we do with the retrodicted Palmer Drought Severity Indices and their resultant maize yields), the second connecting economy to cognition (which we attempt with the utility function).

The most important insight we take from this analysis is the surprising realization that a model of behavior based entirely on the self-interest of the participants—a rational choice model—can explain so much of the variability we see in the settlement record of this portion of the Mesa Verde region between AD 900 and 1300. Villages with their attendant public architecture form more or less when the model predicts they should, and disintegrate when they should, if the chief guiding tenet were the narrow economic self-interest of the participants. The success of methodological individualism (householdism?) in this context flies in the face of the received model of Puebloan adaptations as being highly group-oriented, submerging or effacing the individual.[9] This received model is, of course, based on historic groups, and it may well be correct for historic groups; these results minimally suggest some limits on backwards extrapolation from the "ethnographic present."

If so, it remains a fascinating empirical problem to determine at what point in prehistory or history the strong group-level effects featured in the received model became applicable to Puebloan societies. Certainly the ebb and flow of village life prior to about AD 1275, in comparison with the apparently uninterrupted history of village life after that time, suggest that an important threshold was reached about then. Perhaps with the many new techniques for water harvesting used in the northern Rio Grande, for example (Anschuetz 1992; Moore 1992), the productive conditions after AD 1275 always favored cooperation under a rational choice model: this is an important possibility that should be tested using approaches similar to those developed here for the Mesa Verde area. More likely, we think, is that other ways around the Prisoner's Dilemma were being exploited by this time. One possibility is that the "shadow of the future" loomed larger in late prehistoric times; indeed, in retrospect we know that the probability of future interaction among villagers changed after 1275, but could the founders of an Early Classic village in the Rio Grande have been aware that their villages would endure longer, in some form, than were the inhabitants of a site such as Sand Canyon? Probably not. More likely, what changed was the open commitment of people to a new set of sharing rules that crosscut kin groups and featured village-level activities more strongly than had been the case in the past. Such commitments would have been accompanied by a stronger set of sanctions for potential defectors. Open commitments to a strategy and social sanctions against defection represent powerful methods for by-passing the Prisoner's Dilemma that had, up until about AD 1275/1300, dictated that villages endured only while times were good, and while it was in the best economic interest of all participants to continue to cooperate. Presumably, the katsina cult, the koshare system, and the remapping of kivas and social groups that all appear around this time provided both an attractive and a stable milieu for village life (Adams 1989; Lipe 1989; Steward 1937).

A related but more general way of viewing this transition draws on theory of human action developed by Elster (1989). According to Elster, people decide on various strategies for action—from among those that are practically available to them, within their "opportunity set"—according to either rational choice, or by reference to social norms. The present analysis suggests that rational choice was a dominant force in Anasazi settlement strategies until the late thirteenth century AD, but that after that time, social norms came to dominate rational choice. If so, then this transition offers a remarkable opportunity for understanding how social norms may develop and take over decision areas that were formerly the domain of individual rational choice.

In closing, our approach suggests that disintegration of social groups beyond the level of close kin was a frequent way to cope with resource stress among the northern Anasazi prior to around AD 1275. Our approach also suggests that studies of sharing rules, and how they may change through time, is an area for critical future research. A final important implication of this study is that since the early villages and regional systems we have discussed are built on the backs of surpluses, we must become more sophisticated in our understanding of how these surpluses are mobilized and distributed (Saitta and Keene 1990); this problem is therefore closely connected with the problem of identifying sharing rules.

Acknowledgments

This chapter has benefited from comments on related efforts by Linda Cordell, Richard Duncan, Michelle Hegmon, Keith Kintigh, John H. Miller, Rolf Sinclair, Eric Smith, and Chip Wills. We hope they forgive us for any good suggestions we chose not to follow. We would also like to thank Mark Varien and Angela Schwab of Crow Canyon Archaeological Center for access to unpublished data and manuscripts, and Larry Hammack and Jim Kleidon for recent tree-ring data from Knobby Knee Stockade and Sand Canyon Pueblo. All contributed information, reported more fully in Van West and Kohler (1995), helped make possible the test of the models presented here. William D. Lipe of Washington State University and Crow Canyon improved the clarity of our thinking and approach at several junctures, as did Jeffrey S. Dean of the Laboratory of Tree-Ring Research, University of Arizona. The Wenner-Gren Foundation for Anthropological Research through grant 4799 helped support the research by Van West that produced the annual estimates of prehistoric agricultural productivity on which the present analysis is based. The Graduate School of Washington State University made possible Kohler's professional leave, held at the Santa Fe Institute, providing Kohler with a stimulating environment while revising this chapter for publication.

Notes

1. Their argument is meant to apply only to situations which are not severe enough to endanger the existence of the network itself (Braun and Plog 1982:508). We will try to make the case that systems of interaction, coordination, and sharing exist on several levels of spatial and social generality. Not all of these are subject to collapse, but those at the furthest reaches of social and spatial distance are constantly at risk and do, in fact, collapse frequently.

2. Imagine α as the harvest of some household in one year, and β as the harvest of the same household in the following year. (α could also be the production of a household in year 1, and β the production of another household in the same year.) If—in the first case—the household shares its production with no other household, then after two years, its expected value realized from both harvests will be $[V(\alpha) + V(\beta)]/2$. (Here the value for each year's production is realized on the y-axis and then the average value over two years is calculated.) On the other hand, if the harvest was generally shared by households within a village *before* its value was "realized" (consumed), then the expected value for each household would be the average $V[\bar{x}]$ across all the households in the village. In this case, the mean is calculated on the production axis, and then its value is "realized" on the y-axis.

3. Keeping many plots is another obvious option (Hegmon 1989; 1996). Whereas we do not deny that this may have been of some importance, as it was clearly practiced in historic times, it would be difficult for a household to monitor many plots during the key harvest period. This option is probably more attractive in landscapes with relatively little wild game, as in the Medieval English situation discussed by Winterhalder (1990).

4. These include: (1) the shape of the utility function; (2) that each household seeks to optimize the utility from its consumption; and (3) that local household harvest rates are not perfectly correlated across years.

5. Smith and Boyd (1990) show how an analogous system among hunter/gatherers can profitably be modeled as a two-person or n-person game, having payoffs to participants structured according to the Prisoner's Dilemma (PD). (Sigmund [1993:180–206] provides an excellent introduction to the study of this social dilemma, in which the rational strategy is to defect, but with the undesirable outcome that if both players defect, they receive payoffs that are less than if both cooperate.) For the one-shot PD game, the dominant strategy is to defect (not share). Two obvious ways out of the dilemma, both eminently applicable to the Puebloan situation, are strong social sanctions against defection, and/or an "iterated structure" to the game which makes it very likely that current players will meet again (Axelrod 1984:9–21; see also Gumerman and Dean 1989). In a stable village setting, the future must loom large in this manner. However, we will argue that there may not have been strong social sanctions in place against defection until about AD 1300 throughout much of the northern Southwest. For the system analyzed here, the average payoff for cooperation will be higher than the average payoff for defection only in times of relatively high production (as in the PD); thus, in periods of low production, this is not a PD game.

6. In this chapter we stress climatic-economic reasons why such systems break down, but it is also possible to see, in this formulation, how the system would be endangered if some households *continually* produced more than others. With no expectation for receipt of valuable reciprocal goods in bad years (since there are no bad years), such households could defect (in the absence of very strong social sanctions). This situation is in fact covered by the model, to the extent that we predict sharing to be less valuable in times of little temporal variability in yields. However, the example illustrates that the distribution of yields over time may not be the same for all households in a village, especially given a developed system of land tenure with field ownership at the household, lineage, or clan levels. It also illustrates that defection from a reciprocal system would not necessarily originate in the less-productive households.

7. For the case of three periods with one tie, the median will equal the mode.

8. This disjunction is probably greatest when population levels are relatively low, and farmers have more freedom to select parcels of land that they judge likely to do well in the following year, based on previous experience. When population levels are higher, the variability in achieved production probably more closely mirrors that of the larger landscape.

9. These results also weaken any model that calls for the amassing of surplus by emergent leaders as *prerequisites* for village formation and the related cooperative behaviors we track in Table 24.2. By the model presented here, such general behaviors (although not, perhaps, their specific forms) can be explained more simply by households acting in their own best interests. Hantman (1989: 442)—working on the same problem in the Upper Little Colorado Region—finds no evidence for storage volumes large enough to suggest that "economic surplus was produced and managed prior to the Pueblo IV period" of maximum local aggregation (and even then, may not be large enough to constitute a real "fund of power"). Our result also parallels recent work among insects (reviewed by Deneubourg and Goss 1989; see also Théraulaz and Deneubourg 1992) that demonstrate very complex social behaviors emerging from individuals following a few very simple rules in specific environments, without requiring direction by a dominant individual. We recognize, however, that once established, human villages—particularly as they exist in times of probable surplus—may present fertile environments for the emergence of dominant individuals or groups.

References

Adams, E. C.
1989 Changing Form and Function in Western Pueblo Ceremonial Architecture from AD 1000 to AD 1500. In *The Architecture of Social Integration in Prehistoric Pueblos*, edited by W. D. Lipe and M. Hegmon, pp. 155–160. Crow Canyon Archaeological Center, Cortez.

Adler, M. A.
1990 Population Aggregation and the Anasazi Social Landscape: The View from the Four Corners. Paper presented at the 1990 Southwest Symposium, January, Albuquerque.

Anschuetz, K. F.
1992 Saving a Rainy Day: The Integration of Diverse Agricultural Technologies to Harvest and Conserve Water in the Lower Rio Chama Valley, New Mexico. Paper presented at the New Mexico Archaeological Council Agricultural Symposium, October, Santa Fe.

Axelrod, R.
1984 *The Evolution of Cooperation*. Basic Books, New York.

Braun, D. P., S. Plog
1982 Evolution of "Tribal" Social Networks: Theory and Prehistoric North American Evidence. *American Antiquity* 47:504–525.

Caraco, T., S. Martindale, T. S. Whittam
1980 An Empirical Demonstration of Risk-Sensitive Foraging Preferences. *Animal Behavior* 28:820–830.

Crown, P. L., J. D. Orcutt, T. A. Kohler
1990 Pueblo Cultures in Transition: The Northern Rio Grande. Revised version of paper presented at the conference "Pueblo Cultures in Transition: AD 1150–1350 in the American Southwest." Ms. on file, Crow Canyon Archaeological Center, Cortez.

Deneubourg, J. L., S. Goss
1989 Collective Patterns and Decision-Making. *Ethology, Ecology & Evolution* 1:295–311.

Elster, J.
1989 *Nuts and Bolts for the Social Sciences*. Cambridge University Press, Cambridge, MA.

Floyd, M. L., T. A. Kohler
1990 Current Productivity and Prehistoric Use of Piñon (*Pinus edulis*, Pinaceae) in the Dolores Project Area, Southwestern Colorado. *Economic Botany* 44(2):141–156.

Ford, R. I.
1972 An Ecological Perspective on the Eastern Pueblos. In *New Perspectives on the Pueblos*, edited by A. Ortiz, pp. 1–17. University of New Mexico Press, Albuquerque.

Glassow, M. A.
1977 Population Aggregation and Systemic Change: Examples from the American Southwest. In *Explanation of Prehistoric Change*, edited by J. N. Hill, pp. 185–230. University of New Mexico Press, Albuquerque.

Gumerman, G. J., J. S. Dean
1989 Prehistoric Cooperation and Competition in the Western Anasazi Area. In *Dynamics of Southwestern Prehistory*, edited by L. S. Cordell and G. J. Gumerman, pp. 99–148. Smithsonian Institution Press, Washington, D.C.

Haase, W. R.
1985 Domestic Water Conservation Among the Northern San Juan Anasazi. *Southwestern Lore* 51(2):15–27.

Hantman, J. L.
1989 Surplus Production and Complexity in the Upper Little Colorado Province, East-Central Arizona. In *The Sociopolitical Structure of Prehistoric Southwestern Societies*, edited by S. Upham, K. G. Lightfoot, and R. A. Jewett, pp. 363–388. Westview Press, Boulder.

Hegmon, M.
1989 Risk Reduction and Variation in Agricultural Economies: A Computer Simulation of Hopi Agriculture. *Research in Economic Anthropology* 11:89–121.

1996 Variability in Food Production, Strategies of Storage and Sharing, and the Pithouse-to-Pueblo Transition in the Northern Southwest. In *Evolving Complexity and Environmental Risk in the Prehistoric Southwest*, edited by J. A. Tainter and B. Bagley Tainter pp. 223–250. Addison-Wesley, Reading, Ma.

Kaplan, H., K. Hill, A. M. Hurtado
1990 Risk, Foraging, and Food Sharing Among the Ache. In *Risk and Uncertainty in Tribal and Peasant Economies*, edited by E. H. Cashdan, pp. 107–143. Westview Press, Boulder.

Kohler, T. A.
1992a Fieldhouses, Villages, and the Tragedy of the Commons in the Early Northern Anasazi Southwest. *American Antiquity* 57:617–635.

1992b Prehistoric Human Impact on the Environment in the Upland North American Southwest. *Population & Environment* 13:255–268.

Kohler, T. A., M. H. Matthews
1988 Long-Term Anasazi Land Use and Forest Reduction: A Case Study from Southwest Colorado. *American Antiquity* 53:537–564.

Kohler, T. A., J. D. Orcutt, E. Blinman, K. L. Petersen
1986 Anasazi Spreadsheets: The Cost of Doing Agricultural Business in Prehistoric Dolores. In *Dolores Archaeological Program: Final Synthetic Report*, compiled by D. A. Breternitz, C. K. Robinson, and G. T. Gross, pp. 525–538. USDI, Bureau of Reclamation, Denver.

Krebs, J. R., A. Kacelnik
1991 Decision Making. In *Behavioral Ecology: An Evolutionary Approach*. Third edition, edited by J. R. Krebs and N. B. Davies, pp. 105–136. Blackwell Scientific Publications, Oxford.

Leonard, R. A., H. Reed
1993 Population Aggregation in the Prehistoric American Southwest: A Selectionist Perspective. *American Antiquity* 58:648–661.

Lipe, W. D.
1989 Social Scale of Mesa Verde Anasazi Kivas. In *The Architecture of Social Integration in Prehistoric Pueblos*, edited by W. D. Lipe and M. Hegmon, pp. 53–71. Crow Canyon Archaeological Center, Cortez.

Moore, J. L.
1992 Anasazi Field Systems in the Taos District. Paper presented at the New Mexico Archaeological Council Agricultural Symposium, October, Santa Fe.

Orcutt, J. D., E. Blinman, T. A. Kohler
1990 Explanation of Population Aggregation in the Mesa Verde Region Prior to AD 900. In *Perspectives on Southwestern History*, edited by P. E. Minnis and C. L. Redman, pp. 196–212. Westview Press, Boulder.

Plog, F. T.
1974 *The Study of Prehistoric Change*. Academic Press, New York.

Plog, F., G. J. Gumerman, R. C. Euler, J. S. Dean, R. H. Hevly, T. V. N. Karlstrom
1988 Anasazi Adaptive Strategies: The Model, Predictions, and

Results. In *The Anasazi in a Changing Environment*, edited by G. J. Gumerman, pp. 230–276. Cambridge University Press, Cambridge.

Rima, I. H.
1972 *Development of Economic Analysis*. Richard D. Irwin, Homewood, IL.

Saitta, D. J., A. S. Keene
1990 Politics and Surplus Flow in Prehistoric Communal Societies. In *The Evolution of Political Systems: Sociopolitics in Small-Scale Sedentary Societies*, edited by Steadman Upham, pp. 203–224. School of American Research, Santa Fe, and Cambridge University Press, Cambridge.

Schlanger, S. H.
1988 Patterns of Population Movement and Long-Term Population Growth in Southwestern Colorado. *American Antiquity* 53: 773–793.

Sebastian, L.
1991 Sociopolitical Complexity and the Chaco System. In *Chaco and Hohokam: Prehistoric Regional Systems in the American Southwest*, edited by P. L. Crown and W. J. Judge, pp. 109–134. School of American Research Press, Santa Fe.

Sigmund, K.
1993 *Games of Life: Explorations in Ecology, Evolution, and Behaviour*. Oxford University Press, Oxford.

Smith, E. A.
1988 Risk and Uncertainty in the "Original Affluent Society": Evolutionary Ecology of Resource-Sharing and Land Tenure. In *Hunters and Gatherers 1: History, Evolution, and Change*, edited by T. Ingold, D. Riches, and J. Woodburn, pp. 222–251. Berg, Oxford.

Smith, E. A., R. Boyd.
1990 Risk and Reciprocity: Hunter-Gatherer Socioecology and the Problem of Collective Action. In *Risk and Uncertainty in Tribal and Peasant Economies*, edited by E. Cashdan, pp. 167–191. Westview Press, Boulder.

Speth, J. D., S. L. Scott
1989 Horticulture and Large-Mammal Hunting: The Role of Resource Depletion and the Constraints of Time and Labor. In *Farmers as Hunters: The Implications of Sedentism*, edited by S. Kent, pp. 71–79. Cambridge University Press, Cambridge.

Stephens, D. W.
1990 Risk and Incomplete Information in Behavioral Ecology. In *Risk and Uncertainty in Tribal and Peasant Economies*, edited by E. Cashdan, pp. 19–46. Westview Press, Boulder.

Steward, J.
1937 Ecological Aspects of Southwestern Society. *Anthropos* 32:87–104.

Stone, T. T.
1992 The Process of Aggregation in the American Southwest: A Case Study from Zuni, New Mexico. Unpublished Ph.D. dissertation, Department of Anthropology, Arizona State University, Tempe.

Théraulaz, G., J.-L. Deneubourg
1992 Swarm Intelligence in Social Insects and the Emergence of Cultural Swarm Patterns. In *The Ethological Roots of Culture*, edited by A. Gardner. NATO ASI Series D.

Upham, S.
1984 Ecological and Political Perspectives on Labor Intensive Agriculture and Exchange During the 14th Century. In *Prehistoric Agricultural Strategies in the Southwest*, edited by S. K. and P. R. Fish, pp. 291–307. Anthropological Research Papers 33. Arizona State University, Tempe.

Van West, C. R.
1994 *Modeling Prehistoric Agricultural Productivity in Southwestern Colorado: A GIS Approach*. Reports of Investigations 67. Department of Anthropology, Washington State University, Pullman.

Van West, C. R., T. A. Kohler
1994 A Time to Rend, A Time to Sew: New Perspectives on Northern Anasazi Sociopolitical Development in Later Prehistory. In *Anthropology, Space, and Geographic Information Systems*, edited by M. Aldenderfer and H. D. G. Maschner. Oxford University Press, New York.

White, T. D.
1992 *Prehistoric Cannibalism at Mancos 5MTUMR-2346*. Princeton University Press, Princeton.

Winterhalder, B.
1986 Diet Choice, Risk, and Food Sharing in a Stochastic Environment. *Journal of Anthropological Archaeology* 5:369–392.
1990 Open Field, Common Pot: Harvest Variability and Risk Avoidance in Agricultural and Foraging Societies. In *Risk and Uncertainty in Tribal and Peasant Economies*, edited by E. Cashdan, pp. 67–87. Westview Press, Boulder.

❈ 25 ❈

Conspicuous Consumption as Wasteful Advertising

A Darwinian Perspective on Spatial Patterns in Classic Maya Terminal Monument Dates

Fraser D. Neiman

INTRODUCTION

Stone monuments have played a leading role in the archaeological study of Classic Maya society since archaeologists became aware of its existence. Over a century and a half ago, the remarkable monuments of Copan inspired John Lloyd Stephens to guess that he had discovered the remains of what we would today call a complex society, a place where "orators, warriors, and statesmen, beauty, ambition, and glory had lived and passed away" (1841:105). Inferences similar to Stephens' acquired real historical value when archaeologists learned how to date the monuments by reading the dedicatory dates inscribed on them. This development along with an accurate calibration between the Mayan and Gregorian calendars that was in place at the beginning of this century (Goodman 1905), made it possible to use monuments to chart the efflorescence and demise of Classic Maya civilization. Sylvanus Morley led the way in this endeavor by building histograms of monument dedicatory dates, at both site and regional scales (Morley 1920, 1938, 1946), a tactic still favored today (Jones 1991; Fash 1988; Mathews & Willey 1991). Morley's histograms allowed him to detect the relatively abrupt cessation of dated monument construction that became the famous Maya collapse. Since then region-wide frequency distributions of terminal monument dates—the dedicatory date of the latest known monument constructed at a site—have been used to address possible causes of the collapse (Hamblin & Pitcher 1980; Lowe 1985).

In the wake of the New Archaeology's interest in reconstructing ancient social and political organization, the traditional preoccupation with the distribution of monuments in time has been supplemented by a new interest in their distribution in space (Lowe 1985; Mathews 1985). Interest in the collapse has led to a focus on the spatial distribution of terminal monument dates (Bove 1981; Whitley & Clark 1985; Kvamme 1990).

The common motivation here is the notion that this evidence offers insight into two related topics. The first is the nature of the Classic Maya collapse itself. Different processes responsible for the collapse would appear to have different implications for spatial patterns in the timing of the cessation of dated monument construction. Second, different scenarios concerning the character of what was collapsing, in particular the spatial scale of political competition, yield different expectations about geographical patterns in terminal dates. Researchers have looked to spatial patterns in terminal dates to evaluate competing scenarios in both these areas (Marcus 1993).

My goals in this paper are both empirical and theoretical. On the empirical side, by building on previous work, I hope to advance our understanding of the historical processes responsible for spatial patterns in terminal dates and thereby contribute to our knowledge of ancient Maya history. Achieving this empirical goal requires the introduction of new conceptual tools, drawn from Darwinian theory. Hence I begin by sketching a set of ideas about the mechanistic evolutionary processes that cause conspicuous consumption, manifest in the case at hand by the apparently wasteful energy investment in the construction of dated stone monuments. The result, a Darwinian theory of wasteful social advertising, makes it possible to build a simple model of how historical processes related to the Maya collapse might map onto spatial patterns in terminal dates. It suggests what kinds of phenomena are measured by spatial variation in terminal dates. Expectations about this mapping indicate what sort of patterns we should be looking for in terminal dates and guide the choice of statistical tools likely to uncover them.

I then turn to evidence gleaned from terminal dates from 69 sites in the Maya Lowlands. Local, robust regression (loess) is used to identify the underlying regional trend in terminal dates. The trend offers an unappreciated and important clue to the

---

*Originally published in *Rediscovering Darwin: Evolutionary Theory in Archaeological Explanation*, edited by C. M. Barton and G. A. Clark, pp. 267–290. Archaeological Papers of the American Anthropological Association No. 7, Arlington, VA (1997).

ultimate cause of the cessation of dated monument construction. The residual variation is then examined for spatial pattern using a variogram, a simple spatial-analytic tool borrowed from geostatistics. While the results fall outside initial expectations, the larger theoretical framework indicates what kinds of historical processes created the patterns observed. The resulting inferences support characterizations of Maya polities that have begun to emerge over the last decade, spurred by the decipherment of Mayan hieroglyphic writing (Coe 1993). I conclude by suggesting how they can be evaluated objectively using additional neo-Darwinian theory and independent sources of evidence. The ultimate test of archaeological theory is its ability to deliver objective knowledge about what happened in history from a mute record. To the extent that this paper achieves its empirical goals, it is also an argument for the promise of Darwinian theory in archaeology.

Monuments and Social Complexity

The historical utility of dated stone monuments among Mayanists derives from the assumption that they can be used to monitor ancient phenomena that are variously glossed as "civilization" or "social complexity." Consider some examples, drawn from the recent literature on terminal dates. Justifying their use of terminal dates to document temporal dynamics of an hypothesized peasant revolt, Hamblin and Pitcher (1980:246) observe that dated monument construction requires "literate forms of Classic Maya culture," including hieroglyphic writing, calendrics, mathematics and astronomy. Hence temporal patterns in terminal dates can be regarded as "quantitative indices of the presence of Maya elites, the carriers-producers of the civilization." Working in a similar tradition, but with an eye to spatial patterns, Bove (1981:95) notes that the distribution of stone monuments is "highly correlated with a hierarchical system of settlement, a high degree of specialization, regionalization, and internal specialization" and can be regarded as a "response to increasing sociocultural evolution." Hence the cessation of monument construction is an "indicator of collapse" and the "demise of the elite." In the same vein, Whitley and Clarke (1985:385) regard monuments as "indicators of 'elite' populations and attendant sociopolitical complexity."

This literature lacks a theoretical account of the causal mechanisms that might link the cessation of monument construction with societal "collapse." Instead, the inferential connection is built on a covert empirical generalization that the occurrence of stone monuments on the one hand and "complexity" or "civilization" on the other are correlated across cultural contexts. This simple behavioral correlate is the means by which the distribution of mute monuments in time and space has been given historical meaning since Morley's pioneering efforts. Its use is not, of course, unique to Mayanists. It underlies chronicles of the rise of civilization the world over. Perusal of this wider literature reveals variation in the precise way in which the behavioral meaning of monuments is cast. The most conspicuous source of variation is change in the interpretive fashions of archaeology. Over the last decade, as increased emphasis on social conflict, factionalism, and inequality displaces the functionalist neo-evolutionism of the New Archaeology, inferences tend to invoke "political power," in place of benign "social complexity." However, recent post-processual approaches have failed to supply the missing etiological framework. Their proponents deny the desirability of attempting to build one, preferring instead to rely on their own covert assumptions about human nature (e.g. Hodder 1986; Shanks & Tilley 1987).

Recent work by Trigger (1990) is an apparent exception to archaeologists' inattention to theory in this area. Trigger's account of the link between monumental architecture and "complexity" or "power" begins with the contrast between Zipf's "principle of least effort" (1949) and Veblen's notion of "conspicuous consumption" (1899). Zipf argued that economizing and efficiency govern the production and distribution of goods necessary to sustain human life, while Veblen noted the ubiquity of wasteful spending to enhance social prestige in modern Europe. Trigger connects the two perspectives by defining conspicuous consumption as a violation of the least-effort principle. The definition is important because it raises the possibility that all artifacts and behavior, from pyramids to potlatches, that require significantly more energy than would be expected under the least-effort principle can be understood in the same conceptual framework as manifestations of a single set of causal dynamics.

The crucial question becomes why is conspicuous consumption a universal symbol of power? What is the causal connection between them? Trigger's answer has two parts. The first is that the energy expended often comes from other people's labor. When it does, conspicuous consumption signals the ability to control the behavior of those doing the work. But clearly this is not the crux of the issue. We still have no explanation for why, when people successfully invest energy in placing the labor of others at their disposal, they choose to spend the proceeds of their investment in a wasteful manner. The second part of Trigger's answer is that "monumental architecture and personal luxury goods become symbols of power because they are seen as embodiments of large amounts of human energy and hence symbolize the ability of those for whom they were made to control such energy to an unusual degree" (1989:125). While superficially satisfying, this is no answer at all. It does not explain why the energy expenditure that denotes power is wasteful, that is violates the principle of least effort. Trigger's answer is little more than a reformulation of the original question in the form of an assertion.

The question goes unanswered because Trigger's account, like the empirical generalizations that preceded it, is crafted from common sense and not a larger mechanistic theoretical framework. Should archaeologists interested in figuring out what happened in history, whether in the Maya Lowlands or elsewhere in the world, worry about the lack of an answer? The price we pay for not doing so is the production of inferences about the past that are often impossible to evaluate objectively (Dunnell 1982,

Neiman 1990), a point to which I will return at the end of this paper. It is the promise of objective evaluation that motivates the theoretical development that follows. There is no question that terminal monument dates are correlated with what archaeologists refer to as the "collapse" of Classic Maya civilization. But the foregoing argument does imply the need for greater precision regarding the social phenomena measured by these dates. Greater precision delivers richer expectations about how the historical process associated with the collapse maps onto spatial variation in terminal monument dates. Greater precision also specifies the statistical methods required to isolate relevant empirical patterns and opens the path to objective evaluation by the archaeological community.

A Darwinian Theory of Wasteful Advertising
Temporal Dynamics

What aspects of social organization are we measuring when we look at terminal monument dates? Answering this question requires fundamental theory about the causes of conspicuous consumption. Trigger's treatment contains two clues for us. One is his suggestion that a history of natural selection operating on primate feeding strategies lies behind Zipf's empirical generalization about effort minimization in human behavior. If selection shapes feeding strategies of both human and non-human primates, and the predictive successes of Darwinian theory suggest it does (e.g. Broughton 1994; Kaplan & Hill 1992), then why is it not also responsible for variation in other aspects of human phenotypes, including conspicuous consumption? Placing conspicuous consumption in a Darwinian perspective immediately clarifies the precise sense in which the energy expenditure is wasteful extravagance. It is wasteful because it represents fitness costs that, unlike the time and energy expended on foraging, have no apparent compensatory fitness benefit. Consider dated Maya monuments. They are big chunks of stone in the form of altars, stairs, and stelae. They represent expenditures of resources to assemble and supervise the individuals who quarried, transported, assembled, and carved them. More importantly, they required the training and maintenance of groups of individuals skilled in the use of a complex system of hieroglyphic writing, the accompanying iconographic repertoire, and in carving and painting both on stone. These expenditures represent the diversion of resources that members of the Maya elite might otherwise have invested directly in the maintenance of themselves and their kin, the acquisition of mates, and the production of children. In other words, the individuals responsible for the monuments appear to have suffered significant direct and opportunity fitness costs in order to construct them.

If a Darwinian perspective sharpens the sense in which conspicuous consumption is wasteful, it also renders it even more puzzling: how could selection possibly account for it? The answer I explore in the remainder of this paper was first suggested by the evolutionary biologist Amotz Zahavi (1975, 1987; cf. Kirkpatrick 1986).[1] Graffen (1990a, 1990b) offers several elegant mathematical models that confirm the process envisaged by Zahavi actually leads to the outcome he claimed for it.

If investment in waste does not confer a direct selective advantage on the individual making it, where does the advantage lie? Zahavi's answer was to look for it in the responses waste engenders in other individuals who happen to observe it. Waste is a signal that conveys novel information about some hidden quality of the sender, who incurred a fitness cost to send the signal, to a receiver. Selective advantage to the sender accrues because the signal causes the receiver to act in a way that benefits the sender. It is this beneficial effect that makes sending the signal worthwhile in the first place. Yet it must be in the fitness interest of the receiver to interpret the signal so as to benefit the sender. In this scenario, the investment in waste is smart advertising. It is smart in the technical Darwinian sense that it is engineered by natural selection so that both the sender's and receiver's strategies enhance their fitness, relative to alternative strategies, thus driving the signaling system to fixation.

Although signals may advertise a variety of qualities about the signaler, the one most relevant to Classic Maya dated monuments is competitive ability. By competitive ability, I mean the likelihood that individuals will win political contests with competitors, contests whose outcomes ultimately *do* determine access to mates and resources. An individual's ability to win political contests in turn is a function of a variety of factors, among them physical size, physiological condition, fighting skill, and psychological cleverness, all of which contribute to an individual's ability to build and maintain coalitions comprised of kin and non-kin.

It is not hard to see why signaling competitive ability might be favored by selection on both sender and receiver: if the two parties differ significantly in competitive ability, communication allows both of them to avoid the fitness costs of a contest whose outcome is a foregone conclusion. Less obvious perhaps is the fact that *honest* signaling and accurate interpretation are in the interest of both parties. To see this, we must consider wasteful advertising from the perspective of both the sender and receiver. Senders should send accurate signals. There is little point in investing energy in a deterrent that can be counterfeited by inferior competitors. Hence there is continual selective pressure to invest in wasteful signals at a level that just exceeds the level that inferior competitors can afford. Conversely, too large an investment in a wasteful signal guarantees that you will have nothing left for direct investment in yourself and your offspring.

There is a complementary pressure favoring the accurate interpretation of wasteful signals. Individuals fooled by counterfeit signals from inferior competitors suffer because they forego opportunities to enter political contests that they probably would win. Individuals who use interpretive rules that result in underestimation of the competitive ability of signalers will enter and lose many contests with superior competitors. It is to the signaler's selective advantage to send truthful signals and to the receiver's selective advantage to interpret them accurately.

The result of this coevolutionary selection regime is the fixation of rules for sending and decoding wasteful signals. The rules produce a signaling system with two distinctive properties. First, advertising is on average honest. Individuals of greater competitive ability or quality advertise more and receivers correctly interpret the message. Honesty is a requirement if the signaling rules are to be selectively advantageous for both sender and receiver. Second, advertising is costly. Its costs scale with the amount of advertising. It is the cost of advertising, manifest in apparently wasteful extravagance, that guarantees its honesty. Natural selection is the key link between these two properties. Here, at last, is a causal account of the link between the wasteful extravagance of conspicuous consumption and competitive ability or power.

In our model, the amount of wasteful advertising undertaken in a group of competitors is a function of three factors: the competitive ability of the worst advertiser, the variance in competitive ability in the group, and the payoff to a given investment in advertising. As the competitive ability of the worst competitor increases, so too will the amount of advertising of which she is capable. This in turn will drive up the amount of advertising necessary for superior competitors to distinguish themselves. Second, greater variance in competitive ability leads to more advertising, because there is more variation in political quality to distinguish and individuals higher up the political quality scale must advertise more. The third factor controlling the total amount invested in waste is the payoff to the signal. This is a function of the number of competitors that can be influenced by a given wasteful investment. As the signal reaches a larger audience, the payoff increases. The character of the audience is also an important determinant of payoff, independent of actual audience size. As the proportion of kin to the signaler in the audience increases, the payoff to advertising declines because kin are less likely to be competitors, other things being equal, given a history of selection for kin-based altruism (Hamilton 1964). In fact, selection may favor relatives who cooperate to pool their advertising resources across a nepotistic, kin-selected social coalition. Payoffs to advertising will increase when a larger proportion of the audience is comprised of strangers who have no other means of assessing accurately the competitive ability of signalers. For example, if audience members have a long history of personal acquaintance with the signaler that offers them a means of accurately assessing his competitive ability, or if the determinants of competitive ability are straightforward and available for easy inspection by audience members, then neither signaler nor receiver benefits from wasteful advertising. We can say that variation in kinship levels and prior knowledge of competitive ability alter the "effective" size of the audience.

Monument Design
Were dated Maya monuments the result of selection for wasteful advertising? We can begin to answer this question by looking for design characteristics that should be engineered into the monuments under such a selective regime. The monuments were engineered to be impossible to counterfeit. I have already alluded to the costs of supporting a literate elite, skilled carvers, labor and so forth that were required in order to erect them. These costs guaranteed the signal sent by the monuments was reliable.[2] Some members of the elite paid those costs directly. They were personally engaged in wasteful activities. For example, several sculptors' names, incised into the back of stelae, lintels, and altars carry the title *ahaw,* "lord." The son of the ruler of Naranjo was a painter of ceramic vessels (Stuart 1993). The development and transmission costs of other ancient scripts were paid for by payoffs in the attestation and accounting of resource ownership (e.g., Harrapan and Sumerian writing [Parpola 1994; Schmandt-Besserat 1979]). In contrast, current evidence indicates the Maya hieroglyphic tradition saw much less practical application. This raises the possibility that it owed its existence, although not its precise form, to selection for wasteful advertising.

The semantic content of the monuments, whether conveyed textually or iconographically, offers additional, if more ambiguous, clues. The ambiguity arises because the particular themes chosen do not require a cost that guarantees their truth in a Darwinian signaling system. On the other hand, if the juxtaposition of high manufacturing costs of the monuments with a particular iconographic message made the latter more credible in the brains of receivers, then we can expect those messages to have advertised political quality as well. The major themes of Classic texts and iconography fit this expectation. Among them is time depth in the elite status of individuals whose life histories are catalogued and trans-generational continuity in the elite status of their predecessors. Late Classic inscriptions at Copan trace continuity between powerful rulers back to the time when the site was in fact a tiny village (Mathews 1985:53). This can be seen as an assertion that political quality of the kin-based coalition in question was high, and because temporal variance was low, quality was likely to remain high in the future. A second prominent theme is the participation of elites in self-inflicted bloodletting rituals (Schele & Miller 1986; Stuart 1988). In Darwinian perspective this practice, which has great time depth in Mesoamerica, is an advertisement of political quality. The ability to withstand blood loss without fainting on the spot or more long-term deleterious effects is a function of physiological and nutritional status and hence an accurate indicator of fighting ability and resource access. The emphasis on battles, conquests, and the portrayal of rulers and their captive elite competitors in various postures of submission has a similar, if more obvious, message (e.g., Stuart 1993). These are precisely the kinds of themes that we would expect from political competitors who are advertising their political quality to intimidate rivals.

Our Darwinian theory of advertising gives us some insight into the processes behind the empirical generalizations and covert assumptions that archaeologists have traditionally used to translate conspicuous consumption into "social complexity" and "power." It offers a causal explanation for why monumental architecture should scale with the range in political quality and the payoffs to advertising. It suggests why Mayanists have been right in using terminal monument dates, which mark a downward spike in the amount of wasteful advertising effort in Maya social groups, as an indicator of a "collapse." And finally, it indicates just what "collapse" means: a shrinkage of the range in competitive ability and a shift in audience size and composition, the latter a function of population density and the spatial scale over which political competition occurred within Maya groups.

Spatial Variation in Wasteful Advertising

So far I have outlined the inferences we are entitled to make on the basis of a drop in the amount of energy invested in advertising, signaled by the end of dated stone monument construction, within a single group of individuals. Termination of monument building means a decrease in our control variables: the range of political quality and the effective number of competitors within the group. But what causes spatial variation in the timing of monument termination among multiple groups of individuals distributed continuously across a region like the southern Maya Lowlands?

Consider the following model. We have a population of individuals, each assigned to a randomly chosen location in a region. Individuals sustain themselves and the social coalitions they lead by drawing on resources from a catchment falling within a certain radius of their location. They compete with other individuals for resources that fall within the portions of their catchment that overlap with the catchments of others. Resource catchments are thus also spheres of political competition. Individuals advertise within their catchments, according to the theory described above, in order to deter competitors. Their levels of resource expenditure on advertising are set by their condition, itself determined by resource abundance within their catchment and their past success against competitors, and by the payoff to advertising, which is controlled by the number of competitors they can reach with a given investment in waste.

Now each catchment is comprised of a finite number of resource patches. Resource abundance within each patch is a function of a region-wide parameter plus an error term randomly sampled from some probability distribution with mean 0. This implies that there are no spatial trends in resource abundance across the region and as a result the *expected* abundance of resources within each catchment (and patches within it) is identical. However the actual abundance will be different, because of random variation among patches. The expected difference between any two catchments in resource abundance is therefore a function of the amount of spatial overlap between them.

If we assume that the payoff to advertising is constant across catchments, then the expected difference between catchments in the level of advertising is controlled by the amount of overlap between them as well. We anticipate that difference will increase as a function of the distance between catchment locations, up to a distance equal to the circumference of the catchments. At this distance, catchment pairs no longer overlap and therefore differences in the level of advertising between them attain their maximal value and remain constant as inter-pair distance increases further.

The relevance of this simple model to the question of terminal monument dates becomes apparent if we add a temporal dimension. During successive time periods, we decrease the region-wide expected value for resource abundance. If we further assume that there is a threshold of energy investment in advertising at which construction of wasteful stone monuments is no longer adaptive, then the expected time at which investment falls below this level is the same for all catchments across the region. However, once again the actual times for each catchment will differ, because of random variation among resource patches within catchments. Once again the difference in the timing between pairs of catchments is expected to increase with the distance between them, up to the distance equal to the catchment circumference. The relevance of this simple model to making inferences about the spatial scale on which political competition occurred is apparent. Under the conditions described, an estimate of the distance at which there is no longer a detectable relationship between terminal monument date differences and geographical distance will also be an estimate of the spatial scale of political competition.

Perhaps the least realistic aspect of the model is the assumption of regional homogeneity in expected resource abundance, the timing of decrease in it, and ultimately the expected date at which different locations and their catchments cross the stone-monument building threshold. In the real world of the Classic Maya, there is likely to be a broad spatial trend in the expected crossing times whose contours are themselves determined by the causes of the collapse. Such a trend may confound our attempts to measure the spatial scale of terminal monument date difference that is determined by random variation among patches within catchments. By the same token, a regional trend, if it can be identified, might give us some insight into the causes of the collapse.

It is apparent that if we are to put our model of catchment variation to work with real data, we are going to have to address two methodological issues. The first is the identification of the large-scale trend in the data. Once this trend is identified, we can remove it from the terminal monument dates and examine the local spatial variation that remains. The characterization of that local variation is our second methodological issue. I use robust local regression (loess) to estimate the trend. I then turn to the variogram to examine residual spatial variation in monument dates.

THE SPATIAL TREND AND COLLAPSE DYNAMICS

Data

The data on terminal monument dates used in this study were gleaned from several sources. I started with Bove's (1981) sample of 47 Late Classic sites, and then corrected dates to reflect the

TABLE 25.1. Terminal monument dates from 69 sites used in this analysis. Fitted values are expected terminal dates, computed by loess, as described in the text. The residuals are the deviations of the actual terminal dates from these fitted values. The derivation of the mean annual rainfall estimates is described in Note 4.

Site	Monument	Long Count	Source	Date (AD)	Fitted Value	Residual	Rainfall
Aguacatal	Stela 1	9.16.0.0.0	Houston, n.d.	762	754	8	1600
Aguas Calientes	Stela 1	9.18.0.13.18	Mathews and Willey 1991	791	791	0	2682
Aguateca	New Stela	9.18.3.0.17	Houston, n.d.	793	789	4	2799
Altar de Sacrificios	Stela 15	9.17.0.0.0	Mathews and Willey 1991	771	790	−19	2734
Bonampak	Murals	9.17.15.12.10	Miller 1994	792	792	0	2632
Calakmul	Stela 64	9.19.0.0.0	Morley et al. 1983	810	833	−23	1671
Cancuen	Stela 1	9.18.10.0.0	Morley et al. 1983	800	795	5	3088
Caracol	Stela 17	10.1.0.0.0	Morley et al. 1983	849	810	39	2100
Chinkultic	Stela 1	10.0.15.0.0	Morley et al. 1983	844	855	−11	1462
Comalcalco	Brick	9.19.3.13.12	Houston, n.d.	814	820	−6	2471
Comitan	Stela 1	10.2.5.0.0	Morley et al. 1983	874	875	−1	1003
Copan	Altar L	9.19.11.14.5	Houston, n.d.	822	826	−4	1316
Dos Pilas	Bench 1	9.15.10.17.15	Houston 1993	742	789	−47	2735
El Caribe	Stela 2	9.17.10.0.0	Morley et al. 1983	780	791	−11	2682
El Cayo	New Orleans	9.18.1.12.16	Houston, n.d.	792	787	5	2738
El Chal	Stela 4	9.16.10.0.0	Houston, n.d.	761	804	−43	2501
El Choro	Altar 4	9.17.0.0.0	Mathews and Willey 1991	771	792	−21	2673
El Palmar	Stela 41	10.2.15.0.0	Morley et al. 1983	884	869	15	1652
El Peru	Stela 32	9.18.0.0.0	Houston, n.d.	790	801	−11	2672
Itsimte	Altar 1	9.17.5.0.0	Morley et al. 1983	775	805	−30	2288
Itzan	Stela 6	9.19. 19.16.0	Mathews and Willey 1991	829	791	38	2675
Ixkun	Stela 5	9.18.10.0.0	Morley et al. 1983	800	806	−6	2601
Ixlu	Stela 2	10.2.10.0.0	Morley et al. 1983	879	819	60	2122
Ixtutz	Stela 4	9.17.10.0.0	Morley et al. 1983	780	801	−21	2738
Jimbal	Stela 2	10.3.0.0.0	Morley et al. 1983	889	836	53	1836
Jonuta	—	9.18.0.0.0	Marcus 1976	790	777	13	2589
La Amelia	Panel 1	9.18.17.1.13	Houston, n.d.	807	790	17	2722
La Florida	Stela 7	9.16.15.0.0	Morley et al. 1983	766	784	−18	2886
La Honradez	Stela 4	9.18.0.0.0	Morley et al. 1983	790	863	−73	1790
La Mar	Stela 2	9.18.15.0.0	Morley et al. 1983	805	790	15	2672
La Milpa	Stela 7	9.17.10.0.0	Morley et al. 1983	780	885	−105	1776
La Muneca	Stela 1	10.3.0.0.0	Morley et al. 1983	889	851	38	1617
Lacanha	Lintel 1	9.15.15.0.0	Morley et al. 1983	746	794	−48	2535
Laguna Perdita	Altar 1	9.15.11.2.17	Morley et al. 1983	742	807	−65	2149
Los Higos	Stela 1	9.17.10.7.0	Morley et al. 1983	781	785	−4	1699
Lubaantun	Altar 2	9.18.0.0.0	Bove 1981	790	800	−10	3103
Machaquila	Stela 5	10.0.10.17.5	Mathews and Willey 1991	841	791	50	2844
Morales	Stela 1	9.16.5.0.0	Bove 1981	756	770	−14	2927
Mountain Cow	Altar 1	10.0.5.0.0	Bove 1981	835	820	15	2518
Naachtun	Stela 10	10.16.10.0.0	Morley et al. 1983	761	838	−77	1689
Naj Tunich	Drawing 29	9.17.0.6.3	Houston, n.d.	771	799	−28	3012
Nakum	Stela D	10.1.0.0.0	Morley et al. 1983	849	834	15	1933
Naranjo	Stela 32	9.19.10.0.0	Morley et al. 1983	820	835	−15	2010
Nimli Punit	Stela 14	9.18.10.0.0	Houston, n.d.	800	802	−2	3076
Oxpemul	Stela 7	10.0.0.0.0	Morley et al. 1983	830	833	−3	1593
Palenque	Tablet 96	9.17.13.0.7	Morley et al. 1983	783	798	−15	3681
Piedras Negras	Altar 3	9.19.0.0.0	Morley et al. 1983	810	785	25	2838
Poco Uinic	Stela 3	9.18.0.0.0	Morley et al. 1983	790	828	−38	1530
Polol	Stela 1	9.18.0.0.0	Morley et al. 1983	790	801	−11	2425
Pomona	Stendahl Panel	9.18.0.0.0	Houston, n.d.	790	781	9	3232
Pusilha	Stela E	9.15.0.0.0	Morley et al. 1983	731	794	−63	3217

TABLE 25.1. (cont'd.) Terminal monument dates from 69 sites used in this analysis. Fitted values are expected terminal dates, computed by loess, as described in the text. The residuals are the deviations of the actual terminal dates from these fitted values. The derivation of the mean annual rainfall estimates is described in Note 4.

SITE	MONUMENT	LONG COUNT	SOURCE	DATE (AD)	FITTED VALUE	RESIDUAL	RAINFALL
Quen Santo	Stela 2	10.2.10.0.0	Bove 1981	879	859	20	1625
Quirigua	Temple 1	9.19.0.0.0	Morley et al. 1983	810	792	18	2220
Sacul	Stela 9	9.18.0.0.0	Houston, n.d.	790	805	−15	2667
Seibal	Stalae 18 & 20	10.3.0.0.0	Morley et al. 1983	889	792	97	2663
Tamarindito	Stairway 2	9.16.11.7.13	Mathews and Willey 1991	762	789	−27	2728
Tayasali-Flores	Stela 1	10.2.0.0.0	Morley et al. 1983	869	812	57	2182
Tikal	Stela 11	10.2.0.0.0	Morley et al. 1983	869	831	38	1882
Tila	Stela A	10.0.0.0.0	Morley et al. 1983	830	828	2	2500
Tonina	Monument 101	10.4.0.0.0	Morley et al. 1983	909	824	85	1741
Tortuguero	Stela 2	9.14.0.0.0	Morley et al. 1983	711	820	−109	3615
Uaxactun	Stela 12	10.3.0.0.0	Morley et al. 1983	889	834	55	1831
Ucanal	Stela 4	10.1.0.0.0	Morley et al. 1983	849	817	32	2333
Uxbenka	Stela 15	9.17.10.0.0	Houston, n.d.	780	799	−19	3065
Uxul	–	9.16.0.0.0	Morley et al. 1983	751	823	−72	1715
Xultun	Stela 10	10.3.0.0.0	Morley et al. 1983	889	850	39	1795
Xunantunich	Stela 1	10.1.0.0.0	Morley et al. 1983	849	835	14	2124
Yaxchilan	Lintel 10	9.18.17.13.14	Houston, n.d	808	788	20	2711
Yaxha	Stela 31	9.18.5.16.4	Houston, n.d	796	828	−32	2051

results of more recent research. Terminal dates for an additional 22 sites were culled from several sources, as noted in Table 25.1. Terminal dates in the sample range over a 178-year period, from AD 731 (Tortuguero) to 909 (Tonina). Geographical coordinates for the 69 sites were scaled from two recently published maps at 1:500,000 and 1:1,609,000 (Graham 1982; Garver 1989). Measurement error that affects the resulting estimates of inter-site distance is likely on the order of ±1 km, but this is inconsequential given the scale of spatial pattern investigated here.

Trend Estimation

The question of how best to detect and estimate the broad geographical trend that likely underlies our data is a complex one. The trend in question will represent expected times that each site crossed the monument-building threshold. Actual crossing times will depart from expectation because of random among-patch variation within the political catchment which the site represents. Some clues about how the trend might be estimated can be had by considering just how those site-specific departures (i.e. the error term of the model) are likely to be distributed. We have no reason to believe that the resulting site-specific errors have any particular distribution, let alone a Gaussian or normal one. We can expect unique historical trajectories associated with some political catchments to produce actual terminal dates that greatly depart from expectation, resulting in an error distribution with long tails and perhaps some outliers. In fact, site groups in different areas of the study region may have different error distributions. On the other hand, our model does lead us to expect that the errors for sites that are geographically close to one another will tend to have similar values. This geographical dependence, or spatial autocorrelation, of the errors is precisely one of the phenomena we hope to investigate. These characteristics—stretched-out tails, regional heterogeneity in error variance, and autocorrelation—mean that the traditional parametric trend-surface approach to this problem, based on least-squares estimation, will not yield best unbiased estimates nor will the usual statistical tests be meaningful. Hence we can expect poor results. The problem can also be considered from the perspective of the trend. There is no reason to suspect that the expected crossing times, which after all are a function of historical processes that varied across the region, will be well described by some single parametric surface, i.e., a linear, quadratic, or some higher-order polynomial trend surface globally fit to all the data points. Fitting such a global polynomial is likely to result in spurious inflections in the fitted surface.

What is required then is an estimation procedure that is local, in the sense that its estimates of expected crossing times for a given site are based on sites within a certain local neighborhood. Local estimation is also a way of coping with heterogeneous error distributions across the entire region. Our estimation procedure also needs to be robust so that the fitted surface is not unduly affected by long tails in the error distribution and outliers. One procedure that fits this requirement is loess, a technique of fitting a local regression model with robustness properties. Detailed accounts of loess can be found elsewhere (Cleveland et al. 1991; Cleveland 1993). Two parameters need to be specified to fit a loess model, α and λ. The first of these, α, specifies the size of the local neighborhood to be considered in fitting each point, while the second specifies whether the local fit is to

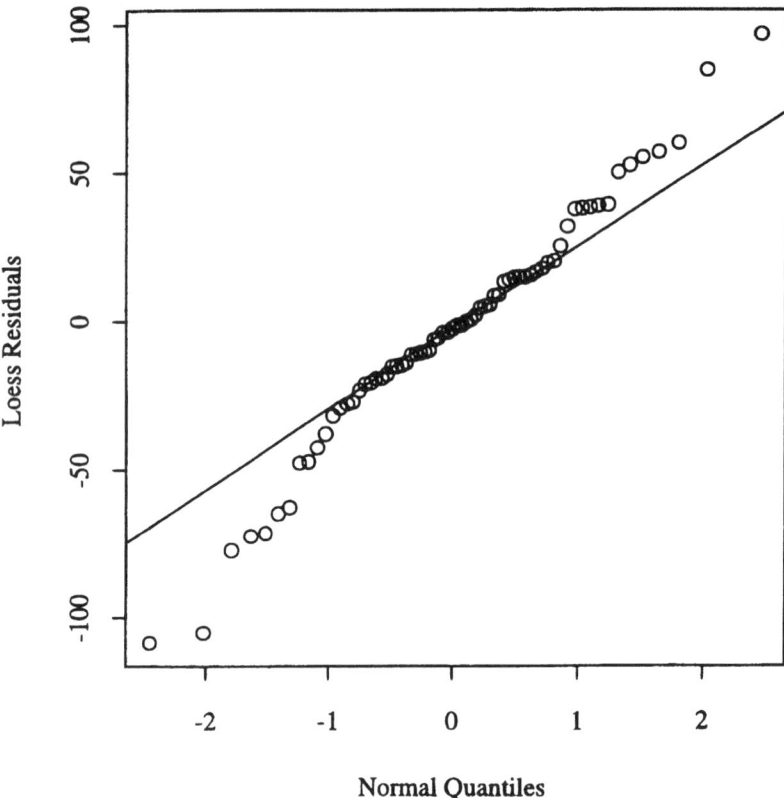

FIGURE 25.1. Normal probability plot of the terminal monument date residuals (actual terminal dates − expected dates, computed from loess). The line passes though the quartiles of the residuals and the reference Gaussian distribution. If the distribution of date residuals were Gaussian, all the points would fall on the line. The observed configuration indicates too much spread in the tails of the distribution. There are also 4 outliers, two at the top (Seibal and Tonina) and two at the bottom of the distribution (Tortuguero and La Milpa).

be a linear or quadratic function of the predictor variable, in this case the site coordinates. Quadratic fitting performs better with curved trends and is used here. The expected terminal date for each is estimated from a weighted regression on its geographical coordinates, with the weights a function of proximity of other sites in the neighborhood to the site in question. Robustness is achieved by iteratively estimating the surface, while re-weighting the influence of each point in the local neighborhood on the basis of its departure from the previous estimate until successive estimates converge.

The key question in using loess is how to choose a value for α. The technique used here is cross validation (Efron & Tibshirani 1993:258–63). For each trial value of α, we proceed as follows. We fit the local regression surface 69 times, leaving one of the 69 sites out in turn. For each of the resulting 69 fits, we compute the expected value for the site we left out, then compute the deviation of the actual from the expected value. The 69 deviations can then be summarized in several ways. Squaring them, adding them up and taking their mean, results in a mean squared prediction error ($MSPE$). I computed this badness of fit measure for successive trial values of α (from .2 to 1.0 in steps of .01). For

the data in question, the minimum $MPSE$ occurs at $\alpha = .90$. I used this value in the analysis reported below.

An examination of the residuals from the expected values makes possible an evaluation of the loess fit. The residuals are estimated in the standard way by subtracting the actual terminal dates from the expected values computed from the loess fit (Table 25.1, Figure 25.1). Plotting them against the corresponding quantiles of a standard normal distribution offers a helpful way to characterize their distribution. If the residuals follow a normal distribution, they all will fall on a single line. Adding a straight line to the plot, placed through the upper and lower quartiles, helps assess how well the residuals conform to this standard. It is obvious that the distribution of residuals has heavy tails at both ends—the departures from the fitted values here are too large. In addition, there appear to be outliers at each end of the residuals distribution. Both the heavy tails and outliers indicate the appropriateness of the loess fit, with its robustness features. There does not appear to be significant long-range dependence of the residuals on absolute spatial location, as demonstrated by conditioning plots of the residuals against their north-to-south and west-to-east spatial coordinates (Fig-

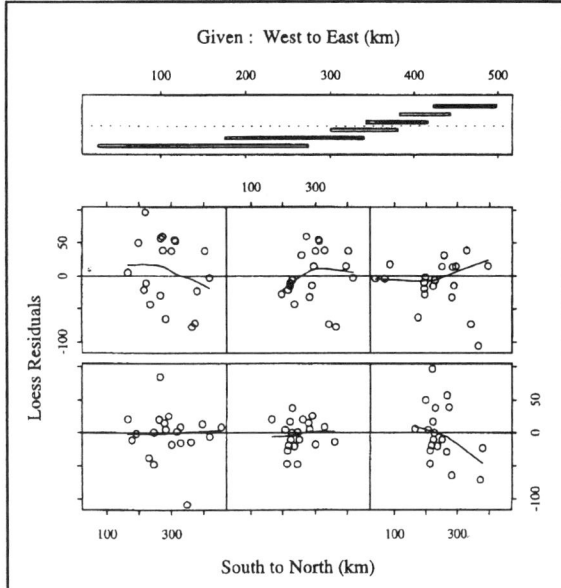
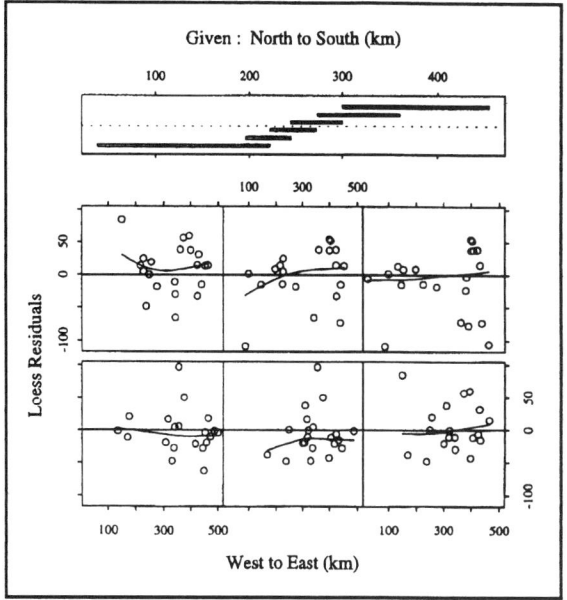

FIGURE 25.2. Conditioning plots of terminal date residuals. *A*. Residuals for six ranges of west-to-east coordinates, shown in the top panel, are plotted against the full range of south-to-north coordinates. *B*. Residuals for six ranges of north-to-south coordinates, shown in the top panel, are plotted against the full range of west-to-east coordinates. In both cases, loess curves fitted to the residuals fail to detect important large-scale trends, indicating that the loess fitted surface (Figure 25.3) captures these.

ure 25.2). Hence the loess surface appears to capture the large-scale trend in the data.

The final loess fitted surface is shown in Figure 25.3. A clear trend is apparent, with expected dates for crossing the monument-construction threshold varying widely across the region from c. AD 760 to 900. This means that the deterministic component behind the shrinkage in political-quality range and audience size, marked by the cessation of monument construction, is a time-transgressive process operating in different places in the region at times separated by nearly a century and a half. The surface is a valley whose floor runs from the Gulf of Mexico southeast to the Gulf of Honduras and whose sides slope up to the northeast and southwest.[3] This result can be used to evaluate collapse scenarios that have been prominent in the literature over the past several decades. There can be little doubt that proximate factors related to the cessation of dated monument construction were many and complexly interwoven. Among these may have been increases in levels of political instability and inter-polity warfare, a possibility that has received increased attention among Mayanists in recent years (e.g., Demarest 1992, 1996). Explanations for the collapse that invoke warfare remain incomplete unless they reference changing ecological or social conditions that changed the costs and benefits of violence. The following discussion covers only ultimate causes whose implications for spatial pattern in terminal dates are reasonably clear. I consider three such causes here: invasion, drought, and long-term population-resource imbalance leading to an ecological disaster.

Invasion

Invasion explanations have enjoyed sporadic popularity over the past 30 years. Willey and Sabloff (1967) proposed that the collapse was caused by an invasion of the groups from the Gulf Coast Lowlands, perhaps ultimately of Mexican derivation. Sabloff and Willey imagined not only that invaders seized Seibal in the Peten, sending raiding parties out from this central base, but that this initial incursion eventually became part of a larger movement of diverse groups originating in the northwest into the Maya area. Invasion hypotheses were popular in the early 1970s. They were featured in the work of Sabloff (1973) and Adams (1973) and were favored among the majority of the participants in the School of American Research conference on the collapse (Culbert 1973). They continue to be proposed today. For example, Sabloff (1992) has resurrected a version of his earlier thesis, in which an incursion of Chontal Maya, now merchants not invaders, traveling up the Usumacinta drainage from the Gulf Coast, had a hand in the collapse. In a similar vein, several workers have proposed incursions from the Yucatan, a region that appears to have come through the collapse relatively unscathed (e.g., Rice 1985; Chase & Chase 1982). In a recent review, Culbert has noted there is "increasing evidence that the southern Lowland Maya may have been under military pressure from the north" (1988:100). Common to the recent versions of the external intrusion hypotheses, is the notion that the intruders were successful because the Maya had already become weak from internal causes (e.g., Culbert 1988:79; Sabloff 1992).

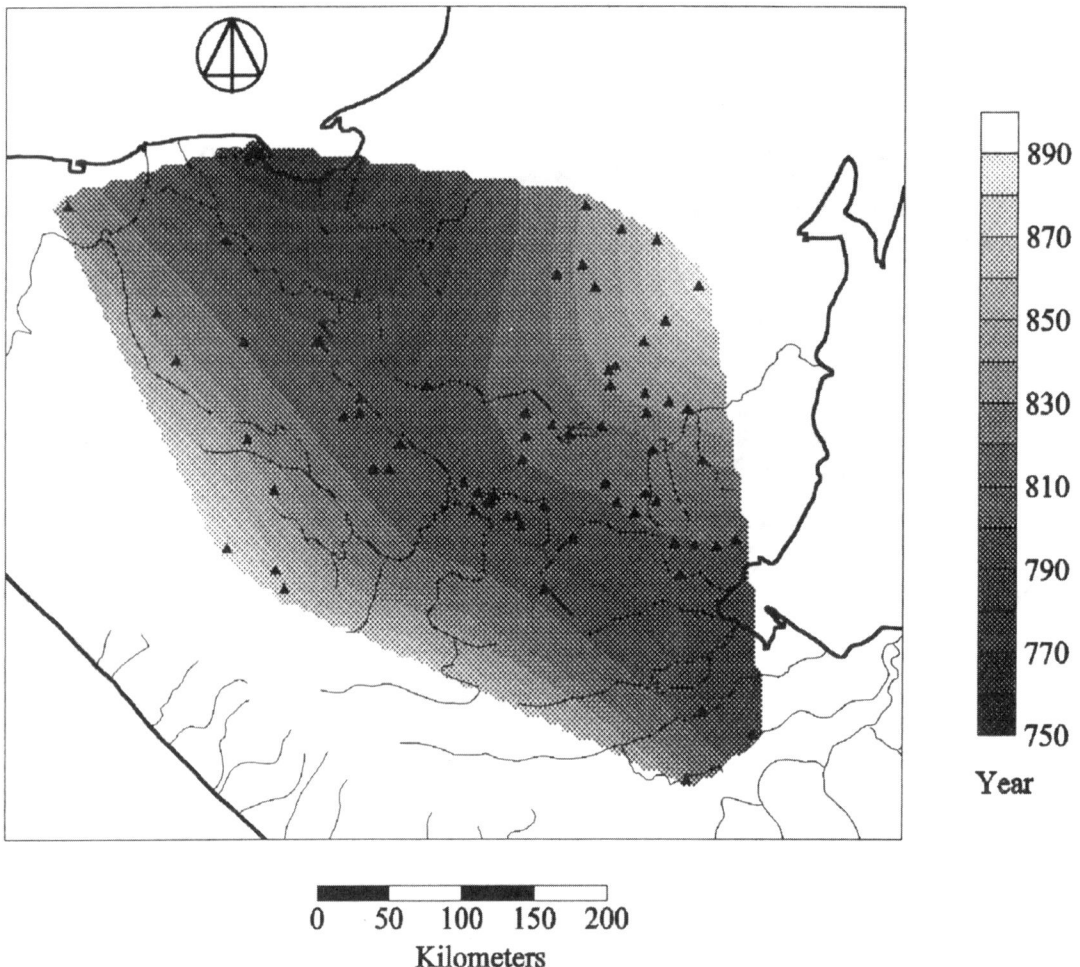

FIGURE 25.3. Contour plot of expected terminal monument dates fitted by loess. The value of the parameter ($a = .90$), which determines the amount of smoothness in the surface, was chosen by cross-validation.

The causal links between an incursion, however conceived, and the cessation of monument construction are poorly specified in the invasion literature (cf. Binford 1968). Under the wasteful advertising hypothesis, they can be sought in a reduction in the variance of political quality within local groups or in a disruption of the transmission of skills required for dated monument construction. If outside intrusions were responsible for the cessation, what sort of spatial patterning in terminal monument dates might result? We can expect a clinal pattern whose configuration would be a function of the route followed by the intruders (Bove 1981). The clinal pattern recovered by loess from the terminal dates fits the pattern expected under Sabloff's various scenarios best. There is a tendency for fitted terminal dates to be earlier in the lower reaches of the Usumacinta drainage. However, the time scale represented by the fitted surface fits neither invasion hypothesis. Both the incursions from the northern Yucatan and from the Gulf Coast are thought to be ninth-century phenomena. But the process mapped here appears to have been underway by the middle of the eighth century.

Drought

The notion that the collapse was the result of climatic forcing has recently received apparent support from evidence derived from sediment coring in Lake Chichancanab in northern Yucatan (Hodell et al. 1995). Temporal trends in oxygen isotopes from shell carbonates and the ratio of calcite to gypsum in sediments suggest that the interval between AD 800 and 1,000 was the driest from the mid-Holocene (c. BC 5000) to the present. The coincidence between this climatic event and the timing of the collapse has led to the suggestion that a two-century long drought triggered the collapse. Although, as with the invasion hypothesis, proponents see internal stresses resulting from "an overall matrix of demographic and environmental pressures" as a necessary prior condition (Sabloff 1995:357).

The connection between drought, specifically the agricultural crisis that it might precipitate, and the cessation of monument construction would be twofold. First we might expect a reduction in the range of political quality among competitors, caused by inability of individuals to produce or extract from others enough food to feed and maintain large cohorts of po-

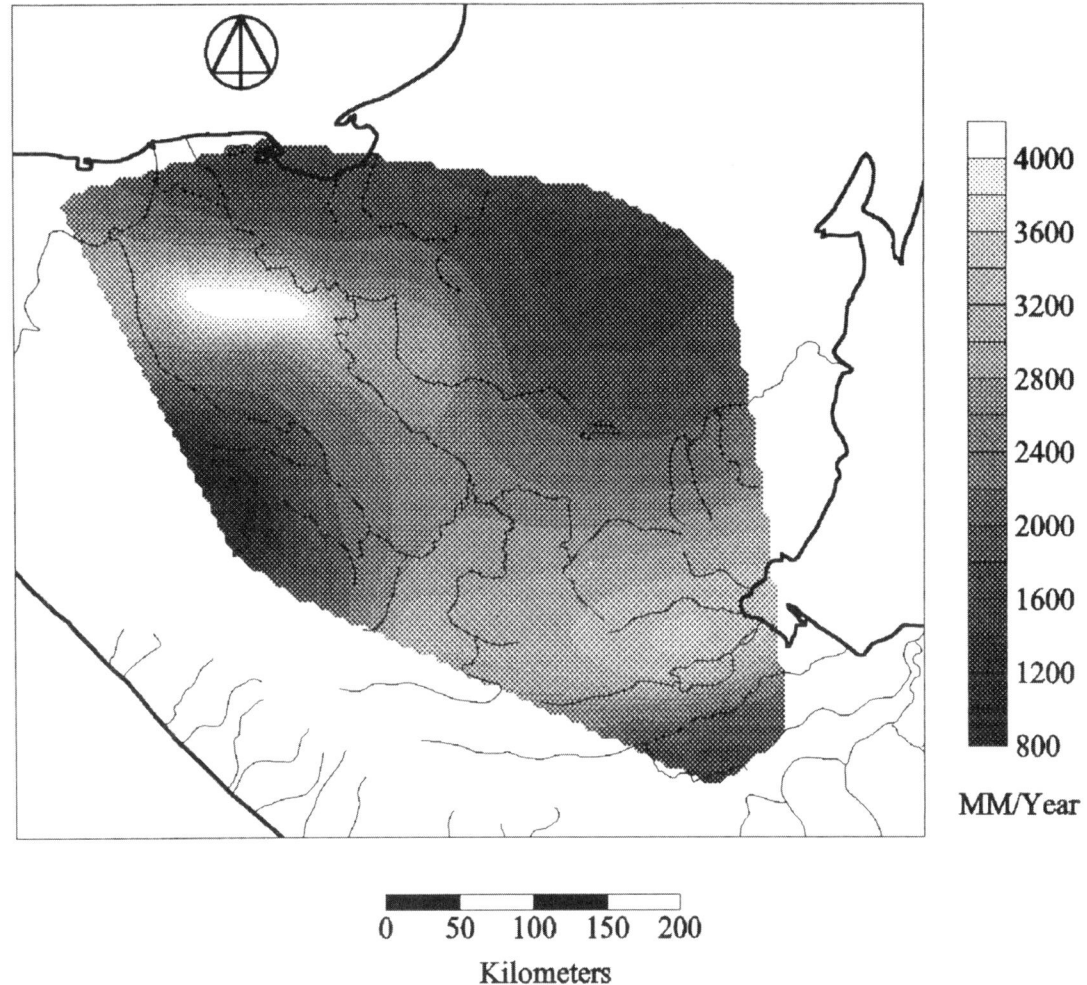

FIGURE 25.4. Contour plot of mean annual rainfall in the Maya Lowlands in millimeters (after Rice 1993).

litical allies and thus maintain former levels of political quality. Second, we might also expect a reduction in population size and thus the number of political competitors, caused by emigration and/or starvation in the face of lowered agricultural productivity.

The fitted surface indicates that the collapse had begun several decades before the onset of this climatic event. In addition, the details of the deterministic component of the spatial pattern identified by loess do not support the drought hypothesis. The southern Lowlands exhibit considerable regional variation in annual rainfall, from 1000 mm to 4000 mm (Rice 1993). If drought were a significant factor in the cessation of monument construction, sites that receive the lowest rainfall should show the earliest effects of the drought and have the earliest fitted terminal dates. However, regional trends in rainfall and in terminal monument dates have precisely the opposite geographical relationship (Figure 25.4). Areas with the *highest* annual rainfall (>2500 mm) have the earliest expected terminal dates. This is the swath extending from the Gulf of Honduras, westward across the southern Peten, into the Pasión and Usumacinta primary drainages, and then northwest down the Usumacinta to its mouth in the Gulf of Mexico. On the other hand, areas on the low end of the rainfall gradient, which bound this swath to the northeast and southwest, have the latest expected terminal dates.

A similar relationship between rainfall and the cessation of monument construction can be identified on a larger spatial scale. Northern Yucatan receives much less rainfall than the southern Lowlands. The mean annual figure for Lake Chichancanab is 1300 mm (Hodell et al. 1995) and this diminishes both to the north and northwest of the lake, where the Post Classic centers of Uxmal, Kabnah, Sayil, and Chichén Itzá lie. Yet these centers began their florescence as the southern Lowlands went into decline, after the initiation of the rainfall hiatus (Chase & Rice 1985; Andrews & Sabloff 1986; Sabloff 1995).

Ecological Disaster
The final collapse scenario considered here invokes a population-resource imbalance leading to an ecological disaster. Evidence in favor of this model has been accumulating from a variety of

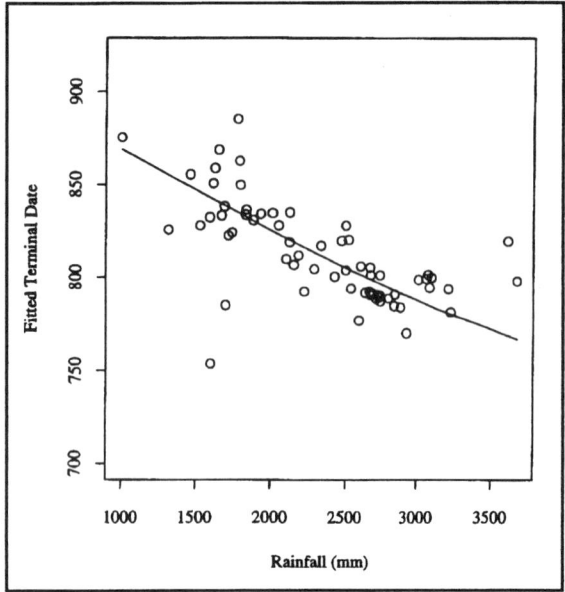 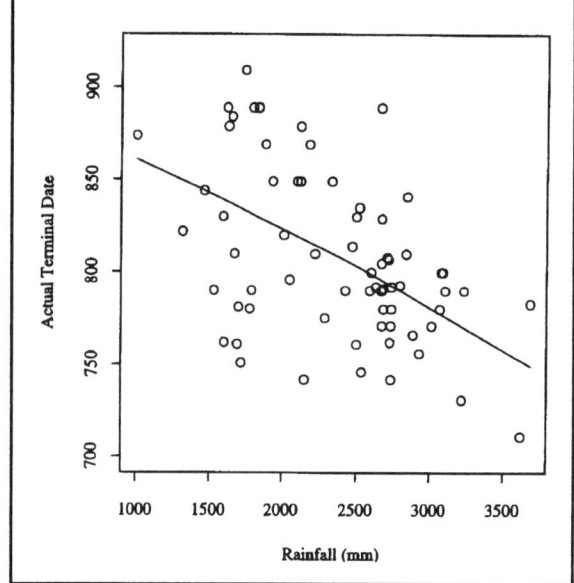

FIGURE 25.5. Scatterplots of fitted terminal dates (A.) and actual terminal dates (B.) as a function of mean annual rainfall. Fitted lines are from loess.

sources over the past two decades. Conspicuous among them is the record of anthropogenic environmental alteration recovered from lake sediment cores (recently reviewed by Rice 1993). The interval from BC 1000 to AD 900 witnessed dramatic increases in the amount of sediment and in silica concentrations, offering convincing evidence for soil erosion on a massive and destructive scale (Deevey et al. 1979; Leyden 1987; Rice et al. 1985). Increased levels of phosphorus attest to soil and ground-cover nutrient leaching. Increases in pollen from grasses and weeds associated with cultivation, and decline in arboreal pollen, point to deforestation wrought by a growing population seeking wood for fuel (Abrams & Rue 1988) and clearing land for swidden agriculture under shorter and ultimately unsustainable fallow cycles (Rice 1993:42). Some archaeologists estimate that by c. AD 900 the Maya had cut down or significantly altered as much as 75% of the lowland forest (Whitmore et al. 1990). In some places, evidence of regrowth cannot be detected until c. AD 1500 (Brenner et al. 1990). That population growth continued unabated into the Late Classic is documented by both archaeological surveys (Culbert 1988; Culbert & Rice 1990) and increased fertility levels in skeletal samples (Whittington 1991). Biological stress, particularly poor nutrition, even among members of the elite, is documented in Classic and Late Classic skeletal samples from Tikal (Haviland 1967), Altar de Sacrificios (Saul 1973), and Copan (Whittington 1989). There is evidence for an increase in childhood morbidity from Late to Terminal Classic times (Wright 1996).[4] Taken together this evidence indicates that by the Late Classic, Maya populations found themselves in an increasingly barren and agriculturally unproductive landscape. That the result was a Malthusian catastrophe is suggested by the fact that Terminal Classic site densities across most of the southern Lowlands range between 15 and 60% of Late Classic levels, with some areas virtually abandoned well into the Post Classic (Culbert 1988:84; 1995:118,127, Culbert & Rice 1990).

Under this scenario, the cessation of monument construction was the predictable outcome of collapse in the range of political quality and the number of competitors, just as in the drought scenario. However in this case, the spatial trends would be determined by the intersection of locally variable population levels, edaphic factors, land use patterns, rainfall levels and the erosion and nutrient depletion that would result from their interaction. An accurate picture of variation in all these variables within the region during the Late Classic is unavailable. However, if local variation in population, soils, and land use were small, spatial variation in annual rainfall would be the major determinant of soil erosion and nutrient depletion rates, with those destructive processes proceeding most rapidly in areas receiving the highest rainfall. This seems unlikely. Yet surprisingly this expectation is met. It is implied in the previous discussion of drought: the southeast to northwest trending swath across the Lowlands with the highest rainfall levels also has the earliest fitted terminal dates. Even more surprising is the strength of the relationship. The rank correlation between rainfall levels and the fitted dates for the 69 sites in our samples is −.69 (Spearman's r).[5] The relationship is so strong that it is even convincingly detectable using the raw terminal dates (Spearman's $r = -.44$) (Figure 25.5). This remarkable and unsuspected result indicates that spatial variation in annual rainfall is the most important determinant of the deterministic component of spatial patterning in terminal monument dates. It provides strong support for the idea that an erosion-driven ecological disaster was the principle, ultimate cause of the cessation of dated-monument construction.

To estimate the probability that these results are an artifact of sampling error, we cannot rely on standard methods that assume observations are independently sampled. Both the fitted date and rainfall values are spatially autocorrelated: sites that are geographically close have similar values. However, recasting the problem in terms of patterns of difference among fitted date and rainfall values makes statistical evaluation possible. By regressing the matrix of fitted date differences on matrices of geographical distances and rainfall differences among the 69 sites, we can partial the effects of distance (and hence spatial autocorrelation) out of the linear relationship between fitted-date and rainfall differences. Null-hypothesis distributions for estimated parameters can be obtained by randomly permuting the rows and columns of the fitted date matrix (Smouse et al. 1986; Smouse and Long 1992).[6] Standardized regression coefficients, which measure effect sizes in standard deviation units, reveal that rainfall has over twice the effect on fitted dates as geographical distance, when each independent variable is held constant ($\beta_{distance} = .19$, $\beta_{rainfall} = .45$). In other words, a unit change in geographical distance brings with it only a .19 unit change in fitted date difference, while a unit change in rainfall difference is associated with a .45 unit change in fitted date difference. Permutation-based testing reveals the probability of observing effects of this size by chance alone is vanishingly small.

THE SPATIAL SCALE OF POLITICAL COMPETITION

The Variogram

Having estimated the deterministic regional trend in terminal monument dates, it now becomes possible to put our model of spatial variation in wasteful advertising to work to infer the average size of spheres of political competition on the eve of the cessation of monument construction. Under that model, the local pattern of residual variation around the expected terminal dates that comprise the loess fitted surface depends on the size of the areas over which elites compete with one another. Our model suggests that the spatial scale or distance over which there is a tendency for the residuals to exhibit statistical dependence will indicate the size of catchments within which Maya elites competed for resources and over which there was a payoff for wasteful advertising of political quality. The job of estimating that scale can be accomplished by computing a variogram for the terminal date residuals.

A variogram is a simple spatial-analytic tool borrowed from geostatistics (Davis 1986; Isaaks & Srivastava 1989). It is a plot of the dissimilarity between the values of a spatially distributed variable at points in space as a function of the distance between those points. In our case the point pairs are pairs of sites, while the spatially distributed variable is the deviation of actual terminal dates from values expected on the basis of the loess fit. The measure of dissimilarity used is called the semivariance. It is half the mean of the squared difference between residual values at all pairs of sites that are a certain distance or range of distances apart. The variogram is a plot of the semivariance against the successive distance classes over which the mean-squared differences were computed.

Local spatial dependence, in the absence of regional trends, will yield a variogram in which dissimilarity increases up to a given inter-site distance class and then remains unchanged over the remaining distance classes. This flat portion of the variogram is called the sill. A nice result from mathematical statistics says that when the process generating the data values contains no deterministic spatial trend, a condition known as stationarity, and only local spatial dependence, then the value of the semivariance at the sill is equal to the total variance in the data values. The distance at which the semivariance reaches its maximal value at the sill represents the range of distances over which terminal dates exhibit statistical dependence on one another or are spatially autocorrelated.

This set of theoretical relationships makes the variogram a useful tool with which to estimate the scale of political competition among monument-building elites on the eve of the collapse. If our model of spatial variation in wasteful advertising approximates competitive dynamics on the eve of the collapse, then the semivariance of our date residuals will increase with increasing distance between sites up to a sill, which is equal to the total variance in the date residuals. After that point, the semivariance will remain flat across ever-larger distance intervals. The distance at which the sill is detected, the range, is the average distance over which wasteful advertising and the political competition of which it was a part took place.

Before we can estimate the variogram, we need first to decide on the width of the distance classes over which we are going to compute the semivariance. Because interest focuses on the shape of the variogram, it is desirable to use relatively narrow distance classes. However, narrow classes are likely to yield a noisy point scatter when the semivariance estimates are plotted against the distance classes over which they were computed. Using wider distance classes may produce a better behaved point scatter, but may obscure interesting details of the semivariance-distance relationship. A way out of this dilemma is to use small distance classes for the semivariance estimates, and then smooth the resulting noisy scatterplot. Loess, with a smoothing parameter α again chosen by cross-validation, recommends itself as a reasonable way to do this.

In computing the variogram, we are assuming that the process that generated the date residuals operated similarly at all sites. Examination of the distribution of residuals suggests this is a reasonable assumption for most of the sites, but perhaps not for the four outliers (Figure 25.1). Hence it seems reasonable to estimate variograms for all the residuals and for a subset of the residuals, with the four outliers removed. I computed the semivariance of the date residuals for successive 6 km distance classes up to 200 km. It is customary to compute the variogram for distances up to half the smallest map dimension. Estimates of the semivariance for distance classes greater than this value depend not only on the distance between pairs of sites, but also on the

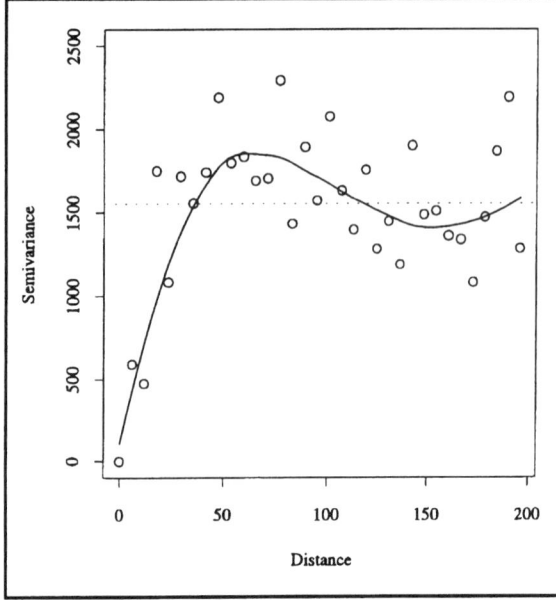 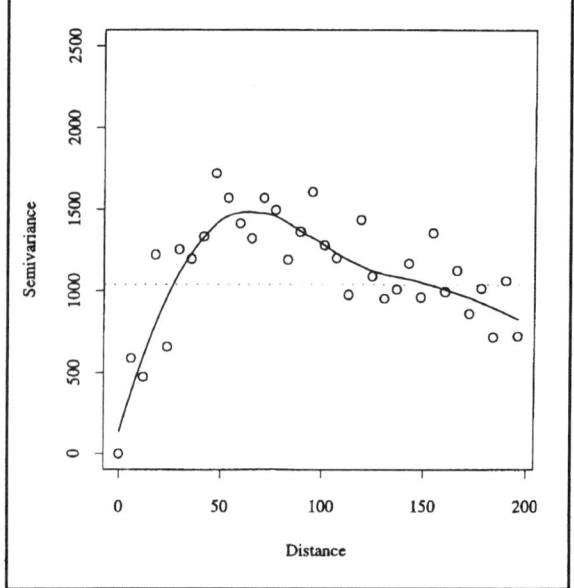

FIGURE 25.6. Variograms of the residuals from expected terminal monument dates. The curves were fitted by loess, with smoothing parameter (α = .65) chosen by cross-validation. The horizontal line represents the total variance of the residuals and is the expected height of the variogram's sill under the stationarity assumption. A. Variogram for all 69 residuals. B. Variogram for 65 residuals, 4 outliers excluded

absolute locations of the sites. For example, in a study area c. 500 km on a side, estimates of the semivariance at distances greater than 250 km will increasingly be dominated by sites that fall close to the map edges. This dependence on location introduces unwanted bias into the semivariance estimates. I used loess to smooth the semivariance-by-distance scatterplots, with α = .65 chosen by crossvalidation. The result is shown in Figure 25.6.

It is clear that for both sets of residuals the variogram, as described by the loess curve, does not behave in the simple way that our model initially led us to expect. Instead of increasing to a flat sill equal to the total variance in the data, the semivariance increases past the total variance, reaching a maximum at a distance of roughly 65 km. Thereafter it declines until it reaches a value that is roughly equal to the total residual variance, at about a distance of about 150 km.[7] This sort of pattern in a variogram is known as the "hole effect" in the geostatistical literature (e.g., Isaaks & Srivastava 1989:156). It means that pairs of sites that are close together do tend to have more similar residual terminal dates than pairs of sites chosen at random, as we expected. However, pairs of sites that are 65 km apart have terminal dates that are less similar than pairs of sites chosen at random. This is a surprise.

What are we to make of this pattern? It indicates that our initial model of spatial variation in terminal monument dates is missing a key component. The Darwinian theory of advertising helps isolate just what this overlooked process might be. Recall that in the model terminal date thresholds are a function of two sets of quantities. The first is the range of political quality, which is determined by resource abundance within each competitor's catchment. This source of variation among sites should have been estimated by the loess fit and it should not appear in the residuals. Even if it does, it is difficult to envisage a process whereby long-term trends in population growth, rainfall, and soil erosion might result in sites that are 65 km apart having unusually different departures from the trend.

The second is the payoff to advertising. Here we find a more likely cause for the tendency of sites 65 km apart to depart from the loess trend in different directions: movement of individuals among resource catchments. As average resource abundance declined across the entire region, random variation among catchments prompted individuals based in areas with lower levels of resources to attempt to move into adjacent catchments where conditions were more favorable. Such movement would result in an increase in the number of elite competitors in recipient catchments and a consequent increase in the effective size of the audience for advertising. Not only would absolute audience numbers increase, audience composition would shift as well, causing further increases in the payoff to advertising. New immigrant competitors would be unlikely to belong to local kin-based coalitions. In addition, immigrants and resident coalition members would be unfamiliar with one another. Hence members of each group would have no means of assessing the competitive ability of members of the other, save through wasteful social advertising. The result would be an increase in the payoff to wasteful advertising for all competitors within recipient political catchments. As resource levels continued to decline across the entire region, this local increase in the payoff to advertising resulted in the recipient catchments crossing the monument-building threshold later than they would have without recent additions to the pool of competitors. On the other hand, departure of emigrant competitors would hasten the crossing of the monument-building threshold in donor catchments by reducing the effective audience size within them. We can expect this movement process would have been accompanied by increased levels of violence, as individuals in political catchments with more favorable resource levels attempted to defend them against the encroachments of outsiders. Thus, spatial heterogeneity in the timing of resource abundance decline and the resulting movement process, documented in the variogram, point the way to a causal account for otherwise unexplained increases in the severity of warfare during the Late Classic that figure prominently in recent fashionable accounts of the collapse (e.g., Demarest 1996).

If the movement process is responsible for the conspicuous "hole effect" in the variogram, then the location of the peak in the semivariance at 65 km represents the mean distance over which such movements took place. In order to achieve the prolongation of monument construction, the immigrant competitors must have been new to the areas into which they moved. Hence 65 km also represents an *upper* limit on the mean diameter of spheres of political competition on the eve of the collapse. Movement during the collapse means that the price of an estimate of the *lower* limit is an additional assumption. The assumption is that the movement process merely altered the spatial distribution of residual variance, shrinking some residuals and stretching others, but leaving the total residual variance unchanged. Among the effects of the movement process on the variogram is an increase in the rate at which the semivariance increases with distance. Hence without the movement process, the slope of the variogram from the origin would be lower. This fact, along with the constant variance assumption, means that the distance at which the loess fit to the variogram equals the total variance provides a lower limit on our estimate of mean size for political catchments—with a lower slope the equality would occur at a greater distance. This distance is roughly 35 km for both variograms. Taking the midpoint between our upper and lower limits on the scale of political competition yields a best estimate of 50 km.

The loess fit for the two variograms is very similar. However, there is much less variation about the curve fit to the 65-site data set from which the four outliers were excluded. Thus we have evidence that these four sites differ from the others, not only in their frequency distribution, but also in their pattern of spatial dependence. These are grounds for concluding that a different process is responsible for the cessation of monument construction at these four sites. For the two positive outliers, Tonina and Seibal, this seems like a reasonable inference, since elites based at these sites continued monument construction long after their adjacent competitors had ceased. Monument construction uniquely ceased at these sites at a time when there was no place else to go. The negative outliers, Tortuguero and La Milpa, may be sites at which the cessation of monument construction was related to ongoing social dynamics that predate the onset of the collapse in their respective areas. It is also possible that the known monuments at these two sites offer especially poor estimates of the timing of the cessation of monument construction.

The "Polity Size" Question

The question of the spatial scale of Classic political competition, discussed traditionally under the rubric of "polity size," is of particular interest because the past 20 years of research have given us, predictably, two opposing answers: "large" and "small."[8] The "large" view has its origins in Joyce Marcus' work in the 1970s and ultimately in the social-systems functionalism of the New Archaeology. Marcus proposed that by the Late Classic Maya polities were hierarchically organized into four or five regional states. Each polity was comprised of a four or five-tiered settlement hierarchy, whose elements Marcus thought could be defined on the basis of site size, patterns of emblem glyph occurrence, and a gloss of the inscriptions on two stelae, at Copan and Seibal respectively, as referring to four regional capitals (Marcus 1976, 1983). Such polities would have spanned 100 to 200 km in diameter. A similar view, based largely on estimates of site size and empirical generalizations about the political correlates of rank-size relationships, is espoused by Richard Adams (1990, Adams & Jones 1981). More recently, Marcus (1993) has suggested that the Classic political hierarchy experienced temporal oscillations, as large regional states formed (c. AD 731), dissolved into their constituent provinces, and then reformed (c. AD 849). Among the initial attractions of Marcus' earlier model was its use of epigraphic evidence. However, the past two decades have seen major advances on the epigraphic front that have rendered Marcus' epigraphic interpretations suspect. Recent epigraphic research renders her reading of the Seibal and Copan monuments unsustainable, casting doubt on her reconstructions of political hierarchies (Houston 1992, 1993:4–8; Stuart 1993).

The epigraphic revolution has produced a competing "small" alternative hypothesis. Its most quantitatively precise rendering is to be found in Peter Mathews' recent work (1985, 1991). It is based on the demonstration that emblem glyphs occur in the context of a personal title, which can be read as "ruler—literally *k'ul ahaw* or holy lord—of place X." This leads to the inference that they refer to political catchments under the control of individual political competitors. Mathews has mapped these units for the Late Classic. The mean nearest-neighbor distance between 32 Late Classic sites with identified emblem glyphs is 52 km (Houston 1993:138). Our best estimate—50 km—of the mean size of political catchments from the variogram fits comfortably with this picture.

The variogram evidence clearly does not support Marcus' original regional state model. However, its relevance to the evaluation of the oscillation model is a little more opaque since the period over which the terminal date evidence accumulated encompassed, on her latest account, more than an entire cycle: from regional states, to fragmentation, to regional states, and then back to fragmentation. Marcus herself thinks the terminal date evidence is germane since she offers Bove's (1981) analysis in support of her more recent conclusions (Marcus 1993). However, the loess fitted surface may be of greater relevance to their evaluation than the variogram. The surface implies that the cessation of monument construction across the region was, as Morley (1920) proposed 75 years ago, a single time and space-transgressive event, and not a palimpsest of two downward trends in an ongoing political cycle.

Conclusion

Summary

How has our exploration of spatial patterning in terminal monuments added to our understanding of Classic Maya history? We have discovered a surprising, and surprisingly strong, relationship between mean annual rainfall and terminal monument dates. This lends support to the idea, warranted from a variety of other sources, that the deterministic component of the cessation of monument construction across the region was ultimately caused by an ecological disaster in which erosion played a leading role. Currently popular alternatives, drought and invasion, seem less likely in the face of the terminal date evidence. In addition, our examination of the scale of spatial autocorrelation in the terminal date residuals indicates that as ecological conditions worsened, groups of individuals moved from lower to higher quality habitats, over an average of 65 km. Finally, we have seen how this pattern of movement fits nicely with the idea, derived from epigraphic sources, that spheres of political competition during the Late Classic averaged about 50 km in diameter.

Caveats

The inferential path leading from terminal monument dates to the causes of the Maya collapse and then to the size of spheres of political competition is a long one. There are obviously many opportunities for missteps. Hence our historical conclusions are provisional. It would be premature to suggest that this is the definitive analysis of terminal monument dates. As new monuments are discovered, we can expect changes in terminal dates for some sites that are included in the current sample. We can also look forward to the addition of new sites to the sample. A glance at the locations of the 69 sites reveals that the western half of the study area is more sparsely sampled than the eastern half. This is a defect that only additional research in this area can remedy.

Evaluation

Our conclusions are also provisional in the sense that they are open to evaluation. Toward the beginning of this paper, to motivate the development of a mechanistic theoretical framework on which the analysis is based, I suggested that the payoff would come in the form of the ability to evaluate the resulting inferences. Over the long term, the inferential process might thereby become less perilous. It is now time to redeem this promissory note and suggest just why we might expect mechanistic theory to deliver this kind of result.

Theory forges the links from which we can model the causal chain that connects monuments to the conditions and mechanisms responsible for them. An explicit mechanistic theoretical framework facilitates model construction—the valid deduction of precise empirical expectations that should hold if our initial ideas about the mechanisms and the historical contexts in which they operated are correct. Precision in empirical expectations is important because it lessens the chance that when there is agreement between them and the evidence, it is the result of chance convergence. Precision also facilitates recognition when agreement is lacking, paving the way for adjustments in our model.

The deductive chain makes manifest the causal role of key theoretical variables in producing empirical variation and motivates the attempt to resolve that variation into the multiple theoretical processes responsible for it. The attempt at resolution means we will probably learn something new. As long as dated monument construction was thought to be a simple correlate of Maya "civilization," it was possible to see the outlines of Maya polities of nearly any sort in terminal dates. Analysis consisted of fitting successively higher order global polynomials until polities emerged from their inflections or their residuals. The absence of theory guaranteed that the polities discovered matched the most *avant garde* interpretation at the time (Marcus 1976). As a result, the terminal date trend was incorrectly identified and its remarkable relationship with mean annual rainfall went unsought and unfound.

The deductive chain also means that when variation does not match initial modeled expectations, there is an audit trail with which we can begin to puzzle out what went wrong. In the case at hand, the possible causes of the departure of the variogram's shape from our initial expectations are clear, easing identification of the most likely overlooked culprit: variation in audience size caused by movement of competitors.

Mechanistic theory makes possible the deduction of multiple sets of expectations implied by the same mechanism's operation in *different* empirical domains. In a Darwinian framework, the domains that recur are the design properties of formal artifact variation, variation in physical traces of past behavior, and the distribution of that variation in time and space (Neiman 1990). In the case at hand, for example, this tactic delivers two complementary sets of expectations under the hypothesis of selection for wasteful advertising. One set details the properties engineered into the monuments and their semantic content (e.g., wastefully high manufacturing cost, low temporal variance in lineage political quality). The second describes the distribution in time and space of the cessation of monument construction under different collapse scenarios (e.g., the relationship with annual rainfall). If we grant that the ecological disaster scenario is the most likely on other grounds and use it to develop expectations regarding spatial patterns for terminal dates, agreement with the observed pattern can be used as independent evidence in favor of a history of selection for smart advertising. If we grant the design arguments establish that the monuments are the result of wasteful advertising, then the spatial distribution evidence can be used to evaluate different collapse scenarios. The second avenue is the one I took here. But both approaches are available and both take advantage of the existence of multiple expectations in different evidentiary domains to evaluate hypotheses about history in an objective fashion.

Another route to evaluation depends on the multidimensional complexity of individual historical trajectories and the existence of multiple theoretical mechanisms to account for different dimensions of variation in a given historical case. We build multiple models, each involving the operation of different mechanisms, to account for different aspects of the same chunk of history. When two (or more) of these theoretically independent models contain a common key parameter, it becomes possible to deduce how variation in a particular parameter value implied by one model, might play out empirically in the context of the second. The two sets of empirical expectations can be made to yield independent estimates of variation in the common parameter. While this powerful line of evaluation has not been pursued here, it is worth considering briefly how it might be in the future.

Consider the question of shifts in audience size caused by movement that were detected in the variogram with the help of the theory of advertising. Darwinian theory offers a second, conceptually independent route to the estimation of audience size through the theory of cultural drift, the mechanism that controls the behavior of stylistic or selectively neutral cultural variation (Dunnell 1978; Neiman 1990). Selective neutrality here means that there are no deterministic forces like selection that favor the transmission or learning of one cultural variant (e.g., design motif) over alternative variants. Elsewhere I have shown that, under the selective neutrality hypothesis, among group variation in the diversity of alternative cultural variants along a dimension of stylistic variation is controlled by the product of the effective size of each group and the rate at which novel variants are introduced into it by immigration from outside (Neiman 1995). Since we know that elite political competitors among the Classic Maya were personally responsible for carving hieroglyphs and iconographic motifs on monuments, the monuments themselves offer a place to look for relevant patterns of diversity. This raises the possibility of using diversity in alternative iconographic motifs (e.g., Proskouriakoff 1950) as a way of measuring variation in the number of immigrants among elites at different sites. Those immigrants represent an increase in the size of the local audience for wasteful advertising. If the movement hypothesis derived from the variogram is correct, then we should find that stylistic diversity among elites at sites with negative residual terminal dates should tend to be lower than diversity among elites at sites with positive ones since political competitors left the former for the latter. It is theoretical clarity on the key role of variation in number of immigrants and in audience size and the overlap between them that makes this evaluation possible.

A similar tactic for evaluation is available when models built from entirely different theoretical frameworks contain common parameters (Kosso 1989). An example is the agreement between the estimate of the size of spheres of political competition based on the variogram and the estimate of polity size derived from Mathews' epigraphic work with emblem glyphs. Agreement lends objective support to both inferences. However, in this case a precise account of the theoretical independence is not possible because the theory that underwrites the mapping of emblem glyphs onto polities is opaque.

The quest for the means to evaluate our ideas about what happened in history is the motivation for building Darwinian models of historical processes in archaeology. Explicit theory about the mechanistic causes of history facilitates the evaluation of inferences about unique historical trajectories. We begin by hunting for agreement between precise, deductively modeled expectations and evidence. Explicit theory means there is an audit trail to follow if initial expectations are not met. Objectivity enters the evaluation process from independence between multiple sources of evidence implicated by the operation of a single mechanism and from independence among estimates of single parameters shared by different mechanisms. Building archaeological inference on simple behavioral correlates or covert, common-sense assumptions about human nature, and both processual and post-processual approaches do precisely that, does not guarantee that a particular inference is wrong. Nor can the use of causal theory guarantee correctness in a particular case. However, working with an explicit mechanistic theory that is the foundation for our scientific understanding of life *uniquely* raises the odds that, over the long term, more accurate inferences, and more powerful models with which to make them, will emerge. This paper is a modest attempt to further this process for ancient Maya history.

Acknowledgments

An initial version of this paper was presented at the Symposium on Mesoamerican Archaeology at the Peabody Museum of Natural History, Yale University, May 1994, in honor of my mentor Michael D. Coe, on the occasion of his retirement. Special thanks to Cary Carson who fundamentally shaped my interest in the problem of conspicuous consumption and my approach to it. I am also grateful for helpful comments on subsequent iterations from Bob Bettinger, Jim Boone, Michael Graves, Martha Hill, Bob Leonard, Garnett McMillan, Ann Ramenofsky, and Patrice Teltser. I owe an enormous debt to Steve Houston for helping me obtain up-to-date estimates of terminal monument dates. Steve also commented extensively on two drafts, enabling me to evade at least some of the pitfalls that await an historical archaeologist poaching in the Maya Lowlands.

Notes

1. Dunnell (1989) has offered a provocative alternative evolutionary model for wasteful energy expenditure. In Dunnell's account waste is a means by which groups achieve reproductive restraint and thus avoid extinction in stochastic environments (cf. Wynne Edwards 1962). His model, recently adumbrated by Graves and Ladefoged (1995), and the apparent group-selection mechanism that runs it deserve a more detailed examination than I can attempt here.
2. Unfortunately, quantitative estimates of these costs are unavailable. Recent attempts to estimate the levels of human effort required to build Classic Maya residential structures (Abrams 1994) offer little guidance since they do not include sculptural and hieroglyphic carving and the skilled labor required to execute it.
3. The surface estimated by loess contrasts with the polynomial trend surfaces, fit by ordinary least squares to these data by previous researchers. Bove (1981) offered linear, quadratic, and cubic polynomial fits. But his interpretive efforts were concentrated on residuals from them, in which he glimpsed four or five polities. Kvamme (1990), on the other hand, fit a single quartic surface and saw the five polities directly in its inflections. Comparison of the location of inflections in Kvamme's fitted surface with actual data values suggests that the former are, to a significant extent, artifacts of the spatial distribution of data points and the global character of the polynomial trend surface approach. Bove's residuals suffer from similar defects, because they are computed from similar surfaces.
4. In her recent study of dental hypoplasias in 160 individuals from the Pasión drainage, Wright argues that "the relative frequency of stress events is shown to have been relatively constant over time," and suggests this finding does not support ecological models of the collapse (1996:242). Her data (Wright 1996:241, Table 2) belie this conclusion. I performed logistic regressions of hypoplastic tooth frequency on temporal period, examining the change from Classic to Late Classic and from Late Classic to Terminal Classic for each tooth separately. Of the 14 teeth examined by Wright, none showed any statistically significant change (using a cutoff of $p < .10$) in hypoplastic frequency during the Classic-to-Late Classic transition. However, significant increases were detected on four teeth during the Late Classic–Terminal Classic transition (upper P3 and I2, lower P3 and P4).
5. Rainfall levels were estimated by digitizing 1000 mm rainfall contours in Rice (1993) and interpolating a 2.5 km grid over them using radial basis functions (Haigh 1992, 1993). The rainfall estimate for each site is the interpolated value for the grid node closest to it.
6. Geographical distance and rainfall are fixed in the usual regression sense. Thus to obtain a proper permutational test, geographical distance and rainfall matrices are held constant, while permuting the fitted date matrix. This protects the interrelations between distance and rainfall (Smouse & Long 1992:192). After 1000 permutations, the 99th percentiles of the null hypothesis distributions for $\beta_{distance}$ and $\beta_{rainfall}$ lie at .14 and .10 respectively. Approximate z-values, obtained by dividing the standard deviation of the permuted null distribution into the parameter estimates, are in line with this result ($Z_{distance} = 3.3$, $Z_{rainfall} = 11.25$). I conducted the analysis on the square roots of all distances and differences to linearize the relationships among them.
7. This result isolates the source of recent controversy over the existence of spatial autocorrelation in terminal dates. Whitley and Clark (1985) divided the Maya lowlands into 17 quadrats, 80 km on a side. They averaged terminal dates within each quadrat, then tested the null hypothesis that the quadrat means were not spatially autocorrelated. They failed to reject this hypothesis. By contrast, Kvamme (1990) dispensed with the 80 km quadrats and examined individual site locations. He definitively rejected the null hypothesis that similarity among terminal dates was unrelated to inter-site distance. The variogram shows why these results diverge. By averaging over 80 km quadrats, Whitley and Clark insured they would not detect spatial dependence on a smaller scale. On the other hand, Kvamme's methods precluded an examination of just what the scale of spatial dependence might have been.
8. Grube and Martin (1994, 1995; Appenzeller 1994) have offered a third alternative, one that posits the existence of two large "superpowers," Tikal and Calakmul, each heading a suite of client states. Grube and Martin see this arrangement lasting from AD 690 to 740. Hence the terminal monument date evidence is too late to be useful in an evaluation of their scenario.

References

Abrams, E. M.
1994 *How the Maya Built Their World: Energetics and Ancient Architecture.* University of Texas Press, Austin.

Abrams, E. M., D. J. Rue
1988 The causes and consequences of deforestation among the prehistoric Maya. *Human Ecology* 16:377–396.

Adams, R. E. W.
1973 Maya collapse: transformation and termination in the ceramic sequence at Altar de Sacrificios. In *The Classic Maya Collapse*, edited by T. P. Culbert, pp. 133–164. University of New Mexico Press, Albuquerque.
1990 Archaeological research at the Lowland Maya city of Rio Azul. *Latin American Antiquity* 1:23–41.

Adams, R. E. W., R. C. Jones
1981 Spatial patterns and regional growth among Maya cities. *American Antiquity* 46:301–322.

Andrews, E. Wyllys, V, J. A. Sabloff
1986 Classic to Postclassic: a summary discussion. In *Late Lowland Maya Civilization: Classic to Postclassic*, edited by J. A. Sabloff & E. W. Andrews, V, pp. 433–456. University of New Mexico Press, Albuquerque.

Appenzeller, T.
1994 Clashing Maya superpowers emerge from a new analysis. *Science* 266:733–734.

Binford, L. R.
1968 Some comments on historical versus processual archaeology. *Southwestern Journal of Anthropology* 24:267–275.

Bove, F. J.
1981 Trend surface analysis and the Lowland Classic Maya Collapse. *American Antiquity* 46:93–112.

Broughton, J. M.
1994 Late Holocene resource intensification in the Sacramento Valley, California: the vertebrate evidence. *Journal of Archaeological Science* 21:501–514.

Brenner, M., B. Leyden, M. W. Binford,
1990 Recent sedimentary histories of shallow lakes in the Guatemalan savannas. *Journal of Paleolimnology* 4:239–252.

Chase, D. Z., A. F. Chase
1982 Yucatec influence in Terminal Classic Northern Belize. *American Antiquity* 47:595–614.

Chase, A., P. M. Rice (editors)
1985 *The Lowland Maya Post Classic: Questions and Answers*. University of Texas Press, Austin.
Cleveland, W. S.
1993 *Visualizing Data*. AT&T Bell Laboratories, Murray Hill, N.J.
Cleveland, W. S., E. Gross, W. M. Shyu
1991 Local regression models. In *Statistical Models in S*, edited by J. M. Chambers & T. Hastie, pp. 309–376. Chapman and Hall, New York.
Coe, M. D.
1993 *Breaking the Maya Code*. Thames and Hudson, New York.
Culbert, T. P.
1988 The collapse of Maya civilization. In *The Collapse of Ancient States and Civilizations*, edited by N. Yoffee & G. L. Cowgill, pp. 69–101. University of Arizona Press, Tucson.
1995 *Maya Civilization*. Smithsonian Books, Washington D.C.
Culbert, T. P. (editor)
1973 *The Classic Maya Collapse*. University of New Mexico Press, Albuquerque.
Culbert, T. P., D. S. Rice
1990 *Precolumbian Population History in the Maya Lowlands*. University of New Mexico Press, Albuquerque.
Davis, J. C.
1986 *Statistics and Data Analysis in Geology*. Wiley, New York.
Deevey, E. S., D. S. Rice, P. M. Rice, H. H. Vaughn, M. Brenner, M. S. Flannery
1979 Maya urbanism: impact on a tropical karst environment. *Science* 206:298–306.
Demarest, A. A.
1992 Ideology in ancient Maya cultural evolution: the dynamics of galactic polities. In *Ideology in Pre-Columbian Civilizations*, edited by A. A. Demarest & G. Conrad, pp. 135–157. School of American Research, Santa Fe.
1996 War, peace, and the collapse of a Native American civilization: lessons for contemporary systems of conflict. In *A Natural History of Peace*, edited by T. Gregor, pp. 215–248. Vanderbilt University Press, Nashville.
Dunnell, R. C.
1978 Style and function: a fundamental dichotomy. *American Antiquity* 43:192–203.
1982 Science, social science, and common sense: the agonizing dilemma of modern archaeology. *Journal of Anthropological Research* 38:1–25.
1989 Aspects of the application of evolutionary theory in archaeology. In *Archaeological Thought in the Americas*, edited by C. C. Lamberg-Karlovksy, pp.35–49. Cambridge University Press, Cambridge.
Efron, B., R. J. Tibshirani
1993 *An Introduction to the Bootstrap*. Chapman and Hall, New York.
Fash, W. L., Jr.
1988 A new look at Maya statecraft from Copan, Honduras. *Antiquity* 62:157–169.
Garver, J. B.
1989 *The Land of the Maya: A Traveler's Map*. Cartographic Division of the National Geographic Society, Washington D.C.
Goodman, G.
1905 Mayan dates. *American Anthropologist* 7:642–647.
Graffen, A.
1990a Sexual selection unhandicapped by the Fisher process. *Journal of Theoretical Biology* 144:473–516.
1990b Biological signals as handicaps. *Journal of Theoretical Biology* 144:517–546.
Graham, I.
1982 *Corpus of Maya Hieroglyphic Inscriptions, Volume 1, Part 3*. Peabody Museum of Archaeology and Ethnography, Harvard University, Cambridge.
Graves, M. W., T. N. Ladefoged
1995 The evolutionary significance of ceremonial architecture in Polynesia. In *Evolutionary Archaeology: Methodological Approaches*, edited by P. Teltser, pp.149–174. University of Arizona Press, Tucson.
Haigh, J. G. B.
1992 Radial basis functions and archaeological surfaces. In *Computer Applications and Quantitative Methods in Archaeology: 1991*, edited by G. Lock & J. Moffett, pp. 157–162. BAR International Series S577, Oxford.
1993 Practical experience in creating digital elevation models. In *Computing the Past: Computer Applications and Quantitative Methods in Archaeology: 1992*, edited by J. Andresen, T. Madsen, & I. Scollar, pp. 67–74. Aarhus University Press, Aarhus.
Hamblin, R. L., B. L. Pitcher
1980 The Classic Maya collapse: testing the class conflict hypothesis. *American Antiquity* 45:246–267.
Hamilton, W. D.
1964 The genetical evolution of social behavior. I, II. *Journal of Theoretical Biology* 7:1–52.
Haviland, W. A.
1967 Stature at Tikal, Guatemala: implications for ancient Maya demography and social organization. *American Antiquity* 32:429–433.
Hodder, I.
1986 *Reading the Past*. Cambridge University Press, Cambridge.
Hodell, D. A., J. H. Curtis, M. Brenner
1995 Possible role of climate in the collapse of Classic Maya civilization. *Nature* 375:391–394.
Houston, S. D.
1992 Classic Maya politics. In *New Theories on the Ancient Maya*, edited by E. C. Danien & R. Sharer, pp. 65–69. University of Pennsylvania Museum, Philadelphia.
1993 *Hieroglyphs and History at Dos Pilas: Dynastic Politics of the Classic Maya*. University of Texas Press, Austin.
n.d. Latest long count dates on monuments in the southern Lowlands. Ms. list in possession of the author.
Isaaks, E. H., R. M. Srivastava
1989 *Applied Geostatistics*. Oxford University Press, New York.
Jones, C.
1991 Cycles of growth at Tikal. In *Classic Maya Political History: Hieroglyphic and Archaeological Evidence*, edited by T. P. Culbert, pp. 102–127. Cambridge University Press, Cambridge.
Kaplan H., K. Hill
1992 The evolutionary ecology of food acquisition. In *Evolutionary Ecology and Human Behavior*, edited by E. A. Smith & B. Winterhalder, pp.167–202. Aldine de Gruyter, New York.
Kirkpatrick, M.
1986 The handicap mechanism of sexual selection does not work. *American Naturalist* 127:222–240.
Kosso, P.
1989 Science and objectivity. *Journal of Philosophy* 86:245–257.
Kvamme, K. L.
1990 Spatial autocorrelation and the Classic Maya collapse revisited:

Leyden, B. W.
1987 Man and climate in the Maya Lowlands. *Quaternary Research* 28:407–414.

Lowe, J. W. G.
1985 *The Dynamics of Apocalypse: A Systems Simulation of the Classic Maya Collapse*. University of New Mexico Press, Albuquerque.

Marcus, J.
1976 *Emblem and State in the Classic Maya Lowlands: An Epigraphic Approach to Territorial Organization*. Dumbarton Oaks, Washington D.C.
1983 Lowland Maya archaeology at the crossroads. *American Antiquity* 48:454–488.
1993 Ancient Maya political organization. In *Lowland Maya Civilization in the Eighth Century AD*, edited by J. A. Sabloff & J. S. Henderson, pp. 111–171. Dumbarton Oaks, Washington D.C.

Martin, S., N. Grube
1994 Evidence for macro-political organization among Classic Maya Lowland States. Ms. in possession of the author.
1995 Maya superstates: how a few powerful kingdoms vied for control of the Maya Lowlands during the Classic Period (AD 300–900). *Archaeology* 48:41–59.

Mathews, P.
1985 Early Classic monuments and inscriptions. In *A Consideration of the Early Classic Period in the Maya Lowlands*, edited by G. R. Willey & P. Mathews, pp. 5–55. Institute of Mesoamerican Studies Publication No. 10, Albany.
1991 Classic Maya emblem glyphs. In *Classic Maya Political History: Hieroglyphic and Archaeological Evidence*, edited by T. P. Culbert, pp. 19–29. Cambridge University Press, Cambridge.

Mathews, P., G. R. Willey
1991 Prehistoric polities of the Pasion region: hieroglyphic texts and their archaeological settings. In *Classic Maya Political History: Hieroglyphic and Archaeological Evidence*, edited by T. P. Culbert, pp. 30–71. Cambridge University Press, Cambridge.

Miller, M. E.
1986 *The Murals of Bonampak*. Princeton University Press, Princeton.

Morley, S. G.
1920 *The Inscriptions of Copan*. Carnegie Institute Publication 219, Washington D.C.
1938 *The Inscriptions of Peten*. 5 vols. Carnegie Institute Publication 437, Washington D.C.
1946 *The Ancient Maya*. Stanford University Press, Stanford.

Morley, S. G., G. W. Brainerd, R. J. Sharer
1983 *The Ancient Maya,* Fourth Edition. Stanford University Press, Stanford.

Neiman, F. D.
1990 An Evolutionary Approach to Archaeological Inference: Aspects of Architectural Variation in the 17th-Century Chesapeake. Ph.D. dissertation. Department of Anthropology, Yale University.
1995 Stylistic variation in evolutionary perspective: inferences from decorative diversity and interassemblage distance in Illinois Woodland ceramics. *American Antiquity* 60:7–37.

Parpola, A.
1994 *Deciphering the Indus Script*. Cambridge University Press, Cambridge.

Proskouriakoff, T.
1950 *A Study of Classic Maya Sculpture*. Carnegie Institute Publication 593, Washington D.C.

Rands, R. L.
1973 The Classic collapse in the southern Maya Lowlands: chronology. In *The Classic Maya Collapse*, edited by T. P. Culbert, pp. 43–62. University of New Mexico Press, Albuquerque.

Rice, D. S.
1985 The Peten Postclassic: perspectives from the central Peten lakes. In *Late Lowland Maya Civilization: Classic to Postclassic*, edited by J. A. Sabloff & E. W. Andrews, V, pp. 301–344. University of New Mexico Press, Albuquerque.

Rice, D. S., P. M. Rice
1993 Physical geography, environment, and natural resources. In *Lowland Maya Civilization in the Eighth Century AD*, edited by J. A. Sabloff & J. S. Henderson, pp. 11–63. Dumbarton Oaks, Washington D.C.

Rice, D. S., P. M. Rice, E. S. Deevey
1985 Paradise Lost: Classic Maya impact on a lacustrine environment. In *Prehistoric Maya Environment and Subsistence Economy*, edited by M. Pohl, pp. 91–105. Peabody Museum of Archaeology and Ethnography, Harvard University, Cambridge.

Sabloff, J. A.
1973 Continuity and disruption during Terminal Late Classic Times at Seibal: ceramic and other evidence. In *The Classic Maya Collapse*, edited by T. P. Culbert, pp. 107–132. University of New Mexico Press, Albuquerque.
1992 Interpreting the collapse of Maya civilization: a case study of changing archaeological perspectives. In *Metaarchaeology: Reflections by Archaeologists and Philosophers*, edited by L. Embree, pp.99–119. Dordrecht: Kluwer Academic Publishers.
1995 Drought and decline. *Nature* 375:357.

Sabloff, J.A, G. R. Willey
1967 The collapse of the Maya civilization in the southern Lowlands. *Southwestern Journal of Anthropology* 23:311–336.

Saul, F. P.
1973 Disease in the Maya Area. In *The Classic Maya Collapse*, edited by T. P. Culbert, pp. 301–324. University of New Mexico Press, Albuquerque.

Schele, L., M. Miller
1986 *The Blood of Kings: Dynasty and Ritual in Maya Art*. Kimball Art Museum, Fort Worth.

Schmandt-Besserat, D.
1979 Reckoning before writing. *Archaeology* 32:22–31.

Shanks, M., C. Tilley
1987 *Re-Constructing Archaeology*. Cambridge University Press, Cambridge.

Smouse, P. E., J. C. Long
1992 Matrix correlation analysis in anthropology and genetics. *Yearbook of Physical Anthropology* 35:187–213.

Smouse, P. E., J. C. Long, R. R. Sokal
1986 Multiple regression and correlation extensions of the Mantel test of matrix correspondence. *Systematic Zoology* 35:627–632.

Stephens, J. L.
1841 *Incidents of Travel in Central America, Chiapas, and Yucatan*. Harper and Brothers, New York.

Stuart, D.
1988 Blood symbolism in Maya iconography. In *Maya Iconography*,

edited by E. Benson & G. Griffin, pp. 175–221. Princeton University Press, Princeton.

1993 Historical inscriptions and the Maya collapse. In *Lowland Maya Civilization in the Eighth Century AD*, edited by J. A. Sabloff & J. S. Henderson, pp. 321–354. Dumbarton Oaks, Washington D.C.

Trigger, B.
1990 Monumental architecture: a thermodynamic explanation of symbolic behavior. *World Archaeology* 22:119–131.

Veblen, T.
1899 *The Theory of the Leisure Class*. Macmillan, New York.

Whitmore, T. M., B. L. Turner, II, B. L. Johnson, R. W. Kates, T. R. Gottschange
1990 Long-term population change. In *The Earth as Transformed by Human Action*, edited by B. L. Turner, pp.25–39. Cambridge University Press, Cambridge.

Whittington, S.
1989 Characteristics of Demography and Disease in Low Status Maya from Copan, Honduras. Ph.D. dissertation, Department of Anthropology, Pennsylvania State University.
1991 Detection of significant demographic differences between subpopulations of prehispanic Maya from Copan. *American Journal of Physical Anthropology* 85:167–185.

Whitley, D. S., W. A.V. Clark
1985 Spatial autocorrelation tests and the Classic Maya collapse: methods and inferences. *Journal of Archaeological Science* 12:377–395.

Wright L. E.
1996 Intertooth patterns of hypoplasia expression: implications for childhood health in the Classic Maya collapse. *American Journal of Physical Anthropology* 102:233–247.

Wynne Edwards, V. C.
1962 *Animal Dispersion in Relation to Social Behavior*. Oliver and Boyd, Edinburgh.

Zahavi A.
1975 Mate selection—a selection for a handicap. *Journal of Theoretical Biology* 53:205–214.
1987 The theory of signal selection and some of its implications. In *Proceedings of the International Symposium of Biological Evolution*, edited by P. V. Delfino, pp. 305–325. Adriatica Editricia, Bari.

Zipf, G. K.
1949 *Human Behavior and the Principle of Least Effort*. Addison-Wesley, Cambridge.

Acknowledgments

C. O. Lovejoy, The Origin of Man. *Science* 211 (1981): 341-350. Reprinted with permission from AAAS.

J. F. O'Connell, Grandmothering and the Evolution of *Homo erectus*. *Journal of Human Evolution* 36 (1999): 461-485. © 1999 by Elsevier, reprinted by permission.

H. Kaplan, A Theory of Human Life History Evolution: Diet, Intelligence, and Longevity. *Evolutionary Anthropology* 9 (2000): 156-185. Reprinted with permission of John Wiley & Sons, Inc.

G. E. Belovsky, An Optimal Foraging-Based Model of Hunter-Gatherer Population Dynamics. *Journal of Anthropological Archaeology* 7 (1988): 329-372. © 1988 by Elsevier, reprinted by permission.

D. K. Grayson and F. Delpech, Changing Diet Breadth in the Early Upper Palaeolithic of Southwestern France. *Journal of Archaeological Science* 25 (1998): 1119-1129. © 1998 by Elsevier, reprinted by permission.

M. C. Steiner, Paleolithic Population Growth Pulses Evidenced by Small Animal Exploitation. *Science* 283 (1999): 190-194. Reprinted with permission from AAAS.

R. L. Kelly, Hunter-Gatherer Foraging and Colonization of the Western Hemisphere. *Anthropologie* 37 (1999): 143-153. Reprinted by permission of the editors of *Anthropologie*.

D. A. Byers, Should We Expect Large Game Specialization in the Late Pleistocene?: An Optimal Foraging Perspective on Early Paleoindian Prey Choice. *Journal of Archaeological Science* 32 (2005): 1624-1640. © 2005 by Elsevier, reprinted by permission.

F. E. Bayham, Effects of a Sedentary Lifestyle on the Utilization of Animals in the Prehistoric Southwest. Occasional Papers of the Archaeological Research Facility, CSU Fullerton, No. 5 (1986), *Agriculture: Origins and Impacts of a Technological Revolution*, pp. 54-78. Reprinted by permission of California State University Fullerton.

K. R. Barlow, Plant Utility Indices: Two Great Basin Examples. *Journal of Archaeological Science* 23 (1996): 351-371. © 1996 by Elsevier, reprinted by permission.

J. M. Broughton, Prey Spatial Structure and Behavior Affect Archeological Tests of Optimal Foraging Models: Examples from the Emeryville Shellmound Vertebrate Fauna. *World Archaeology* 34(2002): 60-83. Permission to reprint courtesy of Taylor and Francis Group, http://www.tanf.co.uk.

M. D. Cannon, A Model of Central Place Forager Prey Choice and an Application to Faunal Remains from the Mimbres Valley, New Mexico. *Journal of Anthropological Archaeology* 22 (2003): 1-25. © 2003 by Elsevier, reprinted by permission.

D. W. Zeanah, Sexual Division of Labor and Central Place Foraging: A Model for the Carson Desert of Western Nevada. *Journal of Anthropological Archaeology* 23 (2004): 1-32. © 2004 by Elsevier, reprinted by permission.

L. Nagaoka, Declining Foraging Efficiency and Moa Carcass Exploitation in Southern New Zealand. *Journal of Archaeological Science* 32 (2005): 1328-1338. © 2005 by Elsevier, reprinted by permission.

S. L. Kuhn, A Formal Approach to the Design and Assembly of Mobile Toolkits. *American Antiquity* 59 (1994): 426-442. Reprinted by permission of the Society for American Archaeology.

C. Beck, Rocks are Heavy: Transport Costs and Paleoarchaic Quarry Behavior in the Great Basin. *Journal of Anthropological Archaeology* 21 (2002): 481-507. © 2002 by Elsevier, reprinted by permission.

J. Bright, A. Ugan, and L. Hunsaker, The Effect of Handling Time on Subsistence Technology. *World Archaeology* 34 (2001): 164-181. Permission to reprint courtesy of Taylor and Francis Group, http://www.tanf.co.uk.

R. G. Elston, Microlithic Technology in Northern Asia: Risk-Minimizing Strategy of the Late Paleolithic and Early Holocene. In *Thinking Small: Perspectives on Microlithization*, eds. R. G. Elston and S. L. Kuhn. Archeological Papers of the American Anthropological Association, No. 12 (2002): 103-116. © 2002 by the American Anthropological Association. All rights reserved. Used by permission.

K. J. Gremillion, Diffusion and Adoption of Crops in Evolutionary Perspective. *Journal of Anthropological Archaeology* 15 (1996): 183-204. © 1996 by Elsevier, reprinted by permission.

S. R. Simms, Bedouin Hand Harvesting of Wheat and Barley: Implications for Early Cultivation in Southwestern Asia. *Current Anthropology* 38 (1997): 696-702. © 2001 by The University of Chicago Press. All rights reserved. Used by permission.

M. S. Alvard, Deferred Harvests: The Transition from Hunting to Animal Husbandry. *American Anthropologist* 103 (2001): 295-311. © 2001 by the American Anthropological Association. All rights reserved. Used by permission.

K. R. Barlow, Predicting Maize Agriculture Among the Fremont: An Economic Comparison of Farming and Foraging in the American Southwest. *American Antiquity* 67 (2002): 65-87. Reproduced by permission of the Society for American Archaeology.

T. A. Kohler, The Calculus of Self-Interest in the Development of Cooperation: Sociopolitical Development and Risk among the Northern Anasazi. In *Evolving Complexity and Environmental Risk in the Prehistoric Southwest*, eds. J. A. Tainter and B. Bagley Tainter, pp. 169-196. © 1996 by Addison-Wesley Publishing Company. Reprinted by permission of Westview PRE, a member of Perseus Books, L.L.C.

F. Neiman, Conspicuous Consumption as Wasteful Advertising: A Darwinian Perspective on Spatial Patterns in Classic Maya Terminal Monument Dates. In *Rediscovering Darwin: Evolutionary Theory in Archaeological Explanation*, eds. C. M. Barton and G. A. Clark, pp. 267-290. © 1997 by the American Anthropological Association. All rights reserved. Used by permission.

Contributors

Chapter 1
Jack M. Broughton is professor of anthropology at the University of Utah.

Michael D. Cannon is a principal investigator with SWCA Environmental Consultants, Salt Lake City, and an adjunct assistant professor in the Department of Anthropology at the University of Utah.

Chapter 2
C. O. Lovejoy
Department of Sociology and Anthropology
Kent State University
Kent, OH 44242
Professor of anthropology, Kent State University; professor of human anatomy at the Northeast Ohio Universities College of Medicine; assistant clinical professor of orthopedic surgery at Case Western Reserve University; research associate, Cleveland Museum of Natural History; and chairman of the Biological Anthropology Area Committee, division of biomedical sciences, Kent State University.

Chapter 3
J. F. O'Connell
Department of Anthropology
University of Utah
Salt Lake City, UT 84112

K. Hawkes
Department of Anthropology
University of Utah
Salt Lake City, UT 84112

N. G. Blurton Jones
Graduate School of Education
Departments of Anthropology and Psychiatry
University of California
Los Angeles, CA

Chapter 4
Hillard Kaplan is a professor in the Department of Anthropology at the University of New Mexico.
hkaplan@unm.edu

Kim Hill is an associate professor at the University of New Mexico.
kimhill@unm.edu

Jane Lancaster is a professor of anthropology at the University of New Mexico.
jlancas@unm.edu

A. Magdalena Hurtado is an associate professor in the Department of Anthropology at the University of New Mexico.
amhurtad@unm.edu

Chapter 5
Gary E. Belovsky
Department of Biology and School of Natural Resources
University of Michigan
Ann Arbor, MI 48109-1115

Chapter 6
Donald K. Grayson
Department of Anthropology and Burke Museum
Box 353010, University of Washington
Seattle, WA 98195

Francoise Delpech
Institut de Préhistoire et de Géologie du Quaternaire
UMR 5808 du CNRS
Université de Bordeaux I
33405 Talence, France

Chapter 7
Mary C. Stiner, Natalie D. Munro, Todd A. Surovell
Department of Anthropology, Building 30
University of Arizona
Tucson, AZ 85721
mstiner@u.arizona.edu

Eitan Tchernov
Department of Evolution, Systematics and Ecology
The Hebrew University of Jerusalem
Jerusalem 91904, Israel

Ofer Bar-Yosef
Department of Anthropology
Peabody Museum, Harvard University
Cambridge, MA 02138

Chapter 8
Robert L. Kelly
Department of Anthropology, Box 3431
University of Wyoming
Laramie, WY 82071
RLKelly@uwyo.edu

Chapter 9
David A. Byers, Andrew Ugan
Department of Anthropology
University of Utah
270 South 1400 East, Salt Lake City, UT 84112
dbyers@uofu.net
andrew.ugan@anthro.utah.edu

Chapter 10
Frank E. Bayham
California State University, Chico

Chapter 11
K. Renee Barlow
College of Eastern Utah
Prehistoric Museum
451 East 400 North
Price, UT 84501

Duncan Metcalfe
Department of Anthropology
University of Utah
Salt Lake City, UT 84112

Chapter 12
Jack M. Broughton
Department of Anthropology
University of Utah
Salt Lake City, UT 84112

Chapter 13
Michael D. Cannon
Department of Anthropology
Southern Methodist University
Box 750336
Dallas, TX 75275-0336
mdcannon@mail.smu.edu

Chapter 14
David W. Zeanah
Department of Anthropology
California State University Sacramento
6000 J Street
Sacramento, CA 95819-6106
zeanah@csus.edu

Chapter 15
Lisa Nagaoka
Department of Geography
PO Box 305279
University of North Texas
Denton, TX 76203-5279
lnagaoka@unt.edu

Chapter 16
Steven L. Kuhn
Department of Anthropology
University of Arizona
Tucson, AZ 85721

Chapter 17
Charlotte Beck
Department of Anthropology
Hamilton College
Clinton, NY 13323
cbeck@hamilton.edu

Amanda K. Taylor
197 Cleveland Drive
Croton-on-Hudson, NY 10520
ataylor@hamilton.edu

George T. Jones
Department of Anthropology
Hamilton College
Clinton, NY 13323
tjones@hamilton.edu

Cynthia M. Fadem
Department of Anthropology
Washington State University
Pullman, WA 99164
cynmfadem@hotmail.com

Caitlyn R. Cook
1115 Morris Avenue
Point Pleasant, NJ 08742
caitlyncook@hotmail.com

Sara A. Millward
90 Butler Lane
Mohnton, PA 19540
sara_millward@hotmail.com

Chapter 18
Jason Bright, Andrew Ugan, and Lori Hunsaker
Department of Anthropology
270 South 1300 East, Room 102
University of Utah
Salt Lake City, UT 84115-0600

Chapter 19
Robert G. Elston
Silver City, NV

P. Jeffrey Brantingham
University of California, Los Angeles

Chapter 20
Kristen J. Gremillion
Department of Anthropology
The Ohio State University
Columbus, OH 43210-1364

Chapter 21
Steven R. Simms and Kenneth W. Russell
Department of Sociology, Social Work, and Anthropology
Utah State University
Logan, UT 84322-0730
ssimms@wpo.hass.usu.edu

Chapter 22
Michael S. Alvard
Department of Anthropology
Texas A&M University
College Station, TX 77843

Lawrence Kuznar
Department Of Sociology and Anthropology
Indiana University–Purdue University at Fort Wayne
Fort Wayne, IN 46805-1499

Chapter 23
K. Renee Barlow
College of Eastern Utah
Prehistoric Museum
451 East 400 North
Price, UT 84501

Chapter 24
Timothy A. Kohler
Department of Anthropology
Washington State University
Pullman, WA
Santa Fe Institute
1399 Hyde Park Road
Santa Fe, NM 87501

Carla R. Van West
Crow Canyon Archaeological Center
Cortez, CO
Statistical Research, Inc.
Tucson, AZ

Chapter 25
Fraser D. Neiman

Index

abandonment, 201, 282, 344, 375, 380, 386, 402, 404
abortion, 85, 92
Acacia, 310–11
acculturation, 66
Ache, 38, 50–55, 59–60, 63, 65–69, 71, 73–74, 364, 386
Acheulean, 37
Acipenser transmontanus, 196, 200
acorns, 154, 195, 251, 353–55, 381, 383, 387–88
actualistic studies/data, 82, 136, 188, 294
Adams, R. E. W., 417, 423
adaptation, 1–3, 9, 17–18, 20–22, 24–25, 67, 71–73, 75, 81, 107, 126, 151, 228, 230–32, 251–52, 271, 275, 288, 320, 335, 345, 374–75, 404–5
Afontova Gora-2 Site, 322
Africa, 13–14, 18, 24, 30–31, 35–37, 40, 67, 72, 98, 133–36, 145–50, 224, 317, 353, 365
agave, 383
agriculture, 8, 31, 95, 98, 153, 155–56, 159–60, 167, 170, 192, 208, 217, 223, 246, 328, 331–36, 340–49, 354–55, 372–76, 378–91, 393, 395, 404, 418–20
Ahler, S. A., 305
Ahmarian, 118
Aiello, L. C., 37
Aldan River Valley, 126
Alectoris, 115
Allenrolfea, 172, 175, 188, 234, 311
Allman, J., 71
allocation, 32–34, 48, 56–58, 60, 70, 73, 135, 193, 331, 335–36, 339, 342–45
allometry, 50, 133, 134, 139, 141, 143, 332, 359, 362, 363
Altar de Sacrificios, 420
Altithermal, 164
altruism, 22, 357, 412
Alvard, M. S., 332–33, 356–57, 360
Alvey Site, 373
Alyawara, 36, 310, 386
Amaranthus, 385
Amazon, 357
Americas, 8, 81, 82, 122, 127, 129, 204
Amerindians, 127
Ammospermophilus, 160, 168
analogy, 18, 122, 128, 228
Anasazi, 372, 374–75, 383, 393–95, 397, 400, 404–5

Anatidae/anatids. *See* geese
Anatolia, 360
anatomically modern humans, 120
Anbarra, 54, 59
Andes, 290, 295, 297, 365, 376
andesite, 290, 295, 297
Anser, 116, 196
antelope, 142, 147, 160–61, 230, 233–34, 236, 313, 376, 382
Antilocapra americana, 135–36, 141, 160, 163, 210, 219, 234, 311. *See also* antelope; pronghorn
Aotinae, 22
apes, 8, 17–18, 24–25, 30–31, 33–34, 63, 67, 69–73, 139, 192, 243, 277–78, 280, 282, 403–4, 406, 411
Apteryx. See kiwis
Araceae, 35
Araya Site, 322
Archaic, 97–98, 151, 153, 156, 159–60, 164–65, 167–68, 170, 173, 186, 193, 282, 289, 290, 297, 374–75, 384, 386
Arctic, 82, 124, 128, 129, 130, 135
Arnhem, 54, 59
artiodactyls, 135, 137–41, 143–48, 153, 160, 163, 167–70, 199, 203–4, 210, 216–17, 219–24
Asia, 24, 120, 122, 272, 320–28, 332–34, 340, 349–54, 360
Ateles paniscus, 359
Atriplex, 234, 384. *See also* shadscale
Aubrey Site, 144
Aurignacian, 104–5, 112, 116, 145
Austad, S., 69
Australia, 36, 40, 81, 86, 122, 145–46, 188–89, 275–76, 283–84, 292, 311, 317, 328, 386
Australopithecus/australopithecines, 6, 13, 17, 31, 34–35, 38, 71–72, 74
Australopithecus afarensis, 13, 17–19, 23
Australopithecus africanus, 18
Australopithecus boisei, 18
Australopithecus robustus, 18

baboon, 21, 71
badger, 313
Baker Village, 372, 385
Bamforth, D., 6, 272
Bantu, 146
baobab, 60

barley, 234, 236, 332, 349–54, 383
Barlow, K. R., 154, 194, 209–10, 212–14, 217, 224, 258, 271–72, 285, 291–93, 304, 332–33
barracouta, 267
Bartram, L. E., Jr., 108
basalt, 290, 295
Basketmaker, 372, 375, 385
basketry, 177, 178, 189, 315
Bayham, F. E., 5, 153–54, 193
Bear River sites, 372, 385
beaver, 135, 144, 358
Beck, C., 271–72
Bedouin, 349, 351
Bedul, 349, 350, 351, 352, 353
Behrensmeyer weathering scale, 264
Belovsky, G. E., 82
beluga whale, 135, 136, 137
Bennett Site, 372, 384
Berelekh Site, 126
Bering Strait, 126
Beringia, 82, 97, 122, 126, 128
Bernoulli, D., 396
berries, 60, 383–86
Bettinger, R. L., 5, 212, 224, 295–96
bifaces/bifacial reduction, 82, 129, 167, 272, 276, 278, 283–86, 290–91, 293–306, 314–16, 320–24, 387
bighorn sheep, 139, 142, 147, 160, 161, 233, 234, 235, 236, 382. *See also Ovis canadensis*
Binford, L. R., 275, 291, 305
bipedality, 4, 13, 17, 18, 19, 22, 25, 31, 71, 72
Bird, D. W., 9, 292
birthrate, 21, 22, 361
birth-spacing, 33
biscuitroot, 36, 234, 236
Bison/bison, 106, 133, 135, 136, 139, 140, 141, 143, 144, 145, 219, 224, 283, 313
bitterroot, 234, 236, 383
bivalves, 54, 116
blackberry, 340
black-tailed deer, 197, 199–200, 202, 204
black-tailed jackrabbit, 160
Blackwater Draw, 138
blades/blade technology, 272, 280, 317, 320–28, 352
Bleed, P., 272
Bliege-Bird, R., 292
blueberry, 340

Bluefish Caves, 127–28
bluegrass, 234, 236, 313, 385, 387
Blumenschine, R. J., 41
Blurton Jones, N. G., 75, 108
body size (of prey in relation to return rate), 5, 153, 193, 208, 216, 313
Boesch, C., 70, 75
Boesch, H., 70, 75
bonobos, 71
Bossert, W. H., 56
Bove, F. J., 410, 413, 424, 426
Boxgrove Site, 66
Brandt's cormorant, 200
Branta, 196
Brantingham, P. J., 272
Braun, D. P., 395, 403
breast-feeding, 124
Bright, J., 272
brittlewort, 188
Bucerotidae, 22
Budyko, M. I., 99
buffalo, 135, 136, 146
bulrush, 234, 236–37, 242, 246–49, 383–85
bushmen, 384
Byers, D. A., 82

caches/caching, 82, 130, 182, 230–31, 249, 275, 384
Calakmul, 426
Caldwell Village, 372, 373, 385
Callahan, E, 297–98
Callista, 115
Callitrichidae/callitrichids, 22, 67
Camassia/camas, 36, 37, 383
camel, 143, 333, 360–61, 363, 365–66
Camelops, 143
Canidae/canids, 22, 23, 25, 163
cannibalism, 402
Canis latrans, 160, 163, 168
Canis lupis, 106
Capra, 145, 366
Capreolus, 366
capybara. See *Hydrochaeris hydrochaeris*
carbohydrate, 35, 36, 37, 38, 178, 180, 351
caribou, 135, 136, 144, 220, 224, 327
carnivores, 14, 19, 40, 63, 74, 105, 106, 129, 163, 167, 168, 262, 263, 264
Carson Desert, 155, 228–35, 239, 241–44, 246–51
Castor, 144
cattail, 233–34, 236, 238, 240–42, 249, 381, 382
cattle, 66, 333, 360–61, 365–66
Cebidae, 359
Cedar Mesa, 385
Celtis. See hackberry
central place foraging, 30–32, 36, 40, 153–55, 164, 166–67, 170, 175–76, 183, 189, 193–95, 208–17, 219–20, 222–24, 228, 232, 244–47, 251, 258–59, 267, 291–94, 296, 299, 304, 305

cervids, 202
Cervus, 106, 110, 135, 144, 196, 219
Chaco Canyon, 374, 386
Chacoan system, 402, 404
Chamberlin, R. V., 188
Charnov, E. L., 33, 56–57, 73, 193, 363–64
Chenopodium, 341, 385
chert, 290, 295, 296, 297, 343
Chesowanja Site, 37
Chiapas, 376, 378, 380, 385, 387
Chichén Itzá, 419
childcare, 231
childhood, 50, 51, 55, 63, 75, 420
chimpanzees, 13, 17–22, 24–25, 31, 33–37, 39–40, 48–51, 55, 58–60, 65–66, 68–71, 73–75
China, 31, 320, 322–24, 326–30
chokecherry, 383
Chontal Maya, 417
Christiansen, P., 134
chub, 234, 236
chukar, 115, 118
Church, R. R., 260
Cienega Creek, 385
Citrullus, 340
Clark, C. W., 332, 359, 362
Clark, W. A. V, 410, 426
Clovis, 82, 124–32, 139
Clutton-Brock, T. H., 21, 70
coevolution, 14, 48, 51, 55–56, 58, 72–75, 336, 412
Colby Site, 130
Cole's equation, 361, 363
collective action, 332, 357
Colobus/colobine or colobus monkeys, 21, 71, 74
colonization, 8, 24, 81–83, 86, 96–99, 122–30, 133, 155, 257
Colorado Plateau, 333, 373, 376, 381
Colorado River, 372
Colubridae, 168
Columbia Plateau, 37
conservation, 192, 204–7, 332, 356–60, 365–71
conspicuous consumption, 409–12
constraints/constraint assumptions, 3, 5, 82, 86–88, 90, 95, 142, 155, 166, 193, 229, 231, 245, 248, 250, 271–73, 276, 296, 332–33, 336, 342–45, 349, 365, 385
Copan, 409, 412, 414, 420, 423
Cordilleran ice sheet, 126
cores/core reduction, 271–72, 275, 277–85, 295–96, 315–16, 322–27
cormorants, 200, 201, 204
corn, 310, 372, 378, 396. See also maize
cottontails, 135, 141–42, 160, 167, 210, 219, 234, 236, 313, 382
cottonwood, 250
cougar, 290
Cova Beneito, 145
Cowboy Cave, 373, 384

Cowboy Rest Creek Quarry, 296–304
Cueva de Malladetes, 145
Culbert, T. P., 417
cultigens, 159, 170, 384, 386
cultivation, 230, 332, 339–40, 342–43, 349–54, 375–76, 378, 383–87, 420
currency, 2–3, 5–7, 67, 208, 277, 295, 323, 336, 338, 360, 398

dacite, 288, 290, 295–97, 303–4
Damuth, J., 139, 147, 363
Danger Cave, 173, 186–87
Darwin, C., 1, 8, 17, 335
Darwinian, 1, 5, 8, 251, 331, 335–36, 409–12, 422, 424–25
Delpech, F., 81
demography, 13, 83, 92, 94–95, 98–99, 124, 366
Demsetz, H., 358
density-mediated attrition, 220–22
Descurainia, 234, 384
"Desert-culture horticulture," 374
Diehl, M. W., 223
diet breadth, 7, 81–82, 104–5, 107–8, 110, 112, 120, 133–34, 139, 141, 144, 147, 217, 230–32, 234, 241, 244, 251–52, 289, 328, 337–38, 344–45, 353
diet breadth model, 2, 104, 139, 142, 208, 234, 244, 252, 382–84, 386
dimorphism, 4, 6, 13, 23, 31, 134, 231
Dinornithiformes, 257
Dioscoreaceae, 35
ditchgrass, 313
Dobe, 66
domesticates, 163, 331, 334, 366
domestication, 333–36, 351–52, 360–61, 365–71
Dordogne River, 104
Dorn, R. I., 127
double-crested cormorant, 200
Drickamer, C., 25
Dryopithecus/dryopithecines, 18, 24
ducks, 141–42, 234, 236, 297, 382
Dunnell, R. C., 426
Dyson-Hudson, R., 357, 365
Dyuktai Cave, 126, 128, 322

ebola, 51
Efe, 69, 135–36, 145
egrets, 23
einkorn, 351
ekwa, 33, 36, 37, 55
eland, 146
elderberry, 340
elephant, 133–40, 145–49
elk, 99, 135, 196–97, 199–200, 203–4, 219
Ellis, C. J., 321–22
Ellsmere Island, 124
Elster, J., 405
Elston, R. G., 272, 294
Elymus triticoides, 234, 239, 383–84

Emeryville Shellmound, 154, 192–205
Eminia, 36
Emlen, S. T., 23
encephalization, 69, 70
endogamy, 366
Ephedra, 187
Epigravettian, 117
Epi-Paleolithic, 83, 115, 116, 117, 120, 353, 354
Equus/equids, 106, 135, 139, 140, 144, 163, 366. See also horses
estrus, 23, 72
Ethiopia, 17
ethnoarchaeology, 108, 198, 332, 349
Eurasia, 24, 41, 81
Europe, 24, 66, 81, 118, 122, 144–45, 204, 276, 283–84, 339–41, 353, 365, 378, 410
Evans Mound, 372, 385
evenness, 107, 110, 112
evolution, 1–9, 13–14, 17–33, 35, 37–49, 51, 55–58, 63, 65, 67, 69, 70–76, 81–82, 115, 120, 122, 130–32, 151, 154, 159–60, 228–29, 243, 250–56, 271–73, 305, 310, 326, 331–37, 339–41, 343–49, 356–57, 359–60, 364–71, 386, 393–95, 409–12, 423, 426
extinction, 20, 82–83, 86, 94–97, 99–100, 124–25, 133, 201, 257, 358–59, 426

Fairview Valley, 249–50, 252
farming, 3–4, 155, 223, 331–35, 341, 343–44, 349, 354, 372, 374–83, 385–88
fat, 22, 72, 135–37, 228–29, 328, 338, 351
fertility, 14, 20, 31, 33–34, 38–41, 48–51, 54, 73, 204, 339, 343, 358, 360, 365, 420
fine-grained search assumption, 3, 5, 154–55, 192–93, 196, 204, 209, 212
Fischer, K. E., 69
Fisher, J. W., Jr., 136
Fisher, R., 357
fishing, 60, 70, 200, 234, 339
fitness, 2–4, 6, 8, 22, 25, 33–34, 39, 57, 71, 85, 96, 98, 204, 228–29, 243, 250, 326, 336–37, 339, 345, 356, 359, 362, 396, 411
Fitzhugh, B., 272
Five-Finger Ridge Site, 372, 385
Flenniken, J. J., 322–23
flintknapping, 296
FLK "Zinj" Site, 40
Floating Island Rockshelter, 187
folivores, 70
Folsom, 130–31, 271, 282
foraging theory, 2–3, 5–8, 81–82, 104, 107, 112, 133, 153, 156, 192–93, 205, 208–9, 212, 216, 243, 251, 257–59, 267, 271–72, 305, 331–33, 354, 356–57, 375–76
Ford, R. I., 400
Formative, 187–88, 372, 374, 378, 380, 384, 386
Fort Ancient, 344
fox, 144. See also *Vulpes*
foxtail, 236, 383
free-riders/free-riding, 357–58, 400

Fremont, 333–34, 372–88
Fremont River, 384
Frison, G., 129, 134, 136, 283
fructan, 36
fructose, 36
frugivory, 35, 70, 71
functionalism, 410, 423

Gadgil, M., 56
Galaz Site, 217–19, 221, 224–25
Galilee, 115
Ganj Dareh, 360
Gardner, P. S., 342
Gazella, 366
geese (Anatidae/anatids), 196, 198–200
gelada baboon, 21
gemsbok, 108
gender, 14, 167, 230, 234, 244–46, 251, 316, 386, 411
generalist/generalization (diet), 21, 23, 129, 147
genetics, 4, 74, 120, 336
George, W., 245
geostatistics, 410, 421–22
Gero, J. M., 316
gestation, 19, 20, 22
gibbons, 23, 71
Gibson, K. R., 70
Gifford, E. W., 196
Gigantopithecus, 18
giraffe, 140, 224
Glycymeris, 115
Goodall, J., 19, 21
Goland, C., 344
Gombe, 19, 49–50, 55, 59
goosefoot, 174, 249, 313
gopher, 139, 141, 143, 163, 168, 236, 313, 382
Gorilla/gorillas, 17, 24, 33, 70–71
Gosiute, 174
Grafen, A., 4, 394, 411
grandmother hypothesis, 13–14, 32, 33–34, 38–39, 41, 49, 73, 75
Grass Valley, 296–97
Gravettian, 145
Grayson, D. K., 6, 81, 294, 305
greasewood, 237, 238, 239, 240
Great Basin, 137, 143, 153–54, 172–75, 177, 180, 184, 186–87, 228, 230, 232, 241, 244, 246, 250–52, 272, 288–91, 296, 310, 315–17, 333, 372, 374, 376, 381, 383, 386
Great Salt Lake, 172, 178, 187, 372, 375
Green River, 372–75, 385
Greenland, 128
Gremillion, K. J., 331–33
Griffiths, D., 193
Grinnell Site, 402
Grotta dei Moscereni Site, 115, 119
Grotta Palidoro Site, 117
Grotta Polesini Site, 117
Grotte XVI Site, 81

ground sloth, 143
groundcherry, 340
group selection, 244, 426
Gwi, 54, 58–59, 67, 122

Haast, J. V., 263, 265
hackberry, 383
Hadar, 17
Hadza, 13, 32–33, 36–38, 50–55, 59–60, 63, 66–67, 73, 75, 108, 137, 140, 146, 224
Hambukushu, 63
Hamilton, W. J., 164–66
handling time/costs, 2–3, 5, 21, 34–35, 38, 40–41, 133, 135–38, 140–43, 154–56, 182, 210–14, 237, 244, 246–47, 252, 259, 272, 310–17, 337, 354
Hantman, J. L., 406
haplorhine primates, 70
Hard, R. J., 387
hares, 81, 115–16, 118–19, 139, 141–42, 144, 230, 317. See also *Lepus*
Harlan, J. R., 351
Harrapan, 412
Harvey, P. H., 70
Haury, E., 159–60
Hawkes, K., 67, 73, 354
Haynes, C. V., 138
Haynes, G., 82
Hayonim Cave, 115–19
hedgehogs, 115–16
Hegmon, M., 397
Heizer, R. F., 230
Helianthus. See sunflower
Henneman, W., 363
herbivores, 63, 71, 89, 129, 139, 140, 143, 147, 362, 363, 365
Herero, 66
herons, 23
Hershkovitz, P., 22
Hildebrandt, W. R., 250–51
Hill, K., 215, 364
Hiwi, 50–54, 59–60, 63–68, 73, 384
Hogup Cave, 173, 186, 372
Hohokam, 153, 159–60, 164, 167–68, 170–71, 193, 372, 374–75
Holliday, V. T., 138
Holloway, R. L., 17
Holmes, W. H., 288
Holocene, 82, 98, 115, 129, 140, 142, 144, 151, 153–57, 161, 192, 195–96, 199, 228, 230, 232–35, 237, 239, 240–44, 247–52, 272, 288–89, 320, 328, 360, 387, 393, 418
Homo erectus, 13, 17–18, 30–35, 37–41
Homo ergaster, 13–14, 31, 40, 66, 71, 74
Homo habilis, 40–41, 98
Homo rudolfensis, 40
Homo sapiens, 13, 17, 30–31, 34, 37–39, 66, 69, 71, 81, 93, 122, 128, 130
hornbills, 22
horses, 135, 139, 143–44, 360

horticulture, 166, 332–34, 357, 374–76, 382–84, 388, 404
Howell, N., 69
Huckell, B. B., 136
Hurtado, A. M., 65, 364
husbandry, 332–34, 356–58, 360–62, 365–66
Hutouliang Site, 322
Hydrochaeris hydrochaeris, 143
hypoplasia (dental/enamel), 35, 241, 426

ibex, 145, 366
Icacinaceae, 35
Illinois River Valley, 387
infanticide, 85, 92
Ingold, T., 357
insectivores, 58
intensification, 21, 23, 151, 154, 156, 159, 164, 167, 170, 264, 272–73, 332–34, 339, 343, 352–54, 366, 385, 387, 395, 404
intentionality, 3, 7–8
interbirth interval, 35, 50–51, 72
Ipomoea, 36, 310, 311
Irwin-Williams, C., 164
Isaac, G. Ll., 32
Isthmus of Panama, 128
Ituri Forest, 135
Iva, 341

jackrabbits, 136–37, 160, 174, 210, 219, 224–25, 234, 236, 313, 376, 382
jaguar, 69
Jakes Valley, 297
Jarmo Site, 360
Jennings, J. D., 374
Jerison, H., 70
Joes Valley Alcove Site, 373
Johanson, D. C., 23
Johnson, J. K., 288, 297–98
Johnson thinning index, 297–304
Jolly, C. J., 17, 18
Jones, K. T., 220, 224
Ju/'hoansi, 63
Judd, N., 374–75
Jumnapai goats, 365
juncos, 396
juniper, 174, 187

Kabnah Site, 419
Kalahari, 58, 108, 122, 384, 386
Kaplan, H., 14, 65, 71, 155, 215
katsina cult, 405
Kebara Cave, 118
Kebaran complex (Geometric), 354
Kelly, R. L., 82, 230–33, 251–52, 293–94, 297
Kibale, 49, 50, 55, 59
kivas, 400, 405
kiwis, 260
Kleiber's rule, 362
Knecht, H., 321
Knudtsen Site, 296–99, 301–4

Kohler, T. A., 393–94
Kongoro, 60
Koobi Fora, 35
Kooyman, B., 260–61, 265
koshare system, 405
Kosoutsy Site, 144
Koster Site, 282
Kozlowski, J, 73
Kua, 108
Kuhn, S., 271
!Kung, 36, 38, 50, 54, 59, 63, 66–67, 69, 73, 86, 146
Kutikina Site, 67
Kuznar, L., 332–33
Kvamme, K. L., 426

La Milpa Site, 423
Labrador, 128
lactation, 13, 22, 72
Laetoli, 18
lagomorphs, 116–17, 120, 135, 140–42, 144–45, 147, 153, 168–70, 197–98, 202
Lake Chichancanab, 418–19
Lancaster, J. B., 25
Laub, R. S., 136
Lazio Site, 115
Le Flageolet Site, 81, 104–14
Leacock, E., 358
learning, 3–4, 8, 14, 20, 25, 31, 48, 55–57, 60, 63, 65–67, 70–75, 275, 425
LeBlanc, S. A., 217
Lee, R., 66
Lee-Thorp, J, 38
Leporidae/leporids, 163, 210, 216, 219–20, 222–23
Lepus spp., 115, 118, 136, 141, 144, 160, 163, 168, 210, 219, 234, 311. See also jackrabbits; lagomorphs
Lese, 135–36
Levallois, 283
Levant/Levantine, 145, 353, 360, 366
Lewisia, 383. See also bitterroot
Lewisville Site, 144
Lewontin, R. C., 304
life expectancy, 38–39, 50, 92–93
Liguria Site, 115
Liliaceae, 35
Limestone Peak Locality I, 297–99, 301–4
Limnopithecus, 18
limpets, 81, 116, 118–19
Linares, O. F., 166
Little Boulder Basin, 272, 311–17
Little Smoky Quarry, 297–304
lizards, 115–17, 119, 159, 311
Lohi sheep, 364
Lomatium spp., 36, 234. See also biscuitroot
Lovejoy, C. O., 13
Loxodonta, 133–34
Lu, T. L. D., 320
Lucy (skeleton), 17

Lycium, 234, 383
Lyman, R. L., 220–21, 260

Macaca/macaques, 19, 25
Machiguenga, 65
macroblades, 322, 328
Madsen, D. B., 375
Mahale, 49–50, 55, 59
maize, 331–33, 335, 339, 341–44, 372–89, 397, 399–400, 402, 405
makalita, 36
Malthusian, 358, 420
mammoths, 82, 97, 130–31, 133–50, 328
Mammut, 133
Mammuthus, 106, 133, 141
Mantaro Valley, 378–79
Marcus, J, 423
marginal value theorem (MVT), 123, 155–56, 193, 209, 258–60, 264
Marlowe, F. W., 75
Marmota/marmots, 115–16, 144, 233–34, 236, 313
marrow, 31, 40, 134, 156, 257, 259–61, 264
mastodons/mastodonts, 97, 133, 143
Mathews, P., 423, 425
Mattocks Site, 217–19, 221, 224
Maxwell, M., 124, 128
Maya, 6, 8, 393–94, 409–26
Maya collapse, 393–94, 409–26
Maynard Smith, J, 4
Mayr, E., 23
McAnally Site, 217–19, 221
McGrew, W. C., 36
McGuire, K. R., 250–51
Meadowcroft Rockshelter, 127–28
Mediterranean Basin, 115, 118
megafauna, 82–83, 86, 96, 97, 133, 137–40, 142–47
Megatherium, 143
Meged Rockshelter, 118
Meltzer, D. J., 82, 144
menopause, 14, 32–34, 38–40, 73
Meriam, 188, 292
Mesa Verde, 383, 395, 405
Mesoamerica, 341, 393, 412, 425
Mesolithic, 151, 156, 323
mesquite, 383
Metcalfe, D., 154, 194, 209–10, 212–14, 217, 220, 224, 258, 271–72, 285, 291–93, 304
microblades, 272, 320–28, 352
microcores, 320, 323, 325–26
microlithic technology/microlithization, 272, 320–28
Milton, K., 70
Mimbres Valley, 155, 208–24
Miocene, 13–14, 17–24
Mio-Pliocene, 18
Mississippi River, 130
Mississippian, 343–44
moas, 155–57, 257–69

mogollon, 208–24, 372, 374–75
Mohave Desert, 127
mollusks, 115–16, 119, 196–97, 200, 204, 233
Mongolia, 328
mongongo nuts, 63, 66
Monodonta, 115
monogamy, 23
Monte Verde Site, 127–28
morbidity, 75, 420
Morley, S., 409–10, 424
Morss, N., 374
Mount Lincoln camp, 245–48
mountain sheep, 135, 137, 313
Mousterian, 282–83
Mubende goats, 365
mule deer, 161, 212, 251
Murray Springs Site, 138
muskrat, 135, 234, 236, 241
mussels, 116, 144, 196
Mytilus, 115, 196

Nagaoka, L., 155–56
Nahal Meged, 115, 118
Naranjo Site, 412
Natufian, 350, 353–54, 366
natural selection, 1, 4, 6–8, 13–14, 17–18, 20–23, 25, 33–34, 38, 48–49, 51, 55–57, 70–71, 73–74, 125, 194, 228–29, 244–45, 250–51, 332, 335–36, 345, 349, 352–54, 364, 376, 411–12, 424–25
Navajo, 365
Nawthis Village, 372, 385
Neandertals, 66, 71, 81, 120
needlegrass, 234, 236, 313
Negev Desert, 351
Neiman, F., 393–94
Nelson, M., 276
Nenana Valley, 128
Neolithic, 116, 120, 320, 328, 349–50, 353–57, 360, 365
Nephi Mounds Site, 372, 385
New Zealand, 153, 155, 257–68
Nichols, J, 127
Nilotic sheep, 364
Nine Mile Canyon, 373, 385
Nukak, 54, 58–59
Numic, 251, 386
Nunamiut, 166, 275

O'Connell, J. F., 6, 8–9, 13–14, 74, 224, 354
oats, 352
obsidian, 290, 295–97
Odell, G. H., 387
Odocoileus spp., 135–36, 141, 160, 163, 168, 197, 210, 218–19, 234, 260. *See also* black-tailed deer; mule deer; white-tailed deer
Ohalo II Site, 353
Oldowan, 37
Olduvai, 35, 40, 74
omnivores, 21, 24

Ondatra, 135, 234
onge, 54, 59
orangutans, 18, 24, 33, 70–71
Orians, G. H., 153, 166, 193, 209–10, 258
Oring, L. W., 23
Oryctolagus, 115
Oryx, 108
Oryzopsis, 385
Ostrea, 115, 196
ostrich, 115–17, 119, 260
otters, 8, 200–204, 222, 313, 315, 317–19, 328, 350, 353, 360, 373, 387
overkill hypothesis, 85–86, 97
Ovibos, 144
Ovis canadensis, 135, 160
ovulation, 23
Owens, I. P. F, 4, 8
ownership, 357, 404, 406, 412
oyster, 196

packrat, 161
Paiute, 174, 189, 245, 251, 374, 378, 384
Paleoarchaic, 272, 288–91, 293, 295–97, 300
Paleoindian, 82–83, 96–99, 122, 128–31, 133–39, 142–50, 276, 282, 285, 288–91
Paleolithic, 6, 40, 81–83, 99, 104, 113, 115–21, 145, 276, 310, 317, 320, 328–30, 353–55, 360
palm, 54, 58, 60, 63, 65, 174, 178, 350, 386, 399, 405, 414
Palmer Drought Severity Index, 399, 405
Paragonah Site, 372
parasites/parasitism, 20, 56, 69
partridges, 81, 115, 116, 118, 119
Pasión drainage, 419, 426
pastoralists, 357, 365–66
patayan, 372, 374
paternity, 13, 22, 25, 39, 228
peach, 234, 236, 335, 340, 341, 345
pearls, 343
Pearson, N. E., 153, 166, 193, 209–10, 258
peccaries, 359–60, 362
Pecos classification, 395
Pedra Furada, 127–28
Perideridia, 383
Perigordian, 104–5, 112
perrisodactyls, 144
persimmon, 340, 341
Peten region, 417, 419
Petra, 349–52
Pharo Village, 372
phenotypic gambit, 4–5
phenotypic/adaptive plasticity, 4, 364
philopatry, 30, 39, 40
pickleweed, 154, 172–89, 236, 383
Pigeon Mountain Basin, 322–24
pigs, 54, 360–61, 365
pine, 154, 172–74, 176–80, 182–91, 212, 230–31, 233–34, 236, 242–43, 376, 381, 383–85, 387
pinnipeds, 74, 194

Pinus, 172, 174, 178, 234, 311, 383
pinyon pine, 154, 172–74, 176–80, 182–91, 212, 230–31, 233–34, 236, 242, 245, 251–52, 295, 383
piranha, 69
Piro, 357, 359–60, 362, 364
pithouses, 217, 219, 221–23, 372, 375, 384, 397
pithouse-to-pueblo transition, 397
Pleistocene, 1, 9, 14, 18, 22, 24–25, 30–32, 37–40, 71–72, 81–83, 86, 96–97, 115, 117, 120, 124, 128–31, 133–40, 142–51, 187, 192, 251–52, 272, 283, 288–89, 320, 328–30, 332–33, 356, 360
Pliocene, 13–14, 18, 20, 24–25, 31, 40
Plio-Pleistocene, 14, 18, 38, 40
Plog, S., 395, 403
plums, 340–41
Poa, 234, 385, 425
polio, 51
polygyny, 21, 23, 194
pongids, 17–19, 24
Pongo, 21
post-encounter return rate, 2–3, 5, 36, 82, 123–24, 129, 135–36, 139–40, 143, 153, 186, 193, 208–10, 215–16, 219, 230, 233–34, 252, 310–11, 381–82
post-processual archaeology, 410, 425
pottery, 8, 222, 313, 315, 317–19, 328, 350, 353, 360, 373, 387
Powers, W. R., 328
pre-adaptation, 21, 71–72, 129
pre-Clovis, 127–28
predator-prey interactions, 115, 118, 151, 170
pregnancy, 19, 33, 38, 72
pretylla, 60
prey model/prey choice model, 2–3, 5, 7–8, 81–82, 104, 133, 144, 154–55, 192–93, 196, 208–9, 212, 215–17, 310, 333, 358, 361
primates, 14, 17–34, 36, 39, 48–50, 55, 58, 63, 67, 69–71, 73–75, 359, 362–63, 411
Prisoner's Dilemma, 405–6
proboscideans, 133–36, 144–50
processual archaeology, 2, 425
Proconsul, 18
prognathism, 18
promiscuity, 23
pronghorn, 135–37, 141, 144, 155, 160–61, 210, 219, 222–26, 234. *See also Antilocapra americana*
prosimians, 69–70
protein, 21–22, 54, 66–67, 86–88, 119, 159, 178, 180, 250, 340, 349, 351
Prunus, 234, 340, 383
ptarmigan, 124
pueblo/puebloan, 217, 219, 223, 374–75, 385, 393–94, 397, 401, 405–7
Pueblo Bonito, 374

Quercus, 383
quinoa, 184

Raab, L. M., 288
rabbits, 81, 115–16, 139, 144, 153, 159–60, 230, 233–34, 317. See also *lepus* spp.; jackrabbits
Ramapithecus, 18
Rangifer, 106, 110, 135–36, 144–45, 220. See also reindeer
Rangwapithecus, 18
rank/ranking of resources, 107–8, 118–20, 125, 133, 140, 143–45, 147, 153, 160, 167, 169–70, 188, 193–94, 200, 204, 230–31, 234, 238–44, 246–47, 249, 251, 257, 259–60, 264, 267, 311, 313, 328, 335, 337–38, 340–41, 344, 354, 375, 381–87
Rapson, D. J., 108–9
ratites, 257, 260, 263
reciprocity, 67, 400, 402
red deer, 110–12, 144–45
reindeer, 110–13, 144, 328, 357
resource depression, 82–83, 107, 151, 153–57, 160, 164–66, 170–71, 192–97, 199–201, 203–10, 216–17, 219–20, 223, 257, 259–60, 262, 372
return rate, see "post-encounter return rate" and "rank/ranking of resources"
Reynolds, V., 24
rheas, 263, 266
rhesus macaques, 19
Rhus, 383
ricegrass, 232–34, 236, 239, 249, 381, 383, 385, 387
richness (taxonomic), 81, 112
Rigaud, J.-P., 104
Rima, I. H., 397
Rio Grande, 386, 405
Riparo Mochi Site, 117–18
risk, 3, 272–73, 322–28, 331, 334, 337–39, 341, 343–45, 366, 393–94, 397–406
Robson, A., 58
Rocky Mountains, 130
rodents, 25, 141, 144, 159, 168, 170, 197–98, 202, 233
roe deer, 366
Rogers, A., 213, 357–58, 364, 366
rookeries, 199–200
Roosevelt, T., 145
roots, 32, 35–37, 51, 54–55, 58–60, 63, 175, 234, 236, 238, 240–41, 245, 366, 381, 383–84, 386
Ross, C., 363
Russell, K. W., 332, 361

Sabloff, J. A., 417–18
sage grouse, 143, 233–34
Sakhalin Island, 327
Salicornia, 188
saltbush, 234, 236–40, 383–84
samphire, 188
San Francisco Bay, 154, 195–97, 199–200
Sand Canyon, 402, 405
Santa Cruz River, 167

savanna, 14, 18–19, 21, 31–32, 37, 72, 97–98
Sayil Site, 419
scavenging, 13, 30–32, 40, 108, 143, 146, 224
Scirpus, 234, 383
Sciuridae, 135, 163
scurvy, 341
seals, 124, 136, 267
search time/costs, 2–3, 21, 65, 98, 133, 137–40, 143, 155–56, 166, 180, 193, 210–12, 214–17, 219–20, 222–23, 234, 237, 244, 252, 295, 304, 310–11, 337–38, 344, 382
seasonality, 13, 18–19, 21–22, 24, 35, 65, 328
Sebastian, L., 397–98, 400
Seibal Site, 417, 423
selectionist, 8, 331
senescence, 33–34, 38, 69, 76, 88, 92
Seong, C., 323
Sevier River, 372
sexual selection, 250
Shabik'eschee Site, 385
shadscale, 234, 236, 384
Shag River Mouth Site, 155, 257–68
Shanidar Site, 360
sharing, 30, 32–33, 39–40, 48, 55, 67–69, 72–73, 75, 120, 229–32, 243, 250, 344, 393, 395, 397–98, 400, 402–8
sheep (domestic), 333, 360–61, 363–66
shellfish, 35, 41, 54, 115–17, 119–20, 154, 188–89, 234, 236, 292–93, 317
Shipman, P., 134
Shott, M., 295
Shuidonggou Site, 328
siamangs, 23
Siberia, 126–28, 320, 322, 326–30, 365
sickles, 320, 332–34, 349, 351–54
signaling, 6–7, 15, 393–94, 409–26
Sillen, A., 38
Silva, M., 147
Simms, S. R., 137–42, 147, 240, 246, 332, 375, 381, 387
simulation modeling, 94, 118–19, 234, 237, 240, 242–44, 246–47, 250, 337–38
Sinagua, 372
Sivapithecus/sivapithecines, 18, 24
skunkbush, 383
Smith, B. H., 71, 74
Smith, E. A., 3–4, 6, 357, 365
snails, 144
Snake Rock Village, 373, 385
snakes, 115, 144
snares, 120, 315
Sonoran Desert, 160
specialist/specialization (diet), 7, 18, 48, 58, 66, 71, 82–83, 133–47, 164, 167, 251, 344, 353
specialization (economic), 67, 229–30, 243, 250, 393, 410
Spermophilus, 141, 160, 163, 168, 234

squirrels, 115–16, 135, 141, 143, 160, 233–34, 236, 313, 382–83

squirreltail grass, 236, 383
starch, 36–37, 54, 60, 63, 65, 241, 315, 343, 386
Steinaker Gap Site, 373
Stephens, J. L., 409
steppe, 37, 41, 328
Steward, J., 174
Stillwater Marsh, 231, 245–49, 252
Stillwater Mountains, 245, 249, 252
Stiner, M. C., 81, 317
storage, 35–38, 48, 56, 67, 71, 74, 82, 130, 164, 166, 172, 174–75, 182, 184, 187, 204, 230–31, 240–46, 249–50, 275, 331, 333, 340–41, 343, 345, 353, 372, 374, 376, 383–84, 386–88, 397, 406–7
strepsirhine primates, 70
sturgeon, 196–97, 199, 200, 204
SU Site, 385
suids, 37
Sumerian, 412
sumpweed, 341
sunflower, 234, 236, 341, 383, 385, 387
Surovell, T., 82
Sus, 144
sustainability, 332, 356–60, 362–63
Swartkrans Cave, 37
Sylvilagus, 135, 141, 160, 163, 168, 210, 219, 234

Taccaceae, 35
Tamar Hat Site, 67
Tanana Valley, 126
tansy mustard, 383–84
Taos Pueblo, 404
taphonomy, 32, 156, 217, 220–22, 224, 262, 264, 266
Tasmania, 67
Tayassu, 359
teleology, 7
temper (pottery), 8, 313, 315, 387
Tepe Ali Kosh Site, 360
Tepe Sarab Site, 360
termites, 22, 60, 70
territoriality, 8, 124, 154, 156, 164, 332, 353, 357–59, 365–66
Testudo, 115, 118
Tewa, 400
Theropithecus/theropiths, 18, 35
Thomas, D. H., 230
Thomomys, 141, 311
Thoms, A., 37, 41
Thule, 128
Tibet, 365
Tierra del Fuego, 127
Tikal, 420, 426
tobacco, 349
Toca da Esperança, 127
Toedokado Paiute, 245, 251
Tonina Site, 415, 423
Topaz Slough Site, 372, 384
Torres Strait Islands, 188, 292

tortoises, 81, 115–20, 317
Tortuguero Site, 415, 423
Tosawihi Quarries, 294–96, 299, 304, 316
tragedy of the commons, 357
traveler-processor model, 251
Trigger, B., 410
Trinkaus, E., 39
tubers, 14–15, 30, 35–38, 40, 60, 72, 88–89, 310–11
Tucson Basin, 384
tule elk, 196, 200, 204
tundra, 41, 97, 328
Turkana, 35, 37
Turner-Look Sites, 373
turtles, 54, 144, 204
Typha spp., 234, 311. *See also* cattail

Udora Site, 144
Uenodaira Site, 322
Ugan, 82
Uintah Basin, 375, 384
Upham, S., 395
Ursus, 106, 135
Ushki Lake Site, 126
USOs (underground storage organs), 35–38
Usumacinta drainage, 417–19
utility, 14, 33, 86, 95, 99, 108, 134, 154, 156, 172–78, 180–84, 187–89, 198–99, 209–10, 212–14, 217, 220–27, 252, 257–69, 271, 276–85, 291–96, 299, 304–8, 311, 315, 323, 326, 328, 338–39, 343, 393, 396–99, 403–6, 410
Uxmal Site, 419

Veblen, T., 410
Ventana Cave, 153, 159–70
Verkhene-Troitskaya Site, 126
vervets, 71
Vickers, W. T., 95, 166
Vigna, 32, 36, 311, 340
vijulla, 60
vole, 313
Vulpes, 106, 144, 163, 168

Wadi Beida, 349, 350, 351, 352
Waguespack, N., 82
Wana, 357
waterfowl, 115, 154, 196, 230–31, 233–34, 344
watermelon, 340
Watt, K. E. F., 164–66
Weigert, R. G., 73
Van West, C. R., 393–94, 399
wetlands, 196, 230–32, 237–42, 244, 246, 249, 251–56, 289, 353, 372
whalers, 359–60
whales, 135–37, 359
wheat, 332, 349–54
Wheeler, C. P., 37
Whipple Site, 144
White, T. D., 402
Whiterocks Village, 372–73
white-tailed deer, 161
Whitley, D. S., 127, 410, 426
wickiups, 374
wildrye. See *Elymus triticoides*
Willard Mounds Site, 372

Willey, G. R., 417
Willig, J. A., 289
Wilson, D. S., 193
Winterhalder, B., 3–4, 91–92, 99, 252, 326, 331, 337–38, 344
wolf, 144
wolfberry, 234, 236, 383
Wolverton, S., 260
woodrats, 233, 234, 236
Wrangham, R. W., 70
Wright, K. I., 352
Wright, L. E., 426

yampa, 383
Yanomamo, 67, 364
Yellen, J., 66
Yora, 65, 69
Younger Dryas, 289, 354, 365
Yucatan, 417–19
Yukon, 127

Zagros Mountains, 360
Zahavi, A., 394, 411
Zawi Chami, 360
Zeanah, D. W., 6, 155, 224
Zeder, M., 366
Zipf, G. K., 410–11
Z-score model, 272, 320, 326–28, 331, 335, 337–38, 341, 344
Zuni, 404